CELL PHYSIOLOGY

MOLECULAR DYNAMICS

second edition

HENRY TEDESCHI
State University of New York at Albany

WCB
Wm. C. Brown Publishers
Dubuque, Iowa • Melbourne, Australia • Oxford, England

Book Team

Editor *Megan Johnson*
Designer *Eric Engelby*
Art Editor *Rachel Imsland*
Permissions Editor *Karen Storlie*

Wm. C. Brown Publishers
A Division of Wm. C. Brown Communications, Inc.

Vice President and General Manager *George Bergquist*
National Sales Manager *Vincent R. Di Blasi*
Assistant Vice President, Editor-in-Chief *Edward G. Jaffe*
Marketing Manager *Carol Mills*
Advertising Manager *Amy Schmitz*
Managing Editor, Production *Colleen A. Yonda*
Manager of Visuals and Design *Faye M. Schilling*
Design Manager *Jac Tilton*
Art Manager *Janice Roerig*
Publishing Services Manager *Karen J. Slaght*
Permissions/Records Manager *Connie Allendorf*

Wm. C. Brown Communications, Inc.

Chairman Emeritus *Wm. C. Brown*
Chairman and Chief Executive Officer *Mark C. Falb*
President and Chief Operating Officer *G. Franklin Lewis*
Corporate Vice President, Operations *Beverly Kolz*
Corporate Vice President, President of WCB Manufacturing *Roger Meyer*

Cover image courtesy of Terry Wagenknecht

Library of Congress Catalog Card Number: 92–72912

ISBN 0–697–16949–9

Printed in the United States of America by Wm. C. Brown Communications, Inc., 2460 Kerper Boulevard, Dubuque, IA 52001

10 9 8 7 6 5 4 3 2 1

This book is dedicated to the memory of Robert Day Allen

Friendship is the sharing

of dreams

Teaching the sharing

of thoughts

Contents

Preface

Cell Physiology: Molecular Dynamics, published in 1974, examined the organization of the cell from two distinct points of view. One section examined the cell in relation to genetic information and its expression, and another examined the cell as a transducer, i.e., an energy converter. The present edition concentrates on this second thread. Obviously, genetic information and gene expression are of paramount importance to cell function, and although not neglected, it has not been used as the primary framework.

As before, the presentation tries to combine the virtues of the conventional coverage of a textbook and the use of selected detailed data from some articles. The references are sufficient to permit exploration of the literature on specific points, and suggested references for study are presented separately in the bibliography of each chapter.

The book is designed to encourage students to read and evaluate original articles and to form views independent of either instructor or textbook. I hope that it will awaken interest in the major issues concerning cell function and enable the readers to follow future progress in this rapidly developing field.

As before, my debt is to many who helped me in many ways, most preciously in their suggestions, criticism, corrections, and patience.

Henry Tedeschi

Introduction

The formulation of new ideas and their testing in the laboratory are exciting tasks. However, these ideas have to rest on knowledge brought to light by others in the endless search of science. How does one acquire the background? At one time (just twenty years ago) it was simple enough to acquire an appropriate reading list, although the availability of well stocked libraries could be and still can be seriously limited. It was entirely possible to read all articles available on one specific topic. At present, however, computer searches with even fairly focused questions yield hundreds of references published in only a few months in periodicals.

How should a neophyte proceed to examine a topic in cell physiology? The first step consists of acquiring a general orientation. Reading part or all of the latest edition of a good textbook is a good start; there are many excellent textbooks in the closely related fields of cell and molecular biology. The present book focuses on the physiology of the cell and provides an orientation so that students may proceed directly to reading the literature on their own. References are provided for further reading.

The second step is a sharper focus and later updates, which can be acquired through review articles that in recent years have become essential and have been honed to provide solid foundations. Specific reviews have been selected in this textbook (see in particular the selected readings preceding the references in each chapter). Scholarly reviews can also be found in *Physiological Reviews, Biological Reviews, Essays in Biochemistry, Progress in Biophysics and Molecular Biology, International Reviews of Cytology, Biochimica and Biophysica Acta,* and the *Annual Reviews* (of cell biology, biochemistry, physiology, biophysics, and biophysical chemistry). These publications review various topics on a rotating basis or when progress has been very rapid. Many reviews have been collected in books on specific topics, and these can also be found readily. The reviews are formidable but need not be forbidding if the reader accepts that he or she may have to check some of the unfamiliar terms or concepts by referring to a textbook or by reading a few of the references.

A recent appearance, the minireview of a few pages sharply focused on key concepts is amazingly useful, particularly to update rusty information and concepts. Again, although the major issues are outlined clearly in most of them, it is impossible to understand them in depth without more reading. But what a joy to know what to look for! Useful short reviews are found in many key journals: *Science, Nature (London), Cell, Cell Regulation, Journal of Biological Chemistry* (minireviews for the past year from this journal are sold in groups of 20 for a nominal amount), *Biochemical Journal, Trends in Cell Biology* (TICB), *Trends in Biochemical Sciences* (TIBS), *Trends in Neurological Sciences* (TINS), *FASEB Journal, News in Physiological Sciences* (NIPS), *FEBS Letters, Experientia, BioTechniques, Current Opinions in Cell Biology,* and many others.

Reviews occupying a position somewhat between a minireview and a complete scholarly review sometimes also appear in the journals listed above. New journals, such as *Seminars in Cell Biology* and *Seminars in Neuroscience,* apparently designed to occupy this intermediate position, have also begun to appear. The list may seem forbidding but occasional perusal of these publications can be very rewarding.

Reading original articles is, however, essential. Learning to read critically is the third step: The actual basis for the conclusions and the limitations and significance of the studies can only be understood well in this way. For this reason, the present textbook critically reviews original articles. Is the reasoning correct? Does the method used actually measure what is intended? Does the evidence unambiguously support the conclusions? What alternative interpretations are possible? These are all questions that have to be asked and the answers require examining the original articles. In this textbook, original data and possible interpretations are presented whenever possible. In addition to providing an overview, it attempts to prepare the reader to ask these questions. Complete bibliographic data for the original articles discussed appear in the reference lists at the end of each chapter, along with those for other recent textbooks and for review articles of the types discussed above.

The most recent articles may be collected by personal examination of key journals or in the weekly listings of *Current Contents* (Institute for Scientific Information, Philadelphia, PA 19104), available in most science libraries and also supplied on floppy disks. Another weekly floppy disk service that provides current titles (and abstracts) is *Reference Update* (Research Information Systems, Inc., Carlsbad, CA 92009). The *Reference Update* database can also be reached directly on a subscriptional basis via a modem. For computer searches of current and past articles, databases providing both titles and abstracts and the necessary software are available through special services. At present, the most useful databases for cell physiology are *Biosis Previews* and *Medline. Medline* searches can be obtained directly from the National Library of Medicine at a modest cost (Medlars Management Section, National Library of Medicine, 8600 Rockville Pike, Bethesda, MD 20844). *Biosis, Medline,* and other important databases are also available from Dialog Retrieval Service, Palo Alto, CA 94304; BRS Information Technology, Latham, NY 12110; and STN International, Columbus, OH 43202. Their databases can be reached through telephone-modem arrangements on a subscription basis or by purchasing compact disks (CDs). Most science libraries have access to these databases and many others. They are equipped to carry out searches on request, usually at reasonable cost. The library at SUNYA, for example, has access to more than 600 databases. Many science libraries also carry the appropriate CDs, which are therefore directly available to their patrons.

Bon appétit!

Chapter 1

Cell Organization: Techniques and a Preliminary View

Since the invention of the light microscope and our first vistas of cells, technology and concepts have gone hand in hand. Hence a review of some of the methodology is useful, although in the chapters that follow, results and techniques are frequently considered together. Our review first examines the techniques more traditionally associated with cell biology: microscopy and the isolation of cell components. These sections are followed by the approaches of molecular biology, such as recombinant DNA technology and immunological techniques, which have revolutionized every aspect of biology and have obvious significance in studies concerning cell function and structure. More traditional biochemical techniques that have become part of biological studies are also included. A preliminary view of cell organization is then presented; other chapters will expand on the themes raised in this section.

I. TECHNIQUES USED IN THE STUDY OF CELLS

A. Microscopic Techniques

The application of video technology and computer-aided analysis of images has allowed studies of cell structure and function that were not possible before. Refinements of more conventional approaches, such as *cytochemistry* and *autoradiography*, have also contributed valuable information. Cytochemistry allows the detection and quantitative estimation of cell components or enzyme reactions, whereas autoradiography detects photographically the presence of radioactive isotopes incorporated into cell components. Similarly, immunological techniques have allowed recognition of the distribution of specific proteins inside cells.

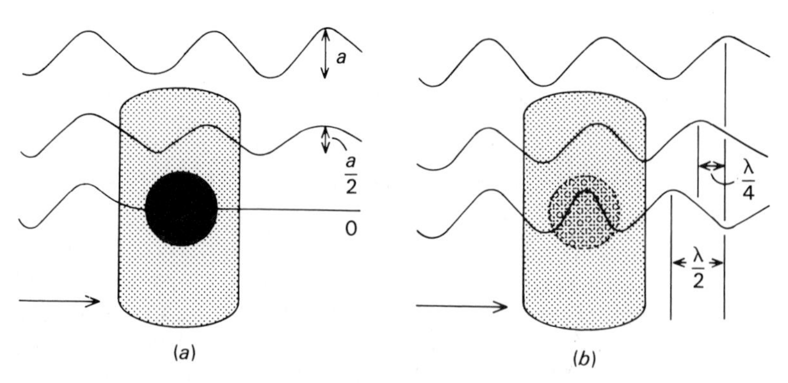

Fig. 1.1 Comparison of absorption and phase interference methods in light microscopy. Based on a diagram from Ross (1967).

Light Microscopy Figure 1.1 illustrates the principles of two of the light microscopic methods currently used in cell biology: conventional light microscopy (Fig. 1.1*a*) and phase and interference microscopy (Fig. 1.1*b*).

With visible light and an ordinary light microscope, the various cell components offer little contrast unless they have been stained or absorb light. Generally, before being examined with the light microscope, a specimen must be fixed, a procedure that immobilizes the macromolecular components either by denaturing or cross-linking them. Then embedding in either plastic or paraffin is needed, generally preceded by dehydration. The material is then sectioned and stained. The absorption or fluorescence of the dyes can be measured quantitatively with a variety of techniques. Figure 1.1*a* illustrates the absorption of light through a stained cell in conventional microscopy.

In light microscopy, immunological techniques have been used by attaching an antibody to a label, a dye with characteristic light absorption or fluorescence (*immunocytochemistry*). Similarly, sites of enzyme activity can be identified by using substrates, sometimes analogs of the natural substrates, which the enzymes convert into insoluble compounds with characteristic absorption or compounds that can be readily precipitated (*enzyme cytochemistry*).

The interference methods illustrated by the special case of phase-contrast microscopy in Fig. 1.1*b* do not require any of the preparatory procedures, so they are best suited for the study of living cells, whose dynamics can be recorded cinematographically or on videotape. These methods take advantage of light interference phenomena. When two coherent beams of light are recombined, they can interfere with each other constructively or destructively to produce a contrast image. The beam passing through an object will be retarded in proportion to the thickness or density of the object. In *phase-contrast microscopy* the beam diffracted by the object is recombined with a reference beam that has passed through the medium or the background material. The reference beam has been advanced or retarded a quarter-wavelength to maximize the interference. The difference in phase between the two beams will give positive or negative interference, the former showing the object as darker and the latter as lighter than the background. In *interference microscopy*, the phase of the reference beam can be varied in relation to the specimen beam, allowing measurement of the retardation of the specimen beam. The method can be used as a quantitative tool to estimate the dry weight of objects.

Nomarski *differential interference contrast microscopy* (DIC) is based on the interference between two closely separated points in the object. The beam passing through the specimen is split by a birefringent plate (a modified Wollaston prism). The image is a gradient of the phase difference between these two adjacent points. Mathematically, the contrast is the derivative of the path differences with respect to distance. The image has a directional contrast resembling a shadow-cast relief map of cellular details. This technique provides a resolution closer to the theoretical limit than any other light microscopic technique. Furthermore, objects above and below the plane of focus are excluded from the image, which provides essentially an optical section. Nomarski optics are ideally suited for observing objects with well-defined boundaries, such as fibers or condensed chromosomes.

Less powerful techniques such as *modulation contrast* (Hoffman) microscopy (Hoffman and Gross, 1975) and other asymmetric illumination methods (Kachar, 1980) provide images that resemble DIC images. Hoffman microscopy has been used recently because of its economy and its advantages in common with DIC such as viewing of optical sections, increased visibility, and contrast of unstained specimens. As in DIC, the images are shadowed and show no halos. However, the principles involved are very different.

Figure 1.2 allows a comparison of images obtained with phase-contrast and Nomarski differential interference microscopy. With Nomarski microscopy (Fig. 1.2*a*) the details are sharp and each cell component is clearly defined. With phase contrast (shown in Fig. 1.2*b* with positive contrast) the details appear more diffuse and, where slightly out of focus, are surrounded by halos.

Generally, under optimal conditions, light microscopy can distinguish as separate images two points that are no less than 0.2 μm apart. This is considered the practical limit of resolution for these techniques. With shorter wavelengths, much closer points can be resolved, and this principle has been exploited in electron microscopy, which uses electrons instead of visible light. The resolution with the electron microscope under optimal conditions approaches atomic dimensions.

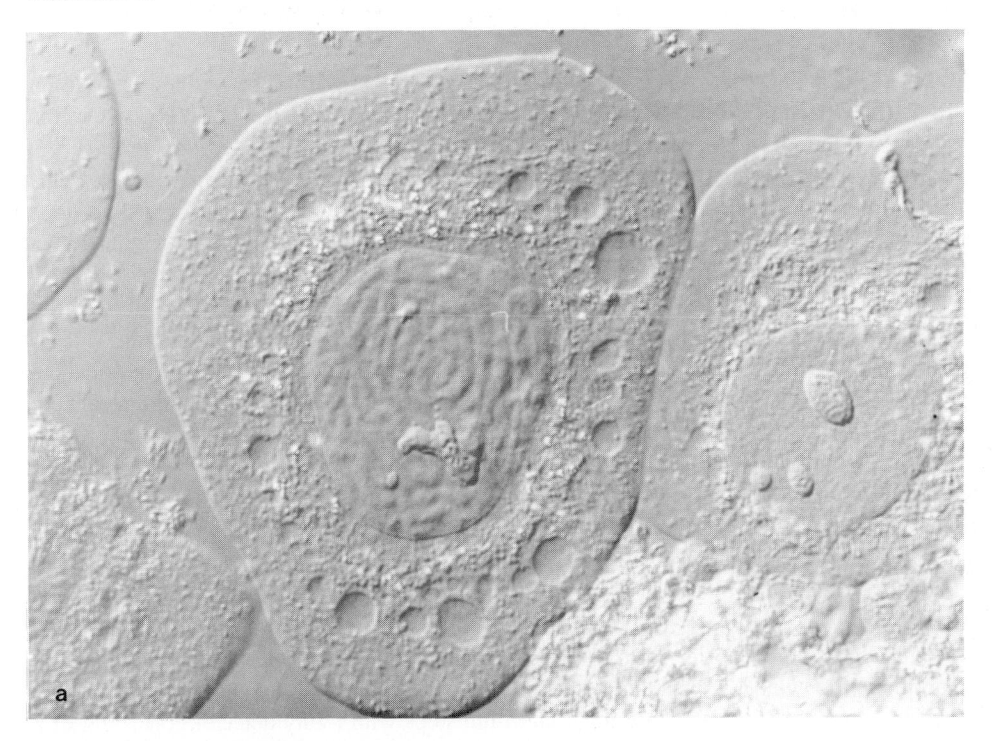

Fig. 1.2 Endosperm cells of *Hemanthus katherinae* shown with (*a*) Nomarski differential interference.

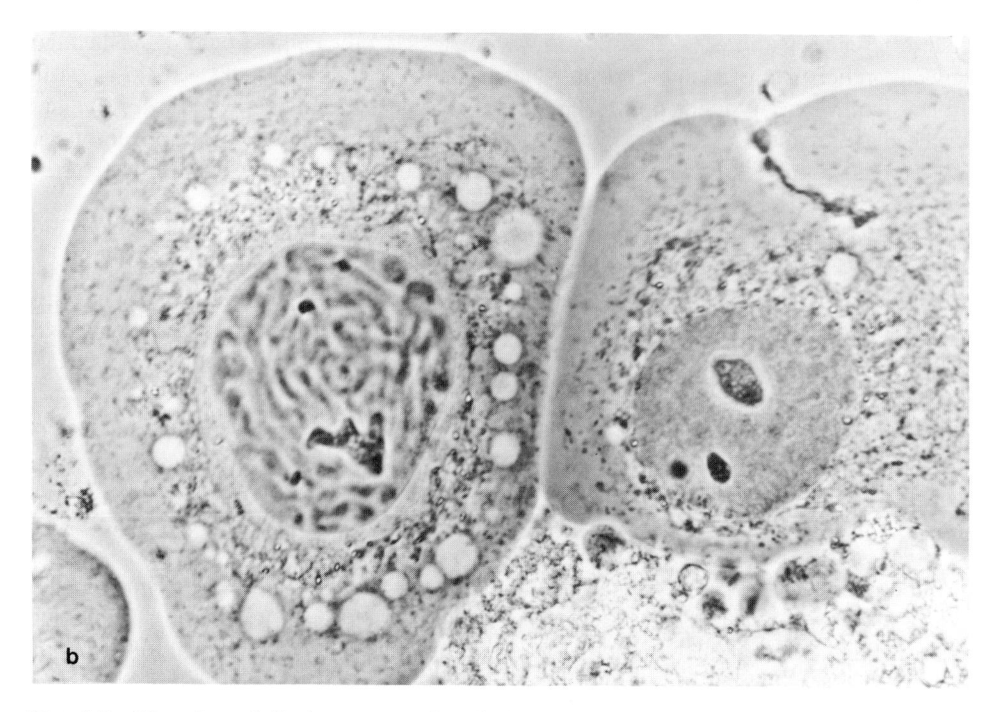

Fig. 1.2 (*Continued*) Endosperm cells of *Hemanthus katherinae* shown with (*b*) positive phase contrast. The figure width corresponds to 212 μm. Photomicrographs by R. D. Allen.

Although light microscopy is limited in resolution, objects below the limits of resolution can still be studied using Nomarski optics and high-performance video cameras. The size of the image, however, does not correspond to that of the object. The video methods enhance the contrast of the objects by subtracting the background and amplifying the remaining signal. This approach allows viewing of single microtubules that are only 25 nm in diameter. Contrast improvements and light intensity measurements have also been provided by digital image processors that filter and average signals.

Confocal microscopes for imaging either fluorescence emission or reflected light have been developed and are now commercially available (White and Amos, 1987; White et al., 1987). Like DIC, they allow the study of optical sections. The instruments contain a computer-controlled system that scans the specimen rapidly in three dimensions with a single small illuminated spot (ideally 0.2 μm in diameter), usually a focused laser beam. Unlike conventional microscopes, they view a region coincident (i.e., confocal) with the illuminated spot. Only the region lying within a narrow depth of focus produces an image that corresponds to a precise location in space. Ideally, out-of-focus areas are not a problem since they are not illuminated. The image can be digitized and stored on computer disks for analysis with conventional programs. Confocal microscopes are ideal for studying the three-dimensional organization of a specimen with fluorescent dyes.

An entirely new concept of microscopy developed by M. Isaacson and A. Lewis (Pool, 1988) promises to extend the resolving power of the light microscope to as little as 50 nm. This microscope scans the illuminated specimen in 15-nm steps with a tiny detector consisting of a tube with an aperture as small as 50 nm in inner diameter.

Electron Microscopy Electron microscopy has brought the study of cell structure to the so-called *submicroscopic* or *ultrastructural* level, where even macromolecules can now be observed. The electron microscope (EM) depends on the fact that electron beams can be bent and focused by electromagnetic fields and used in much the same way as light in light microscopy. The electron beam originates from a filament and, after passing through an evacuated microscope column, is focused on a fluorescent screen or a photographic plate.

Transmission Electron Microscopy With the *transmission electron microscope* (TEM), the electron beam passes through the specimen, which is generally placed on a plastic-coated metal grid. The denser areas in the specimen scatter more electrons and the corresponding regions in the image appear darker. Electron stains of high density are usually used to enhance the contrast. The specimen preparation generally used for TEM resembles in principle the fixation, embedding, sectioning, and staining used for the light microscope but, as might be expected, must be carried out under much more rigidly controlled conditions. Many fixatives such as OsO_4 also serve as stains. Others, such as glutaraldehyde, do not add to the contrast, and additional treatment with electron stains is needed. Generally, the sections have to be much thinner than those used for light microscopy, which must be embedded in plastic. In part, the thinner sections must be used because of the limited penetrating power of conventional TEM; in part, they are needed to avoid the superposition of features in the image above or below the structures of interest.

High-voltage or intermediate-voltage electron microscopes are now in use. Because of the increased penetration of the electrons provided by the higher voltages, intact cells or much thicker sections can be used. Together with tilting of the EM stage, these techniques provide three-dimensional information and a wealth of detail. However, because of the overlap of the information in these images, they are difficult to interpret. Developments in computer analysis of such images are providing greater information about the three-dimensional organization of cells.

Isolated cell components, macromolecules, and macromolecular assemblies such as ribosomes and viruses have been studied with the TEM using *negative staining*. With this technique the particles are suspended, generally in a solution of an electron-dense compound (e.g., phosphotungstate or uranyl salts). The preparation is sprayed or spread in some manner on a grid and then dried. Structures that are not permeated by the salt appear light, whereas those containing the salt—for example, the previously hydrated areas—appear dark.

The scanning transmission electron microscope (STEM) has been used to study the structure of the freeze-dried ciliary and flagellar protein dynein (discussed in Chapters 17 and 18) without the use of stain or fixative (Johnson and Wall, 1983). This technique, which uses a very low electron dosage, also allows the determination of mass from the electron scattering intensities over a particle.

The newest advance in TEM is the imaging of cells and macromolecules that are embedded in amorphous ice at very low (liquid nitrogen) temperatures. These specimens are unfixed and unstained, so the contrast arises directly from differences in density of the cell components (proteins and nucleic acids appear dark, lipids and water light). Instrumental or computer techniques are usually needed to enhance the low inherent contrast of these frozen-hydrated images (see Fig. 1.3*e*).

Figure 1.3 illustrates application of electron microscopy to a membrane channel. Figure 1.3*a* is an electron micrograph of a negatively stained mitochondrial outer membrane, and Fig. 1.3*e* is a reconstruction of a preparation embedded in vitreous ice without stain. Other details of this figure are discussed below in relation to computer-aided reconstruction (p. 7).

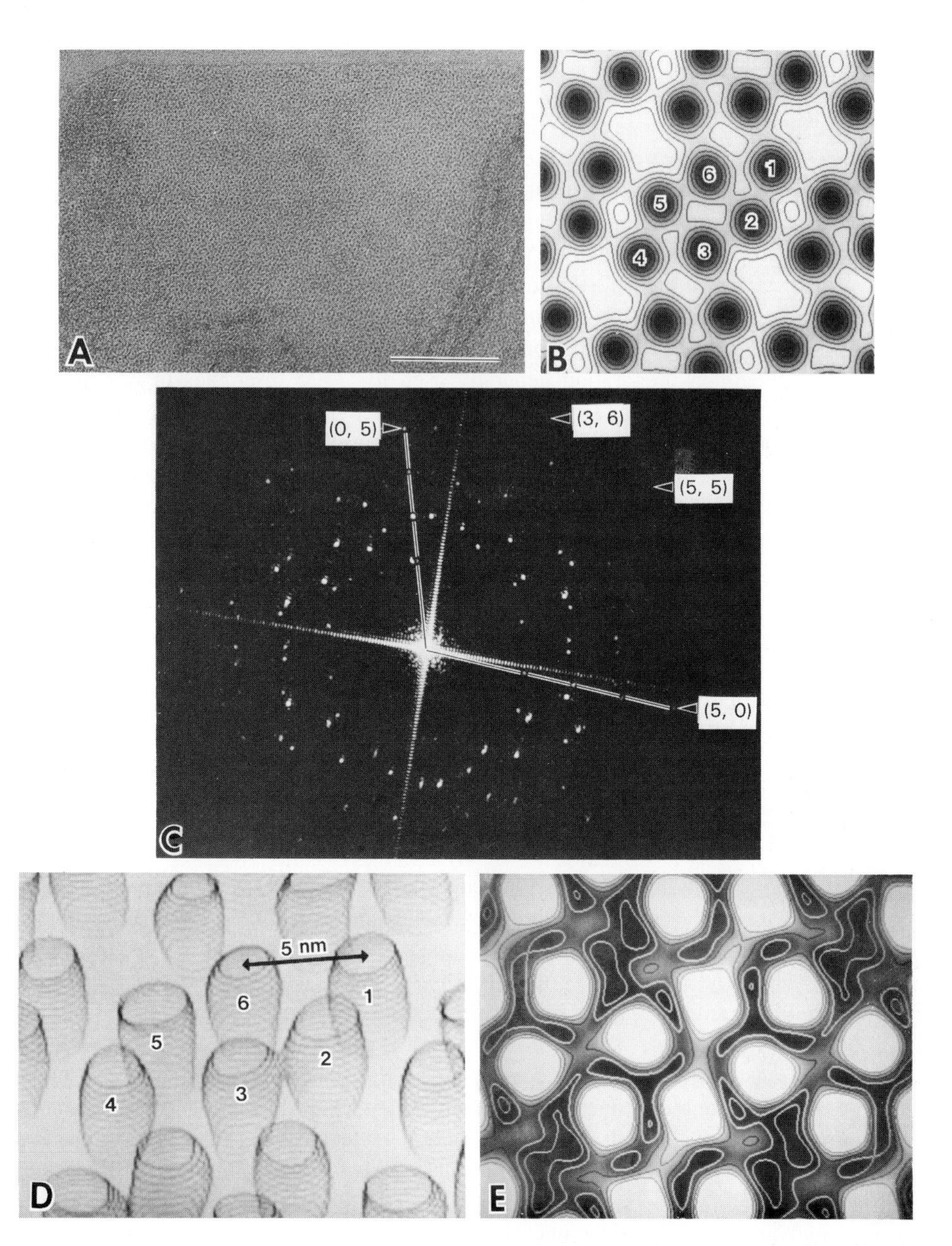

Fig. 1.3 Electron crystallography of a membrane channel. (*a*) Electron micrograph of a negatively stained mitochondrial outer membrane from the fungus *Neurospora crassa* (scale bar, 100 nm). (*b*) The two-dimensional repeat motif of one of the membrane arrays. (*c*) Optical diffraction of micrograph taken with coherent illumination from a laser. (*d*) Three-dimensional reconstruction of the array. (*e*) The projected density of the same membrane array embedded in vitreous ice without stain. (*a*) and (*c*) are reproduced from *The Journal of Cell Biology,* (1982), vol. 94, pp. 680–687, by C. A. Mannella, by copyright permission of Rockefeller University Press; (*b*), (*d*), and (*e*) courtesy of C. A. Mannella.

Scanning Electron Microscopy In the *scanning electron microscope* (SEM), a beam of electrons is brought to a focus and scanned across the specimen in a process similar to that used to produce television images. The microscope collects and displays electrons scattered from the surface of the object (secondary electrons). Generally, the specimen has to be dried and coated with metal. The metal increases the electrons scattering and also acts as a conductor to avoid accumulation of charges.

The SEM produces images with a stereoscopic appearance because curved portions of the object reflect more electrons than flat areas. The SEM is very useful in observing three-dimensional objects. However, it is typically restricted in resolution to above 5 to 10 nm.

Computer Processing Computer processing of TEM images permits reconstruction of three-dimensional structures. The analysis is more direct for periodically repeating units, such as those in two-dimensional crystals formed in membranes, which allow the application of crystallographic methods. The images of untilted and tilted specimens are first digitized. The three-dimensional reconstruction of the two-dimensional crystal is done with Fourier transforms (FTs). The three-dimensional FT of a two-dimensional crystal (in the *xy* plane in real space) is an array of lines normal to the plane of the crystal (Fig. 1.4) (Amos et al., 1982). Variations in the specimen density in the *z* direction in real space result in modulation of intensity along each *z* line. The two-dimensional FT of a projection image of the crystal represents a central section, that is, a slice through the origin of the three-dimensional FT with the angle of the slice equal to the tilt angle of the specimen in the EM. By combining many projection transforms (corresponding to different tilt and azimuthal angles), the three-dimensional Fourier volume is filled in, except for a missing cone defined by the maximum tilt angle (typically 50–60°). The projection transforms are brought to a common phase origin (equivalent to aligning the images in real space) and the inverse transformation is performed, yielding the three-dimensional volume in real space.

Figure 1.3*a* (Mannella, 1982) shows an electron micrograph of a two-dimensional crystal produced by concentrating the proteins in an outer membrane preparation by partially removing the lipid with phospholipase. The optical diffraction pattern from the micrograph is shown in Fig. 1.3*c*. The crystals are formed by the so-called mitochondrial porin or voltage-dependent anionic channel (VDAC). The pores are black when filled with stain (Fig. 1.3*b*) and white when filled with water (Fig. 1.3*c*). Figure 1.3*d* is a three-dimensional reconstruction of the negatively stained crystals showing the channel structure.

Unordered molecules on an EM grid, which are viewed generally with negative staining or after freezing, pose a different challenge, since the analysis is necessarily much more complex. The EM images are recorded with and without tilting of the specimen to provide three-dimensional information, as done for the two-dimensional crystals. The computer three-dimensional reconstruction of the nonperiodic structures is done in real space by *tomography*. Again, multiple projection images are used; these may be a "conical series," as shown in Fig. 1.5 (Radermacher et al., 1986), in which there is a constant large tilt angle and the complete range of azimuthal angles is covered. The reconstruction requires back-projecting the density at each point in each image into appropriate strips in the three-dimensional volume (Fig. 1.6) (Frank et al., 1985) after careful alignment of the images and prefiltering in Fourier space to enhance high-resolution details. A flow diagram of the reconstruction procedure is shown in Fig. 1.7 (Radermacher et al., 1986).

Figure 1.8*a* and *b* show 50S ribosomal subunits tilted by 50° and untilted, respectively. A reconstruction of the ribosomes and various ribosomal subunits is shown in Fig. 1.9 (Frank et al., 1988).

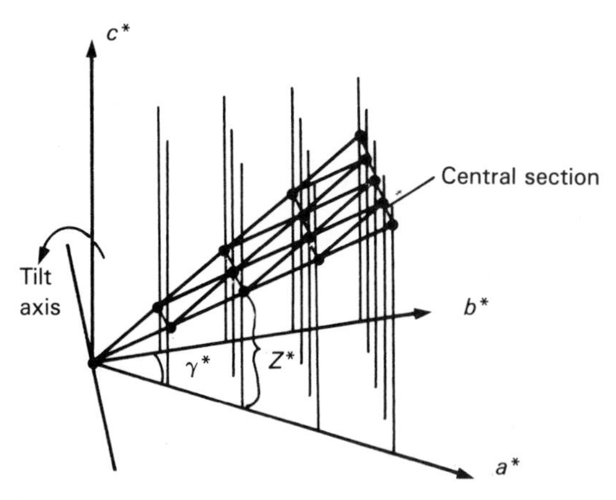

Fig. 1.4 Schematic diagram showing the transform of a two-dimensional crystal. The transform takes the form of a number of lattice lines extending perpendicular to the plane of the crystal. Each micrograph contains an image that, when Fourier-transformed, gives the values of amplitude and phase at points along the lattice lines where the central section (perpendicular to the viewing direction) intersects them (Amos et al., 1982). Reprinted from *Progress in Biophysics and Molecular Biology, 39,* L. A. Amos, R. Henderson, and P. N. T. Unwin, 3–D structure determination by electron microscopy of 2–D crystals, Copyright © 1982, Pergammon Journals Ltd.

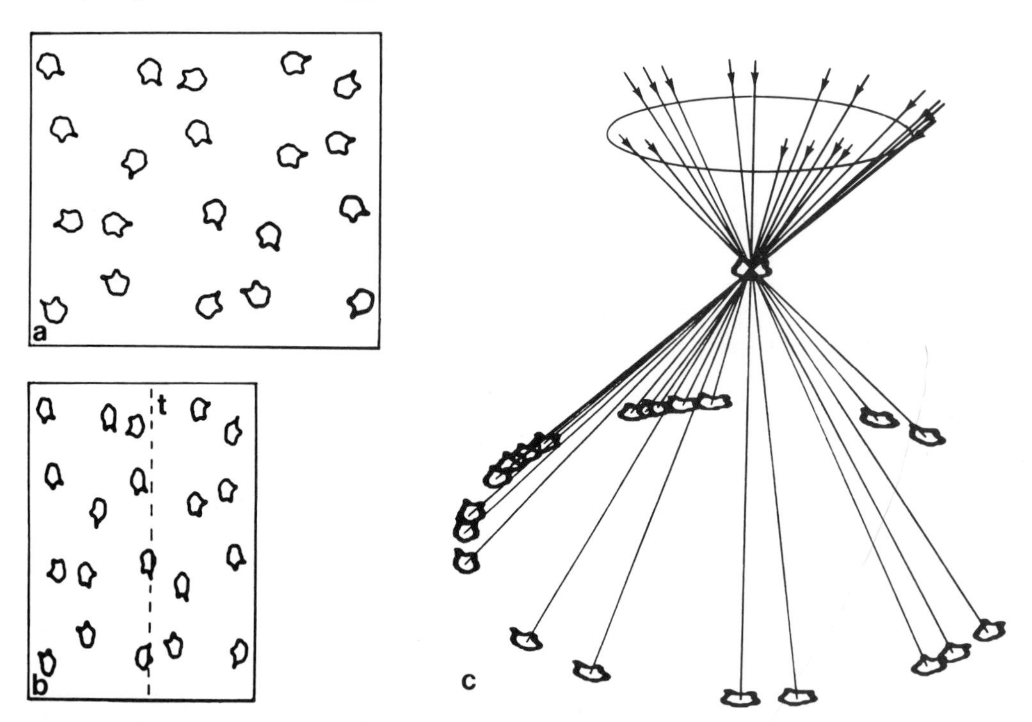

Fig. 1.5 Basic principle of the reconstruction scheme. (*a*) View of a specimen with randomly oriented 50S particles lying flat in the plane of the specimen. (*b*) Projection of the specimen in (*a*), tilted by 50°. The images that can be extracted from the tilted image (*b*) form the conical tilt series shown in (*c*), equivalent to a tilt series of a single particle with random projection directions, all lying on the surface of a cone. Reprinted with permission from M. Radermacher, et al., *Journal of Microscopy,* 146:113–136. Copyright © 1986 Royal Microscopical Society, Oxford, England.

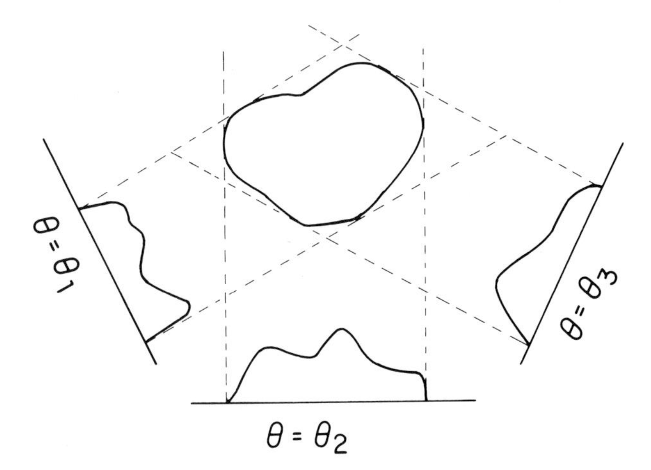

Fig. 1.6 Illustration of the back-projection method of three-dimensional reconstruction. In any plane perpendicular to the tilt axis, the optical density profiles measured along corresponding lines of the projections are smeared out into the direction of projection and are added up to form a slice of the three-dimensional object to be reconstructed. Reprinted with permission from J. Frank, et al., *New Technologies in Studies of Protein Configuration.* Copyright © 1985 Van Nostrand Reinhold, New York, NY.

Scanning Tunneling Microscopy The scanning tunneling microscope (STM) has been applied to the study of biological surfaces to a very limited extent, since it is a relatively new application. However, its potential is considerable since it allows observations with atomic resolution (see Zadzinski, 1989).

The method requires very thin molecular layers attached to a conductive substrate or alternatively coated with a conducting surface. A metal tip is brought close to the surface and operates at moderate voltages (in contrast to those in electron microscopes), in the range of 2 mV to 2 V. Electrons tunnel between the tip and the conducting surface. A feedback system maintains the current constant by changing the height of the tip (alternatively, the current can be allowed to vary and the distance maintained constant). The image is a map of the tip height in relation to the lateral position of the probe.

B. Isolation of Cell Components

The isolation of cell components has also led to enormous progress in our understanding of the molecular organization of cells. Basically, this approach depends on rupturing the cell in a medium that is compatible with preservation of the integrity of the organelle under study, followed by physical separation of the various cell components. Generally, the medium is a solution that is isosmotic or hyperosmotic in relation to the cell interior; chilled sucrose solutions are often used.

Cells and tissues are usually broken down mechanically. Bacteria and plants present unique problems because of their protective capsules or walls. The capsules or walls can be broken down by enzymatic digestion, or can be prevented from forming by use of special conditions (e.g., using a mutant with a defective capsule). Without such coverings, homogenization can be relatively gentle. For example, with the homogenization device shown in Fig. 1.10, up-and-down motion of the plunger forces the cells to squeeze between plunger and vessel wall. The homogenizer can be maintained in a ice bath to dissipate any of the heat generated and to minimize the effect of hydrolytic enzymes.

The most widely used methods for large-scale separations involve centrifugation. The

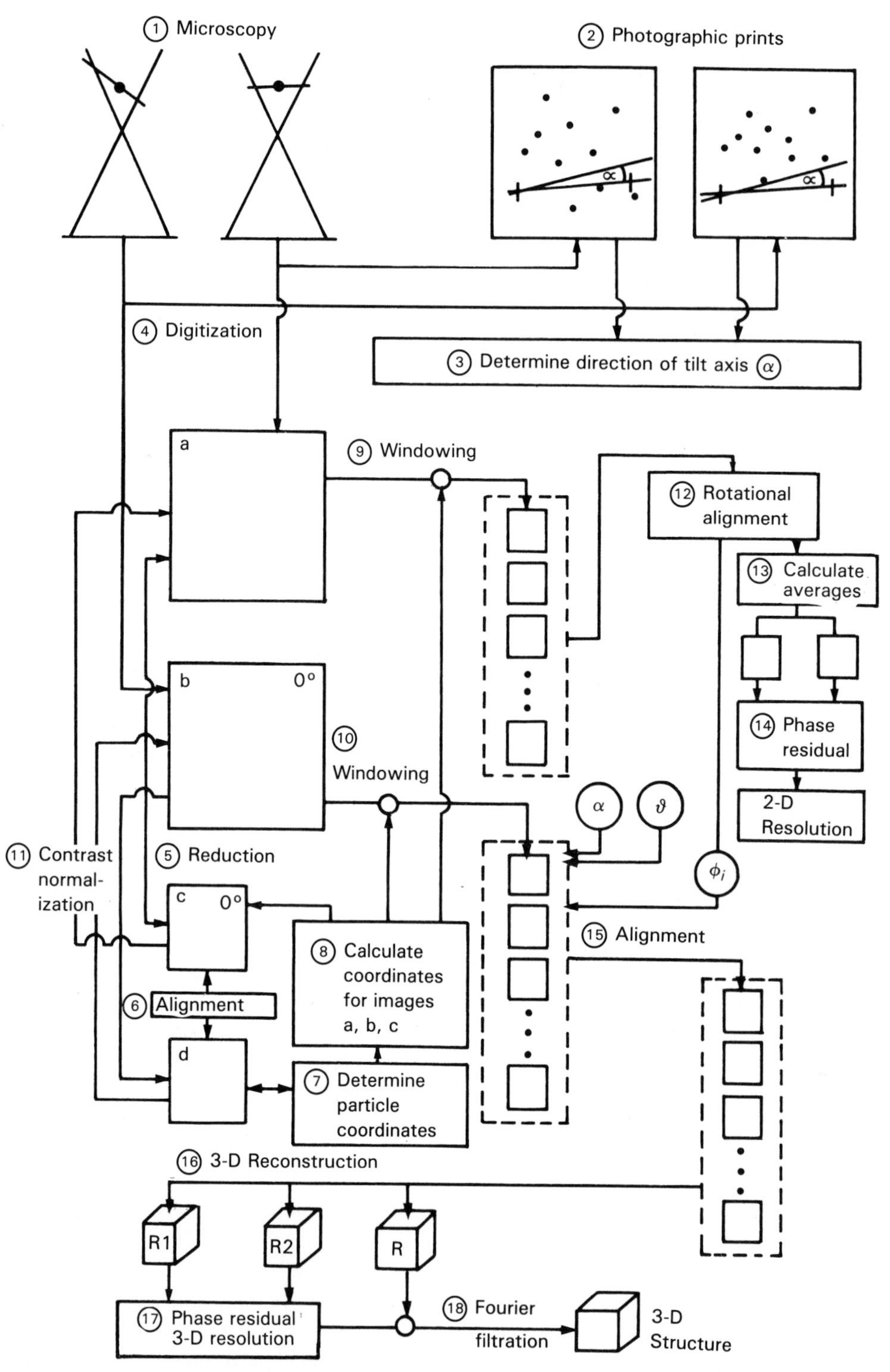

Fig. 1.7 Flow diagram outlining the complete reconstruction procedure. Reprinted with permission from M. Radermacher, et al., *Journal of Microscopy,* 146:113–136. Copyright © 1986 Royal Microscopical Society, Oxford, England.

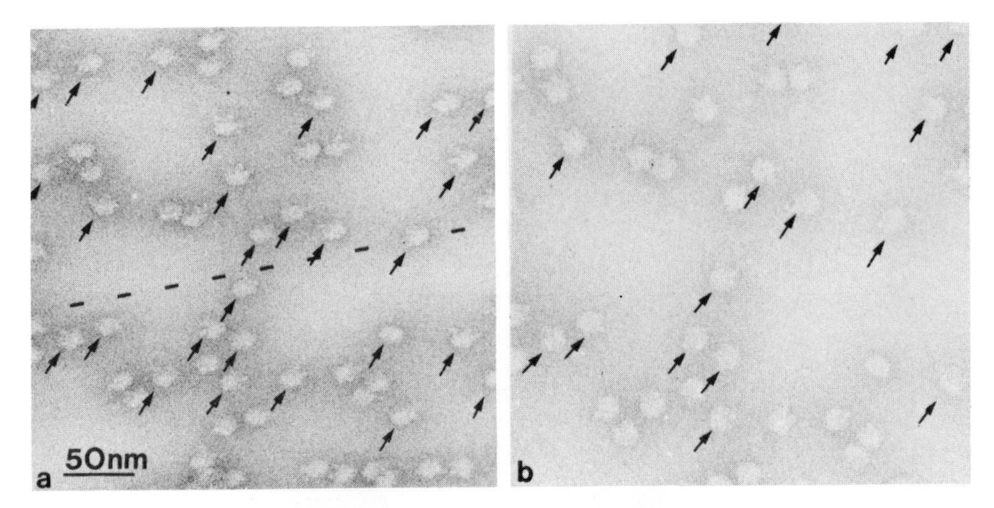

Fig. 1.8 Portion of the particle fields. (*a*) Projection of the specimen tilted by 50°; the particles initially selected are marked at the lower left corner. (*b*) The same field without tilt; the particles finally used are marked. Only the area common to both micrographs and lying in the underfocus range in the tilted view (as determined by optical diffraction) was used for evaluation to avoid errors due to the sign change of the transfer function at low spatial frequencies. Average densities calculated in areas showing only carbon foil were used to normalize the contrast of the images. Reprinted with permission from Radermacher et al. (1986).

homogenate is placed in a plastic or heavy-walled test tube and then centrifuged. The rotation generates a centrifugal force, which increases with the speed of rotation and the radius of the rotor. The rate of sedimentation of a particle is given by Stokes' law, shown in Eq. (1.1):

$$\frac{dx}{dt} = \frac{KG(d_p - d_m)r^2}{\eta} \tag{1.1}$$

In this equation, x is the displacement in units of length, t is time, G is the force exerted on the particle in gravitational units, d_p is the density of the particle and d_m that of the medium, r is the radius of the particle, and η is the viscosity of the medium. For a sphere and when distance is expressed in centimeters, η in poises, density in grams per cubic centimeter, and time in seconds, K corresponds to 2/9. The centrifugal force is a function of the rotations per minute (rpm) of the rotor as represented in Eq. (1.2), where L is the radius of the rotor in cm

$$G = 11.17L\left(\frac{\text{rpm}}{1000}\right)^2 \tag{1.2}$$

Equation (1.1) shows that particles differing in r and d_p could be isolated by sedimenting them sequentially by increasing the speed with each centrifugation, as done in *differential centrifugation*. In contrast, when d_p is lower than the density of the medium ($d_p < d_m$), dx/dt would be negative and the particle would be displaced upward.

In differential centrifugation, after each centrifugation the fraction is collected from the

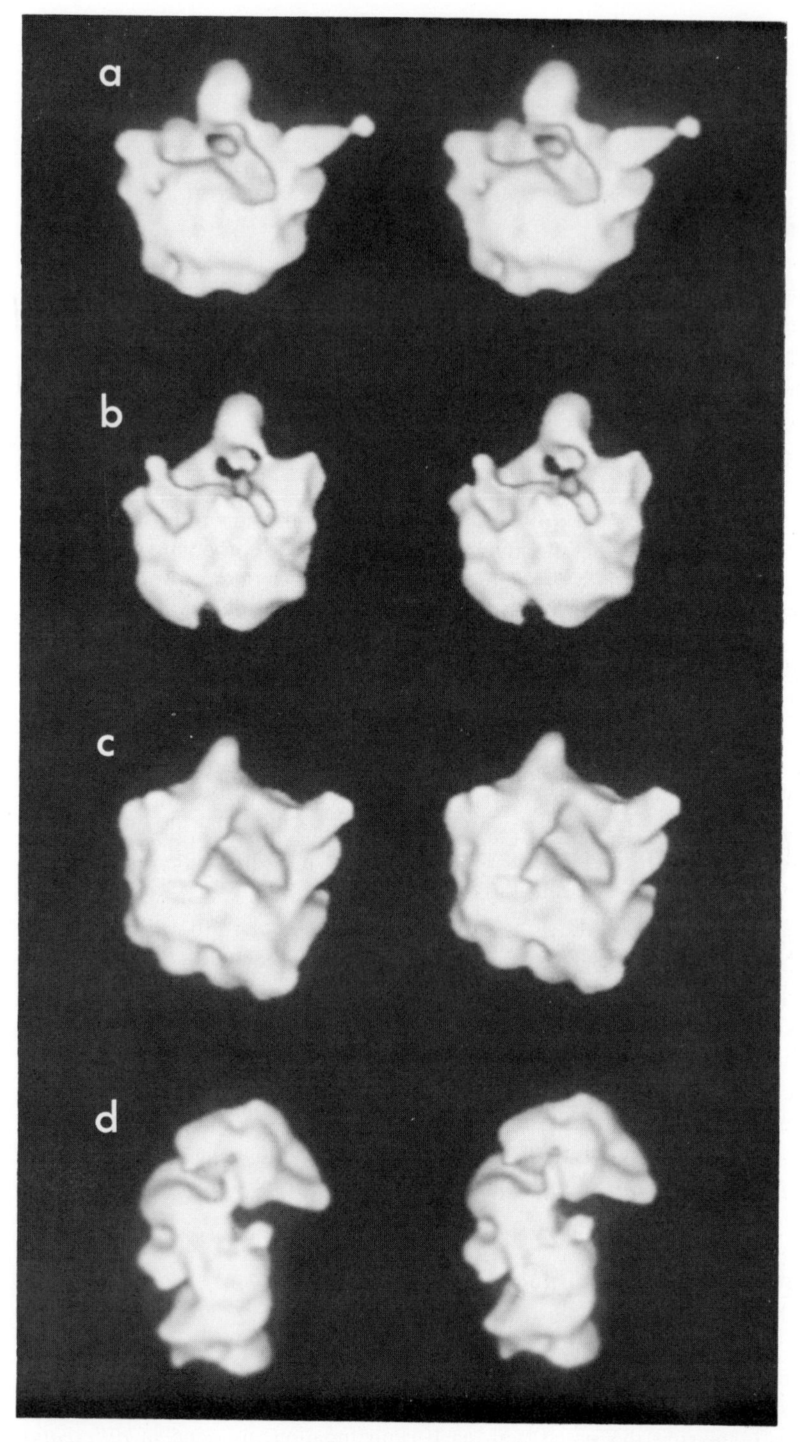

Fig. 1.9 Gallery of ribosomes and ribosomal subunits reconstructed with the new technique, presented as stereo pairs (scale bar, 10 nm). (*a*) 50S ribosomal subunit of *E. coli*, oriented with its interface surface toward the viewer. (*b*) 50S ribosomal subunit of *E. coli* depleted of L7/L12 proteins. Orientation as in (*a*). (*c*) 70S monosome of *E. coli* oriented so that the 50S domain matches position with (*a*) and (*b*). (*d*) 40S ribosomal subunit from rabbit reticulocytes in a lateral-view orientation. Reprinted with permission from Frank et al. (1988).

sedimented material (*pellet*) and the material that remains in suspension (*supernatant*) is recentrifuged at a higher speed. The larger or denser particles sediment in earlier centrifugations. This process is illustrated in Fig. 1.10. Differential centrifugation cannot separate cell components into pure fractions in a single series of centrifugations, since particles will sediment according to their position in the tube rather than just size or density. However, repeated resuspensions and resedimentations can produce a satisfactory level of purity. The effect of position in the tube on isolation can be avoided by layering the homogenate on top of a layer of a denser solution. This process approximately equalizes the distance of travel for all particles. Layers or gradients can be produced by varying the concentration of sucrose or of a polymer (e.g., dextran or Ficoll) in the layers, with the denser layers in the bottom.

Separation with high resolution can be obtained by centrifuging through layers or gradients, as in *density gradient centrifugation*. As a particle approaches a layer of identical density ($d_p = d_m$), the rate of sedimentation decreases [Eq. (1.1)] and eventually, at equilibrium, sedimentation stops. The separation will depend on density to varying degrees, depending on the method chosen. With equilibrium centrifugation it will depend exclusively on the density, not the size of the particle. Not surprisingly, density gradient centrifugation has been found most useful for separating organelles of approximately the same size but different densities—for example, separating lysosomes from mitochondria.

Centrifugation with a two-phase system has also been found useful. Particles lighter than the bottom layer are rejected and remain at the interface.

Centrifugation techniques require refrigeration, since the centrifugation can generate a good deal of heat. In addition, to reach high speeds the friction has to be decreased by use of a partial vacuum around the rotor.

Subcellular particles can also be separated by taking advantage of their surface properties. Many aqueous solutions containing two polymers, e.g., methylcellulose and dextran, separate out into phases. If the solutions have been mixed with subcellular components, viruses, or macromolecules, these substances will tend to favor one of the two phases or the interface between the two.

The isolation techniques have made available large quantities of material for study by standard biochemical techniques. However, in some cases, lack of purity and possible disruption of native organelles still pose a problem.

The microscopic and biochemical techniques discussed have individual advantages and disadvantages and have been used together to produce a complex picture of cellular organization.

C. Cell Cultures

The use of cell cultures opened a new phase in the study of cells. There are very few sections in this book in which at least some of the experiments cited were not carried out with cells in culture.

Cells to be cultured are first dissociated from the tissue, most commonly by using proteolytic enzymes to detach them from the protein material that anchors them to other cells or extracellular matrix. Alternatively, fibroblasts can be allowed to grow out of a piece of tissue, which is subsequently removed. The cells are placed on plates and provided with appropriate conditions, such as temperature, nutrients, and protein fractions derived from serum, which probably provide

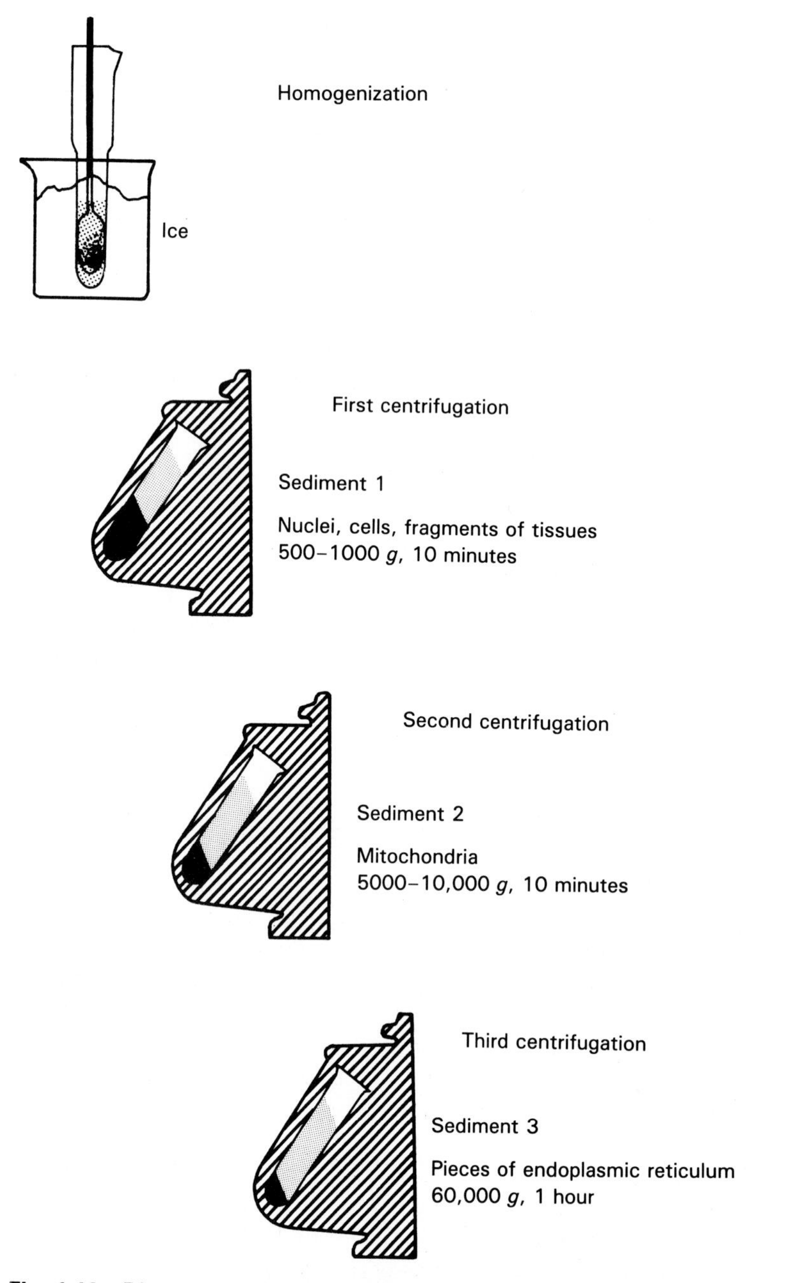

Fig. 1.10 Diagrammatic representation of differential centrifugation.

growth factors. The isolated cells display many of the characteristics of their differentiated state. Generally, in order to grow and divide they must have an attachment surface. Cells attached to an appropriate surface may divide until they become confluent. The contact between cells arrests their replication (*contact inhibition*), and a monolayer of cells forms.

Cultures directly derived from a tissue are referred to as *primary cultures*. When *subcultured,* that is, removed from the original culture and cultured in a new medium at a lower concentration, the cells continue to divide. The cells can be maintained by repeated subculturing to form a cell line. However, all normal vertebrate cells can undergo only a finite number of divisions. In contrast, transformed or malignant cells do not have this limitation and, if maintained under the appropriate conditions, continue to proliferate to a much higher concentration than normal cells, since they do not exhibit contact inhibition. Transformed cells will also grow in suspension without an attachment surface. Cells can be transformed by infection with oncogenic viruses or treatment with certain chemicals. Alternatively, malignant cells can be cultured directly from tumors. One of the most commonly used cell lines, the HeLa cell line, was originally derived from cervical cancer cells from a single patient.

Cells can be stored frozen at liquid nitrogen temperature in the presence of cryoprotective agents.

Cells from recognized clones can be obtained from the American Type Culture Collection (Rockville, Maryland) or the Human Genetic Mutant Cell Repository, located at the Coriell Institute for Medical Research (Camden, New Jersey) and have been cataloged (e.g., Catalog of Cell Lines and Hybridoma 1988, American Type Culture Collection).

II. TECHNIQUES OF MOLECULAR BIOLOGY

A. Recombinant DNA Techniques

The study of the macromolecules involved in cell function has been revolutionized by the development of recombinant DNA technology. Combinations of unrelated genes can be created, cloned, and amplified. Furthermore, the production of complementary DNA (cDNA, discussed below), its cloning to produce large amounts of material, and subsequent sequencing have allowed the study of proteins that could not be readily studied before.

Constructing New DNA A novel piece of DNA can be constructed by cleaving the DNA or a piece of DNA (which we will call *donor* DNA) with a *restriction enzyme*. Restriction enzymes or endonucleases are found in a variety of prokaryotes. They recognize specific base sequences in double-helical DNA and cleave them in such a way that the new ends are single stranded and complementary to each other; that is, two complementary strands are cut in slightly different positions, producing staggered ends (see Fig. 1.11*a*). The same restriction enzyme can be used to cut *vector* DNA, creating staggered ends with precisely the same sequence so that they can be annealed to the donor DNA. Vectors are phages or plasmids—independently replicating nonchromosomal DNA. Treatment with DNA ligase will then join the two kinds of DNA, the vector and the donor piece (Fig. 1.11*b*). Similarly, a small, artificially synthesized DNA piece can be covalently joined to either vector or donor DNA.

The covalent joining of a DNA fragment to a DNA vector permits *cloning* of the fragment, that is, the production of many identical copies. DNA can be introduced into cells by a variety of techniques, including infection with a virus containing the donor DNA. The modified DNA replicates in host cells. A commonly used host for cloning is *Escherichia coli,* the intestinal bacterium that has already provided us with a wealth of genetic and biochemical information.

The cells containing the donor DNA can be selected by various procedures. Most simply, the DNA is fused to a vector containing a marker, such as antibiotic resistance; then, in the presence of the antibiotic, only the cells expressing the marker survive. Alternatively, cells from each clone can be tested for the desired DNA sequence by Southern blotting (using a labeled complementary DNA; see below) or by assaying for a protein coded for by the new DNA, either by enzymatic assay if the protein is an enzyme or with a specific antibody (Western blotting; see below).

Selected genes can be cloned from the DNA representing the complete genome of a cell

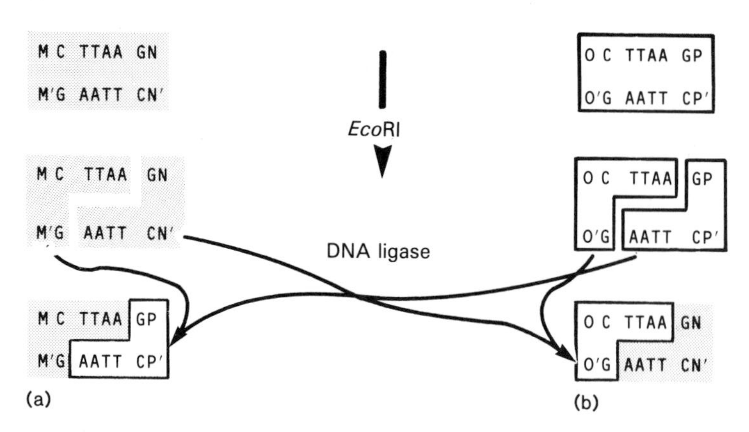

Fig. 1.11 Production of a novel DNA. Only the staggered ends are shown. (*a*) Production of staggered ends by endonuclease. (*b*) Joining of the two distinct DNAs by DNA ligase.

type. The DNA is fragmented by shearing or by use of restriction endonucleases and is separated out by electrophoresis to select pieces about 20 kilobases (kb) long. When the DNA pieces are linked to vectors such as lambda phage DNA and the infective phage is reconstituted in vitro, they can be used to infect *E. coli* and replicate repeatedly to form a so-called *genomic library*.

Infection and subsequent lysis of *E. coli* cells on a plate by a dilute suspension of phage produce plaques, each corresponding to the progeny of a single phage. When a sheet of nitrocellulose is applied to the plate, some of the DNA from lysed cells sticks to the sheet and forms a replica. After denaturation of the DNA with OH^-, specific DNA sequences can be identified by hybridization to radioactive DNA or RNA *probes* (Southern or Northern blotting, see next section). The hybridization sites can be recognized by autoradiography and the corresponding plaques still present in the original plate can be picked out and grown to form millions of DNA clones.

Amplification by synthesis of DNA using the *polymerase chain reaction* (PCR) with a selected DNA serving as a template provides an alternative to cloning (Saiki et al., 1988). Single genomic sequences as large as 2 kb have been amplified more than 10 million times with a high-specificity, heat-resistant DNA polymerase extracted from a thermophilic bacterium. A thermostable polymerase is advantageous, since newly synthesized strands have to be separated (generally done by heat) before the next round of replication.

DNA probes can be produced from the corresponding messenger RNA (mRNA). The mRNA can be isolated from cells that produce predominantly a single protein (e.g., reticulocytes, which synthesize hemoglobin). Alternatively, the mRNA can be extracted from polysomes isolated by precipitation with antibodies against the protein in the process of being synthesized. Then cDNA can be synthesized using reverse transcriptase and the mRNA as a template. Reverse transcriptase is an enzyme found in retroviruses, in which it synthesizes DNA complementary to genomic RNA during genomic replication. The cDNA, after attachment to appropriate vectors, can be efficiently replicated in *E. coli* to produce the probe molecule in large amounts or can be expressed to produce the corresponding protein, a technique that has allowed the production of large quantities of specific proteins. Small oligonucleotide pieces synthesized to correspond to the protein's amino acid sequence can also serve as probes.

Site-specific mutagenesis has obvious advantages over conventional mutagenesis, since the latter is a random rather than a directed process. Altering single nucleotides in isolated DNA permits synthesis of the modified DNA. In practice, this can be accomplished by first synthesizing a small oligonucleotide piece complementary to a short sector of the modified DNA containing the nucleotide change. The rest of the strand can then be synthesized by DNA polymerase, which uses the small oligonucleotide as a primer. Site-specific mutagenesis has been used to modify proteins in a controlled and systematic manner in the study of mechanisms of catalysis and the relationship between protein structure and function.

In *transfection,* the genetic information contained in a vector is incorporated in the host's DNA. When the cell is a fertilized egg, the transfer can produce *transgenic* organisms that will pass the newly acquired gene from generation to generation.

Recombinant DNA techniques have revolutionized genetics, the study of gene expression, and our understanding of the functioning of protein molecules by making it possible to produce or modify genes, change their genetic environment, and change proteins. Used to produce large amounts of proteins and other compounds, recombinant DNA has entirely changed the study of their physiological role and therapeutic use.

Recognition Techniques DNA and fragments obtained by restriction endonuclease treatment of DNA can be separated out by polyacrylamide or agarose gel electrophoresis (see below). The positions of the bands can be found by using autoradiography when the DNA has been labeled, usually with ^{32}P or alternatively by staining with dyes such as the fluorescent ethidium bromide.

Specific base sequences can be recognized by hybridization techniques in which double-stranded DNA is denatured and the resulting single strand is annealed to a *probe*. Such a probe is a specific sequence of DNA, usually radiolabeled so that its presence, and hence that of complementary sequences, can be recognized by autoradiography. In the laboratory, the transfer of denatured DNA bands onto nitrocellulose sheets before annealing proved most practical. This technique is known as *Southern blotting*. The same technique applied to the recognition of RNA sequences is referred to as *Northern blotting*. Recognition of a protein by using a specific antibody as a probe is referred to as *Western blotting*.

B. Recognition and Separation of Proteins

Recognition and separation of proteins have played an important role in the study of cells.

Use of Antibodies Injection of an antigen into an animal elicits the production of antibodies originating from different antibody-producing cells, each responding to a different determinant in the antigen molecule. However, a single cell or progeny of that cell (i.e., a clone) will synthesize only a *monoclonal* antibody, specific for one or a few closely related determinants (epitopes). Techniques that take advantage of these properties have been developed using *myeloma* cells. *Multiple myeloma* is a malignancy of antibody-producing cells. Many cells (a clone) can be generated from the proliferation of a single myeloma cell. In practice, antibody-producing cells are fused to myeloma cells. Lymphocytes and plasma cells are obtained from the spleen of a mouse that has been immunized against an antigen. They are then fused to myeloma cells by special procedures. The hybrid cells (*hybridoma cells*) proliferate in vitro, in contrast to normal cells such as nonfused cells from the spleen of the immunized mouse. The hybrid cells can be selected from the unfused malignant cells if the myeloma cells used in the fusion had a biochemical defect so that only the fused cells containing the wild-type genome of the mouse cells can survive. In the standard techniques, the defect is absence of the enzyme hypoxanthine-guanine phosphoribosyltransferase (HGPRT).

Monoclonal antibodies can be used to recognize specific groups in proteins and, when incorporated in affinity columns, to aid in the isolation of a specific protein from a mixture (see below).

In studying the location of specific components of cells, either heterogeneous or monoclonal antibodies are conjugated to markers. For light microscopy, a fluorescent dye is frequently used. Submicroscopic markers such as ferritin or colloidal gold are used with electron microscopy. Immunological techniques have provided valuable information about cell structure and function.

Other Protein Probes In addition to the DNA and antibody probes already discussed, *affinity* probes can be used in the study of binding sites to specific ligands in enzymes or other proteins. Affinity labels resemble the natural ligand, but they bind covalently. *Photoaffinity* labels can be used in a similar manner. They are generally analogues of the natural ligand, usually azido compounds that are stable in the absence of light. When photolyzed by exposure to light of the appropriate wavelength, they form highly reactive groups that bind to proteins.

Separation Techniques Many methods have been used to isolate, identify, and purify proteins. Some of these stand out for their usefulness or frequency of use.

Antibodies can be used to precipitate (*immunoprecipitate*) specific proteins. On incubation of a mixture with a specific antibody, the antigen is bound. Multivalent antigens and multivalent antibodies form large complexes, which precipitate. With the exception of monoclonal antibodies, serum contains a variety of antibodies for different determinants in the antigen molecule. Precipitation can also be produced by adding *Staphylococcus* A cells. These cells contain at their surface a protein that binds the constant region of most antibodies, producing very large aggregates. Obviously, specific antibodies cannot be produced unless a pure protein is available first to serve as a specific antigen.

The techniques used for the purification of proteins are varied. Some of the classical approaches depend on the solubility of proteins under a variety of conditions and concentrations of salts (such as ammonium sulfate) or binding of proteins to columns.

In ion-exchange chromatography, proteins are separated according to their charge. An ion exchanger, either an organic cation or anion, is bound to supportive material. The proteins separate out by binding differentially to the ion exchanger. Since the charge of the protein depends on its degree of ionization, the protein charge and hence the degree of separation depend on pH.

Many other techniques are based on separation of proteins by size. Among many others, they include gel filtration, density gradient centrifugation, and sodium dodecyl sulfate-polyacrylamide gel electrophoresis (SDS-PAGE).

In gel filtration, separation of proteins depends on passage of the soluble mixture through a column containing hydrated carbohydrate beads. The smaller the protein, the larger the portion of the water in which it will be distributed, so the larger molecules will migrate faster through the gel. The rate of passage of various protein molecules of known size allows a molecular weight calibration that can be used to size other proteins.

In electrophoresis, proteins are separated in an electric field because of their charge. When a polyacrylamide gel is used, it acts as a restrictive matrix that also separates proteins by their size. In SDS-PAGE, the proteins are denatured and reduced (to eliminate sulfhydryl bridges) so that the component polypeptides are separated. Furthermore, they are coated with the detergent SDS, which is negatively charged. Ideally, this coating is evenly distributed, so that the proteins will separate electrophoretically by size only. The protein bands are stained or otherwise made visible (e.g., by autoradiography if they were radioactively labeled).

Affinity chromatography is potentially one of the most powerful techniques. In this technique a specific ligand (e.g., a cofactor, substrate, or antibody) is attached to a matrix. When an extract containing proteins is passed through the column, only the targeted protein remains in the column and can be subsequently eluted (e.g., by free ligand).

C. Sequencing of Proteins and Computer Analysis

Techniques are available for sequencing proteins; in fact, the methodology has been automated. In practice it is easier to sequence DNA because of the capacity to multiply the DNA either through cloning or PCR so that the amino acid sequence is most frequently deduced from the nucleotide sequence. The methodology has been standardized to the extent that it is possible to commercially contract for deriving sequencing information from purified compounds. Conversely, it is possible to synthesize peptides or polynucleotides.

A known sequence of a protein can be compared with others by the use of software and data bases presently available (e.g., see Devereux et al., 1984). Similarly, it is possible to search for particular motifs in proteins (or nucleic acids) through computer searches. One of the interesting applications of this approach is the deduction of function by identifying amino acid sequence domains that are in common with proteins of known function.

III. ORGANIZATION OF THE CELL: A PRELIMINARY VIEW

A photomicrograph of a spread-out living fibroblast obtained with a phase-contrast microscope is shown in Fig. 1.12 (Buckley and Porter, 1967). The technique shows a good deal of cellular structure but without much detail. The nucleus is clearly visible, and lamellar elements that are known to be associated with the Golgi apparatus can also be distinguished. There are various cell inclusions or organelles, filamentous mitochondria, spherical lysosomes, and fibers called stress fibers present in focal contacts (see Chapter 3, V, D; Chapter 17, IV, D). Many variations on this basic plan occur. Some cells have specialized surfaces and specialized ends. For example, secretory cells such as the acinar cells of the pancreas, which we will have occasion to discuss later, are polar—that is, they have distinct cell surfaces. The luminal side corresponds to the

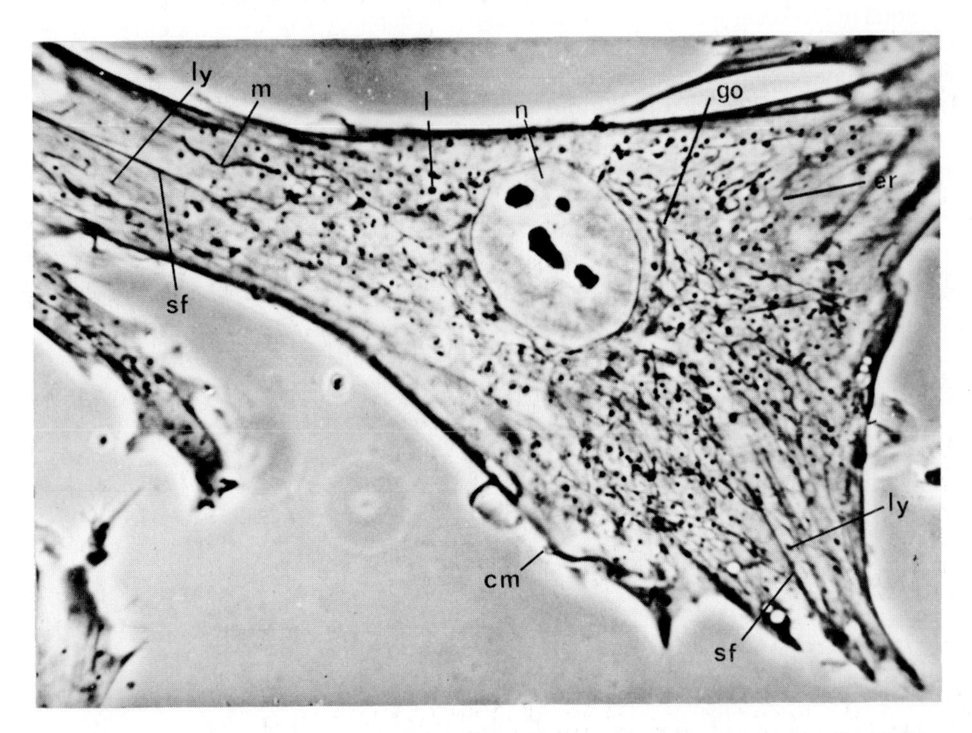

Fig. 1.12 Phase-contrast photomicrograph of living cultured rat embryo cell with its nucleus (n) surrounded by Golgi apparatus (go), endoplasmic reticulum (er), mitochondria (m), lipoprotein inclusions (l), and small vacuoles, which are probably lysosomes (ly). In several places around the cell margins, the cell membrane (cm) is engaged in undulating movements, which at the lower cell process are associated with the formation of pinocytotic vacuoles. A small number of phase dark lines, the intracytoplasmic stress fibers (sf), are seen within the cell processes (×675). Reprinted with permission from Buckley and Porter (1967).

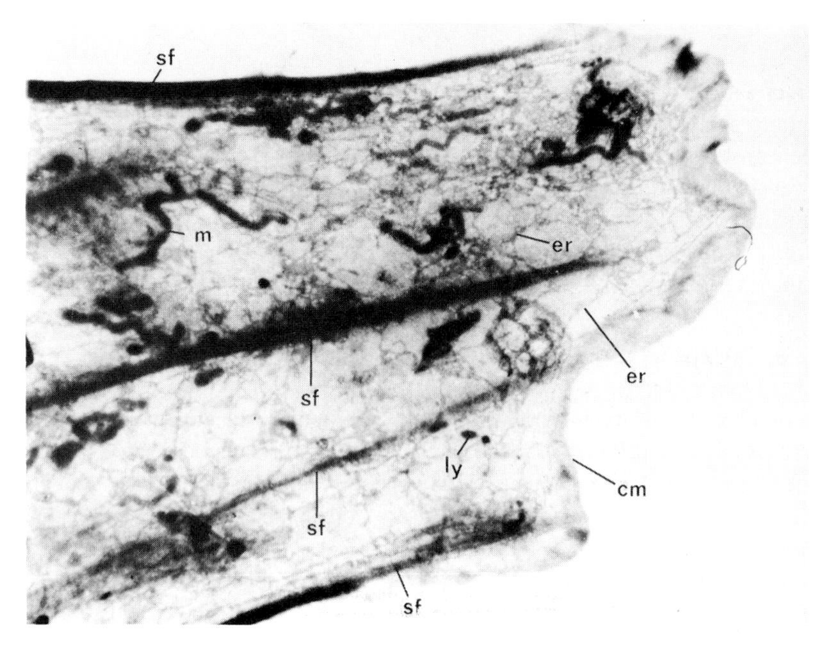

Fig. 1.13 Electron micrograph of projecting process of an unsectioned glutaraldehyde-osmium–fixed cultured rat embryo cell showing, within the cell margin (cm), filamentous mitochondria (m), lysosomes (ly), the network of the endoplasmic reticulum (er), and dense linear structures, the stress fibers (sf). Note that the images of the stress fibers here are similar to the phase-contrast images of stress fibers in the living cell (×4900). Reprinted with permission from Buckley and Porter (1967).

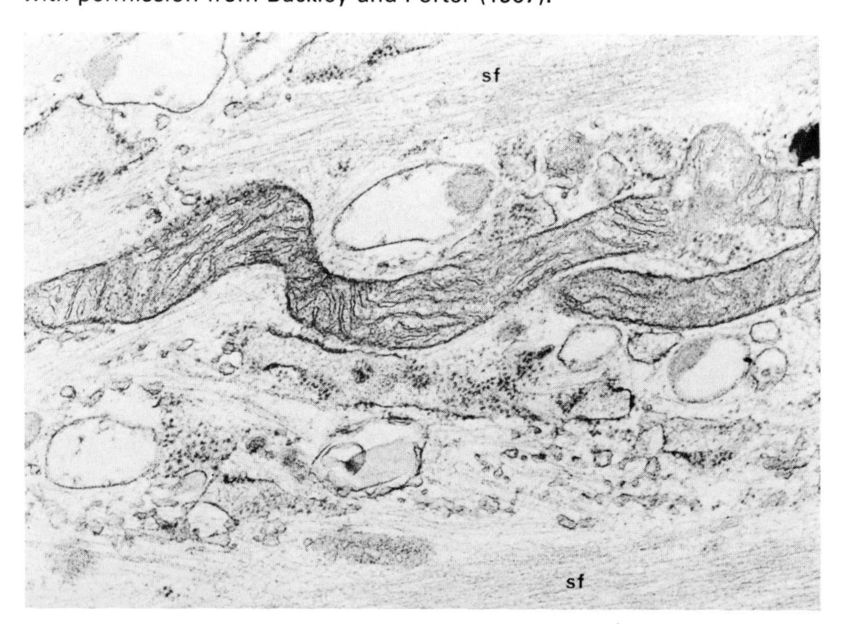

Fig. 1.14 Section through cultured rat embryo cell showing cytoplasmic structures in greater detail. Centrally, a branching mitochondrion is surrounded by profiles of the endoplasmic reticulum, some of which are studded with ribonucleoprotein particles and contain a somewhat granular material. Flanking this cluster of organelles are portions of sectioned stress fibers (sf), which appear as bundles of near-parallel filaments of about 7.5 nm diameter (×26,000). Reprinted with permission from Buckley and Porter (1967).

Table 1.1 Functional Distribution in the Eukaryotic Cell

Metabolic reaction or composition	Location	Role in the cell	Chapter
Glycolysis and its regulation	Cytoplasm	Major role	8
	Nucleus	Minor role	
Oxidative metabolism of carbohydrates, lipids, and amino acids; oxidative phosphorylation; fatty acid synthesis	Mitochondria		10, 12
Photosynthesis	Chloroplasts (chromatophores in bacteria)		11, 12
Oxidative shunt	Peroxisomes	In plants it may have a role in the synthesis of special compounds and in photorespiration	10
Fat metabolism in plants	Glyoxysomes	Fatty acid metabolism in germinating seeds	10
Oxidation of drugs, pesticides, and some metabolites	Endoplasmic reticulum	Unsaturation of fatty acids, disposal of drugs	10
Lipid synthesis	Endoplasmic reticulum	Major role	
Genetic information and presence of DNA	Nucleus	Major role	
	Mitochondria and chloroplasts	Minor role	
Synthesis of macromolecules and their control	Nucleus, organelles, cytoplasm		9
	Endoplasmic reticulum		4
Intracellular digestion	Lysosomes, phagosomes	Major role	4
Communication inside cell			2, 5
Communication between cells	Cell surface		2, 3
Transport of secretory products and posttranscriptional modification	Endoplasmic reticulum, Golgi cisternae		4
Movement	Cilia, flagella, cytoplasm		17, 18
Passage of solutes through membranes			2, 13–15

surface at which the secretory products are discharged. In addition to cell polarity, the kinds of organelles present or their relative proportions may differ from one cell type to another or from one organism to another.

Transmission electron micrographs of spread-out cells (Fig. 1.13) (Buckley and Porter, 1967) without sectioning add little to the phase micrographs. However, sectioning and magnification provide considerable detail (Fig. 1.14). The vacuoles are clearly vesicles, and the mitochondria show both an inner and an outer membrane component.

Cell structures have been presented in this section very sketchily. They will be considered in more detail later. Table 1.1 summarizes some of the biochemical compartments of the cell.

SUGGESTED READING

Light Microscopy

Allen, R. D. (1985) New observations on cell architecture and dynamics with video-enhanced contrast optical microscopy. *Annu. Rev. Biophys. Chem.* 14:265–290.

Allen, R. D., and Stromgen Allen, N. (1983) Video-enhanced microscopy with a computer frame memory. *J. Microsc.* 129:3–17.

Inoué, S. (1986) *Video Microscopy*. Plenum, New York. See Chapter 5, pp. 93–148; Chapter 6, pp. 149–186; Chapter 11, pp. 393–418.

Salmon, T., Walker, R. A., and Pryer, N. K. (1989) Video-enhanced differential interference contrast light microscopy. *BioTechniques* 7:624–633.

Shuman, H., Murray, J. M., and Di Lullo, C. (1989) Confocal microscopy: an overview. *BioTechniques* 7:154–163.

Spencer, M. (1982) *Fundamentals of Light Microscopy*, p. 93. Cambridge University Press, London.

Wick, R. A. (1989) Photon counting imaging: applications in biomedical research. *BioTechniques* 7:262–269.

Transmission and Scanning Electron Microscopy

Meek, G. E. (1977) *Practical Electron Microscopy for Biologists*. Wiley, New York.

Nagatani, T. (1989) The ultra-high resolution scanning electron microscope and some applications to biological studies. *BioTechniques* 7:270–275.

Cryoelectron Microscopy

Dubochet, J., Adrian, M., Lepault, J., and McDowall, A. W. (1985) Cryo-electron microscopy of vitrified biological specimens. *Trends Biochem. Sci.* 10:143–146.

High-Voltage Electron Microscopy

King, M. V., Parsons, D. F., Turner, J. N., Chang, B. B., and Ratkowski, A. J. (1980) Progress in applying the high-voltage electron microscope to biomedical research. *Cell Biophys.* 2:1–95.

Image Processing

Frank, J. (1989) Three-dimensional imaging techniques in electron microscopy. *BioTechniques* 7:164–173.

Frank, J., McEwen, B. F., Radermacher, M., Turner, J. N., and Rieder, C. L. (1987) Three-dimensional tomography reconstruction in high voltage electron microscopy. *J. Electron Microsc. Tech.* 6:193–205.

Frank, J., Verschoor, A., Wagenknecht, T., Radermacher, M., and Carazo, J.-M. (1988) A new non-crystallographic image-processing technique reveals the architecture of ribosomes. *Trends Biochem. Sci.* 13:123–127.

Stewart, M. (1986) Computer analysis of ordered microbiological objects. In *Ultrastructure Techniques for Microorganisms* (Aldrich, H. C., and Todd, W. J., eds.), pp. 333–364. Plenum, New York.

Scanning Tunneling Microscopy

Zadzinski, J. A. N. (1989) Scanning tunneling microscopy with applications to biological surfaces. *BioTechniques* 7:174–187.

Cell Culture

Adams, R. L. P. (1982) *Cell Culture for Biochemists*, pp. 1–292. Elsevier Science Publishers, New York.

Recombinant DNA Techniques

Berg, P. (1981) Dissection and reconstruction of genes and chromosomes. *Science* 213:296–303.

Berger, S. L., and Kimmel, A. R. (1987) Guide to molecular cloning. *Methods Enzymol.* vol. 152.

Botstein, D. and Shortle, D. (1985) Strategies and application of *in vitro* mutagenesis. *Science* 229:1193–1201.

Glover, D. M., ed. (1985) *DNA Cloning,* 3 vols. IRL Press, Oxford and Washington D.C.

Innis, M. A., Gelfand, D. H., Sninsky, J. J., and White, T. J., eds. (1990) *PCR Protocols, A Guide to Methods and Applications.* Academic Press, San Diego, Calif.

Sambrook, J., Fritsch, E. F., and Maniatis, T., eds. (1989) *Molecular Cloning—A Laboratory Manual,*

Watson, J. D., Gilman, N., Witowski, J., and Zoller, M. (1992) *Recombinant DNA,* Second Edition, Scientific American Books (distributed by W. H. Freeman and Co., New York). See Chapters 5–14.

Immunological Techniques

Milstein, C. (1980) Monoclonal antibodies. *Sci. Am.* 243(3):66–74.

Campbell, A. M. (1984) *Monoclonal Antibody Technology.* Elsevier, Amsterdam.

Isolation of Cell Components

Fleischer, S., and Packer, L., eds. (1974). *Methods in Enzymology,* Vol. 31, Part A. Academic Press, New York.

McNamee, M. G. (1989) Isolation and characterization of cell membranes. *BioTechniques* 7:466–475.

REFERENCES

Amos, L. A., Henderson, R., and Unwin, P. N. T. (1982) 3-D determination by electronmicroscopy of 2-D crystals. *Prog. Biophys. Mol. Biol.* 39:183–231.

Benz, R. (1985) Porin from bacterial and mitochondrial outer membranes. *CRC Crit. Rev. Biochem.* 19:145–190.

Buckley, I. K., and Porter, K. R. (1967) Cytoplasmic fibril in living cultured cells. A light and electron microscope study. *Protoplasma* 64:349–390.

Devereux, J., Haeberli, P., and Smithies, O. (1984) A comprehensive set of sequence analysis programs for VAX. *Nucleic Ac. Res.* 12:387–395.

Frank, J., Verschoor, A., and Wagenknecht, T. (1985) Computer processing of electron microscopic images of single macromolecules. In *New Methodologies in Studies of Protein Configuration* (Wu, T. T., ed.), p. 36. Van Nostrand Reinhold, New York.

Frank, J., Verschoor, A., Wagenknecht, T., Radermacher, M., and Carazo, J.-M. (1988) A new non-crystallographic image-processing technique reveals the architecture of ribosomes. *Trends Biochem. Sci.* 13:123–127.

Hoffman, R., and Gross, L. (1975) Modulation contrast microscopy. *Appl. Opt.* 14:1169–1176.

Johnson, K. A., and Wall, J. S. (1983) Structure and molecular weight of the dynein ATPase. *J. Cell Biol.* 96:669–678.

Kachar, B. (1980) Asymmetric illumination contrast: a new method of image formation in video light microscopy. *Science* 227:766–768.

Mannella, C. A. (1982) Structure of the outer mitochondrial membrane: ordered arrays of porelike subunits in outer-membrane fractions from *Neurospora crassa* mitochondria. *J. Cell Biol.* 94:680–687.

Pool, R. (1988) Near-field microscopes beat the wavelength limit. *Science* 241:25–26.

Radermacher, M., Wagenknecht, T., Verschoor, A., and Frank, J. (1986) Three-dimensional reconstruction from a single-exposure, random conical tilt series applied to 50S ribosomal subunit of *Escherichia coli. J. Microsc.* 146:113–136.

Ross, K. F. A. (1967) *Phase Contrast and Interference Biology for Cell Biologists*. St. Martin's Press, New York.

Saiki, R. K., Gelfand, D. H., Stoffel, S., Scharf, S. J., Higuchi, R., Horn, G. T., Mullis, K. B., and Ehrlich, H. A. (1988) Primer directed enzymatic amplification of DNA with thermostable DNA polymerase. *Science* 239:487–491.

White, J. G., and Amos, W. B. (1987) Confocal microscopy comes of age. *Nature* (*London*) 328:183–184.

White, J. G., Amos, W. B., and Fordham, M. (1987) An evaluation of confocal versus conventional imaging of biological structures by fluorescence microscopy. *J. Cell Biol.* 105:41–48.

Chapter 2

The Cell Membranes

The earliest views of the cell membrane were accurate within the technological restraints of the time, which were in part compensated by well-thought-out experiments. New information collected over many years forced some changes in these ideas, and although the interpretations of those early times were not incorrect, not surprisingly many of the early perspectives were not accurate.

The cells of multicellular organisms were considered to be independent units, grouped to form specialized tissues. The membranes at the surface of the cells were thought to be partially responsible for this independence by regulating exchanges between the cells and their environment. Lipids were considered to play the most significant role in membrane structure and function, whereas proteins were thought to be of minor importance. Furthermore, the cell membranes were considered not to participate in the dynamic changes or the biochemical events of the cells. As we will see in the discussion that follows, these views had to be modified in fundamental ways.

Section I describes some of the experiments leading to these early concepts and then proceeds to more recent developments. In Section II the organization of the proteins in the cell membrane is discussed. Section III focuses on how cells communicate through gap junctions, and Sections IV and V examine in more detail the organization and some of the properties of the membrane. Finally, Section VI examines some of the properties of the membranes enclosing the intracellular compartments.

I. THE PLASMA MEMBRANE: THE LIPID BILAYER

The early concept that the lipid at the periphery of the cell is organized predominantly as a bilayer is still valid. This model can be represented as shown in Fig. 2.1. The lines represent hydrophobic chains of the phospholipid molecules and the circles represent the polar of hydrophilic portions. The lateral forces holding the structure together correspond to the van der Waals forces of hydrophobic bonding. More important, water molecules tend to squeeze the nonpolar hydrocarbon chains together because of the high affinity of water molecules for each other. This property favors arrangement of the phospholipids in lamellae or micelles. Depending on the nature of the molecules, the charges of the polar groups may also be involved. This model was originally based mostly on the results of two kinds of experiments, one concerning the permeability of certain cells to nonelectrolytes, the other concerning the composition of the membrane of mammalian red blood cells. Mammalian red blood cells have been very useful for the study of membranes. Aside from their ready availability, uniform population, and several other ideal features, they have no internal organelles and they lyse osmotically to form the so-called red blood cell ghosts, from which the almost pure plasma membranes are readily isolated by simple procedures.

Other cells, such as internodal cells of the freshwater alga *Chara,* were also found very useful for studying the permeability of the plasma membrane. These multinucleated cells, as long as 1 cm, are relatively easy to manipulate and analyze because of their size. As we shall see later, they have also been useful in the study of electrical events and cell movement. Data on the permeability of *Chara* to nonelectrolytes (Collander, 1949) have been analyzed as shown in Fig. 2.2 (Davson, 1970). In this figure the permeability of the cells multiplied by the square root of the molecular weight ($PM^{1/2}$) is shown on the ordinate. The abscissa represents the oil-water partition coefficient, a measure of the solubility of the nonelectrolyte in oil divided by its solubility in water. These values can be obtained experimentally by equilibrating a water phase containing the nonelectrolyte with an identical volume of oil (or in some cases organic solvents) and then determining analytically the concentration of the nonelectrolyte in the two phases. A ratio of the two concentrations corresponds to the partition coefficient. Each point in Fig. 2.2 corresponds to the permeability constant (ordinate) and partition coefficient (abscissa) of a nonelectrolyte. The product $PM^{1/2}$ was used rather than the permeability constant P because a smaller molecule is expected to enter the plasma membrane faster for steric reasons. This method of plotting corrects for differences in molecular size, following a theoretical treatment that is based on several assumptions (Davson and Danielli, 1952). In this treatment, a molecule of solute is considered to detach itself from the water phase and move into the lipid phase. It travels through the hydrophobic layer in steps, each with a characteristic energy of activation. Then it must enter the water phase again from the lipid.

Figure 2.2 shows that there is a correlation between the permeability and the partition coefficient, indicating that substances that are more soluble in oil than in water penetrate the plasma membrane more readily. Similar data are available for the red blood cell. The correlation between the permeability constant and the solubility in oil suggests that the solute molecules have to dissolve in a lipid layer in order to pass through.

The significance of these observations is more readily apparent when the results (in this case for the red blood cell) are compared to the values calculated for a hypothetical cell enclosed

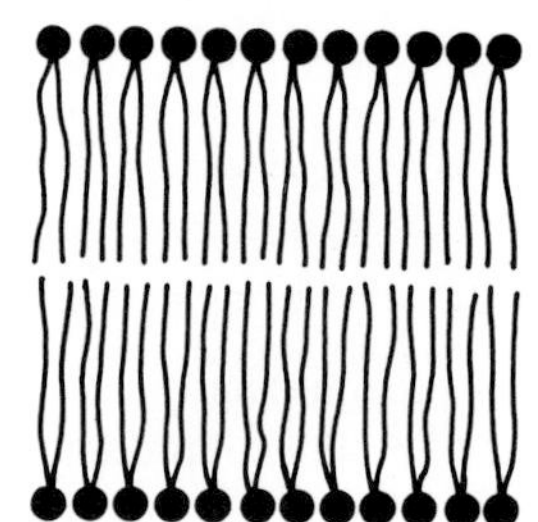

Fig. 2.1 Diagrammatic representation of phospholipid molecules in a bilayer arrangement.

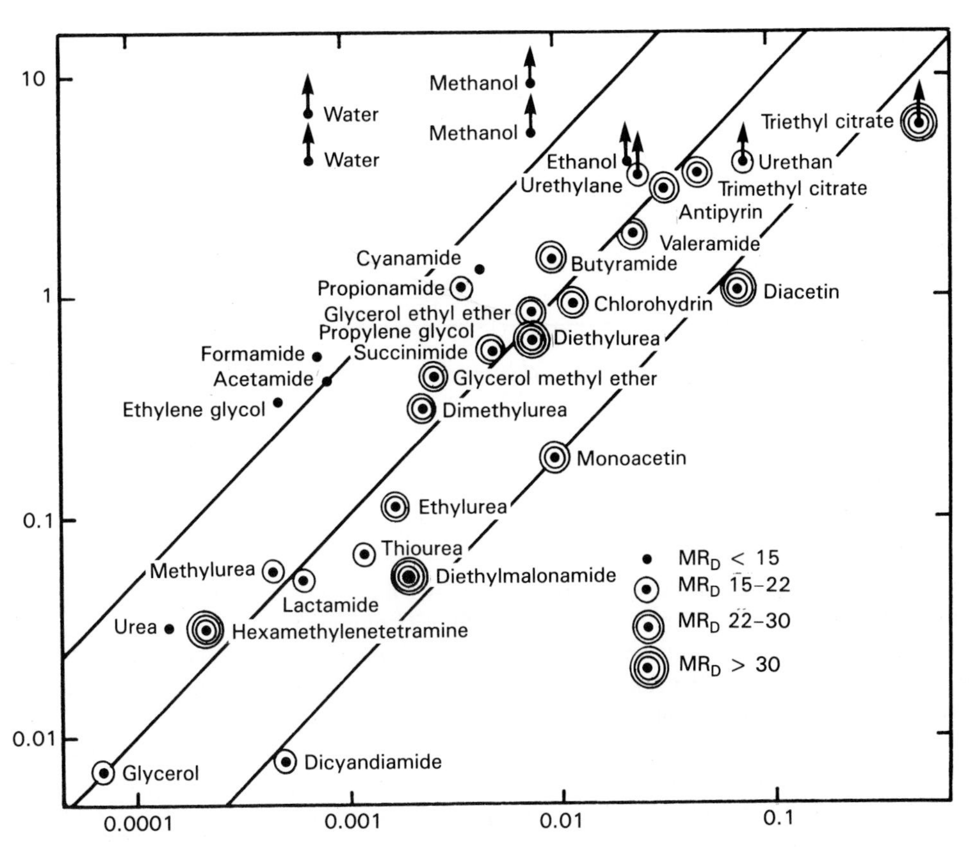

Fig. 2.2 Permeability of *Chara* cells plotted against oil-water partition coefficient. Ordinate: permeability (cm/h) × $M^{1/2}$. Abscissa: oil-water partition coefficient. Reprinted with permission from H. Davson, *A Textbook of General Physiology,* Vol. 1, 4th ed. Copyright © 1970 Churchill Livingston, England.

by a water layer of approximately the thickness of the plasma membrane (Table 2.1) (Jacobs, 1952). In Table 2.1 the permeability is expressed as the time to reach 90% of equilibrium, a parameter that is inversely proportional to the permeability constant. Columns 5 and 6 give the times required for 90% equilibration. In contrast to observations with the red blood cells (column 6), equilibration in the model system (column 5) is almost instantaneous and the system is unable to distinguish between one solute molecule and another.

In addition to the permeability data, the early conclusion that the phospholipids are present in the membrane as a bilayer was based in part on the known properties of lipids—in particular their behavior at water-air interfaces—and in part on the results of a pioneering study by Gorter and Grendel (1925). Gorter and Grendel used a Langmuir trough, which is represented in Fig. 2.3. When the phospholipids are placed in the trough, they arrange themselves so that their hydrophilic ends are immersed in the water and their hydrophobic chains stick out in the air. When the lipid molecules are close to each other, they essentially form a continuous monomolecular lipid sheath. The surface of the trough available per lipid molecule can easily be calculated from the number of lipid molecules added to the surface and the dimensions of the trough. The force exerted on the system by compression can be measured. Data obtained by varying the position of the movable end of the trough can be plotted as shown in Fig. 2.4, where the pressure is shown on the ordinate and the surface available to each molecule on the abscissa. As the

Table 2.1 Calculations of the Time Required for 90% Equilibration of the Penetrants Listed in Column 1[a]

(1) Solute	(2) *k* in water (cm^2/s)	(3) *k′* in model (cm/s)	(4) *k′* in cells (cm/s)	(5) Times for 90% equilibrations (s) Model	(6) Cells
Urea	1.18×10^{-5}	11.8	1.94×10^{-4}	0.76×10^{-5}	0.5
Glycerol	0.83×10^{-5}	8.3	1.50×10^{-6}	1.08×10^{-5}	67
Mannitol	0.58×10^{-5}	5.8	0	1.55×10^{-5}	>2 days
Sucrose	0.43×10^{-5}	4.3	5.50×10^{-20}	2.09×10^{-5}	∞

Based on Jacobs (1952).
[a] The calculations assume the membrane thickness to be 10 nm, the red blood cell volume to be 30×10^{-12} cm^3, and the surface area to be 77×10^{-8} cm^2.

pressure is increased, the surface area per lipid molecule decreases until a limiting value is reached. At this point no further compression is possible and the monolayer buckles (not shown). The limiting value shown in the abscissa corresponds to the area occupied by one lipid molecule when the monolayer is maximally compressed.

Gorter and Grendel extracted the lipids from a known number of red blood cells and placed them in a Langmuir trough. The minimal area of the film at its limiting value was then determined as described above. The area occupied by the surfaces of the red blood cells from which the lipid has been extracted can be calculated at least approximately from the dimensions of one cell and a count of the number of cells. The total area of the compressed film over the area of the cells was found to be approximately 2, indicating that enough lipid was extracted for two monomolecular layers, that is, a bilayer. The results are at least qualitatively in agreement with the notion that the lipid component of the membrane corresponds approximately to a bilayer.

II. PROTEINS IN THE PLASMA MEMBRANE

The plasma membrane contains proteins as well as lipids. A current view of how proteins are distributed in the plasma membrane of the red blood cell is shown in Fig. 2.5 (Fowler, 1986). Some of the evidence on which this model is based will be discussed later. In this diagram, the lipid bilayer is represented by the ball and stick double layer. Several proteins extend across the membrane and are arranged so that characteristic portions are on either the outer or the inner face. The hairlike processes shown in the diagram on the outer surface represent the carbohydrate portion of the glycoproteins. Although only two kinds of proteins are shown traversing the membrane, several others not shown in the diagram are present in lesser amounts. Proteins that are embedded in the bilayer are known as *integral* proteins; they can be removed from the membrane only when the lipid is disrupted and replaced by detergents. The anionic transporter

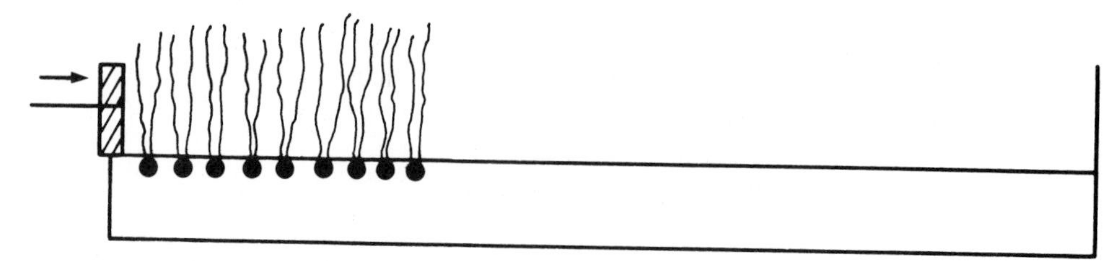

Fig. 2.3 Langmuir trough.

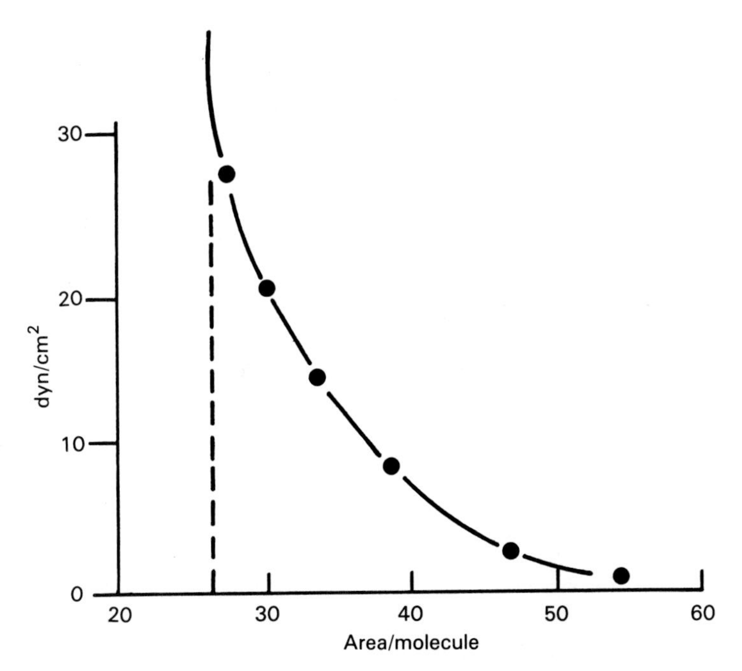

Fig. 2.4 Representation of data obtained from compression of the lipid layer in a Langmuir trough.

or anion channel (also known as band 3) will be discussed briefly in Chapter 15; it represents the major protein of the membrane. Generally, the proteins are numbered following the corresponding bands on sodium dodecyl sulfate (SDS) gel electrophoresis.

At least part of the band 3 protein is anchored to the network of filaments and fibers under the plasma membrane by *ankyrin,* a protein with a molecular mass of 210,000 daltons (Da). The network is distributed through the cytoplasm and is known as the *cytoskeleton*. It can be isolated after the cell membrane is disrupted by detergents. In this case the filaments contain, among other proteins, *spectrin* and *actin*. Actin, discussed several times in this book, is part of the contractile apparatus of muscle and, generally in combination with myosin, is also involved in movement in other cells. In fact, the red blood cell contains myosin, and strong arguments (Fowler, 1986) have been made for a role of actomyosin in the contraction of this cell.

The pattern of the cytoskeletal elements closely associated with the red blood cell membrane has been known for some time from extraction and binding studies. A clear demonstration of the undisturbed structural pattern has been shown in experiments in which the meshwork of fibers has been isolated intact from the ghosts after detergent treatment (Liu et al., 1987). An electron microscopic examination of the spread-out network after negative staining reveals the organization shown in Fig. 2.6 (Liu et al., 1987). The individual components labeled in the figure were identified by selective extraction of the isolated network. The globules containing ankyrin are likely also to contain band 3 protein, since they are generally extracted together. As shown, the fibers form a hexagonal lattice (Fig. 2.6*b* and *c*). The network at lower magnification suggests that it is a continuous structure lining the membrane surface.

The arrangement of membrane proteins at the surface as depicted in Fig. 2.5 offers the possibility of specific interactions between cells or between the cell surface and extracellular components, for which there is considerable information. A class of compounds termed neural cell adhesion molecules (NCAM) are thought to play a crucial role in neural differentiation

(Rutishauser, 1984). Furthermore, the presence of proteins traversing the membrane may make it possible for the cell to respond readily to outside stimuli. The interconnection between membrane proteins and cytoskeletal elements suggests that cell membrane elements can produce changes in shape of the cell and perhaps even movements at the surface (e.g., see Jinbu et al., 1984).

The details of the molecular architecture are likely to differ in different cells, but the same general plan is thought to apply to all cells (Bennett and Davis, 1984). *Filamin,* which is found in several different kinds of cells, is similar to spectrin and may play a similar role. Ankyrin has been found in several cell types.

The location of protein molecules across the plasma membrane suggests that at least in some cases they may operate as channels for the passage of water or certain solutes. In contrast to the permeability model already discussed, these molecules would not have to pass through the hydrocarbon layer of the phospholipid. There are indeed indications that certain specialized transport processes take place through such protein channels. Protein channels are likely to play a role in the passage of water through the membrane (Macey, 1984; Solomon et al., 1983). The cell membrane has a much higher permeability to water than do phospholipid bilayers. Passage of water in response to an osmotic gradient is much more rapid than diffusion of water, as would be expected if channels are present. Furthermore, mercurial sulfhydryl reagents block water transport in the cell membrane so that it becomes as impermeable to water as phospholipid bilayers. An effect that depends on the presence of sulfhydryl groups is most likely to involve proteins. The transport of several nonelectrolytes also seems to be sensitive to reagents that are

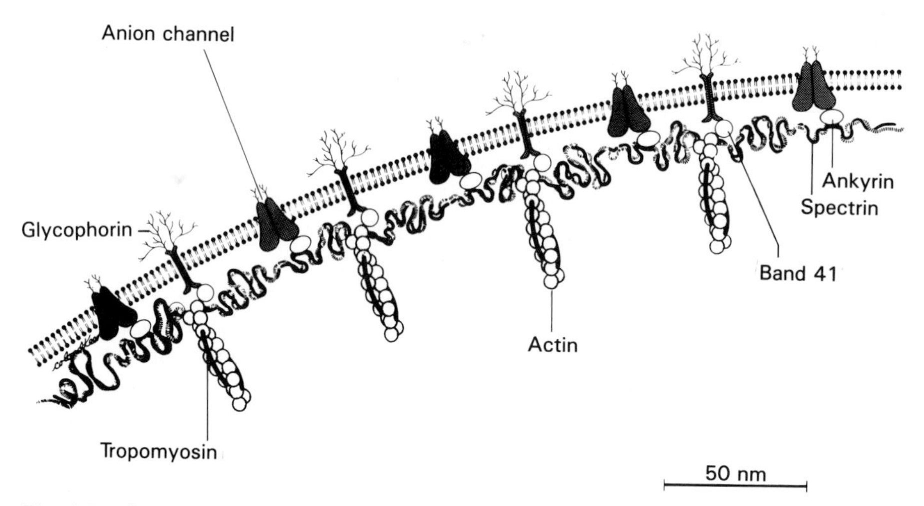

Fig. 2.5 Organization of the erythrocyte cytoskeletal filaments. Only one spectrin tetramer connects any two adjacent actin filaments in the cross section of the membrane shown here, but each actin filament would be attached to four to six spectrin tetramers in a plane perpendicular to the plane of the paper. The 60-nm spacings between the actin filaments and the excess length of the spectrin tetramers linking them are based on direct observations of the arrangement of the spectrin-actin network in negatively stained preparations of artificially spread membrane skeletons. The linear end-to-end length of a spectrin tetramer is about 194 nm. The actin filaments shown here are intended to be oriented with their barbed ends attached to the spectrin molecules. Linear dimensions of the cytoskeletal structures are to scale, but the actual sizes of the individual molecules are enlarged for purposes of illustration. Reprinted with permission from V. M. Fowler, *Journal of Cell Biochemistry,* 31:1–9. Copyright © 1986 Alan R. Liss, Inc.

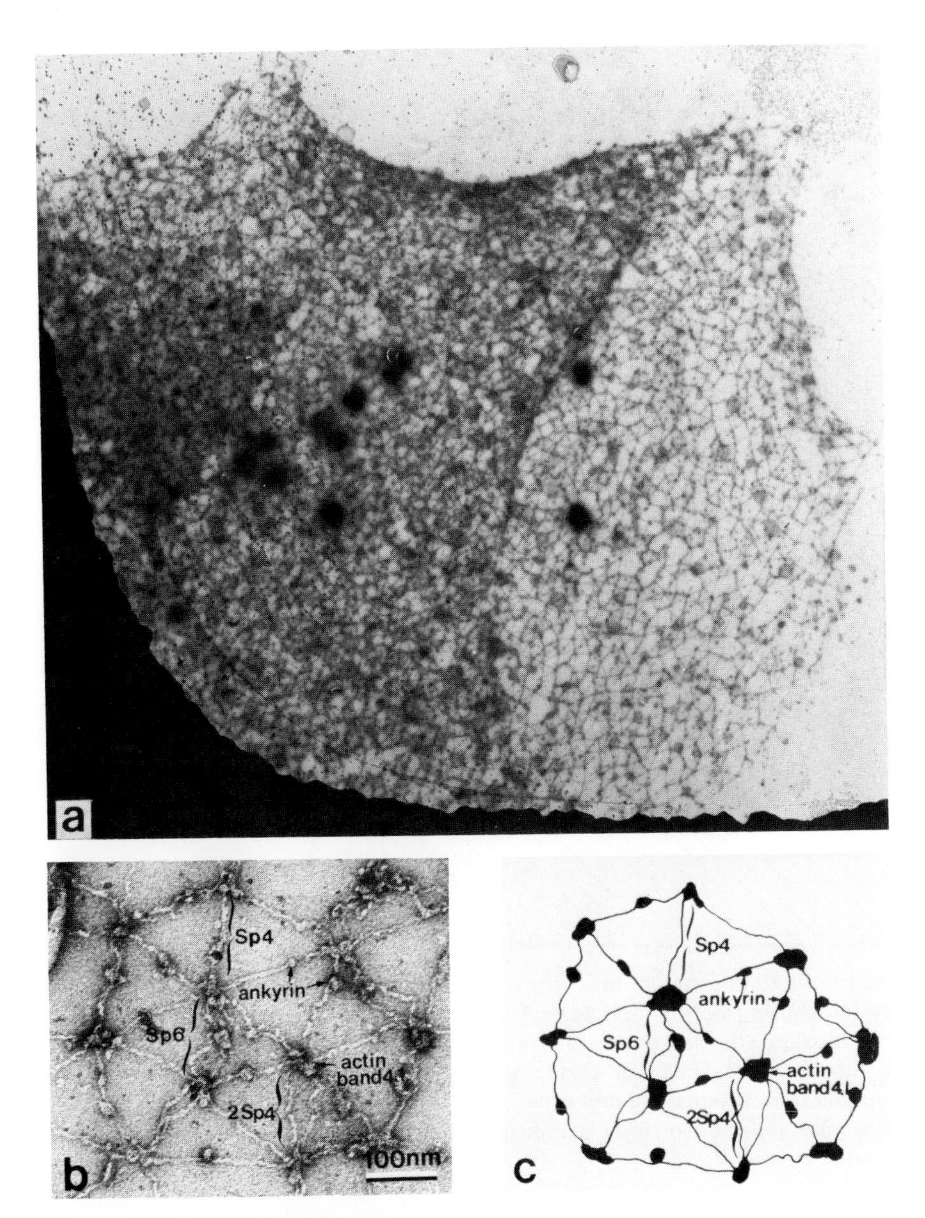

Fig. 2.6 Spread membrane skeleton examined by negative-staining electron microscopy. Membrane skeletons derived from Triton-treated (2.0%) normal red cell ghosts were isolated by sucrose density gradient centrifugation under hypotonic conditions. The skeletons were applied to thin carbon-coated grids, fixed with glutaraldehyde, stained with uranyl acetate, air dried, and examined by transmission electron microscopy. (*a*) A large area of a spread meshwork revealing the marginal region of the exposed bottom layer of the skeleton. (*b*) A higher magnification of the spread meshwork reveals the hexagonal lattice of junctional complexes, presumably containing short F-actin and band 4.1, cross-linked by spectrin tetramers (*Sp4*), three-armed spectrin molecules (*Sp6*), and double spectrin filaments (*2 Sp4*). Globular structures of ankyrin (or ankyrin-containing complexes) are attached to spectrin filaments at the ankyrin-binding site, i.e., 80 nm from the distal end of spectrin. (*c*) Tentative assignment of these structural elements in a schematic diagram. From Liu et al. (1987).

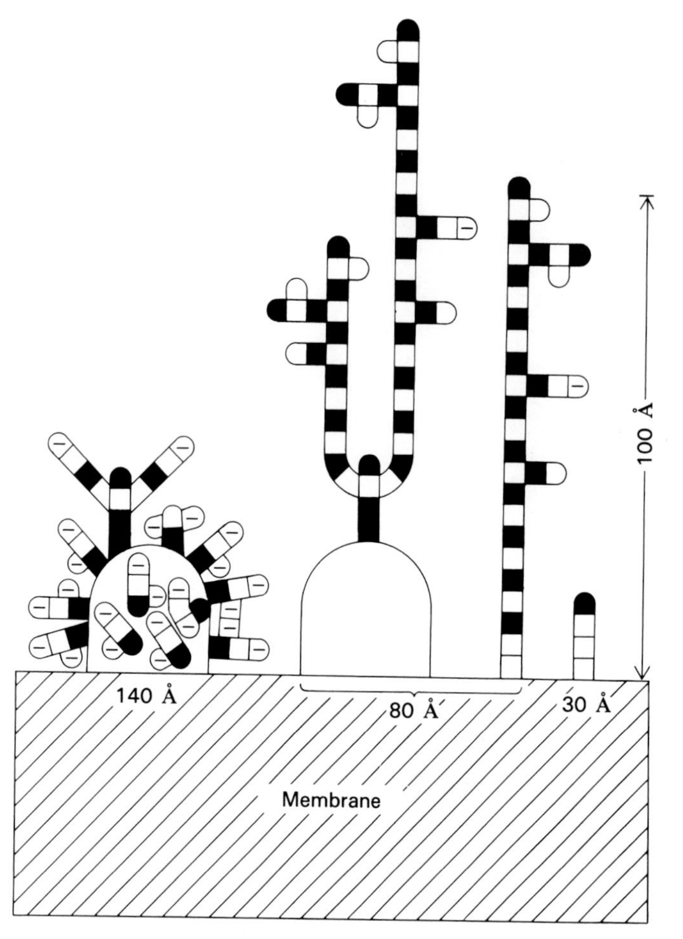

Fig. 2.7 Four carbohydrate components of the red blood cell surface. The first two represent the integral proteins glycophorin A and band 3. The second two represent polylactosamine ceramide and globoside. The blackened areas represent hexosamines (*N*-acetylglucosamine and *N*-acetylgalactosamine). The blank areas represent hexoses (fucose, galactose, glucose, and mannose). The minus signs indicate sialic acid. The scale of the projections is shown on the right. Distances indicated on the *x*-axis correspond to the calculated average distances between like molecules.

likely to react with proteins. A role of channels in the proteins involved in ion transport is discussed in Chapter 15.

The carbohydrate components, whether attached to the lipid or to the protein, are probably significant in biological interactions (e.g., see Chapter 3). They form a special domain that can interact with the extracellular environment.

Information about the carbohydrate components is most complete in the case of the red blood cell (Viitala and Jarnefelt, 1985). A summary for purposes of illustration is shown in Fig. 2.7, which represents two of the glycoproteins and two of the glycolipids that project from the bilayer into the cell's exterior. The globular structures represent the part of the proteins that juts out of the bilayer. The chains represent the carbohydrate present. In the diagram, the black portions represent hexosamines (*N*-acetylglucosamines and *N*-acetylgalactosamines). The open portions represent hexoses such as fucose, galactose, mannose, or glucose. Alternatively, the open portions represent sialic acid when a minus sign (indicating its negative charge) is shown. The scale on the right indicates the reach of the arms from the membrane surface, and the scale along the *x* axis indicates the calculated distance between like molecules. Clearly, the surface is densely populated by carbohydrate structures, some of them highly charged, and this coat must alter significantly the properties of the surface.

III. COUPLING BETWEEN CELLS

Electrical techniques and the microinjection of fluorescent molecules have demonstrated that cells in tissues are interconnected.

When a current is passed through a microelectrode inserted in one cell, a current is detected in the adjoining cell, as shown later in the records of Fig. 2.12. The signal would not be detectable at these current levels if the two electrodes were in the bathing medium or the cells were not interconnected. When one of the cells is damaged, this electrical coupling can no longer be observed, suggesting that the cells have been uncoupled. Several experiments indicate that cells are uncoupled by increases in the concentration of Ca^{2+} or H^+ (Peracchia and Peracchia, 1980a, b), which are normally at very low levels. Calcium ion may play a role in response to cell injury. The Ca^{2+} concentration of the extracellular fluid is much greater than that inside the cell, so an increase in the internal concentration could readily occur when the membrane is damaged. The low levels of internal H^+ or Ca^{2+} also raise the possibility that there is a subtle interplay in the regulation of the opening of the channels even in uninjured cells. At least in *Chironomus* salivary glands, the openings have also been found to be sensitive to changes in electrical potential (Obaid et al., 1983).

As might be predicted from the electrical findings, a sufficiently small fluorescent molecule microinjected into a cell will spread into adjacent cells. Microinjection of uncharged proteins of different sizes provided further information (Schwarzmann et al., 1981). The probes were obtained by varying degrees of proteolysis of the same glycoprotein. Probes above a certain critical size failed to pass from one coupled cell to the other, suggesting that a pore controls the exchanges. In mammalian cells the size is approximately 1.6 to 2 nm and in insect salivary cells 2 to 3 nm.

Coupling has been found at *gap junctions*. Whereas the membranes of confluent cells may be tightly apposed, in gap junctions the electron microscope reveals regular clear spaces or gaps traversed by structures at regular intervals as shown in Fig. 2.8 (Peracchia and Dulhunty, 1976).

In tangential views of the gap junctions (Fig. 2.9) (Peracchia and Dulhunty, 1976), a regular array of structures is also shown with either negative staining (a technique in which dense material penetrates the water spaces) or freeze fracture. (In freeze fracture the specimens are frozen rapidly at low temperatures and fractured; the structures are subsequently exposed by sublimation and then coated so that a replica is examined.) These structures, the *connexons,* appear in hexagonal patterns and are thought to correspond to the structures bridging the gap junctions. The regular array of the connexons, shown by the membrane faces, seems to be characteristic of the uncoupled state. When coupled, the connexons appear to be in a more disordered arrangement. As is visible in some of the electron micrographs, the connexons themselves are made up of particles in a regular hexagonal arrangement. SDS gel electrophoresis of isolated gap junctions reveals one major protein of approximately 26,000 to 27,000 Da, about the right size to correspond to one of the six subunits of the connexons (Finbow et al., 1980).

Because of the regularity of the structures, packed pellets of the membranes containing gap junctions can be studied with x-ray diffraction techniques. Figure 2.10 corresponds to the interpretation of combined x-ray diffraction and electron microscopy studies (Makowski et al.,

1977). Each connexon appears to be formed by two basic units, one from each membrane and each formed by six subunits. The close juxtaposition of the units from the two membranes suggests the possibility of a continuous and patent central channel.

If the connexons are the structures responsible for the coupling between cells, it is reasonable to expect structural differences that would correspond to the two physiological states. Gap junctions isolated in the presence and the absence of Ca^{2+} should correspond to the two states. A careful analysis of electron micrographs (Unwin and Zampighi, 1980) suggests that in the coupled state there is a central channel and that this channel can be closed by a rotation and sliding of the subunits of the connexons in relation to each other, as shown in Fig. 2.11.

Lowenstein et al. (1978) observed the electrical resistance between cells during coupling and uncoupling of junctions in embryos of the African frog *Xenopus laevis*. Coupling occurred when two cells formed contacts, with the result that the voltage recorded from the adjacent cell increased. Uncoupling could be elicited by introducing Ca^{2+}, which was delivered electrophoretically in the vicinity of the junction. The results are shown in Fig. 2.12. The increase in voltage began 1.7 minutes after the formation of the contact (see arrow at *a*) and it took place in steps. The uncoupling accompanying the delivery of Ca^{2+} (arrow at *b*) also occurred in steps, suggesting that individual channels or groups of channels open one at a time, a response that has been referred to as quantal.

The implications of cell coupling for regulatory phenomena are far reaching, although at the moment they are difficult to evaluate precisely. However, cell coupling is thought to have a role in embryonic development and metabolic regulation.

Transfer of metabolites between cells may play an important role in metabolism. A possible role is shown by intracellular injections of glycolytic intermediates into confluent cultured

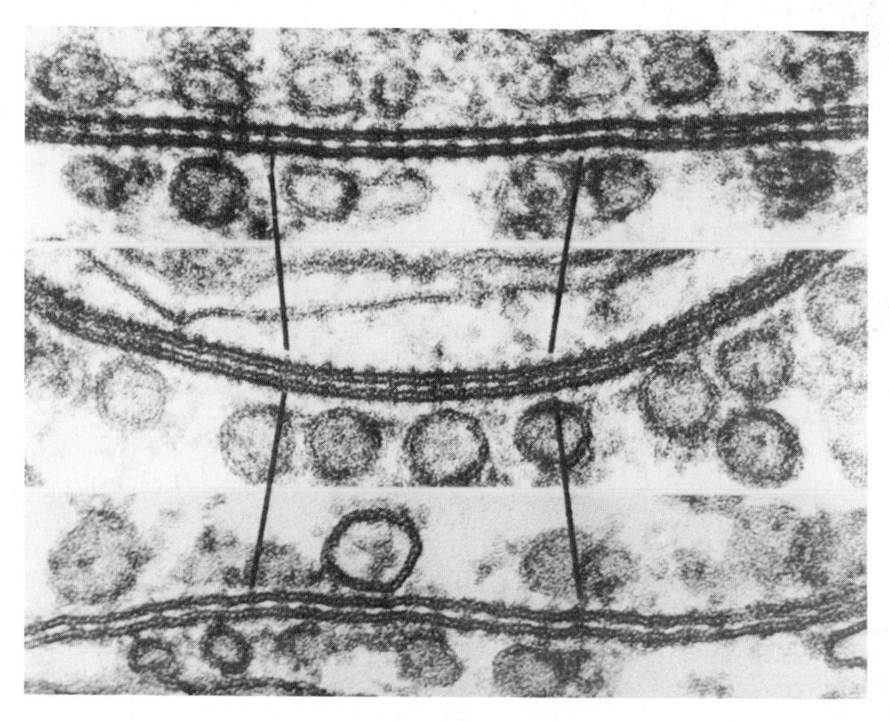

Fig. 2.8 Electron micrograph showing the "gap" seen between membranes of apposing cells at the site of a gap junction. From Peracchia and Dulhunty (1976).

Fig. 2.9 Comparison of freeze-fracture and negative staining of junctions from control preparations. Face P of a fractured junction is shown. Most of the particles have been fractured away with the external membrane leaflet, leaving a fairly regular hexagonal array of pits. A few pits are occupied by particles. (Inset) Fragment of an isolated PTA negatively stained junction. The unit cell dimension of the arrays is about 20 nm in the freeze-fractured junction and 15 to 15.5 nm in the negatively stained one. Notice the difference between the two hexagonal areas, which contain the same number of particles at the same magnification. Reprinted with permission from Peracchia and Dulhunty (1981).

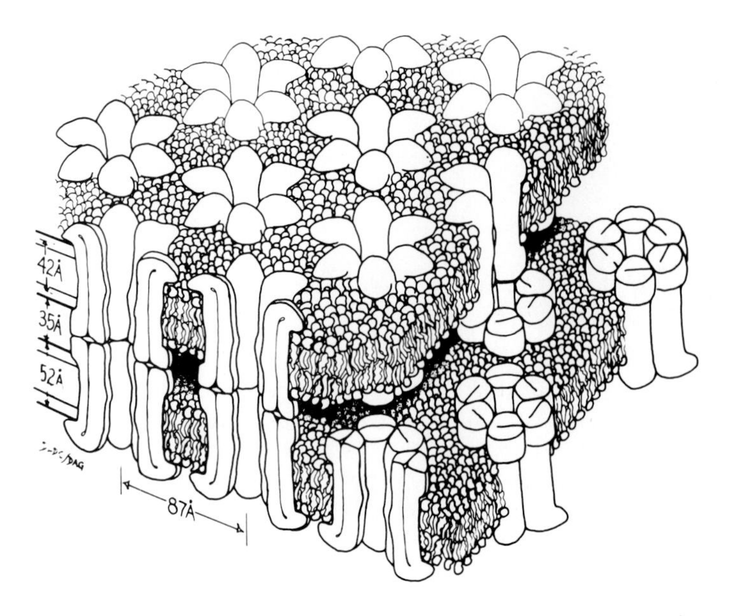

Fig. 2.10 Schematic representations of connexons in a gap junction based on x-ray diffraction and electron microscopy studies (Makowski et al., 1977). Reproduced from *The Journal of Cell Biology,* 1977, vol. 74, pp. 629–645 by copyright permission of The Rockefeller University Press.

pancreatic cells. Injection into adjacent cells can reduce the oxidized pyridine nucleotides of the connected cells (Kohen et al., 1979), indicating that metabolites generated by one cell can be used by another. Studies of metabolic regulation may therefore have to consider the metabolite pool of several cells.

The possible implications of cell coupling for embryonic development are shown by experiments in which different types of cells are cocultured. When the cells form connections they become capable of responding to cyclic AMP-mediated hormonal signals to which they normally cannot respond. For example, myocardial cells normally respond to catecholamines by increasing

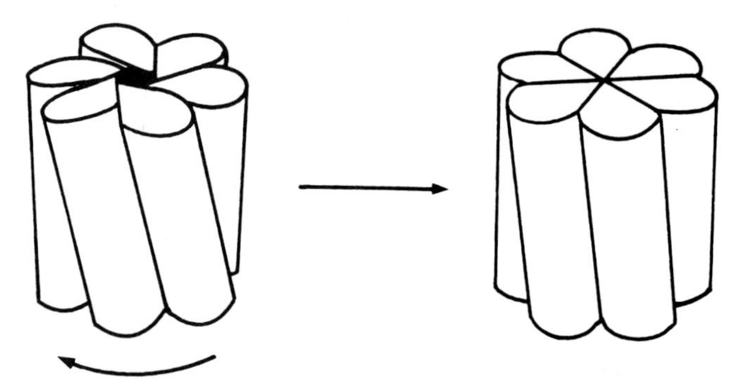

Fig. 2.11 Model of the connexon, depicting the transition from the "open" to the "closed" configuration. It is proposed that the closure on the cytoplasmic face (uppermost) is achieved by the subunits sliding against each other, decreasing their inclination and hence rotating in a clockwise sense, at the base (Unwin and Zampighi, 1980). Reprinted by permission from *Nature* 283:545–549 copyright 1980 Macmillan Magazines Ltd.

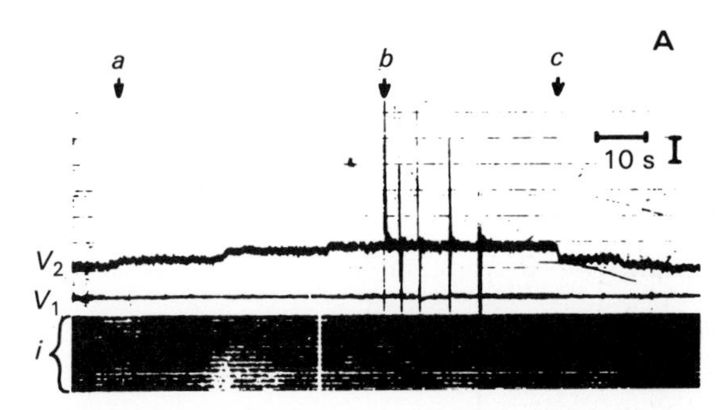

Fig. 2.12 Oscillograph record of quantal steps in transfer resistance during channel formation and abolition. The high-resolution V_2 record shows a series of three upsteps beginning at *a*, 1.7 min after cell contact. At *b*, five successive Ca^{2+} pulses are delivered into cell 1, in the vicinity of the junction, causing three downsteps, which begin at *c*. (The large spikes in the V_2 trace are capacitive artifacts due to the solenoid that controls hydraulic injection.) Both up- and downsteps are quantal in this experiment (Lowenstein et al., 1978). Reprinted by permission from *Nature* 274:133–136, copyright © 1978 Macmillan Magazines Ltd.

their beat frequency in a mechanism mediated by the cytoplasmic cAMP (see Chapter 5). They have receptors at the cell surface that bind the neurohormone. Since these cells have no receptors for *follicle-stimulating hormone* (FSH), they do not respond to FSH. This hormone, which is secreted by the *anterior pituitary* (the *adenohypophysis*), affects ovarian cells such as granulosa cells as well as cells elsewhere. When myocardial cells are cocultured with granulosa cells, which contain FSH receptors at the cell surface but not catecholamine receptors, the myocardial cells respond to FSH by increasing their beat frequency (Lawrence et al., 1980). This is shown in Fig. 2.13, in which the lower curve represents myocardial cells alone, in the absence of granulosa cells, and the upper curve represents myocardial and granulosa cells that have made contact. The results are expressed as beats per minute as a function of time after the addition of FSH. Although the presence of the granulosa cells decreases the baseline of myocardial cell activity before FSH is added (compare upper curve to lower curve at time zero), there is no question that the addition of FSH causes a dramatic increase in beat frequency. In contrast, in the absence of the granulosa cells (lower curve), FSH has no effect. Phosphodiesterase, which hydrolyzes cAMP, did not affect the beat frequency when added to the cell suspension, indicating that the increase in beat frequency is not the result of cAMP secretion. The results show that coupling can alter the pattern of responses of cells to hormones.

Coupling of cells does play a significant role in differentiation, as demonstrated with antibodies to the connexon protein (Warner et al., 1984). Antibodies to the major protein of rat liver gap junctions were injected into early embryos of *Xenopus*. Injection into cells in the eight-cell stage interfered with either electrical coupling or dye coupling of the injected cell and its progeny, resulting in substantial developmental defects. Although these experiments show unequivocally that gap junctions play a role, the actual mechanism by which they exert their effect is still unknown.

IV. ASYMMETRY OF THE MEMBRANE

A. The Bilayer

The lipid bilayer of the plasma membrane could conceivably be formed of two identical lipid leaflets. However, all indications are that the two leaflets differ in composition; that is, the bilayer is asymmetric.

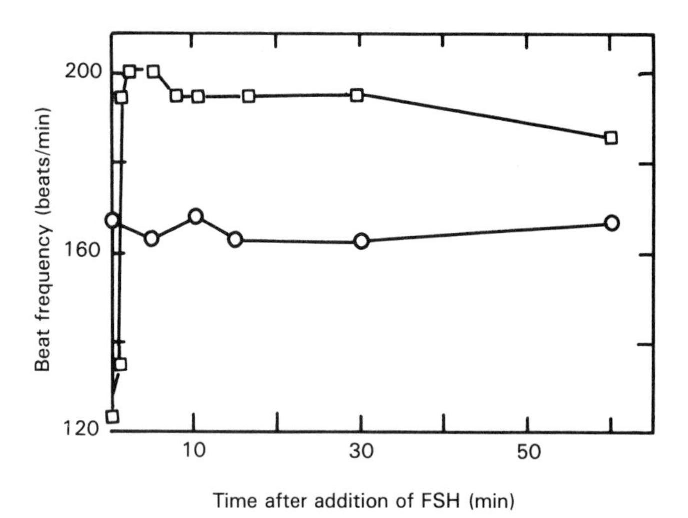

Fig. 2.13 Effect of FSH on myocardial cells and granulosa cocultures. Myocardial cells (5×10^4) were plated into adjacent wells of a Falcon multiwell plate and 24 to 72 h later 3×10^5 granulosa cells were added to one of the two wells. The medium was removed 4 h later; then the wall between the wells was removed and L-15 supplemented with 10% calf serum was added. FSH (10 μg/ml) was added to the cultures 14 to 18 h later. The myocardial cells cultured separately (◯) were unaffected, while there was a rapid increase in the beat frequency of the myocardial cells cocultured in the adjacent well (□). From Lawrence et al. (1980). Reprinted by permission from *Nature* 272:501–506, copyright 1980 Macmillan Magazines Ltd.

The location of reactive groups in the two faces of the cell membrane can be recognized by using one of several possible approaches. Certain reagents can react with sites in the membrane. Such reagents can be synthesized so that, in addition to the reactive group, they contain polar groups. Although the polar groups affect the reactivity only minimally, they confer important properties to the reagents. They will enter the membrane of intact cells or vesicles only with great difficulty (see discussion of permeability, above). Therefore, when the reagents are used under appropriate conditions (e.g., short exposure time), they will react only with the sites on the external surface of the membrane. Alternatively, in the absence of polar side groups, all reactive sites in the membrane are susceptible. Hence, sites that react only with the second kind of reagents must be internal to the cell surface. In addition, hydrolytic enzymes such as lipases or proteases can be used to determine location. Because of their high molecular weights, they will not enter the cell and therefore will generally digest only surface components.

Chemical labeling or enzyme treatment provides further information when used with membrane vesicles in which the membrane has reversed polarity, that is, has been turned inside out. The red blood cell membrane can be broken up into sealed vesicles that have the same orientation as intact cells; they are right side out (RO vesicles), or they have been turned inside out (IO vesicles). A similar IO preparation can be produced from other cells by taking advantage of the phenomenon of *endocytosis*. Latex particles present with cultured cells are ingested and sequestered in vesicles (the *phagosomes*). The cell membrane invaginates (see Fig. 2.14) so that the phagosomes form inside-out sealed pieces of cell membrane. Rupture of the cells and isolation of the phagosomes, a process made easier by the fact that they are heavy because of the internal latex, provides a preparation analogous to the red cell IO vesicles.

Results of the digestion of red blood cells with lipases are shown in Table 2.2 (Rennoij et

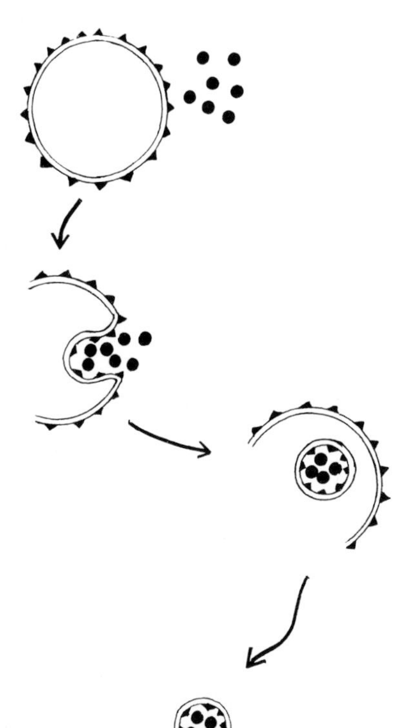

Fig. 2.14 Formation of vesicles with membranes of reverse polarity from that of intact cells by endocytosis.

al., 1976). The hydrolytic attack of phospholipases and sphingomyelinase is shown in Fig. 2.15. Since the enzymes attack the polar ends of the lipids, any lipid in the outer leaflet will be exposed to hydrolytic attack. However, since the reaction may liberate fatty acids, any free fatty acid must be removed to avoid secondary effects on the membrane. Generally this is done by adding albumin to the medium. In the experiments of Table 2.2, the phospholipid population indicated in column 1 is shown as a percentage of the total in column 2. The results of the digestion with phospholipase (column 3), sphingomyelinase (column 4), or both (column 5) are shown as

Table 2.2 Nonhemolytic Degradation of Phospholipids by Phospholipases in Intact Rat Erythrocytes

(1)	(2)	(3)	(4) Degradation by	(5)
Phospholipid	Phospholipid composition (%)	Phospholipase A_2 (%)	Sphingomyelinase (%)	Phospholipase A_2 + sphingomyelinase (%)
Lysolecithin	5	—	0	—
Sphingomyelin	12	0	100	100
Lecithin	42	48	0	62
Phosphatidylserine[a]	16	0	0	6
Phosphatidylethanolamine	25	8	0	20
Total phospholipid	100	22	12	44

Reprinted with permission from W. Rennoij, et al., *European Journal of Biochemistry,* 61:53–58.

[a] Including phosphatidylinositol.

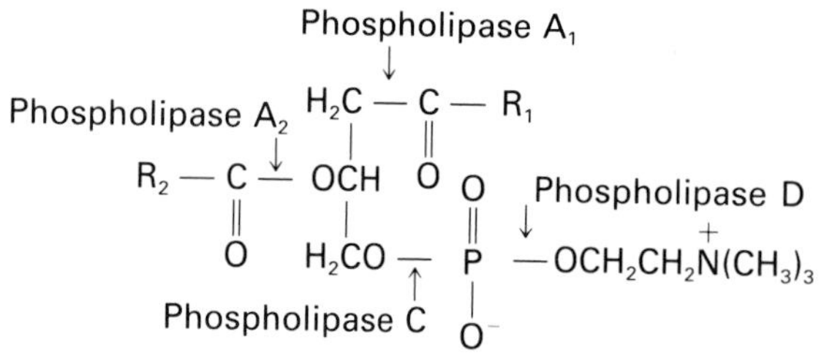

Phospholipase A_1

Phospholipase A_2 $H_2C — C — R_1$

$R_2 — C — OCH$ O O Phospholipase D

O $H_2CO — P — OCH_2CH_2\overset{+}{N}(CH_3)_3$

Phospholipase C O^-

Phosphatidylcholine

$HO—P—OCH_2—CH_2—\overset{+}{N}(CH_3)_3$ Phosphorylcholine

Sphingomyelinase

CH_2

H—C— NH

H—C—OH C=O

HC $(CH_2)_{14}$ Ceramide

CH CH_3

$(CH_2)_{12}$

CH_3

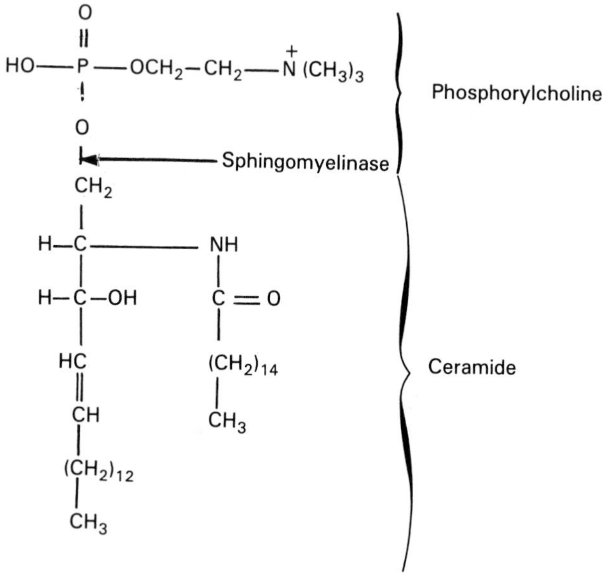

Sphingomyelin

Fig. 2.15 Hydrolytic points of attack of enzymes.

percentages of the individual phospholipid hydrolyzed. The results indicate that all of the sphingomyelin is at the external surface since it is all hydrolyzed. A substantial portion, about 50%, of the lecithin [phosphatidylcholine (PC)] also appears to be hydrolyzed, and more of it is available for hydrolysis (about 60%) when sphingomyelinase is present, suggesting that the hydrolysis of sphingomyelin exposes more PC to the enzyme. Only small portions of the phosphatidylserine and phosphatidylethanolamine (PS and PE, respectively) are hydrolyzed, which indicates that they are predominantly in the inner leaflet.

Results obtained with red blood cell ghosts and inside-out vesicles are entirely comparable to those just discussed. The bilayers of other eukaryotic cells also appear asymmetric. However, the composition and distribution of the lipids may differ considerably (Sandra and Pagano, 1978).

There are indications that the maintenance of the different composition of the leaflets depends on the presence of ATP in the cytoplasm (Seigneuret and Devaux, 1984) and in red blood cells is related in some manner to spectrin (Haest et al., 1978), since cross-linking of the latter eliminates the asymmetry.

$$\begin{array}{c} NH_2 \\ \| \\ CH_3 - C - OR \end{array}$$

Isethionyl acetimidate (IAI)
$R = -CH_2-CH_2-SO_3^-$ Polar, lipid insoluble

$R = -CH_2-CH_3$ (EAI)
Ethylacetimidate nonpolar, lipid soluble

Fig. 2.16 A polar and a nonpolar probe used to localize proteins in relation to the two faces of the plasma membrane.

B. The Membrane Proteins

The membrane is also asymmetric in relation to proteins. Generally, integral proteins extend across the cell membrane. Others are present only at one of the two interfaces. Conceivably, the proteins that span the membrane could be oriented randomly in the membrane; that is, 50% of the time one end of a particular molecule would be either in the internal or the external face. All evidence indicates, however, that the orientation of the membrane proteins is invariably the same. This not only indicates that the membrane is asymmetric in relation to the proteins but also demonstrates that the intrinsic proteins do not flip-flop from one lipid leaflet to the other. This is an important consideration in the mechanism of ion transport (see Chapter 15).

One kind of experiment demonstrating protein asymmetry used two related reagents, isethionyl acetimidate (IAI) and ethylacetimidate (EAI) (Whiteley and Berg, 1974). The formulas for these compounds are represented in Fig. 2.16. As shown in the diagram, IAI contains an SO_3^- group and is therefore polar and water soluble. In contrast, IAI is hydrophobic and therefore lipid soluble and easily permeates into the cell. The reactive site of either compound can react with amino groups as shown in Eq. (2.1),

$$\begin{array}{c} NH_2 \\ \| \\ {}^*CH_3 - C - OR \end{array} + MNH_2 \rightarrow \begin{array}{c} NH_2^+ \\ \| \\ {}^*CH_3 - C - NHM \end{array} + HOR \qquad * = {}^3H \text{ or } {}^{14}C \quad (2.1)$$

where M represents any component of the membrane with an amino group. This compound could correspond to a phospholipid (PS or PE) or a protein. If the reagent is labeled (by incorporating either ^{14}C or ^3H) the membrane product will also be labeled.

Table 2.3 (Whiteley and Berg, 1974) shows the pattern of labeling expressed as percentage

Table 2.3 Percentage of Amino Groups Modified at Saturation

(1) Preparation	(2) Reagent	(3) Total amino groups (%)	(4) Protein amino groups (%)	(5) Lipid amino groups (%)
Intact cells	IAI	1.0	1.5	≤0.3
Ghosts	IAI	30	28	31
Intact cells	EAI	52	[a]	[a]
Ghosts	EAI	54	62	46

Reprinted with permission from N. M. Whiteley and H. C. Berg, *Journal of Molecular Biology,* 87:541–561. Copyright © 1974 Academic Press, London.

[a] Not assayed.

of the total lipid or protein when the polar IAI is in the presence of intact cells. Few of the amino groups have reacted. However, both the lipids and the proteins of the red blood cell ghosts are labeled by the reagent much more extensively. This demonstrates that most of the membrane amino groups are internal. In the case of lipids, these results confirm the previously discussed studies using hydrolytic enzymes which show that PE and PS are predominantly in the inner leaflet. EAI, the nonpolar reagent, labels a much larger proportion of the proteins and lipids of ghosts.

The effects of IAI with ghosts are, unfortunately, somewhat ambiguous. The preparatory procedure necessary to make ghosts may have altered the membrane structure enough to expose sites that normally are not available to IAI or EAI. Results of an experiment designed to examine this question are shown in Table 2.4 (Whiteley and Berg, 1974). In this experiment the cells were first treated with saturating concentrations of unlabeled EAI. Subsequently intact cells (item a) or ghosts (item b) were exposed to labeled EAI and ghosts (item c) were exposed to labeled IAI. Since the labeling of the cells and that of the ghosts are comparable in magnitude (third column), it follows that no new groups were exposed by the preparatory procedures. For comparison, item d represents the direct labeling with EAI of ghosts of cells not pretreated with unlabeled EAI. These findings validate the conclusion that the distribution of amino groups in both proteins and lipids is asymmetric.

Considerably more information can be gained from experiments in which the various proteins are separated out after the cells are labeled, first with IAI (which labels the groups on the external interface) and then with EAI (which labels the remaining amino groups). One reagent can be labeled with ^{14}C and the other with ^{3}H. Since the two radioactive isotopes can be counted separately even when they are present in the same solution, it is possible to distinguish between these two kinds of sites. Among various techniques used to separate molecules, SDS acrylamide gel electrophoresis, which separates out polypeptides primarily by molecular weight, has been found very useful (see Chapter 1). In this technique as applied to membrane proteins, the detergent SDS is added to the membranes to disrupt them. Generally, the proteins are then denatured (so that they will be present in an extended form) and chemically reduced to disrupt any disulfide bonds holding the component peptides together. After this treatment, the extended polypeptides in the presence of SDS become coated by the detergent; 1.4 SDS are bound per unit weight of protein. Since the sulfate groups of SDS are anionic, the polypeptides will migrate toward a positive electrode in an acrylamide gel. The gel acts as a sieve, with the higher molecular weight polypeptides moving more slowly. It is possible to calibrate the gels with proteins of known molecular weight, so that the molecular weights of the unknowns can be determined. In Fig. 2.17 the radioactivity is expressed in relation to the molecular weight determined directly by the use of standards rather than position in the gel.

The lower half of Fig. 2.17 (Whiteley and Berg, 1974) shows the results of first labeling intact cells with [^{14}C]IAI, which reacts with the external amino groups (line and closed circles).

Table 2.4 Labeling of Membranes of Cells Saturated with Unlabeled Ethyl Acetimidate

Substrate	Reagent	Specific activity of membrane protein (counts/min per μg)
a. Cells saturated with EAI	[^{3}H]EAI	228±11
b. Ghosts from cells saturated with EAI	[^{3}H]EAI	214±11
c. Ghosts from cells saturated with EAI	[^{3}H]IAI	144±7
d. Ghosts from unmodified cells	[^{3}H]EAI	4960±250

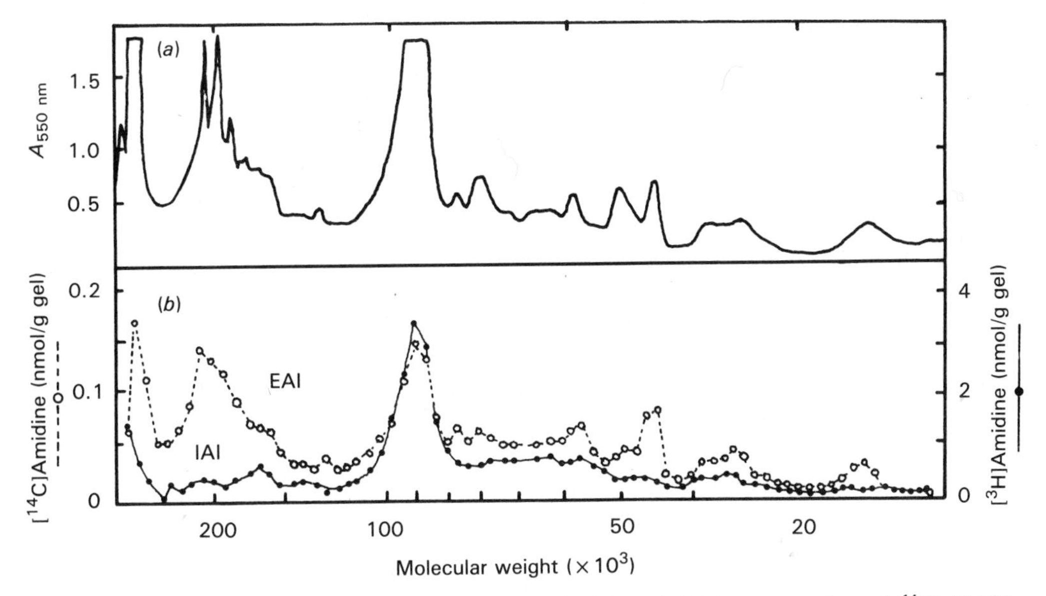

Fig. 2.17 Gel A. (*a*) Absorbance of stained protein (25 μg). (*b*) Incorporation of [^{14}C]IAI (●) and [^{3}H]EAI (○) per 15 μg of total protein. Reprinted with permission from N. M. Whiteley and H. C. Berg, *Journal of Molecular Biology,* 87:541–561. Copyright © 1974 Academic Press, London.

This is followed by labeling with [^{3}H]EAI (dashed line and open circles). The membranes are then isolated, solubilized, and subjected to SDS gel electrophoresis. The ordinate in the upper curve represents the light absorption of the gel after staining with Coomassie Blue. This stain serves to indicate the position of the proteins. The ordinates of the lower curves represent radioactivity, on the left hand for ^{14}C (indicating IAI) and on the right for ^{3}H (indicating EAI). The abscissa represents molecular weight, which has been derived from the position of the radioactivity in the gel. Each of the peaks shown in the upper graph represents a protein or several proteins of the same molecular weight. Most of these peaks correspond to a radioactivity peak for EAI (open circles). This indicates that these proteins are present only inside the membrane or, alternatively, on the internal face. Notable among these is a protein of about 210,000 Da that corresponds to spectrin, which we discussed earlier. One and perhaps a few other peaks are labeled by both EAI (open circles) and IAI (closed circles), most notably one of approximately 90,000 Da (corresponding to band 3). The observation suggests that this molecule (or molecules) spans the membrane and has been exposed sequentially to both reagents. This conclusion can be checked by digesting the protein in the intact cell with pronase. Figure 2.18 shows that the label associated with both IAI and EAI and originally corresponding to the 90,000-Da protein is digested and displaced to a lower molecular weight peak (Fig. 2.18, arrow). If there were two separate proteins of equal molecular weight, one labeled with IAI and the other with EAI, only one (that labeled at the external surface by IAI) would be digested.

The use of reagents to pinpoint the location of integral proteins or protein segments in relation to the membrane has been referred to as *vectorial* labeling. In addition to the use of chemicals to label proteins present in the two faces of the membrane, the hydrophobic membrane-spanning segments can be labeled with hydrophobic *photoactivated probes.* These can be incorporated into the bilayer first and then activated by light. These chemicals can be derivatives of phospholipids generally and can photogenerate nitrenes or carbenes.

Some insights into the arrangement of plasma membrane components can be gained by

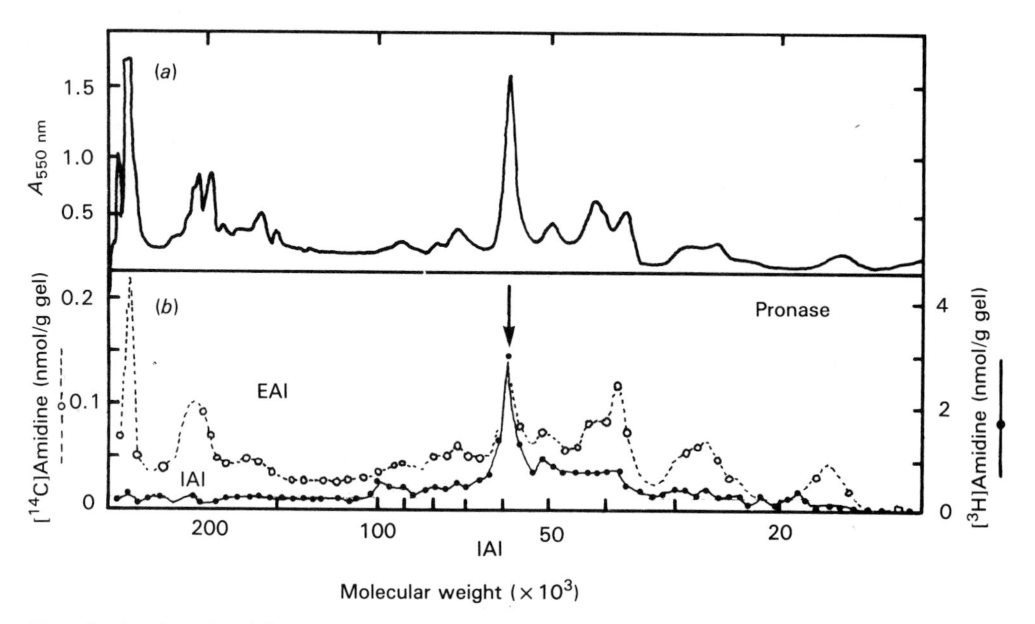

Fig. 2.18 Gel B. (*a*) Absorbance of stained protein (10 μg). (*b*) Incorporation of [^{13}C] IAI (●) and [^{3}H]EAI (○) per 15 μg of total protein. Reprinted with permission from N. M. Whiteley and H. C. Berg, *Journal of Molecular Biology,* 87:541–561. Copyright © 1974 Academic Press, London.

chemical treatment and enzyme digestion as already discussed. The details of this arrangement are generally referred to as the *topology* of a membrane protein. Our knowledge of the topography of many integral proteins has also been augmented by finding the location of binding sites and active groups of integral proteins involved in ion transport (see Jennings, 1989). These approaches have recently been supplemented by pinpointing sites with monoclonal antibodies, that is, antibodies that react with specific amino acid sequences (Ovchinnikov, 1987). Increased knowledge of the properties of proteins has also made it possible to predict their arrangement in the membrane.

The technique of cloning the DNA (see Chapter 1) that codes for either the Na^+,K^+-ATPase (Schull et al., 1985) or the Ca^{2+}-ATPase (MacLennan et al., 1985) has permitted sequencing of the DNA and hence deduction of the amino acid sequence of these two proteins (see Cantley, 1986). Knowledge of the hydrophilic and hydrophobic properties of the amino acid side chains allows use of a scale, the *hydropathy* index, in which positive values indicate hydrophobicity. Plotting this index averaged for polypeptide segments as a function of their position in the chain permits the construction of models (Kyte and Doolittle, 1982). These models, supplemented by other information such as the position of the phosphorylated amino acid or the ATP binding site, predict the location of the segments in the phospholipid bilayer and in relation to its two faces.

Profiles for the Na^+,K^+-ATPase are shown in Fig. 2.19 (Shull et al., 1985). In these representations, the polypeptides are shown from the amino terminal on the left to the carboxyl terminal on the right. Hydrophobicity is plotted upward. The shaded parallelograms indicate areas of *homology* (i.e., amino acid sequence similarity), P* is the phosphorylation site, and A* is the ATP binding site. Clearly, the two profiles are very similar and suggest that the two proteins resemble each other, as also shown by their biochemical properties. A tentative reconstruction of the Na^+,K^+-ATPase is presented in Fig. 2.20 (Cantley, 1986). The numbered

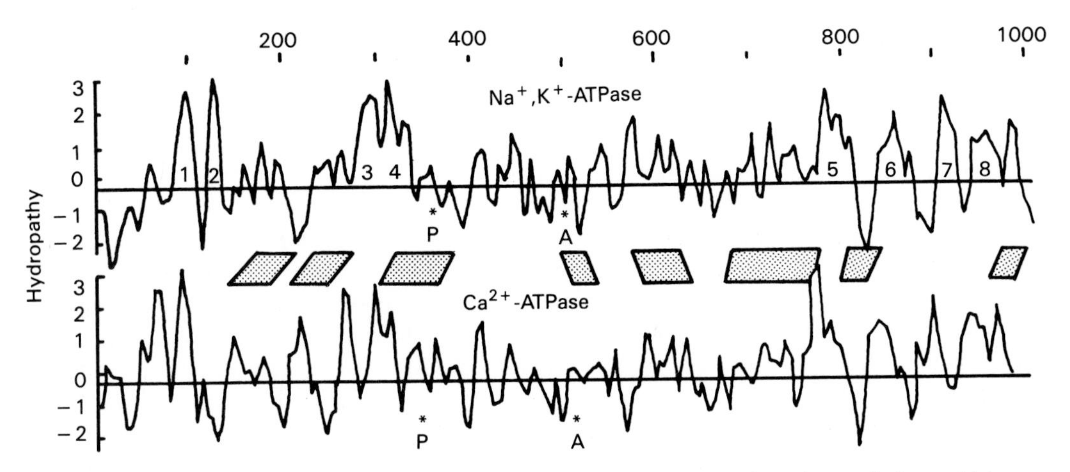

Fig. 2.19 Comparison of amino acid homology and hydropathy plots of sheep kidney Na^+,K^+-ATPase α-subunit and rabbit cardiac Ca^{2+}-ATPase (Shull et al., 1985). Reprinted by permission from *Nature* 316:691–695, copyright © 1985 Macmillan Magazines Ltd.

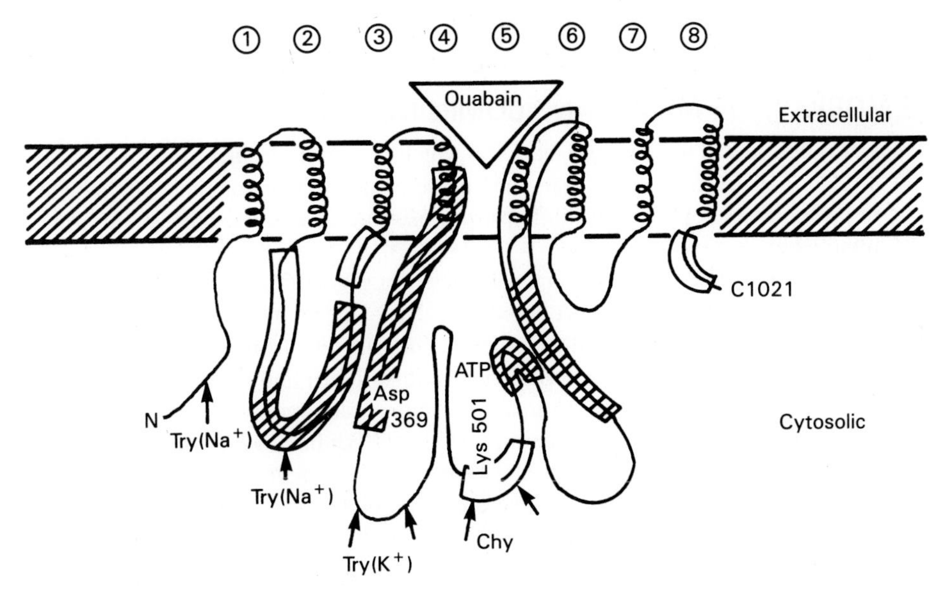

Fig. 2.20 A possible model for folding of the Na^+,K^+-ATPase catalytic subunit across the plasma membrane. Regions in the sequence where trypsin (Try) and chymotrypsin (Chy) cleave the native protein from the cytosolic side of the membrane are indicated by arrows. The sites indicated by (Na^+) and (K^+) are the primary trypsin cleavage sites when either sodium or potassium is bound to the transport sites. The hydrophobic regions proposed to be transmembrane are designated by the circled numbers one through eight. The aspartate residue at position 369 in the sheep kidney α sequence accepts phosphate as an intermediate in ATP hydrolysis and lysine 501 has been implicated in the ATP binding site by affinity labeling. Regions with sequence homology to sarcoplasmic reticulum Ca^{2+}-ATPase are indicated by open boxes and regions with homology to both Ca^{2+}-ATPase and bacterial K^+-ATPase are indicated by shaded boxes. The triangle indicates the binding site for ouabain, an inhibitor of Na^+,K^+-ATPase. Reprinted with permission from L. Cantley, *Trends in Neuro-Sciences,* 9:1–3. Copyright © 1986 Elsevier Science Publishers, London.

hydrophobic portions in Fig. 2.20 correspond to those shown in Fig. 2.19 and are thought to traverse the bilayer (indicated by the shaded portion) as α-helices.

Figure 2.20 also shows sites of proteolytic cleavage that indicate the side of the membrane on which the polypeptide segment is located. The cation indicated (Na^+ or K^+) denotes that the proteolytic effect depends on whether Na^+ or K^+ is bound to the transport sites. As discussed later (Chapter 15), the conformation of the transport proteins changes during transport. In the case of Na^+,K^+-ATPase, the conformational changes require Na^+ or K^+ binding to specific groups.

As suggested by this example, in order to determine the topology of a membrane protein, the information provided by hydrophobicity properties must be supplemented by other data that pinpoint the location of specific segments. Hydrophobicity alone can be misleading, since hydrophobic segments exist in soluble proteins (in the interior of the folded polypeptide).

Integral proteins have been classified on the basis of how many times the polypeptide spans the membrane. *Monotopic* proteins, although associated hydrophobically with the membrane, do not span the membrane entirely. They are relatively rare. *Bitopic* proteins cross the membrane once, and *polytopic* proteins cross it more than once (such as the protein represented in Fig. 2.20).

Our knowledge of the structure and function of integral proteins will certainly be refined with more information. The possibility of producing mutations in the cDNA and therefore amino acid substitutions in the transport proteins should play an important role in future progress.

V. THE MOBILITY OF MEMBRANE COMPONENTS

The lipid bilayer is considered to be a fluid matrix in which protein molecules are able to move (Singer and Nicolson, 1972). In addition to moving in the plane of the membrane, components could rotate in position and conceivably flip-flop from one interface to the other or from one lipid leaflet to the other.

That membrane fluidity does allow movement has been most graphically demonstrated using two different antibodies labeled with distinct fluorescent dyes. One antibody is prepared against the membrane antigens of mouse cells and the other against antigens from human cells. When one of the antibodies is added to the appropriate cells, it distributes uniformly, indicating that the antigens are evenly distributed. After fusion of the human and mouse cells, the antigens, each in its corresponding domain, rapidly intermix (Frye and Edidin, 1970).

The rate of diffusion in the membrane has also been studied through the use of spin-labeled components. Synthetic lipids containing a radical are incorporated into membranes. Since an electron of the radical is unpaired, the labeled molecules are paramagnetic. The *electron paramagnetic resonance* (EPR) spectrum of the spin label indicates the fluidity of the layer (Marsh, 1975) and in this case is consistent with fluidity of the lipid membrane. The intensity of the spectrum is also a function of the concentration of the label: because of spin-spin interactions, the higher the concentrations, the less distinct the lines. This fact leads to experiments in which the diffusion of components in the plane of the membrane is followed solely by using EPR. A spot of labeled lipid is placed in a model system (such as a multilayer of lipid). Diffusion of the label away from the concentrated spot increases the sharpness of the lines, permitting calculation of diffusion coefficients (Devaux and McConnell, 1972). The results obtained with this and other techniques are listed in Table 2.5 (Marsh, 1975) and Table 2.6 (Edidin, 1974; Marsh, 1975). The diffusion coefficients are generally higher than 1×10^{-8} cm^2/s (for comparison, note that the diffusion coefficient of ribonuclease in water is about this value). From the values of the diffusion coefficient, the viscosity of the membrane can be calculated; it corresponds to about 1 to 6 poises (similar to motor oil).

These experiments show that diffusion of the lipids in the plane of the membrane takes

Table 2.5 Lipid Lateral Diffusion Coefficients in Phospholipid Bilayers and Membranes (Determined by Spin Label Measurements)

Bilayer/membrane	D (cm²/s) × 10^8
Dipalmitoyl PC, 50°C	3
Egg PC, 25°C	2
Egg PC: cholesterol (4:1), 40°C	12
Sarcoplasmic reticulum membrane, 37°C	6
E. coli membrane, 40°C	3
Liver microsomes, 30°C	11

Source: Marsh (1975). Reprinted by permission from *Essays in Biochem.* 11:139–180, copyright © 1975 The Biochemical Society, London.

place at a fairly high rate. What can be said about the rate of flip-flop of a lipid from one leaflet of the bilayer to the other? When the spin label is reduced, it is no longer paramagnetic and therefore cannot be detected. This observation suggests an experimental design. Since ascorbate does not penetrate liposomes (artificial vesicular structures made up of lipids), it can only reduce the label present in the outer leaflet of the bilayer. After an initial reduction of the label, any subsequent reduction must be the result of a flip-flop from the inner layer. Therefore, the rate of disappearance of the label after the initial reduction of the outer leaflet serves as a measure of the rate of flip-flop. In model systems this rate has generally been found to be extremely low, on the order of hours (Kornberg and McConnell, 1971; Rothman and Dawidowicz, 1975). However, experiments with vesicles derived from *Electrophorus electricus* using the same

Table 2.6 Diffusion of Lipids in Natural Membranes and Their Extracted Lipids

Membrane	D (cm²/s) × 10^8	Temperature (°C)	Method
E. coli	1.8	31	NMR
Sciatic nerve	0.5	31	NMR
Sciatic nerve lipids	0.8	31	NMR
Sarcoplasmic reticulum	0.6	8	NMR
	1.8	25	NMR
	1.0	50	NMR
	2.5	25	ESR[a]
	6.0	40	ESR[a]
Liver microsomes	9.5	20	ESR[b]
	11.0	30	ESR[b]
	13.7	40	ESR[b]
Sarcoplasmic reticulum lipids	0.4	31	NMR
	10.0	40	ESR[a]
Electroplax membrane	≧0.1	33	NMR

Plasma membrane	D (cm²/s) × 10^9	Method	Label
Cultured rat muscle fiber	1–2	Spread of a fluorescent spot	Fluorescent antibody fragment
Amphibian disks	4–5	Randomization of bleached rhodopsin molecules	Retinol

Source: Edidin (1974). Reproduced, with permission, from the *Annual Review of Biophysics and Bioengineering,* Volume 3, © 1974 by Annual Reviews Inc.

[a] Labeled lecithin.

[b] Labeled fatty acid.

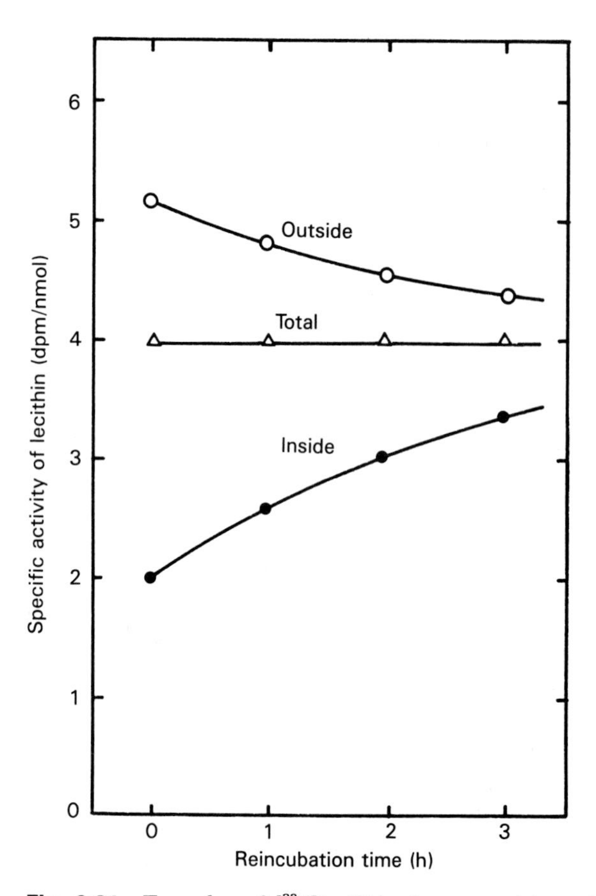

Fig. 2.21 Transfer of [^{32}P]lecithin from outside to inside of rat erythrocyte membranes. Asymmetric labeling was achieved by incubating nonradioactive red cells in ^{32}P-labeled plasma for 2 h at 37°C. Then the plasma is removed by washing and the labeled erythrocytes are reincubated in a salt buffer for several hours at 37°C. Reprinted with permission from W. Rennoij, et al., *European Journal of Biochemistry,* 61:53–58. Copyright © 1976 Springer-Verlag, New York, NY.

method suggest very rapid flip-flop, with a half-time of a few minutes (McNamee and McConnell, 1972). Such rapid flip-flop would make it extremely difficult to understand the asymmetry of the lipids in biological membranes just discussed. The results could also be explained by assuming that the ascorbate enters the *Electrophorus* vesicles in an unexpected manner.

Another study using an entirely different approach with rat red blood cell membranes has reached very different conclusions. The lipids of the outer leaflet are labeled with ^{32}P, and then their fate and location are followed over time by incubating the membranes in an unlabeled salt solution. Incorporation of label in the outer leaflet is accomplished by incubation with plasma containing ^{32}P-labeled phospholipids. After the labeling, the distribution of the label at any time is estimated by digesting the lipids with phospholipases, which digest only the lipids of the outer leaflet. The remaining radioactivity corresponds to that of the internal leaflet. The results are presented in Fig. 2.21 (Rennoij et al., 1976), which shows the radioactivity of phosphatidylcholine (lecithin) in the membranes of the cells as a function of incubation time. The total specific activity (and in fact the radioactivity) remains the same over time. However, the labeled lecithin appears increasingly in the internal leaflet, suggesting that flip-flop takes place with a half-time of 1 to 2 h. Although the rate is much lower than that reported for the *Electrophorus* vesicles, it is still much higher than that of synthetic systems (Rennoij et al., 1976). Experiments studying the flip-flop in human red blood cells suggest the involvement of an enzyme, an amino phospholipid translocase that transfers the amino lipids to the inner membrane leaflet (see review of Devaux, 1990). The enzyme would be responsible for the phospholipid asymmetry.

The motion of proteins in the membrane has also been studied, usually by techniques that depend on photobleaching. When pigments or chromophore-labeled proteins are exposed to light of sufficient intensity, the energy absorbed produces chemical changes so that the molecule is bleached. The recovery of the system from photobleaching can be monitored with time by measuring the pigment's absorption of a nondamaging light beam (either of low intensity or of short duration). The earlier studies involved rhodopsin, the visual pigment in the rods of the retina (Cone, 1972). Rhodopsin in situ is dichroic when the electric vector of the light is parallel to the surface of the disk membranes of the rods, indicating that the chromatophores are parallel to the surface of the disk membranes. This is so because 11-*cis*-retinal, the chromatophore of rhodopsin, is dichroic. Dichroism is a consequence of absorbing light differently depending on its polarization, in this case most strongly when the electric vector is parallel to the long conjugated chain of the molecule. However, when the rods are viewed end-on they are not dichroic, indicating that there is no preferred orientation of the chromatophore in relation to the long axis of the rod. When the rods are bleached with a beam of end-on polarized light, the rhodopsin becomes dichroic since it is bleached differentially. Disappearance of the dichroism should indicate the rate of rotation. The results are shown in Fig. 2.22 (Cone, 1972). The upper line indicates the dichroism ratio (absorbance of a parallel beam/absorbance of a perpendicular beam) of the rods. The lower line represents identical experiments in which the rhodopsin was immobilized by the fixative glutaraldehyde, which cross-links the molecules. The unfixed rods recover rapidly with a recovery time of about 20 μs, whereas the fixed rods do not recover.

Similarly, the diffusion of rhodopsin can be measured by bleaching it with a large spot of light illuminating the retinal rod (Poo and Cone, 1974). The return of the rhodopsin to the bleached spot (which can be monitored by following the absorption of light of the appropriate wavelength in the same location) provides a value for its diffusion coefficient, which was found to be approximately 3–5 $\times$ 10^{-9} cm^2/s.

The idea that proteins diffuse freely in the plane of the membrane cannot be entirely right.

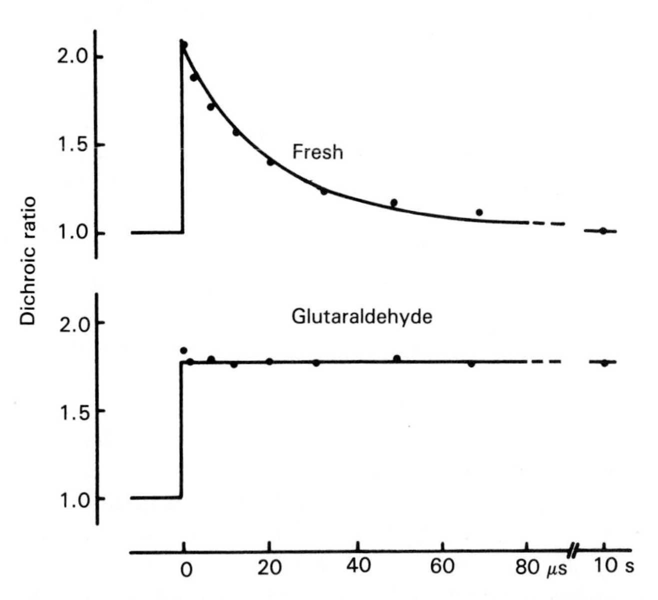

Fig. 2.22 Dichroic ratios plotted as functions of time after the actinic flash (Cone, 1972). Reprinted by permission from *Nature* 236:39–43, copyright © 1972 Macmillan Magazines Ltd.

Receptors at the cell surface are known to be at specific locations or patches. For example, at chemical synapses or at neuromuscular junctions we would expect a concentration of acetylcholine receptors (Axelrod et al., 1976). In fact, at least a fraction of a number of proteins have been found to be essentially immobile (e.g., Schlessinger, 1983). A number of mechanisms could account for the immobility (e.g., cross-linking to other surface components to form a meshwork). Surface interactions are clearly important, since the external domain of the glycoproteins has been shown to slow down their lateral diffusion (Wier and Edidin, 1988). However, it is most likely that the attachment of the intrinsic proteins to cytoskeletal elements is the primary factor, as represented in Fig. 2.5 for the case of the red blood cell. In this figure, ankyrin, spectrin, and actin are shown to be involved in holding the band 3 protein (anion channel) immobile (Schlessinger, 1983). There is good evidence for this model and for the conclusion that glycophorin specifically binds the band 4.1 protein (Anderson and Lovrien, 1984).

VI. INTERNAL MEMBRANES

In Chapter 1 we saw that the cell's compartments perform separate complex tasks. The membranes surrounding these compartments are responsible for maintaining the internal environment of the organelles and actively controlling the accessibility of the compartment. Furthermore, the internal membranes of mitochondria or the thylakoid lamellae of chloroplasts are involved in energy transduction (see Chapters 10–12). In this section we will primarily discuss the role of the nuclear and mitochondrial membranes in controlling exchanges with the cytoplasm.

A good deal is known about the permeability of the various cellular inclusions such as lysosomes (Lloyd, 1969) and endoplasmic reticulum (Nilsson et al., 1973), which will not be discussed here.

A. The Nuclear Envelope and the Nucleopore Complex

By virtue of its location, enclosing the chromosomes, the nuclear envelope must play a central role in the development and physiology of the cell. The envelope is important in controlling the traffic of RNA produced by transcription and the passage in or out of the nucleus of the macromolecules involved in the control of genetic expression.

Recent studies carried out with cells in culture show that the transport through the nuclear envelope depends on the phase of cellular activity (Feldherr and Akin, 1990). Furthermore, the location of certain proteins, whether in the nucleus or in the cytoplasm, is thought to reflect the developmental stage of cells. This is indicated by experiments (Dreyer et al., 1982) in which the fates of four proteins present in the oocyte nuclei of *Xenopus*, the African frog, were followed in early development, from cleavage to neurula, by using fluorescently labeled monoclonal antibodies. In the early stages the proteins became isolated in the cytoplasm and shifted to a nuclear location at stages specific for the individual proteins. These experiments suggest a significant role of these proteins and of the transport mechanism in development. Since RNA is predominantly synthesized in the nucleus, whereas protein synthesis occurs predominantly in the cytoplasm, it follows that the nuclear envelope must also play a role in the outward passage of newly synthesized RNA.

The nuclear envelope is made up of two membranes, joined at the openings of the numerous *nucleopore complexes*. In these complexes (Unwin and Milligan, 1982) eight spokes form a diaphragm extending into the lumen of the pore. A large central particle approximately 35 nm in diameter is frequently present at the center of the lumen and is thought to have a role in transport (see below), and particles of approximately 22 nm are frequently attached to the cytoplasmic rim of the pore. The structural arrangement is shown in Fig. 2.23 and Fig. 2.24 (Akey, 1989; Unwin and Milligan, 1982).

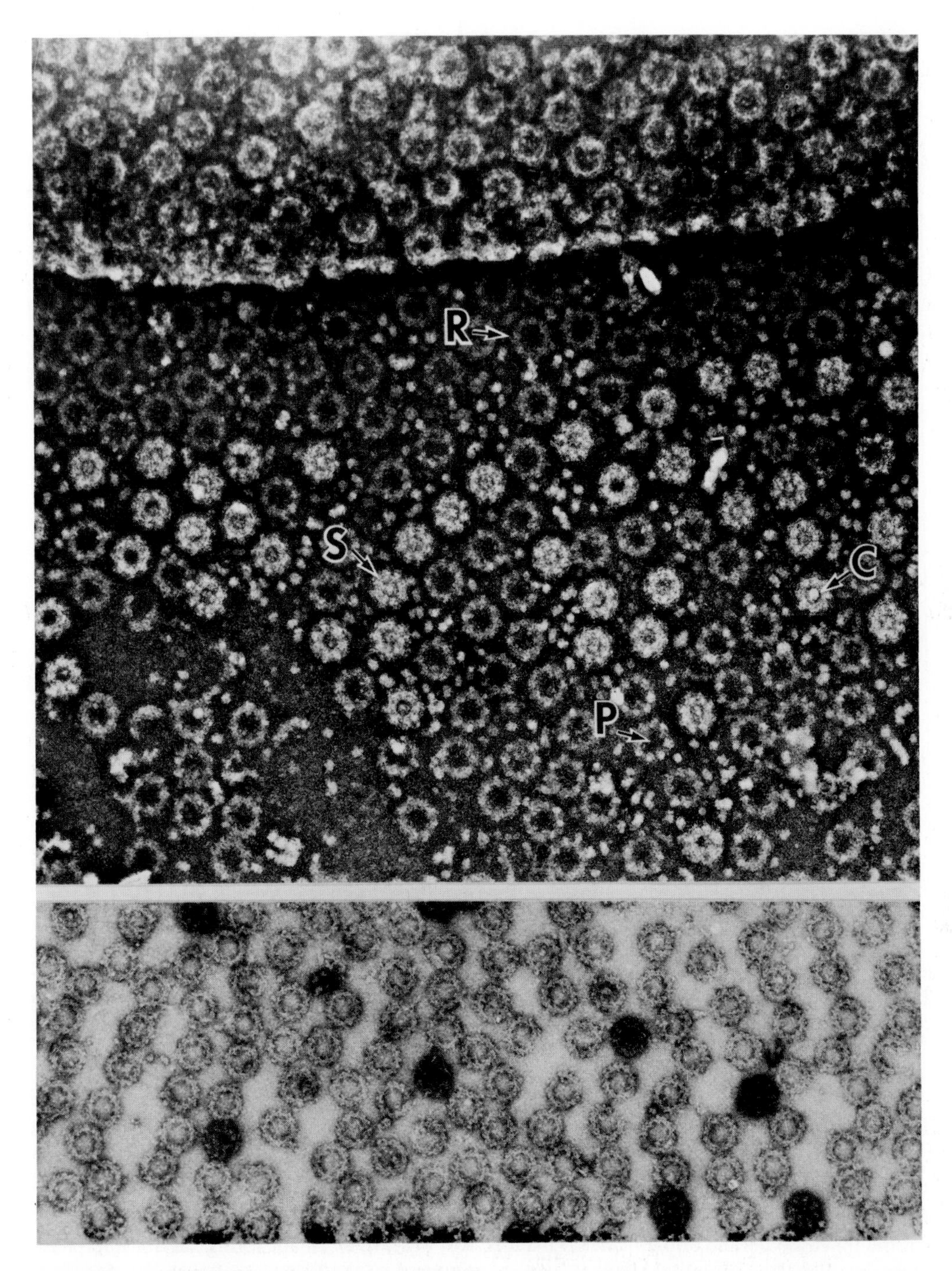

Fig. 2.23 Constituents of the nuclear envelope released after its partial detachment from a polylysine-treated carbon film in the presence of 0.1% Triton X-100. The constituents most obviously related to the pore complex are rings (R), central plug (C), spokes (S), and particles (P), occasionally observed around the rings. As the lower micrograph shows, the rings are sometimes obtained in large numbers by themselves. Uranyl acetate stain (×60,000). Reprinted with permission from Unwin and Milligan (1982).

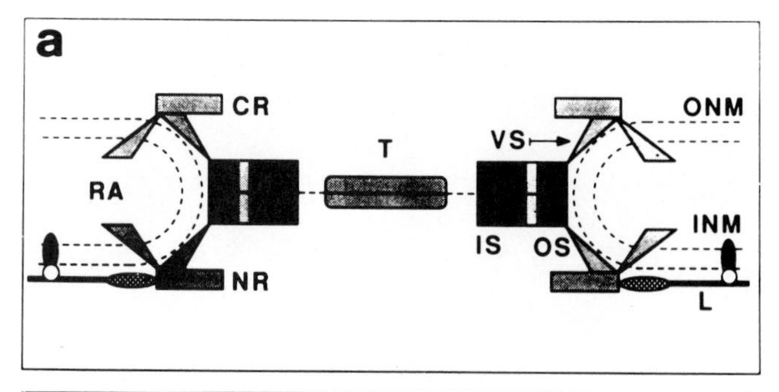

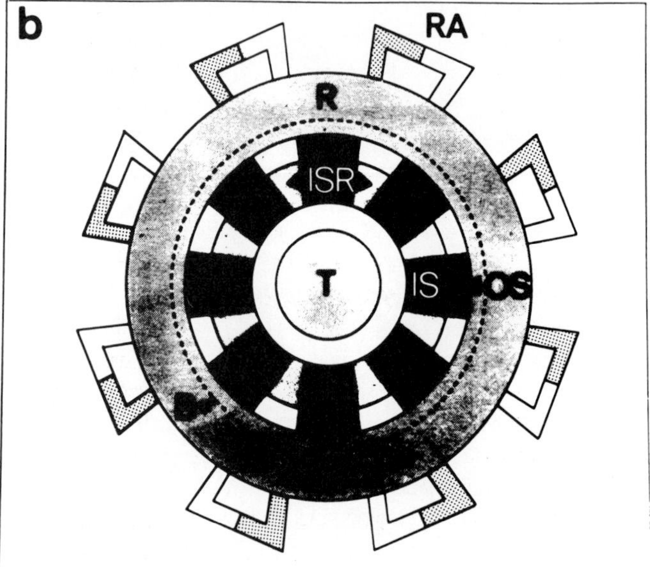

Fig. 2.24 Diagram of a nuclear pore complex (*a*) in central cross section and (*b*) seen from above. The major structural domains are indicated: inner spokes (IS), outer spokes (OS), vertical supports (VS), cytoplasmic coaxial ring (CR), nucleocytoplasmic coaxial ring (NR), and radial arms (RA). T indicates the transporter, which presumably has a central channel (not shown), and L the lamina attachment (Akey and Goldfarb, 1989). Reproduced from the *Journal of Cell Biology,* 1989, vol. 109, pp. 955–970 by copyright permission of the Rockefeller University Press.

Most eukaryotic nuclei contain a distinct electron-dense layer, the *lamina,* between the inner membrane and adjacent chromatin elements. In some other eukaryotic cells the lamina is not prominent but the presence of analogous elements is suspected. The laminae are likely to disassemble reversibly during mitosis, possibly in a process regulated by phosphorylation-dephosphorylation cycles (Gerace et al., 1984). The lamina is thought to serve as a framework for the inner membrane of the nuclear envelope and to serve as an anchoring site for interphase chromosomes (Gerace et al., 1978). Similarly, the fibers of the lamina anchor the nuclear pore complexes (Aaronson and Blobel, 1975).

The pore complexes have most commonly been isolated from rat liver and *Xenopus* oocytes. Extraction of the isolated complexes with high salt concentrations and detergent has revealed relatively few major proteins. A 190-kDa glycoprotein (gp 190) that binds concanavalin A has been isolated, and antibodies to this protein, covalently attached to ferritin, label the cytoplasmic surface of the complex and suggest that 25 copies are present per pore (Gerace et al., 1982).

Another protein of 62 kDa (Davis and Blobel, 1986), which binds wheat germ agglutinin (WGA), has also been identified as part of the complex. Ferritin-labeled WGA also locates it on the cytoplasmic surface (Davis and Blobel, 1986). In rat liver, 15 WGA are bound per pore, suggesting that there are many copies of the protein per pore.

The effect of WGA on nucleocytoplasmic exchanges suggests a role of the 62-kDa protein in the translocation of macromolecules through the nucleopores. WGA blocks the passage of ^{14}C-prelabeled ribonucleoproteins from isolated rat liver nuclei (Baglia and Maul, 1983) and the influx of fluorescently labeled nucleoplasmin (a nuclear protein, which is discussed later) in a similar preparation (Finlay et al., 1987).

Much of our knowledge of how macromolecules are transferred through the nuclear envelope has been obtained by microinjecting macromolecules into either the cytoplasm or nucleus and tracing their passage with the help of fluorescent labels (for light microscopy), radioactive labels (for autoradiography or other techniques), or colloidal gold (for electron microscopy). The large size of the amphibian oocyte and its nucleus, such as that of *Xenopus*, has made this cell the choice for micromanipulations, although the same principles seem to apply when other cells such as somatic mammalian cells are used.

Passive Transport The microinjection of radioactively labeled dextrans into the cytoplasm shows that the size of the molecule is significant in controlling the influx into the nucleus. Dextrans of increasing size have increasing difficulty in entering, as shown in Fig. 2.25 (Paine et al., 1975). The diameter of a pore with the permeability properties shown corresponds to 4.5 nm, approximately the size of a globular protein of 60 kDa. Figure 2.25 shows the concentration of probe in the nucleus over that in the cytoplasm as a function of time for three probes of different sizes. The results are consistent with entry occurring passively through the nucleopores. After equilibration, the apparent concentration of the probe molecules is higher in the nucleus, probably because of binding by the nuclear components.

In addition to this sievelike effect, it has become obvious that special mechanisms are also present, since much larger particles can enter. Microinjection of colloidal gold coated with *nucleoplasmin* (Fig. 2.26) shows that the particles enter rapidly through the nucleopores (Feldherr et al., 1974). Figure 2.26*a* shows the colloidal gold in the cytoplasm just after injection, and Fig. 2.26*b* and *c* show the entry into the nucleus through the nucleopore complex (part *b*). Nucleoplasmin is a protein of 165 kDa extracted from the nucleus of *Xenopus* oocytes.

Active Translocation The inhibition of protein macromolecular transport by WGA, which binds the 62-kDa protein of the nucleopore complex, and the demonstration of the passage of large particles coated with a nuclear protein suggest a special and complex role of the nucleopore in nuclear-cytoplasmic transport and interactions.

Other experiments show that this is in fact the case (Dingwall et al., 1982; Dingwall and Laskey, 1986). Nucleoplasmin can be systematically cleaved with proteolytic enzymes. The protein is a pentamer, each peptide of which possesses a tail segment (see Fig. 2.27). Digestion can produce a core pentamer segment missing some or all of the tail segments and, of course, the corresponding free tail segments. Any portion containing a tail, including a single tail segment, is capable of being rapidly transferred into the nucleus. The naked core segment, however, is not transferred in either direction. This has been demonstrated in experiments in which the protein is radioactively labeled and, after the appropriate hydrolysis with a protease, injected into the cytoplasm or the nucleus of the cells. The procedure and the results are shown in Fig. 2.27 (Dingwall and Laskey, 1986). These results suggest that the nucleoplasmin has a specific amino acid sequence that serves as a signal, analogous to the signal sequences that determine the transfer of nascent polypeptide chains into the cisternae of the endoplasmic

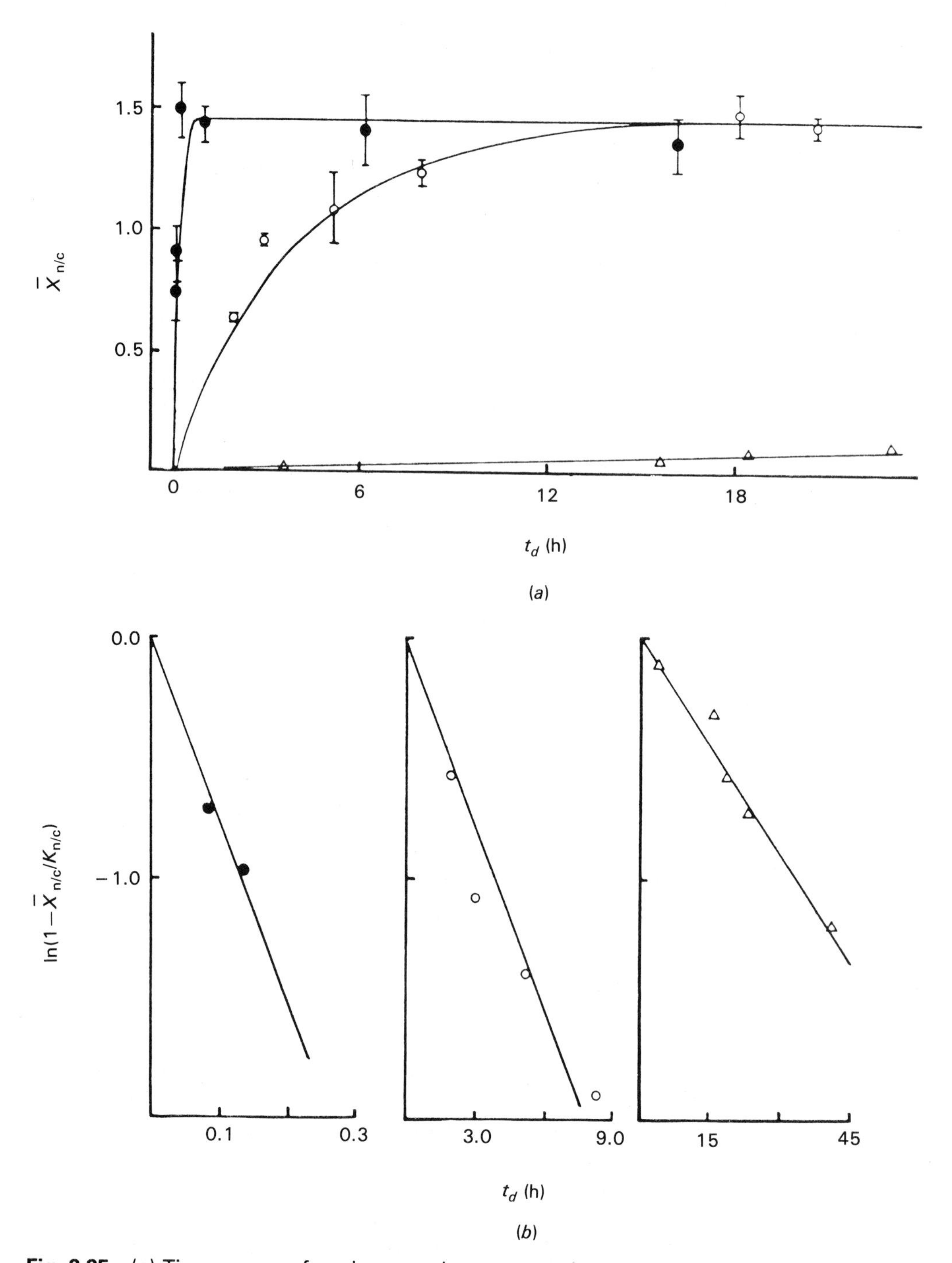

Fig. 2.25 (*a*) Time course of nuclear envelope permeation, expressed as average nuclear/cytoplasmic grain density $\overline{X}_{nc}$ as a function of diffusion time after microinjection, t_d, for 1.2 (●), 2.33 (○), and 3.55 (△) nm ^{3}H-labeled dextrans. Vertical bars show standard error of the mean. (*b*) First-order exponential entry kinetics illustrated by plots for 1.2-, 2.33-, and 3.55-nm ^{3}H-labeled dextrans, respectively. K_{nc} is assumed to be 1.45. Note the different time scales on the abscissae (Paine et al., 1975). Reprinted by permission from *Nature* 254:109–114, copyright © 1975 Macmillan Journals Ltd.

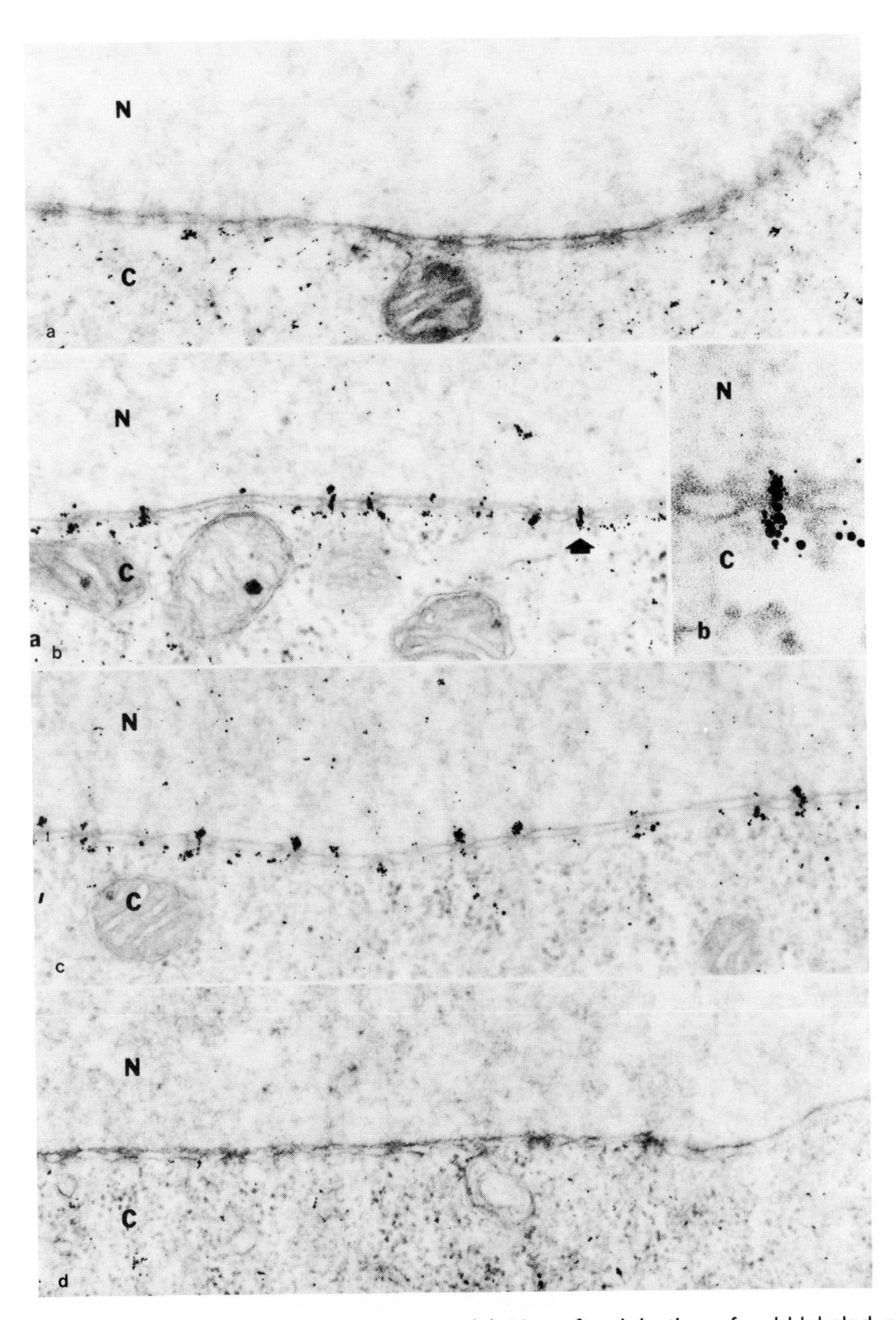

Fig. 2.26 Nucleoplasmin gold injection (*a*) 10 s after injection of gold-labeled nucleoplasmin. The gold is present only in the cytoplasm (C), ×60,000. (*b*) At 15 min after injection, gold is present in the cytoplasm and the nucleus (N). The accumulation can be seen next to and in the pore, ×60,000 (inset ×200,000). (*c*) At 1 h after injection, the concentration in the nucleus is higher, ×60,000. (*d*) Control: experiment in which the nucleoplasmin has been hydrolyzed. Gold is present only in the cytoplasm, ×60,000. Reprinted with permission from Feldherr et al. (1984).

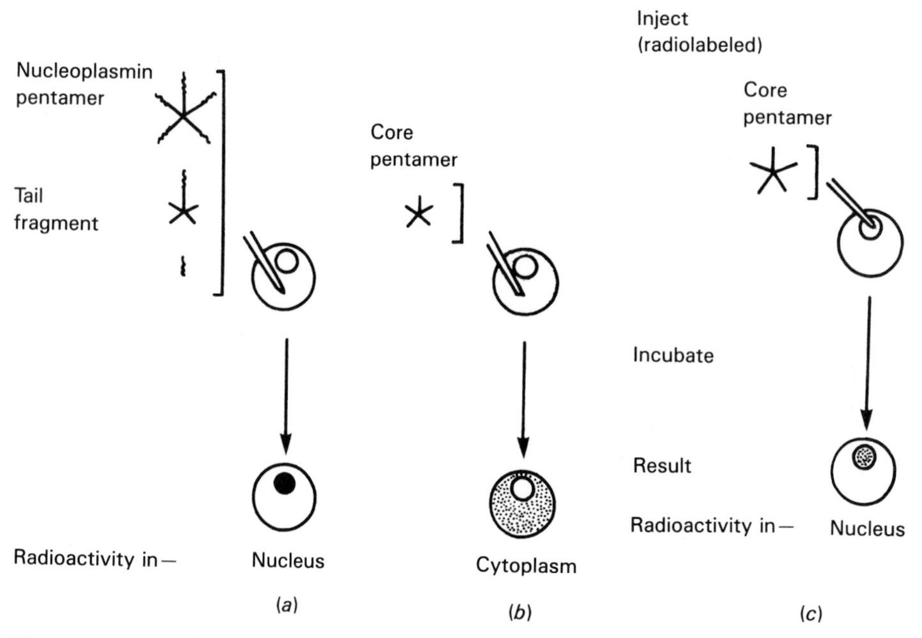

Fig. 2.27 Diagram illustrating the transport of nucleoplasmin molecules into the *Xenopus* oocyte nuclei. (*a*) The intact nucleoplasmin pentameter, a pentameter with a single intact subunit, and the isolated "tail" fragment can accumulate in the nucleus after microinjection in the cytoplasm. (*b*) The nucleoplasmin "core" molecule cannot enter the nucleus after microinjection in the cytoplasm. (*c*) The core cannot reach the cytoplasm when microinjected in the nucleus (Dingwall and Laskey, 1986). Reproduced, with permission, from the *Annual Review of Cell Biology* 2, © 1986 by Annual Reviews Inc.

reticulum or into the membrane structure itself. Many other experiments have demonstrated a role of signal sequences, although apparently many different sequences can act in this fashion (Dingwall et al., 1986; Silver et al., 1989). These sequences have been referred to as *nuclear localization sequences* (NLSs). In some cases, unlike nucleoplasmin, the NLSs are in the middle of the polypeptide chain (e.g., Nelson and Silver, 1990). In other cases, it has been shown that translocation requires two independent NLSs (see Underwood and Fried, 1990).

How do NLSs provide the necessary information to trigger translocation? As a first step it would seem likely that the NLSs bind to receptor proteins. These receptors could be on the nucleopore complex. Alternatively, there is evidence for cytoplasmic receptors that bind the NLSs before the complex of the two binds to the nucleopore components (Adam and Garace, 1991).

A seven-amino-acid signal sequence has been identified and synthesized. The entry of the protein can be recognized by microinjection into the amphibian oocyte test system. When the sequence is covalently attached to a nonnuclear protein, it directs the entry of the protein into the nucleus. This has been observed even for a protein such as ferritin, of 465 kDa and 9.4 nm diameter (Goldfarb et al., 1986; Lanford et al., 1986), or coated colloidal gold particles as large as 28 nm in diameter. Individual pores are capable of recognizing and transporting proteins that contain different nuclear targeting sequences (Dworetzky et al., 1988).

Similar principles seem to regulate the nuclear efflux of RNAs. Gold particles coated with RNA and microinjected into *Xenopus* oocytes were found to be transported through the centers of the nucleopores (Dworetzky and Feldherr, 1988). Transfer RNA (tRNA), 5S RNA, and polyadenosine were equally effective in a saturable transport process. Nonphysiological polynucleotides such as polyinosine or polydeoxyadenosine were also transported. Several studies have examined the efflux of initiation tRNA, and genetic studies using mutant tRNA have been able to pinpoint the tRNA sequences that seem to be needed for the transport. The results are summarized in Figs. 2.28 and 2.29 (Tobian et al., 1985). The principles governing the passage of RNA seem to be similar to those controlling protein entry. The signal recognized by the nucleopore complex translocation system, however, may not be in the RNA. The RNA could bind specifically to a protein containing the NLS.

The transfer of the small nuclear riboproteins (snRNPs) is particularly interesting. snRNPs are involved in the splicing of pre-mRNA. Since they are assembled in the cytoplasm, they must be translocated into the nucleus where the mRNA is processed. The protein portion alone cannot be transported. Therefore, the NLS must reside in the RNA portion or result from an snRNA-protein interaction. The snRNAs contain a m_3G cap (m_3G = 2, 2, 7-trimethylguanosine) with the sequence $m_3G_{ppp}G$. Injection of modified snRNAP into the *Xenopus* oocyte system indicates that the cap has a signaling role (Fischer and Lührmann, 1990; Hamm et al., 1990). However, the situation is more complex since for optimal translocation a protein component was needed (Hamm et al., 1990).

The ability of nuclei to transfer RNA and protein specifically poses an important question. Are the pores specialized, some being responsible for the transfer of proteins and perhaps others for RNA? The transfers of nucleoplasmin and RNA can be followed simultaneously with the electron microscope when the two are labeled with gold particles of different sizes. The two transfers appear to proceed through the same pore even though they proceed in opposite directions (Dworetzky and Feldherr, 1988). In fact, a monoclonal antibody of nuclear pore complex proteins, when injected into *Xenopus* oocytes, inhibits the active import of proteins and the export of RNA. However, it has no effect on passively transported proteins (Featherstone et al., 1988).

A number of studies using colloidal gold coated with test molecules have implicated the nucleopores in the transport. The analysis has been carried one step further in a recent study (Akey and Goldfarb, 1989). Probe-coated gold particles were injected into oocytes and the nuclei examined at different times using cryoelectron microscopy and image processing (see Chapter 1). The probes consisted of an antibody to a nuclear pore protein, WGA, and nucleoplasmin. The probes bind to similar sites. The results obtained with nucleoporin suggest the time sequence illustrated in Fig. 2.30. First, the probe molecule seems to be predominantly bound to the periphery (part 1). This is then followed by positioning at the center of the central particle thought to have a role in transport (and therefore called the transporter in this model) (part 2, docking). Finally, the probe molecule passes through the transporter, which is hypothesized to open (part 3), although currently there is no evidence for the latter mechanism.

Detailed studies of the molecular mechanisms require in vitro systems. These studies have begun using isolated nuclei from rat liver maintained in the presence of *Xenopus* egg extracts, and it has already become evident that ATP is required for the entry of labeled nucleoplasmin (Newmeyer et al., 1986).

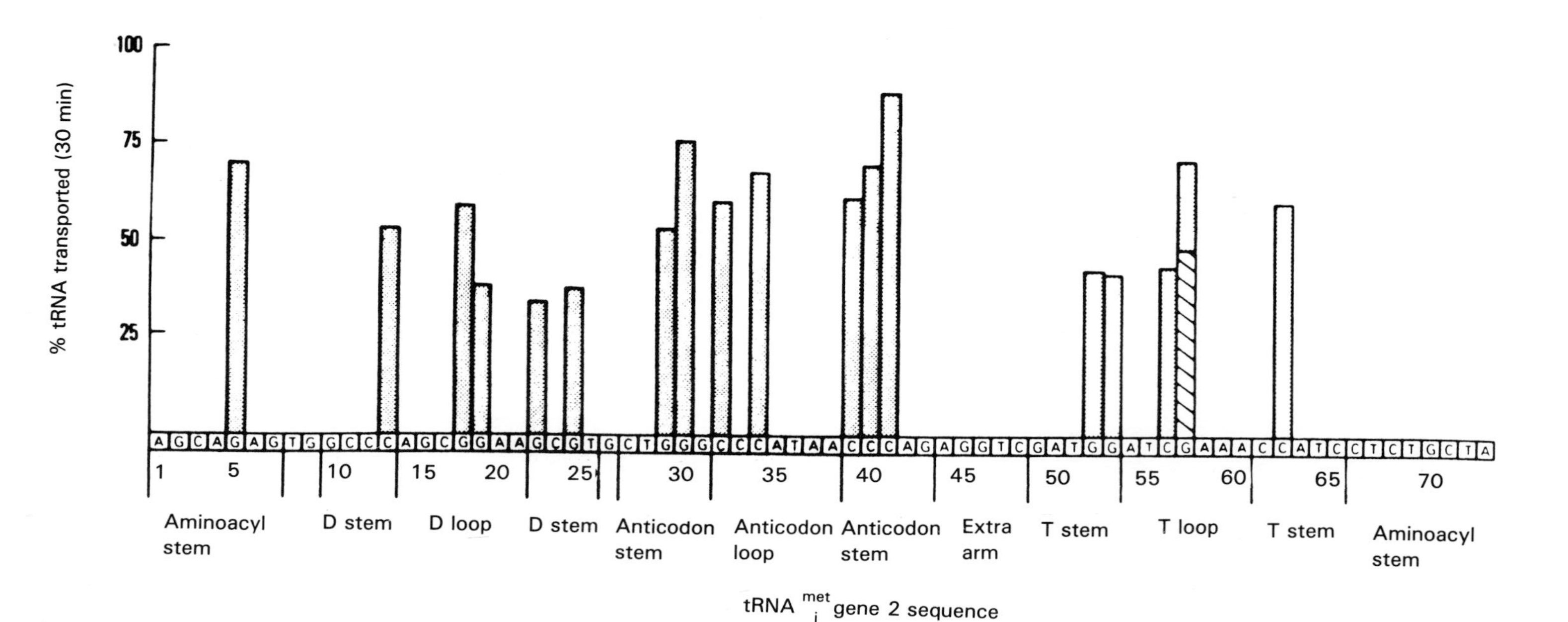

Fig. 2.28 Summary of transport kinetics: percent of mature tRNA transported at 30 min. The tRNA$_i^{met}$ gene sequence is shown at the bottom of the graph with position 17 deleted (to conform to the conventional tRNA numbering scheme). Bars represent percent tRNA transported to the cytoplasm 30 min after intranuclear injection of gel-purified ^{32}P-labeled tRNA with a mutation at that position. At position 57, the hatched bar indicates the transport properties of the natural variant that has a G-to-T transversion and the stippled bar represents the transport properties of the mutant that has a G-to-A transition. At 30 min, 80% of the wild-type tRNA$_i^{met}$ has been transported to the cytoplasm. Reprinted with permission from Tobian et al. (1985), copyright © 1985 by Cell Press.

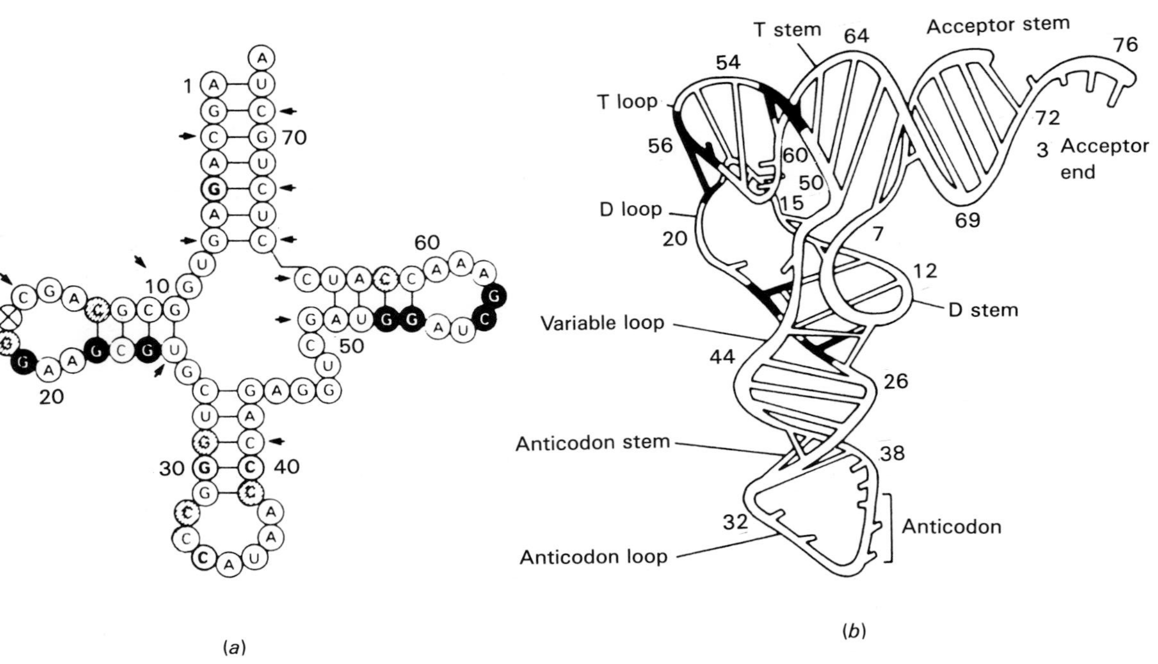

Fig. 2.29 Secondary and tertiary folding of tRNA$_i^{met}$: localization of mutations resulting in transport-defective phenotypes. (*a*) Secondary structure indicating mutations resulting in transport-defective tRNAs. (Solid black circle) 30 to 50% transported at 30 min; (hatched circle) 50 to 65% transported at 30 min; (stippled circle) 65 to 70% transported at 30 min. (Arrows) Positions of mutations that yielded almost wild-type or wild-type transport values in the initial screening and in some cases in kinetic experiments. (*b*) Tertiary structure indicating positions of mutations that produce severely defective tRNAs (corresponding to mutations represented by black crosses in *a*). Reprinted with permission from Tobian et al. (1985), copyright © 1985 by Cell Press.

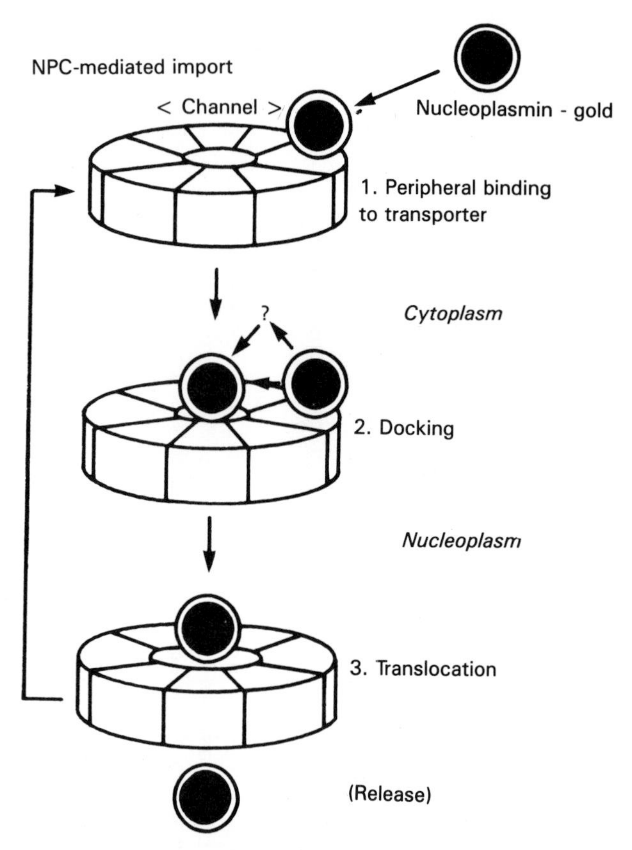

Fig. 2.30 Possible three-step mechanism of nuclear import at the level of the nuclear pore complex (NPC) transporter. The simple model of nuclear import involves (1) specific binding of transport substrates at the perimeter of the transporter (radius 10 to 12.5 nm), possibly involving O-linked *N*-acetylglucosamine-containing nucleoporins; (2) docking of substrate over the center of the closed transporter; and (3) opening of the transporter resulting in translocation, possibly followed by a release step. The translocation step is shown schematically as a dilation of the transporter; however, the actual mechanism is not known (Akey and Goldfarb, 1989). Reproduced from the *Journal of Cell Biology,* 1989, vol. 109, pp. 955–982, by copyright permission of the Rockefeller University Press.

Table 2.7 Mitochondrial Metabolite Transporters

Category	Name	Physiological substrates	Inhibitors: Specific	Inhibitors: Sulfhydryl reagent
Electroneutral proton compensated	Phosphate	Phosphate, arsenate		Organic mercurials, *N*-ethylmaleimide
	Glutamate	Glutamate	Avenaciolide	*N*-Ethylmaleimide
	Pyruvate	Monocarboxylic acids, ketone bodies, branched-chain ketoacids	α-Cyano-3-hydroxycinnamate	Organic mercurials, *N*-ethylmaleimide
	Ornithine	Ornithine, citrulline, lysine		
Electroneutral anion exchange	Dicarboxylate	Phosphate, malate, succinate, oxaloacetate	Butylmalonate, bathophenanthroline, iodobenzylmalonate, phenylsuccinate, phthalonate	Organic materials
	α-Ketoglutarate	Malate, α-ketoglutarate, succinate, oxaloacetate	Phthalonate, bathophenanthroline, phenylsuccinate, butylmalonate	
	Tricarboxylate	Citrate, isocitrate, phosphoenolpyruvate, malate, succinate	1,2,3-Benzene tricarboxylate, α-cetylcitrate, bathophenanthroline	
Neutral	Carnitine	Carnitine, acylcarnitine	Sulfobetaines	Organic mercurials, *N*-ethylmaleimide
	Neutral amino acids	Neutral amino acids		Organic mercurials
	Glutamine	Glutamine		Organic mercurials
Electrophoretic	Glutamate-aspartate	Glutamate/aspartate	Glisoxepide (nonspecific)	
	Adenine nucleotide	ADP, ATP	Atractyloside, carboxyatractyloside, bongkrekate, long-chain acyl CoA, α-cetylcitrate	

Source: Schoolwerth and LaNoue (1985). Reproduced with permission, from the *Annual Review of Physiology,* Volume 47, © 1985 by Annual Reviews Inc.

B. The Mitochondrial and Chloroplast Membranes

Mitochondria are bounded by a double set of membranes. The outer membrane contains channels that allow passage of rather large molecules (Colombini, 1987). The function of these channels is still not well understood.

The internal membranes of mitochondria, studied by a variety of techniques, appear to be a barrier to diffusion similar to the plasma membrane. The permeabilities of cell and mitochondrial inner membranes to nonelectrolytes depend similarly on the oil-water partition coefficient (Tedeschi and Harris, 1955). ATP synthase and the dehydrogenases involved in oxidative phosphorylation are on the mitochondrial matrix side of the inner mitochondrial membrane. Therefore, a relatively impermeable barrier would block the passage of metabolites as well as ADP and ATP, since they are charged and would find it difficult to pass through an ordinary lipid bilayer. In mitochondria as in cells, a system of transport proteins seems to be responsible for these exchanges. The ADP/ATP transporter, which is needed to provide the energy generated by oxidative phosphorylation to the rest of the cell, is prominent among these. The various known transport systems are listed in Table 2.7 (Schoolwerth and LaNoue, 1985).

The inner envelope of the chloroplast encloses the thylakoid vesicles, which carry out photosynthesis, a function analogous to oxidative phosphorylation. This membrane is also provided with many transporter proteins in accordance with the biochemical needs of the plant cells (Heber and Heldt, 1981). In contrast, as in mitochondria, the outer envelope is highly permeable to many low molecular weight substances.

SUGGESTED READING

Devaux, P. F. (1990) The aminophospholipid translocase: a transmembrane lipid pump-physiological significance, *News Physiol. Sci.* 5:53–58.

Edidin, M. (1987) Rotational and lateral diffusion of membrane proteins and lipids: phenomena and function. *Curr. Top. Membr. Transp.* 29:91–127.

Fraser, S. E. (1985) Gap junctions and cell interactions during development. *Trends Neurosci.* 8:3–4.

Gennis, R. B. (1989) *Biomembranes: Molecular Structure and Function*, Chapters 1 to 5. Springer-Verlag, New York.

Goldfarb, D., and Michaud, N. (1991) Pathway for the nuclear transport of proteins and RNAs. *Trends in Cell Biol.* 1:20–24.

Guidotti, G. (1986) Membrane proteins: structure, arrangement, and disposition in the membrane. In *Membrane Physiology*, 2d ed. (Andreoli, T. E., Hoffman, J. F., Fanestil, D. D., and Schulz, S. G., eds.), pp. 45–58. Plenum Medical Book Co., New York.

Hanover, J. A. (1992) The nuclear pore: At the crossroads, *FASEBJ.* 6:2288–2295.

Kleinfeld, A. M. (1987) Current views of membrane structure. *Curr. Top. Membr. Transp.* 29:1–27.

Schlessinger, J. (1983) Mobilities of cell membrane proteins: how are they modulated by the cytoskeleton? *Trends Neurosci.* 6:360–363.

Stein, W. D., and Lieb, W. R. (1986) *Transport and Diffusion Across Cell Membranes*, Chapters 1–3. Academic Press, New York.

Thompson, T. E., and Huang, C. (1986) Composition and dynamics of lipids in biomembranes. In *Membrane Physiology*, 2d ed. (Andreoli, T. E., Hoffman, J. F., Fanestil, D. D., and Schultz, S. G., eds.), pp. 25–44. Plenum Medical Book Co., New York.

Unwin, P. N. T. (1987) Gap junction structure and the control of cell to cell communication. *CIBA Found. Symp.* 125:78–91.

Other Articles of General Interest

Anderson, D. J. (1984) New clues to protein localization in neurons. *Trends Neurosci.* 7:355–357.

Bennett, M. V. L., Spray, D. C., Harris, A. L., Ginzberg, R. D., Decarvalho, A. C., and White, R. L. (1984) Control of intercellular communication by way of gap junctions. *Harvey Lect.* 78:23–57.

Bennett, V., and Davis, J. Q. (1984) Proteins closely related to spectrin and ankyrin general components of cell membranes. In *The Red Cell, Sixth Ann Arbor Conference* (Brewer, G. J., ed.), pp. 457–468. Alan R. Liss, New York.

Branton, D., Cohen, C. M., and Tyler, J. (1981) Interaction of cytoskeletal proteins on human erythrocyte membranes. *Cell* 24:24–32.

Gall, W. E., and Edelman, G. M. (1981) Lateral diffusion of surface molecules in animal cells and tissues. *Science* 213:903–905.

Garcia-Bustos, J., Heitman, J., and Hall, M. N. (1991) Nuclear protein localization. *Biochim. Biophys. Acta* 1071:83–101.

Houslay, M. D., and Stanley, K. K. (1982) *Dynamics of Biological Membranes*, p. 325. Wiley, New York. See Chapters 1–4 and 7.

Jennings, M. L. (1989) Topography of membrane proteins. *Annu. Rev. Biochem.* 58:999–1027.

Koeppel, D. E., and Sheetz, M. P. (1981) Fluorescence photobleaching does not alter the lateral mobility of erythrocyte membrane glycoprotein. *Nature (London)* 293:159–161.

Revel, J.-P. Nicholson, B. J., and Yancey, S. B. (1984) Molecular organization of gap junctions. *Fed. Proc.* 43:2672–2677.

Rutishauser, U. (1984) Developmental biology of a neural cell adhesion molecule. *Nature (London)* 310:549–554.

Schwartzmann, B., Wiegandt, H., Rose, B., Zimmerman, A., Ben-Haim, D., and Lowenstein, W. R. (1981) Diameter of cell-to-cell junctional membrane channel as probed with neutral molecules. *Science* 213:551–554.

Viitala, J., and Jarnefelt, J. (1985) The red cell surface revisited. *Trends Biochem. Sci.* 10:392–395.

Warren, G. (1981) Membrane proteins: structure and assembly. In *Membrane Structure* (Finnean, J. B., and Mitchell, R. H., eds.), pp. 215–257. Elsevier North-Holland, Amsterdam.

Yeagle, P. (1987) *The Membrane of Cells*, p. 292. Academic Press, New York. See Chapters 1–7.

REFERENCES

Aaronson, R. P., and Blobel, G. (1975) Isolation of nuclear pore complexes in association with lamina. *Proc. Natl. Acad. Sci. U.S.A.* 72:1007–1011.

Adam, S. A., and Gerace, L. (1991) Cytosolic proteins that specifically bind nuclear location signals are receptors for nuclear import. *Cell* 66:837–847.

Akey, C. W. (1989) Interactions and structure of the nuclear pore complex revealed by cryo-electron microscopy. *J. Cell Biol.* 109:955–970.

Akey, C. W., and Goldfarb, D. S. (1989) Protein import through the nuclear pore complex is a multistep process. *J. Cell Biol.* 109:971–982.

Anderson, R. A., and Lovrien, R. E. (1984) Glycophorin is linked by band 4.1 protein to the human erythrocyte membrane skeleton. *Nature (London)* 307:655–658.

Axelrod, D., Radvin, P., Koeppel, D. E., Schlessinger, J., Webb, W. W., Elson, E. L., and Podleski, T. R. (1976) Lateral motion of fluorescently labeled acetylcholine receptors in membranes of developing muscle fibers. *Proc. Natl. Acad. Sci. U.S.A.* 73:4594–4598.

Baglia, F. A., and Maul, G. G. (1983) Nuclear ribonucleoprotein release and nucleoside triphosphatase activity are inhibited by antibodies directed against one nuclear matrix glycoprotein. *Proc. Natl. Acad. Sci. U.S.A.* 80:2285–2289.

Bennett, V., and Davis, J. Q. (1984) Proteins closely related to spectrin and ankyrin are general components of cell membranes. In *The Red Cell* (Brewer, G. J., ed.), pp. 457–458. Alan R. Liss, New York.

Cantley, L. (1986) Ion transport systems sequenced. *Trends Neurosci.* 9:1–3.

Collander, R. (1949) Die Verteilung Organischer Verbindunger zwischen Ather und Wasser. *Acta Chem. Scand.* 3:717.

Colombini, M. (1987) Regulation of the mitochondrial outer membrane channel, VDAC. *J. Bioenerg. Biomembr.* 19:309–320.

Cone, R. A. (1972) Rotational diffusion of rhodopsin in visual receptor membrane. *Nature New Biol.* 236:39–43.

Davis, L. I., and Blobel, G. (1986) Identification and characterization of a nuclear pore complex protein. *Cell* 45:699–709.

Davson, H. (1970) *A Textbook of General Physiology,* Vol. 1. Churchill Livingston, Edinburgh (see Fig. 219).

Davson, H., and Danielli, J. F. (1952) *The Permeability of Natural Membranes,* 2d ed. Cambridge University Press, London.

Devaux, P. F. (1990) The aminophospholipid translocase: a transmembrane lipid pump-physiological significance, *News Physiol. Sci.* 5:53–58.

Devaux, P., and McConnell, H. M. (1972) Lateral diffusion in spin-labeled phosphatidylcholine multilayers. *Am. Chem. Soc. J.* 94:4475–4481.

Dingwall, C., Sharnick, S. V., and Laskey, R. A. (1982) A polypeptide domain that specifies the migration of nucleoplasmin into the nucleus. *Cell* 30:449–458.

Dingwall, C., Burglin, T. R., Kearsey, S. E., Dilworth, S., and Laskey, R. A. (1986) Sequence features of the nucleoplasmin tail region and evidence of selective entry mechanism for transport into the cell nucleus. In *Nucleocytoplasmic Transport* (Peters, R., and Trendelburg, M. R., eds.), pp. 159–169. Springer-Verlag, New York.

Dingwall, C., and Laskey, R. A. (1986) Protein import into the cell nucleus. *Annu. Rev. Cell Biol.* 2:367–390.

Dreyer, C., Scholtz, E., and Hausen, P. (1982) The fate of oocyte nuclear proteins during early development of *Xenopus laevis. Wilhelm Roux' Arch. Dev. Biol.* 191:228–233.

Dworetzky, S. I., and Feldherr, C. M. (1988) Translocation of RNA-coated gold particles through the nuclear pores of oocytes. *J. Cell Biol.* 106:575–584.

Dworetzky, S. I., Lanford, R. E., and Feldherr, C. M. (1988) The effects of variations in the number and sequence of targeting signals on nuclear uptake. *J. Cell Biol.* 107:1279–1287.

Edidin, M. (1974) Rotational and translational diffusion in membranes. *Annu. Rev. Biophys. Bioeng.* 3:179–201.

Featherstone, C., Darbly, M. K., and Gerace, L. (1988) A monoclonal antibody against the nuclear pore complex inhibits nucleocytoplasmic transport of protein and RNA *in vivo. J. Cell Biol.* 107:1289–1297.

Feldherr, C. M., and Akin, D. (1990) The permeability of the nuclear envelope in dividing and nondividing cells. *J. Cell. Biol.* 111:1–8.

Feldherr, C. M., Kallenbach, E., and Schulz, N. (1984) Movement of a karyophilic protein through the nuclear pores of oocytes. *J. Cell Biol.* 99:2216–2222.

Finbow, M., Yancey, S. B., Johnson, R., and Revel, J. P. (1980) Independent lines of evidence suggesting a major gap junctional protein with a molecular weight of 26,000. *Proc. Natl. Acad. Sci. U.S.A.* 77:970–974.

Finlay, D. R., Newmeyer, D. D., Price, T. M., and Forbes, D. J. (1987) Inhibition of *in vitro* nuclear transport by a lectin that binds to nuclear pores. *J. Cell Biol.* 104:189–214.

Fischer, U., and Lührmann, R. (1990) An essential signaling role for m_3G cap in the transport of U1 snRNP to the nucleus. *Science* 249: 786–790.

Fowler, V. M. (1986) An actomyosin contractile mechanism for erythrocyte shape transformations. *J. Cell Biochem.* 31:1–9.

Frye, L. D., and Edidin, M. J. (1970) The rapid intermixing of cell surface antigens after the formation of mouse-human heterokaryons. *J. Cell Sci.* 7:319–335.

Gerace, L., Blum, A., and Blobel, G. (1978) Immunochemical localization of the major polypeptides of the nuclear pore complex—lamina fraction. *J. Cell Biol.* 79:546–566.

Gerace, L., Comeau, C., and Benson, M. J. (1984) Organization and modulation of nuclear lamina structure. *J. Cell Sci. Suppl.* 1:137–160.

Gerace, L., Ottaviano, Y., and Kondor-Koch, C. (1982) Identification of a major polypeptide of the nuclear pore complex. *J. Cell Biol.* 95:826–837.

Goldfarb, D. S., Gariepy, J., Schoolnik, G., and Kornberg, R. D. (1986) Synthetic peptides as nuclear localization signals. *Nature (London)* 322:641–644.

Gorter, E., and Grendel, F. (1925) On bimolecular layers of lipids on the chromocytes of the blood. *J. Exp. Med.* 41:439–443.

Haest, C. W. M., Plasa, G., Kamp, D., and Deuticke, B. (1978) Spectrin as a stabilizer of the phospholipid asymmetry in the human erythrocyte membrane. *Biochim. Biophys. Acta* 509:21–32.

Hamm, J., Darzynkiewicz, E., Tahara, S. M., and Mattaj, I. W. (1990) The trimethylguanosine cap structure of U1 snRNA is a component of a bipartite nuclear targeting signal. *Cell* 62:569–577.

Heber, U., and Heldt, H. W. (1981) The chloroplast envelope: structure, function and role in leaf metabolism. *Annu. Rev. Plant Physiol.* 32:139–168.

Jacobs, M. H. (1952) The measurement of cell permeability with particular reference to the erythrocyte. In *Modern Trends in Physiology and Biochemistry* (Barron, E. S. G., ed.), pp. 149–171. Academic Press, New York.

Jinbu, Y., Sato, S., Nakao, T., Nakao, M., Tsukita, S., Tsukita, S., and Ishikawa, H. (1984) The role of ankyrin in shape and deformability change of human erythrocyte ghosts. *Biochim. Biophys. Acta* 773:237–245.

Kohen, E., Kohen, C., Thorell, B., Mintz, D. H., and Rabinovitch, A. (1979) Intracellular communication in pancreatic islet monolayer cultures: a microfluorometric study. *Science* 204:862–865.

Kornberg, R. D., and McConnell, H. M. (1971) Inside-outside transitions of phospholipids in vesicle membranes. *Biochemistry* 10:1111–1120.

Kyte, J., and Doolittle, R. F. (1982) A simple method for displaying the hydropathic character of protein. *J. Mol. Biol.* 157:105–132.

Lanford, R. E., Kanda, P., and Kennedy, R. C. (1986) Induction of nuclear transport with a synthetic peptide homologous to the SV40 T antigen transport signal. *Cell* 46:575–582.

Lawrence, T. S., Beers, W. H., and Gilula, N. B. (1980) Transmission of hormonal stimulation by cell-cell communication. *Nature (London)* 272:501–506.

Liu, S.-C., Derick, L. H., and Palek, J. (1987) Visualization of the hexagonal lattice in the erythrocyte membrane skeleton. *J. Cell Biol.* 104:527–536.

Lloyd, J. E. (1969) Studies on the permeability of rat liver lysosomes to carbohydrates. *Biochem. J.* 115:703–707.

Lowenstein, W. R., Kanno, Y., and Socolar, S. J. (1978) Quantum jumps of conductance during formation of membrane channels at cell-cell junctions. *Nature (London)* 274:133–136.

Macey, R. I. (1984) Transport of water and urea in red blood cells. *Am. J. Physiol.* 246:C195–C203.

MacLennan, D. H., Brandl, C. J., Konczak, B., and Green, N. M. (1985) Amino-acid sequence of a Ca^{2+} and Mg^{+}-dependent ATPase from rabbit muscle sarcoplasmic reticulum deduced from its complementary DNA sequence. *Nature (London)* 316:696–700.

Makowski, L., Caspar, D. L. D., Phillips, W. C., and Goodenough, D. A. (1977) Gap junction structures. II. Analysis of the x-ray diffraction data. *J. Cell Biol.* 74:629–645.

Marsh, D. (1975) Spectroscopic studies of membrane structure. In *Essays in Biochemistry* (Campbell, P. N., and Aldridge, W. N., eds.), Vol. 11, pp. 139–180. Academic Press, New York.

McNamee, M., and McConnell, H. (1972) Transmembrane potentials and phospholipid flip-flop in excitable membrane vesicles. *Biochemistry* 12:2951–2957.

Nelson, M., and Silver, S. (1989) Context affects nuclear protein localization in *Sacharomyces* Cerevisae, *Mol. Cell Biol.* 9:384–389.

Newmeyer, D. D., Finlay, D. R., and Forbes, D. J. (1986) *In vitro* transport of a fluorescent nuclear protein and exclusion of non-nuclear proteins. *J. Cell Biol.* 103:2091–2102.

Nilsson, R., Peterson, E., and Dallner, G. (1973) Permeability of microsomal membranes isolated from rat liver. *J. Cell Biol.* 56:762–776.

Obaid, A. L., Socolar, S. J., and Rose, B. (1983) Cell-to-cell channels with two independently regulated gates in series: analysis of functional conductance modulation by membrane potential, calcium and pH. *J. Membr. Biol.* 73:69–89.

Ovchinnikov, Y. A. (1987) Probing the folding of membrane proteins. *Trends Biochem.* 12:434–438.

Paine, P. L., Moore, L. C., and Horowitz, S. B. (1975) Nuclear envelope permeability. *Nature (London)* 254:109–114.

Peracchia, C., and Dulhunty, A. F. (1976) Low resistance junctions in crayfish. *J. Cell Biol.* 70:419–439.

Peracchia, C., and Peracchia, L. L. (1980a) Gap junction dynamics: reversible effects of divalent cations. *J. Cell Biol.* 87:708–718.

Peracchia, C., and Peracchia, L. L. (1980b) Gap junction dynamics: reversible effects of hydrogen ions. *J. Cell Biol.* 87:719–737.

Poo, M., and Cone, R. A. (1974) Lateral diffusion of rhodopsin in the photoreceptor membrane. *Nature (London)* 247:438–441.

Rennoij, W., Van Golde, L. M. G., Zwaal, R. F. A., and VanDeenen, L. L. M. (1976) Topical asymmetry of phospholipid metabolism in rat erythrocyte membranes. *Eur. J. Biochem.* 61:53–58.

Rothman, J. E., and Dawidowicz, E. A. (1975) Asymmetric exchange of vesicle phospholipids catalyzed by the phosphatidylcholine exchange protein. Measurement of inside-outside transitions. *Biochemistry* 14:2809–2816.

Rutishauser, U. (1984) Development biology of neural cell adhesion molecule. *Nature (London)* 310:549–554.

Sandra, A., and Pagano, R. E. (1978) Phospholipid asymmetry in LM cell plasma membrane derivatives: polar head group and acyl chain distributions. *Biochemistry* 17:332–338.

Schlessinger, J. (1983) Mobilities of cell-membrane proteins: how are they modulated by cytoskeleton? *Trends Neurosci.* 6:360–363.

Schoolwerth, A. C., and LaNoue, F. F. (1985) Transport of metabolic substrates in renal mitochondria. *Annu. Rev. Physiol.* 47: 143–171.

Schwartzmann, G., Wiegandt, H., Rose, B., Zimmerman, A., Ben-Haim, D., and Lowenstein, W. R. (1981) Diameter of cell-to-cell junctional membrane channel as probed with neutral molecules. *Science* 213:551–553.

Seigneuret, M., and Devaux, P. F. (1984) ATP-dependent asymmetric distribution of spin-labeled phospholipids in the erythrocyte membrane: relation to shape changes. *Proc. Natl. Acad. Sci. U.S.A.* 81: 3751–3755.

Shull, G. E., Schwartz, A., and Lingrel, J. B. (1985) Amino-acid sequence of the catalytic subunit of the (Na^+ + K^+) ATPase deduced from a complementary DNA. *Nature (London)* 316:691–695.

Silver, P., Sadler, I., and Osborne, M. A. (1989) Yeast proteins that recognize nuclei localization sequences. *J. Cell Biol.* 109:983–989.

Singer, S. J., and Nicolson, G. L. (1972) The fluid mosaic model of the structure of cell membrane. *Science* 175:750–751.

Solomon, A. K., Chasan, B., Dix, J. A., Lukacovic, M. F., Toon, M. R., and Verkman, A. S. (1983) The aqueous pore in red cell membrane—band 3 as a channel for anions, cations, non-electrolytes and water. *Ann. N.Y. Acad. Sci.* 414:97–124.

Tedeschi, H., and Harris, D. L. (1955) The osmotic behavior and permeability to non-electrolytes of mitochondria. *Arch. Biochem. Biophys.* 58:52–67.

Tobian, J. A., Drinkard, L., and Zasloff, M. (1985) tRNA nuclear transport: defining critical regions of human $tRNA_i^{met}$ by point mutagenesis. *Cell* 43:415–422.

Underwood, M. R., and Fried, H. (1990) Characterization of nuclear localizing sequences derived from the yeast ribosomal protein L29. *EMBO J.* 9:91–100.

Unwin, P. N. T., and Milligan, R. A. (1982) A large particle associated with the perimeter of the nuclear pore. *J. Cell Biol.* 93:63–75.

Unwin, P. N. T., and Zampighi, G. (1980) Structure of the junction between communicating cells. *Nature (London)* 283:545–549.

Viitala, J., and Jarnefelt, J. (1985) The red cell surface revisited. *Trends Biochem. Sci.* 10:392–395.

Warner, A. E., Guthrie, S. C., and Gilula, N. B. (1984) Antibodies to gap junctional communication in early amphibian embryo. *Nature (London)* 311:127–132.

Whiteley, N. M., and Berg, H. C. (1974) Amidination of the outer and inner surfaces of the human erythrocyte membrane. *J. Mol. Biol.* 87:541–561.

Wier, M., and Edidin, M. (1988) Constraint of the translational diffusion of a membrane glycoprotein by its external domain. *Science* 242:412–414.

Chapter 3

Multifaceted Roles of the Cell Surface

As discussed in Chapter 2, the cell membrane is a dynamic structure and its components, such as integral proteins or lipids, can move in the plane of the membrane. Furthermore, membrane proteins are associated, again dynamically, with components of the cytoskeleton. Chapters 13 to 15 will discuss special integral membrane proteins that are engaged in the transport of solutes through the membrane. The present chapter is concerned in part with events taking place at the cell surface and in part with events that follow these interactions.

Many membrane proteins in contact with the surrounding medium are involved in interactions with extracellular elements, such as peptide hormones and growth factors, foreign proteins, the surfaces of other cells, or the extracellular matrix. As a consequence, knowledge of the cell surface and its various receptors is important for our understanding of the responses of cells to a variety of stimuli, the regulation of cell growth, immunological responses, differentiation, morphogenesis, and malignancy. The receptors involved in endocytosis have been studied extensively and will be discussed first.

I. ENDOCYTOSIS

In *endocytosis,* cells ingest extracellular materials by trapping them in invaginations of the cell membrane, which then pinch off to form membrane-lined intracellular vesicles. In protozoans, endocytosis has a role in feeding. In specialized phagocytic cells of multicellular organisms such as macrophages, endocytosis functions in the removal of foreign particulate material and its delivery to the lysosomal system, where it is digested. However, endocytosis is not limited to specialized cells. All animal cells are thought to be capable of endocytotic uptake, which typically proceeds at a prodigious rate. In mouse cell fibroblasts, the amount of surface membrane taken up by endocytosis has been estimated to be 50% per hour (Pearse, 1975).

Endocytosis can occur without demonstrable receptors. *Receptors* are proteins that specifically bind a ligand with high affinity. Presumably, in the absence of receptors the trapping of extracellular protein is nonselective. In contrast, in receptor-mediated endocytosis, proteins present in the medium are taken up after binding specific surface receptors with a high degree of specificity. Our discussion will be primarily concerned with endocytosis involving receptors.

The purpose of receptor-mediated endocytosis is not always clear. The *low-density lipoprotein* (LDL) receptor system functions in the processing of cholesterol by the cell. The uptake of *nerve growth factor* (NGF) is likely to have a role in transporting the NGF to its target. However, this case is probably an exception, and the endocytotic uptake of receptor and ligand probably plays a role in regulating the surface concentration of receptor required to provide its physiological response. The failure of certain drugs such as alkylamines or the antibiotic bacitracin to block the mitogenic activity of *epithelial growth factor* (EGF) (Maxfield et al., 1979), while greatly interfering with the movement of receptors to coated pits, supports this view. This displacement is required for endocytosis (see next section). Nevertheless, the physiological roles of peptide hormones and growth factors and endocytosis are unavoidably linked since the two functions involve binding to the same receptors.

A. Receptor-Mediated Endocytosis

The receptors involved in the endocytotic uptake of ligands are transmembrane glycoproteins with the carbohydrate moiety attached to the amino terminal region. In principle, the arrangement is similar to that of other intrinsic proteins discussed in Chapter 2. All have residues such as phosphate, palmitate, or oligosaccharides that have been added posttranslationally. Some aspects of their processing and intracellular transport are discussed in Chapter 4.

In receptor-mediated endocytosis, the uptake of ligand occurs within minutes of binding to the receptor. In some cases, the receptors are preferentially located in coated pits; in others the receptors migrate to the coated pits after binding the ligand. The coated pits are indentations in the membrane surface lined with a fuzzy coat (see Fig. 3.1) (Anderson et al., 1977a). Most generally, the proteins taken up by endocytosis are in coated vesicles, which lose their coat when they form *endosomes*. In many cases, the proteins are eventually digested by the lysosomal enzymes (however, see C). *Lysosomes* are vesicles that contain a full array of hydrolytic enzymes capable of digesting materials taken up by endocytosis. The ways in which the endosomal contents and the lysosomal enzymes interact is presently under discussion (see E, below).

The LDL receptor system, which functions in the transport of cholesterol from the plasma into cells, was one of the first to be examined in detail and will be the first subject of our discussion.

B. The LDL Receptors

In mammals, cholesterol is transported in the bloodstream in spherical LDL particles 22 nm in diameter, which originate in the liver. The particles are complexes containing a core of 1500 cholesterol molecules esterified to long-chain fatty acids covered by phospholipids, cholesterol, and protein (Goldstein and Brown, 1977).

The receptors, which correspond to a lipoprotein of molecular mass 164 kDa (Schneider et al., 1982), are synthesized in the endoplasmic reticulum when cholesterol is required. After binding and internalization, the receptor-LDL particle complexes are delivered to the lysosomes, where the protein component of the LDL particles is hydrolyzed and cholesterol is regenerated from the cholesteryl esters. Cholesterol in the cytoplasm controls the synthesis of the receptor

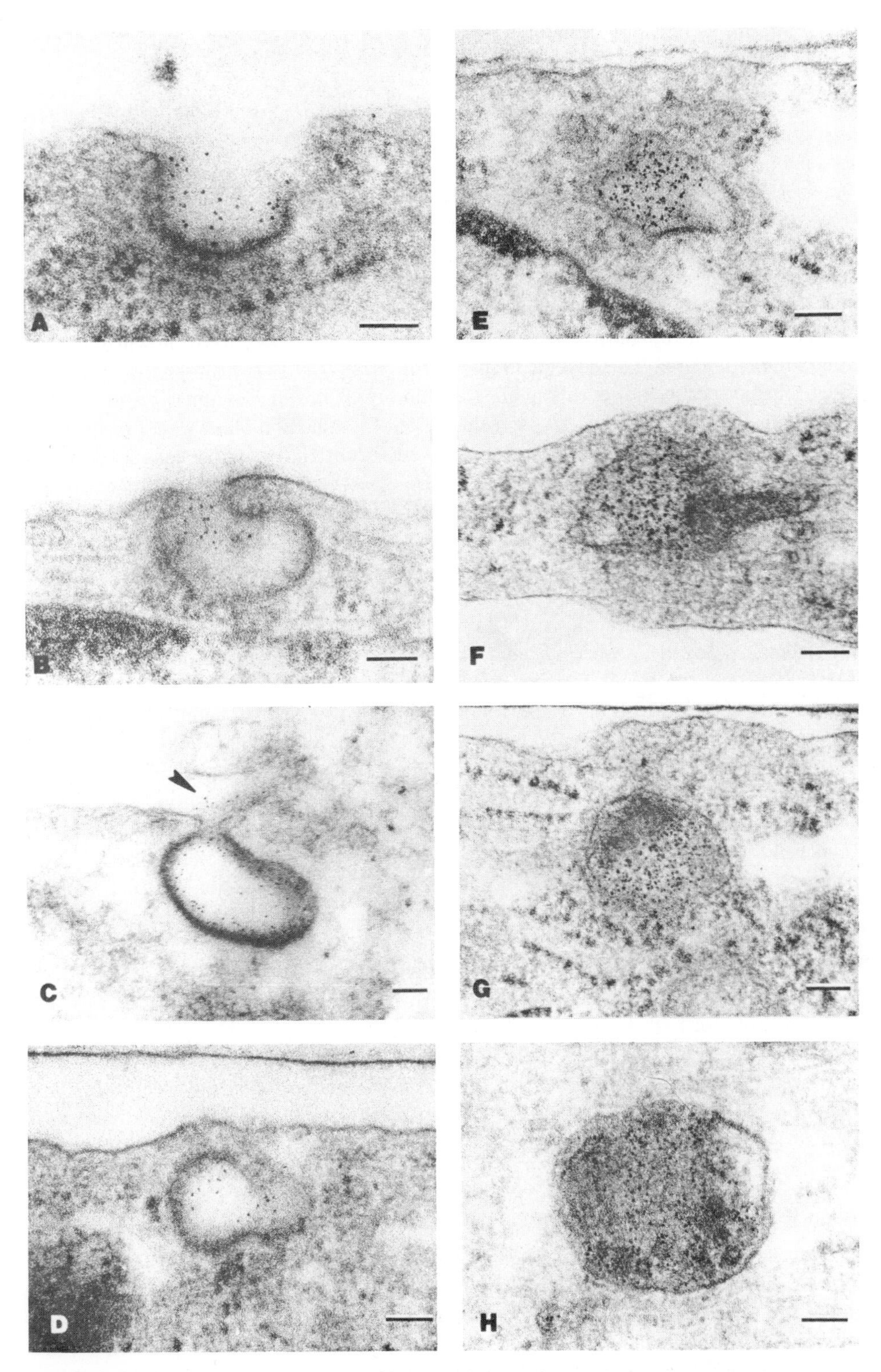

Fig. 3.1 Role of the coated endocytotic vesicle in the uptake of receptor-bound low-density lipoprotein in human fibroblasts. The bars represent 100 nm. See text. Reprinted with permission from Anderson et al. (1977a), copyright 1977 by Cell Press.

molecules and thereby prevents an excessive accumulation (Goldstein and Brown, 1977). Apparently, the receptors return to the surface, where they cluster again in the coated pits (Anderson et al., 1977b; Goldstein et al., 1976).

Between 50 and 80% of the receptors are clustered in 2% of the cell surface and have been shown to be present in coated pits by electron microscopy of cells allowed to take up LDL coupled to ferritin (Anderson et al., 1977a; Orci et al., 1978). Since the ferritin contains iron, it is visible with the electron microscope by virtue of its high density. In contrast to other systems, the LDL receptors are present in coated pits before the addition of LDL. This has been demonstrated by adding the ferritin-labeled LDL to cells that have been fixed, in which presumably migration is impossible (Anderson et al., 1976).

The probable sequence of events is illustrated in Fig. 3.1, which summarizes the electron microscopic observations with the LDL- ferritin marker. Cultured fibroblasts were first incubated at 4°C for 2 h and then, after extensive washing, were incubated for various lengths of time in the presence of ferritin-labeled LDL at 37°C. Figure 3.1A–C show that after incubation for 1 min the various configurations of early endocytosis are already present in coated pits. Figure 3.1D shows a coated vesicle that is beginning to lose its coat after a 2-min incubation and Fig. 3.1E shows a vesicle or endosome without a coat, also after a 2-min incubation. Figure 3.1F–H correspond to configurations in the formation of secondary lysosomes, that is, lysosomes formed by fusion of endosomes and newly formed primary lysosomes. In Fig. 3.1F and G the amount of ferritin per vesicle seems to be greater, possibly indicating a fusion of various vesicles; Fig. 3.1H corresponds to a fully formed lysosome containing ferritin.

The kinetics of endocytosis have been followed directly by two other approaches (Goldstein et al., 1976), which support the interpretation of the electron microscopy studies. The binding and incorporation of LDL were followed by labeling LDL with ^{125}I. The actual endocytotic uptake and the binding were distinguished from each other because heparin releases surface LDL, so the ^{125}I-labeled LDL that could be released after heparin treatment could be assumed to be attached to the LDL receptor at the surface (see Fig. 3.2). The release of the LDL from the receptor was not the result of damage to the receptors. After incubation in heparin and its removal, the binding of LDL was the same as that of untreated cells. The proteolysis of LDL was followed by measuring the radioactivity of trichloroacetic acid (TCA)-soluble cell extracts. Protein is insoluble in TCA, in contrast to the hydrolytic products, that is, amino acids and low molecular weight peptides. An increase in the TCA-soluble radioactivity of the extract then serves as a measure of LDL hydrolysis (see Fig. 3.4).

Figure 3.2 represents the radioactivity released in the medium after incubation in labeled LDL followed by washing and shows that there is considerable initial nonspecific binding. However, most of the labeled LDL is removed by washing. In normal cells (filled circles), a portion that does not come off even after eight washes is released by the addition of heparin. Interestingly, cells from a patient with a genetically determined receptor deficiency (open circles) can bind LDL readily with low affinity, but the heparin-sensitive binding is absent.

Figure 3.3 shows that the surface-bound ^{125}I-labeled LDL is taken up very rapidly, most within 5 min. However, the heparin-insensitive uptake of LDL continues linearly with time at a lower rate, suggesting that it corresponds to rapid uptake into cytoplasmic vesicles.

Figure 3.4 represents the degradation of the LDL at 37°C after an initial binding at 4°C. The radioactivity of the bound protein disappears (curve 1) and most of it appears in a TCA-soluble form (curve 2).

The location of coated pits coincides with the position of the underlying cytoplasmic stress fibers (Anderson et al., 1978), which have been shown in other studies to contain actin. In fact,

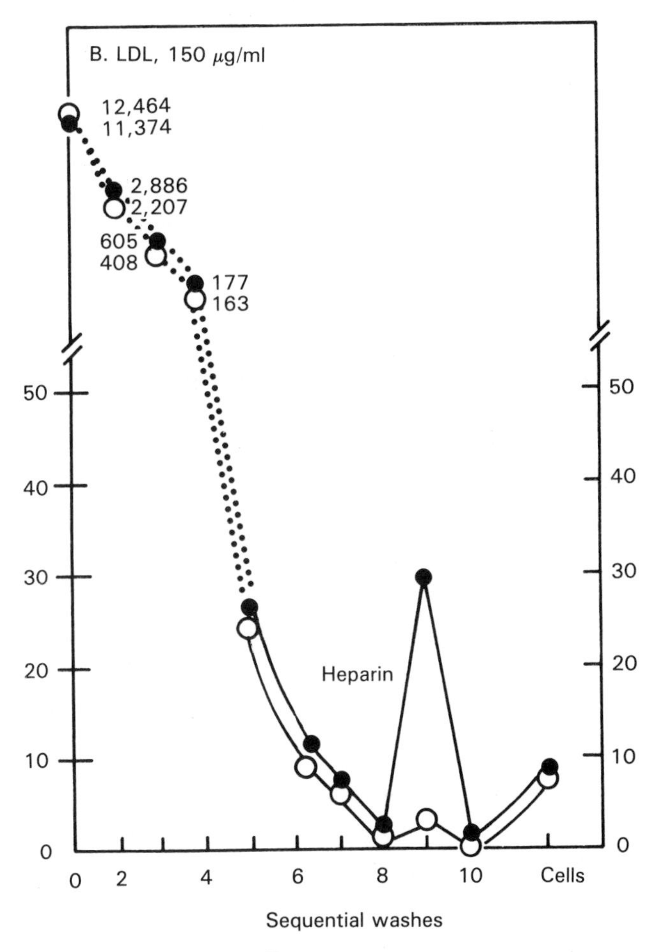

Fig. 3.2 Amount of ^{125}I-labeled LDL released from normal (●) in ordinate (ng per dish per wash) and FH (O) fibroblast by sequential washes. Reprinted with permission from J. L. Goldstein, et al., *Cell,* 7:85–95. Copyright 1976 by Cell Press.

clathrin, the major protein of coated vesicles, can bind actin. Since actin fibers are associated with movement, it is possible that they have a role in the internalization of vesicles.

Studies similar to those described for LDL have also been carried out for EGF (Gaspodarowicz, 1985).

Accumulation in the coated pits and binding of the ligand are two separate functions of the LDL receptors, since they are affected by different mutations. This conclusion is based on observations of patients with familial hypercholesterolemia (FH), a disease characterized by very high levels of cholesterol in the blood. One kind of FH can be traced to a mutation in which the receptors are unable to bind LDL; another kind involves a mutation in which the receptors are unable to be incorporated into coated pits. These observations suggest that the receptors have two separate sites: the site for binding the LDL facing the medium and the site that interacts with the coated pits, which is presumably cytoplasmic. As discussed in Chapter 4, some of the receptor molecules are likely to have other specialized sites responsible for their sorting out into various cellular compartments.

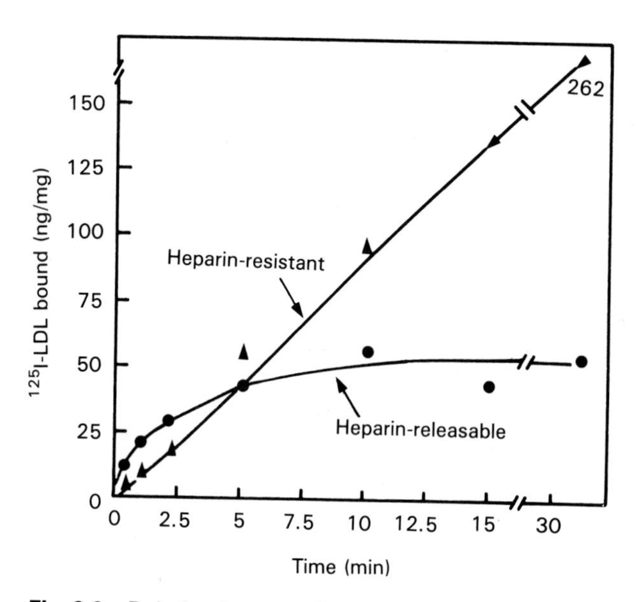

Fig. 3.3 Relation between heparin-releasable and heparin-resistant ^{125}I-labeled LDL binding at 37°C at early time points. Reprinted with permission from J. L. Goldstein, et al., *Cell,* 7:85–95. Copyright 1976 by Cell Press.

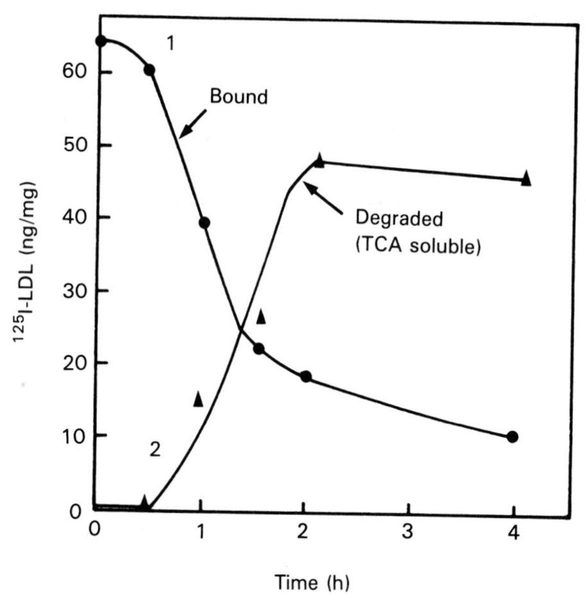

Fig. 3.4 Proteolytic degradation at 37°C of ^{125}I-labeled LDL previously bound to normal fibroblasts at 4°C. Reprinted with permission from J. L. Goldstein, et al., *Cell,* 7:85–95. Copyright 1976 by Cell Press.

C. Other Receptors

Involvement in Endocytosis Various proteins are taken up specifically by cells, as shown in Table 3.1 (Goldstein et al., 1979), and so far 25 specific receptors involved in endocytosis have been recognized.

Table 3.2 (Brown et al., 1983) lists some of the cell surface receptors that have been purified and characterized. The receptors have been generally identified by their ability to bind

Table 3.1 Systems for Receptor-Mediated Endocytosis of Proteins

Protein	Cell type	Internalization via coated pits and coated vesicles	Fate of internalized protein: Degraded in lysosomes	Fate of internalized protein: Other
Transport proteins				
LDL	Fibroblasts, smooth muscle cells, endothelial cells, adrenocortical cells, lymphocytes	Yes	Yes; cholesterol retained by cells	—
Yolk proteins (phosvitin, lipovitellin)	Oocytes (chicken, mosquito)	Yes	No	Delivered to yolk granules
Transcobalamin II	Kidney cells, hepatocytes, fibroblasts	Data not available	Yes, vitamin B_{12} retained by cells	—
Transferrin	Erythroblasts, reticulocytes	Yes	Data not available	Iron retained by cells
Protein hormones				
Epidermal growth factor	Fibroblasts, 3T3 cells	Yes	Yes	—
Nerve growth factor	Sympathetic ganglion cells	Data not available	Data not available	Carried in vesicles retrograde up the axon
Insulin	Hepatocytes, hepatoma cells, lymphocytes, adipocytes, 3T3 cells	Data not available	Yes	Also delivered to Golgi apparatus and nuclei
Chorionic gonadotropin	Leydig tumor cells, ovarian luteal cells	Data not available	Yes	—
β-Melanotropin	Melanoma cells	Data not available	Data not available	Delivered to Golgi apparatus and melanosomes
Other proteins				
Asialoglycoproteins	Hepatocytes	Data not available	Yes	—
Lysosomal enzymes	Fibroblasts	Data not available	No	Delivered to lysosomes and Golgi-associated structures; enzymes remain active for many days
α_2-Macroglobulin	Fibroblasts, macrophages, 3T3 cells	Yes	Yes	—
Maternal immunoglobulins (IgG)	Fetal yolk sac, neonatal intestinal epithelial cells	Yes	No	Transferred intact in coated vesicles to basal surface of cells, where IgG is discharged into fetal or neonatal circulation

Source: From Goldstein et al. (1979). Reprinted by permission from *Nature* 279:679–685, copyright © 1979 Macmillan Magazines Ltd.

Table 3.2 Structures of Cell Surface Receptors That Concentrate in Coated Pits

Receptor	Source	Estimated molecular mass (kDa) Subunit	Holoreceptor	Residues added posttranslationally
LDL	Bovine adrenal cortex Human fibroblasts	160	160	O-linked oligosaccharide; sulfate on N-linked oligosaccharides
Transferrin	Human leukemia cells	90	180 (disulfide-linked dimer)	Palmitate; phosphate on tyrosine and serine
Epidermal growth factor	Human A-431 cells	170	170	Phosphate on tyrosine and serine
Insulin	Rat adipose tissue Rat liver Human placenta	90 125	~350 (two subunits of 90 kDa and two subunits of 125 kDa, disulfide-linked)	Phosphate on tyrosine and serine
Lysosomal enzyme (mannose-6-phosphate)	Bovine liver Rat chondrosarcoma	215	215	
Asialoglycoproteins	Chicken liver	26	26	Phosphate on serine
	Rat liver	43 54 64	110 (?dimer)	
Fibroblast growth factor	Pituitary	15		

From M. S. Brown, et al., *Cell,* 32:663–667. Copyright 1983 by Cell Press, reprinted with permission.

the appropriate ligand. The LDL receptors (Orci et al., 1978) were extracted from cell membrane preparations of bovine adrenal cortex. Adrenal cells are particularly rich in these receptors because they use cholesterol in the synthesis of steroid hormones. It has been estimated that there are 100,000 receptor molecules per adrenal cell. Binding of ^{125}I-labeled LDL was used as the assay of the receptor through the various steps of the fractionation procedures.

In many of the cases listed in Table 3.1, coated pits and coated vesicles have been demonstrated, and in other cases their involvement is strongly suspected. However, some of the details of the uptake and processing differ. We saw that the LDL receptors are located in the coated pits in the absence of LDL and are continuously internalized. Various receptors follow this pattern. However, others such as the EGF receptors are distributed throughout the surface and cluster only after binding the ligand (Hagler et al., 1978). As we saw in Chapter 2, membrane components can diffuse two-dimensionally inside the membrane, and the diffusional movement of receptors toward the coated pits is sufficiently rapid to account for the clustering (Bretscher and Pease, 1984; Hopkins, 1985).

Fate of Ligand and Receptor Many of the systems recycle; for instance, the LDL receptor is rapidly returned to the surface and has a half-life as long as 15 h (Brown et al., 1981). As might be expected, blocking protein synthesis with cycloheximide does not have an immediate effect on LDL endocytosis. In contrast, the EGF receptor is degraded in most tissues (Schlessinger et al., 1978), although it is recycled in the liver (Dunn and Hubbard, 1984).

Many of the proteins bound to the receptors are degraded, as we saw for LDL. In contrast, yolk proteins are accumulated in yolk granules and NGF, which enters by endocytosis at the tip of the axon, is accumulated in the cell body (Bradshaw, 1978). Some of the insulin taken up by endocytosis is degraded and some remains intact inside the cell.

Both receptor and ligand may recycle, as is the case for *transferrin* (Bleil and Bretscher, 1982), a protein that functions in the transport of iron in organisms. After uptake of transferrin by endocytosis, the iron is removed from the transferrin in the endosome; the apotransferrin (i.e., transferrin stripped of iron) remains attached to the receptor and both are returned to the cell surface (Geutze et al., 1984).

Four possible pathways for the receptor-ligand complex are shown in Fig. 3.5 (Goldstein et al., 1985).

Mechanisms: Hydrolysis, Recycling, and Sorting As we saw for LDL, after endocytosis the coated vesicles shed their coat and fuse to form smooth larger vesicles, the endosomes. These in turn fuse with the lysosomes, where digestion can take place. In this process the interior of the endosome becomes acidic, as shown by the endocytosis of α_2-macroglobulin conjugated to fluorescein (Tycko and Maxfield, 1982). Fluorescein is a dye whose fluorescence varies with pH. After 20 min of endocytosis induced by the labeled macroglobulin, the fluorescence of the

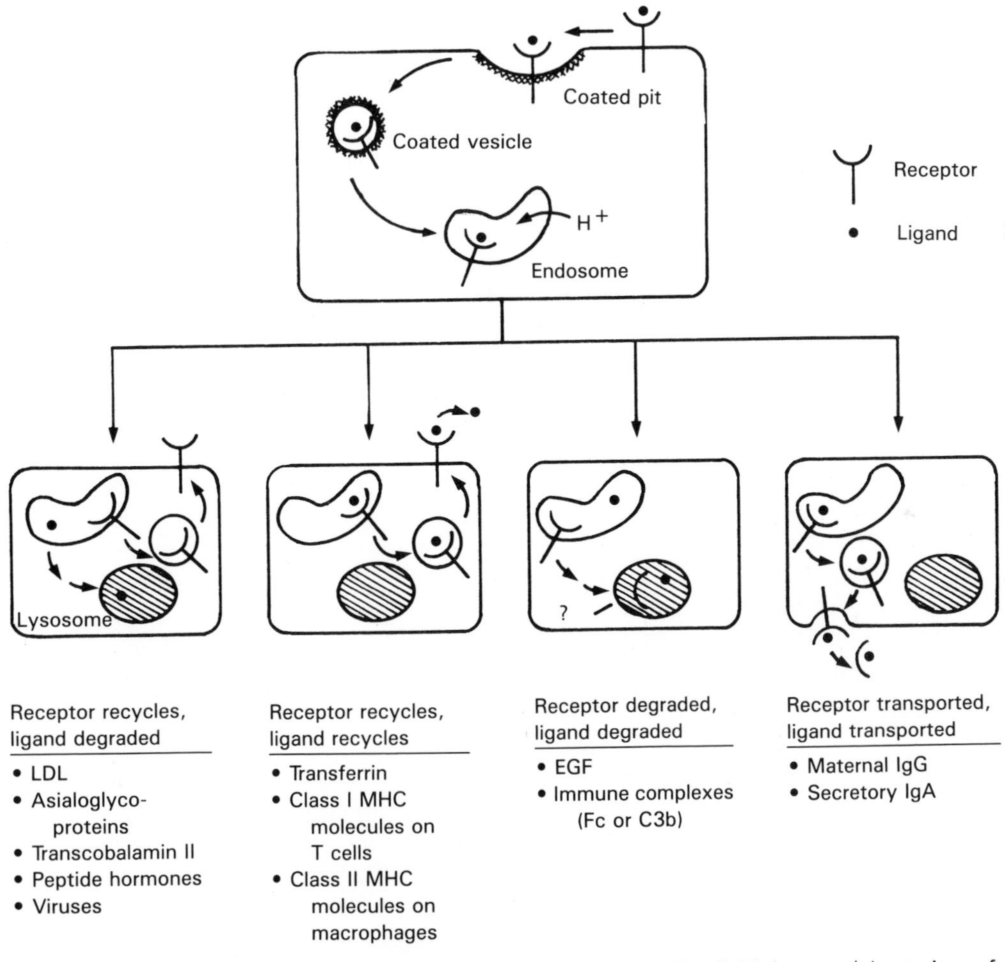

Fig. 3.5 Four pathways of receptor-mediated endocytosis. The initial steps (clustering of receptors in coated pits, internalization of coated vesicles, and fusion of vesicles to form endosomes) are common to the four pathways. After entry into acidic endosomes, a receptor-ligand complex can follow any of the four pathways shown. From Goldstein et al. (1985). Reproduced, with permission, from the *Annual Review of Cell Biology,* Volume 1, © 1985 by Annual Reviews Inc.

dye in the vesicles indicated a pH of approximately 5. The internal acidification is the result of an H^+-pump powered by ATP hydrolysis, as demonstrated in experiments with isolated vesicles (Galloway et al., 1983). Apparently, the acidity of the endosome has a role in detaching the ligand from the receptors. In at least some cases the ligands and receptors have separate fates, and it is difficult to imagine how this can take place unless there is some special mechanism for separating them and conveying them to different compartments.

The asialoglycoprotein receptor system has been studied with the electron microscope after labeling with separate antibodies for ligands, receptors, and clathrin. *Asialoglycoproteins* are abnormal plasma glycoproteins that have been stripped of the sialic acid residue that normally covers the terminal galactose. The antibodies used in this study could be visualized and distinguished from one another by coupling them to colloidal gold particles of different sizes. The results indicate that after separation, ligands and receptors are segregated in a vesicle (*compartment of uncoupling receptor and ligand,* CURL) containing tubular extensions. The receptors attach to the membranes of the tubules, whereas the ligands remain in the lumen (Geutze et al., 1983). Presumably, budding of the tubules could produce smaller vesicles, which could return the receptors to the surface. Figure 3.6*a* shows that both ligand (coupled to the 5-nm-diameter particle) and receptor (coupled to the 8-nm particle) are present in a vesicle close to the cell surface. Figure 3.6*b* and *c* show how the ligand remains in the lumen of the CURL, whereas the tubules favor the receptors. In *b* the receptor is labeled with the 8-nm gold particle and in *c* the labeling is reversed.

The movement of receptor or receptor-ligand complex through several compartments, with accurate sorting in each, suggests that the receptor may have many functional domains for interaction with the ligand and the various macromolecular species capable of redirecting it to a new target.

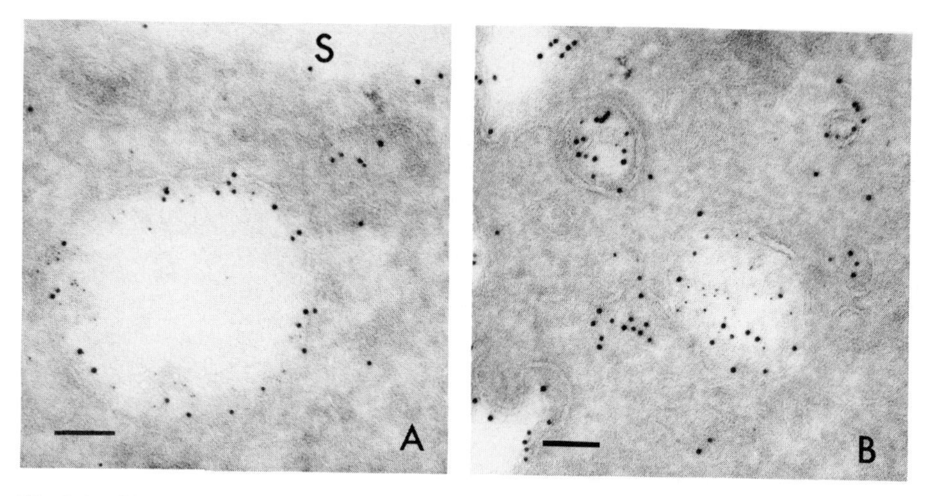

Fig. 3.6 Simultaneous demonstration of receptor and ligand in CURL and in a multivesicular body. The bars correspond to 100 nm. (*a*) Vesicle just beneath plasma membrane at the sinus (S), with ligand (5-nm gold) associated with receptor (8-nm gold). (*b*) Coated pit with receptor (8-nm gold) and ligand (5-nm gold) at upper left. The slightly tangential view of early endosomes shows a heterogeneous distribution of receptor in the vesicular portion and abundant receptor in associated tubules.

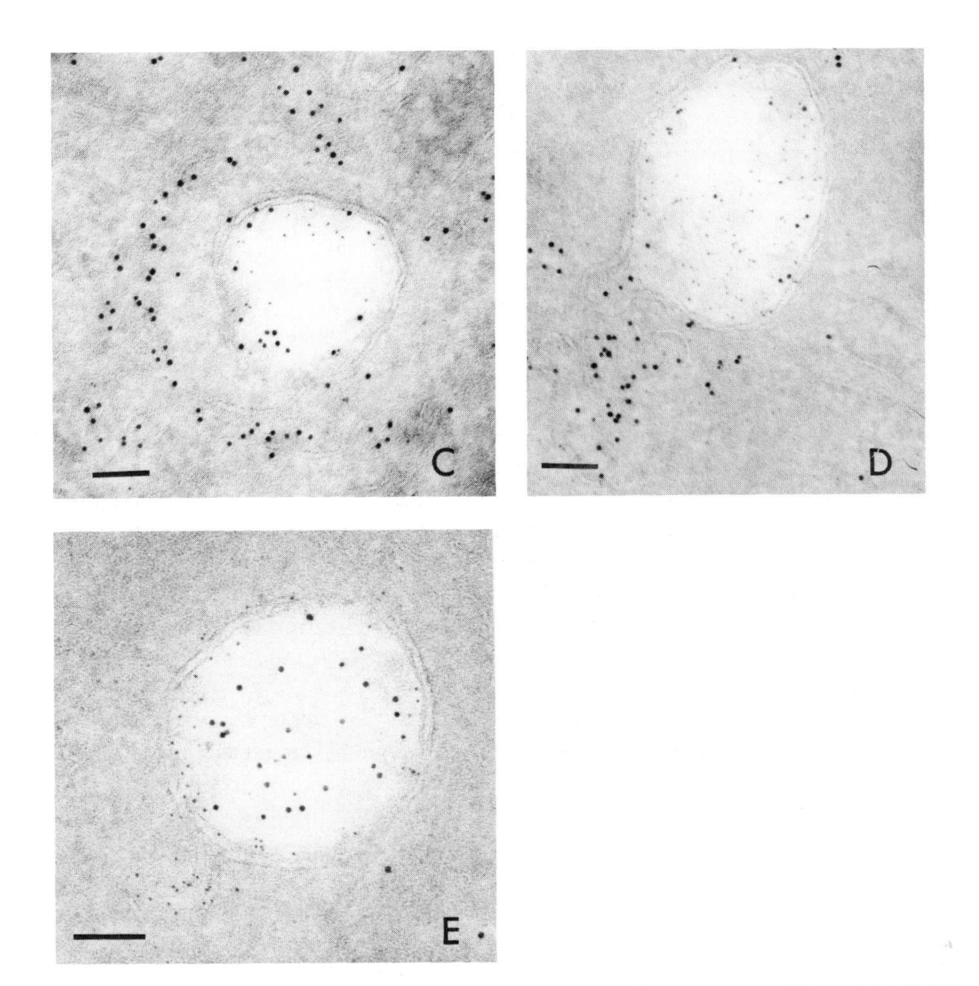

Fig. 3.6 (*Continued*) Simultaneous demonstration of receptor and ligand in CURL and in a multivesicular body. (*c*) Early endosomes profile shows peripheral ligand (5-nm gold) and heterogeneous labeling of receptor (8-nm gold). Intense receptor labeling is present over the tubules adjacent to the vesicular portion of early endosome. (*d*) Free ligand (5-nm gold) can be seen in the lumen of the vesicular portion of early endosome, which also shows scarce and heterogeneous receptor (8-nm gold) labeling. Receptor labeling is intense over the connecting tubules. (*e*) Early endosome profile in which the receptor (5-nm gold) is located predominantly at the pole, where a tubule with heavy labeling of receptor is connected. Most of the ligand (8-nm gold) is present free in the vesicle lumen. Reprinted with permission from Geuze et al. (1983), copyright 1983 by Cell Press.

D. Clathrin

Coated pits and coated vesicles have been found in virtually all nucleated animal cells. The coated pits have a fuzzy cytoplasmic coat that, at higher resolution, can be shown to correspond to periodically spaced fibers.

While coated pits have been found to be involved in the uptake of proteins from the medium, coated vesicles have been implicated in intracellular membrane transport. Accordingly, they have been found in the Golgi region and they appear to transfer material to the lysosomes (Friend

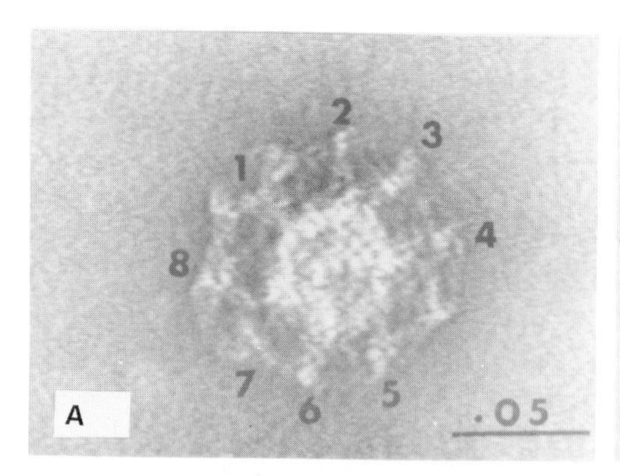

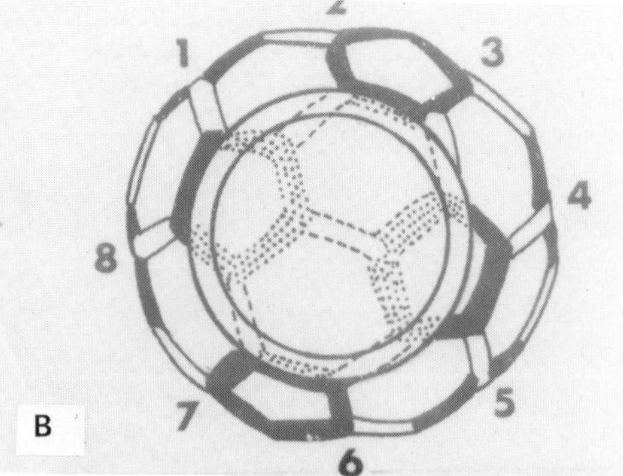

Fig. 3.7 (*a*) Isolated and negatively stained clathrin basket (×340,000). (*b*) Reconstruction of a vesicle and its hexagonal basket. Reprinted with permission from Kaneseki and Kadota (1969).

and Farquhar, 1967; Holzman et al., 1967) and to be involved in secretion (Franke et al., 1977). The introduction of anticlathrin antibody into living cells resulted in inhibition of both receptor-mediated endocytosis and fluid endocytosis (Doxsey et al., 1987). However, the antibody did not block the transport accompanying secretion, suggesting that the transport may involve different idiotypes of clathrin.

The details of endocytosis were examined using horseradish peroxidase, a protein taken up by the cells of the *vas deferens* (Friend and Farquar, 1967). The protein serves as a tracer, since its enzymatic activity can be identified by the insoluble products of diaminobenzidine oxidation seen with the electron microscope. At 20 to 45 min after exposure, the peroxidase was still present in the large smooth vesicles. However, the peroxidase also appeared in multivesicular and dense bodies, identified as lysosomes.

During the endocytotic uptake of peroxidase, small coated vesicles not containing peroxidase increased in numbers in the Golgi region of the cell and later in the apical cytoplasm, toward the lumen. Phosphatase reactions—detected through phosphate precipitates seen with the electron microscope and characteristic of the lysosomes—were found localized in the Golgi cisternae, small coated vesicles of the Golgi, and multivesicular and dense bodies. Presumably, the coated vesicles are involved in the formation of the lysosomes.

Isolated coated vesicles correspond to lipid vesicles, which appear to be surrounded by a basketlike arrangement of protein (Kanaseki and Kadota, 1969), predominantly *clathrin* of 180 kDa and a smaller polypeptide of about 35 kDa (Pearse, 1973, 1975, 1978). Characteristic structures seen with the electron microscope and reconstructed from the images are shown in Fig. 3.7 (Kanaseki and Kadota, 1969).

Clathrin can be readily removed from coated vesicles, which bind it noncovalently. Under special conditions, free clathrin spontaneously assembles into a network of hexagons and pentagons similar to the basketlike arrangement present in coated vesicles (Keen et al., 1979; Schook et al., 1979; Woodward and Roth, 1979).

E. Endosomes and Lysosomes: Interactions

The discussion of the fate of ligands and receptors after endocytosis (sections A and C) included the digestion of material present in endosomes by lysosomal hydrolytic enzymes. The properties of either endosomes or lysosomes differ depending on their stage during their cycle. Early endosomes close to the cell's surface differ from late endosomes which interact with lysosomes. It has been assumed that these differences are the result of a maturation process and that the late lysosomes and endosomes interact by fusion.

Recently, a view has gained favor that endosomes and lysosomes form part of systems which include stationary organelles akin to the Golgi stacks and tubular elements discussed in Chapter 4 (III). All transfers would be carried out by transport vesicles (e.g., Storrie et al. 1988; Griffiths and Gruenberg, 1991). The stationary elements would act as the relay stations; for example, in the case of the endosomal system, it would receive endocytotic transport vesicles and deliver the contents to the lysosomal system where digestion would take place. In part the proposals find support from structural observations. The early endosomes of many kinds of cells are constituted of cisternae, tubules, and large vesicles (Wall et al., 1980; Hopkins, 1983; Geuze et al., 1983; Griffiths et al. 1989) exhibiting coated buds (Geuze et al., 1983; Van Deurs and Linausen, 1982). Similarly, a complex lysosomal system has been described (e.g., Swanson et al., 1987; Knapp and Swanson, 1990). Kinetic data suggest that the transfer of material taken up by endocytosis (e.g., horse radish peroxidase) is discontinuous and is mediated by transport vesicles (e.g., Griffiths et al., 1989; Bomsel et al., 1990) possibly transported along microtubules. In addition, the properties of the vesicles derived from these systems are in harmony with the general concept. Early endosomes can fuse with each other in vitro (e.g., Bomsel et al., 1990), a fusion that is regulated by a GTP-binding protein (Gorvell et al., 1991) that has been implicated in intracellular transport (see Chapter 4, III, B). Direct fusion between early and late endosomes does not take place in vitro. However, the putative endosomal-transport vesicles fuse with late endosomes (Bomsel et al., 1990).

II. POLYPEPTIDE GROWTH FACTORS

The proliferation and survival of normal cells are controlled by a variety of substances, including a group of polypeptide hormones termed *polypeptide growth factors* (PGFs). The PGFs are a large family of regulatory peptides of which EGF, discussed above, is one (Kris et al., 1985).

The PGFs promote the proliferation (Lillien and Claude, 1981) and differentiation of various cell types. Typically, they increase the size and number of the cells they regulate. During embryonic development, when growth is complete the cells have become differentiated. PGFs also play an important role in repair and in tissues that are maintained by continued or intermittent cell turnover, such as epithelium and hematopoietic tissues. The PGFs exert their effect after binding to specific plasma membrane receptors that are also involved in endocytosis. Each growth factor seems to have a specific target cell or tissue. The amino acid sequences of several PGFs are known, either from amino acid sequence analysis or from cDNA studies.

We have already seen that EGF binds to receptors that seem to move freely in the plane of the membrane. After binding, the receptors cluster, mainly over clathrin-coated pits (Hagler et al., 1978, 1979; Hopkins et al., 1981; Schlessinger et al., 1978). Within 1 or 2 min, the occupied receptors appear in the endosomes (Schlessinger et al., 1978) and then in the lysosomes, where in most tissues they are degraded (Carpenter and Cohen, 1976). As already mentioned, in liver they are recycled and not degraded (Dunn and Hubbard, 1984).

Like many other PGFs, EGF is a tyrosine autokinase; that is, it phosphorylates its own tyrosine residue (Hunter and Cooper, 1981). In addition, it induces the phosphorylation of the serine and threonine residues of the receptor in reactions that are probably catalyzed by a separate enzyme, kinase C (Cochet et al., 1984; Iwashita and Fox, 1984). EGF increases phosphatidylinositol turnover, diacetylglycerol production, and Ca^{2+} influx. All these effects have been implicated in signal transduction for many hormones and are responsible for the biological effects of the factors. These will be discussed in more detail in Chapter 5.

Nerve growth factor was the first PGF studied (Levi-Montalcini, 1954). It affects central nervous system, sensory, and sympathetic neurons as well as nonneuronal cells such as chromaffin and mast cells. NGF has an essential maintenance role in early embryonic stages, as shown by the massive cell death that follows its removal by, for example, exposing the cells to NGF antibody. However, NGF also has a supportive role in fully differentiated cells. In addition, it directs growth and regeneration along its concentration gradient (Gundersen and Barrett, 1979).

NGF binds to the receptors present in the cell bodies and the growth cone of target cells. The growth cone is the axonal structure from which the axon grows during development and regeneration. The binding results in a change in the affinity of the ligand for its receptor (Landreth and Shooter, 1980), followed by clustering (Levi et al., 1980) and endocytotic uptake (Calissano and Shelanski, 1980). When taken up by the neuronal terminals, NGF is transported to the cell body (Hendry et al., 1974). Interference with this retrograde transport kills the cell. In the cytoplasm, some of the NGF receptors are associated with the cytoskeletal elements and others are present in the nucleus.

The NGF gene codes for a precursor protein containing 307 amino acids. The NGF is made up of two nonidentical monomers 178 amino acids long.

The mechanism of action of NGF is not known. It may stimulate the expression of genes coding for proteins that affect cytoskeletal elements, such as tau (Drubin et al., 1985) or the *microtubular* or *microfilament-associated proteins* (MAPs), which function in the assembly of the corresponding cytoskeletal structures. NGF might be involved in the activation of MAPs, perhaps through its phosphorylation (Burnstein et al., 1985). Alternatively, the effect could be through the activation of oncogenes or perhaps much more indirect mechanisms such as control of the transport of some essential nutrient.

Platelet-derived growth factor (PDGF) has also been intensively studied (Ross and Vogel, 1978; Ross et al., 1986). Discovered in platelets, PDGF is produced by a variety of other cells as well, such as megakaryocytes, endothelium, or smooth muscle. PDGF is thought to have a role at sites very close to its release, since in vivo it has a very short half-life.

PDGF from human platelets has a molecular weight of approximately 30,000. Various peptide species ranging in molecular weight between 14,000 and 17,000 and held together by disulfide bridges make up the native PDGF molecule.

PDGF stimulates general protein synthesis and collagen synthesis in responsive cells and also stimulates the production of enzymes active in the hydrolysis of these proteins. So far, PDGF receptors have been demonstrated only in connective tissue components such as fibroblasts, vascular smooth muscle, glial cells, and chondrocytes, where PDGF stimulates cell proliferation. PDGF also has many indirect effects, such as enhancement of erythropoiesis and production of vasoconstriction.

III. GROWTH FACTORS AND ONCOGENES

New information about growth factors and oncogenes has opened the door to our understanding of not only genetic control of cell growth but also a variety of problems involving malignancy (Aaronson et al., 1985; Kris et al., 1985).

The mechanisms of cell transformation are central to our understanding of the control of growth and the role of oncogenes. Transformation involves the acquisition of a number of new characteristics, including rounding up of the cells, a varied karyotype, and an increased capacity to produce tumors when injected into test mice. In culture, the transformed cells are more efficient in forming colonies when grown in soft agar. This growth produces proliferation around foci representing the original transformed parent cells. This kind of growth permits easy recognition when screening for transformation.

Various tumors and virally transformed cells secrete *transforming growth factors* (TGFs), which induce cell division and transformation (Assoian et al., 1983; Delarco and Todaro, 1978; Marquadt et al., 1983; Roberts et al., 1981). Some of these TGFs have significant similarities to EGF (see Delarco and Todaro, 1978).

The fact that growth factors and oncogenes are related was first shown for PDGF. The product of the transforming gene of simian sarcoma virus p28 v-*sis* closely resembles one of the two polypeptide chains of PDGF (Delarco and Todaro, 1978; Waterfield et al., 1983). Both virally transformed cells and various tumor cells expressing the *sis* oncogenes secrete growth factors that resemble PDGF (Devel et al., 1983). Similarly, the product of the *erb-B* oncogene resembles the EGF receptor (Herschman, 1985; Ulrich et al., 1984). These considerations suggest that the oncogenes may act through the cell's machinery that normally responds to the growth factors.

Active oncogenes have been identified in a bioassay in which DNA isolated from human tumor cells or normal cells is tested for its ability to transform NIH 3T3 mouse cells to form colonies (Cooper et al., 1980; Shih and Weinberg, 1982). The NIH 3T3 cells have been transformed by several viruses and the viral DNA integrated into the cells' genome. This approach has allowed the identification of more than 23 oncogenes.

However, oncogenes per se are not responsible for transformation or malignancy. DNAs cloned from human oncogene DNA were used to probe for the presence and location of oncogenes in the chromosomes of normal cells, and as a result it has become apparent that oncogenes are widely distributed throughout the genome (Aaronson et al., 1985). Apparently, the oncogenes are activated to produce tumor cells in at least one of two ways. Gene amplification of the oncogenes could produce an excess of oncogene mRNA. Alternatively, alteration of the expression of oncogenes could follow chromosomal rearrangements, presumably by putting the genes under new regulatory controls.

Oncogenes are thought to function in normal cells in the regulation of cell growth. Accordingly, they could affect a number of mechanisms. Some oncogenes code for proteins related to growth receptors. Among these, the V-*erb-B* oncogene codes for the cytoplasmic domain of the EGF receptor (Ulrich et al., 1984). The products of *src* oncogenes have tyrosine kinase activity, which is thought to have a regulatory effect on the interaction between receptors and growth factors.

Other oncogenes code for growth factors; for instance, one of the polypeptide chains of PDGF is coded by the *sis* oncogene (Doolittle et al., 1983; Waterfield et al., 1983).

The *ras* gene family codes for proteins in the inner face of the plasma membrane (the G proteins, discussed in Chapter 13) and regulates the effects of some growth factors. These genes are thought to mimic the effect of growth factor receptors by affecting the coupling between the receptor and the internal signal (Der et al., 1982; Paladail et al., 1982; Shih and Weinberg, 1982).

The *myb* family of genes encodes proteins in the nucleus of cells (Gonda and Bishop, 1983; Van Beveren et al., 1982), and in normal cells they are presumably involved in the control of gene expression during development. Two of these genes code for proteins that are expressed in normal cells in response to growth factors.

Our knowledge of neoplastic (tumor-related) growth has been aided by studies with acutely transforming retroviruses. *Retroviruses* are different from other viruses in having a genome coded by RNA rather than DNA. Some transforming retroviruses rapidly transform cells both in vivo and in vitro (Bishop, 1983) by virtue of their possession of specific genes that are considered oncogenes. The viral oncogenes are thought to have been incorporated in the viral genome from the transcript of normal genes involved with the regulation of growth (Carpenter and Cohen, 1976). Supporting this view, the retroviral *onc* sequence used as a DNA probe detects homologous sequences in the DNAs of a wide variety of vertebrate species (Aaronson et al., 1985).

Slow-acting retroviruses do not have oncogenes but transform cells by integrating the viral promoter adjacent to a cellular oncogene (Hayward et al., 1981) and thereby altering the regulation of the oncogene in an otherwise normal cell.

The account of the regulation of growth factors and oncogenes is fascinating, bringing together very fundamental concepts of cell regulation. However, it may well be only half the story. Other data, still less detailed, suggest the presence of suppressor genes that seem to have a role opposite to that of the oncogenes. These genes have been detected by fusing normal and malignant cells and finding a reversion from a transformed to a nonmalignant phenotype (Klein, 1987). The role of the suppressor genes is likely to be as widespread as that of the oncogenes themselves.

IV. RECEPTORS AND IMMUNITY

The cell surface also has a role in the production of antibodies in a manner somewhat related to its involvement in the action of growth factors.

Specific antibodies are produced by the immune system in response to proteins recognized as foreign, the *antigens*. *Antibodies* are proteins that can specifically bind a chemical moiety of the antigen, the *epitope*.

At present, there is considerable evidence that an antibody is produced by a process of clonal selection and activation. The theory of clonal selection assumes that each lymphocyte of an immunocompetent animal possesses a receptor at the cell surface that can bind only to the epitope. Each cell would have only one of these special receptors, so only a small portion of the lymphocytes could respond to one antigen. The binding of the antigen to the receptor stimulates the cell to divide, producing a clone of cells with the same receptor specificity. The progeny of this clone includes cells that synthesize and secrete antibody molecules of the same specificity as the receptor molecules.

The fact that the receptor to a given antigen is present in only a few cells before exposure to the antigen has been demonstrated in two different kinds of experiments. One kind selected for destruction cells with the appropriate receptor. The other marked these cells so that they could be separated from the rest of the lymphocyte population. In the first kind of experiment, cells from mouse spleen were exposed to a ^{125}I-labeled antigen, *Salmonella* flagellin. Only 1 in 5000 cells could bind the antigen, and after incubation for 16 to 20 h the high radioactivity of the label destroyed them. The remaining cells were tested for immunological competence by injecting them into mice whose immunological systems had been destroyed by x-rays. Mice injected with the flagellin were incapable of producing the antibody, although they could produce an antibody to flagellin from a related strain of *Salmonella* (Ada and Byrt, 1969). The results of this experiment are therefore consistent with the theory of clonal selection in that only a portion of the cells could bind the antibody and the rest of the cells were unable to produce the appropriate antibody. The results are shown in Table 3.3. The two different *Salmonella* strains are distinguished as SW1338 and SL871. Column 2 lists the pretreatment with the antigen, and columns 3 and 4 show the amount of antibody produced by the test mice.

The positive selection experiments were carried out with antigen labeled with a fluorescent dye. When the antigen was added to the spleen cells, some of the cells were able to bind the antigen. The cells were then sorted out by a cell sorter. The sorter functions by passing the

Table 3.3 Antibody Titers of X-Irradiated Mice Injected with Two Serologically Distinct Flagellar Antigens 1 Day After Receiving Syngeneic Cells Pretreated with ^{125}I-Labeled or Unlabeled Antigen

(1)	(2)	(3)	(4)
		Mean antibody titer ($\log_2$) from groups of 10 mice after challenge with	
Experiment	Antigenic pretreatment (per 1.3×10^7 cells)	SW1338	SL871
1	0.5 μg SW1338, labeled	0.5	4.2
	0.5 μg SW1338, unlabeled	2.7	5.1
	5 μg SW1338, labeled	<0.5	5.4
	5 μg SW1338, unlabeled	2.9	4.3
	No antigen	2.3	4.7
2	5 μg SL871, labeled	0.82	<0.5
	5 μg SL871, unlabeled	1.6	4.8
	No antigen	1.1	4.2
3	5 μg SL871, labeled	2.0	1.1
	5 μg SL871, unlabeled	3.3	4.2
	No antigen	2.4	4.9

Source: Ada and Byrt (1969). Reprinted by permission from *Nature* 222:1291–1292, copyright © 1969 Macmillan Magazines Ltd.

suspension through a vibrator, which produces liquid droplets containing individual cells. A sensitive photocell can detect the fluorescent label when the cells are illuminated by a laser beam. The sorter imparts a charge to the fluorescent droplets so that labeled and unlabeled cells can be collected at metal plates with opposite charge. In this experiment (Julius et al., 1972), 0.1 to 3% of the mouse spleen cells were found to contain the fluorescent antigen. The sorter was able to collect cells that were fluorescent and more capable than other cells of producing immunity to the antigen when tested in the irradiated mice. The collection of labeled cells is shown in Table 3.4.

The formation of an antibody actually involves a complex interaction between at least two kinds of lymphocytes. The events are thought to begin with binding of the antigen by macrophages. After binding to surface receptors, the antigen is taken up by endocytosis (Pernis, 1985; Pernis and Axel, 1985) and partially hydrolyzed to shorter peptide fragments. The proteolytic fragments are transported back to the cell surface, where they become available to the well-characterized antigen receptor of neighboring T lymphocytes. The remaining steps in the production of antibody are complex, involve interactions between B and T lymphocytes, and are still not well understood.

V. INTERACTION WITH THE EXTRACELLULAR MATRIX

Interactions between the cell surface and the extracellular matrix (ECM) play an important role in cell differentiation, development, and morphogenesis. All indications are that the effect of the matrix components is mediated by binding at receptor sites on the cell surface.

The importance of the interaction between cells and the extracellular matrix is illustrated by the inability of nontransformed cells to proliferate without contact with a solid support (Stoker et al., 1968), even in the presence of growth factors (Otsuka and Moskowitz, 1975). Accordingly, DNA synthesis depends on attachment to a substrate (O'Neill et al., 1986). An example of this

Table 3.4 Enrichment of Antigen Binding Cells

Experiment	Strain	Priming antigen[a]	Staining technique	Before separation	Undeflected	Deflected	Enrichment factor
1	BALB/cN	KLH	Direct	0.1	<0.1	40	400
2	BAB/14	KLH	Direct	0.1	<0.1	52	520
3	CWB	H Albumin	Indirect	3.5	1	65	18.5
4	CWB	H Albumin	Indirect	1.3	<0.1	55	42
5	BALB/cN	Human gamma globulin	Indirect	1	<0.1	62	62

Reprinted with permission from M. H. Julius, et al., Proceedings of the National Academy of Sciences, 69:1934–1938, 1972.
[a]KLH, Keyhole limpet hemocyanin; H albumin, human serum albumin.

dependence is shown in Fig. 3.8, which represents the number of cells that are labeled in an [^{3}H]thymidine medium when plated on a surface of palladium on poly(hydroxyethyl methacrylate). As indicated, the rate of incorporation (representing cell divisions) is a function of the area of contact between the cells and the substrate.

Other experiments show a correlation between attachment, cell shape, and proliferation (Ben-Ze'ev, 1985). When cells are placed in a liquid medium after a period of growth, mRNA [detected as poly(A) RNA], rRNA, and DNA synthesis stops almost immediately. Eventually, protein synthesis also decreases. If the same cells are then plated on a solid medium, all these activities resume relatively rapidly. The correlation between mean cell shape and DNA synthesis is excellent, as shown in Figs. 3.9 and 3.10 (Ben Ze'ev et al., 1980). In these experiments, cell spreading on a solid medium is decreased by coating with different concentrations of poly(2-hydroxyethyl methacrylate) [poly(HEMA)], as shown in the micrographs and representation of mean cell diameter of Fig. 3.9. In turn, the DNA synthesized correlates well with the concentration of poly(HEMA) as shown in Fig. 3.10.

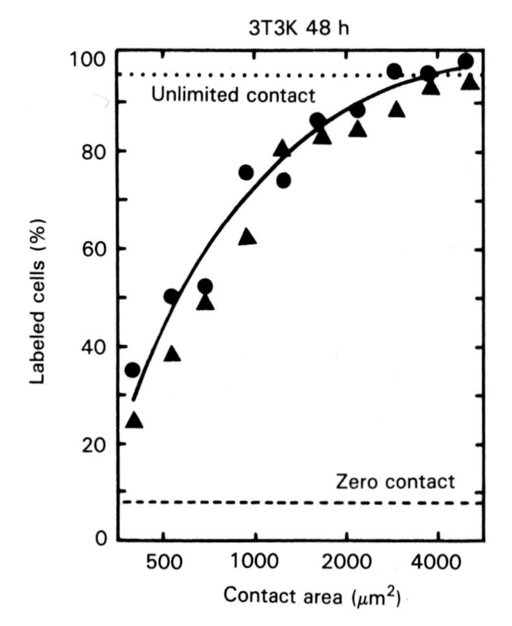

Fig. 3.8 Effect of attachment area on DNA synthesis in 3T3 cells. Reprinted with permission from C. O'Neill, et al., *Cell,* 44:489–496. Copyright 1986 by Cell Press.

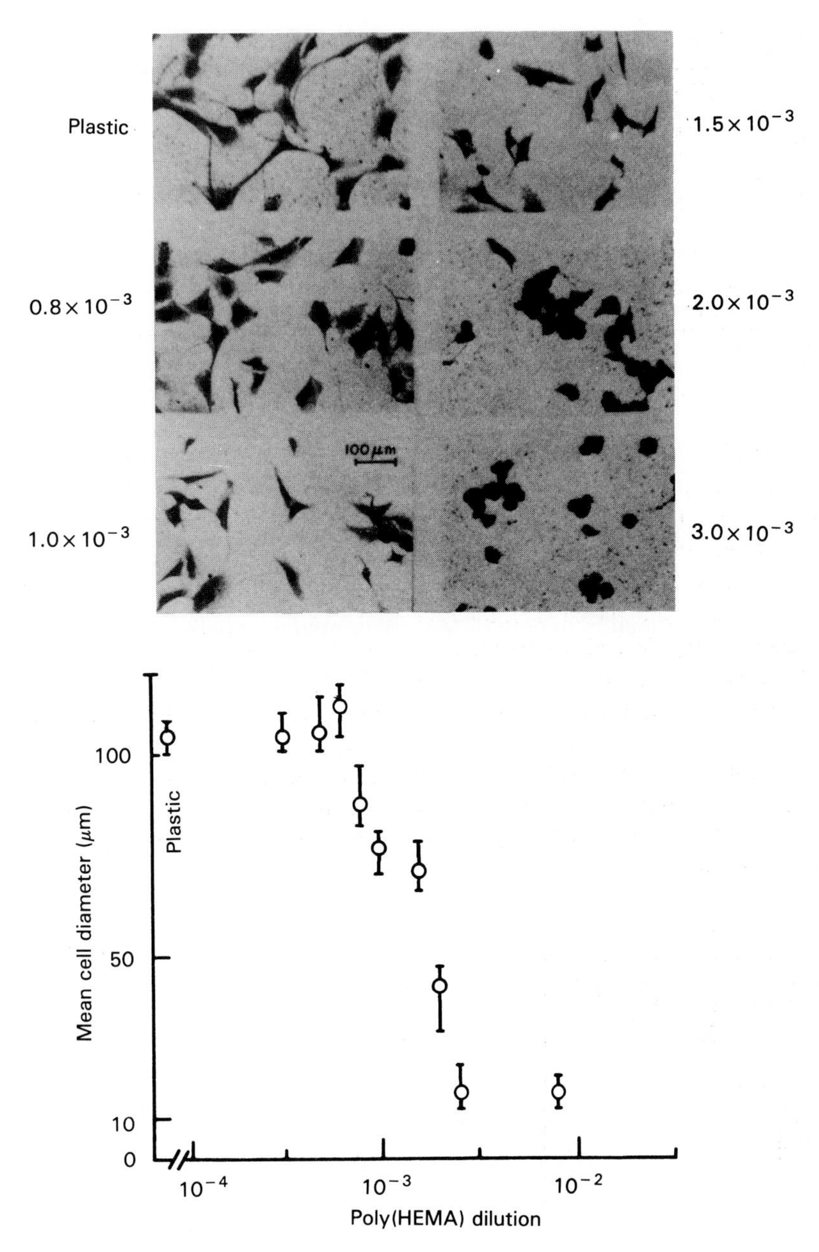

Fig. 3.9 Control of cell spreading by coating the culture dish with different poly-(HEMA) concentrations. Reprinted with permission from A. Ben-Ze'ev, et al., *Cell,* 21:365–372. Copyright 1980 by Cell Press.

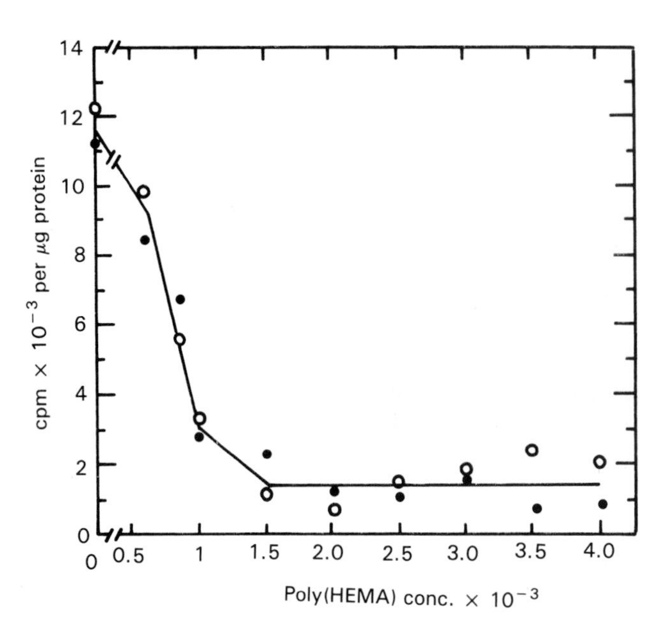

Fig. 3.10 DNA synthesis as a function of poly(HEMA). Reprinted with permission from A. Ben-Ze'ev, et al., *Cell,* 21:365–372. Copyright 1980 by Cell Press.

A role for microtubular and microfilament systems, including formation of the structural framework of the nuclear matrix that allows the cell to respond to growth factors or hormones, has been proposed by many (Ben-Ze'ev, 1985; Ben-Ze'ev et al., 1979). The possible linkage between shape and expression of the cytoskeletal protein genes may serve as a model for understanding how shape may affect other biological activities.

Microtubules are disrupted by a variety of agents. Colchicine depolymerizes microtubules so that there is an increase in the concentration of free tubulin. In contrast, vinblastine, which also disrupts microtubules, induces aggregation into paracrystals; under these conditions, the concentration of free tubulin decreases. In a classical feedback response, the increase in free tubulin results in a decrease in tubulin synthesis, whereas a decrease in tubulin concentration results in an increase in synthesis. Apparently, the control is at the level of the gene, since the mRNA coding for tubulin mirrors the changes in synthetic rate. The mRNA has been identified with an in vitro translational system (Ben-Ze'ev, 1985). It is likely that changes in shape are accompanied by changes in the polymerization of microtubular and microfilament elements and, as in the tubulin case, these changes could have regulatory effects on genes required for the action of the growth factors.

An association of cell shape and cell activity is illustrated by results of experiments with mammary epithelial cells. When the cells are grown on a floating collagen gel they are cuboidal and secrete milk proteins (Emerman and Pitelka, 1977; Lawler and Hynes, 1986). However, they fail to become cuboidal or to secrete milk proteins when grown on floating collagen that has been cross-linked with glutaraldehyde or on collagen attached to the dishes (Lee et al., 1984)

The experiments with artificial substrates may suggest a nonspecific role of attachment. However, it is entirely possible that the cells first secrete extracellular matrix proteins and that these have the mitogenic effect, the synthetic substrate only supplying a suitable support.

The role of extracellular matrix is undoubtedly complex. There is evidence, however, that part of the action of its components may be through a mechanism analogous to that of growth factors. The sequencing of the cDNA of extracellular proteins, such as laminin (Sasaki et al., 1987), an 800-kDa protein associated with the basal membrane; cartilage matrix protein, a 148-kDa protein (Argraves et al., 1987); or thrombospondin (Lawler and Hynes, 1986), a 420-kDa extracellular glycoprotein secreted by cells in culture, has revealed in these molecules amino acid sequences homologous to those of EGF (see V, B, below).

Transmission of a signal from the cell surface to the cell's interior is likely to require interaction of the signal with an intrinsic membrane protein and in many ways resembles the action of hormones and growth factors which bind to surface receptors (see V, C).

A. The Extracellular Matrix

The ECM contains collagen, glycoproteins, proteoglygans and glycosaminoglycans secreted by cells and interwoven into an organized network (Hay, 1984), which includes hyaluronic acid, fibronectin, laminin, entactin, and elastin.

Collagen With a few possible exceptions, collagen molecules are rodlike structures 30 nm long and 1.5 nm in diameter (see Linsenmayer, 1991). Ninety-five percent of the molecule forms a triple helix, in which three α chains of about 1000 amino acids, each forming a left-handed helix, are wrapped around one another. The conformation is stabilized by hydrogen bonding, and the peptide bonds are buried in the interior of the molecule. Figure 3.11 (Linsenmayer, 1981, 1991) shows these features for type I collagen.

Every third residue of the α-helices corresponds to glycine, which is responsible for holding the chains in the proper configuration. The other two amino acids of this triplet, glycine-X-Y, can be any amino acids, but frequently X is proline or hydroxyproline (100 sites per chain). At the NH_2 and COOH terminals of the α chains there are nonhelical extensions of about 20 amino acids. These ends can link in forming collagen fibrils. The hydroxyprolines originate posttranslationally from prolines, a reaction catalyzed by prolyl hydroxylase. Two glycosylation reactions can take place at hydroxylated sites to form galactosylhydroxylysine (GAL, OH-LYS) and glycosylgalactosylhydroxylysine (GLU, GAL, OH-LYS). The amounts of hydroxylation and glycosylation differ in different organisms and even in different tissues of the same organism.

Lysine and hydroxylysine are modified extracellularly to their reactive aldehyde forms, allysine and hydroxyallysine, preferentially at the terminal extensions and usually after collagen fibrils have already formed. These sites can form stable cross-bridges with adjacent fibrils.

Collagen is formed intracellularly as procollagen, which has a higher molecular weight than the extracellular collagen because it contains extensions at both the NH_2 and COOH terminals. Furthermore, a still higher molecular weight procollagen is produced by the translational system containing collagen mRNA.

There are various types of collagen (for a review see van der Rest and Garrone, 1991). Type I is present in many adult connective tissues such as skin, bone, tendon, and cornea. Type I collagen forms striated fibrils (in which the triple-helical arrangements are in register) 20 to 100 nm in diameter. Two distinct kinds of α chains form the triple helix.

Type II collagen is present in cartilage and is synthesized in chondrocytes. The fibrils are generally 10 to 20 nm in diameter. The α chains of type II collagen are identical. This type of collagen has more hydroxylysine and glycosylation than type I.

Type III collagen is present primarily in loose connective tissue such as derma, blood vessel walls, and placenta and in the 50-nm fibers of reticular connective tissue. This collagen is composed of a single type of α chain, and the three chains making up the triple helices are held together by disulfide bonds. Accordingly, the amino acid composition is relatively rich in half cystines. Type III collagen has a higher glycine concentration than other collagens.

Types IV and V are collagens of the basement membrane. Type IV probably corresponds to two different α chains rich in hydroxyproline and cysteine and with a good deal of glycosylation.

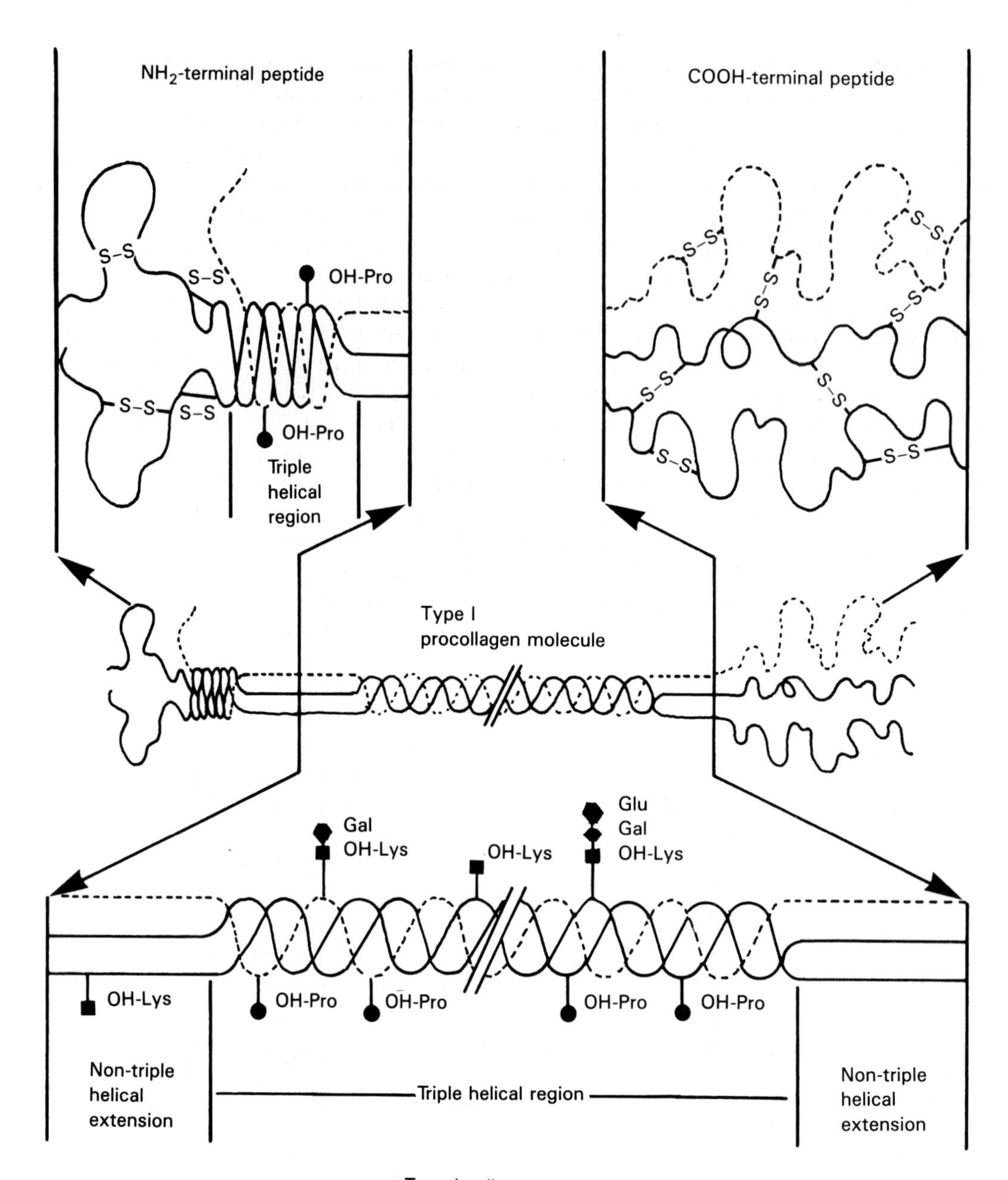

Type I collagen molecule

Fig. 3.11 Diagram of the major characteristics of the type I collagen molecule and its procollagen form. OH-PRO, hydroxyproline; OH-LYS, hydroxylysine; –S–S–, disulfide bonds; GAL, galactose; GLU, glucose. The NH_2-terminal propeptide and COOH-terminal propeptide show the two pieces that are cleaved from procollagen during its processing into a collagen molecule. Reprinted with permission from T. F. Linsenmayer, *Cell Biology of Extracellular Matrix.* Copyright © 1981 Plenum Publishing Corporation.

Proteoglycans Proteoglycans (see Wight et al., 1991) are complex macromolecules containing a protein core covalently bound to one or more glycosaminoglycan (GAG) chains. Many have complex structures with domains similar to those of other protein families. These domains may confer special functional properties. Some of the proteoglycans (such as aggrecan) are matrix proteins. Others (such as the syndecans) are integral membrane proteins with large extracellular domains. The matrix proteoglycans create a water-filled matrix which is freely available to low molecular solutes but excludes large molecules. Their multiple negative charges attract cations, not only increasing hydration but also serving as a selection device.

In cartilage proteoglycan the protein is 250,000 in molecular weight and 300 nm in length. This core protein constitutes only 5 to 10% of the total mass, the rest being carbohydrate. Figure 3.12 shows an electron micrograph of a negatively stained cartilage proteoglycan. A diagram of the structure is shown on the right side of the figure. More details are shown in Figure 3.13. In the extracellular matrix, the proteoglycan complexes attach along hyaluronic acid (HA) strands. Hyaluronic acid is a polymer in which the repeating unit is a disaccharide made up of *D*-glucuronic acid and *N*-acetyl-*D*-glucosamine. The binding site on the proteoglycan has a high affinity for long HA strands, the decasaccharide or longer. However, generally the binding is even tighter, since it is mediated by one of two linking proteins. An electron micrograph of this configuration is shown in Figure 3.14 (p. 91), which also contains an explanatory diagram. In this case the chondroitin sulfate side chains are collapsed, in contrast to those shown in Figure 3.12. A network of collagen and proteoglycan is shown in a section of cartilage in Figure 3.15 (p. 92).

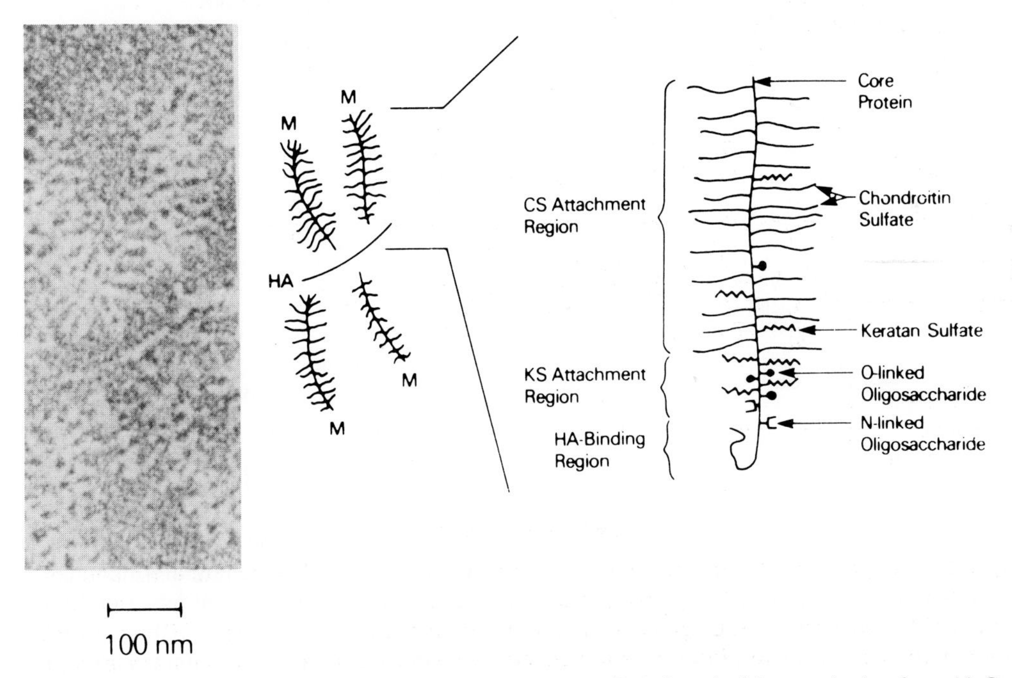

Fig. 3.12 Cartilage proteoglycan, monomer structure. Reprinted with permission from V. C. Hascall and G. K. Hascall, *Cell Biology of Extracellular Matrix.* Copyright © 1981 Plenum Publishing Corporation.

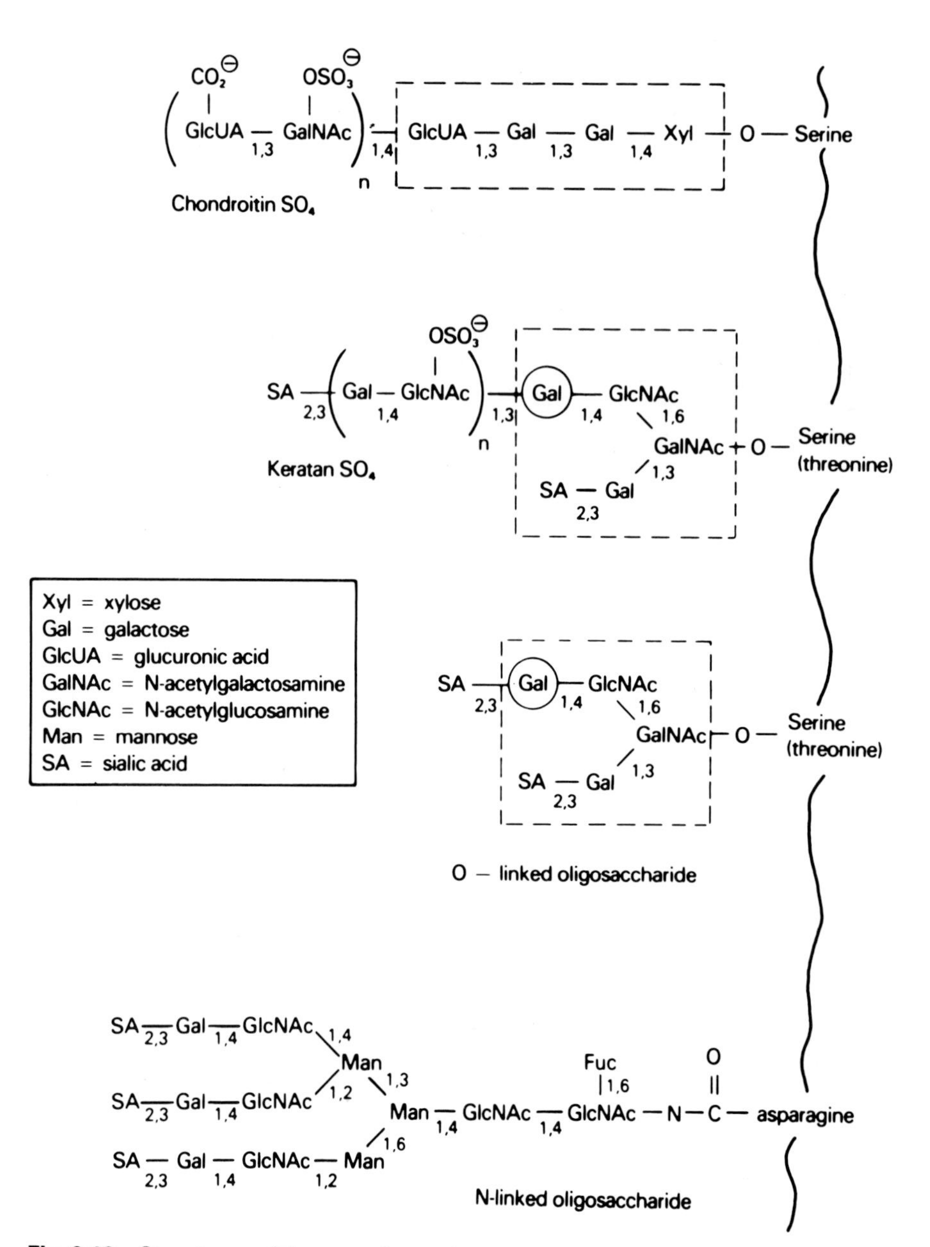

Fig. 3.13 Structures of the complex carbohydrates attached to cartilage proteoglycans. The repeating disaccharide structures of chondroitin sulfate and keratan sulfate are indicated, along with the specialized attachment regions by which the chains are covalently bound to the core protein (shown in the boxes). The O-linked oligosaccharides are related to the linkage region structure for keratan sulfate as indicated by the circled, similarly located galactose residues. Reprinted with permission from V. C. Hascall and G. K. Hascall, *Cell Biology of Extracellular Matrix.* Copyright © 1981 Plenum Publishing Corporation.

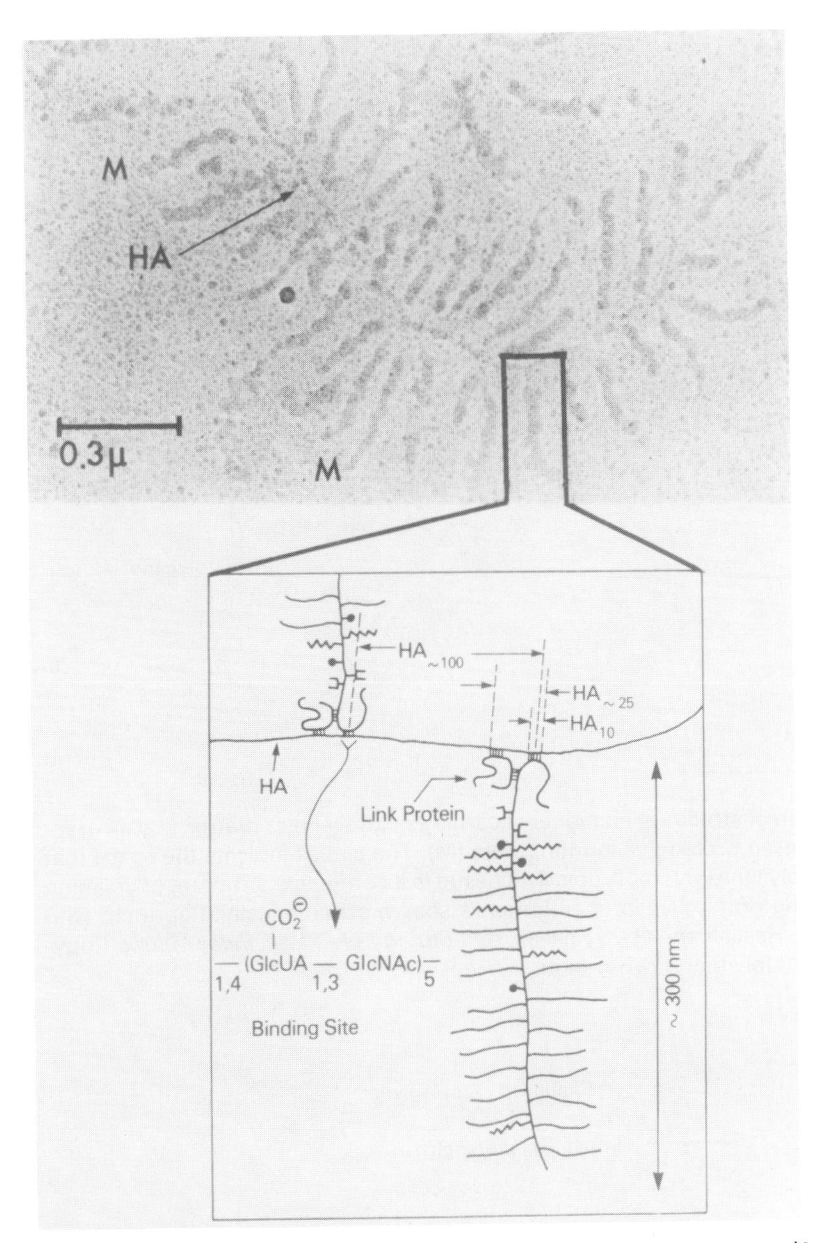

Fig. 3.14 Structure of cartilage proteoglycan aggregates. Monomers (1,4) are aligned on a central filament of HA. Reprinted with permission from V. C. Hascall and G. K. Hascall, *Cell Biology of Extracellular Matrix.* Copyright © 1981 Plenum Publishing Corporation.

The proteoglycans of the aorta are very similar to those in collagen. They also combine with HA and link proteins. Furthermore, they are likely to interact with collagen. However, the proteoglycans are very different in other tissues. Some of these proteoglycans are represented in Figure 3.16 (p. 92).

Proteoglycans have many functions. Select sequences within the molecular structure are modulated during varying cellular responses, suggesting very important physiological roles. One of the recent exciting findings is the binding of growth factors by proteoglycans which are now thought to play a significant role in the regulation of their activity (Ruoslahti and Yamaguchi, 1991).

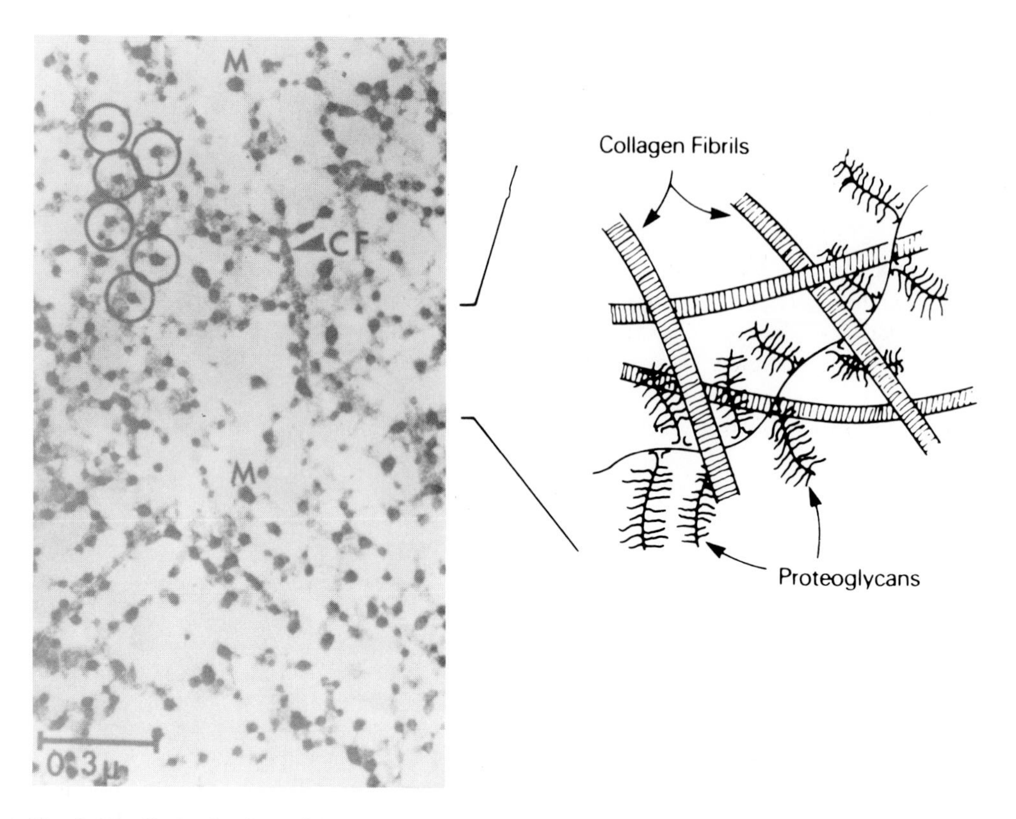

Fig. 3.15 Organization of structural elements of cartilage extracellular matrix. Matrix granules represent condensed proteoglycan monomers (M). The circles indicate the space that the expanded proteoglycans would occupy. The tissue is a composite structure of collagen fibrils with intertwining proteoglycan aggregates as shown schematically. Reprinted with permission from V. C. Hascall and G. K. Hascall, *Cell Biology of Extracellular Matrix.* Copyright © 1981 Plenum Publishing Corporation.

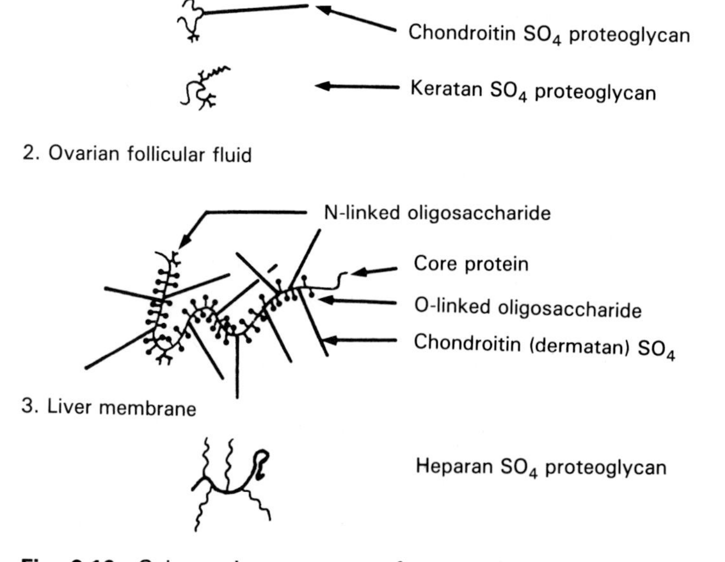

Fig. 3.16 Schematic structures of proteoglycans from different tissues. Reprinted with permission from V. C. Hascall and G. K. Hascall, *Cell Biology of Extracellular Matrix.* Copyright © 1981 Plenum Publishing Corporation.

Arrangement of Cells in the Matrix Connective tissue cells, such as chondrocytes and dermal fibroblasts, are completely surrounded by the extracellular matrix. Endothelial cells may be separated from connective tissue by a specialized sheet, the basement membrane or lamina. Each cell may be completely surrounded by a basal membrane or the membrane may be on only one side of the cell.

The basement membranes primarily contain type IV collagen; heparan sulfate proteoglycans and laminin are their major noncollagen components, and minor components may include fibronectin and entactin (Kefalides et al., 1979). The heparan sulfate proteoglycans in the basement membranes provide an anionic barrier to block the passage of proteins, which are generally negatively charged (Kamwar et al., 1980). Laminin is produced on one side of epithelial cells and is therefore localized on one side of these cells.

B. Cell Adhesion Proteins

The formation of tissues during development, their physiology and differentiation are crucially influenced by the interactions between the ECM and cells and between cells and other cells.

Association of cells with the ECM Cells bind to the ECM by interacting with matrix-molecules. The ability of some specialized glycoproteins to bind cells to other elements of the ECM are so striking that they are referred to as ECM-adhesion molecules. The adhesion glycoproteins affect cellular activity and regulate the formation of the matrix. These proteins can generally be shown to correspond to families of related proteins such as the fibronectins, chondronectins, thrombospondins, laminins, and tenascins—each group with a more or less distinct tissue distribution and activity toward certain cell types. These proteins have dramatic effects on cell migration, morphology, and metabolic activity and play an important role in development, cell organization, and repair. The importance of the cell-adhesion ECM molecules and their receptors can be illustrated in the case of the unc-6 gene which is involved in directing cell migration during development in the nematode *Caenorhabditis elegans* (Hedgecock et el., 1990). Mutations at this site produce a defective circumferential movement of axons. This gene probably encodes for a laminin B2 receptor. Furthermore, isoforms of laminin (Beck et al., 1990) correspond to distinct stages of organ development in mammals (Sanes et al., 1990; Klein et al., 1990). Schematic drawings of the ECM-adhesion glycoproteins are shown in Fig. 3.19. Their structure and various domains are represented in Fig. 3.20 (Engel, 1991).

The fibronectins (FNs) are a family of glycoproteins composed of two subunits 220–250 kDa with an acidic isoelectric point (pI = 5.5–6.0) which are joined by disulfide bridges. A protein is considered an FN if, in addition to these characteristics, it binds gelatin and cross-reacts immunologically with other FNs. The significance of the FNs is indicated by the fact that their quantity, location, and in some cases structure vary with embryonic development (Wartiavaaren and Vaheri, 1980), disease (Akiyama and Yamada, 1983), and aging (Chanda-sekharis et al., 1983).

Fibronectin is present in significant amounts at locations of tissue remodeling and cell migration and in basement membranes. Experiments in which the effect of fibronectin is blocked with antifibronectin suggest that it influences the distribution of collagen (McDonald et al., 1982).

Fibronectin exists in a soluble form in plasma. Cellular fibronectin is secreted by fibroblasts and possibly other cells; it forms a fibrillar network at the cell surface and in the extracellular matrix. Although the two kinds of fibronectins are very similar, they differ significantly in structure and activity.

Table 3.5 Biological Activities of Fibronectin

Cell-to-substrate adhesion
Attachment and spreading of cells on plastic or glass
Cell attachment to collagen or fibrin
Cell-to-cell adhesion
Cell morphology
Maintenance of flattened, fibroblastic shape and minimal numbers of cell surface microvilli
Alignment of confluent fibroblasts in parallel arrays
Promotion of actin microfilament bundle organization
Cell migration
Stimulation of cell motility
Haptotaxis
Stimulation or inhibition of cell differentiation
Nonimmune opsonic activity for macrophage phagocytosis

From K. M. Yamada, et al., *The Role of Extracellular Matrix in Development.*

Table 3.5 (Yamada et al., 1984) summarizes some of the known activities of fibronectin. The mechanism of these effects is still not clear. However, specific regions of the molecule, the binding domains, can be shown to bind specifically to various components, and this capacity may help explain how fibronectin acts. The presence of different domains allows the fibronectin to act as a bridge, so that one domain could attach to a component such as a cell, and another domain of the same fibronectin molecule could attach to a collagen fiber or some other element.

The affinity of each domain can be demonstrated after limited digestion of the molecule by trypsin or another protease that leaves intact large fragments, which can then be tested. The approach is summarized in Figure 3.17 (Schneider et al., 1982). The binding information derived in these experiments is summarized in the fibronectin structure depicted in Figure 3.18. In this diagram, the proteolytic fragments are represented by boxes and the affinity of the domain for a specific component is written under the boxes. As shown in the diagram, fibronectin consists of two similar subunits held together by disulfide bonds. Corresponding domains of the two subunits are the same, although the structures differ slightly.

The FNs have a major role in the adhesion of fibroblasts, hepatocytes, and endothelial cells to the ECM. The binding occurs at specific cell surface receptors (discussed also in the next section). The receptors proved elusive and were implicated using antisera to cell surface components. The antibodies were tested for their ability to block FN binding when added to cells. When found effective they could be used to identify the cell surface protein. Proteolysis of a cell surface component was used with the same rationale (e.g., Urushihara and Yamada, 1986). When the proteolysis prevented attachment, the protein involved in the binding would correspond to the one or more proteins missing (e.g., as seen in a PAGE gel, see Chapter 1, II, B) from the treated preparation. In addition to these procedures, chemical cross-linking of cell surface proteins to FN (Aplin et al., 1981) was used to reveal the receptors. These experiments implicated a protein or proteins of 45 to 47 kDa. Other experiments using antisera and, in particular, two monoclonal antibodies, identified a family of glycoproteins of 115 to 165 kDa (e.g., Hasegawa et al., 1985; Knudsen et al., 1985). Their solubility properties suggested that they were integral membrane proteins. These are the *integrins* discussed in detail in the next section.

Another group of cell adhesion molecules, the *laminin* family, has been found in the basal lamina of many tissues (for a review, see Beck et al., 1990). Basement membranes are highly organized thin extracellular matrix structures that separate epithelial cells from underlying

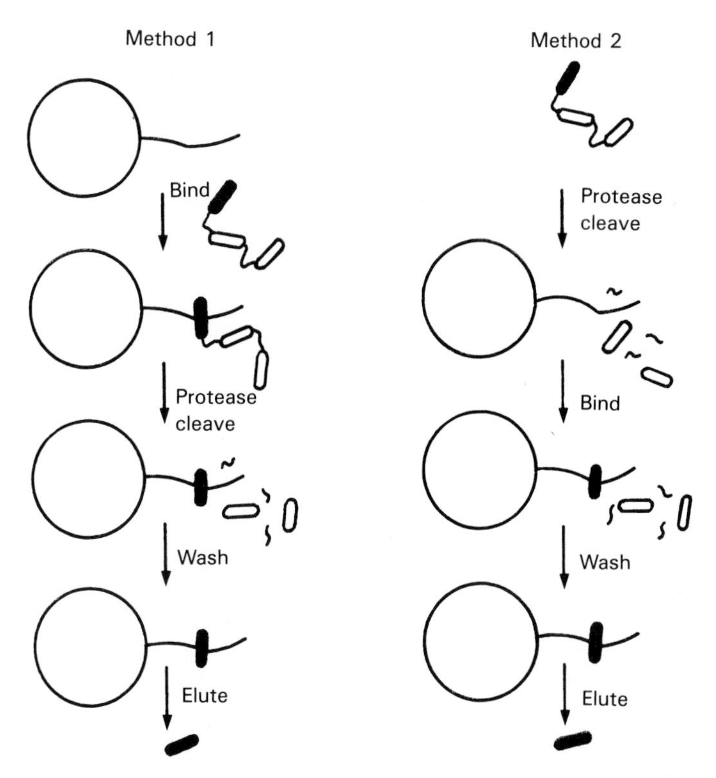

Fig. 3.17 Method for the isolation of functional domains from a binding protein. Circles at the left indicate agarose beads, with one of the attached ligand molecules depicted as extending to the right. Three protease-resistant domains are shown as black or shaded rodlike structures separated by flexible regions of polypeptide chain; the domain that binds specifically to the ligand is shown in black. From K. M. Yamada, et al., *The Role of Extracellular Matrix in Development.* Copyright © 1984 Alan R. Liss, Inc.

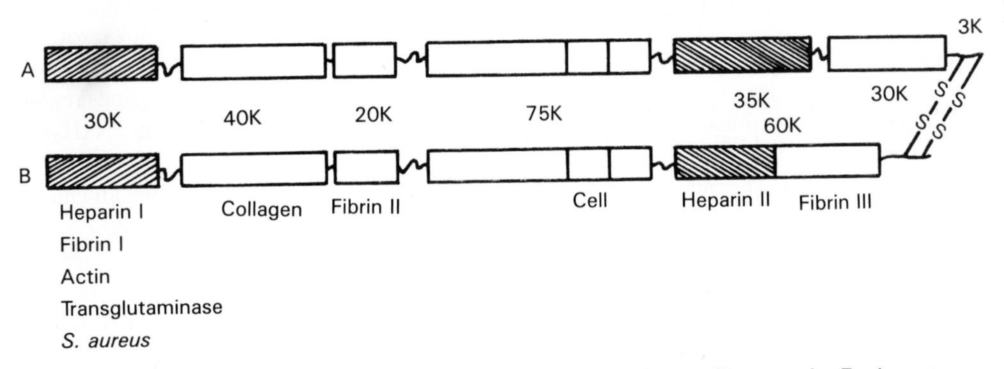

Fig. 3.18 Early map of the functional domains of human plasma fibronectin. Each rectangular box represents a protease-resistant functional domain of the molecule. The domains of the A and B subunits differ in at least one site near the carboxyl terminus. Sizes of the domains are indicated by the numbers; for instance, 30K means an apparent molecular weight of 30,000. The binding activities of each domain are listed underneath; some of these domains can bind to the same ligand and are numbered from amino terminus (left) to carboxyl terminus (right). From K. M. Yamada, et al., *The Role of Extracellular Matrix in Development.* Copyright © 1984 Alan R. Liss, Inc.

stromal tissues. The basal membranes, viewed with the electron microscope, have an electron-dense region (lamina densa) and a more electron- transparent region (lamina lucida). The dense region contains a network of filaments (Kefalides et al., 1979). Laminin is composed of three B chains of 200 kDa and one A chain of 400 kDa, held together by disulfide bonds and arranged as a cross (Timpl, 1982).

Among other activities, laminin promotes the extension of axons (e.g., Edgar et al., 1984). The activity of laminin overlaps some of those of fibronectin, although some of their activities are distinct. It is clear, however, that individual cell types can interact with more than one kind of adhesion molecule.

Laminins are the first known ECM-adhesive molecules to appear in embryogenesis where they have been detected at the two-cell stage. The laminins have been shown to promote cell attachment, growth, and differentiation. Laminin has a cruciform structure that can be observed with electron microscopy. Two short arms are 36 nm in length, another short arm is 48 nm, and a long arm is 77 nm in length (Bruch et al., 1989) (see Fig. 3.19). Mouse EHS tumor laminin has three different polypeptide chains: A (440 kDa), B1, and B2 (about 220 kDa each), linked by disulfide bonds. The glycosylation is uneven, so the actual molecular weight is variable.

Laminins can self-associate and also bind to heparin, cellular elastin, and collagen. Several cell surface receptors have been implicated in laminin binding. These include a 67 kDa protein found in metastatic cells (von der Mark and Kühl, 1985), and a 66 kDa protein from skeletal muscle, aspartactin (Hall et al., 1988). In addition, members of the ubiquitous integrin family (discussed below) seem to serve as receptors for laminin (Gehlsen et al., 1988).

Tenascin is one of the ECM adhesive proteins thought to play a role in the development of the nervous system. The six identical subunits of this glycoprotein are approximately 190–320 kDa (depending on the species) and are linked by disulfide bridges. Tenascin binds to ECM molecules such as FN, heparin, and proteoglycans (see Chiquet-Ehrismann, 1990). It also binds to cell surface receptors such as the proteoglycan, syndecan (Salmivirta et al., 1991), and integrins (e.g., Bourdon et al., 1989).

Many of the amino acid sequences of ECM adhesive proteins have been deciphered through DNA technology (see Chapter 1, II, A). Typically, ECM proteins are found to have repeat sections as well as regions of homology with other ECM proteins. Tenascin has a heptad repeat region (see Fig. 3.20), EGF and fibronectin-like repeats, and a region of homology with the globular domain of fibrinogen.

Despite knowledge of sequences, it is very difficult to assign function to the various regions of the molecules. Functional information is available only for some of the domains in FN (Hynes, 1990), laminin (Beck et al., 1990), and tenascin (Chiquet-Ehrismann, 1991).

An arg-gly-asp sequence in one of the FN domains is needed for the attachment and spreading of cells and the binding to its integrin receptor ($\alpha 5\beta 1$, see below). Furthermore, the binding is inhibited by the synthetic peptide with this sequence, which also inhibits the attachment of other ECM adhesive proteins (see Ruoslhati, 1988). However, many other binding domains in other ECM proteins do not depend on this sequence. In some cases single chains in random coil conformation are inactive (Deutzmann et al., 1990) suggesting that secondary and tertiary structure may play a significant role.

Some of the domains in ECM proteins have homology to EGF and transforming growth factor α (TGF-α) (see Fig. 3.19, 3.20). A growth-promoting effect has been reported for laminin, thrombospondin, and tenascin (Engel, 1989), and in the case of laminin, this function was localized to a specific sector with 25 EGF-like repeats (Panayotou et al., 1989).

Specific binding sites for growth factors have also been identified in several ECM proteins, e.g., TGF-β in fibronectin and proteoglycans (Fava and McClure, 1987; Ruoslhati and Yamaguchi, 1991). Their function is still not clear. However, the storage of FGF in the ECM has

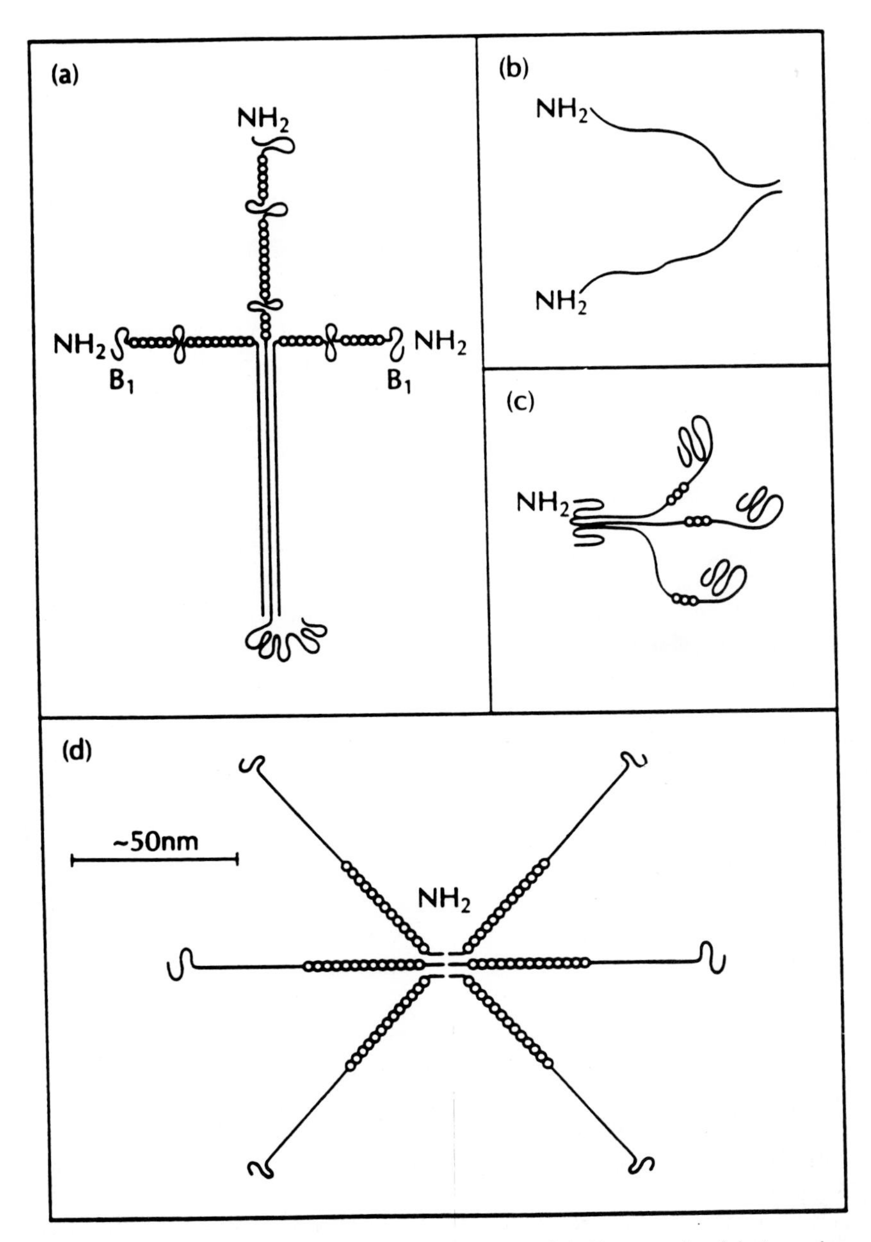

Fig. 3.19 Schematic drawing of (a) laminin, (b) fibronectin, (c) thrombospondin and (d) tenascin. The molecules are drawn approximately to scale. The circles indicate EGF-like domains. For other details see the more extensive diagram of Fig. 3.20. Reproduced from J. Engel, *Current Opinion in Cell Biology,* 3:779–785. Copyright © 1991 Current Biology Ltd., Philadelphia, PA.

been detected. It has been proposed that the storage and the release of the factor fulfills a regulatory role (Vlodasky et al., 1991). Many more domains may well play a significant role. There are domains in ECM proteins which have homologies with proteases (e.g., Engel, 1991). Although specific functions have not been found yet, these findings suggest possible important physiological roles.

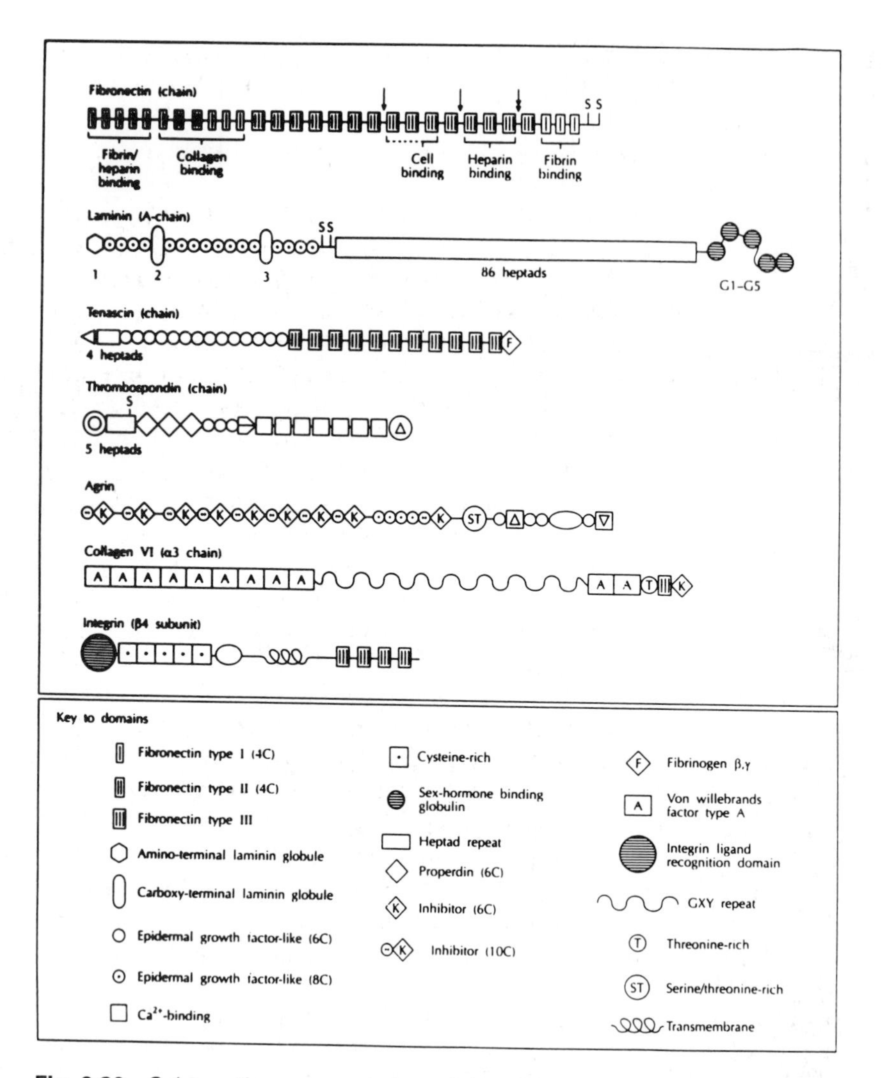

Fig. 3.20 Schematic representation of the various domains of some ECM-proteins, represented as single chains. Heptad repeats are repeats of seven amino acids which tend to assume a coil-coil conformation. See Engel, 1991 for more details. Reproduced from J. Engel, *Current Opinion in Cell Biology,* 3: 779–785. Copyright © 1991 Current Biology Ltd., Philadelphia, PA.

Some of the ECM proteins have anti-adhesive effects. Tenascin, which promotes cell adhesion (but not spreading), blocks the effect of fibronectin (Spring et al., 1989).

As seen from these few examples, the interactions between the various elements of this system are of great complexity which had only been suspected until now.

Cell-to-cell adhesion Cells can bind other cells. The significance of cell-to-cell adhesion is illustrated by the classical finding that cells isolated from any one particular tissue can be sorted out when mixed with those of other tissues (e.g., Townes and Holtfreter, 1955), a finding explainable by the differential adhesion hypothesis (Steinberg, 1963). This hypothesis proposes that each cell type has a specific adhesive capacity that allows it to attach only to like cells. Presumably this capacity could be mediated by cell-to-cell adhesion molecules. The *cadherins* are adhesion molecules that mediate such attachments. The appearance of cadherin is developmentally regulated, the expression of specific cadherins coinciding with morphogenic events (see Takeichi, 1991). Adhesion between cells also has obvious implications in malignancy and metastasis, which require the disruption of connection between cells in order to disperse them to other tissues.

The proteins required for cell-to-cell adhesion have been classified into several families. Some of these bind like molecules (*homophilic binding*); others bind molecules unlike themselves (*heterophilic binding*). The adhesion molecules include cadherins, immunoglobulins (Ig), some integrins (which generally attach cells to extracellular matrix), and selectins (in lymphocytes).

Cadherins constitute a family of transmembrane glycoproteins which mediate Ca^{2+}-dependent cell-to-cell adhesion and are responsible for determining cell adhesion specificity for the majority of cell types (see Takeichi, 1991). Since these proteins were discovered independently in a number of laboratories, they were named repeatedly. Uvomorulin, L-CAM, cell-CAM 120/80 are all cadherins. All cell types that form solid tissues have cadherins, each with a specific homophilic binding capacity. The predominance of the role of cadherins is demonstrated in experiments in which cell layers are treated with cadherin antibodies. This treatment was found to induce cell dispersion (see Takeichi, 1990) presumably by competing with the cells for the binding of cadherin molecules. Conversely, cadherin-deficient cells acquire a Ca^{2+}-dependent cadherin-mediated cell-to-cell adhesion when transfected with cDNA coding for cadherin (see Takeichi, 1990).

Different cadherins have different binding specificities. When cells expressing different cadherins are mixed, they reaggregate separately (Takeichi et al., 1981). Similarly, weakly aggregating cells transfected with different cDNAs, and therefore producing different kinds of cadherins, segregate out when mixed. Cells generally coexpress multiple classes of cadherins producing a very large number of possible specificities.

An amino acid sequence of 113 amino acids at the amino-terminal of the cadherins determines their binding specificities (Nose et al., 1990). The intracellular domain that includes the carboxy-terminal is the most conserved region. Partial or complete deletion of this domain blocks cell-to-cell adhesion (e.g., Fujimori et al., 1990), implying that the conformation of the external portion depends on the cytoplasmic domain.

Junctional regions between cells have a high concentration of cadherins. These plaques also contain vinculin, α-actinin, radixin, and actin filaments (Geiger, 1989). These cadherins were found in contact sites containing cortical actin bundles as shown by immunofluorescence techniques (e.g., Matsuzaki et al., 1990). Solubilized cadherins are associated with cytoplasmic proteins, *catenin,* of which there are three recognized kinds: α, β, and γ. α-catenin has been purified and its cDNA cloned (Nagafuchi et al., 1991). It has been found to be similar to vinculin, a component of focal adhesions and adherens junctions (see V, D below for a discussion of the junctions). β-catenin resembles another desmosomal protein, pakoglobin (McCrea et al., 1991). The cadherin domain which binds catenins coincides with the intracellular region needed for adhesion. Furthermore, at least catenin associated with E-cadherin can bind to globular actin.

The use of *Drosophila melonagaster* holds much promise for the study of cell-adhesion molecules and the understanding of their function in cell interaction and embryogenesis. The embryology and genetics of *Drosophila* are well known. The adhesion molecules isolated from this system have many similarities to those of vertebrates and have similar functions. Furthermore, the study of the molecular biology of cell-adhesion molecules and their surface receptors has been given a significant boost by the use of a line of cells (Schneider-2, S-2) (see Hortsch and Bieber, 1991) that grow predominantly unattached and rounded because of lack of adhesion molecules and are therefore ideal for testing the presence of an exogenous adhesion protein. The study can be carried out by transfecting the cells with cDNA coding for adhesion molecules. Several convenient vectors are available.

Among the *Drosophila* proteins that have been identified using this system are homophilic adhesion glycoproteins fasciclins I, II, and III. Fasciclin I is anchored covalently to the glycosyl-phosphatidylinositol moiety of membrane lipid. Fasciclin II and III are integral glycoproteins. Fasciclin II (Harrelson and Goodman, 1988) and another adhesion protein, neuroglian (Bieber et al., 1989) are related to the vertebrate NCAM and are members of the Ig superfamily.

Heterophilic cell-to-cell interactions have also been studied. *Notch(N)* and *Delta (D1)* belong to a group of genes responsible for the decision of pre-epidermal cells to either become epithelial or neuronal (Fehon et al., 1990). They code for two transmembrane proteins with EGF repeats in their extracellular domain (Wharton et al., 1965; Vässin et al., 1987).

C. Cell Surface Receptors

Interactions between extracellular elements (or other signals) and the cell's interior must also involve molecules bridging the plasma membrane, i.e., they must involve integral proteins. At the external face, they would bind extracellular elements to act as receptors. At the inner membrane face, they would transmit a chemical signal or bind to elements of the cytoskeleton. We saw that the binding domain of some extracellular elements resembles motifs present in growth factors. The binding of extracellular fibrous elements transmitted to the cell's interior can eventually affect gene expression (Spiegelman and Ginty, 1983; Werb et al., 1989) as detected, for example, by the synthesis of specific mRNAs. In many cases the interactions may well be mediated by the second messenger system discussed in more detail in Chapter 5.

Several cell surface receptors that bind extracellular matrix components directly have been isolated. Among these, anchorin CII, which binds to collagen II, has been isolated from plasma membranes of chondrocytes. When incorporated into artificial lipid vesicles (liposomes), the protein was found to bind with high affinity and specificity (Mollehauer and von der Mark, 1983). A similar protein was detected in fibroblasts and myoblasts.

Another collagen binding protein, of 68 kDa, has been isolated from platelets (Chiang and Kang, 1982), and a laminin binding protein of similar size, anchorin ML, has been reported in other cells (Rao et al., 1983; Terranova et al., 1983)

The formation of tissue patterns during embryogenesis may result from qualitative and quantitative differences in cell adhesion in time and space (Steinberg, 1970). The *neural cell adhesion molecule* (NCAM) has a number of properties that suggest it has a role in this process in neural tissue.

NCAM is an integral protein, a single polypeptide chain. A site at the amino terminal allows one NCAM molecule to interact with another NCAM molecule. NCAM can also interact with the extracellular matrix components, fibronectin in particular. The carboxyl terminal portion goes through the cell membrane. A middle portion is attached to carbohydrate that contains polysialic acid of variable length, which modulates the binding: a decrease in length increases adhesion (Rutishauser and Goridis, 1986). There are 100,000 NCAMs per neuron.

NCAM has been found to vary in quantity and quality within individual cells and from tissue to tissue. Furthermore, these variations occur spatially and chronologically with embryogenesis, suggesting that they may play a role in these events. However, the correlation between NCAM changes and the events of embryogenesis still has not led to the elucidation of specific mechanisms (Rutishauser, 1986).

Integrins are thought to correspond to the most ubiquitous family of bridging molecules. Integrins are involved in binding to the fiber systems of the cell's interior (see below). They are also thought to be involved in the activations of certain ion translocases such as the Na^+/H^+ transport system (Schwartz et al., 1991), possibly by triggering second messengers (Banga et al., 1986) and to have a role in Ca^{2+}-homeostasis, possibly by an effect on Ca^{2+}-channels (Brass, 1985). FNs and, by implication, integrins, have been found to have effects on cell shape and cytoskeletal organization and also to affect gene expression (Spiegelman and Ginty, 1983; Werb et al., 1989). Presumably the role of the integrins could be transmitted through a conformational change in the protein upon binding a ligand. Conformational change of the platelet integrin complex IIb/IIIa induced by ligand binding has been demonstrated in vitro (Parise et al., 1987).

Twenty distinct integrin heterodimers formed by noncovalent association of α and β subunits are now known (Springer, 1990; Quaranta, 1990). There are at least 11 different kinds of α chains and 6 β chains. Generally, one kind of α chain associates only with one kind of β chain; however, there are many exceptions. The class of chain is indicated by a number (sometimes as a subscript) following the Greek letter. The sequences of some of the integrin chains have been determined (e.g., Tamura et al., 1991) using cDNA techniques (see Chapter 1, II, A). The extracellular ligands of integrin include matrix glycoproteins such as laminin, fibronectin, and collagen as well as the cell adhesion molecules 1CAM-1, 1CAM-2, and VCAM-1.

One of the significant connections bridged by integrins is the basal membranes or basal laminae separating epithelial cell sheets and tubes from other ECM components, or located around muscle, fat, and Schwann cells (see p. 93). Laminins have a binding domain for integrins (see Mecham, 1991).

D. Junctional Structures

Interactions between external matrix and elements in the cytoplasm frequently involve specialized structures. These structures are connected to either one of two major classes of filament elements of cytoskeletal filaments, i.e., actin and intermediate size filaments which are associated with the inner face of the plasma membrane. The attachment may be between adjacent cells (*intercellular junctions*) or between cells and extracellular matrix (*asymmetric junctions*).

Desmosomes are intercellular junctions that provide anchorage for intermediate filaments. Asymmetric junctions involving the bottom surface of basal epithelial cells and the extracellular basal lamina also involve intermediate filaments and are referred to as *hemidesmosomes* (Schwarz et al., 1990).

Actin molecules are part of some cell-to-cell junctions and some asymmetric junctions, the *focal contacts*. The two kinds of junctions have been grouped together and referred to as the *adherence-type junctions* (see Geiger and Ginsberg, 1991). Major proteins forming part of these junctions are listed in Table 3.6 (Geiger and Ginsburg, 1991). A reconstruction of focal contact using presently available information that is still preliminary is shown in Fig. 3.21 (Burridge and Fath, 1989).

The cadherins referred to in the table are so-called *desmosomal cadherins* or *desmocollins* (DGI to DGIII). They have some similarities to the cadherins involved in cell-to-cell contacts (Wheeler, 1991); however, the cytoplasmic domains are very different. This is to be expected since they have to provide a connection which ultimately will bind intermediate filaments rather than actin.

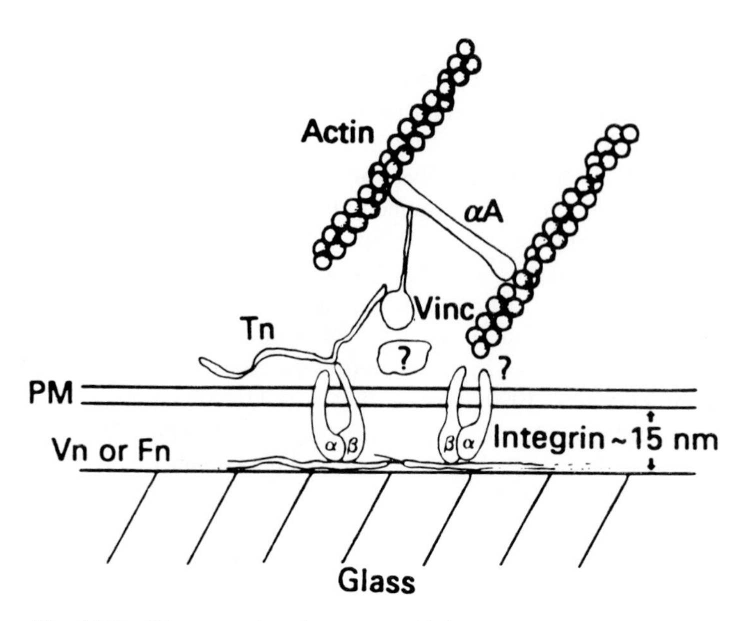

Fig. 3.21 Diagram showing some of the components at a focal contact and their probable arrangement. Abbreviations are as follows: PM, plasma membrane; Vn, vitronectin; Fn, fibronectin; Tn, talin; vinc, vinculin; αA, α-actinin. From K. Burridge and K. Fath, *Bioessays,* 10:104–108. Copyright © 1989 Company of Biologists Ltd., England.

Table 3.6 Major Components of Adherens-Type Junctions[a]

Molecule	Polypeptide MW (dDA)	Presence in	
		C-C[d]	C-S[d]
Cadherins(s)[b]	120–140	+	–
Integrins(s)[b]	100–200[c]	–	+
Actin	43	+	+
α-Actinin	100	+	+
Catenins α, β, γ	102,88,80	+	–
FC-1	60	–	+
Fimbrin	63	–	+
Paxillin	68	–	+
Plakoglobin	83	+	–
Radixin	82	–	+
Talin	215–235	–	+
Tensin	200,150	–	+
Tenuin	400	+	+
Vinculin	116	+	+
Zyxin	82	+	+
200 K	200	+	+
30B6 antigen	175	+	+
70 kD	70	–	+
$p60^{src}$	60	+	+
$p120^{v\text{-}gag\text{-}abl}$	120	?	+
$p80,p90^{v\text{-}gag\text{-}yes}$	80,90	+	+
Protein kinase C	82	?	+
CDPII	80	–	+

[a]From B. Geiger and D. Ginsberg, *Cell Motility and the Cytoskeleton,* 20: 1–6. Copyright © 1991 Wiley-Liss, Inc., division of John Wiley & Sons, Inc.
[b]Cell-specific members of multigene families.
[c]Heterodimers of diverse α and β chains.
[d]C-C: Cell-to-Cell; C-S: Cell-to-Substratum.

SUGGESTED READING

Endocytosis

Bretscher, M. S., and Pearse, B. M. F. (1983) Coated pits in action. *Cell* 38:3–4.
Goldstein, J. L., Brown, M. S., Anderson, R. G. W., Russel, D. W., and Schneider, W. J. (1985) Receptor-mediated endocytosis. *Annu. Rev. Cell Biol.* 1:1–39.
Morris, S., Ahle, S., and Ungewickell, E. (1989) Clathrin coated vesicles. *Current Opin. Cell Biol.* 1:684–690.
Pearse, B. M. F., and Cowther, R. A. (1987) Structure and assembly of coated vesicles. *Annu. Rev. Biophys. Biophys. Chem.* 16:49–68.

Polypeptide Growth Factors

Deuel, T. F. (1987) Polypeptide growth factors: roles in normal and abnormal cell growth. *Annu. Rev. Cell Biol.* 3:443–492.

Oncogenes and Oncogene Suppressors

Adamson, E. D. (1987) Oncogenes in development. *Development* 99:449–471.
Kris, R. M., Liberman, T. A., Avivi, A., and Schlessinger, J. (1985) Growth-factor receptors and oncogenes. *Rev. Biotechnol.* 3:135–140.
Sager, R. (1989) Tumor suppressor genes: The puzzle and promise. *Science* 246:1406–1412.

Interaction between Cells

Takeichi, M. (1991) Cadherin cell adhesion receptors as a morphogenic regulator. *Science* 251:1451–1455.

Interaction between Cells and the Cell Matrix

Beck, K., Hunter, I., and Engel, J. (1990) Structure and function of laminin: anatomy of a multidomain glycoprotein. *FASEB J.* 4:148–160.
Ben-Zee'v, A. (1991) Animal cell shape changes and gene expression. *BioEssays* 13:207–211.
Hardingham, T. E., and Fosang, A. J. (1992) Proteoglycans: many forms and many functions. *FASEB J.* 6:861–870.
Hynes, R. O. *Fibronectins,* Springer-Verlag, New York, Berlin. Chapter 5: Interactions of fibronectins; Chapter 6: Structure of fibronectin; Chapter 8: Cellular adhesion and cell surface receptors.
Hynes, R. O. (1992) Integrins: Versatility, modulation and signaling in cell adhesion, *Cell* 69:11–25.

Specialized Junctions

Burridge, K., and Fath, K. (1989) Focal contacts: transmembrane links between the extracellular matrix and the cytoskeleton. *Bioessays* 10:104–108.
Geiger, B., and Ginsberg, D. (1991) The cytoplasmic domain of adherens-type junctions. *Cell Motility and Cytosk.* 20:1–6.
Ruoslahti, E. (1989) Proteoglycans and cell regulation. *J. Biol. Chem.* 264:13369–13372.

REFERENCES

Aaronson, S. A., Tronick, S. R., and Robbins, K. C. (1985) Oncogenes and the pathway to malignancy. In *Control of Animal Cell Proliferation,* Vol. 1 (Boynton, A. L., and Leffert, H. L., eds.), pp. 3–24. Academic Press, Orlando, Fla.

Ada, G. L., and Byrt, P. (1969) Specific inactivation of antigen reactive cells with ^{125}I-labelled antigen. *Nature (London)* 222:1291–1292.

Akiyama, S. K., and Yamada, K. M. (1983) Fibronectin in disease. In *Connective Tissue Diseases* (Wagner, B. M., Fleishmajer, P., and Kaufman, N., eds.) pp. 55–96. Williams & Wilkins, Baltimore.

Anderson, R. G. W., Goldstein, J. L., and Brown, M. S. (1976) Localization of low density lipoprotein receptors on the plasma membrane of normal human fibroblasts and their absence in cells from a familial hypercholesterolemic homozygote. *Proc. Natl. Acad. Sci. U.S.A.* 73:2434–2438.

Anderson, R. G. W., Brown, M. S., and Goldstein, J. L. (1977a) Role of the coated endocytotic vesicle in the uptake of receptor bound low density lipoprotein in human fibroblasts. *Cell* 10:351–364.

Anderson, R. G. W., Goldstein, J. L., and Brown, M. S. (1977b) A mutation that impairs the ability of lipoprotein receptors to localize in coated pits in the cell surface of human fibroblasts. *Nature (London)* 270:695–699.

Anderson, R. G. W., Vasile, E., Mello, R. J., Brown, M. S., and Goldstein, J. L. (1978) Immunochemical visualization of coated pits and vesicles in human fibroblasts: relation to low density lipoprotein receptor distribution. *Cell* 15:919–933.

Aplin, J. D., Hughes, R. C., Jaffe, C. L., and Sharon, L. (1981) Reversible cross-linking of cellular components of adherent fibroblasts to fibronectin and lectin-coated substrata. *Exp. Cell Res.* 134:488–494.

Assoian, R. K., Komoriya, A., Meyers, C. A., Miller, D. M., and Sporn, M. B. (1983) Transforming growth factor-beta in human platelets. *J. Biol. Chem.* 258:7155–7160.

Banga, H. S., Simons, E. R., Brass, L. F., and Rittenhouse, S. E. (1986) Activation of phospholipase A and C in human platelets exposed to epinephrine: role of glycoproteins IIb/IIIa and dual role of epinephrine. *Proc. Natl. Acad. Sci. U.S.A.* 83:9197–9201.

Beck, K., Hunter, I., and Engel, J. (1990) Structure and function of laminin: anatomy of a multidomain glycoprotein. *FASEB J.* 4:148–160.

Ben-Ze'ev, A. (1985) Cell shape, the complex cellular networks and gene expression: cytoskeletal protein genes as a model system. *Cell Muscle Motil.* 6:23–53.

Ben-Ze'ev, A., Farmer, S. R., and Penman, S. (1979) Mechanisms of regulating tubulin synthesis in cultured mammalian cells. *Cell* 17:319–325.

Ben-Ze'ev, A., Farmer, S. R., and Penman, S. (1980) Protein synthesis requires cell-surface contact while nuclear events respond to cell shape in anchorage-dependent fibroblasts. *Cell* 21:365–372.

Bieber, A. J., Snow, P. M., Hortsch, M., Patel, N. H., Jacobs, J. R., Traquinma, Z. R., Schilling, J., and Goodman, C. S. (1989) *Drosophila neuroglian*: a member of the immunoglobin superfamily with extensive homology to vertebrate neural adhesion molecule L1. *Cell* 59:447–460.

Bishop, J. M. (1983) Cellular oncogenes and retroviruses. *Annu. Rev. Biochem.* 52:301–354.

Bleil, J. D., and Bretscher, M. S. (1982) Transferrin receptor and its recycling in HeLa cells. *EMBO J.* 1:351–355.

Bomsel, M., Parton, R., Kuznetsov, S. A., Schroer, T. A., and Gruenberg, J. (1990) Microtubule- and motor-dependent fusion in vitro between apical and basolateral endocytotic vesicles. *Cell* 62:719–731.

Bourdon, M. A., and Ruoslahti, E. (1989) Tenascin mediates cell attachment through an RGD-dependent receptor. *J. Cell Biol.* 108:1149–1155.

Bradshaw, R. A. (1978) Nerve growth factor. *Annu. Rev. Biochem.* 47:191–217.

Brass, L. F. (1985) Ca^{2+} transport across the platelet plasma membrane: a role for membrane glycoproteins IIb and IIIa. *J. Biol. Chem.* 260:2231–2236.

Bretscher, M. S., and Pease, B. M. F. (1984) Coated pits in action. *Cell* 38:3–4.

Brown, M. S., Anderson, R. G., Basu, S. K., and Goldstein, J. L. (1981) Cell surface receptors. Observations from the LDL receptor. *Cold Spring Harbor Symp. Quant. Biol.* 46:713–721.

Brown, M. S., Anderson, R. G. W., and Goldstein, J. L. (1983) Recycling receptors: the round trip itinerary of migrant membrane proteins. *Cell* 32:663–667.

Bruch, M., Landwehr, R., and Engel, J. (1989) Dissection of laminin by cathepsin G into its long arm and short structures and localization of regions involved in calcium dependent stabilization and self-association. *Eur. J. Biochem.* 185:271–279.

Burnstein, D. E., Seeley, B. J., and Greene, L. A. (1985) Lithium ion inhibits nerve growth factor-induced neurite outgrowth and phosphorylation of nerve growth factor-modulated microtubule associated protein. *J. Cell Biol.* 101:862–870.

Burridge, K., and Fath, K. (1989) Focal contacts: transmembrane links between the extracellular matrix and the cytoskeleton. *Bioessays* 10:104–108.

Burridge, K., Fath, K., Kelly, T., Nuckolls, G., and Turner, C. (1988) Focal adhesions: transmembrane junctions between the extracellular matrix and the cytoskeleton. *Annu. Rev. Cell Biol.* 4:487–525.

Calissano, P., and Shelanski, M. L. (1980) Interaction of nerve growth factor with tight binding and pheochromocytoma cells. Evidence for sequestration. *Neuroscience* 5:1033–1039.

Carpenter, G., and Cohen, S. (1976) I^{125} labeled human epidermal growth-factor. Binding, internalization, and degradation in human fibroblasts. *J. Cell Biol.* 71:159–171.

Chandrasekharis, S., Sorrentino, J. A., and Millis, A. J. T. (1983) Interaction of fibronectin with collagen: age specific defect in biological activity of human fibroblast fibronectin. *Proc. Natl. Acad. Sci. U.S.A.* 80:4747–4751.

Chiang, T. M., and Kang, A. H. (1982) Isolation and purification of collagen $\alpha 1$ (I) receptor from human platelet membrane. *J. Biol. Chem.* 257:7581–7586.

Chiquet-Erismann, R. (1990) What distinguishes tenascin from fibronectin? *FASEB J* 4:2598–2604.

Chiquet-Erismann, R. (1991) Anti-adhesive molecules of the extracellular matrix. *Curr. Opin. Cell Biol.* 3:800–804.

Cochet, C., Gill, G. N., Meisenhelder, J., Cooper, J. A., and Hunter T. C. (1984) C-kinase phosphorylates the epidermal growth factor receptor and reduces its epidermal growth factor-stimulated tyrosine protein kinase activity. *J. Biol. Chem.* 259:2553–2558.

Cooper, G. M., Okenquist, S., and Silverman, L. (1980) Transforming activity of DNA of chemically transformed and normal cells. *Nature* 284:418–421.

Delarco, J. E., and Todaro, G. J. (1978) Growth factors from murine sarcoma virus-transformed cells. *Proc. Natl. Acad. Sci. U.S.A.* 75:4001–4005.

Der, C. J., Krontiris, T. G., and Cooper, G. M. (1982) Transforming genes of human bladder and lung carcinoma cell lines and homologous to the *ras* genes of Harvey and Kirsten sarcoma viruses. *Proc. Natl. Acad. Sci. U.S.A.* 79:3637.

Deutzmann, R., Aumailley, M., Wiedemann, H., Pysny, W., Timpl, R., and Edgar, D. (1990) Cell adhesion, spreading and neurite stimulation by laminin fragment E8 depends on maintenance of secondary and tertiary structure in its rod and globular domain. *Eur. J. Biochem.* 191:513–522.

Devel, T. F., Huang, J. S., Huang, S. S., Stroobant, P., and Waterfield, M. D. (1983) Expression of a platelet-derived growth factor-like protein in simian sarcoma virus transformed cells. *Science* 221:1348.

Doolittle, R. F., Hunkapiller, M. W., Hood, L. E., Devare, S. G., Robbins, K. C., Aaronson, S. A., and Antoniades, H. N. (1983) Simian sarcoma virus one gene, v-*sis,* is derived from the gene (or genes) encoding a platelet-derived growth factor. *Science* 221:275–277.

Doxsey, S. J., Brodsky, F. M., Blank, G. S., and Helenius, A. (1987) Inhibition of endocytosis by anti-clathrin antibodies. *Cell* 50:453–463.

Drubin, D. G., Feinstein, S. C., Shooter, E. M., and Kirschner, M. W. (1985) Nerve growth factor induced neurite outgrowth in PC 12 cells involves the coordinate induction of microtubule assembly and assembly-promoting factors. *J. Cell Biol.* 101:1799–1801.

Dunn, W. A., and Hubbard, A. L. (1984) Receptor-mediated endocytosis of epidermal growth factor by hepatocytes in the perfused rat liver: ligand and receptor dynamics. *J. Cell Biol.* 98:2148–2159.

Edgar, D., Timpl, R., and Thoenen, H. (1984) The heparin-binding domain of laminin is responsible for the effects on neurite outgrowth and neuronal survival. *EMBO J.* 3:1463–1468.

Emerman, J. T., and Pitelka, D. R. (1977) Maintenance and induction of morphological differentiation in dissociated mammary epithelium on floating collagen membranes. *In Vitro* 13:316–328.

Engel, J. (1989) EGF-like domains in extracellular matrix proteins: localized signals for growth and differentiation? *FEBS Lett.* 251:1–7.

Engel, J. (1991) Common structural motif in proteins of the extracellular matrix. *Curr. Opin. Cell Biol.* 3:779–785.

Fava, R. A., and McClure, D. B. (1987) Fibronectin associated transforming growth factor. *J. Cell. Physiol.* 131:184–189.

Fehon, R. G., Kooh, P.J., Rebay, I., Regan, C. L., Xu, T., Muskavitch, M. A. T., and Antavaris-Tsokonas, S. (1990) Molecular interactions between the protein products of the neurogenic loci *Notch* and *Delta,* two EGF-homologous genes in *Drosophila. Cell* 61:523–534.

Franke, W. W., Lüder, M. R., Kartenbeck, J., Zerban, H., and Deenan, T. W. (1977) Involvement of vesicle coat material in casein secretion and surface regeneration. *J. Cell Biol.* 69:173–195.

Friend, D. S., and Farquhar, M. G. (1967) Functions of coated vesicles during protein absorption in the rat *vas deferens. J. Cell Biol.* 35:357–376.

Fujimori, T., Miyatani, S., and Takeichi, M. (1990) Ectopic expression of N-cadherin perturbs histogenesis in *Xenopus. Dev. Biol.* 110:97–104.

Galloway, C. J., Dean, G. E., Marsh, M., Rudnick, G., and Mellman, I. (1983) Acidification of macrophage and fibroblast endocytotic vesicles *in vitro. Proc. Natl. Acad. Sci. U.S.A.* 80:3334–3338.

Gaspodarowicz, D. (1985) Epidermal and fibroblastic growth factor. In *Control of Animal Cell Proliferation,* Vol. 1 (Boynton, A. L., and Leffert, H. L., eds.), pp. 61–90. Academic Press, Orlando, Fla.

Gehlsen, K. R., Dillmer, L., Engvall, E., and Ruoslahti, E. (1988) The human laminin receptor is a member of the integrin family of cell adhesion receptors. *Science* 241:1228–1229 (correction, *Science* 245:342–343).

Geiger, B. (1989) *Curr. Opin. Cell Biol.* 1:103.

Geiger, B., and Ginsberg, D. (1991) The cytoplasmic domain of adherens-type junctions. *Cell Motility and Cytosk.* 20:1–6.

Geutze, H. J., Slot, J. W., Strous, G. J. A. M., Lodish, H. F., and Schwartz, A. L. (1983) Intracellular sites of asiologlycoprotein receptor-ligand uncoupling: double-label immunoelectronmicroscopy during receptor-mediated endocytosis. *Cell* 32:277–287.

Geutze, H. J., Slot, J. W., Strous, G. J., Peppard, J., von Figura, K., Hasilik, A., and Schwartz, A. L. (1984) Intracellular receptor sorting during endocytosis: comparative immunoelectronmicroscopy of multiple receptors in rat liver. *Cell* 37:195–204.

Goldstein, J. L., and Brown, M. S. (1977) The low-density lipoprotein pathway and its relation to atherosclerosis. *Annu. Rev. Biochem.* 46:897–930.

Goldstein, J. L., Basu, S. K., Brunschede, G. Y., and Brown, M. S. (1976) Release of low density lipoprotein from its cell surface receptor by sulfated glycosaminoglycans. *Cell* 7:85–95.

Goldstein, J. L., Anderson, R. G. W., and Brown, M. S. (1979) Coated pits, coated vesicles, and receptor-mediated endocytosis. *Nature (London)* 279:679–685.

Goldstein, J. L., Brown, M. S., Anderson, R. G. W., Russel, D. W., and Schneider, W. J. (1985) Receptor-mediated endocytosis: concepts emerging from the LDL receptor system. *Annu. Rev. Cell Biol.* 1:1–39.

Gonda, T. J., and Bishop, J. M. (1983) Structure and transcription of the cellular homolog (c-*myb*) of the avian myeloblastosis virus transforming gene (v-*myb*). *J. Virol* 46:212–220.

Gorvell, J.-P., Chavrier, P., Zerial, M., and Gruenberg, J. (1991) rab5 controls early endosome fusion in vitro. *Cell* 64:915–925.

Griffiths, G., and Gruenberg, J. (1991) The arguments for pre-existing early and late endosomes. *Trends in Cell Biol.* 1:5–9.

Griffiths, G., Back, R., and Marsh, M. (1989) A quantitative analysis of the endocytotic pathway in baby hamster kidney cells. *J. Cell Biol.* 109:2703–2770.

Gundersen, R. W., and Barrett, J. N. (1979) Neuronal chemotaxis: chick dorsal-root axons turn toward high concentrations of nerve growth factor. *Science* 206:1079–1080.

Hagler, H., Ash, J., Singer, S. J., and Cohen, S. (1978) Visualization by fluorescence of binding and internalization of epidermal growth factor in human carcinoma cells A-431. *Proc. Natl. Acad. Sci. U.S.A.* 75:3317–3321.

Hagler, H. T., McKanna, J. A., and Cohen, S. (1979) Direct visualization of the binding and internalization of a ferritin conjugate of epidermal growth factor in human carcinoma cells A-43. *J. Cell Biol.* 81:382–395.

Hall, D. E., Frazer, K. A., Hann, B. C, and Reichardt, L. F. (1988) Isolation and characterization of a laminin-binding protein from rat and chick muscle. *J. Cell Biol.* 107:687–697.

Harrelson, A. L., and Goodman, C. S. (1988) Growth cone guidance in insects: fasciclin II is a member of the immunoglobulin superfamily. *Science* 242:700–708.

Hascall, V. C., and Hascall, G. K. (1981) Proteoglycans. In *Cell Biology of Extracellular Matrix* (Hay, E. D., ed.), pp. 39–63. Plenum, New York.

Hasegawa, T., Hasegawa, E., Chen, W. T., and Yamada, K. M. (1985) Characterization of membrane-associated glycoprotein complex implicated in cell adhesion to fibronectin. *J. Cell Biochem.* 28:307–318.

Hay, E. D. (1984) Cell-matrix interaction in embryo: cell shape, cell surface, and cell skeleton and their role in differentiation. In *The Role of Extracellular Matrix in Development* (Trelstad, R. L., ed.), pp. 1–31. Liss, New York.

Hayward, M. S., Neel, B. G., and Astrin, S. M. (1981) Activation of a cellular *onc* gene of promoter insertion in AIV-induced lymphoid leukosis. *Nature* 290:475–480.

Hedgecock, E. M., Culotti, J. G., and Hall, D. H. (1990) The unc-5, unc-6 and unc-40 genes guide circumferential migrations of pioneer axons and mesodermal cells in the epidermis in *C. elegans*. *Neuron* 4:61–85.

Hendry, I. A., Stockel, K., Thoenen, H., and Iversen, L. L. (1974) The retrograde axonal transport of nerve growth factor. *Brain. Res.* 68:103–121.

Herschman, H. R. (1985) The EGF receptor. In *Control of Animal Cell Proliferation,* Vol. 1 (Boynton, A. L., and Leffert, H. L., eds.), pp. 169–199. Academic Press, Orlando, Fla.

Holzman, E., Novikoff, A. B., and Villaverde, H. (1967) Lysosomes and GERL in dermal and chromatolytic neurons of the rat ganglion nodosum. *J. Cell Biol.* 33:419–435.

Hopkins, C. R. (1983) Intracellular routing of transferrin and transferrin receptors in epidermoid carcinoma A431 cells. *Cell* 35:321–330.

Hopkins, C. R. (1985) The appearance and internalization of transferrin receptors at the margins of spreading human tumor cells. *Cell* 40:199–208.

Hopkins, C. R., Boothroyd, B., and Gregory, H. (1981) Early events following the binding of epidermal growth factor to surface receptors on ovarian granulosa cells. *Eur. J. Cell Biol.* 24:259–265.

Hortsch, M., and Bieber, A. J. (1991) Sticky molecules in non-sticky cells. *Trends in Biochem. Sci.* 16:283–287.

Hunter, T., and Cooper, J. A. (1981) Epidermal growth factor induces rapid tyrosine phosphorylation of proteins in A431 human tumor cells. *Cell* 24:741–752.

Hynes, R. O. (1990) *Fibronectins,* Springer-Verlag, New York.

Iwashita, S., and Fox, C. F. (1984) Epidermal growth factor and potent phorbol tumor promoters induce epidermal growth factor receptor phosphorylation in a similar but distinctively different manner in human epidermoid carcinoma A431 cells. *J. Biol. Chem.* 259:2559–2567.

Julius, M. H., Masuda, T., and Herzenberg, L. A. (1972) Demonstration that antigen binding cells are precursors of antibody producing cells after purification using a fluorescence activated sorter. *Proc. Natl. Acad. Sci. U.S.A.* 69:1934–1938.

Kamwar, Y. S., Linker, A., and Farquhar, M. G. (1980) Increased permeability of the glomerular basement membrane to ferritin after removal of glycosaminoglycans (heparan sulfate) by enzyme digestion. *J. Cell Biol.* 86:688–693.

Kanaseki, T., and Kadota, K. (1969) Vesicles in a basket. A morphological study of coated vesicles isolated from nerve endings of guinea pig brain with special reference to mechanism of membrane movement. *J. Cell Biol.* 42:202–220.

Keen, J. H., Willingham, M. C., and Pastan, I. H. (1979) Clathrin-coated vesicles. Isolation, dissociation and factor-dependent reassociation of clathrin baskets. *Cell* 16:303–312.

Kefalides, N. A., Alper, R., and Clark, C. C. (1979) Biochemistry and metabolism of basement membranes. *Int. Rev. Cytol.* 61:167–228.

Klein, G. (1987) The approaching era of the tumor repressor genes. *Science* 238:1539–1545.

Knapp, P. E., and Swanson, J. A. (1990) Plasticity of the tubular lysosomal compartment in macrophages. *J. Cell Sci.* 95:433–439.

Knudsen, K. A., Horwitz, A. F., and Buck, C. A. (1985) A monoclonal antibody identifies a glycoprotein involved in cell-substratum, adhesion. *Exp. Cell Res.* 157:218–228.

Kris, R. M., Liberman, T. A., Avivi, A., and Schlessinger, J. (1985) Growth-factor receptors and oncogenes. *Biotechnology* 3:135–140.

Landreth, G. E., and Shooter, E. M. (1980) Nerve growth factor receptors on PC 12 cells: ligand-induced conversion from low to high-affinity states. *Proc. Natl. Acad. Sci. U.S.A.* 77:4751–4755.

Lawler, J., and Hynes, R. O. (1986) The structure of human thrombospondin, an adhesive glycoprotein with multiple calcium binding sites and homologies with several different proteins. *J. Cell Biol.* 103:1635–1648.

Lee, E. Y. -H., Parry, G., and Bissell, M. J. (1984) Modulation of secreted proteins of mouse mammary epithelial cells by collagenous substrata. *J. Cell Biol.* 98:146–155.

Levi, A., Shechter, Y., Neufeld, E. J., and Schlessinger, J. (1980) Mobility, clustering, and transport of nerve growth factor in embryonal sensory cells and sympathetic neuronal cell. *Proc. Natl. Acad. Sci. U.S.A.* 77:3469–3473.

Levi-Montalcini, R. (1954) Effect of mouse tumor transplantation in a mouse system. *Ann. N.Y. Acad. Sci.* 55:330–343.

Lillien, L. E., and Claude, P. (1981) Nerve growth factor is a mitogen for cultured chromaffin cells. *Nature (London)* 317:632–634, 1985.

Linsenmayer, T. F. (1991) Collagen. In *Cell Biology of Extracellular Matrix* (Hay, E. D., ed.), pp. 7–44. Plenum, New York.

Marquadt, H., Hunkapiller, M. W., Hood, L. E., Twardzik, D. R., DeLarco, J. E., Stephenson, J. R., and Todaro, G. J. (1983) Transforming growth factors produced by retrovirus-transformed rodent fibroblasts and human melanoma cells: amino acid sequence homology with epidermal growth factor. *Proc. Natl. Acad. Sci. U.S.A.* 80:4684–4688.

Matsuzaki, F., Mège, R.-M., Jaffe, S. H., Friedlander, D. R., Gallin, W. J., Goldberg, J. I., Cunningham, B. A., and Edelman, G. M. (1990) cDNAs of cell adhesion molecules of different specificity induce changes in cell shape and border formation in cultured S180 cells. *J. Cell. Biol.* 110:1239–1252.

Maxfield, F. R., Davies, P. J. A., Klempner, L., Willingham, M. C., and Pastan, I. (1979) Epidermal growth factor stimulation of DNA synthesis is potentiated by compounds that inhibit its clustering in coated pits. *Proc. Natl. Acad. Sci. U.S.A.* 76:5731–5735.

McCrea, P., Tuck,. C. W., and Gumbiner, B. (1991) A homologue of *armadillo* protein in *Drosophila* (plakoglobin) associates with E-cadherin. *Science,* 254:1359–1361.

McDonald, J. A., Kelley, D. G., and Broekelmann, T. J. (1982) Role of fibronectin in collagen deposition: Fab′ to the gelatin-binding domain of fibronectin inhibits both fibronectin and collagen organization in fibroblast extracellular matrix. *J. Cell Biol.* 92:485–492.

Mecham, R. P. (1991) Laminin receptors. *Annu. Rev. Cell Biol.* 7:71–91.

Mollehauer, J., and von der Mark, K. (1983) Isolation and characterization of a collagen-binding glycoprotein from chondrocyte membranes. *EMBO J* 2:45–50.

Nagafuchi, A., Takeichi, M., and Tsukita, S. (1991) The 102 kd cadherin-associated protein: similarity to vinculin and posttranslational regulation of expression. *Cell* 65:849–857.

O'Neill, C., Jordan, P., and Ireland, G. (1986) Evidence for two distinct mechanisms of anchorage in freshly explanted and 3T3 Swiss mouse fibroblasts. *Cell* 44:489–496.

Orci, L., Carpentier, J. -L., Perrelet, A., Anderson, R. G. W., Goldstein, J. L., and Brown, M. S. (1978) Occurrence of low density lipoprotein receptors within large pits on the surface of human fibroblasts as demonstrated by freeze etching. *Exp. Cell Res.* 113:1–13.

Otsuka, H., and Moskowitz, M. (1975) Arrest of 3T3 cells in G1 phase in suspension culture. *J. Cell Physiol.* 87:213–220.

Paladail, F., Tabin, C. J., Shih, C., and Weinberg, R. A. (1982) Human E5 bladder carcinoma oncogene is homologue of Harvey sarcoma virus ras gene. *Nature (London)* 297:474–478.

Palade, G. E. (1956) The endoplasmic reticulum. *J. Cell Biol.* 2(Suppl.):85–98.

Panayotou, G., End, P., Aumailley, M., Timpl, R., and Engel, J. (1989) Domains of laminin with growth-factor activity. *Cell* 56:93–101.

Parise, L. V., Helgerson, S. L., Steiner, B., Nannizzi, L., and Phillips, D. R. (1987) Synthetic peptides from fibrinogen and fibronectin change the conformation of purified platelet glycoprotein IIb-IIIa. *J. Biol. Chem.* 262:12597–12604.

Pearse, B. M. F. (1973) Clathrin: a unique protein associated with intracellular transfer of membrane by coated vesicles. *Proc. Natl. Acad. Sci. U.S.A.* 73:1255–1259.

Pearse, B. M. F. (1975) Coated vesicles from pig brain: purification and biochemical characterization. *J. Mol. Biol.* 97:92–98.

Pearse, B. M. F. (1978) Structure and functional components of coated vesicles. *J. Mol. Biol.* 126:803–812.

Pernis, B. (1985) Internalization of lymphocyte membrane components. *Immunol. Today* 6:45–49.

Pernis, B., and Axel, R. (1985) A one and a half receptor mode for MHC-restricted antigen recognition by T lymphocytes. *Cell* 45:13–16.

Quaranta, V. (1990) Epithelial integrins. *Cell Diff. Dev.* 32:361–366.

Roberts, A. B., Anzano, M. A., Cambil, C., Smith, J. M., and Sporn, M. B. (1981) New class of transforming growth factors potentiated by epidermal growth factor: isolation from non-neoplastic tissues. *Proc. Natl. Acad. Sci. U.S.A.* 78:5339–5343.

Ross, R., and Vogel, A. (1978) The platelet-derived growth factor. Review. *Cell* 14:203–210.

Ross, R., Raines, E. W., and Bowen-Pope, D. F. (1986) The biology of platelet-derived growth factor. *Cell* 46:155–169.

Ruoslahti, E. (1988) Fibronectin and its receptors. *Annu. Rev. Biochem.* 57:375–413.

Ruoslahti, E., and Yamaguchi, Y. (1991) Proteoglycans as modulators of growth factor activity. *Cell* 64:867–869.

Rutishauser, U. (1986) Differential cell adhesion through spatial and temporal variations of NCAM. *Trends Neurosci.* 9:374–378.

Rutishauser, U., and Goridis, C. (1986) NCAM. The molecule and its genetics. *Trends Genet.* 2:72–76.

Salmivirta, M., Elenius, K., Vainio, S., Hofer, U., Chiquet-Ehrismann, R., Thesleff, I., and Jalkenen, M. (1991) Syndecan from embryonic tooth mesenchyme binds tenascin. *J. Biol. Chem.* 266:7733–7739.

Sanes, J. R., Engvall, E., Butkowski, R., and Hunter, D. D. (1990) Molecular heterogeneity of basal laminae: isoforms of laminin and collagen IV at the neuromuscular junction and elsewhere. *J. Cell Biol.* 111:1685–1699.

Sasaki, M., Kato, S., Kohno, K., Martin, G. R., and Yamada, Y. (1987) Sequence of the cDNA encoding laminin B1 chain reveals a multidomain protein containing cysteine-rich repeats. *Proc. Natl. Acad. Sci. U.S.A.* 84:935–939.

Schlessinger, J., Shechter, Y., Willingham, M. C., and Pastan, I. (1978) Direct visualization of binding, aggregation, and internalization of insulin and epidermal growth factor on living fibroblastic cells. *Proc. Natl. Acad. Sci. U.S.A.* 75:2659–2663.

Schneider, W. J., Beisiegel, U., Goldstein, J. L., and Brown, M. S. (1982) Purification of the low density lipoprotein receptor, an acidic lipoprotein of 164,000 molecular weight. *J. Biol. Chem.* 257:2664–2673.

Schook, W., Puskin, S., Bloom, W., Ores, C., and Kachwa, S. (1979) Mechanochemical properties of brain clathrin: interaction with actin, alpha-actininin and polymerization in basketlike structures or filaments. *Proc. Natl. Acad. Sci. U.S.A.* 76:116–120.

Schwartz, M. A., Lechene, C. P., and Ingber, D. E. (1990) Activation of cytoplasmic signal by integrin $\alpha5\beta1$. *J. Cell Biol.* 111:263a.

Schwarz, M. A., Owaribe, K., Kartenbeck, J., and Franke, W. W. (1990) Desmosomes and hemidesmosomes: constitutive molecular components. *Annu. Rev. Cell Biol.* 6:461–491.

Spring, J., Beck, K., and Chiquet-Ehrismann, R. (1989) Two contrary functions of tenascin: dissection of the active sites by recombinant tenascin fragments. *Cell* 59:325–334.

Shih, C., and Weinberg, R. A. (1982) Isolation of a transforming sequence from a human bladder carcinoma line. *Cell* 29:161–169.

Sonnenberg, A., Calafat, J., Janssen, H., Daams, H., van der Raaj-Helmer, L. M. H., Falcioni, R., Kennel, S. J., Applin, J. D., Baker, J., Loizidou, M., and Garrod, D. (1991) Integrin $\alpha6/\beta4$ complex is located in hemidesmosomes, suggesting a major role in epidermal cell-basement membrane adhesion. *J. Cell Biol.* 113:907–917.

Spiegelman, B. M., and Ginty, C. A. (1983) Fibronectin modulation of cell shape and lipogenic gene expression in 3T3 adipocyte. *Cell* 35:657–666.

Springer, T. A. (1990) Adhesion receptors of the immune system. *Nature* 346:425–434.

Steinberg, M. S. (1963) Reconstruction of tissues by dissociated cells. *Science* 141:401–408.

Steinberg, M. S. (1970) Does differential adhesion govern self-assembly processes in histogenesis? Equilibrium configurations and emergence of hierarchy among population of embryonic cells. *Exp. Zool.* 173:395–434.

Stoker, M., O'Neill, C., and Berrymans, S. (1968) Anchorage and growth regulation in normal and virus transformed cells. *Int. J. Cancer* 3:683–693.

Storrie, B. (1988) Assembly of lysosomes: perspectives from comparative molecular cell biology. *Int. Rev. Cytol.* 111:53–105.

Swanson, J., Bushnell, A., and Silverstein, S. C. (1987) Tubular lysosomes morphology and distribution within macrophages depend on the integrity of cytoplasmic microtubules. *Proc. Natl. Acad. Sci.* 84:1921–1925.

Takeichi, M., Atsumi, T., Yoshida, C., Uno, K., and Okada, T. S. (1981) Selective adhesion of embryonic carcinoma cells and differentiated cells by Ca^{2+}-dependent sites. *Dev. Biol.* 87:340.

Takeichi, M. (1990) Cadherins: a molecular family important in selective cell-cell adhesion. *Annu. Rev. Biochem.* 59:237–252.

Takeichi, M. (1991) Cadherin cell adhesion receptors as a morphogenic regulator. *Science* 251:1451–1455.

Tamura, R. N., Rozzo, C., Starr, L., Chambers, J., Reichardt, L. F., Cooper, H. M., and Quaranta, V. (1990) Epithelial integrin $\alpha_6\beta_4$: complete primary structure of α_6 and variant forms of β_4. *J. Cell Biol.* 111:1593–1604.

Terranova, V. P., Rao, C. N., Kalebic, T., Margulies, I. M., and Liotta, L. A. (1983) Laminin receptor on human breast carcinoma cells. *Proc. Natl. Acad. Sci. U.S.A.* 80:444–448.

Timpl, R. (1982) Antibodies to collagens and procollagens. *Methods Enzymol.* 82:472–498.

Townes, P. L., and Holtfreter, J. (1955) Directed movement and selective adhesions of embryonic amphibian cells. *J. Exp. Zool.* 128:53–120.

Tycko, B., and Maxfield, F. R. (1982) Rapid acidification of endocytotic vesicles containing macroglobulin. *Cell* 28:640–651.

Ulrich, A., Coussens, L., Haytlick, J. S., Dull, T. J., Gray, A., Tam, A. W., Lee, J., Yarden, Y., Libermann, T. A., Schlessinger, J., Downward, T., Mayes, E. L. U., Whittle, N., Waterfield, M. D., and Seeburg, P. H. (1984) Human epidermal growth factor receptor cDNA sequence and aberrant expression of the amplified gene in A431 epidermoid carcinoma cells. *Nature* 309:418–425.

Urushihara, H., and Yamada, K. M. (1986) Evidence for involvement of more than one class of glycoprotein in cell interactions with fibronectin. *J. Cell Physiol.* 126:323–332.

Van Beveren, C., Curran, T., Muller, R., and Verma, I. M. (1982) Analysis of FBJ-MuSV provirus and *c-fos* (mouse) gene reveals that viral and cellular *fos* gene-products have different carboxy termini. *Cell* 32:1241–1245.

van der Rest, M., and Garrone, R. (1991) Collagen family of proteins. *FASEB J.* 5:2814–2823.

Van Deurs, B., and Nilausen, K. (1982) Pinocytosis in mouse L-fibroblast: ultrastructural evidence for a direct membrane shuttle between the plasma membrane and the lysosomal compartment. *J. Cell Biol.* 94:279–286.

Vässin, H., Bremer, K. A., Knust, E., and Campos-Ortega, J. A. (1987) The neurogenic gene Delta of *Drosophila melanogaster* is expressed in neurogenic territories and encodes putative transmembrane protein with EGF-like repeats. *EMBO J.* 6:3431–3440.

Vlodavsky, I., Bar-Shavit, R., Ishai-Michaeli, R., Bashkin, P., and Fuks, Z. (1991) Extracellular sequestration and release of fibroblast growth factor: a regulatory mechanism. *Trends Biochem. Sci.* 16:268–271.

von der Mark, K., and Kühl, U. (1985) Laminin and its receptor. *Biochim. Biophys. Acta* 823:47–160.

Wacholtz, M. C., Patel, S. S., and Lipsky, P. E. (1989) Leukocyte function associated antigen-1 is an activation molecule for human T-cells. *J. Exp. Med.* 170:431–448.

Wall, D. A., Wilson, G., and Hubbard, A. L. (1980) The galactose specific recognition system of mammalian liver: the route of ligand internalization in rat hepatocytes. *Cell* 21:79–93.

Wartiavaaren, J., and Vaheri, A. (1980) Fibronectin in early mammalian embryogenesis. *Dev. Mamm.* 4:233–266.

Waterfield, M. D., Scrace, G. T., Whittle, N., Stroobant, P., Johnsson, A., Wasteson, A., Westermark, B., Heldin, C. H., Huang, J. S., and Deuel, T. F. (1983) Platelet-derived growth factor is structurally related to the putative transforming protein $p28^{sis}$ of simian sarcoma virus. *Nature* 304:35–39.

Werb, Z., Tremble, P. M., Beherendtsen, O., Crowley, E., and Damsky, C. H. (1989) Signal transduction through the fibronectin receptor induces collagenase and stromelysin gene expression. *J. Cell Biol.* 109:877–889.

Wharton, K. A., Johansen, K. M., Xu, T., and Artavanis-Tsokonas, S. (1985) Nucleotide sequence from the neurogenic locus Notch implies a gene product that shares homology with proteins containing EGF-like repeats. *Cell* 43:567–581.

Wheeler, J. N., Buxton, R. S., Arnemann, J., Reese, D. A., King, I. A., and Magee, A. I. (1991) Desmosomal cadherins I, II, and III: novel members of the cadherin superfamily. *Biochem. Soc. Trans.* 19: 1060–1064.

Wight, T. N., Heinegård, D. K., and Hascall, V. C. (1991) Proteoglycans: Structure and function, in *Cell Biology and Extracellular Matrix* (Hay, E. D. ed.) pp. 45–77, Plenum, New York and London.

Woodward, M. P., and Roth, T. F. (1979) Coated vesicles: characterization, selective dissociation and reassembly. *Proc. Natl. Acad. Sci. U.S.A.* 75:4394–4398.

Yamada, K. M., Hayashi, M., Hirano, H., and Akiyama, S. K. (1984) Fibronectin and cell surface interactions. In *The Role of Extracellular Matrix in Development* (Trelstad, R. L., ed.), pp. 89–121. Liss, New York.

Chapter 4

The Intracellular Biosynthetic Transport System

No discussion of cell structure can do justice to its dynamics. Cell components are in continuous motion and change. Even when cells appear to be in a steady state, membrane-enclosed vesicles are continuously taken up and processed by cells, as we examined in some detail in Chapter 3. In addition, new components, including membrane elements, are produced and used up. These biosynthetic events and the transport of newly synthesized proteins are the central topics of this chapter. Since the transport involves a variety of intracellular vesicles and compartments, this account also concerns membranes.

All cells are engaged in biosynthetic protein transport. After their synthesis and segregation in vesicles, the proteins are transported to various destinations. One of these destinations may be a storage secretory granule that discharges on receipt of the appropriate physiological signal, frequently a neurotransmitter (*regulated secretion*). The secretory granules serve as stores in which the concentration of secretory products is greater than in the Golgi cisternae from which they originated, as much as 10 times greater in exocrine secretion and 200 times greater in endocrine secretion. Other proteins may be destined to remain in intracellular organelles such as lysosomes, while still others may be destined for the cell surface, to form new plasma membrane or to be continuously discharged to the outside by exocytosis (*constitutive secretion*). The transport of integral membrane proteins and constitutive secretion are facets of the same process, since in *exocytosis* the vesicular membrane becomes continuous with the plasma membrane and discharges its contents (Breckenbridge and Almers, 1987). In fact, it has been shown, using immunoelectron microscopy with individual antibodies attached to gold particles of different sizes, that the membrane proteins and constitutive secretion products are present in the same vesicle (Strous et al., 1983). The regulated secretory system was the first to be studied in detail and has influenced the study of all other intracellular biosynthetic protein transport. For this reason it is presented in the first section of this chapter.

I. INTRACELLULAR TRANSPORT AND PROTEIN SYNTHESIS

The fate of a newly synthesized macromolecule can be traced by pulse-chase techniques using electron microscopy (EM) and radioautography. The radioactivity is recorded by placing a photographic emulsion next to a tissue section. Generally, the radioautograph has to be stored in the dark for weeks. Where a radioactive disintegration has taken place, a silver grain will appear. Sharp localization can be obtained using transmission electron microscopy (TEM) with components labeled with ^{3}H. The β radiation of ^{3}H is of low energy and therefore cannot travel very far, permitting localizations within 0.1 to 0.2 μm. The amount of incorporation can be estimated by counting the grains. Alternatively, the radioactivity of various cell fractions can be measured after they are isolated.

The chronological sequence of events can be followed by a pulse-chase procedure. Travel of the marker with time, observed by sampling the cells or tissue at various times, corresponds to a migration of the labeled material. In pulse-chase, the incubation of the cells or tissue with a radioactively labeled percursor for a short time period (*pulse*) is followed by the introduction of a very large excess of unlabeled precursor (*chase*). In essence, the chase excludes from observation any incorporation of the radioactive precursor that occurs after the pulse incubation.

Experiments in which [^{14}C]leucine was injected into guinea pigs and various cell fractions were isolated showed that the synthesis takes place in the polysomes associated with the endoplasmic reticulum, since the highest specific radioactivity (radioactivity per milligram of enzyme) was detected in this fraction (i. e., polysomes attached to vesicles) at the earliest possible sampling time and not in the detached polysomes (Siekevitz and Palade, 1960).

Radioautographic pulse-chase studies have been carried out with slices of guinea pig pancreas. The experiments with the radioautographic technique using L-[^{3}H]leucine are illustrated in Fig. 4.1 (Jamieson and Palade, 1967). The electron micrograph of an acinar cell shows the vesicles of the *rough endoplasmic reticulum* (RER) (i.e., endoplasmic reticulum with polysomes), mitochondria, the nucleus, and secretory granules—the dense spherical inclusions on the left side of the figure. The radioautograph corresponds to an incubation of the pancreatic slice for 3 minutes, and the grains are predominantly on the RER. A summary of the radioautographic data as a function of time is shown in graphical form in Fig. 4.2. Curve 1 represents the percentage of grains over the RER. The radioactivity is incorporated into the condensing vacuoles of the Golgi complex (curve 2) and eventually the secretory granules (curve 3) are released into the acinar lumen (not shown). Incorporation into peripheral Golgi vesicles precedes that into the condensing vacuoles and is shown by the small dark dots on the graph marked with arrows. For clarity, a curve has not been drawn through the points. Figure 4.3 (Farquhar, 1985) summarizes a model that incorporates this and other information that will be discussed later.

A qualitatively similar chronology was found for the incorporation of [^{3}H]leucine in monocytes, in which the primary product is sequestered in lysosomes (Cohn et al., 1966). This finding is part of the evidence that the initial steps of biosynthetic protein transport are common to various distinct processes.

As indicated by the pulse-chase experiments just discussed, newly synthesized proteins are transferred from the RER to the Golgi apparatus. From the Golgi, various vesicles are sorted out to form either storage secretory vesicles, lysosomes, or vesicles involved in constitutive secretion. The divergent fates of the various components suggest the sorting out of the proteins or the vesicles as already indicated in Chapter 3 for the endocytotic pathway. The Golgi system

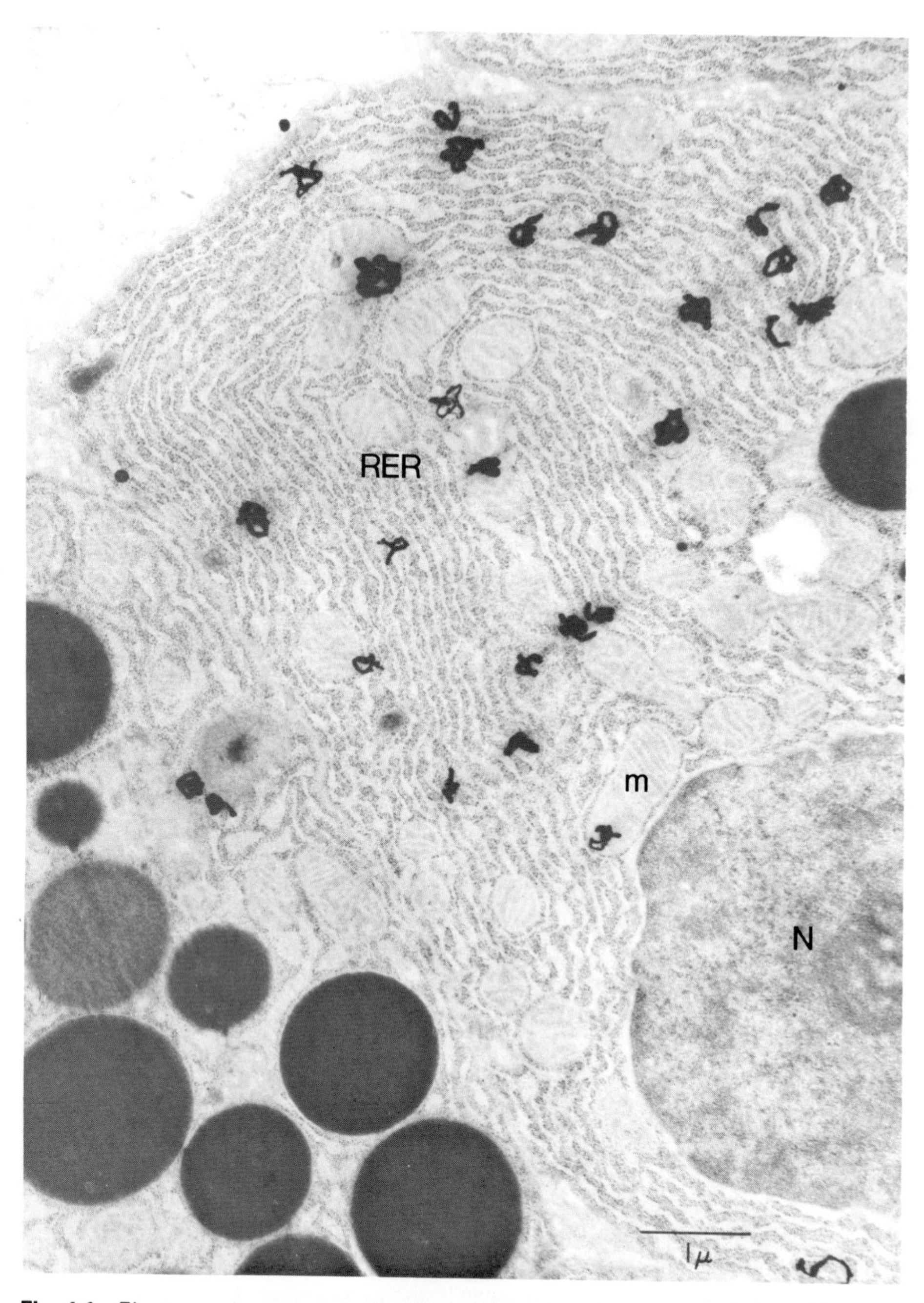

Fig. 4.1 Electron microscopic radioautograph of an acinar cell at the end of pulse labeling for 3 min with L-[^{3}H]leucine. The radioautographic grains are located almost exclusively over elements of the rough ER (RER). A few grains partly overlie mitochondria but, for reasons discussed in the text, most likely label the adjacent rough ER. m, Mitochondrion; N, nucleus. ×17,000. Reprinted with permission from Jamieson and Palade (1967).

seems to have a central role in this sorting. The various processes of sorting are discussed in the rest of this chapter.

II. ROLE OF THE ENDOPLASMIC RETICULUM

The experiments described in the previous section showed that the polysomes of the endoplasmic reticulum (ER) are responsible for the synthesis of proteins destined for secretion as well as for the enzymatic complement of lysosomes. Similarly, the synthesis of plasma membrane integral proteins is also associated with these polysomes.

The information for the localization of newly synthesized or nascent polypeptides resides in a discrete segment of the polypeptide, the *signal sequence* or *leader sequence*. Actual delivery of the protein requires interaction of the signal sequence with receptors in the cytoplasm or in the ER membrane. The receptors have a role in targeting the protein to the ER and may have a role in its translocation into the ER lumen. We have examined a mechanism that resembles the leader sequence–translocation process in the passage of specific proteins and RNA through the nuclear envelope (Chapter 2). Signal sequences represented by discrete segments of targeted proteins and the corresponding receptors are also thought to play a role in the targeting of protein to other membranes, such as the mitochondrial membranes, which are in part formed from proteins synthesized in the free polysomes of the cytoplasm.

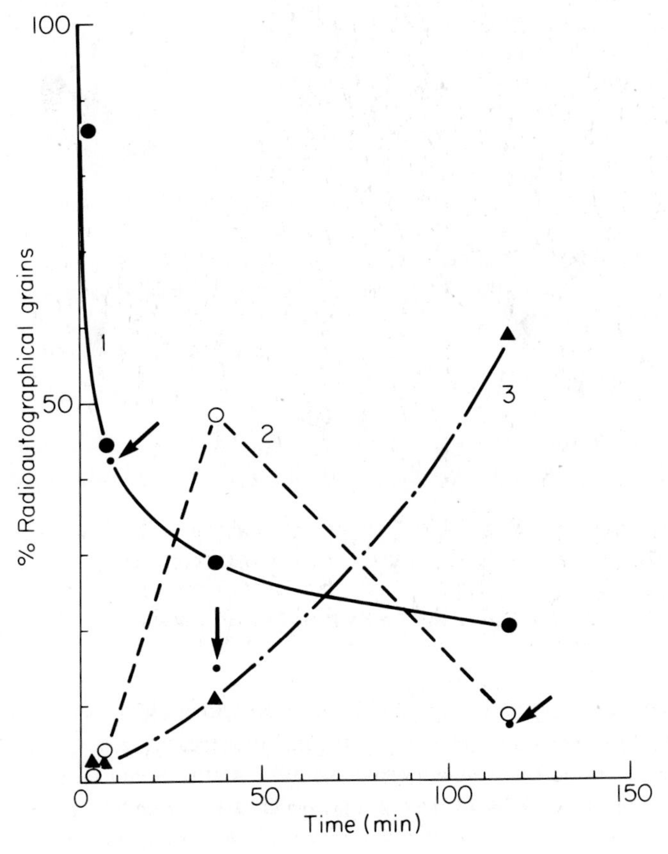

Fig. 4.2 Pulse-chase of pancreatic acinar cell. See text. From data of Jamieson and Palade (1967).

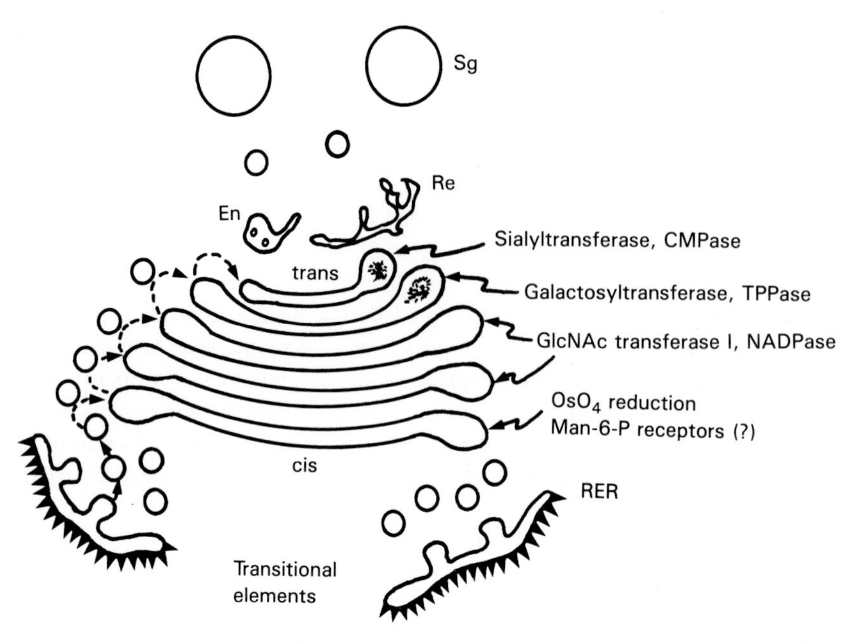

Fig. 4.3 Stationary cisternae model of the Golgi complex. The membrane components that have been localized in situ, together with their most frequent localization in either cis, middle, or trans cisternae, are indicated on the *right.* Note that exceptions and functional modulations in these distributions have also been reported. The flow of biosynthetic products through the Golgi complex is diagrammed on the *left.* The main features of this model are that (1) each cisterna represents a separate subcompartment with a distinctive membrane composition and internal milieu; (2) products move vectorially from rough ER to transitional elements (part rough, part smooth) located on the cis side of the Golgi complex, and then unidirectionally across the stacks (cis to trans), traversing the cisternae one by one; (3) transport along the route is effected by vesicular carriers; and (4) the main flow of traffic is to the dilated rims of the cisternae. *Sg,* Secretory granules; *En,* endosome; *Re,* reticular element (Farquhar, 1985). Reproduced, with permission, from the Annual Review of Cell Biology, Volume 1, copyright © 1985 by Annual Reviews Inc.

When an integral protein is translocated through a membrane, the insertion requires a *stop-transfer sequence* that arrests the translocation so that the protein remains in the phospholipid bilayer.

The translocation reactions for secretory (Blobel and Dobberstein, 1975a, 1975b), lysosomal (Erickson et al., 1983), and some integral proteins (Katz et al., 1977) have been studied in isolated systems and are discussed in some detail in the rest of this section.

A. Signal Sequences

The interaction between polysomes, ER membranes (forming *microsome* vesicles in isolated preparations), and the nascent polypeptide chain has been studied in many cell types. One of the systems studied was that of murine myeloma cells that engage in the synthesis and secretion of immunoglobulin (Blobel and Dobberstein, 1975a, 1975b). The mRNA for the light chain of immunoglobulin was found exclusively in membrane-bound polysomes. When this mRNA was used in a microsome-free translation system, the product was a protein larger than the secreted light chains (Blobel and Dobberstein, 1975a). In contrast, completion of chains contained by

RER vesicles produced only chains of normal length. These experiments suggest that the vesicle components are responsible for the processing needed to produce shorter mature proteins by cleavage of the signal peptide with a signal peptidase. Experiments using mRNA, the translational system, and various concentrations of microsomes stripped of polysomes are in agreement with this conclusion. Table 4.1 shows the results of experiments by Blobel and Dobberstein (1975b) in which the two proteins of distinct size were characterized using sodium dodecyl sulfate–polyacrylamide gel electrophoresis (SDS-PAGE). The first column indicates the addition of a preparation of stripped microsomes, and the second and third columns show the production of processed (Li) and unprocessed (PLi) light IgG chains, respectively. In the absence of stripped microsomes (row a) there is no synthesis of processed light chains; only the unprocessed proteins are produced. However, processing does take place (rows b–e) when the stripped microsomes are added to the mixture. The processing is stopped by heating the membranes at 55°C before the incubation (row f), as expected if the membranes play an active role in the protein processing. Increasing the concentration of microsomes beyond optimal values (rows d and e) decreases the synthesis of either processed or unprocessed protein, possibly because of nonspecific binding of needed components at the higher concentrations of membranes.

The experiment of Fig. 4.4 provides more information. A translation mixture containing intact microsomal vesicles is first incubated in the presence of radioactive amino acids, and then the vesicles are disrupted with detergent. Synthesis of processed protein (curve 1) continues after the disruption, probably from the pool of polypeptides not yet completed when the vesicles were disrupted. At about the time when the synthesis of processed polypeptides ceases, the unprocessed peptides begin to make their appearance (curve 2) and continue being produced thereafter. These experiments show that the processing by the microsomal membranes is cotranslational and the cleavage of the signal peptide takes place before the polypeptide is completed, since processed nascent chains continue to be produced even after the vesicles are removed. However, when the synthesis is initiated in the absence of membranes, the polypeptides remain unprocessed.

The essential features of signal sequences are still not clear, since they show no amino acid

Table 4.1 Synthesis of Processed (Li) and Nonprocessed (PLi) Light Chains of IgG in an Initiation System Containing Light Chain mRNA and Either No Added EDTA-Stripped Microsomes (0 μl RM-EDTA), Increasing Amounts of RM-EDTA (5, 10, 25, 50 μl), or Heat-Inactivated RM-EDTA (25 μl)

RM-EDTA (μl)	Li	PLi
(a) 0[a]	0.0	3.4
(b) 5	4.9	0.8
(c) 10	4.8	0.9
(d) 25	2.8	0.4
(e) 50	0.8	0.0
(f) 25 (55°C)	0.0	3.2

From G. Blobel and B. Dobberstein, "Transfer of Proteins Across Membranes" in *Journal of Cell Biology,* 67:852–862. Copyright © 1975 Rockefeller University Press, New York, NY.

[a]25-μl aliquots of incubation mixture were used for SDS-PAGE, while others (not marked with an *a*) were derived from 50-μl aliquots of the incubation mixture.

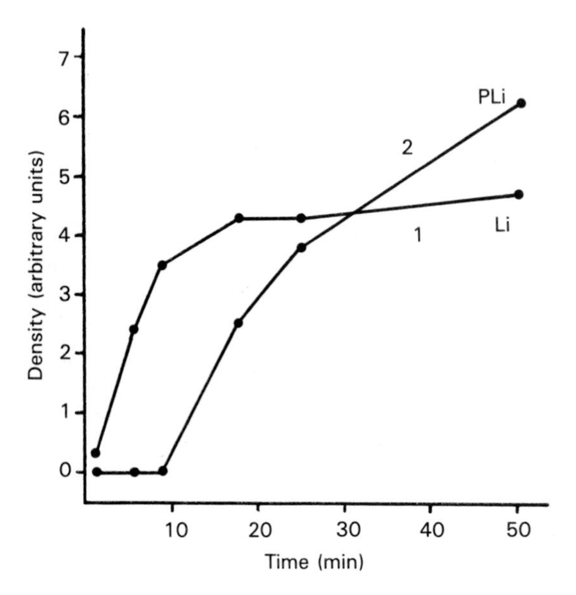

Fig. 4.4 Quantitation by densitometry of the autoradiograph. PLi and Li designate the unprocessed and the processed light chains of IgG, respectively. See text. Reprinted with permission from G. Blobel and B. Dobberstein, "Transfer of Proteins Across Membranes" in *Journal of Cell Biology,* 67:852–862. Copyright © 1975 Rockefeller University Press, New York, NY.

sequence homology (von Heijne, 1985). All signal sequences have in common a variable stretch of hydrophobic amino acids that are essential for function. Bacteria have a similar system for transport of proteins between the inner and outer membrane. Point mutations in the hydrophobic segment abolish function, indicating that this segment is essential for transfer of the protein through the membranes (Lee and Beckwith, 1986). Furthermore, cDNA encoding the signal sequence for bacterial β-lactamase, fused to globin DNA, contains information for the translocation of this protein into ER vesicles (Lingappa et al., 1984). Normally, globin is produced and remains in the cytoplasm. In the study by Lingappa et al., hybrid genes were produced by fusing in a plasmid the DNA coding for the amino terminal portion of bacterial β-lactamase and codons for the carboxyl terminal of α-globin. The genetic characteristics of the plasmids are shown in Fig. 4.5. Figure 4.5*a* shows the plasmid containing the two fused portions; the arrows show the construction of plasmids with deleted portions of the lactamase fragment. Figure 4.5*b* shows the proteins translated by the deleted plasmids in a cell-free system, and Fig. 4.5*c* is the genetic map of the region obtained with restriction enzymes; the deletions are indicated by the lines.

In Fig. 4.6, the proteins translated in a membrane-free system by the transcription products of the vectors outlined in Fig. 4.5*b* are separated by SDS-PAGE. These proteins have been immunoprecipitated with antibodies against the globin (g) and the lactamase (l). The control (n) was treated with normal rabbit serum, and no proteins appear in the gel. The proteins were further characterized by their sensitivity to trypsin (Fig. 4.7). Trypsin will hydrolyze only proteins that remain unprotected outside the vesicles. All of the proteins produced reacted with the globin antibody, whereas pGB14 (with only the signal sequence portion of the lactamase remaining) and pGM/N1 (with only two codons of the signal sequence remaining) do not react with the lactamase antibody. When the proteins were translated by a system containing microsomal vesicles (Fig. 4.7), the proteins containing the signal sequences coded by pG2 and pGB14 were shortened by a length corresponding to the signal sequence portion and were insensitive to the added trypsin, indicating that they had been translocated into the vesicles. In

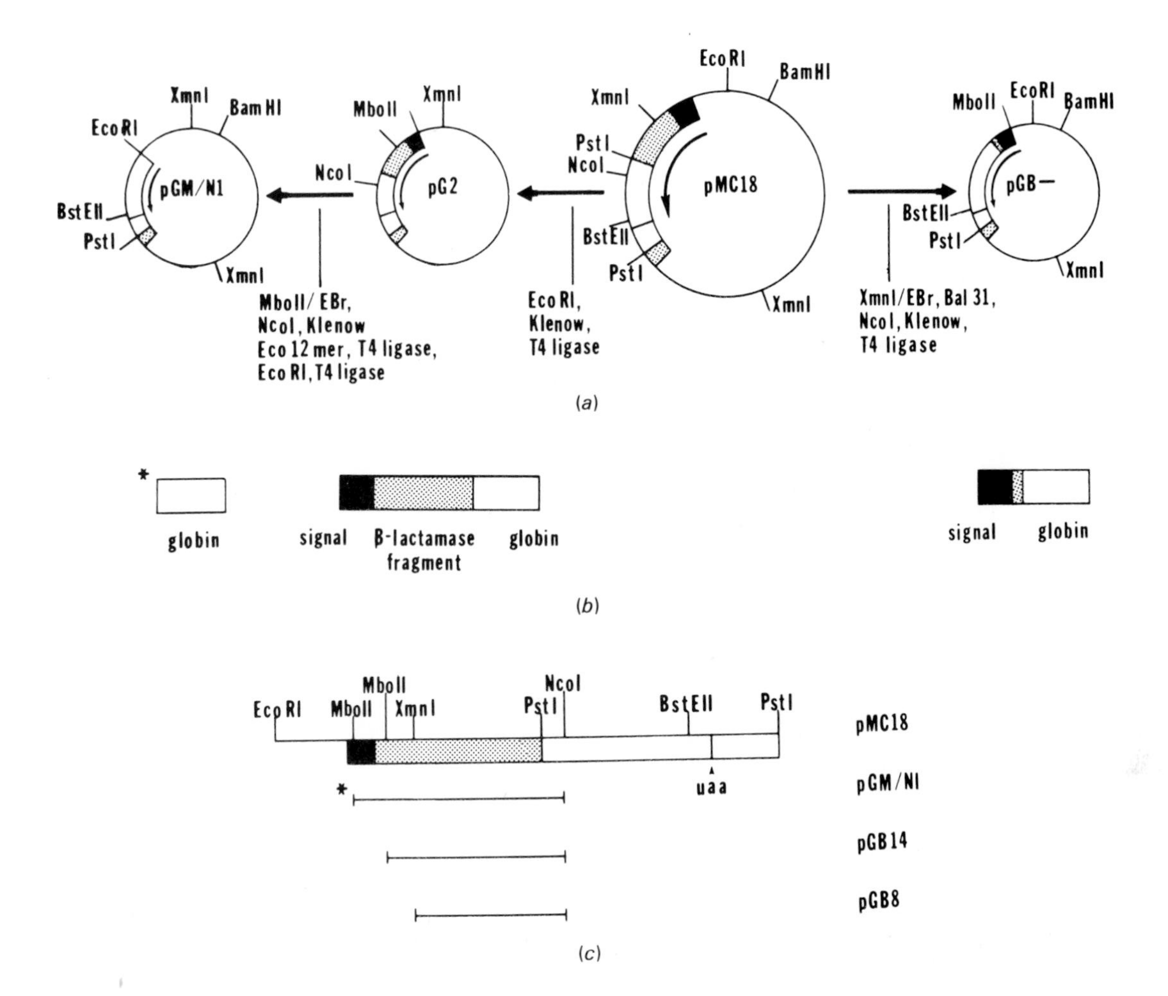

Fig. 4.5 Structure of pMC18 and deletion plasmids. (*a*) Plasmid pMC18, showing the β-lactamase gene (stippled bars) interrupted by chimpanzee α-globin cDNA sequences (white bars). Arrows on the plasmid diagrams indicate direction of transcription. The β-lactamase signal sequence is indicated (black bars). Leftward arrows indicate the steps involved in constructing pGM/N1, in which all but the two amino-terminal codons of the β-lactamase signal sequence are deleted. The *Eco*RI site was "filled-in" to generate an *Xmn*I site by sequential treatment with *Eco*RI DNA polymerase (Klenow fragment) plus dNTPs, then T4 ligase. The resultant plasmid, pG2, was linearized with *Mbo*II in the presence of ethidium bromide (EBr) at 5 μg/ml and treated with *Nco*I, which cuts at the start codon of the α-globin gene. After treatment with DNA polymerase (Klenow fragment), the terminus generated at the *Mbo*II site following the first two codons of the β-lactamase gene was fused to the *Nco*I-generated terminus at the start codons of globin with an intervening *Eco*RI 12-base-pair linker. The rightward arrow shows the scheme used to generate deletion plasmids such as pGB14 and pGB8. pMC18 was linearized by *Xmn*I in the presence of ethidium bromide (5 μg/ml), treated with exonuclease *Bal*31, and religated after treatment with DNA polymerase (Klenow fragment). (*b*) Representation of the proteins produced by the β-lactamase-globin fusion genes (not to scale). The asterisk indicates the extra six amino acids preceding the normal globin start codon encoded by the first two codons of β-lactamase and the *Eco*RI linker. (*c*) Restriction map of the relevant region of pMC18 with deletions indicated by the lines below. The asterisk indicates the insertion of the *Eco*RI linker. In pGM/N1, the *Eco*RI site was changed to an *Xmn*I site as described above. The UAA terminator codon for globin is shown. From V. R. Lingappa, et al., *Proceedings of the National Academy of Sciences,* 81:456–460. Copyright © 1984 Lingappa, et al.

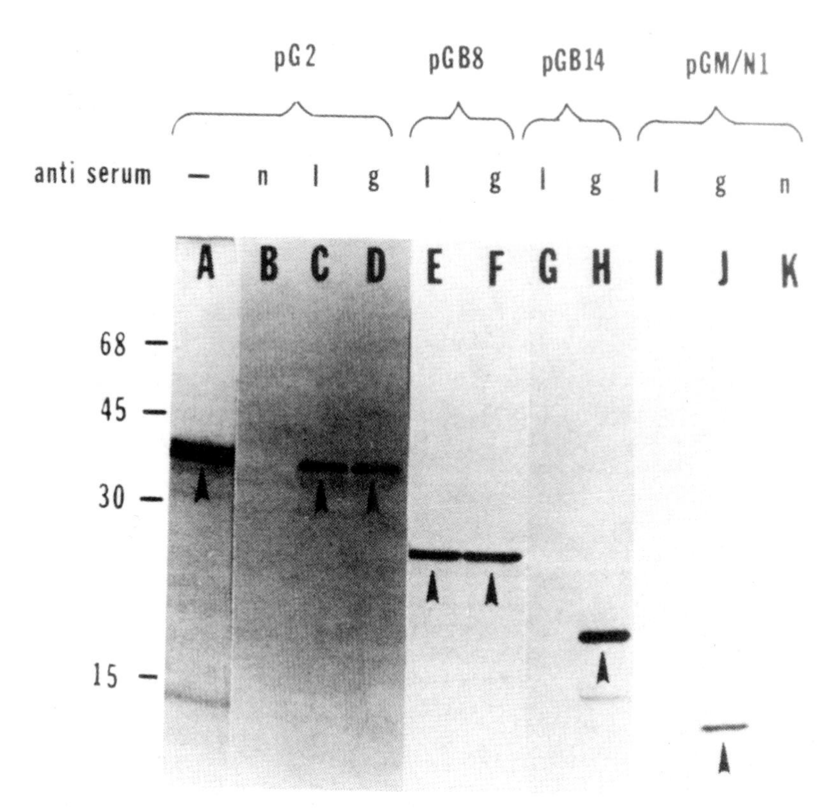

Fig. 4.6 Protein products of β-lactamase-globin fusions. Reprinted with permission from Lingappa et al. (1984).

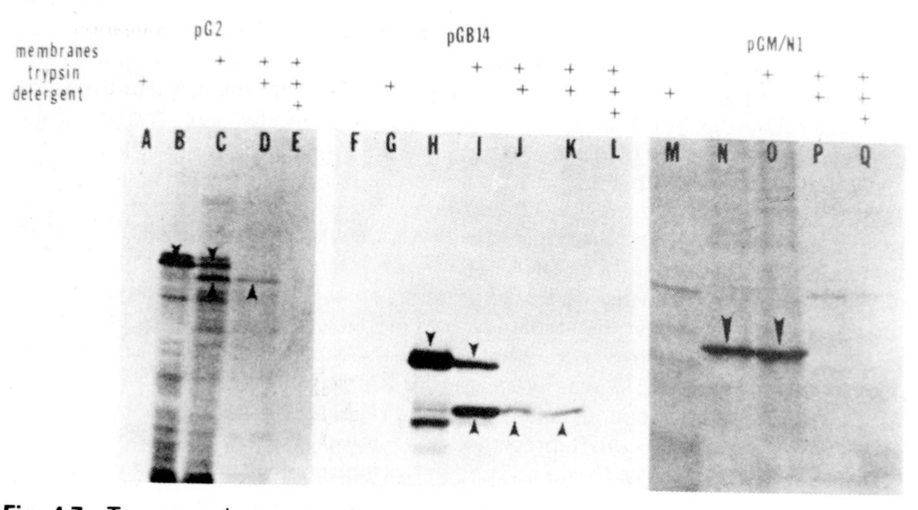

Fig. 4.7 Transmembrane translocation of β-lactamase-globin fusion proteins. Reprinted with permission from Lingappa et al. (1984).

the absence of membranes, there was no processing or protection from added protease. Protein pGM/N1, which lacks virtually all the signal sequence, is neither processed (compare lanes N and O) nor protected (lanes M and Q). These experiments showed that the 23 codons of the signal sequence (pGB14) and a small portion of the β-lactamase (5 codons, pG2) are sufficient to signal for translocation into the ER. Interestingly, the code for the signal sequence need not precede that of the peptide for translocation to occur (Simon et al., 1987).

B. Protein Processing

As the proteins gain access to the lumen of the ER, they are modified. The signal peptide is cleaved, disulfide bonds form, and glycosylation of the amino terminal takes place. However, cleavage by the signal peptidase is not required for translocation, as indicated, for example, by secreted proteins such as ovalbumin which have a hydrophobic signal portion that is not removed during processing (Palmiter et al., 1978).

C. Targeting

In salt-extracted canine pancreatic microsomes, recognition of the nascent peptide by the membrane system requires addition of the *signal recognition particle* (SRP) (Walter and Blobel, 1980, 1981). This particle includes a 300-nucleotide 7S RNA (Walter and Blobel, 1982a) and six nonidentical polypeptides organized in four SRP proteins.

Reconstitution experiments indicate that the SRP attaches to ribosomes. When a nascent polypeptide emerges, the SRP-ribosome complex is targeted to the membrane of the ER by an interaction of the SRP with its receptor (*docking protein*) (Gilmore and Blobel, 1983). The receptor is an ER integral protein (Hortsch et al., 1985). The ribosome–nascent polypeptide complex remains attached to the ER membrane, forming a ribosome-membrane junction where the translocation of the nascent chain takes place. The SRP and the docking protein are released to enter a new cycle. In the presence of SRP but in the absence of ER membranes, the translation is arrested, showing that synthesis of the entire polypeptide requires the complete system (Meyer et al., 1982a; Walter and Blobel, 1982b). The SRP is involved in the arrest of elongation. The complete cycle is represented in Fig. 4.8 (Walter et al., 1984).

The SRP receptor was first implicated in the translation-translocation system of the ER when it was found that proteolytic digestion of ER membranes blocked translocation and the activity could be reconstituted by addition of an extract solubilized by partial protease treatment (Meyer and Dobberstein, 1980; Walter et al., 1979). The active factor was subsequently shown to be a 52-kDa fragment of a 69-kDa integral membrane protein of the ER, as demonstrated using immunological and peptide mapping techniques (Gilmore et al., 1982a, 1982b). The 69-kDa protein was isolated using SRP-Sepharose affinity chromatography (Meyer et al., 1982b).

The 69-kDa docking protein (referred to as the α subunit) is thought to be part of a complex with a 30-kDa protein (the β subunit), since the two have been found tightly bound in most preparations (Tajima et al., 1986).

The SRP interacts directly with the signal sequence, since in isolated translating systems the SRP binds to the signal peptide. This has been shown for a signal peptide rich in lysine, where the attachment of SRP is blocked by the lysine analog β-hydroxyleucine (Walter and Blobel, 1981). In addition, in the case of the peptide hormone precursor preprolactin, photoactivated cross-linking reagents were shown to be incorporated into the amino region of the polypeptide and to cross-link to the SRP (Kurzchalia et al., 1986).

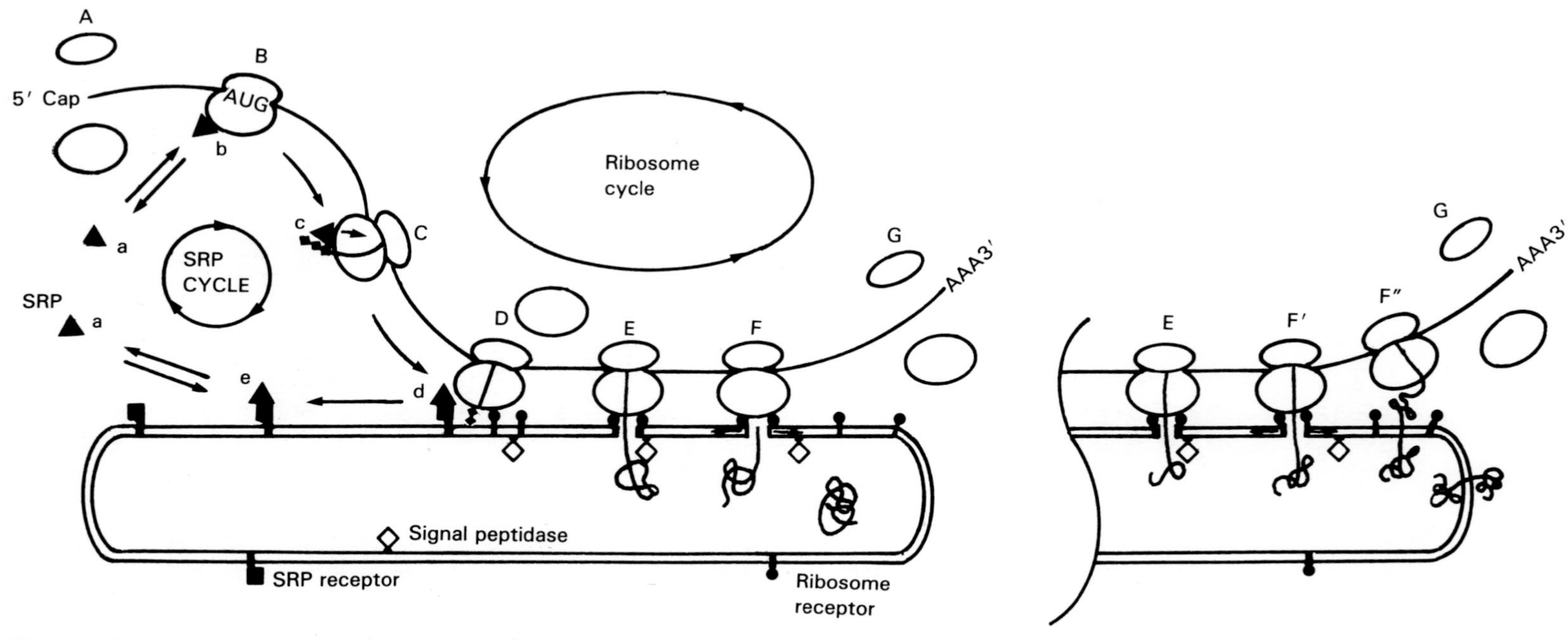

Fig. 4.8 Protein translocation across or into the membrane of the endoplasmic reticulum. Reprinted with permission from Walter et al. (1984), copyright 1984 by Cell Press.

D. Translocation

The precise events of translocation and the involvement of membrane components are still not entirely clear. Although in the mammalian in vitro systems translocation is cotranslational, this seems to be unrelated to the mechanism of translocation. In yeast the translocation is posttranslational. Furthermore, the potential for posttranslational translocation has been demonstrated in mammalian systems. The glucose transporter protein is transferred into vesicles even when it is presynthesized and the microsomal vesicles are added only subsequently (Mueckler and Lodish, 1986). Similarly, the newly formed proteins remained attached to the ribosomes following transcription and translation in the absence of microsomes from cDNA from which the termination codons were deleted. However, addition of the stripped microsomes elicited their transfer into the vesicles (Perara et al., 1986). These findings suggest that the translocation machinery is part of the ER membrane and the mechanism is not part of the translational process itself. The posttranslational transfer and presumably the normal transfer were found to require an energy source, in these experiments supplied by ATP, GTP, and phosphocreatine in the presence of creatine phosphokinase (Perara et al., 1986).

Two integral proteins termed *ribophorins* (Kreibich et al., 1978) are suspected to play a role in translocation, but this is still to be established. Furthermore, the signal peptidase (Evans et al., 1986) that is needed to cleave the signal sequence to form the mature protein has six subunits, and it has been suggested that this enzyme may have a more complex role than its peptidase activity, perhaps being involved in translocation. So far there is evidence for the involvement of nine separate proteins in the translocation (Rapoport, 1991).

III. SORTING PROTEINS IN THE ENDOPLASMIC RETICULUM AND THE GOLGI COMPLEX

In tracing the pathway of biosynthetic transport, the study of cells infected by animal viruses has been very useful. The virus uses the synthetic machinery of infected cells. During completion of the synthesis of new virus, the viral envelope containing its own characteristic integral proteins is generated from the plasma membrane of the host cell by a process of budding. The fate of the newly synthesized viral coat proteins can therefore be used as a model for the pathway followed by plasma membrane integral proteins. Vesicular stomatitis virus (VSV) and Semliki Forest virus (SFV) have been particularly useful. Semliki Forest virus, which infects mosquitoes, is related to yellow fever virus and is named after a forest in Uganda. Vesicular stomatitis virus is a mild pathogen of cattle. Generally, cells in culture such as Chinese hamster ovary (CHO) cells and baby hamster kidney (BHK) fibroblasts have been used. An entirely different approach has been used to study the biosynthetic transport system of the yeast *Saccharomyces cerevisiae*. The yeast cells have a biosynthetic transport system analogous to that of mammalian cells. The central vacuole corresponds functionally to the lysosome, and the yeast also secretes proteins into its periplasmic space. The study of yeast permits the use of the powerful recombinant DNA techniques (see Chapter 1) in cells whose genetics is generally well understood. All of these approaches and materials will make several appearances in the account that follows.

Many proteins are partially glycosylated in the ER, and both glycosylated and unglycosylated proteins are transported out of the ER at variable rates. The Golgi complex is the intracellular site where biosynthetic, endocytotic, and recycling membrane traffic converge. The Golgi system has a role in the posttranscriptional *N*-glycosylation of integral membrane proteins and in sorting the various proteins.

The Golgi complex has three to eight flattened cisternae, which exhibit functional and topographic polarity; that is, the displacement of newly synthesized proteins with time is accompanied by stepwise processing by enzymes that occupy specific locations in the system in an organization analogous to an assembly line (Dunphy et al., 1981; Rothman, 1981). Depending on their position, the Golgi cisternae are referred to as cis, medial, and trans components, where cis corresponds to the elements closest to the RER (see Fig. 4.3). Although the morphological sidedness is not always obvious, the enzymatic and hence cytochemical polarity is generally demonstrable. A number of other complexities have also been described. Large *Golgi-associated vacuoles* (GAVs) have been recognized in the cis side of the system (Katz et al., 1977), and a network of tubular vesicles, the *trans Golgi network* (TGN), has been observed in the trans side of the system (Griffiths and Simon, 1986). In analogy with the trans system, the term *cis Golgi network* (CGN) has been adopted more recently to include all cis elements including GAV (see Duden et al., 1991a).

The Golgi cisternae are held in a central position in the cell by their tendency to move along the microtubules in the direction of their minus end, toward the microtubular organizing center (MTOC) (see Kreis, 1990). In almost all mammalian cells, the major MTOC corresponds to the *centrosome* (consisting of two centrioles), the organelle from which the microtubules begin to assemble, and which is generally located to one side of the nucleus. The cisternae do not separate when the cell membrane is disrupted or after micromanipulation, suggesting that they are held together by some adhesive molecules.

All proteins processed by the Golgi complex, regardless of eventual destination, can be found throughout the cisternae, as shown, for example, by immunocytochemistry. Sorting must therefore occur when the proteins leave the Golgi at the trans end. The TGN has been proposed to play a special role in the sorting (Griffiths and Simon, 1986). At least part of the transport is thought to take place by *bulk flow,* a nonspecific translocation that is discussed later in this chapter.

Primarily three types of vesicles have been observed to form from Golgi stacks. Their contents have been identified using immunocytochemistry (Brown and Farquhar, 1984). The secretory storage vesicle has a partial clathrin coat, contains densely packed secretion products, and originates from the trans cisternae (Orci et al., 1984). Small clathrin-coated vesicles contain acid hydrolases and mannose-6-phosphate receptors (e.g., Campbell et al., 1983; Schulze-Lohoff et al., 1985) and presumably form lysosomes. Clathrin is probably present only in the trans face of the Golgi, although some of the data are contradictory (Griffiths et al., 1985; Orci et al., 1985). The bulk carrier, involved in a form of intracellular transport discussed in Section III, B, is likely to correspond to non-clathrin-coated vesicles that are found at all levels of the Golgi and contain VSV G protein in infected cells (Orci et al., 1986). Studies with Semliki virus membrane glycoproteins (Saraste and Hedman, 1983) suggest the involvement of at least two vesicles, one coated and of approximately 50 nm and a smooth vesicle of about 80 nm.

Vesicles are thought to be the vehicle of transport between the various compartments. They appear to be formed by budding and to deliver their contents in a process in which their membranes fuse with those of the target compartment. Morphological observations suggest that the vesicles are formed from the rims of the Golgi cisternae, as indicated in Fig. 4.3. Electron microscopic studies with isolated Golgi stacks of CHO cells infected with VSV virus also showed the formation of vesicles containing G protein of the virus from the rims of the Golgi stacks. These vesicles did not contain clathrin, supporting the notion that non-clathrin-coated vesicles are involved in the bulk flow. The vesicle formation required ATP and cytosol (Balch et al., 1984b; Orci et al., 1986).

Golgi derived non-clathrin-coated vesicles have been isolated and studied recently using a new strategy. These vesicles accumulate in a cell-free system in the presence of the non-hydrolyzable nucleotide analog, GTPγS (Malhotra et al., 1989; Serafini et al., 1991). As discussed below (III, B), GTP-binding proteins, needed for vesicular transport, require for function the presence of GTP (by an unfortunate duplication of names these proteins are also called G proteins but are unrelated to the viral coat protein). When blocked by the GTP analog, the vesicles accumulate and therefore can be isolated in quantity and examined for their protein content. These vesicles have four major proteins thought to be coat proteins (α, β, γ, and δ-coating proteins, COP), ranging in molecular weight between 60 and 160 kDa, and a few smaller proteins. β-COP has been localized with immunofluorescence and immuno-electron microscopy in the CGN and TGN and has also been found in soluble cytosolic complexes (Duden et al., 1991b; Waters et al., 1991). Presumably, the cytosolic β-COP is the source of the vesicle's coat protein.

The rest of this section concerns the sorting functions of the system.

A. Export from the ER

The ER plays a role in sorting by controlling the duration of residence of the various proteins (Fries et al., 1984; Scheele and Tartakoff, 1985). This role could be explained by a model in which these proteins are bound to receptors and can be transported only after their release.

Some evidence presently available indicates that proteins have to be folded and assembled before leaving the ER. For example, the retinol-binding protein cannot be transported unless it binds its ligand (Ronne et al., 1983), and similarly the heavy chain of IgM accumulates in the ER unless it is able to bind to the light chains. These observations suggest that the sorting signal for leaving the ER corresponds to patches in the protein molecule that are conformation dependent. Unfolded or unassembled chains appear to be retained in the ER until they are assembled.

Unfolded proteins such as mutant proteins of influenza hemagglutinin have been found in some cases to be associated with a 77-kDa protein, heavy chain-binding protein (BiP), in a pattern that suggests an association during an intermediate step of transport (Gething et al., 1986). It has been suggested that unfolded and unassembled proteins have a tendency to form aggregates that cannot proceed through transport unless solubilized. The BiP and protein disulfide isomerase would facilitate this solubilization (Pfeiffer and Rothman, 1987). BiP is released from immobilized immunoglobulin heavy chains on addition of ATP (Munro and Pellham, 1956). This suggests a possible energy-requiring cycle that disaggregates proteins so that transport can proceed.

BiP, protein disulfide isomerase (PDI), needed for correct disulfide bond formation, and glucose regulating protein (GRP94), and other soluble proteins residing in the ER have a role in the initial steps of maturation of secretory proteins (see Pelham, 1990). These proteins must be present in a functional form and be properly folded, yet they are retained; therefore, they must be distinguished from the proteins which are rapidly transported through the Golgi complex. For these cases, recognition sequences are definitely involved. Comparison of the amino acid sequences (computer comparisons can be carried out using currently available databanks, see Chapter 1, II, C) revealed common or similar sequences at the carboxy-terminal corresponding to a tetrapeptide (usually Lys-Asp-Glu-Leu, KDEL; in yeast, His-Asp-Glu-Leu, HDEL). When expressed in monkey COS cells, BiP lacking this sequence was secreted (Munro and Pelham, 1987) (COS cells are transformed kidney cells from the African green monkey). In this and other experiments discussed below, the proteins were modified by genetic manipulation using vectors containing the appropriate DNA sections (Chapter 1, II, A). The proteins themselves were frequently identified using Western blots (see Chapter 1, II, B). Addition of the last 6 amino acids of BiP to secretory, lysosomal, or vacuolar proteins caused retention in the ER (e.g., Munro and Pelham, 1987). Similar results were obtained with other mammalian cells, plants, and yeast.

Mammalian liver esterases of the ER have a number of exceptions to the KDEL and HDEL requirement. Although some have a similar signal tetrapeptide, some have entirely different ones (e.g., Long et al., 1988).

Although retention signals have in these cases a primary role, conformation of the protein can also be important, presumably by controlling the accessibility of the tetrapeptide to receptors (e.g., Pelham et al., 1988). Some integral membrane proteins are thought to be retained by their association with large aggregates that cannot enter the transport vesicles. However, at least some of them have been shown to have retention signals (see Pelham, 1990).

How are the ER soluble proteins retained? The proteins could be immobilized by binding to receptors excluded from the nascent transport vesicles as discussed above for other proteins. However, in at least some cases the process appears to be more complicated, involving the delivery to a post-ER compartment (probably part of the Golgi complex) followed by retrieval. This has been demonstrated ingeniously by adding the KDEL retention sequence to the lysosomal enzyme cathepsin D. Lysosomal enzymes are processed in the Golgi stacks (see III, B), in this case by the addition of GlcNAc-1-phosphate. The cathepsin was found in the ER with the post-ER addition (Pelham, 1988). Similar results were obtained in yeast (Dean and Pelham, 1990).

These experiments suggest that a membrane receptor for the retained protein is activated in a post-ER component, and the complex is then returned to the ER. Mutants of yeast unable to retain the HDEL signal have been isolated, and a putative receptor has been recognized (Semenza et al., 1990).

Cell-free preparations from VSV-infected CHO cells were used to study the transfer of G protein of the virus from ER to Golgi (Balch et al., 1987). The trimming of high-mannose oligosaccharide from G protein by mannosidase I, which occurs in the Golgi, was used to monitor the transport. The transport from ER to Golgi requires ATP and cytosolic factors.

B. Passage Through the Golgi System and Transfer to the Cell Surface

Sequential Events and Sequential Sites The passage of proteins through the Golgi system can be traced by initiating the synthesis of a protein at a specific time and following its passage. This can be accomplished with viral infection. The viral glycoproteins enter the Golgi complex at the cis face and exit through the trans face (Bergmann and Singer, 1983; Saraste and Hedman, 1983), as demonstrated in the study discussed below.

BHK fibroblasts infected with a temperature-sensitive SFV strain were unable to transport the viral proteins from the ER membrane at 39°C, the restrictive temperature. When the cells were shifted to 28°C, the permissive temperature, transport resumed. Antibodies to the glycoproteins were used to reveal their location (Saraste and Hedman, 1983) and were visualized with peroxidase covalently attached to coat protein A of *Staphylococcus aureus*. Protein A specifically attaches to the Fc portion of the immunoglobulin molecules. The peroxidase reaction was detected cytochemically from the precipitated products of the appropriate reaction.

An uninfected cell (control) treated in this way is shown in Fig. 4.10*c*. Figure 4.9*a* and *b* show the localization 2.5 min after the shift to 28°C, and Fig 4.9*c* shows the results after 5 min. Figure 4.10*a* and *b* correspond to 3.5 h and 30 min, respectively. After 2.5 min the label is in the large GAV and after 5 to 30 min in the cisternae stacks—the cis cisternae after 5 min and the more distal stacks after 15 to 30 min.

A finished glycoprotein requires several posttranscriptional stepwise enzymatic additions to arrive at its mature structure. The specific locations of the necessary enzymes, mirroring the

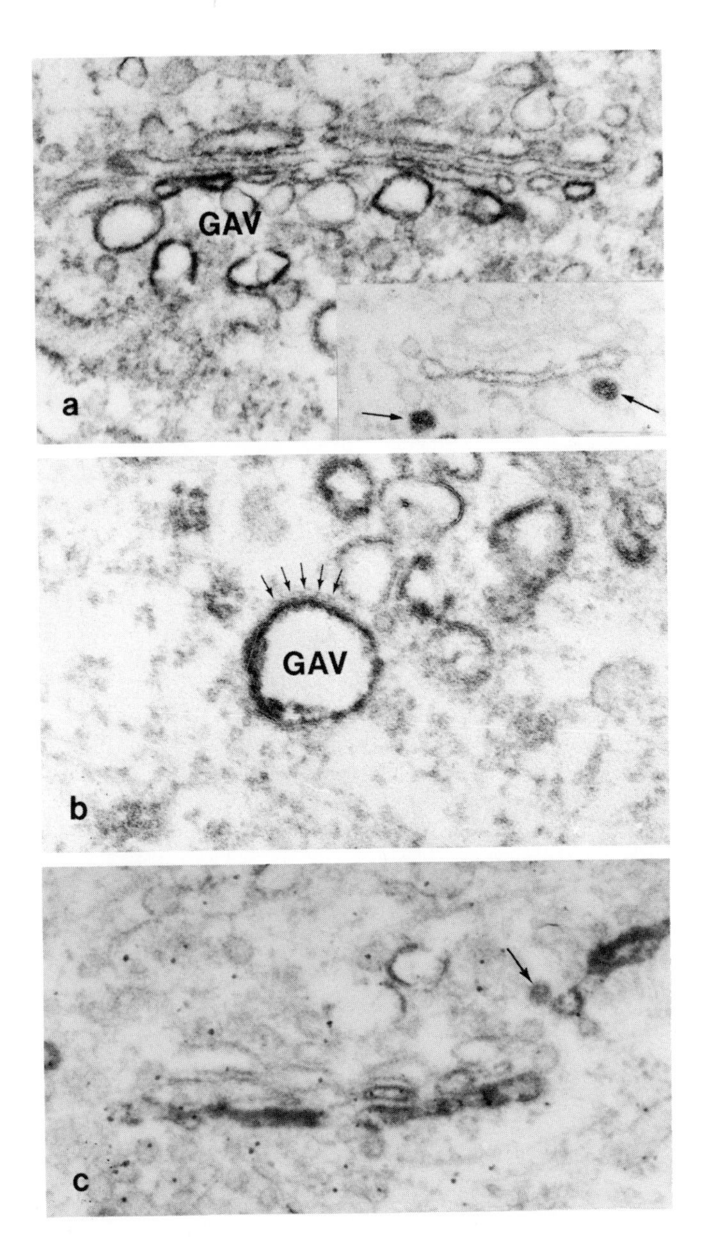

Fig. 4.9 (*a*) A cell fixed at 2.5 min after the shift. Several labeled GAVs are seen in the vicinity of cisternal elements, which contain little label (×38,000). The inset shows two labeled 50-nm vesicles and a labeled Golgi cistern (×60,000). (*b*) Cell fixed at 2.5 min after shift. Peroxidase-stained GAV with a coated region on its cytoplasmic surface (thin arrows) (×70,000). (*c*) Golgi labeling at 5 min after shift. Peroxidase stain is deposited in one or two proximal cisternae of the stack, in a 50-nm vesicle (arrow), and in the ER (×60,000). Reprinted with permission from Saraste and Hedman (1983).

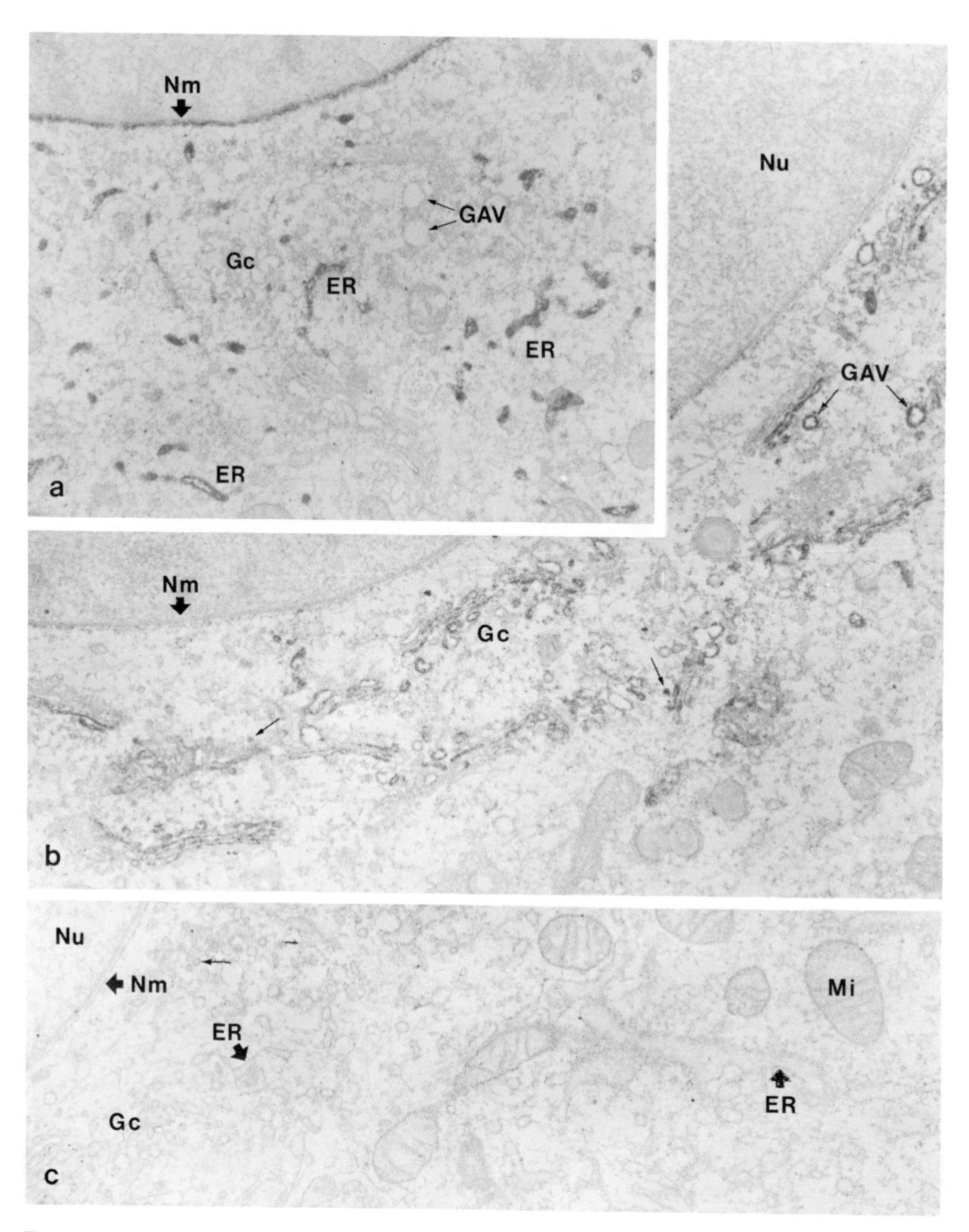

Fig. 4.10 (*a*) Immunoperoxidase staining of SFV virus membrane glycoproteins in ts-1 mutant infected cells grown at the restrictive temperature (39°C) for 3.5h Peroxidase stain is deposited at the ER and at the nuclear membrane (Nm), whereas the Golgi complex (Gc), including the cisternal stacks, large GAV, and small vesicles, is unstained (×15,000). (*b*) Ts-1 infected cells fixed at 30 min after shift to the permissive temperature (28°C). Peroxidase reaction product is deposited at the cisternal stacks on the Golgi complexes (Gc), at membranes of large GAV, and in vesicles of the Golgi region (arrows). The nuclear membrane (Nm) and the ER are relatively unstained. Nu = nucleus (×20,000). (*c*) Control showing an uninfected cell labeled for SFV membrane proteins by the peroxidase procedure. All structures including ER, the Golgi complex, and vesicles of the Golgi region (arrows) are unstained. Mi = mitochondrion (×25,000). Reprinted with permission from Saraste and Hedman (1983).

order of the necessary steps, are evidence of vectorial processing akin to an assembly line (e.g., Dunphy et al., 1985). Our present understanding of the distribution of enzymes in the ER and the Golgi system is shown in Fig. 4.11 (Goldberg and Kornfeld, 1983). The evidence for this distribution comes in part from the experiments in which the components of CHO cells were fractionated in sucrose gradients and the glycoprotein-processing enzymes were found in different fractions. Enzymes acting earlier in the processing pathway were found in the heavier fractions. This was confirmed in more detail using mouse lymphoma cells, in which the order of fractionation in the sucrose gradients coincided with the order of processing. Although this evidence shows that the location is vectorial it is somewhat harder to correlate the presence of an enzyme with an actual location inside the cell. Immunocytochemistry has confirmed the location of some of these enzymes in the appropriate Golgi compartment (Farquhar, 1985).

Experiments involving a complementation assay that required cell fusion confirmed that proteins are transferred from one Golgi stack to another in a unidirectional manner (Rothman et al., 1984a, 1984b). In these experiments VSV-infected CHO cells were fused to noninfected cells. The Golgi complexes of the two types of cells maintained their integrity, as shown in the electron micrograph of Fig. 4.12. Three clones of cells were used. In clone 13 the VSV G protein can acquire *N*-acetylglucosamine (GlcNAc) terminals but cannot add either galactose (Gal) or sialic acid. Clone 1021 can add GlcNAc and Gal but not sialic acid. Clone 15B is competent in these three activities. The fusion of the various types of cells is accomplished by

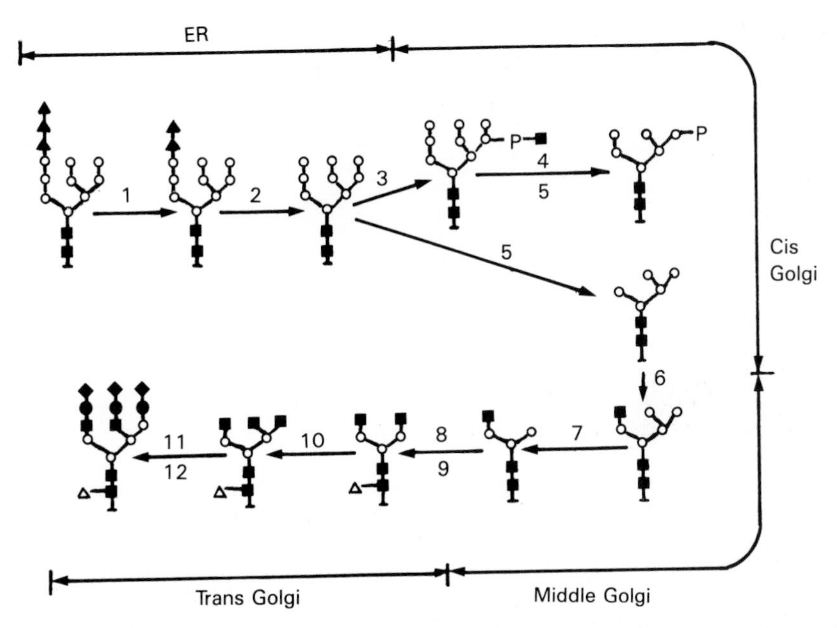

Fig. 4.11 Steps in the processing of asparagine-linked oligosaccharides and their presumptive intracellular site. Steps for addition of M6P to lysosomal enzymes are indicated in the side branch (3–5). 1; Glucosidase I; 2, glucosidase II; 3, lysosomal enzyme, *N*-acetylglucosaminylphosphotransferase; 4, lysosomal enzyme, phosphodiester glycosidase; 5, mannosidase I; 6, GlcNAc transferase I; 7, mannosidase II; 8, GlcNAc tranferase II; 9, fucosyltransferase; 10, GlcNAc transferase IV; 11, galactosyltransferase; 12, sialyltransferase. (▲) Glucose; (■) GlcNAc; (○) mannose; (●) galactose; (△) fucose; (◆) sialic acid; P, phosphate. Note: ER mannosidase is not indicated in the diagram. Reprinted with permission from Goldberg and Kornfeld (1983).

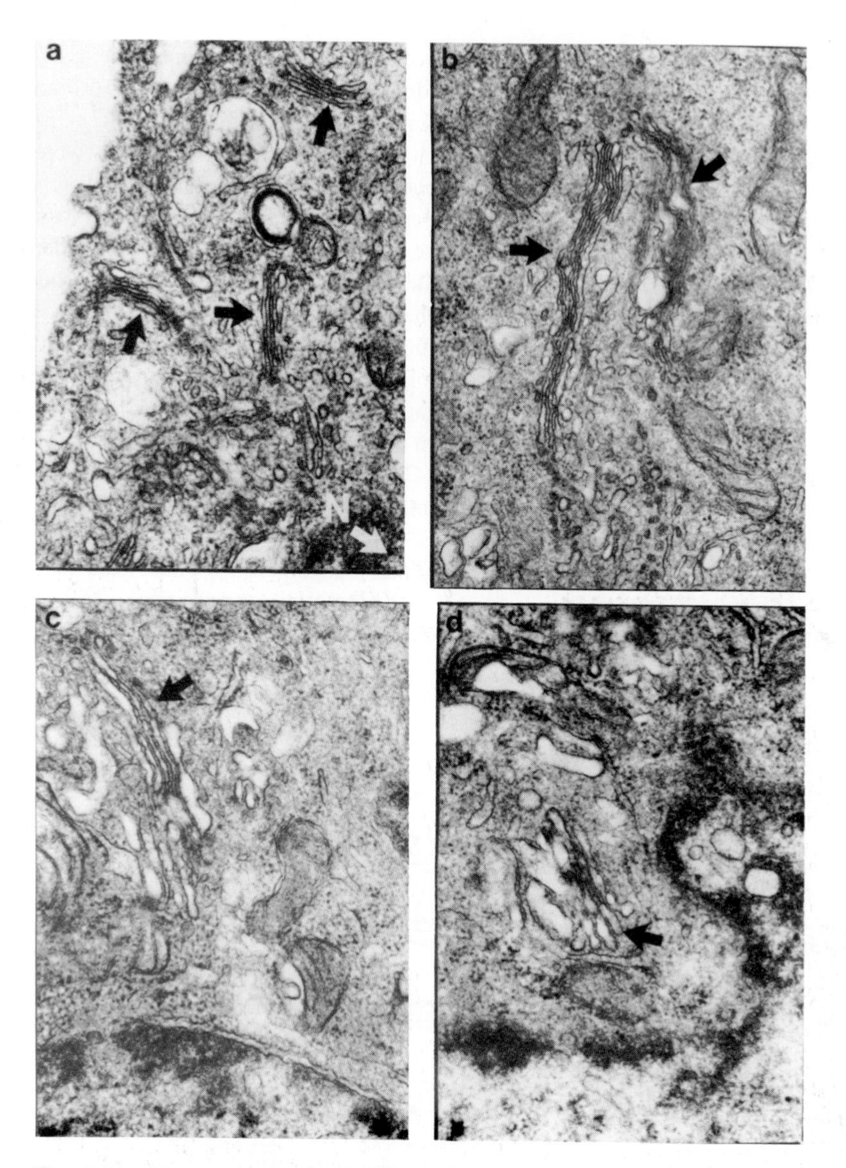

Fig. 4.12 Morphology of Golgi stacks (a) before and (b–d) after cell fusion. The nucleus (N) is at the lower right corner of (a), just outside this photograph. Portions of nucleus are shown in (b) and (d). About ×40,000 (Rothman et al., 1984b). Reproduced from *The Journal of Cell Biology* (1984) 99:248–259, by copyright permission of the Rockefeller University Press.

short exposure of the cells to low pH. The Gal-containing G protein was identified using ^{3}H-labeled G proteins. The G protein was first immunoprecipitated and then digested with a proteolytic enzyme, pronase. The Gal-containing fragments were isolated using ricin, a lectin that is specific for Gal. Similarly, a slug lectin was used for sialic acid. A *lectin* is a protein capable of binding a specific sugar residue in carbohydrate chains such as those of glycoproteins.

The fusion between clone 13 cells (no ability to incorporate Gal), pulse-labeled (see p. 113) with [^{3}H]GlcNAc, and clone 15B cells is represented in Fig. 4.13 (Rothman et al., 1984a). The results of this experiment, represented in Fig. 4.14 on p. 132 (Rothman et al., 1984b), show that the galactosylation does take place so that a transfer from the Golgi of clone 13 cells to 15B cells must have taken place. The rate of incorporation is about the same as that with 15B cells alone, with or without fusion and before or after pulsing. However, there is a progressive loss of activity when the chase period, corresponding to the time span before fusion, becomes longer. These results are shown in Fig. 4.15. This finding is in agreement with the assembly-line model; it suggests that the transfer is vectorial and that when the G protein is transferred to another compartment it can no longer accept Gal, since the latter compartment does not contain the appropriate enzyme.

Experiments with a somewhat different design have been used to study the transfer more directly. VSV-infected clone 1021 cells (deficient in ability to add sialic acid) were labeled with either [^{3}H]Gal or [^{3}H]GlcNAc and then fused to 15B cells (which can add either Gal or sialic acid). This experiment is represented in Fig. 4.16. Vectorial transfer of Gal-labeled G protein represents a transfer from a later stack of the Golgi than the transfer of [^{3}H]GlcNAc, since the step occurs later biochemically. Table 4.2 shows the results expressed as the percent of total G

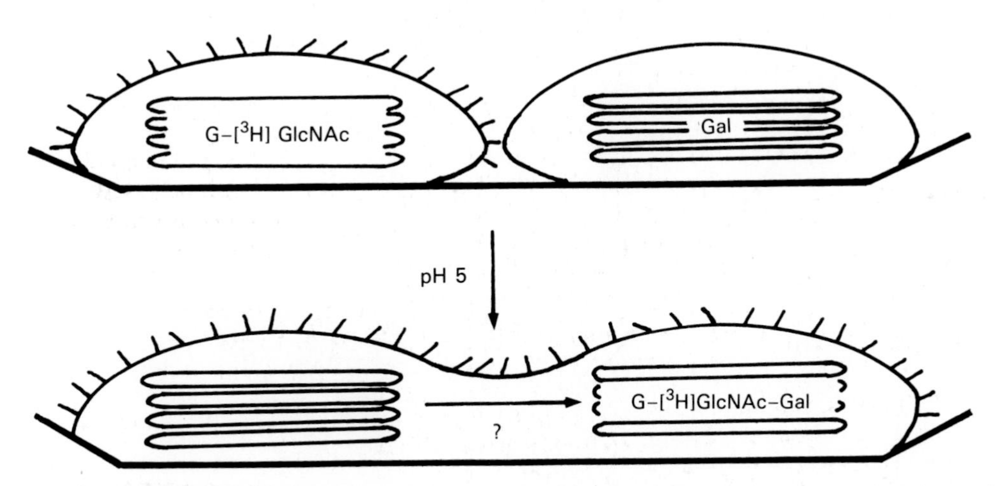

Fig. 4.13 Design of a cell-fusion experiment to detect transfers between two Golgi populations. A mixed monolayer is formed containing VSV-infected CHO clone 13 cells and uninfected CHO clone 15B cells. VSV G protein is labeled in the Golgi complexes of clone 13 cells with [^{3}H]glucosamine (GlcNH$_2$), incorporated as peripheral *N*-acetylglucosamine (GlcNAc). The clone 13 Golgi is unable to add galactose (Gal). The clone 13 cells containing [^{3}H]GlcNAc-labeled G protein are then fused to neighboring clone 15B cells (via a brief exposure to pH 5) whose Golgi complexes are able to add Gal. Transfer of G protein from the clone 13 to the clone 15B Golgi complexes is monitored by the addition of Gal to [^{3}H]GlcNAc-labeled G protein after fusion (Rothman et al., 1984a). Reproduced from *The Journal of Cell Biology* (1984) 99:260–271, by copyright permission of the Rockefeller University Press.

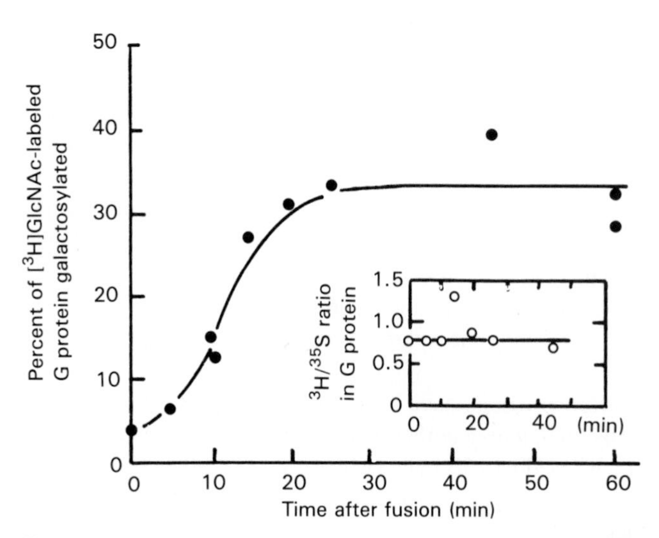

Fig. 4.14 Kinetics of galactosylation of G protein labeled with [^{3}H]GlcNAc in VSV-infected clone 13 cells after fusion to uninfected clone 15B cells (Rothman et al., 1984a). Reproduced from *The Journal of Cell Biology* (1984) 99:260–271, by copyright permission of the Rockefeller University Press.

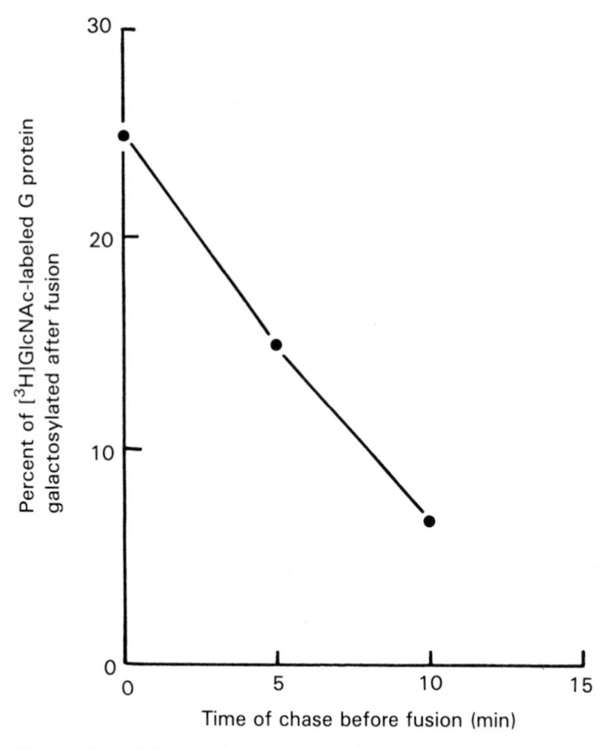

Fig. 4.15 Effect of a period of chase before fusion of [^{3}H]GlcNH$_2$-labeled VSV-infected clone 13 cells with uninfected clone 15B cells. Plotted is the extent of galactosylation of [^{3}H]GLcNAc-labeled G protein 1 h after fusion versus the time at which fusion was initiated. A variable period of chase (0, 5, or 10 min) intervened between the 3-min pulse and the initiation of fusion (Rothman et al., 1984b). Reproduced from *The Journal of Cell Biology* (1984) 99:260–271, by copyright permission of the Rockefeller University Press.

VSV-infected clone 1021 cell

- Will add [^{3}H]GlcNAc and [^{3}H] Gal to G protein in its Golgi
- Won't add sialic acid to G protein
- Pulse label with either [^{3}H]Gal or [^{3}H]$GlcNH_2$

Uninfected clone 15B cell

- Will add Gal and sialic acid to G protein following transfer to its trans Golgi

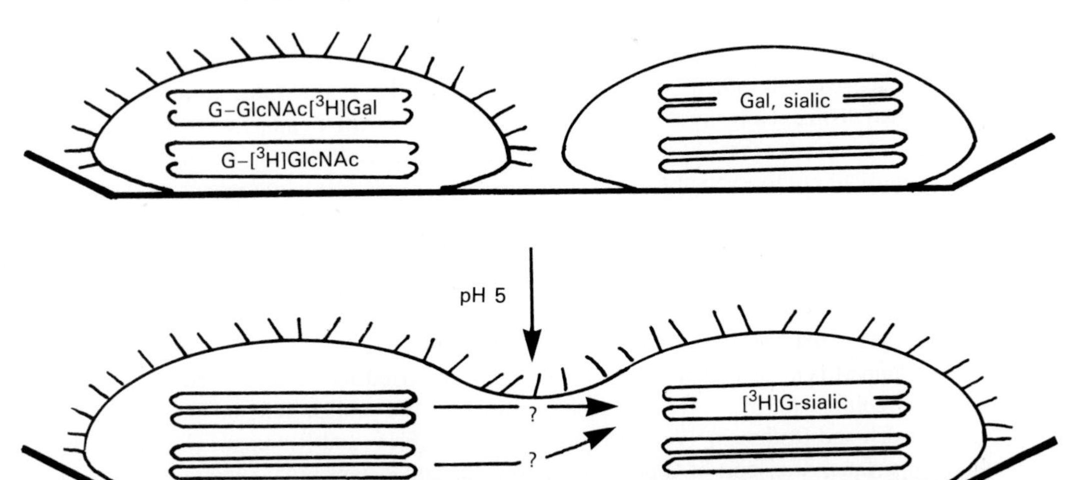

Fig. 4.16 Design of a cell fusion experiment to measure the relative efficiency with which G protein present in two different Golgi complex subcompartments (in which GlcNAc and Gal, respectively, are added) is transferred to an exogenous Golgi population. A mixed monolayer is formed containing VSV-infected CHO clone 1021 cells and uninfected clone 15B cells. VSV G protein is labeled in the Golgi complex of clone 1021 cells either with [^{3}H]$GlcHN_2$ (incorporated as peripheral GlcNAc) or with [^{3}H]Gal. Clone 1021 is able to incorporate both of these sugars, but not sialic acid. The clone 1021 cells, now harboring ^{3}H-G protein in their Golgi complexes, are then fused to neighboring clone 15B cells (via a brief exposure to pH 5) whose Golgi complex are able to add sialic acid. Transfer of G protein from the clone 1021 to the clone 15B Golgi complex is monitored by the addition of sialic acid to the ^{3}H-G protein upon arrival. In the case of the [^{3}H]GlcNAc-labeled G protein, the addition of Gal (presumably in the 15B Golgi complex) must occur before sialic acid can be added (Rothman et al., 1984a). Reproduced from *The Journal of Cell Biology* (1984) 99:248–259, by copyright permission of the Rockefeller University Press.

Table 4.2 Addition of Sialic Acid to G Protein after Fusion of VSV-Infected Clone 1021 Cells to Uninfected Clone 15B Cells

Protocol	^{3}H incorporated into	[^{3}H]G protein sialylated[a] (percent of total)
(a) Pulse, fuse	GlcNAc	10.3 ± 0.5 (7)
(b) Pulse, fuse	Gal	3.7 ± 0.1 (4)
(c) Pulse, don't fuse	GlcNAc	1.6 ± 0.2 (2)
(d) Pulse, don't fuse	Gal	1.9 ± 0.8 (2)
(e) Pulse, fuse to clone 1021	GlcNAc	0.8 ± 0.3 (2)
(f) Fuse, then pulse ("prefuse")	GlcNAc	32.3 ± 1.9 (4)
(g) Fuse, then pulse ("prefuse")	Gal	30.3 (1)

Source: Rothman et al. (1984b). Reproduced from *The Journal of Cell Biology* (1984) 99:260–271, by copyright permission of the Rockefeller University Press.

[a]Number of samples is shown in parentheses.

protein that is sialylated. After pulsing and fusion, the cells transfer sialic acid to the GlcNAc-labeled proteins [(row a) in Table 4.2] to a greater extent than to the Gal-labeled proteins (row b). The results in row b are comparable to those obtained without fusion (rows c and d) or with fusion of two sets of cells not competent to add sialic acid (row e, 1021 cells fused to 1021 cells). However, when the cells were already fused when pulsed with either Gal or GlcNAc, the efficiency of the two processes was the same, so the results cannot be explained by a difference in efficiency between the attachment of galactose and that of sialic acid. The results show that galactose is added before sialic acid vectorially. Furthermore, the rate of transfer between two distinct Golgi systems is the same as that occurring in intact cells (Rothman et al., 1984a). This indicates that physical proximity is not a significant factor. The mechanism of the transfer most likely to explain these results involves vesicle formation where the vesicles are the vehicles for the sequential transfer from one Golgi stack to another.

Experiments with cell-free systems of CHO cells provide similar evidence. When stacks from the mutant CHO cells not containing UDP-GlcNAc transferase I and infected with VSV are complemented with those of the wild type (Balch et al., 1984a; Braell et al., 1984; Fries and Rothman, 1980) the G protein is appropriately processed. Therefore, the G protein from the cis compartment of mutants must have been transferred to the medial compartment. The transport requires the presence of cytosol extract, ATP, and protein in the surface of the Golgi (Balch and Rothman, 1985; Balch et al., 1984a; Fries and Rothman, 1980). In general, intercompartment transport of G protein requires sequential steps probably involved in budding and fusion (Balch et al., 1984b; Wattenberg et al., 1986). ATP is required for each step, and cytosol is required for all except the last.

Proteins Involved in Transport, the G Proteins The use of yeast temperature-sensitive mutants together with the study of mammalian cell-free systems has allowed the dissection of each stage of transfer in the secretory pathway of transport. At least 12 proteins needed for the secretory pathway have been recognized, and several genes have been identified and studied in yeast (see Balch, 1990). The SEC genes have been found to be involved primarily in vesicle formation or alternatively in vesicle fusion to membranes. For example, one of them (SEC 7) is involved in intra-Golgi transfers, another (SEC 4) in targeting and fusion to the plasma membrane. However, each step requires the coordinate functioning of more than one protein.

In yeast, temperature-sensitive secretory mutants were selected. Cells blocked in the later stages of transport could be selected because of their higher density at the restrictive temperature (e.g., Novick et al., 1980). Cells blocked in the early stages could be selected by growing the cells in ^{3}[H] mannose. Since the glycoside portion of the glycoproteins is added early in the sequence of reactions, cells defective in the earlier steps would have a greater chance of survival than those blocked in later steps (Newman et al., 1987). Mutants were found to be blocked at four different steps: entry into the ER, leaving the ER, entering the Golgi apparatus, and finally, between the Golgi and exocytosis. Several genes were identified from the ability of a vector containing the appropriate sector of yeast wild-type DNA to reestablish function. The DNA of the gene could then be cloned and sequenced and the protein identified (e.g., Salminen and Novick, 1987).

These steps in intracellular transport were found to require a novel group of GTP-binding proteins coded by genes belonging to the *ras* family, the G proteins (e.g., see Hall, 1990). The *ras* genes are oncogenes that were discussed in Chapter 3, section III.

In mammalian cells the G-proteins were studied using partially disrupted cells and blocking the transport with GTPγS (e.g., Beckers and Balch, 1989). These results implicated the same steps as the genetic studies and found Ca^{2+} as a necessary factor. In addition, synthetic polypeptides homologous to the domain thought to correspond to the effector domain of the *ras* proteins were found to block the transport from the ER to the Golgi (Plutner et al., 1990).

In mammalian systems an *N-ethylmaleimide-sensitive fusion protein* (NSF) (e.g., Beckers et al., 1989), a homotetramer of 76 kDa subunits, is required for the fusion of vesicles in the transport from ER to Golgi and intra-Golgi transport. NSF is dissolved in the cytoplasm or attached to the Golgi as a peripheral protein. Release of the vesicles from the membranes requires the hydrolysis of ATP. The binding depends on the presence of an integral membrane receptor and other soluble proteins.

Although an outline of the molecular events of transport is beginning to emerge, at this time little is known about the mechanism of action of the various proteins.

Retention of Resident Proteins in the Golgi Stacks The individual components of the Golgi system differ in composition, including the lipid elements (see e.g., Machamer, 1991). The characteristic composition of the Golgi stacks could be accounted for in part as the result of a steady state of products flowing through the system. However, at least the enzymatic machinery, that is, the glycosidases and glycosyl transferases and some of the structural components (such as the scaffolding thought to maintain their shape) must be retained by some special mechanism. All the resident proteins of the Golgi examined so far have been found to be associated with the membranes and not in solution in the lumen.

Retention has been studied using proteins of the cis, trans, or TGN compartments. At least in the cases analyzed, the key to retention appears to be a signal sequence with polar amino acids in transmembrane domains. An avian coronovirus, infectious bronchitis virus (IBV) has been used to define the retention signal of the cis compartment. When expressed from cDNA in animal cells, this glycoprotein is trapped in the cis Golgi membranes (Machamer et al., 1990). IBV E1 is an integral protein which spans the membrane three times; the retention information is thought to be in the first span (Machamer and Rose, 1987). The uncharged polar residues Asn, Thr, and Gln, appear to be the significant feature of the retention signal, and when added to proteins normally transferred to the plasma membrane, they remain in the cis Golgi (Swift and Machamer, 1991).

The targeting signal of endogenous trans Golgi proteins, glycosyltransferases, and glycosidases was studied through examination of the cDNAs. All of these proteins have the N-terminus in the cytoplasm and an uncleaved signal sequence as a membrane spanning domain (e.g., Paulson and Colley, 1989). This spanning domain is needed for retention. If substituted with a cleavable signal sequence, the protein is secreted.

Therefore, both cis and trans proteins seem to be retained by a signal which spans the membrane. In contrast, early indications are that the TGN requires the cytoplasmic C-terminus (Luzio et al., 1990).

These findings indicate that unlike the retention signals of the ER, which pertain to soluble proteins, the retention sequences of the Golgi are retained in the membrane, suggesting that there are specific receptor sites in the membrane, possibly proteins or perhaps even lipid microenvironments.

Bulk Flow and Sorting Signals The transported proteins share the same pathway until their final packaging in the appropriate vesicle. What determines their destination? The different proteins could be coded by separate *sorting signals* analogous to the signal peptides initially responsible for the entry of the protein in the ER (Kelly, 1983). Each sorting signal could then mark the destination of the protein. Alternatively, all proteins not containing a sorting signal could be transported through the various compartments culminating with arrival at the cell surface. A sorting signal would control a destination to lysosomes or storage secretory vesicles. Recognition of the signal is likely to require interaction between the signal (i.e., some specific sequence or configuration of the protein molecule) and the receptor. Therefore the packaging system should be saturable and, when overproduced, even a protein containing a signal would follow the nonselective route or *bulk flow* (Pfeiffer and Rothman, 1987).

Evidence from studies of a variety of cells supports a mechanism involving bulk flow, with a sorting signal only for proteins destined for storage in lysosomes or secretory vesicles. In yeast, overproduction of the vacuolar enzyme carboxypeptidase Y (CPY) results in constitutive secretion to the periplasmic space. The vacuolar compartment corresponds to the lysosomes of other cells, although, in contrast to lysosomes (see below), the recognition signal is not the carbohydrate portion of the glycoprotein. Before reaching the vacuole, CPY is present in the cell as proCPY, since processing to the mature form of the enzyme takes place in the vacuole. In experiments testing the effect of overproduction, the CPY structural gene (PRC1) was cloned to produce plasmids containing multiple copies. The cloned portion of the genome is shown in Fig. 4.17*a* (Stevens et al., 1986). After a pulse-chase with $[^{35}S]SO_4^{2-}$, the protein was precipitated with CPY antibody and fractionated. The results of SDS-polyacrylamide gel radioautograph are shown in Fig. 4.17*b*. Lane I corresponds to internal CPY, presumably vacuolar. Lanes P and M correspond, respectively, to protein released from the periplasm or present in the medium. The results are summarized in Table 4.3. The increased gene dosage overproduced CPY (in the form of proCPY), presumably saturating the processing system and resulting in constitutive secretion. This constitutive secretion is blocked when a late step in the process in inhibited (Patzak and Winkler, 1986). Furthermore, deleting the NH_2 terminus of proCPY, which in this case is a sorting signal, leads to constitutive secretion (Johnson et al., 1987; Valls et al., 1986).

Fusion of sequences coding for the NH_2-terminal portion of CPY to the gene coding for the secretory enzyme invertase (Inv) has allowed study of the sorting signal (Johnson et al., 1987). The results are summarized in Fig. 4.18. Three types of combinations are shown. In class I, the CPY-Inv hybrid proteins are all delivered to the vacuole. Class II hybrids have less than the needed sequence and are partially or totally secreted. Class III hybrids, with only a small portion of the signal sequence remaining, remain in the cytoplasm.

Fifty amino acids of the CPY NH_2 terminal appear to be sufficient to code for the transfer. Twenty of these correspond to the signal peptide, so the 30 remaining are likely to contain the vacuolar sorting signal. In fact, deletion of this segment in the CPY gene produces missorting.

The results obtained with animal cells are in agreement with those obtained with yeast. With bulk flow, all proteins going to the surface would share a common mechanism and should be transferred at the same rate, faster than other proteins. In fact, some secretory proteins traverse the Golgi system at the same high rate (Lodish et al., 1983). The model predicts that

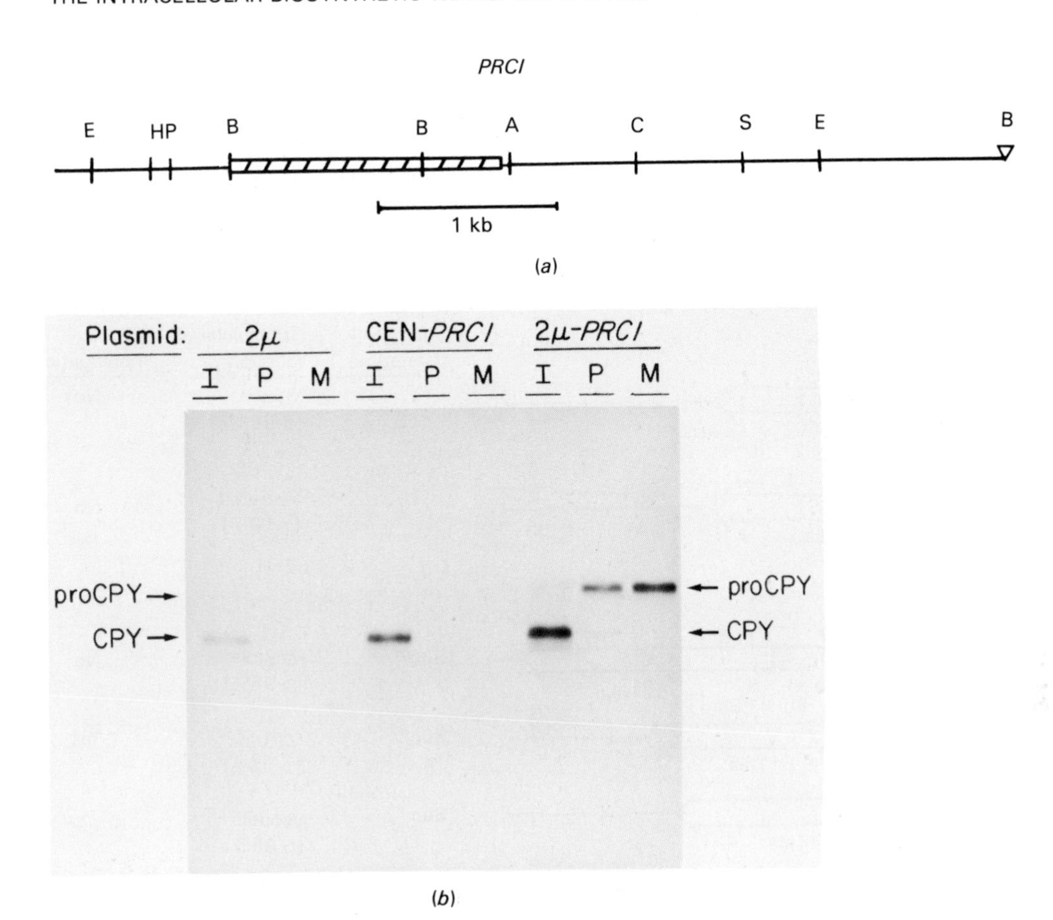

Fig. 4.17 (*a*) *PRC1* insert cloned from the YEp13 yeast library. The 2.6-kb II-*Cla*I fragment is the smallest subclone that complements PRC 1–1 mutation. A, *Acc*I; B, *Bam*HI; C, *Cla*I; E, *Eco*RI; H *Hin*dIII; P, *Pvu*II; S, *Sal*I. The sequence encoding CPY is represented by the hatched box. (*b*) Secretion of CPY in a strain carrying *PRC1*-containing plasmids. Reproduced from the *Journal of Cell Biology,* 102:1551–1557, 1986, by copyright permission of the Rockefeller University Press.

Table 4.3 Overproduction of CPY Results in Secretion of the Protein

Plasmid	Estimated *PRC1* gene copy number	Relative CPY synthesis	CPY secreted (%)
2μ	1	1	6
CEN-*PRC1*	2	2	12
2μ-*PRC1*	>5	6–8	50–55

Reproduced from the *Journal of Cell Biology,* 102:1551–1557, 1986, by copyright permission of the Rockefeller University Press.

any inert substance should be transported at the higher rate. This premise was tested with a synthetic acyl tripeptide labeled with ^{125}I, in which the acyl moiety corresponded to fatty acid chains ranging from 2 to 10 carbons (Wieland et al., 1987). The acyl group is added to impart lipid solubility to the molecule and facilitate its diffusion through cell membranes (see Chapter 2). CHO or hepatoma cells in culture were used. The tripeptide contained the Asn-X-Ser terminal, which is glycosylated by the ER.

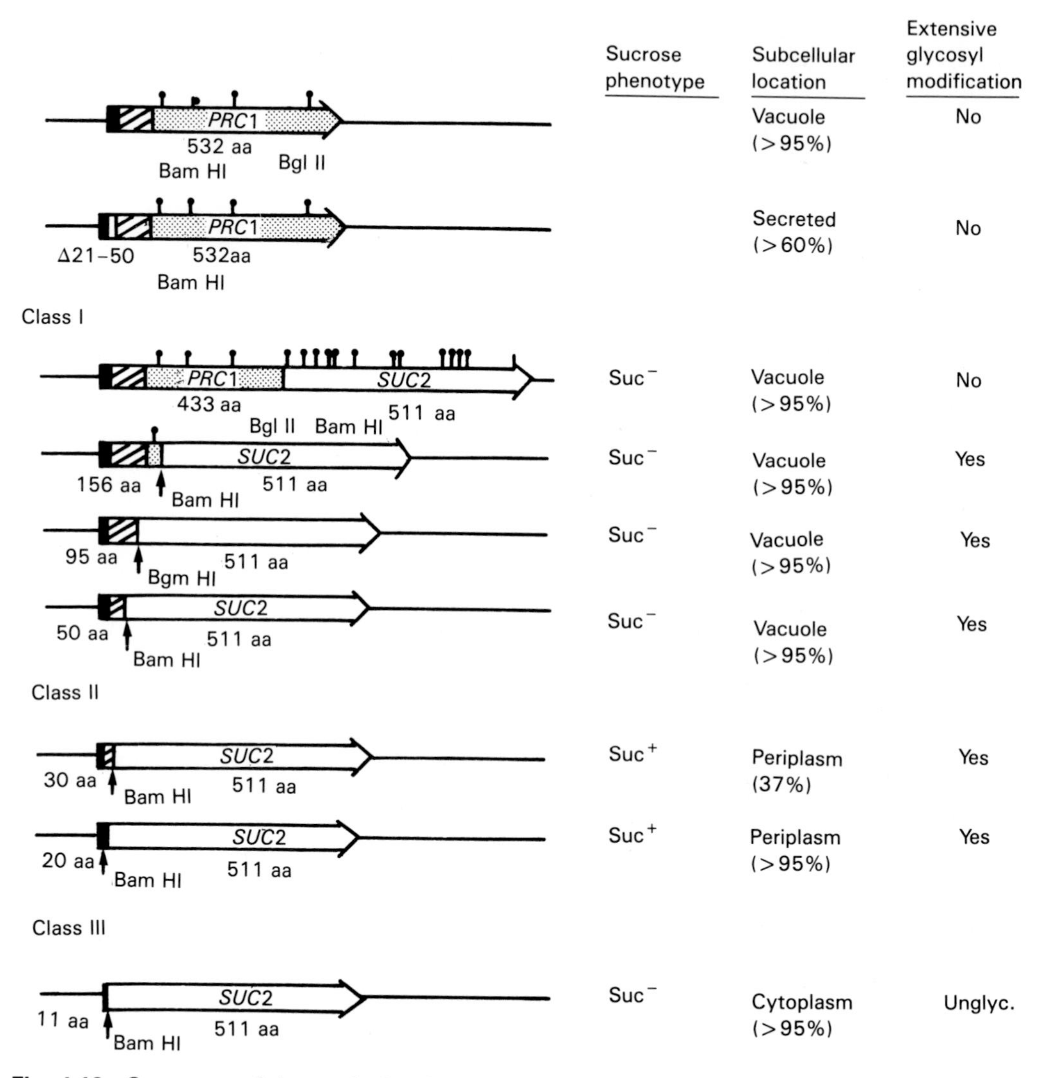

Fig. 4.18 Summary of the analysis of CPY-Inv hybrid proteins. At the top is a schematic representation of the gene encoding CPY (*PRC1*). The signal sequence is indicated by a solid black box, the pro-segment by hatched lines, and the mature sequences by dots. Four oligosaccharide addition sites present in CPY are shown by lollipops. The deletion of amino acids 21 to 50 in *PRC1* is shown immediately below the wild-type *PRC1* gene. All the gene fusions contain the same *SUC2* gene fragment, though only the first fusion shows the position of the 13 predicted invertase oligosaccharide addition sites. Reprinted with permission from Johnson et al. (1987), copyright 1987 by Cell Press.

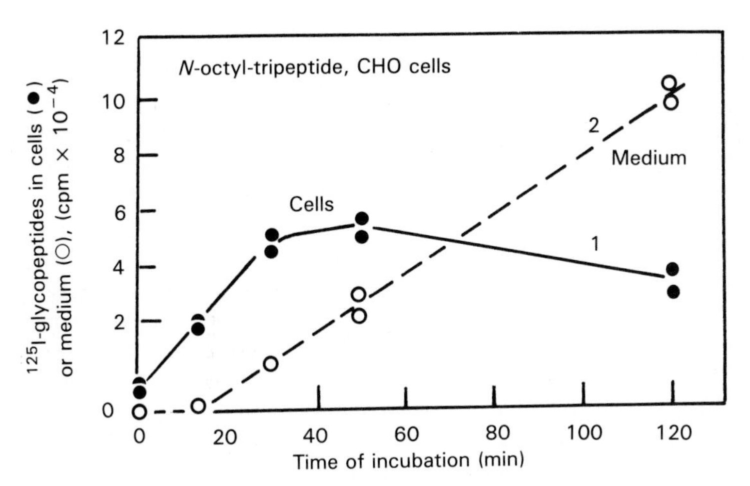

Fig. 4.19 Rate of appearance of ^{125}I-labeled octylglycotripeptides in CHO cells and their medium. Reprinted with permission from Wieland et al. (1987), copyright 1987 by Cell Press.

Figure 4.19 shows the appearance of radioactivity in the cell and in the medium after extraction. To ensure that the radioactivity was not from a breakdown product, the glycosylated tripeptide was first selected by binding to the lectin concanavalin A, conjugated to Sepharose beads. As shown in curve 1, the cells immediately proceeded to glycosylate the peptide without a lag period, indicating that the acyl tripeptides diffused freely into the cell and were trapped by glycosylation. The delay in the appearance of label in the medium, shown in curve 2, corresponds to the time for the transport of the glycopeptide to the surface. Figure 4.20 shows

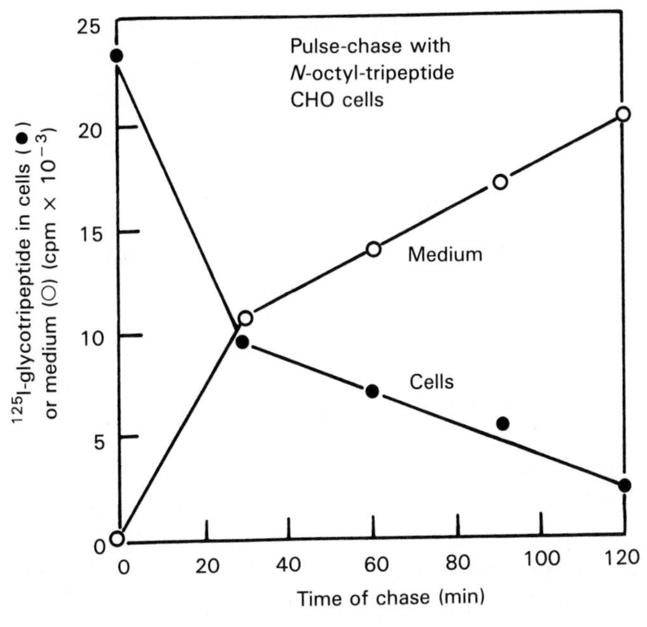

Fig. 4.20 Secretion of ^{125}I-labeled octylglycopeptides from CHO cells in a pulse-chase experiment. Reprinted with permission from Wieland et al. (1987), copyright 1987 by Cell Press.

the results obtained using the same technique with a long pulse (1 h) followed by a chase, corrected for the radioactivity present in the medium at zero time. This experiment also agrees with very fast transport. The half-time of the transport from the ER to the cell surface was less than 20 min, a rate comparable to or higher than that observed for other secreted proteins (Lodish et al., 1983). The low molecular weight of the tripeptide and the fact that oligosaccharides do not serve as signals in this transport pathway make it unlikely that it contained a sorting signal.

In animal cells, as in yeast, overproduction of proteins that require packaging, in this case in lysosomes and storage secretory vesicles, produces a spill of the excess through the bulk flow pathway. For example, constitutive secretion of the lysosomal enzymes takes place in I cells, cultured animal cells with a lysosomal storage disease (see below), which cannot tag lysosomal hydrolases with mannose-6-phosphate (M6P) or in which the M6P receptor is deficient (Gonzalez-Noriega et al., 1980; Hasilik and Neufeld, 1980). Similarly, treatment of AtT-20 cells, a line derived from pituitary cells, with chloroquinone, which blocks the storage route, causes constitutive secretion of the ACTH precursor (Moore et al., 1983). Integral membrane proteins and constitutively secreted proteins, presumably transported by bulk transport, are in the same vesicle (Strous et al., 1983), and in yeast the two are biochemically linked (Holcomb et al., 1987) so that a mutation blocking secretion also blocks membrane growth (Tschopp et al., 1984).

In summary, the evidence suggests that for constitutive secretion or the transport of plasma membrane proteins, there is no signal for the transfer from the Golgi to the cell surface and the transport proceeds through a nonspecific bulk pathway. Proteins destined for lysosomes or storage secretory vesicles must therefore possess a sorting signal (see later sections) analogous to the yeast sorting sequence, which has been demonstrated in the case of a protein targeted to the vacuole.

Transport to the Cell Surface Transport to the cell surface has been studied with a cell-free system from cells infected with influenza virus (Woodman and Edwardson, 1986). In this study the transfer of the viral neuraminidase to the plasma membrane was followed. To detect its arrival, an acceptor fraction was prepared by binding [^{3}H]sialic acid-labeled Semliki Forest virus to cell surfaces before homogenization. The observed production of free [^{3}H]sialic acid served as an assay of the fusion of exocytotic vesicles containing the enzyme with the labeled acceptor preparation. This reaction requires ATP and a variety of proteins.

C. Formation of Lysosomes and Secretory Storage Vesicles

Lysosomes The study of the transport system in the formation of lysosomes received great impetus from the recognition of at least 30 human lysosomal storage disorders. In one of these, I-cell disease, a deficiency in lysosomal enzymes results from a failure in the recognition marker needed for targeting. Study of the I-cell mutant cells has permitted the identification of the recognition marker, M6P residues, and the M6P receptor, MPR, a 215-kDa protein (Sahagian et al., 1981). The receptor spans the membrane, and 10 kDa of its COOH terminal protrudes into the cytosol (Sahagian and Steer, 1985).

Lysosomal enzyme precursors seem to be transported through the common pathway, as indicated by immunocytochemical experiments (Geutze et al., 1984). However, processing continues on arrival in the cis Golgi (see Fig. 4.21), where lysosomal hydrolases are recognized by GlcNAc-phosphotransferase, which adds GlcNAc-phosphate to α-1,2-mannose residues of the hydrolases. The phosphotransferase probably recognizes a signal patch, since recognition is very sensitive to conformation changes. After this modification, M6P residues are exposed by removal of the *N*-acetylglucosamine and are recognized by MPR. The interaction between receptors and M6P residues is responsible for the lysosomal enzyme segregation (Kornfeld,

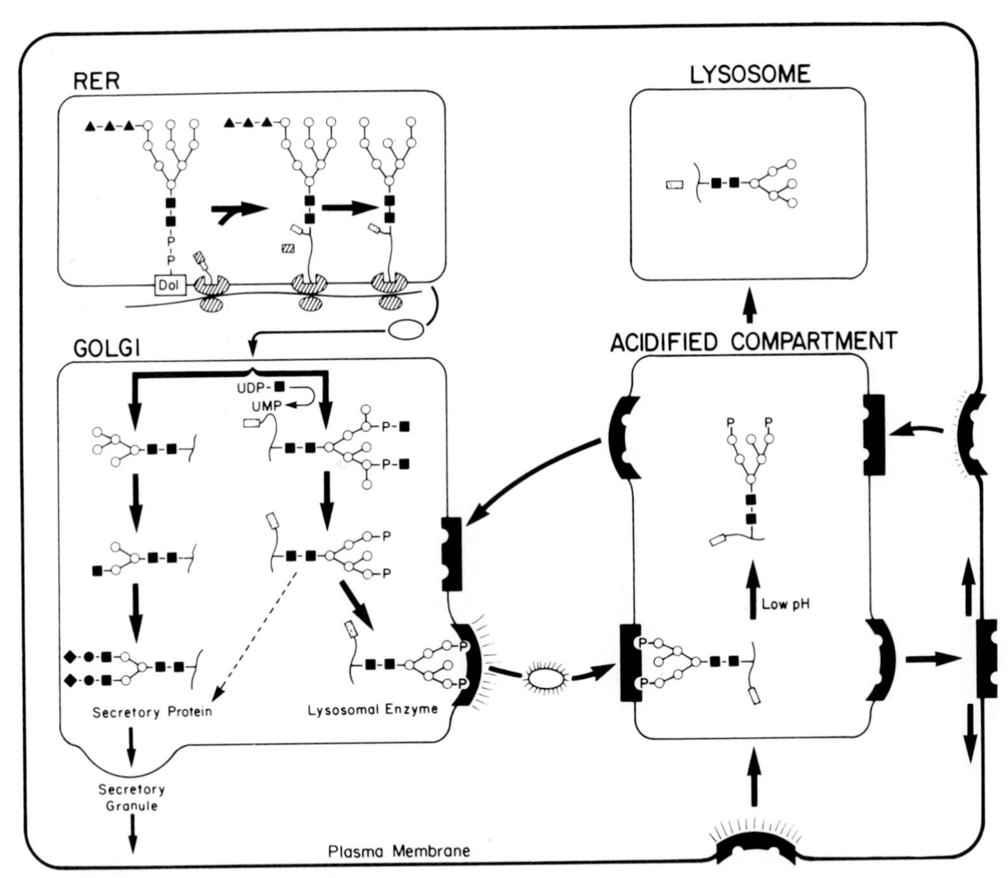

Fig. 4.21 Schematic pathway of lysosomal enzyme targeting to lysosomes. Lysosomal enzymes and secretory proteins are synthesized in the rough endoplasmic reticulum (RER) and glycosylated by the transfer of a preformed oligosaccharide from dolichol-P-P-oligosaccharide (Dol). In the RER, the signal peptides (hatching) are excised. The proteins are translocated to the Golgi, where the oligosaccharides of secretory proteins are processed to complex-type units and the oligosaccharides of lysosomal enzymes are phosphorylated. Most of the lysosomal enzymes bind to mannose 6-phosphate receptors (MPRs) ( ) and are translocated to an acidified prelysosomal compartment where the ligand dissociates. The receptors recycle back to the Golgi or to the cell surface, and the enzymes are packaged into lysosomes where cleavage of their propieces is completed ( ). The P_i may also be cleaved from the mannose residues. A small number of the lysosomal enzymes fail to bind to the receptors and are secreted along with secretory proteins (---→). These enzymes may bind to surface MPRs in coated pits ( ) and be internalized into the prelysosomal compartment. (■) *N*-Acetylglucosamine; (○) mannose; (▲) glucose; (●) galactose; (◆) sialic acid. Reprinted with permission from S. Kornfeld, *Federation of American Societies for Experimental Biology Journal,* Vol. 1, No. 6: 463, 1987.

1987; von Figura and Hasilik, 1986). The two can be detected together in buds and coated vesicles in the TGN (Geutze et al., 1985; Griffiths et al., 1985) which eventually form lysosomes.

In addition to the processing of the oligosaccharides, lysosomal enzymes are proteolytically cleaved to their mature form (Gieselman et al., 1983).

Before the primary lysosomes are formed, the MPRs are recycled to the trans Golgi (Kornfeld, 1987; von Figura and Hasilik, 1986). Recovery of receptor probably follows its dissociation from the ligands brought about by lowering the pH, as in the case of endocytotic vesicles discussed in Chapter 3. A small portion of the lysosomal enzymes are secreted and are thought to be recovered by endocytotic uptake after binding MPRs present at the surface (Willingham et al., 1981).

The current view of the production and segregation of lysosomal enzymes is shown in Fig. 4.21.

Secretory Storage Granules The mechanism for selection of the contents of the secretory storage vesicles which occurs in the trans stacks is not known; however, the process is very discriminating, as shown in cells which exclude a portion of the secretory proteins they produce from some of the vesicles. On the other hand, they can package a variety of different proteins in some vesicle, suggesting that they recognize a signal patch rather than a specific amino acid sequence. A process of self-aggregation of the proteins that takes place in the trans Golgi could be important in this recognition.

Other experiments using cDNA encoding for trypsinogen show that a sorting signal sequence resembling a conventional signal sequence is not involved (Burgess et al., 1987). Trypsinogen is a regulated exocrine secretion of the pancreas. In these experiments the cDNA was introduced in AtT-20 cells, anterior pituitary cells that secrete ACTH, also a regulated endocrine product. The trypsinogen was found sorted in the same vesicles as the ACTH, as shown by immunoelectron microscopy with gold particles of two different sizes, each size labeling one of the antibodies. Replacing the trypsinogen signal peptide with that of immunoglobulin κ light chain, a constitutively secreted protein, or eliminating 12 amino acids from the NH_2 terminal did not alter the pathway.

Various experiments suggest a pH-dependent receptor-ligand dissociation and recycling of receptor (Wagner et al., 1986), as examined in the case of lysosomes (see above) and receptors involved in endocytosis (Chapter 3).

As already discussed, vesicles are produced by budding of Golgi cisternae. Segregation of secretory products or lysosomal enzymes in specific domains in the cisternae could be accomplished if the receptors were integral proteins segregated in some membrane domain. Budding from these membrane domains could result in specific packaging. We saw that MPR, an integral protein, binds lysosomal enzymes. A similar mechanism could take place with storage secretory vesicles, and this view is supported by the demonstration that the precursor of insulin, proinsulin, is bound to Golgi membranes (Munro and Pellham, 1986).

In contrast to the normal lysosomal pathway, the storage secretory vesicles are discharged to the outside by exocytosis on receipt of a physiological signal (for example, a neurohormone).

D. Recognition of Cell Surfaces and Transcytosis

In an organized tissue many cells are polarized—that is, the cytoplasm and the cell surface have specialized functions—and the polarity is frequently preserved when the cells are cultured on a solid medium. The acinar cells of the pancreas (shown, for example, in Fig. 4.1) discharge the regulated secretory granules primarily on their apical side, not at the surfaces facing other cells (the basolateral sides) or the surface in contact with the basal lamina. Epithelial cells in general have a similar polarity; the apical and basolateral surfaces of polarized epithelial cells have distinct protein compositions forming membrane domains (Simons and Fuller, 1985). Accordingly, newly synthesized integral membrane proteins are targeted to specific cell surface domains (Fuller et al., 1985; Rindler et al., 1984). The most convenient model system for these studies is virally infected Madine-Darby canine kidney (MDCK) cells. The viral coat proteins share the transport pathway through the Golgi system, as shown by immunological EM methods using colloidal gold markers of different sizes in doubly infected cells (Rindler et al., 1984). Simian virus 5 (SV5) and influenza hemagglutinin are targeted to the apical surface, whereas VSV is directed to the basolateral surface. The delivery of the G protein of VSV to the basolateral surface (Pfeiffer et al., 1985) and of the hemagglutinin of influenza virus to the apical surface is direct (Matlin and Simon, 1984; Matlin et al., 1983). For the hemagglutinin, this has been shown (Matlin and Simon, 1984) by growing cells on Millipore filters, which makes the cells accessible to treatment at either apical or basolateral sides. The location of the proteins was recognized by their sensitivity to trypsin or binding to antibodies. In addition, in the case of the apical transfer of hemagglutinin, continuous exposure of the basolateral surfaces to either trypsin or antibody did not interfere with the appearance of the markers in the apical surface, evidence that the proteins arrive directly at their target and are not initially distributed at random or to the opposite surface.

The G protein accumulated in the cytoplasm when the transport was arrested by placing the cells at 20°C (Pfeiffer et al., 1985). Despite this accumulation, when the transport was reactivated by raising the temperature to 37°C, the protein appeared mostly on the basolateral surface. If the protein had to be transferred in two steps, arriving first at the apical surface, the procedure would also be expected to increase the G protein at the apical surface.

In rat hepatocytes the sorting mechanism is different (Bartles et al., 1987). Hepatocytes are also polar, and newly synthesized membrane proteins are also sorted to either an apical (bile canicular) or a basolateral domain. The destination of the targeted proteins was followed by pulse labeling them with [^{35}S]methionine injected in rats. After isolation of various cellular fractions, several membrane proteins were estimated by immunoprecipitation. When vesicles formed from basolateral membranes were isolated after a 45-min chase, they were found to contain both apical and basolateral (sinusoidal) proteins. However, with longer chase periods, the apical proteins began to appear in vesicle preparations that originated from the apical membrane. These experiments suggest that in hepatocytes both kinds of proteins are first delivered to the basolateral surface and the apical proteins are subsequently rerouted to the apical domain.

Although some cells release secretory products around their perimeter, as mentioned above, many polar cells release them only in specific regions of the membranes. Thus the same kind of targeting has to be conserved in secretion. In the MDCK cells, constitutively secreted endogenous proteins are released in the apical domain (Gottlieb et al., 1986; Kondor-Koch, 1985), suggesting that this is the destination of the bulk flow.

A role of the cell adhesion molecule, uvomorulin (a protein of the cadherin family, see Chapter 3, V, B), in the distribution of Na^+, K^+-ATPase to the basolateral surface of epithelial cells has been suggested (Nelson and Hammerton, 1989).

Some cells carry out receptor-mediated endocytosis (Chapter 3), in which the ligand is generally degraded in the lysosomes and the receptor is either degraded or recycled to the surface. However, in *transcytosis,* which is ubiquitous in epithelial cells, the endosomes transfer receptor and ligand to the surface of opposite polarity, so that the material traverses the cell.

The transcytosis of polymeric immunoglobulin (poly-Ig) is probably the best understood of the various known cases. Poly-Ig is produced by plasma cells and transported through epithelial cells by transcytosis. In this process, the epithelial cell adds a polypeptide, the secretory component (SC) or secretory piece, part of the receptor molecule, to the poly-Ig. The poly-Ig receptor is originally incorporated in the basolateral surface, where it binds poly-Ig. After transcytosis, the endocytotic vesicle discharges its contents at the apical surface by exocytosis and the receptor is cleaved.

Despite the flow of membranes from one pole of the cell to the other, the two surface domains remain distinct. One way in which this could happen is illustrated by G protein, which is normally in the basolateral surface of the infected MDCK cells. When artificially inserted into the apical surface by fusing it to viral coats at low pH, the G protein is endocytosed and delivered to the basolateral surface (Matlin and Simon, 1984).

E. Recycling of the Plasma Membrane

The process of exocytosis adds a considerable amount of material to the plasma membrane (Tajima et al., 1986). Much of this material is recycled through endocytosis, so the turnover time of the membranes of granules is much longer than that of soluble proteins (Meldolesi, 1974).

The availability of antibodies that bind to the luminal side of the secretory vesicle membrane has permitted the demonstration of exocytosis and the fate of the vesicles during recycling (Patzak and Winkler, 1986) in the regulated secretion of catecholamine granules. Glycoprotein III (gpIII) appears in coated pits and vesicles in the first 5 min after exocytosis. Then it passes through the smooth ER and reappears in the trans Golgi network and in dense-core secretory granules within 30 to 45 min. The protein was never found in the cisternal lumen, indicating that the membrane itself is being recycled.

SUGGESTED READING

Balch, W. E. (1990) Molecular dissection of early stages of the eukaryotic secretory pathway. *Current Opin. Cell Biol.* 2:634–641.

Dunphy, W. G., and Rothman, J. E. (1985) Compartmental organization of Golgi stacks. *Cell* 42:13–21.

Gething, M.-J., and Sambrook, J. (1990) Transport and assembly processes in the endoplasmic reticulum. *Seminars Cell Biol.* 1:65–72.

Griffiths, G., and Simone, K. (1986) The trans Golgi network: sorting at the exit site of the Golgi complex. *Science* 234:438–443.

Kelly, R. B. (1983) Pathways of protein secretion in eukaryotes. *Science* 230:25–32.

Kornfeld, S. (1987) Trafficking of lysosomal enzymes. *FASEB J.* 1:462–468.

Lodish, H. F. (1988) Transport of secretory and membrane glycoproteins from the rough endoplasmic reticulum to the Golgi: a rate limiting step in protein maturation and secretion. *J. Biol. Chem.* 263:2107–2110.

Morris, S. A., Ahle, S., and Ungewickell, E. (1989) Clathrin coated vesicles. *Current Opin. Cell Biol.* 1:684–690.

Mostov, K. E., and Simister, N. E. (1985) Transcytosis. *Cell* 43:389–390.

Rapoport (1991) Protein transport across the endoplasmic reticulum: facts, models, mysteries. *FASEB J.* 5:2792–2798.

Verner, K., and Schatz, G. (1988) Protein translocation across membranes. *Science* 241:1307–1313.

Walter, P., Gilmore, R., and Blobel, G. (1984) Protein translocation across the endoplasmic reticulum. *Cell* 38:5–8.

Waters, M. G., Serafini, T., and Rothman, J. E. (1991) 'Coatamer': a cytosolic protein complex containing subunits of non-clathrin-coated Golgi transport vesicles. *Nature* 349:248–251.

Minireviews

Balch, W. E. (1989) Biochemistry of interorganelle transport. A new frontier in enzymology emerges from versatile *in vitro* model systems. *J. Biol. Chem.* 264:16965–16968.

Dahms, N., Lobel, P., and Kornfeld, S. (1989) Mannose 6-phosphate receptors and lysosomal enzyme targeting. *J. Biol. Chem.* 264:12115–12118.

Kelly, R. B. (1990) Microtubules, membrane traffic, and cell organization. *Cell* 61:5–7.

Pfeffer, S. R. (1992) GTP-binding proteins in intracellular transport. *Trends in Cell Biol.* 2:41–46.

Rothman, J. H., Yamashiro, C. T., Kane, P. T., and Stevens, T. H. (1989) Protein targeting to the yeast vacuole. *Trends Biochem. Sci.* 14:347–350.

Wilson, D. W., Whiteheart, S. W., Orci, L., and Rothman, J. E. (1991) Intracellular membrane fusion. *Trends in Biochem. Sci.* 16:334–337.

Recent Reviews

Fleischer, R. (1988) Functional topology of Golgi membranes. In *Protein Transfer and Organelle Biogenesis* (Das, R. C., and Robbins, P. W., eds.), pp. 289–317. Academic Press, New York.

Hanover, J. A., and Dickson, R. B. (1988) Organelles of endocytosis and exocytosis. In *Protein Transfer and Biogenesis* (Das, R. C., and Robbins, P. W., eds.), pp. 401–463. Academic Press, New York.

Moore, H. -P., Orci, L., and Oster, G. F. (1988) Biogenesis of secretory vesicles. In *Protein Transfer and Biogenesis* (Das, R. C., and Robbins, P. W., eds.), pp. 521–562. Academic Press, New York.

Parent, J. B. (1988) Role of carbohydrate in glycoprotein traffic and secretion. In *Protein Transfer and Organelle Biogenesis* (Das, R. C., and Robbins, P. W., eds.), pp. 51–108. Academic Press, New York.

Perara, E., and Lingappa, V. R. (1988) Transport of proteins into and across the endoplasmic reticulum membrane. In *Protein Transfer and Organelle Biogenesis* (Das, R. C., and Robbins, P. W., eds.), pp. 3–50. Academic Press, New York.

REFERENCES

Bacon, R. A., Salminen, A., Ruohola, H., Novick, P., and Ferro-Novick, S. (1989) The GTP binding protein YPT1 is required for transport *in vitro*: the Golgi apparatus is defective in ypt1 mutants. *J. Cell Biol.* 109:1015–1022.

Balch, W. E. (1990) Molecular dissection of early stages of the eukaryotic secetory pathway. *Cell Biol.* 2:634–641.

Balch, W. E., and Rothman, J. E. (1985) Characterization of protein transport between successive compartments of the Golgi apparatus: asymmetric properties of donor and acceptor activities in a cell-free system. *Arch. Biochem. Biophys.* 240:413–425.

Balch, W. E., Dunphy, W. G., Braell, W. A., and Rothman, J. E. (1984a) Reconstitution of the transport of protein between successive compartments of the Golgi measured by the coupled incorporation of *N*-acetylglucosamine. *Cell* 39:405–416.

Balch, W. E., Glick, B. S., and Rothman, J. E. (1984b) Sequential intermediates in the pathway of intercompartmental transport in a cell free system. *Cell* 39:525–536.

Balch, W. E., Wagner, R. R., and Keller, D. S. (1987) Reconstitution of transport vesicular stomatitis virus G protein from the endoplasmic reticulum to the Golgi complex using a cell free system. *J. Cell Biol.* 104:749–760.

Balch, W. E. (1990) Molecular dissection of early stages of the eukaryotic secretory pathway. *Curr. Opin. Cell Biol.* 2:634–641.

Bartles, J. R., Feracci, H. M., Stieger, B., and Hubbard, A. L. (1987) Biogenesis of the rat hepatocyte plasma membrane in vivo: comparison of the pathways taken by apical and basolateral proteins using subcellular fractionation. *J. Cell Biol.* 105:1241–1251.

Beckers, C. J. M., and Balch, W. E. (1989) Calcium and GTP: essential components in vesicular trafficking between the endoplasmic reticulum and Golgi apparatus. *J. Cell Biol.* 108:1245–1256.

Beckers, C. J. M., Block, M. R., Glick, B. S., Rothman, J. E., and Balch, W. E. (1989) Vesicular transport between the endoplasmic reticulum and the Golgi stack requires NEM-sensitive protein. *Nature* 339:397–398.

Bergmann, J. E., and Singer, S. J. (1983) Immunoelectron microscopic studies of the intracellular transport of the membrane glycoprotein (G) of vesicular stomatitis virus in infected Chinese hamster ovary cells. *J. Cell Biol.* 97:1777–1784.

Blobel, G., and Dobberstein, B. (1975a) Transfer of proteins across membranes. I. Presence of proteolytically processed and unprocessed nascent immunoglobulin light chains on membrane-bound ribosomes of murine myeloma. *J. Cell Biol.* 67:835–851.

Blobel, G., and Dobberstein, B. (1975b) Transfer of proteins across membranes. II. Reconstitution of functional rough microsomes from heterologous components. *J. Cell Biol.* 67:852–862.

Braell, W. A., Balch, W. E., Dobbertin, D. C., and Rothman, J. E. (1984) The glycoprotein that is transported between successive compartments of the Golgi in a cell free system resides in stacks of cisternae. *Cell* 39:511–524.

Breckenbridge, L. J., and Almers, W. (1987) Final steps in exocytosis observed in a cell with giant secretory granules. *Proc. Natl. Acad. Sci. U.S.A.* 84:1945–1949.

Brown, W. J., and Farquhar, M. G. (1984) The mannose-6-phosphate receptor for lysosomal enzymes is concentrated in the *cis* Golgi cisternae. *Cell* 36:295–307.

Burgess, T. L., Craik, C. S., Matsuuchi, C., and Kelly, R. B. (1987) In vitro mutagenesis of trypsinogen: the role of the amino terminus in intracellular protein targeting to secretory granules. *J. Cell Biol.* 105:659–668.

Campbell, C. H., Fine, R. E., Squicciarino, J., and Rome, L. H. (1983) Coated vesicles from rat and calf brain contain cryptic mannose-6-phosphate receptors. *J. Biol. Chem.* 258:3006–3011.

Cohn, Z. A., Fedorko, M. E., and Hirsch, J. C. (1966) The *in vitro* differentiation of mononuclear phagocytes. V. The formation of macrophage lysosomes. *J. Exp. Med.* 123:757–766.

Dean, N., and Pelham, H. R. B. (1990) Recycling of proteins from Golgi compartment to ER in yeast. *J. Cell Biol.* 111:369–377.

Duden, R., Allan, V., and Kreis, T. (1991a) Involvement of β-COP in membrane traffic through the Golgi complex. *Trends Cell Biol.* 1:14–19.

Duden, R., Griffiths, G., Frank, R., Argos, P., and Kreis, T. E. (1991b) β-COP, a 110 kd protein associated with non-clathrin-coated vesicles and Golgi complex, shows homology to β-adaptin. *Cell* 64:649–665.

Dunphy, W. G., Fries, E., Urbani, L. J., and Rothman, J. E. (1981) Early and late functions associated with the Golgi apparatus reside in distinct compartments. *Proc. Natl. Acad. Sci. U.S.A.* 78:7453–7457.

Dunphy, W. G., Brands, R., and Rothman, J. E. (1985) Attachment of terminal *N*-acetylglucosamine to asparagine-linked oligosaccharide occurs in central cisternae of Golgi stack. *Cell* 40:463–472.

Erickson, A. H., Walter, P., and Blobel, G. (1983) Translocation of a lysosomal enzyme across the microsomal membrane requires signal recognition particle. *Biochem. Biophys. Res. Commun.* 115:275–280.

Evans, E. A., Gilmore, R., and Blobel, G. (1986) Purification of a microsomal signal peptidase as a complex. *Proc. Natl. Acad. Sci. U.S.A.* 83:581–585.

Farquhar, M. G. (1985) Progress in unraveling pathways of Golgi traffic. *Annu. Rev. Cell Biol.* 1:447–488.

Fries, E., and Rothman, J. E. (1980) Transport of vesicular stomatitis virus glycoprotein in cell-free extract. *Proc. Natl. Acad. Sci. U.S.A.* 77:3870–3874.

Fries, E., Gustaffson, L., and Petersen, P. (1984) Four proteins synthesized by hepatocytes are transported from the endoplasmic reticulum to Golgi complex at different rates. *EMBO J.* 3:147–152.

Fuller, S. D., Bravo, R., and Simons, K. (1985) An enzymatic assay reveals that proteins destined for the apical or basolateral domains of an epithelial cell line share the same late Golgi compartments. *EMBO J.* 4:297–307.

Gething, M. -J., McCammon, K., and Sambrook, J. (1986) Expression of wild type and mutant forms of influenza hemagglutinin, the role of folding in intracellular transport. *Cell* 46:939–950.

Geutze, H. J., Slot, J. W., Strous, J. A. M., Hasilik, A., and von Figura, K. (1984) The ultrastructural localization of the mannose 6-phosphate receptor in rat liver. *J. Cell Biol.* 98:2047–2054.

Geutze, H. J., Slot, J. W., Strous, J. G., Hasilik, A., and von Figura, K. (1985) Possible pathway for lysosomal enzyme delivery. *J. Cell Biol.* 101:2253–2263.

Gieselman, V., Pohlmann, R., Hasilik, A., and von Figura, K. (1983) Biosynthesis and transport of cathepsin D in cultured human fibroblasts. *J. Cell Biol.* 97:1–5.

Gilmore, R., and Blobel, G. (1983) Transient involvement of signal recognition particle and its receptor in the microsomal membrane prior to the protein translocation. *Cell* 35:677–685.

Gilmore, R., Blobel, G., and Walter, P. (1982a) Protein translocation across the endoplasmic reticulum. I. Detection in the microsomal membrane of a receptor for the signal recognition particle. *J. Cell Biol.* 95:463–469.

Gilmore, T., Walter, P., and Blobel, G. (1982b) Protein translocation across the endoplasmic reticulum. II. Isolation and characterization of the signal recognition particle receptor. *J. Cell Biol.* 95:470–477.

Goldberg, D. E., and Kornfeld, S. (1983) Evidence for extensive subcellular organization of asparagine-linked oligosaccharide processing and lysosomal enzyme phosphorylation. *J. Biol. Chem.* 258:3159–3165.

Gonzalez-Noriega, A., Grubb, J. H., Talkad, V., and Sly, W. S. (1980) Chloroquine inhibits lysosomal enzyme pinocytosis and enhances lysosomal enzyme secretion by empairing receptor recycling. *J. Cell. Biol.* 85:839–852.

Gottlieb, T. A., Beaudry, G., Rizzolo, L., Coleman, A., Rindler, M., Adesnik, M., and Sabatini, D. D. (1986) Secretion of endogenous and exogenous proteins from polarized MDCK cell monolayers. *Proc. Natl. Acad. Sci. U.S.A.* 83:2100–2104.

Griffiths, G., and Simon, K. (1986) The trans Golgi network: sorting at the exit site of the Golgi complex. *Science* 234:438–443.

Griffiths, G., Pfeiffer, S., Simon, K., and Matlin, K. (1985) Exit of newly synthesized membrane proteins from the trans cisternae of the Golgi complex to the plasma membrane. *J. Cell Biol.* 101:949–964.

Hall, A. (1990) The cellular functions of small GTP-binding proteins. *Science* 249:635–640.

Hasilik, A., and Neufeld, E. F. (1980) Biosynthesis of lysosomal enzymes in fibroblasts. Synthesis as precursor of a higher molecular weight. *J. Biol. Chem.* 255:4937–4950.

Holcomb, C. L., Hansen, W. B., Etcheverry, T. E., and Shekman, R. (1987) Plasma membrane protein intermediates are present in secretory vesicles of yeast. *J. Cell Biochem. Suppl.* 11A:247.

Hortsch, M., Griffiths, G., and Meyer, D. I. (1985) Restriction of docking protein to the rough endoplasmic reticulum: immunochemical localization in rat liver. *Eur. J. Cell Biol.* 38:271–279.

Jamieson, J. D., and Palade, G. E. (1967) Intracellular transport of secretory proteins in the pancreatic exocrine cell. II. Transport to condensing vacuoles and zymogen granules. *J. Cell Biol.* 34:597–615.

Johnson, L. M., Bankaitis, V. A., and Emr, S. D. (1987) Distinct sequence determinants direct intracellular sorting and modification of yeast vacuolar protease. *Cell* 48:875–885.

Katz, F. N., Rothman, J. E., Lingappa, V. R., Blobel, G., and Lodish, H. F. (1977) Membrane assembly *in vitro:* synthesis, glycosylation, and asymmetric insertion of a transmembrane protein. *Proc. Natl. Acad. Sci. U.S.A.* 74:3278–3282.

Kelly, R. B. (1983) Pathways of protein secretion in eukaryotes. *Science* 230:25–32.

Kondor-Koch, C., Bravo, R., Fuller, S. D., and Garoff, H. (1985) Exocytotic pathways exist to both apical and basolateral cell surface of the polarized epithelial cell MDCK. *Cell* 43:297–306.

Kornfeld, S. (1987) Trafficking of lysosomal enzymes. *FASEB J.* 1:462–468.

Kreibich, G., Freinstein, C. M., Pereyra, B. N., Ulrich, B. L., and Sabatini, D. D. (1978) Proteins of the rough microsomal membranes related to ribosome binding. II. Cross linking of bound ribosomes to specific membrane proteins exposed at the binding sites. *J. Cell Biol.* 77:488–506.

Kreis, T. E. (1990) Role of microtubules in the organization of the Golgi apparatus. *Cell Motility Cytosk.* 15:67–70.

Kurzchalia, T. V., Wiedmann, M., Girshovich, A. S., Bochkareva, E. S., Bielka, H., and Rapoport, T. A. (1986) The signal sequence of nascent preprolactin interacts with the 54 K polypeptide of the signal recognition particle. *Nature* 320:634–636.

Lee, C., and Beckwith, J. (1986) Cotranslation and posttranslation protein translocation in prokaryotic systems. *Annu. Rev. Cell Biol.* 2:315–336.

Lingappa, V. R., Chaidez, J., Yost, C. S., and Hedgepeth, J. (1984) Determinants for protein localization: β-lactamase signal sequence directs globin across membranes. *Proc. Natl. Acad. Sci. U.S.A.* 81:456–460.

Lodish, H. F., Kong, N., Snider, M., and Strous, G. J. A. M. (1983) Hepatoma secretory proteins migrate from rough endoplasmic reticulum to Golgi at characteristic rates. *Nature* 304:80–83.

Long, R. M., Satoh, H., Martin, B. M., Kimura, S., Gonzalez, F. J., and Pohl, L. R. (1988) Rat liver carboxyesterase: cDNA cloning, sequencing and evidence for multigene family. *Biochem. Biophys. Re. Comm.* 156:866–873.

Luzio, J. P., Brake, B., Banting, G., Howell, K. E., Braghetta, B., and Stanley, K. K. (1990) Identification, sequencing and expression of an integral membrane protein of the trans Golgi network (TGN38). *Biochem. J.* 270:97–102.

Machamer, C. E. (1991) Golgi retention signals: do membranes hold the key? *Trends in Cell Biol.* 1:141–144.

Machamer, C. E., Mentone, S. A., Rose, J. K., and Farquhar, M. G. (1990) The E1 glycoprotein of an avian coronavirus is targeted to the *cis* Golgi complex. *Proc. Natl. Acad. Sci.* 87:6944–6948.

Machamer, C. E., and Rose, J. K. (1987) A specific transmembrane domain of coronavirus E1 glycoprotein is required for its retention in the Golgi region. *J. Cell Biol.* 105:1205–1214.

Malhotra, V., Serafini, T., Orci, L., Shepherd, J. C. and Rothman, J. E. (1989) Purification of a novel class of coated vesicles mediating biosynthetic protein transport through Golgi stacks. *Cell* 58:329–336.

Matlin, K., and Simon, K. (1984) Sorting of a plasma membrane glycoprotein occurs before it reaches the cell surface in cultured epithelial cells. *J. Cell Biol.* 99:2131–2139.

Matlin, K., Bainton, D. F., Pesonen, M., Louvard, D., Genty, N., and Simons, K. (1983) Transepithelial transport of viral membrane glycoprotein implanted into the apical plasma membrane of Madin-Darby canine kidney cells. I. Morphological evidence. *J. Cell Biol.* 97:627–637.

Meldolesi, J. (1974) Dynamics of cytoplasmic membranes in guinea pig pancreatic acinar cells. I. Synthesis and turnover of membrane proteins. *J. Cell Biol.* 61:1–13.

Meyer, D. I., and Dobberstein, B. (1980) A membrane component essential for vectorial translocation of nascent proteins across the endoplasmic reticulum: requirements for its extraction and reassociation with the membrane. *J. Cell Biol.* 87:498–502.

Meyer, D. I., Krause, E., and Dobberstein, B. (1982a) Secretory protein translocation across membranes—the role of the docking proteins. *Nature* 297:647–650.

Meyer, D. I., Louvard, D., and Dobberstein, B. (1982b) Characterization of molecules involved in protein translocation using a specific antibody. *J. Cell Biol.* 92:579–583.

Moore H. -P., Gumbiner, B., and Kelly, R. B. (1983) Chloroquine diverts ACTH from a regulated to a constitutive secretory pathway in AtT-20 cells. *Nature* 302:434–436.

Mueckler, M., and Lodish, H. F. (1986) The human glucose transporter can insert posttranslationally into microsomes. *Cell* 44:629–637.

Munro, S., and Pelham, H. R. B. (1986) An HSP70-like protein in the ER: identity with the 78 kd glucose-regulated protein and immunoglobulin heavy chain binding protein. *Cell* 46:291–300.

Munro, S., and Pelham, H. R. B. (1987) A C-terminal signal prevents secretion of luminal ER proteins. *Cell* 48:899–907.

Nelson, W. J., and Hammerton, R. W. (1989) A membrane-cytoskeleton complex containing Na^+, K^+-ATPase, ankyrin and fodrin in Madin-Darby canine kidney (MDCK) cells: implications from the biogenesis of epithelial cell polarity. *J. Cell Biol.* 108:893–902.

Newman, A. P., and Ferro-Novick, S. (1987) Characterization of new mutants in the early part of the yeast secretory pathway isolated by a ^{3}[H]mannose suicide selection. *J. Cell Biol.* 105:1587–1594.

Novick, P., Field, C., and Shekman, R. (1980) Identification of a 23 complementation group required for post-translational events in the yeast secretory pathway. *Cell* 21:205–215.

Orci, L., Halbam, P., Amherdt, M., Ravazzola, M., Vassali, J. -D., and Perrelet, A. (1984) A clathrin-coated, Golgi related compartment of the insulin secreting cell accumulates proinsulin in the presence of monensin. *Cell* 39:39–47.

Orci, L., Ravazzola, M., Amherdt, M., Louvard, D., and Perrelet, A. (1985) Clathrin-immunoreactive sites in the Golgi apparatus are concentrated at the trans pole in polypeptide hormone secreting cells. *Proc. Natl. Acad. Sci. U.S.A.* 82:5385–5389.

Orci, L., Glick, B. S., and Rothman, J. E. (1986) A new type of coated vesicular carrier that appears not to contain clathrin: its possible role in protein transport within the Golgi stacks. *Cell* 46:171–184.

Palmiter, R. D., Gagnon, J., and Walsh, K. A. (1978) Ovalbumin: a secreted protein without a transient hydrophobic leader sequence. *Proc. Natl. Acad. Sci. U.S.A.* 75:94–98.

Patzak, A., and Winkler, H. (1986) Exocytotic exposure and recycling of membrane antigens of chromaffin granules: ultrastructural evaluation after immunolabeling. *J. Cell Biol.* 102:510–515.

Paulson, J. C., and Colley, K. J. (1989) Glycosyltransferase. Structure, localization, and control of cell type specific-localization. *J. Biol. Chem.* 264:17615–17618.

Pearse, B. M. F., and Robinson, M. S. (1990) Clathrin, adaptors, and sorting. *Annu. Rev. Cell. Biol.* 6:151–172.

Pelham, H. R. B. (1988) Evidence that luminal ER proteins are sorted from secreted proteins in a post-ER compartment. *EMBO J.* 7:913–918.

Pelham, H. R. B. (1990) The retention signal for soluble proteins of the endoplasmic reticulum. *Trends in Biochem. Sci.* 15:483–486.

Pelham, H. R. B., Hardwick, K. G., and Lewis, M. J. (1988) Sorting of soluble ER proteins in yeast invertase fusion gene. *EMBO J.* 7:1757–1762.

Perara, E., Rothman, R. E., and Lingappa, V. R. (1986) Uncoupling translocation from translation: implications for transport of protein across membranes. *Science* 232:348–352.

Pfeiffer, S. R., and Rothman, J. E. (1987) Biosynthetic protein transport and sorting by the endoplasmic reticulum and Golgi. *Annu. Rev. Biochem.* 56:829–852.

Pfeiffer, S., Fuller, S. D., and Simons, K. (1985) Intracellular sorting and basolateral appearance of the G protein of vesicular stomatitis virus in Madin-Darby canine kidney cells. *J. Cell Biol.* 101:470–476.

Plutner, H., Schwaninger, R., Pind, S., and Balch, W. S. E. (1990) Synthetic peptides of the *rab* domain inhibit vesicular transport through the secretory pathway. *EMBO J.* 9:2375–2383.

Rapoport, T. A. (1991) Protein transport across the endoplasmic reticulum membrane; facts, models, mysteries. *FASEB J.* 5:2792–2798.

Rindler, M. J., Ivanov, I. E., Plesken, H., Rodriguez-Boulan, E., and Sabatini, D. D. (1984) Viral glycoproteins destined for apical or basolateral plasma membrane domains traverse the Golgi apparatus during the intracellular transport in doubly infected Madine-Darby canine kidney cells (MDCK). *J. Cell Biol.* 98:1304–1319.

Ronne, H., Ocklind, C., Wiman, K., Rask, L., Obring, B., and Peterson, P. A. (1983) Ligand dependent regulation of intracellular protein transport: effect of vitamin A on the secretion of the retinol binding protein. *J. Cell Biol.* 96:907–910.

Rothman, J. E. (1981) The Golgi apparatus: two organelles in tandem. *Science* 213:1212–1219.

Rothman, J. E., Urbani, L., and Brands, R. (1984a) Transport of protein between cytoplasmic membranes of fused cells: correspondence to process reconstituted in cell-free systems. *J. Cell Biol.* 99:248–259.

Rothman, J. E., Miller, R. L., and Urbani, L. J. (1984b) Intercompartmental transport in the Golgi complex is a dissociative process: facile transfer of membrane proteins between two Golgi populations. *J. Cell Biol.* 99:260–271.

Sahagian, G. G., and Steer, C. J. (1985) Transmembrane orientation of a mannose-6-phosphate receptor in isolated clathrin coated vesicles. *J. Biol. Chem.* 260:9838–9842.

Sahagian, G. G., Distler, J., and Jourdian, G. W. (1981) Characterization of a membrane-associated receptor from bovine liver that binds phosphomannosyl residues of bovine testicular β-galactosidase. *Proc. Natl. Acad. Sci. U.S.A.* 78:4289–4293.

Saliminen, A., and Novick, P. J. (1987) The *ras*-like protein is required for a post-Golgi event in yeast secretion. *Cell* 49:527–538.

Saraste, J., and Hedman, K. (1983) Intracellular vesicles in the transport of Semliki Forest virus membrane proteins to the cell surface. *EMBO J.* 2:2001–2006.

Scheele, G., and Tartakoff, A. (1985) Exit of non-glycosylated secretory proteins from the RER is asynchronous in exocrine pancreas. *J. Biol. Chem.* 260:926–931.

Schulze-Lohoff, E., Hasilik, A., and von Figura, K. (1985) Cathepsin D precursors in clathrin coated organelles from human fibroblasts. *J. Cell Biol.* 101:824–829.

Semenza, J. C., Hardwick, K. G., Dean, N., and Pelham, H. R. B. (1990) *ERD2*, a yeast gene required for the receptor mediated retrieval of luminar ER proteins from the secretory pathway. *Cell* 61:1349–1357.

Serafini, T., Stenbeck, G., Brecht, A., Lottspeich, F., Orci, L., Rothman, J. E., and Wieland, F. T. (1991) A coat subunit of Golgi-derived non-clathrin vesicles with homology to the clathrin-coated vesicle protein β-adaptin. *Nature* 349:215–220.

Siekevitz, P., and Palade, G. E. (1960) A cytochemical study on the pancreas of guinea pig. V. *In vivo* incorporation of leucine-1-C^{14} into chymotrypsinogen of various cell fractions. *J. Biophys. Biochem. Cytol.* 7:619–630.

Simons, K., and Fuller, S. D. (1985) Cell surface polarity in epithelium. *Annu. Rev. Cell Biol.* 1:243–288.

Simons, K., Perara, E., and Lingappa, V. R. (1987) Translocation of globin fusion proteins across the endoplasmic reticulum membrane in *Xenopus laevis* oocytes. *J. Cell Biol.* 104:1165–1172.

Stevens, T. H., Rothman, J. H., Payne, G. S., and Shekman, R. (1986) Gene dependent secretion of yeast vacuolar carboxypeptidase. *J. Cell Biol.* 102:1551–1537.

Strous, G. J. A. M., Willemsen, R., van Kerkkof, P., Slot, P. W., Geutze, H. J., and Lodish, H. F. (1983) Vesicular stomatitis virus glycoprotein, albumin and transferrin are transported to the cell surface via the same Golgi vesicles. *J. Cell Biol.* 97:1815–1822.

Swift, A. M., and Machamer, C. E. (1991) A Golgi retention signal in a membrane-spanning domain of coronavirus E1 protein. *J. Cell Biol.* 115:19–30.

Tajima, S., Lauffer, L., Rath, V. L., and Walter, P. (1986) The signal recognition particle receptor is a complex that contains two distinct polypeptide chains. *J. Cell Biol.* 103:1167–1178.

Tschopp, J., Esmon, P. C., and Shekman, R. (1984) Defective plasma membrane assembly in yeast secretory mutants. *J. Bacteriol.* 160:966–970.

Valls, L. A., Hunter, C. P., Rothman, J. H., and Stevens, T. H. (1987) Protein sorting in yeast: localization determinant of yeast vacuolar carboxypeptidase Y resides in propeptide. *Cell* 48:887–897.

von Figura, K., and Hasilik, A. (1986) Lysosomal enzymes and their receptors. *Annu. Rev. Biochem.* 55:167–193.

von Heijne, G. (1985) Signal sequences. The limits of variation. *J. Mol. Biol.* 184:99–105.

Wagner, D. D., Mayadas, T., and Marder, V. J. (1986) Initial glycosylation and acidic pH in the Golgi apparatus are required for multimerization of von Willebrand factor. *J. Cell Biol.* 102:1320–1324.

Walter, P., and Blobel, G. (1980) Purification of a membrane-associated protein complex required for protein translocation across the endoplasmic reticulum. *Proc. Natl. Acad. Sci. U.S.A.* 77:7112–7116.

Walter, P., and Blobel, G. (1981) Translocation of proteins across the endoplasmic reticulum. II. Signal recognition protein (SRP) mediates selective binding to microsomal membranes of in-vitro assembled polysomes synthesizing secretory protein. *J. Cell Biol.* 91:551–556.

Walter, P., and Blobel, G. (1982a) Signal recognition particle contains a 7S RNA essential for protein translocation across the endoplasmic reticulum. *Nature* 299:691–698.

Walter, P., and Blobel, G. (1982b) Translocation of proteins across the endoplasmic reticulum. III. Signal recognition protein (SRP) causes signal sequence dependent and site specific arrest of chain elongation which is released by microsomal membranes. *J. Cell Biol.* 91:557–561.

Walter, P., Ibrahim, I., and Blobel, G. (1981) Translocation of proteins across the endoplasmic reticulum. I. Signal recognition protein (SRP) binds in-vitro assembled polysomes synthesizing secretory protein. *J. Cell Biol.* 91:545–550.

Walter, P., Jackson, R. C., Marcus, M. M., Lingappa, V. R., and Blobel, G. (1979) Tryptic dissection and reconstitution of translocation activity for nascent presecretory proteins across microsomal membranes. *Proc. Natl. Acad. Sci. U.S.A.* 76:1795–1799.

Walter, P., Gilmore, R., and Blobel, G. (1984) Protein translocation across the endoplasmic reticulum. *Cell* 38:5–8.

Warren, G. (1987) Signal and salvage sequences. *Nature* 327:17–18.

Waters, M. G., Serafini, T., and Rothman, J. E. (1991) 'Coatamer': a cytosolic protein complex containing subunits of non-clathrin-coated Golgi transport vesicles. *Nature* 349:248–251.

Wattenberg, B. W., Balch, W. E., and Rothman, J. E. (1986) A novel prefusion complex formed during protein transport between Golgi cisternae in a cell free system. *J. Biol. Chem.* 261:2202–2207.

Wieland, F. T., Gleason, M. L. Serafini, T. A., and Rothman, J. E. (1987) The rate of bulk flow from the endoplasmic reticulum to the cell surface. *Cell* 50:289–300.

Willingham, M. C., Pastan, I. H., Sahagian, G. G., Jourdian, G. W., and Neufeld, E. F. (1981) Morphologic study of the internalization of a lysosomal enzyme by mannose-6-phosphate receptor in cultured Chinese hamster ovary cells. *Proc. Natl. Acad. Sci. U.S.A.* 78:6967–6971.

Woodman, P. G., and Edwardson, J. M. (1986) A cell free assay for the insertion of viral glycoprotein into the plasma membrane. *J. Cell Biol.* 103:1829–1835.

Chapter 5

Second Messengers

As discussed in Chapter 3, a variety of chemical signals, neurotransmitters and hormones, act by binding cell surface receptors. These signals frequently require the mediation of internal signals, the second messengers. In the cytoplasm, the second messenger in turn induces the molecular changes that are responsible for the physiological action of the signal.

As discussed in various chapters, changes in the intracellular concentration of Ca^{2+} serve as a signal for responses involving secretion (e.g., see neurotransmitters in Chapter 16) and other biochemical functions and induce contraction in muscle and other contractile systems (Chapters 17 and 18). As an activator of secretion, Ca^{2+} enters cells through channels in the plasma membrane. In the activation of contractile systems, the Ca^{2+} is released from the sarcoplasmic reticulum. Calcium as an internal messenger, a so-called *second messenger,* is discussed in the first section of this chapter. The next section discusses the possible involvement of an Na^+–H^+ exchange system in mediating the action of growth factors. Recent studies have implicated nitric oxide as a second messenger. Two other second-messenger systems have been studied intensively. One acts through the production of cyclic AMP (cAMP); the other, the inositol phospholipid system, acts through the production of diacylglycerol and inositol-1,4,5-triphosphate (IP_3), produced by the degradation of phosphatidylinositol-4,5-biphosphate (PIP_2), and also through the release of Ca^{2+} from internal stores. These two systems are the subject of the remaining sections of the chapter.

The cAMP system has been implicated in glycogenolysis and lipolysis and in the nervous system in mediating the effects of opiates (Drummond, 1983), control of electrical activity (Drummond, 1983), formation of synapses between nerve and muscle cells, and initiation of simple behavioral patterns (Kendel and Schwartz, 1982). *Opiates* found in the central nervous system, such as *endorphins* and *enkephalins,* are natural compounds that have an analgesic effect comparable to that of opium derivatives such as morphine.

The inositol phospholipid system also acts in response to a variety of chemical signals and is responsible for various functional responses (Nishizuka, 1984).

I. CALCIUM AND CALMODULIN

Calcium serves as a messenger for a whole class of extracellular stimuli and couples them to specific cellular responses (Rasmussen and Barrett, 1984). Generally, Ca^{2+} acts through receptor proteins such as calmodulin or directly on calcium-activated enzymes. Various interactions of Ca^{2+} and the proteins that mediate its effect are listed in Table 5.1.

For Ca^{2+} to serve as a signal, its concentration must be kept low, approximately 10^{-7} M. When Ca^{2+} is used as a signal, its concentration increases rapidly to a value as high as 10^{-5} M. In order to maintain the low resting level, several transport systems are operative, including an Na^{+}–Ca^{2+} exchange that uses the Na^{+} gradient as an energy source (Schatzmann, 1985) and the Ca^{2+}–ATPase system (Penniston, 1983). In addition, organelles such as the endoplasmic reticulum or the sarcoplasmic reticulum of muscle accumulate Ca^{2+} against a gradient (see chapters 14, 15, and 18). The mitochondrial transport system is also thought to play a role at higher Ca^{2+} concentrations.

Calcium is involved in the functioning of other second messengers, so generalizations about its role are oversimplifications. As shown in Table 5.1, one of the signaling systems involving Ca^{2+} is mediated by calmodulin, a single polypeptide of 16.7 kDa. At elevated Ca^{2+} concentrations, each molecule of calmodulin binds four Ca^{2+} ions. The binding produces a conformational change in the calmodulin such that it can interact with various calmodulin-regulated enzyme systems (see Klee et al., 1986; Schulman and Low, 1989).

The role of calmodulin has been illustrated by the experiment shown in Fig. 5.1 (Teo and Wang, 1973). The activity of cAMP phosphodiesterase, the enzyme that hydrolyzes cAMP to 5′-AMP, is shown as a function of calmodulin concentration. Curve 1 represents results obtained in the absence of added Ca^{2+}, and curves 2 and 3 show results after the addition of two different concentrations of Ca^{2+}. Although some activity is present in the absence of either Ca^{2+} or calmodulin, the presence of optimal concentrations of both increases the activity considerably. Calcium has no effect in the absence of the calmodulin, which is absolutely required.

Table 5.1 Calcium Receptor Proteins

I. True receptor proteins
- A. Soluble homologous class (cytosol)
  1. Calmodulin—all cells
  2. Troponin C—skeletal and cardiac muscle
  3. Parvalbumin—skeletal muscle
  4. Myosin light chain
- B. Membrane bound
  1. Calmodulin

II. Calcium-activated enzymes without specific calcium receptor subunit
- A. Bound to mitochondrial membrane
  1. Glyceraldehyde phosphate dehydrogenase
  2. Mitochondrial substrate transport protein
- B. Mitochondrial matrix
  1. Pyruvate dehydrogenase
  2. α-Ketoglutarate dehydrogenase
- C. Cytosol
  1. Calcium-activated, phospholipid-dependent protein kinase

From H. Rasmussen and P. G. Barrett, *Physiological Review,* 64:940.

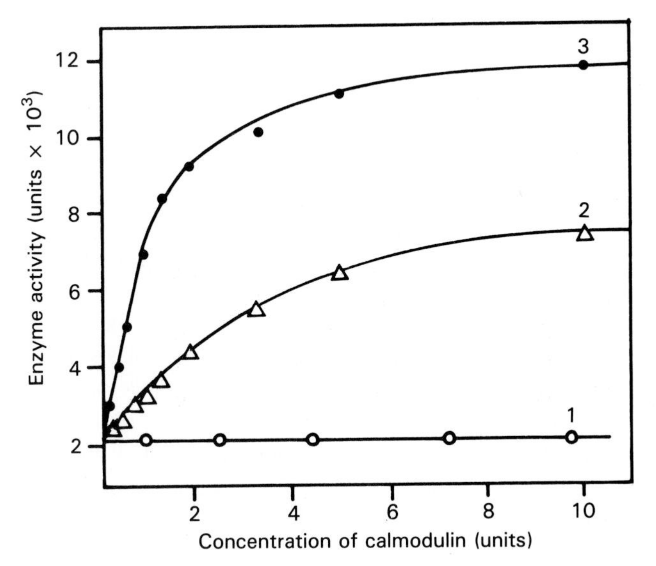

Fig. 5.1 Activation of cAMP phosphodiesterase by calmodulin plus Ca^{2+} was 0, 4 μM, and 100 μM in curves 1, 2, and 3, respectively. Reprinted with permission from T. S. Teo and J. H. Wang, *Journal of Biological Chemistry,* 248:5950–5955. Copyright © 1973 American Society of Biological Chemists, Inc.

This example also illustrates the complexity of second-messenger regulation, since cAMP itself, hydrolyzed by this enzyme, is also a second messenger.

II. SODIUM INFLUX

Fertilization or the binding of growth factors to their appropriate receptors activates influxes of Na^+ that are thought to stimulate the growth of cells. These growth factors include epidermal growth factor (EGF), platelet-derived growth factor (PDGF), nerve growth factor (NGF), fibroblast growth factor (FGF), thrombin, and insulin. Although the influx and the growth have been strongly linked (see below), it is not at all clear that Na^+ can be considered a second messenger.

Since the influx proceeds by an Na^+–H^+ exchange catalyzed by an antiporter protein, the cytoplasm becomes alkaline, as shown in several experiments by using pH-sensitive dyes (Cassel et al., 1983).

The naturally occurring growth factors, also referred to as *mitogens* (i.e., mitotic stimulators), acting on the G_0 or G_1 stage of cell replication (i.e., the arrested state or the period preceding DNA synthesis) stimulate an Na^+ influx that has been found to be inhibited by amiloride (Leffert and Koch, 1985). The Na^+ influx ranges from 5 to 20 mol/10^6 cells per minute, a 1.5- to 5-fold elevation from the basal level. Mitogen-activated Na^+ influx bursts have been observed seconds after the addition of the factor, and the uptake persists 2 to 60 min. There is a good correlation between incorporation of tritiated thymidine and initial Na^+ influx rates. Furthermore, amiloride blocks both the mitogen-activated Na^+ influx and the DNA synthesis. However, the time courses of the two events are so different, minutes in the case of Na^+ and hours in the case of DNA synthesis, that it is difficult to consider them as being directly linked.

This observation supports the idea that the Na^+ influx directly or indirectly links the growth factors to their physiological effects. However, there is no indication that the intracellular Na^+ concentration is changed, and the effect may be indirect.

III. NITRIC OXIDE

Nitric Oxide (NO) was recognized only recently as the active agent responsible for the activity of the so-called endothelial-derived relaxing factor of vascular smooth muscle. Since NO can diffuse rapidly through cell membranes, it can act not only as second messenger but also as a hormone. The signal generated by one cell can be transmitted to many other unconnected cells, although the action is limited by its short half-life (estimated to be 1 to 5 s, Ignarro, 1990).

NO is synthesized by endothelial cells, macrophages, and brain neuronal cells (for the latter, see Bredt et al., 1990). NO is produced by NO synthase which converts arginine to citrulline and NO (Moncada et al., 1989). The neural and epithelial enzymes are activated by the Ca^{2+}-calmodulin system (Bredt and Snyder, 1990). The enzymes in macrophages, fibroblasts, and smooth muscle cells are induced by γ-interferon (Werner-Felmayer et al., 1990), a macrophage activator important during microbial infections, and necrosis factor-α (Marsden and Ballermann, 1990), an inflammatory cytokine which functions as an immunostimulant and mediator of host resistance.

In the nervous system, NO is likely to play a role in the efficacy of neuronal communication at synapses and may have a role in neurotransmitter release.

IV. CYCLIC AMP

The binding of a variety of neurotransmitters or hormones (H) to specific receptors (R) at the cell surface leads to the activation of adenylate cyclase (C) on the inner face of the membrane. C catalyzes the formation of the second messenger, cAMP, from ATP. In contrast, binding to an inhibitory receptor leads to inhibition of adenylate cyclase.

In this section we discuss prostaglandins, epinephrine, and norepinephrine, which in common with other hormones activate C. *Prostaglandins,* released by certain tissues, have a role in the control of local circulation. *Adrenergic receptors* physiologically respond to *norepinephrine* or *epinephrine,* an effect that is mimicked by *sympathomimetic* drugs such as isoproterenol. These receptors have been classified as α or β, depending on their response to the sympathomimetic drugs.

Cyclic AMP activates a protein kinase. As indicated by its name, the kinase phosphorylates proteins, which presumably are then responsible for the biological response. Phosphorylation is one of the mechanisms controlling the activity of enzymes. (Table 7.5 in Chapter 7 lists a variety of enzymes regulated by phosphorylation.)

The formation of cAMP in response to the ligand binding actually involves three separate units, as discussed in the rest of this section.

A. R and C Are Separate Units

The results of many experiments indicate that R is a separate entity from C. One such experiment introduced a new R into the plasma membrane of erythroleukemia (Fc) cells containing the cyclase system normally activated by prostaglandin E_1 (PGE_1). This membrane combination was accomplished by fusing the Fc cells to the turkey red blood cells lacking C, in this case because of chemical treatment. These cells (indicated in the diagram as ENEM) normally possess a system activated by β-adrenergic activators such as isoproterenol. Cell fusion takes place when the surface properties of cells are altered, in this case by introducing Sendai virus.

The results of the fusion experiment are shown in Fig. 5.2 (Schulster et al., 1978). Cyclic AMP production was examined as a function of time after fusion of the cells and in the presence

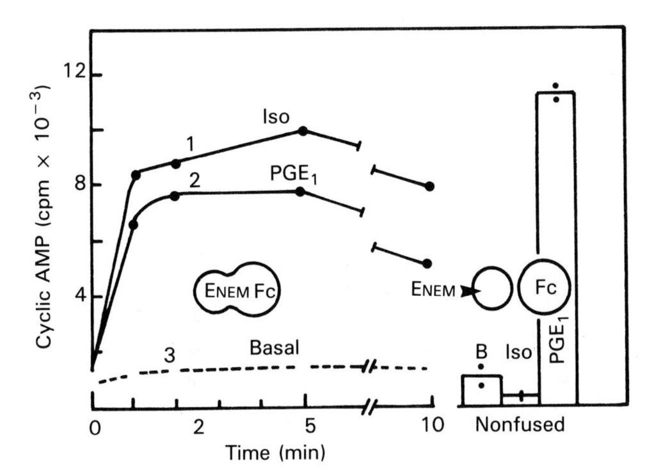

Fig. 5.2 Cyclic AMP production after fusion (graph) or in the absence of fusion. Reprinted with permission from D. Schulster, et al., *Journal of Biological Chemistry,* 253:1201–1206. Copyright © 1978 American Society of Biological Chemists, Inc.

of either PGE_1 (curve 2) or isoproterenol (curve 1) or without either activator (curve 3). For comparison, the activity of a mixture of cells not exposed to the virus, and therefore not fused, is shown in the histograms on the right. Although neither set of nonfused cells can be activated by isoproterenol, clearly the hybrid cells can respond to either PGE_1 or isoproterenol. Therefore the R of one cell has been combined with the rest of the system from another cell, indicating that R and C are separate entities.

Complementation experiments using extracts, but otherwise similar to the experiments depicted in Fig. 5.2, also demonstrate that the agonist-sensitive (i.e., sympathomimetic) cyclase activity involves at least two components (Ross and Gilman, 1977a). Cell membranes of B82 cells, a mouse fibroblast line, contain C but have no β-adrenergic receptors. They are therefore unable to express β-adrenergic cyclase activity. On the other hand, a variant clone of S49 lymphoma cells (AC^-) has no cyclase activity, since it lacks the regulatory protein G (discussed later) but contains the β receptors. With a combination of AC^- membranes and detergent extracts of B82, conventional β-adrenergic cyclase activity can be reestablished. The results of the reconstitution experiments are shown in Fig. 5.3. The various curves represent adenylate cyclase activity as a function of amount of AC^- membrane added to the B82 extracts. Curves 1 and 2 represent the results obtained in the presence of NaF or guanylylimidodiphosphate, Gpp(NH)p. Activation by these two agents bypasses the receptor step. Curve 3 shows the effect of isoproterenol in the presence of GTP, and curve 4 shows the effect of GTP alone. Clearly, the AC^- membranes together with the B82 extracts can reconstitute the whole system so that it is able to respond to isoproterenol. The dashed line, to which the scale on the right refers, represents the increase in enzyme activity to a sixfold maximum from the control level produced by the addition of isoproterenol.

B. Presence of a Third Component

The experiments represented in Figs. 5.2 (Schulster et al., 1978) and 5.3 (Ross and Gilman, 1977a) demonstrate that R and C are two separate units. Other experiments, however, reveal the need for a third component that mediates the response from R to C. This third unit is referred to as G, since it requires GTP or analogs of GTP.

The detergent extracts of cells lacking R but with cyclase activity can be separated into two protein fractions. Neither fraction alone can express the catalytic activity, which requires the

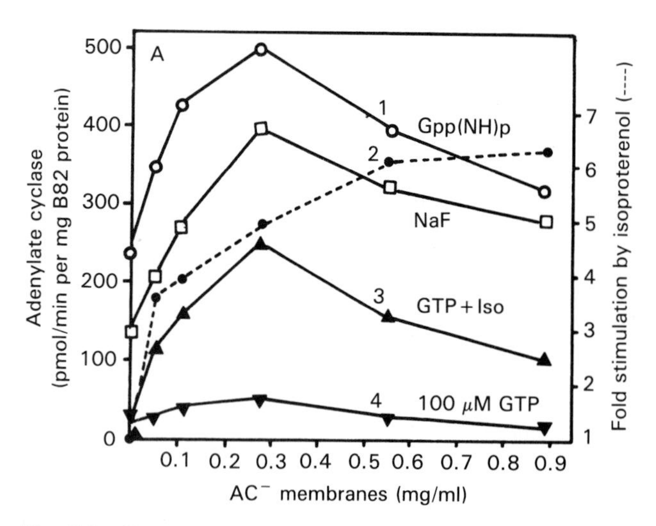

Fig. 5.3 Dependence of cyclase activity on concentration of donor extract or membranes. Reprinted with permission from E. M. Ross and A. G. Gilman, *Proceedings of the National Academy of Sciences,* 74:3715–3719, 1977a.

presence of both (Ross and Gilman, 1977b). One fraction is thermolabile, is inactivated by incubation at 37°C, and is sensitive to *N*-ethylmaleimide. The other fraction is less sensitive to either treatment and has been referred as thermostable. In these experiments the cyclase activity was assayed in the presence of Gpp(NH)p, which bypasses the need for R. Figure 5.4 (Ross and Gilman, 1977b) shows the adenylate cyclase activity of an extract of wild-type (WT) S49 lymphoma cells after treatment at 37°C and subsequent mixing in various proportions with extracts of AC⁻ cells. When a fixed concentration of either extract alone was tested, no significant activity was found. Cyclase activity was reconstituted when the two were mixed together. In curve 1, the lymphoma extract, WT, is kept constant and the amount of AC⁻ extract is varied. Curve 2 represents the reverse design. The results indicate that, aside from R, a minimum of two other proteins are needed for cyclase activity. The shape of the curves of Fig. 5.4 suggests

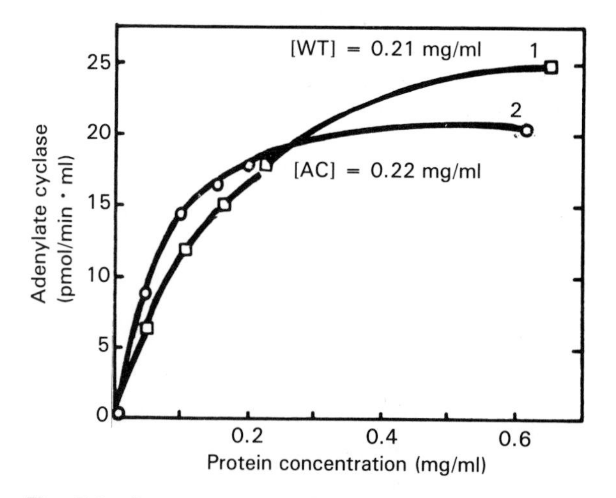

Fig. 5.4 Dependence of adenylate cyclase activity on heated wild-type extract (WT) and AC⁻ extract assayed in the presence of Gpp(NH)p. Reprinted with permission from E. M. Ross and A. G. Gilman, *Journal of Biological Chemistry,* 252:6966–6969. Copyright © 1977 American Society of Biological Chemists, Inc.

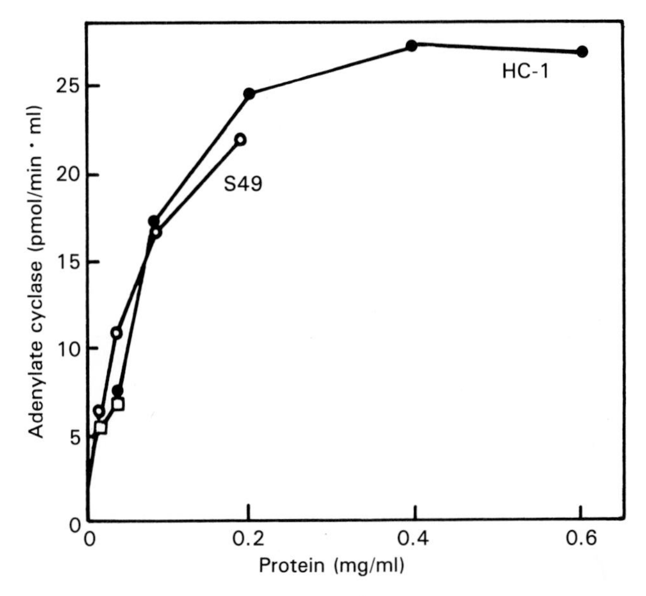

Fig. 5.5 Reconstitution of Mg^{2+}-dependent adenylate cyclase from extracts of AC^- and HC-1 membranes using Gpp(NH)p as an activator. The wild-type (S49) protein was heated. Reprinted with permission from E. M. Ross, *Journal of Biological Chemistry,* 253:6401–6412. Copyright © 1978 American Society of Biological Chemists, Inc.

saturation with increasing protein concentration, consistent with the idea that the two components combine in a fixed stoichiometry. The presence of at least two separate proteins is also supported by experiments using a different combination of cell lines: the AC^- lymphoma line, which we saw was deficient in cyclase activity and lacks the thermostable component, and a hepatoma cell line (HC-1), which contains only the thermostable activity. The reconstitution is shown in Fig. 5.5 (Ross et al., 1978). The experiment was carried out in a manner analogous to the experiment of Fig. 5.4.

Hormonally stimulated adenylate cyclase requires the presence of GTP, which is normally hydrolyzed by the cyclase complex. GDP and a variety of analogs that are not hydrolyzed, such as Gpp(NH)p, are also effective. Table 5.2 (Rodbell et al., 1971) lists various nucleotides and their effect in activating adenylate cyclase and shows that either GTP, GDP, or the analog 5′-guanylyl-diphosphonate (GMP-PCP) is effective and that in their absence there is only a trace of activity (Rodbell et al., 1971).

Table 5.2 Specificity of Action of Nucleotides on Glucagon-Stimulated Adenylate Cyclase Activity

Addition	Cyclic AMP formed (pmol/mg protein)
None	21
GTP (0.01 mM)	235
GDP (0.01 mM)	237
GMP-PCP (0.1 mM)	201
UTP (0.1 mM)	75
CTP (0.1 mM)	53

From J. Rodbell, et al., *Journal of Biological Chemistry,* 246:1877–1882. Copyright © 1971 American Society for Biological Chemists, Inc.

Table 5.3 Composition of Guanyl Nucleotides Released from Turkey Red Blood Cells Pretreated with Isoproterenol and [^{3}H]GTP

Nucleotide	(a) Membrane bound	(b) Released at 37° counts/min − Isoproterenol	(c) Released at 37° counts/min + Isoproterenol	(d) b − c Increased by isoproterenol
GTP	650	70	75	5
GDP	2250	340	1040	700
GMP	60	55	85	30
Total	2960	465	1200	735

Reprinted with permission from D. Cassel and D. Selinger, *Proceedings of the National Academy of Sciences,* 75:4155–4159, 1978.

The nucleotide binding component distinct from C was also demonstrated with extracts (Pfeuffer, 1977).

The considerations just discussed agree with the presence of a minimum of three proteins, together corresponding to a complex (R, G, and C). The fact that G and C seem to bind stoichiometrically suggests that they are actually attached. However, exogenous receptors can be inserted in a cell membrane to activate C, which suggests that R is free to diffuse in the membrane.

When the complex is activated by a hormone, GTP is required and is hydrolyzed during the activation (Cassel and Selinger, 1977) so that eventually the cyclase activity stops. However, when G is activated by Gpp(NH)p or some other analog that is not hydrolyzed, the activation persists with time.

The interaction of the entire complex with GTP has been studied in detail (Cassel and Selinger, 1978) in turkey erythrocyte membranes incubated in a medium containing [^{3}H]GTP in the presence of isoproterenol. After extensive washing to remove the unreacted labeled GTP and the isoproterenol, some of the radioactivity remained in the membranes. This radioactivity was released when the membranes were incubated with the isoproterenol. As shown in Table 5.3 (Cassel and Selinger, 1978), most clearly in column (d), the released radioactivity, expressed in counts per minute released by isoproterenol, is in the form of [^{3}H]GDP.

The results indicate that the inactive cyclase complex tightly binds the GDP produced by the hydrolysis of GTP. In addition, GDP is released because, after binding H, R changes the affinity of G for GDP. Figure 5.6 summarizes the observations. Cholera toxin blocks the hydrolysis by ADP-ribosylating G_s and thereby enhances the cyclase activity (Cassel and Selinger, 1977).

The mechanism of this complex effect involving all three components is thought to take place as shown in Fig. 5.7 (Schramm and Selinger, 1984). In this figure the circles and attached sticks represent the lipid component of the membrane, R and C are appropriately labeled, and G is the protein binding either GTP or GDP. In this scheme, H binds to R, which is in contact with G, and allows the latter to activate C. Simultaneously, H is released because the affinity of R for H is decreased. The interaction of HR with G also facilitates the release of GDP and the binding of GTP. The G with the bound GTP activates C by complexing with it. The GTP hydrolyzes at the G site, resulting in the dissociation of G and C and cessation of cyclase activity.

G can act either as an activator or as an inhibitor. To distinguish between the two, G_s is generally used to denote a G involved in activation and G_i for inhibition.

The G_s, β-adrenergic receptors, and C have all been purified. When these purified constituents were incorporated into phospholipid vesicles, the hormonally activated synthesis of cAMP was reconstituted (May et al., 1985). The G proteins are discussed in more detail in Section V.

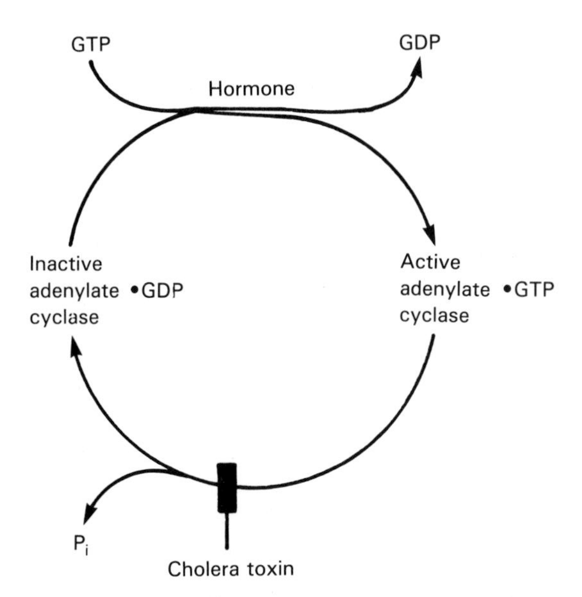

Fig. 5.6 The regulatory GTPase cycle of turkey erythrocyte adenylate cyclase. Reprinted with permission from D. Cassel and D. Selinger, *Proceedings of the National Academy of Sciences,* 75:4155–4159, 1978.

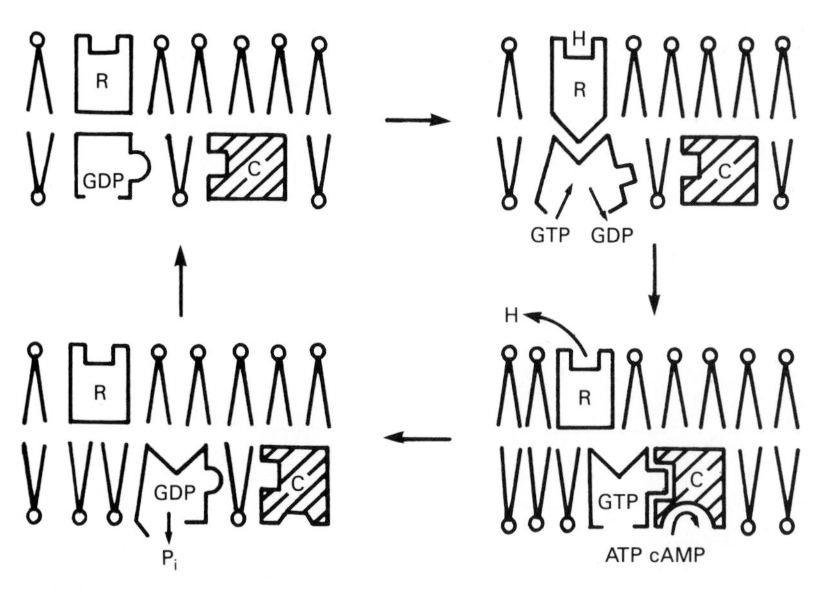

Fig. 5.7 Scheme of component interactions in the activation and inactivation of the adenylate cyclase system. A section of the cell membrane shows the three components (upper left). Phospholipid is represented by the polar heads with a pair of fatty acid tails interspersed between the components. When the neurotransmitter or hormone, H, binds to R, HR interacts with the binding protein G to release GDP and to facilitate the tight binding of GTP (upper right). The G binding protein thus activated by GTP associates with the catalytic unit C to form the active enzyme complex GC (lower right). Hydrolysis of the GTP at the G protein site, with release of inorganic phosphorus (P_i), results in dissociation of G from C and cessation of enzyme activity (lower and upper left areas). R is arbitrarily drawn as limited to the outer layer of the cell membrane. There appears to be no evidence as yet that R transverses the entire membrane. Reprinted with permission from M. Schramm and Z. Selinger, *Science,* 225:1350–1356. Copyright 1984 by the AAAS.

V. THE INOSITOL PHOSPHOLIPID SYSTEM

In contrast to the system that produces cAMP, the inositol phospholipid system involves the phospholipid of the plasma membrane. The attachment of hormone to the receptor induces the hydrolysis of PIP_2, probably catalyzed by phospholipase C, to produce 1,2-diacylglycerol (DG) and IP_3 (Abdel-Latif et al., 1977). Present evidence indicates the biochemical reactions shown in Fig. 5.8 (Nishizuka, 1984). Apparently, the production of IP_3, induced by activation of the receptor, in turn activates the release of Ca^{2+} from intracellular stores. The simultaneous production of DG stimulates protein kinase C, a multifunctional enzyme that phosphorylates a variety of proteins. DG acts by increasing the affinity of this enzyme for Ca^{2+} (Kishimoto et al., 1980). Since DG is a membrane component, the protein kinase must also be located in the membrane. These two effects, the increase in Ca^{2+} and the activation of the kinase, mediate the physiological effects of the chemical signal.

In this section, the effect of DG in activating protein kinase C is discussed first, followed by the release of Ca^{2+}.

A. Diacylglycerol

The physiological response of platelets, which release a variety of compounds, is mediated by the inositol phospholipid system and coincides with the phosphorylation of two proteins, of molecular weight 20,000 and 40,000. The smaller, 20 kDa protein corresponds to the light chain

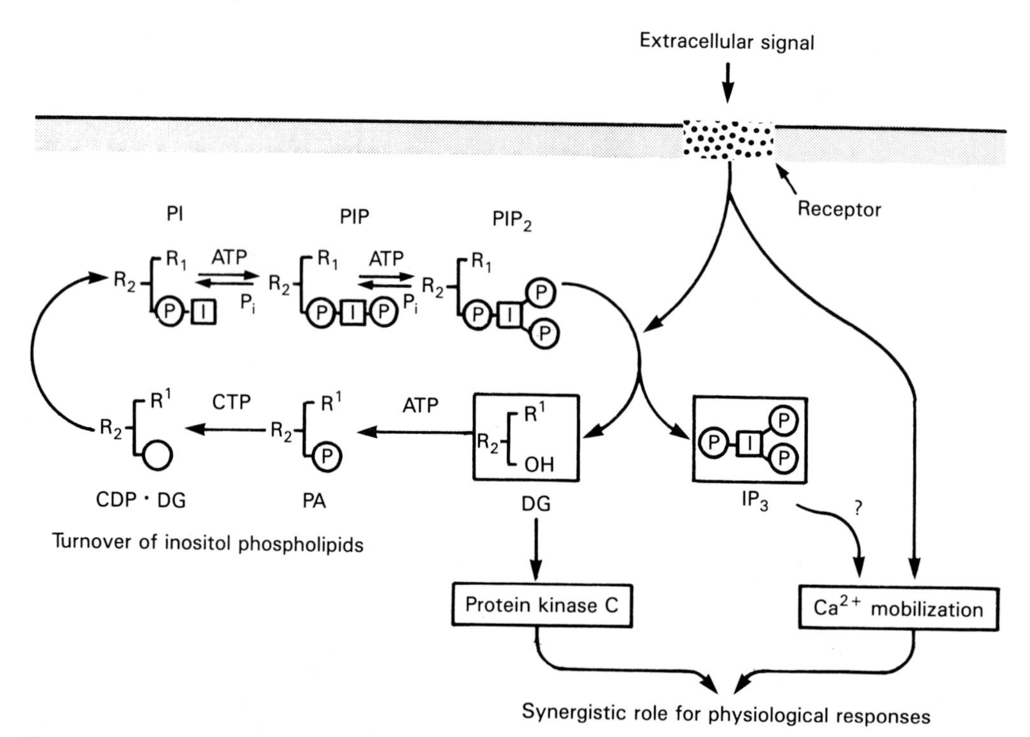

Fig. 5.8 Turnover of inositol phospholipids and signal transduction. Abbreviations: PI, phosphatidylinositol; PIP_2, phosphatidylinositol 4,5-biphosphate; DG, diacylglycerol; IP_3, inositol triphosphate; PA, phosphatidic acid; PIP, phosphatidylinositol 4-phosphate. Reprinted with permission from Z. Nishizuka, *Science,* 225:1365–1370. Copyright 1984 by the AAAS.

of myosin, and the enzyme carrying out the phosphorylation is unknown. In contrast to protein kinase C, this enzyme is activated by Ca^{2+} in a calmodulin-dependent process (Hathaway and Adelstein, 1979). As discussed in Section I of this chapter, calmodulin is a protein that combines with Ca^{2+} and in this form can activate a variety of enzymes. The phosphorylation of the 40-kDa protein is catalyzed by protein kinase C (Kawahara et al., 1980), which can be inhibited by a variety of drugs acting on the membrane phospholipid.

Platelets can be activated by several signals. The platelet activator thrombin is a plasma protein that is formed during clotting and catalyzes the formation of fibrin. Figure 5.9 (Sano et al., 1983) shows how thrombin induces the transient production of DG (curve 1), followed by the phosphorylation of the 40-kDa protein (curve 2) and eventually the release of serotonin, the physiological response to the thrombin (curve 3). Accordingly, as shown in Fig. 5.10, PI, which serves as a source of DG, decreases rapidly, while (Fig. 5.10) PA, of which DG is a precursor, increases as expected. In these experiments the phosphorylation of the 40-kDa protein was followed after a preliminary incubation of the platelets with $^{32}P_i$; phospholipids were labeled by incubation in [^{3}H]arachidonic acid, and [^{14}C]serotonin was also present.

In the inositol phospholipid signal system only a small fraction of the total turnover of the phospholipids is affected, and there is no change in turnover when it is mimicked by the addition of compounds with detergentlike action.

The effect of DG on the activity of protein kinase C is shown in Fig. 5.11 (Kishimoto et al., 1980), which shows the activity of the purified kinase as a function of Ca^{2+} concentration. At zero Ca^{2+}, any trace of free calcium has been removed by addition of the chelator ethylene glycol (aminoethyl ether) tetraacetate (EGTA). Curves 3, 4, and 5 show the effect of adding

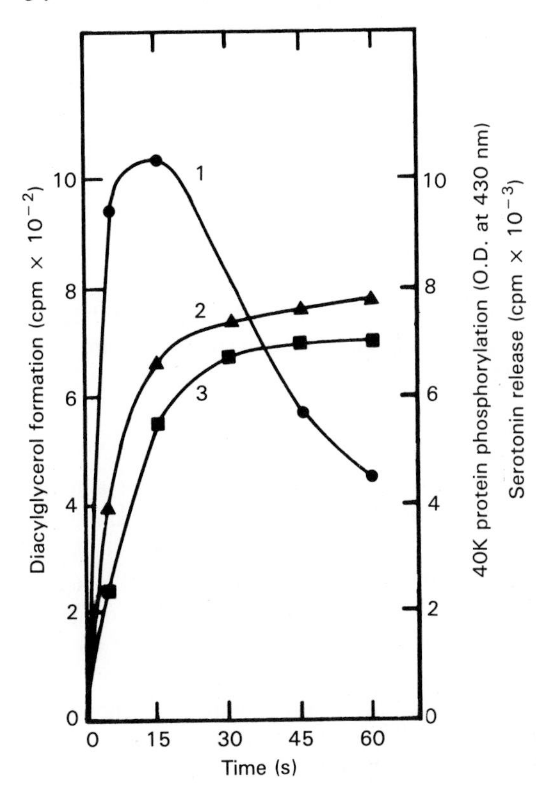

Fig. 5.9 Time courses for diacylglycerol formation (curve 1), 40-kDa protein phosphorylation (curve 2), and serotonin release in platelets (curve 3) stimulated by thrombin. Reprinted with permission from K. Sano, et al., *Journal of Biological Chemistry,* 258: 2010–2013. Copyright © 1983 American Society of Biological Chemists, Inc.

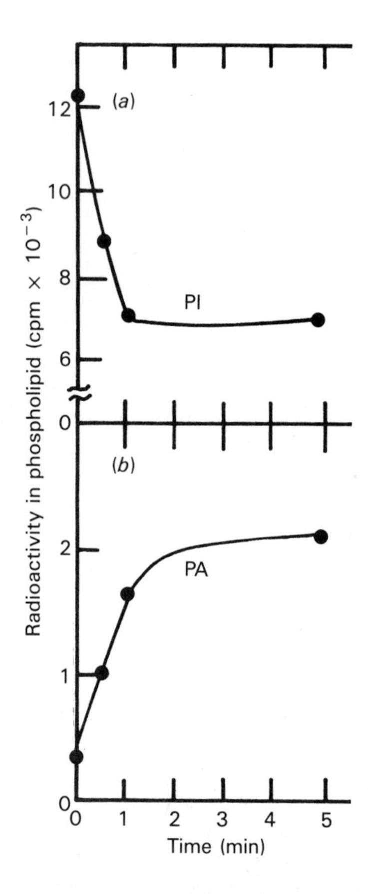

Fig. 5.10 Effects of thrombin on phosphatidylinositol and phosphatidic acid in platelets. Reprinted with permission from K. Sano, et al., *Journal of Biological Chemistry,* 258:2010–2013. Copyright © 1983 American Society of Biological Chemists, Inc.

each component alone: phospholipid, diolein, or neutral lipid, respectively. The effect is small, and phospholipid has the most effect. However, the effect is much greater when phospholipid is combined with the diolein, as shown by curve 1. The diolein can be replaced by neutral lipid, as shown by curve 2, although the effect is somewhat less in the latter case. The various DGs can be regarded as acting by modifying the sensitivity of the kinase to Ca^{2+}.

Phosphatidylserine (PS) is required for the activation of the kinase by DG. Phosphatidylethanolamine (PE) enhances this effect, whereas phosphatidylcholine (PC) and sphingomyelin (SM) decrease it (Kaibuchi et al., 1981).

B. Inositol Triphosphate

In hepatocytes the breakdown of PIP_2 to form DG and PIP_3 is induced by α-adrenergic agonists and vasoactive peptides. The probable involvement of PIP_3 in the release of Ca^{2+} from internal stores has been shown by the addition of exogenous PIP_3 after hepatocytes were made leaky by treatment with the detergent digitonin. The concentration of Ca^{2+}, $[Ca^{2+}]$, can be followed by using Ca^{2+}-sensitive electrodes or by monitoring the light absorption or fluorescence emission of a Ca^{2+}-sensitive dye. In the experiments discussed here, either the fluorescence of the dye Quin 2 or the electrode method was used.

The results are shown in Figs. 5.12–5.15 (Joseph et al., 1984). Figure 5.12 shows the Ca^{2+} concentration measured with the Ca^{2+}-sensitive electrode, expressed as pCa^{2+} (which

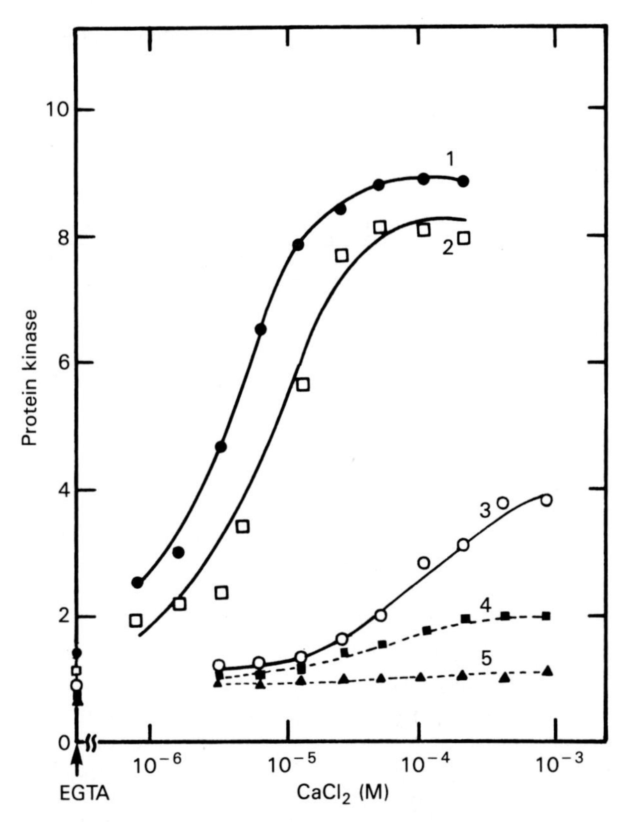

Fig. 5.11 Effects of neutral lipid and diolein on reaction velocity of protein kinase C at various concentrations of Ca^{2+}. Curve 1, diolein and phospholipid added; curve 2, neutral lipid and phospholipid added; curve 3, phospholipid alone; curve 4, diolein alone; curve 5, neutral lipid alone. Reprinted with permission from A. Kishimoto, et al., *Journal of Biological Chemistry,* 255:2273–2276. Copyright © 1980 American Society of Biological Chemists, Inc.

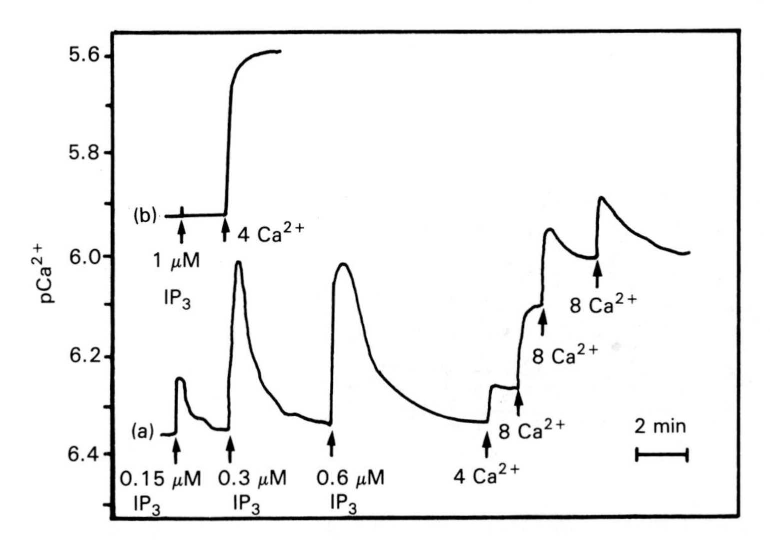

Fig. 5.12 Release of Ca^{2+} induced by IP_3 in saponin-permeabilized hepatocytes. Reprinted with permission from S. K. Joseph, et al., *Journal of Biological Chemistry,* 259:3077–3081. Copyright © 1984 American Society of Biological Chemists, Inc.

corresponds to $-\log_{10}[Ca^{2+}]$) as a function of time. ATP and a system containing phosphocreatine and creatine kinase to regenerate ATP were present in the medium to provide energy. At the times indicated by the arrows, IP_3 was added in various concentrations. After each addition there was a Ca^{2+} release. The release was only temporary, as shown later (see Fig. 5.15), because IP_3 is hydrolyzed by the system. Insert (b) in Fig. 5.12 simply shows that the system does not respond to IP_3 in the absence of hepatocytes. As shown at the right side of the figure, Ca^{2+} was added at various concentrations at the end of the experiment to provide the scale shown in the ordinate. The numbers shown at the arrows indicate the added Ca^{2+} in nanomoles. Experiments such as this allow the construction of a curve showing Ca^{2+} release as a function of IP_3 concentration (Fig. 5.13).

Figure 5.14 shows the uptake and release of Ca^{2+} estimated from the changes in the fluorescence emission of the calcium-sensitive dye Quin 2. Again, each addition is indicated by an arrow. In parts A and B, succinate is added first, followed by Mg-ATP. The mitochondrial use of the Mg-ATP was blocked by adding oligomycin, which inhibits the mitochondrial ATPase (Chapters 10 and 12). Succinate is primarily a mitochondrial substrate (Chapter 10). Although succinate did not support the uptake of Ca^{2+}, Mg-ATP supported considerable uptake, presumably in vesicles other than mitochondria. Addition of IP_3 resulted in release of Ca^{2+}. The mitochondrial uncoupler of oxidative phosphorylation, 1799, did not induce a release as long as Mg-ATP was present (Fig. 5.14A and B). However, addition of the Ca^{2+} ionophore Ionomycin resulted in considerable release, presumably by making internal vesicles leaky to Ca^{2+}. The results of these experiments were taken as evidence for a role of vesicles other than mitochondria, probably endoplasmic reticulum, in the release and uptake of Ca^{2+}. The small role of mitochondria is confirmed in Fig. 5.14C, where only succinate is added. Under these conditions Ca^{2+} was not taken up or released. However, at higher Ca^{2+} concentrations (micromolar, not shown), mitochondria seem to have a significant role in both uptake and IP_3-induced release.

As shown by the dashed line in Fig. 5.15 (Joseph et al., 1984), the cells hydrolyze IP_3. In this experiment IP_3 was labeled with ^{32}P, and the radioactivity remaining in the IP_3 was measured

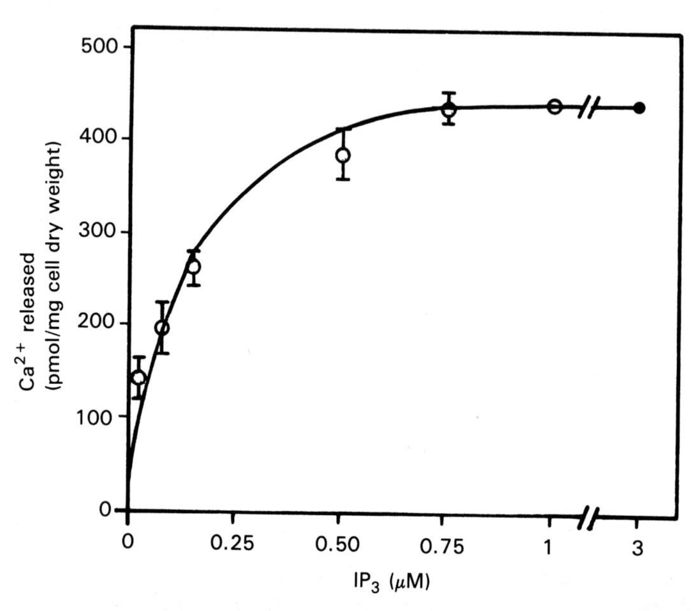

Fig. 5.13 Relationship between the amount of Ca^{2+} released and the concentration of added IP_3. Reprinted with permission from S. K. Joseph, et al., *Journal of Biological Chemistry,* 259:3077–3081. Copyright © 1984 American Society of Biological Chemists, Inc.

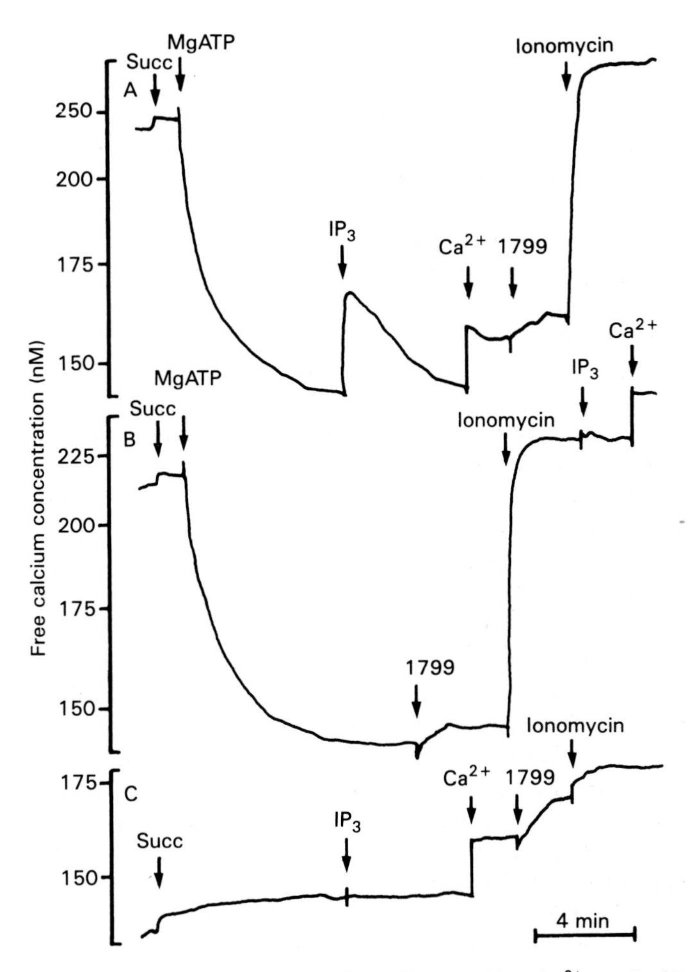

Fig. 5.14 Characterization of the IP_3-sensitive Ca^{2+} pool of isolated hepatocytes. Reprinted with permission from S. K. Joseph, et al., *Journal of Biological Chemistry,* 259:3077–3081. Copyright © 1984 American Society of Biological Chemists, Inc.

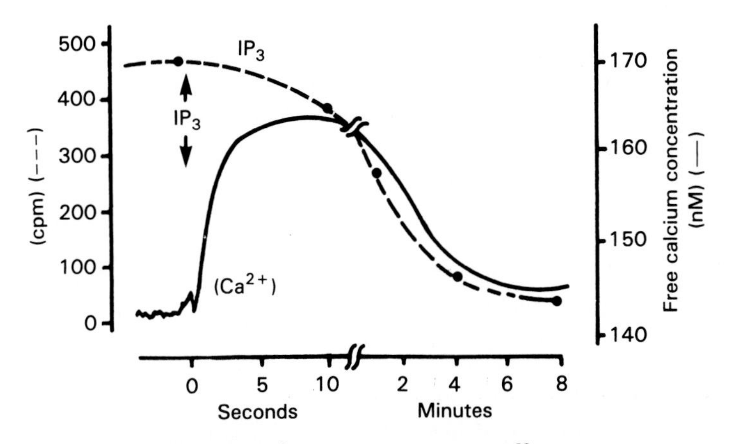

Fig. 5.15 Uptake of Ca^{2+} and degradation of ^{32}P-labeled IP_3 in saponin-permeabilized hepatocytes. Reprinted with permission from S. K. Joseph, et al., *Journal of Biological Chemistry,* 259:3077–3081. Copyright © 1984 American Society of Biological Chemists, Inc.

with time after separating the IP_3 from the mixture. The $[Ca^{2+}]$ in the medium was monitored by measuring the fluorescence of Quin 2. The uptake of Ca^{2+} parallels closely the disappearance of IP_3. Similar demonstrations have been carried out with pancreatic acinar cells (Streb et al., 1983).

In summary, the results with permeabilized cells indicate that IP_3 releases Ca^{2+} from internal stores, primarily nonmitochondrial at low $[Ca^{2+}]$. Calcium itself can act as an activator of a variety of enzyme systems.

The responses of the inositol phospholipid system to extracellular signals—that is, the activation of protein kinase C and the mobilization of Ca^{2+}—together are probably essential for the action of the corresponding hormones (Karbuchi et al., 1983). However, except for a few cases, such as platelets, little is known about the target of the phosphorylation and its immediate consequences.

Where is the Ca^{2+} stored? Where does IP_3 act? Three approaches have been used to define these sites using antibodies to proteins involved in Ca^{2+} release or storage. The drug ryanodine blocks Ca^{2+} release from the sarcoplasmic reticulum. From all indications, all biological chemicals we have examined so far act by binding a receptor. Receptors have been identified for both IP_3 and ryanodine, and they have been sequenced by cDNA cloning (Furuichi et al., 1989, and Takeshima et al., 1989) (see Chapter I, II, A). Interestingly, they are related proteins and both act as Ca^{2+} release channels. Ca^{2+} is generally sequestered by specialized proteins in vesicles such as *calsequestrin, calreticulin,* or related proteins (see Treves et al., 1990; Krause et al., 1990). Why not then search for the Ca^{2+} organelle by using antibodies to these three kinds of proteins? Antibodies labelled with colloidal gold have been used to localize the proteins with electron microscopy. Results of these experiments have led to the concept that a specialized smooth vesicle, the *calciosome,* is responsible for the accumulation of Ca^{2+} (Volpe et al., 1988; Hashimoto et al., 1988; Treves et al., 1990). Results obtained with all three probes using cerebellar Purkinje cells provide a complex picture (see Burgoyne and Cheek, 1991). Some membranes respond to one probe, some to another, and a third group of membranes binds to all three.

VI. THE G PROTEINS

The G proteins, which were discussed briefly in Section III in relation to their role in the cAMP system, are a family of signal-coupling proteins that mediate signals in both the adenylate cyclase and inositol phospholipid second-messenger systems. Their broad involvement in physiological functions is shown in Table 5.4 (Stryer and Bourne, 1986), which includes hormonal activation and also sensory transitions of eukaryotes. In the latter role they control the opening of channels and thereby affect the membrane potential of the receptor.

The two groups of G proteins—G_s, associated with activation, and G_i, associated with inhibition—have been purified.

Pertussis toxin, also known as islet-activating protein, blocks the inhibition of adenylate kinase produced by inhibitory hormones by binding to G_i (Bokoch et al., 1983, 1984; Codina et al., 1983, 1984) and catalyzing the ADP-ribosylation of a 41-kDa polypeptide found in the plasma membrane. This polypeptide is a subunit of G_i (Katada and Ui, 1982). This property permitted a strategy for the recognition of G_i that led to its isolation.

The G_s protein is a heterotrimer of subunits α, 45 to 52 kDa; β, 35 to 36 kDa; and γ, approximately 8 kDa. The β and γ components are required for attachment of the α polypeptide, which is

Table 5.4 Examples of Physiological Processes Mediated by G Proteins

Stimulus	Receptor	G protein	Effector	Physiological response
Epinephrine	β-Adrenergic receptor	G_S	Adenylate cyclase	Glycogen breakdown
Serotonin	Serotonin receptor	G_S	Adenylate cyclase	Behavioral sensitization and learning in *Aplysia*
Light	Rhodopsin	Transducin	cGMP phospho-diesterase	Visual excitation
IgE-antigen complexes	Mast cell IgE receptor	G_{PLC}	Phospholipase C	Secretion
fMet peptide	Chemotactic receptor	G_{PLC}	Phospholipase C	Chemotaxis
Acetylcholine	Muscarinic receptor	G_K	Potassium channel	Slowing of pacemaker activity

Source: Stryer and Bourne (1986). Reproduced, with permission, from the Annual Review of Cell Biology, Volume 2, © 1986 by Annual Reviews Inc.

soluble even in the absence of detergent (Schulster et al., 1978). The α subunit alone can activate the cyclase (Cerione et al., 1984; May et al., 1985). However, at least for activation of the acetylcholine receptor, the β and γ subunits open the K^+ channel (Logothetis et al., 1987).

Activation of G_s with GTP analogs dissociates α from the other two subunits. The cycle of activation and inactivation of G_s is shown in Fig. 5.16 (Stryer and Bourne, 1986).

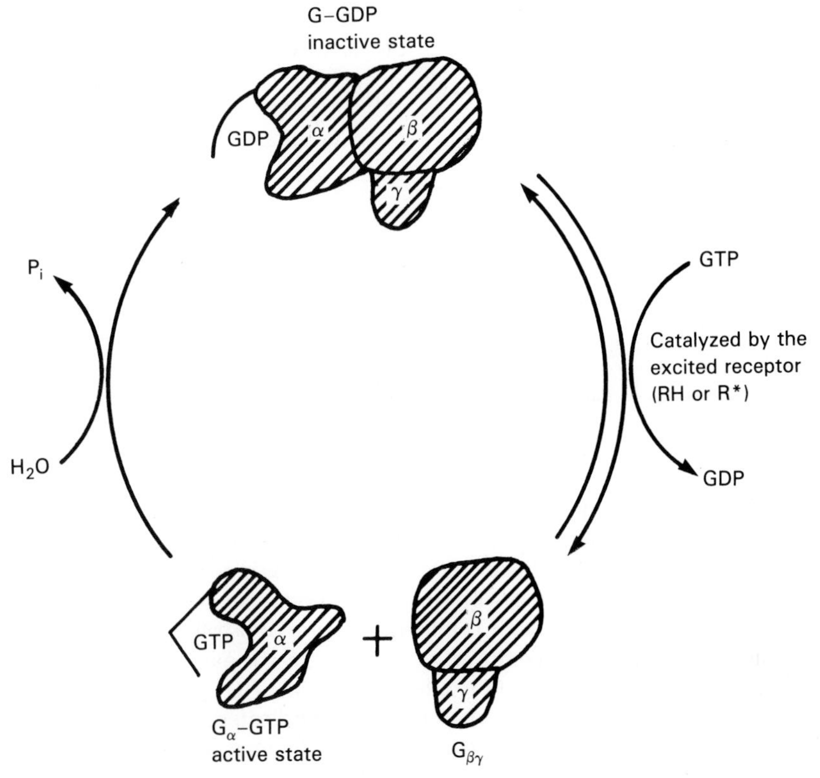

Fig. 5.16 Cycle of activation and inactivation of G_s (Stryer and Bourne, 1986). Reproduced, with permission, from the Annual Review of Cell Biology, Volume 2, © 1986 by Annual Reviews Inc.

Table 5.5 Functions and Properties of Purified G Proteins

Functional and structural parameters	G_s	G_i	G_o	Transducin
Signal detector	β-Adrenergic receptor, glucagon receptor, and many others	Muscarinic receptor, opiate receptor, and many others	Unknown	Rhodopsin
Effector protein	Adenylate cylcase	Adenylate cyclase	Unknown	cGMP phosphodiesterase
Function	Stimulation	Inhibition	—	Stimulation
Subunit masses (kDa)				
α	45 and 52	41	39	39
β	35 and 36	35 and 36	35 and 36	36
γ	8	8	8	8
Toxin susceptibility	Cholera	Pertussis	Pertussis	Pertussis and cholera
Location	Nearly all cells	Nearly all cells	Brain	Retinal rod outer segments

Source: Stryer and Bourne (1986). Reproduced, with permission, from the Annual Review of Cell Biology, Volume 2, © 1986 by Annual Reviews Inc.

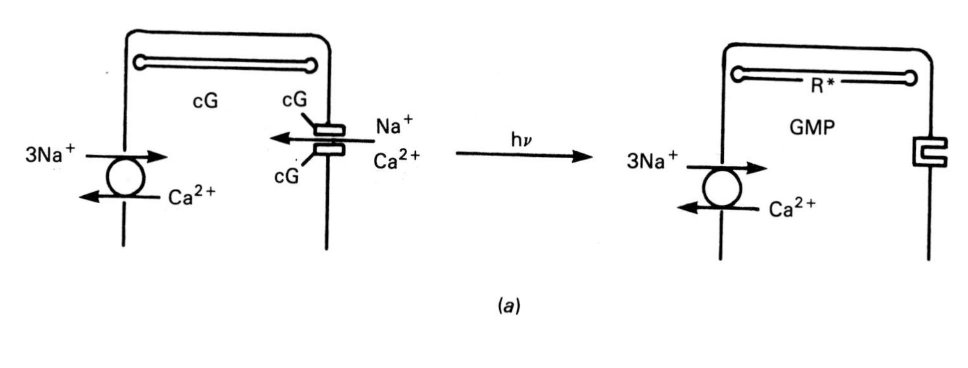

(*a*)

R → R* → T-GTP → PDE* → cG + H_2O (open channels) / 5′ GMP + H^+ (closed channels)

(*b*)

Fig. 5.17 Cyclic GMP controls sodium channels in the plasma membrane. (*a*) In the dark, a high level of cGMP in the cytosol opens sodium channels in the plasma membrane. Na^+ and Ca^{2+} enter the outer segment through these channels. Ca^{2+} is extruded in exchange for Na^+ by an Na^+/Ca^{2+} exchanger. On illumination, photoexcited rhodopsin triggers a cascade that results in the hydrolysis of cGMP to GMP. The lowered level of cGMP closes sodium channels. Ca^{2+} continues to be extruded from the outer segment by the exchanger. (*b*) Flow of information in the cGMP cascade R* is photoexcited rhodopsin. T-GTP is the activated form of transducin, and PDE* is the activated form of the phosphodiesterase (Stryer, 1986). Reproduced, with permission, from the Annual Review of Neuroscience, Volume 9, © 1986 by Annual Reviews Inc.

The G_i protein also contains an α subunit of 41 kDa and β and γ subunits and functions analogously to G_s. Table 5.5 (Stryer, 1986) lists the various G proteins.

VII. ROLE OF G PROTEINS IN RECEPTION

Perhaps one of the most exciting functional roles of G proteins is that of *transducin* (T) in visual reception in the retinal rods.

Retinal rod cells function as single-photon receptors through the light-induced cis-trans isomerization of a single molecule of rhodopsin, which, by a cascade amplification mechanism, can block the entry of 10^6 Na^+. At rest, the Na^+ channels of rod cells are kept open by the presence of cyclic GMP (cGMP), which results in a characteristic resting potential. In excitation, the cGMP is hydrolyzed by cGMP phosphodiesterase activated by a T-mediated system. The resulting block of the Na^+ channels produces a hyperpolarization that serves as a signal to the cell synapse. These events are summarized in Fig. 5.17 (Stryer, 1986).

In this mechanism, the photoexcited *trans*-rhodopsin serves as the activated receptor, R, which catalyzes the GTP-GDP exchange in the T-GTP. In turn, the T-GTP activates the phosphodiesterase. To permit rapid recovery, the activated rhodopsin is rendered ineffective by phosphorylation followed by combination with a molecule of arrestin (A). The complete cycle of this mechanism is represented in the scheme of Fig. 5.18 (Stryer and Bourne, 1986).

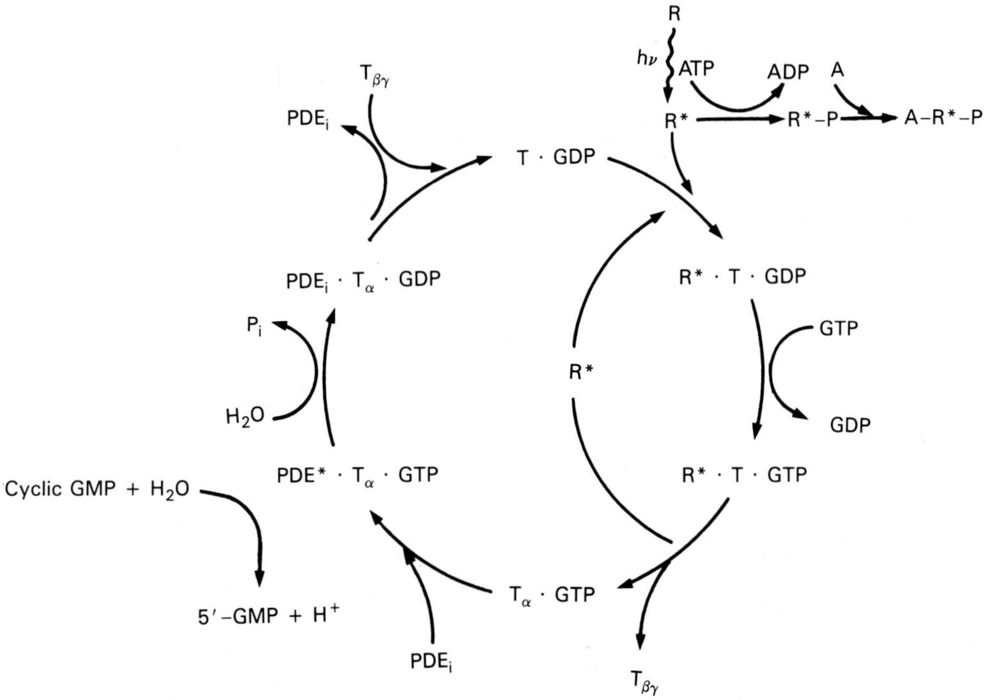

Fig. 5.18 Light-triggered transducin cycle of vertebrate photoreceptors. A, arrestin; PDE_i and PDE*, inhibited and activated forms of cGMP phosphodiesterase; R and R*, unexcited and photoexcited rhodopsin; T, transducin (Stryer and Bourne, 1986). Reproduced, with permission, from the Annual Review of Cell Biology, Volume 2, © 1986 by Annual Reviews Inc.

The fact the cGMP keeps the channel open has been shown by patch clamping of an excised piece of rod outer segment (Fesenko et al., 1985).

SUGGESTED READING

Berridge, M. J., and Irvine, R. F. (1989) Inositol phosphate and cell signalling. *Nature* 341:197–205.

Freissmuth, M., Casey, P. J., and Gilman, A. G. (1989) G proteins control diverse pathways of transmembrane signaling. *FASEB J.* 3:2125–2131.

Leffert, H. L., and Koch, K. S. (1985) Growth regulation by sodium influxes. In *Control of Animal Cell Proliferation* (Boynton, A. L. and Leffert, H. L., eds.), pp. 367–413. Academic Press, New York.

Nishizuka, Y. (1984) Turnover of inositol phospholipids and signal transduction. *Science* 225:1365–1370.

Rasmussen, H., and Barrett, P. Q. (1984) Calcium messenger systems, an integrated view. *Physiol. Rev.* 64:938–984.

Schramm, M., and Selinger, Z. (1984) Message transmission: receptor controlled adenylate cyclase system. *Science* 225:1350–1356.

Stryer, L. (1986) Cyclic GMP cascade of vision. *Annu. Rev. Neurosci.* 9:87–119.

Stryer, L., and Bourne, H. R. (1986) G proteins: a family of signal transducers. *Annu. Rev. Cell Biol.* 2:391–420.

Other Reviews

Berridge, M. J. (1987) Inositol triphosphate and diacylglycerol: two interacting second messengers. *Annu. Rev. Biochem.* 56:159–193.

Gilman, A. G. (1987) G proteins: tranducers of receptor-generated signals. *Annu. Rev. Biochem.* 56:615–649.

REFERENCES

Abdel-Latif, A. A., Akhtar, R. A., and Hawthorne, J. N. (1977) Acetylcholine increases the breakdown of triphosphoinositide of rabbit iris muscle prelabelled with (^{32}P) phosphate. *Biochem. J.* 162:61–73.

Agranoff, B. W., Murthy, P., and Seguin, E. B. (1983) Thrombin-induced phosphodiesteratic cleavage of phosphatidylinositol biphosphate in human platelets. *J. Biol. Chem.* 258:2076–2078.

Berridge, M. J., and Irvine, R. F. (1989) Inositol phosphate and cell signalling. *Nature* 341:197–205.

Bokoch, G. M., Katada, T., Northup, J. K., Hewlett, E. L., and Gilaman, A. (1983) Identification of the predominant substrate for ADP-ribosylation by islet activating proteins. *J. Biol. Chem.* 258:2072–2075.

Bokoch, G. M., Katada, T., Northup, J. K., Ui, M., and Gilman, A. G. (1984) Purification and properties of the inhibitory guanine nucleotide-binding regulatory component of adenylate cyclase. *J. Biol. Chem.* 259:3560–3567.

Bredt, D. S., and Snyder, S. H. (1990) Isolation of nitric oxide synthetase, a calmodulin requiring enzyme. *Proc. Natl. Acad. Sci. U.S.A.* 87:682–685.

Bredt, D. S., Hwang, P. M., and Snyder, S. H. (1990) Localization of nitric oxide synthase indicating a neural role for nitric oxide. *Nature* 347:768.

Burgoyne, R. D., and Cheek, T. R. (1991) Locating intracellular calcium stores. *Trends in Biochem. Sci.* 16:319–320.

Cassel, D., and Selinger, D. (1976) Catecholamine-stimulated GTPase activity in turkey erythrocyte membranes. *Biochim. Biophys. Acta* 452:538–551.

Cassel, D., and Selinger, D. (1977) Activation of turkey erythrocyte adenylate cyclase and blocking of the catecholamine-stimulated GTPase by guanosine 5′(-thio) triphosphate. *Biochem. Biophys. Res. Commun.* 77:868–873.

Cassel, D., and Selinger, D. (1978). Mechanism of adenylate cyclase activation through the β-adrenergic receptor catecholamine-induced displacement of bound GDP by GTP. *Proc. Natl. Acad. Sci. U.S.A.* 75:4155–4159.

Cassel, D., and Selinger, D. (1977). Mechanism of adenylate cyclase activation by cholera toxin: inhibition of GTP hydrolysis at the regulatory site. *Proc. Natl. Acad. Sci. U.S.A.* 74:3307–3311.

Cassel, D., Rothenberg, P., Zhuang, Y., Deuel, T. F., and Glaser, L. (1983) Platelet derived growth factor stimulates Na^+/H^+ exchange and induces cytoplasmic alkalinization in NR6 cells. *Proc. Natl. Acad. Sci. U.S.A.* 80:6224–6228.

Cerione, R. A., Sibley, D. R., Codina, J., Benovic, J. L., Winslow, J., Neer, E. J., Birnbaumer, L., Caron, M. G., and Lefkowitz, R. J. (1984) Reconstitution of a hormone-sensitive adenylate cyclase system. *J. Biol. Chem.* 259:9979–9982.

Codina, J., Hilderbranch, J., Tyengar, R., Birnbaumer, L., Sekura, R. D., and Manclark, C. R. (1983) Pertussis toxin substrate, the putative Ni component of adenyl cyclases, is an heterodimer regulated by guanine nucleotide and magnesium. *Proc. Natl. Acad. Sci. U.S.A.* 80:4276–4280.

Codina, J., Hildebranch, J., Sunyer, T., Sekura, R. D., Manclark, C. R., Iyengar, R., and Birnbaum, L. (1984) Mechanisms in the vectorial receptor-adenylate cyclase signal transduction. *Adv. Cyclic Nucleotide Res.* 17:111–125.

Drummond, G. I. (1983) Cyclic nucleotides in the nervous system. *Rev. Cyclic Nucleotide Res.* 15:373–494.

Fesenko, E. E., Kolesnikov, S. S., and Lyubarsky, A. L. (1985) Induction by cyclic GMP of cationic conductance in plasma membrane of retinal rod outer segment. *Nature* 313:310–313.

Furuichi, T., Yoshikawa, S., Miyawaki, A., Wada, K., Maeda, N., and Mikoshiba, K. (1989) Primary structure and functional expression of the inositol 1,4,5-triphosphate-binding protein P_{400}. *Nature* 342:32–38.

Hashimoto, S., Bruno, B., Lew, D. P., Pozzan, T., Volpe, P., and Meldolesi, J. (1988) Immunocytochemistry of calcisomes in liver and pancreas. *J. Cell Biol.* 107:2523–2531.

Hathaway, D. R., and Adelstein, R. S. (1979) Human platelet myosin light chain kinase requires the calcium-binding protein calmodulin for activity. *Proc. Natl. Acad. Sci. U.S.A.* 76:1653–1657.

Ignarro, L. J. (1990) Nitric oxide. A novel signal transduction mechanism for intracellular communication. *Hypertension* 16:477–483.

Joseph, S. K., Thomas, A. P., Williams, R. J., Irvine, R. F., and Williamson, J. R. (1984) *Myo*-inositol 1,4,5-trisphosphate: a second messenger for the hormonal mobilization of intracellular Ca^{2+} in liver. *J. Biol. Chem.* 259:3077–3081.

Kaibuchi, K., Takai, Y., and Nishizuka, Y. (1983) Synergistic functions of protein mobilization in platelet activation. *J. Biol. Chem.* 258:6701–6704.

Kaibuchi, K., Takai, Y., and Nishizuka, Y. (1981) Cooperative roles of various membrane phospholipids in the activation of calcium-activated, phospholipid-dependent protein kinase. *J. Biol. Chem.* 256:7146–7149.

Katada, T., and Ui, M. (1982) Direct modification of the membrane adenylate cyclase system by islet-activating protein due to ADP-ribosylation of a membrane protein. *Proc. Natl. Acad. Sci. U.S.A.* 79:3129–3133.

Kawahara, Y., Takai, Y., Minakuchi, R., Sano, K., and Nishizuka, Y. (1980) Phospholipid turnover as a possible transmembrane signal for protein phosphorylation during human platelet activation by thrombin. *Biochim. Biophys. Res. Commun.* 97:309–317.

Kendel, E. R., and Schwartz, J. H. (1982) Molecular biology of learning-modulation of transmitter release. *Science* 218:433–443.

Kishimoto, A., Takai, Y., Mori, T., Kikkawa, U., and Nishizuka, Y. (1980) Activation of calcium and phospholipid-dependent protein kinase by diacylglycerol, its possible relation to phosphatidylinositol turnover. *J. Biol. Chem.* 255:2273–2276.

Klee, C. B., Newton, D. L., Ni, W.-C., and Hiech, J. (1986) Regulation of the calcium signal by calmodulin. In *Calcium and the Cell. CIBA Found. Symp.* 112:162–170.

Krause, K.-H., Simmerman, H. K. B., Jones, L. R., and Campbell, K. P. (1990) Sequence similarity of calreticulin with a Ca^{2+}-binding protein that copurifies with an Ins(1,4,5) P_3-sensitive Ca^{2+} store in HL-60 cells. *Biochem. J.* 270:545–548.

Leffert, H. L., and Koch, K. S. (1985) Growth regulation by sodium ion fluxes. In *Control of Animal Cell Proliferation* (Boynton, A. L., and Leffert, H. L. eds.), Vol. 1, pp. 367–413. Academic Press, New York.

Logothetis, D., Kurachi, Y., Galper, J., Neer, E. J., and Calpham, D. E. (1987) The subunits of GTP-binding proteins activate the muscarinic K^+ channel in heart. *Nature* 325:321–326.

Marsden, P. A., and Ballermann, B. J. (1990) Tumor-necrosis-factor-α activated soluble guanylate cyclase in bovine mesangial cell via an L-arginine dependent mechanism. *J. Exp. Med.* 172:1843–1852.

May, D. C., Ross, E. M., Gilman, A. G., and Smigel, M. D. (1985) Reconstitution of catecholamine-stimulated adenylate cyclase activity using three purified proteins. *J. Biol. Chem.* 260:15829–15833.

Moncada, S., Palmer, R. M. J., Higgs, E. A. (1989) Biosynthesis of nitroxide from L-arginine—a pathway for the regulation of cell function and communication. *Biochem. Pharm.* 38:1709–1715.

Nishizuka, Y. (1984) Turnover of inositol phospholipids and signal transduction. *Science* 225:1365–1370.

Penninston, J. T. (1983) Plasma membrane Ca^{2+} ATPases as active Ca^{2+}-pumps. *Calcium and Cell Function* 4:100–149.

Pfeuffer, T. (1977) GTP-binding proteins in membranes and the control of adenylate cyclase activity. *J. Biol. Chem.* 252:7224–7234.

Rasmussen, H., and Barrett, P. Q. (1984) Calcium messenger systems, an integrated view. *Physiol. Rev.* 64:938–984.

Rodbell, M., Birnbaumer, L., Pohl, S. L. and Krans, H. M. J. (1971) The glucagon-sensitive adenyl cyclase system in plasma membranes of rat liver. *J. Biol. Chem.* 246:1877–1882.

Ross, E. M., and Gilman, A. G. (1977a) Reconstitution of catecholamine-sensitive adenylate cyclase activity: interaction of solubilized components with receptor-replete membranes. *Proc. Natl. Acad. Sci. U.S.A.* 74:3715–3719.

Ross, E. M., and Gilman, A. G. (1977b) Resolution of some components of adenylate cyclase necessary for catalytic activity. *J. Biol. Chem.* 252:6966–6989.

Ross, E. M., Howlett, A. C., Ferguson, K. M., and Gilman, A. G. (1978) Reconstitution of hormone-sensitive adenylate cyclase activity with resolved components of the enzyme. *J. Biol. Chem.* 253:6401–6412.

Sano, K., Takai, Y., Yamanashi, J., and Nishizuka, Y. (1983) A role of calcium-activated phospholipid-dependent protein kinase in human platelet activation. *J. Biol. Chem.* 258:2010–2013.

Schatzmann, H. J. (1985) Calcium extrusion across the plasma membrane by the Ca^{2+}-pump and the Ca^{2+}-Na^{+} exchange. In *Calcium and Cell Physiology* (Marmé, D., ed.), pp. 18–52. Springer-Verlag, Berlin.

Schramm, M., and Selinger, Z. (1984) Message transmission: receptor controlled adenylate cyclase system, *Science* 225:1350–1356.

Schulman, H., and Lou, L. L. (1989) Multifunctional Ca^{2+}/calmodulin dependent protein kinase: domain structure and regulation. *Trends Biochem. Sci.* 14:62–66.

Schulster, D., Orly, J., Seidel, G., and Schramm, M. (1978) Intracellular cyclic AMP production enhanced by a hormone receptor transferred from a different cell. *J. Biol.Chem.* 253:1201–1206.

Sternweis, P. C. (1986) The purified subunits of G_o and G_i from bovine brain required for association with phospholipid vesicles, *J. Biol. Chem.* 261:631–637.

Streb, H., Irvine, R. F., Berridge, M. J., and Schulz, I. (1983) Release of Ca^{2+} from a nonmitochondrial intracellular store in pancreatic acinar cells by inositol-1,4,5-trisphosphate. *Nature* 306:67–69.

Stryer, L. (1986) Cyclic GMP cascade of vision. *Annu. Rev. Neurosci.* 9:87–119.

Stryer, L., and Bourne, H. R. (1986) A family of signal transducers. *Annu. Rev. Cell Biol.* 2:391–420.

Takeshima, H., Nishimura, S., Matsumoto, T., Ishida, H., Kangawa, K., Minamino, N., Matsuo, H., Ueda, M., Hanaskam, M., Hirose, T., and Numa, S. (1989) Primary structure and expression from complementary DNA of skeletal muscle ryanodine receptor. *Nature* 339:439–445.

Teo, T. S., and Wang, J. H. (1973) Mechanism of activation of a cyclic adenosine 3′:5′-phosphate diesterase. *J. Biol. Chem.* 248:5950–5955.

Treves, S., De Mattei, M., Lanfredi, M., Villa, A., Green, M., MacLennan, D. H., Meldolesi, J., and Pozzan, T. (1990) Calreticulin, a candidate for a calsequestrin-like function in Ca^{2+}-storage compartments (calcisomes) of liver and brain. *Biochem. J.* 272:473–480.

Volpe, P., Krause, K.-H., Hashimoto, S., Zorzato, F., Pozzan, T., Meldolesi, J., and Lew, D. P. (1988) "Calciosome," a cytoplasmic organelle, the inositol, 1,4,5-triphosphate-sensitive Ca^{2+} store of nonmuscle cells. *Proc. Natl. Ac. Sci. U.S.A.* 85:1091–1095.

Werner-Felmayer, G., Werner, E. R., Fuchs, D., Hausen, A., Reibnegger, G., and Wachter, H. (1990) Tetrahydropterin-dependent formation of nitrite and nitrate in murine fibroblasts. *J. Exp. Med.* 172:1599–1607.

Chapter 6

Energy and Biological Systems

Energy is central to cell function. Energy is needed for carrying out the tasks of the cell, the displacement of mass in biological movement, the fluxes of solutes through membranes against electrochemical gradients (concentration and electrical potential gradients), the emission of light (bioluminescence), and the synthesis of macromolecules to replace cell components that are broken down. Ultimately, the energy must come from the environment: from the light absorbed by photosynthetic pigments and from the substances that serve as substrates in the cell's metabolism. In a sense, cells or cell organelles may be regarded as transducer systems, i.e., devices that convert one form of energy into another. Two transducer systems familiar in everyday life are shown schematically in Fig. 6.1*a* and *b*. Figure 6.1*a* shows the conversion of mechanical work into sound as, for example, in a record player. Figure 6.1*b* shows the conversion of light into an electrical current, as in a photocell. The present chapter reviews very briefly some of the principles of bioenergetics and is intended to serve as a framework for the discussion of later chapters.

I. FREE ENERGY

The Gibbs free energy change, ΔG, expresses the maximal amount of work that can be performed by a system or, conversely, the minimal amount of energy input required for work.

The definition of ΔG is such that net work can be performed by a reaction when ΔG is less than zero; i.e., the reaction is exergonic. Unless coupled to exergonic reactions, endergonic reactions ($\Delta G > 0$) take place to a very limited extent. When $\Delta G = 0$, the system is at equilibrium. For reactions taking place at constant temperature, ΔG can be expressed as

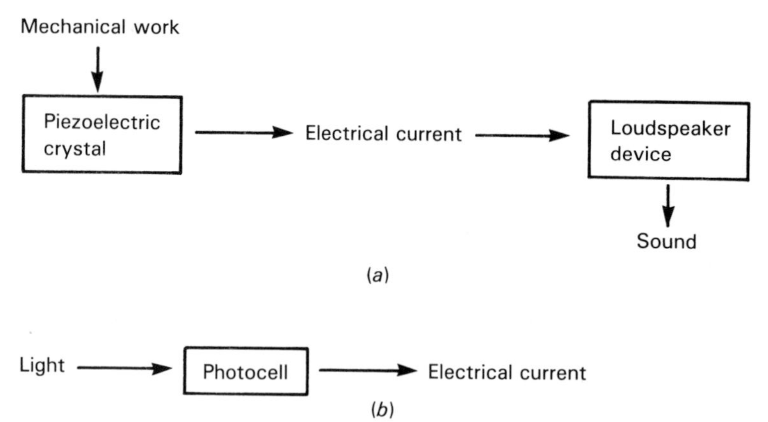

Fig. 6.1 Familiar transducer systems.

$$\Delta G = \Delta H - T\Delta S \tag{6.1}$$

ΔH is the heat transferred between the system under discussion and the surrounding at constant temperature. In an exothermic reaction heat is released by the system ($\Delta H < 0$), whereas in an endothermic reaction heat is absorbed by the system ($\Delta H > 0$). The entropy of the system, ΔS, is defined as shown in Eq. (6.2), where dQ is the change in the heat of the system.

$$dS = -\frac{dQ}{T} \tag{6.2}$$

In practical terms, the entropy is related to the organization of the system. In an isolated system (i.e., one that does not exchange heat or matter with its environment) ΔS must be positive.

Two examples of exergonic reactions are shown below in Eqs. (6.3) and (6.4). The superscript ° (e.g., in $\Delta G°$, $\Delta H°$) means that the quantity has been determined under standard temperature (usually 298 K), pressure (1 atmosphere), and concentration (1 molal). Since biological reactions do not take place under these standard conditions, we should be concerned with ΔG, but for simplicity the system will be assumed to be at standard conditions. The difference between ΔG and $\Delta G°$ and the meaning of these two parameters are discussed in Sections IV and V. The oversimplification of equating ΔG to $\Delta G°$ has been vigorously challenged (Banks and Vernon, 1970). The ΔG value changes as a function of concentration, and in the case of active transport of ions ΔG is also a function of membrane potential. This aspect will be taken up in Sections IV and VI.

$$\begin{aligned} \text{ATP} \rightarrow \text{ADP} + \text{P}_i \qquad \Delta H° &= -7.4 \text{ kcal} \\ -T\,\Delta S° &= \underline{-2.6 \text{ kcal}} \\ \Delta G° &= -10.0 \text{ kcal} \end{aligned} \tag{6.3}$$

Equation (6.3) represents the hydrolysis of the terminal phosphate of adenosine triphosphate (ATP) to form adenosine diphosphate (ADP) and inorganic orthophosphate (P_i). In this reaction $\Delta H°$ and $\Delta G°$ are both less than zero. However, $\Delta H°$ need not always be less than zero for a reaction to take place, as shown for the activation of NADH dehydrogenase, represented by E in Eq. (6.4).

$$\text{E} + \text{AMP} \rightarrow \text{E—AMP} \qquad \begin{aligned} \Delta H^\circ &= 12.5 \text{ kcal} \\ -T\,\Delta S^\circ &= -17.3 \text{ kcal} \\ \hline \Delta G^\circ &= -4.8 \text{ kcal} \end{aligned} \tag{6.4}$$

The units used for energy vary in part for convenience, in part for historical reasons. The calorie (or kilocalorie = 1000 cal) has been frequently used, and kilocalories can easily be converted to electrical potential units by dividing by the Faraday constant, F (23 kcal/V). For a direct comparison of ΔG values to the experimentally obtained redox potentials, see Section III. In muscular contraction the joule (1×10^7 ergs, 0.24 cal) is frequently used.

II. COUPLED REACTIONS

Chemical reactions with $\Delta G > 0$ can take place to a significant extent only if coupled to another reaction in which $\Delta G < 0$. The coupling requires the product of the first reaction to be the reactant in the subsequent reaction. If the reactions were not linked in this manner, they would occur independently of each other and therefore no energy could be transferred between these molecules by ordinary means at constant temperature and pressure. A hypothetical set of coupled reactions is shown in Eq. (6.5).

$$\text{A} + \text{B} \rightleftharpoons \text{C} + \text{D} \qquad \Delta G_1^\circ = 5 \text{ kcal} \tag{6.5a}$$

$$\text{C} \rightleftharpoons \text{X} + \text{Y} \qquad \Delta G_2^\circ = -7 \text{ kcal} \tag{6.5b}$$

$$\text{A} + \text{B} \rightleftharpoons \text{D} + \text{X} + \text{Y} \qquad \Delta G^\circ = -2 \text{ kcal} \tag{6.5c}$$

The hypothetical product of reaction (6.5a), C, is a reactant in the second reaction, (6.5b). Consequently, in the overall reaction (6.5c), C does not appear at all, since no net change in C occurs. The ΔG of the overall reaction (6.5c) is less than zero since the ΔG of the coupled reactions is additive. Equation (6.6a) gives the formal description of an actual biochemical reaction that requires an energy input in order to proceed to a significant extent. The energy-yielding reaction is represented by Eq. (6.6b). Equations (6.6a) and (6.6b) correspond to an outline. The component needed to link the two reactions is ignored. In fact, the molecular details need not be known; only the overall reaction is necessary.

$$\text{glutamate} + \text{NH}_4^+ \rightleftharpoons \text{glutamine} \qquad \Delta G^\circ = 3.7 \text{ kcal} \tag{6.6a}$$

$$\text{ATP} \rightleftharpoons \text{ADP} + \text{P}_i \qquad \Delta G^\circ = -7.4 \text{ kcal} \tag{6.6b}$$

$$\text{NH}_4^+ + \text{glutamate} + \text{ATP} \rightleftharpoons \text{glutamine} + \text{P}_i + \text{ADP} \qquad \Delta G^\circ = -3.7 \text{ kcal} \tag{6.6c}$$

The reaction actually takes place by the mechanism shown in Eqs. (6.6d) and (6.6e). An intermediate glutamyl — P — E in the first reaction is a participant in the second reaction.

$$\text{glutamate} + \text{ATP} + \text{E} \rightleftharpoons \text{glutamyl — P — E} + \text{ADP} \tag{6.6d}$$

$$\text{glutamyl — P — E} + \text{NH}_4^+ \rightleftharpoons \text{glutamine} + \text{P}_i + \text{E} \tag{6.6e}$$

In Eqs. (6.6d) and (6.6e), E represents an enzyme molecule. The energy for reaction (6.6a) is said to proceed from the hydrolysis of ATP, reaction (6.6b), and the actual common intermediate (glutamyl — P — E) need not be known to carry out calculations involving the energy balance of the reactions, i.e., ΔG°.

The synthesis of ATP from ADP and P_i in so-called substrate-level phosphorylation takes place by a coupling similar to that discussed in these reactions. They involve as an intermediate a phosphorylated substrate. In contrast, in most cells, the reactions responsible for most of the synthesis of ATP involve oxidation-reduction reactions of the cytochrome system in a process referred to collectively as oxidative phosphorylation. These reactions are likely to occur by a distinct mechanism (see later discussion). In eukaryotic cells the cytochrome system is in the mitochondria.

An example of substrate-level phosphorylation is the oxidation of glyceraldehyde 3-phosphate to form 3-phosphoglycerate in one of the reactions of glycolysis. In this step, ADP is concomitantly phosphorylated to form ATP. Equations (6.7a) and (6.7b) represent the process schematically.

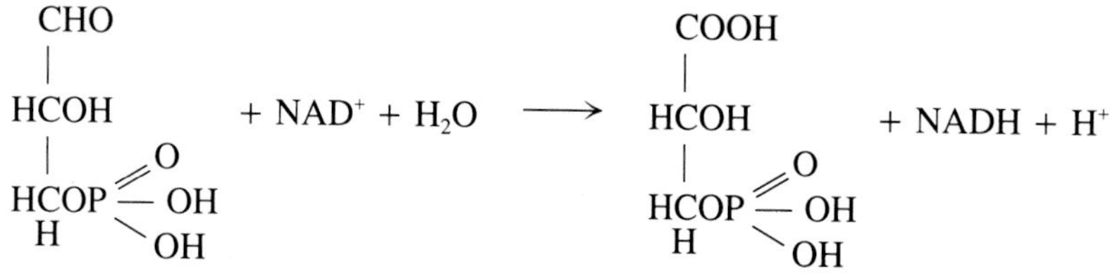

$$\begin{array}{l} \text{CHO} \\ | \\ \text{HCOH} \\ | \\ \text{HCOP(=O)(OH)OH} \\ \text{H} \end{array} + NAD^+ + H_2O \longrightarrow \begin{array}{l} \text{COOH} \\ | \\ \text{HCOH} \\ | \\ \text{HCOP(=O)(OH)OH} \\ \text{H} \end{array} + NADH + H^+ \qquad (6.7a)$$

Glyceraldehyde 3–phosphate 3–Phosphoglycerate

$$ADP + P_i \longrightarrow ATP \qquad (6.7b)$$

The oxidation of glyceraldehyde 3-phosphate has a ΔG° of less than −10 kcal, whereas the phosphorylation of ADP is endergonic and has a ΔG° of about 10 kcal. In effect, the energy yielded by one reaction is trapped by the synthesis of ATP. A later event, hydrolysis of the ATP, can yield the energy necessary for other processes.

The details of the reactions show that in one reaction, Eq. (6.8a), a product, 1,3-phosphoglycerate, is formed that is used in the subsequent reaction depicted by Eq. (6.8b). The coupling scheme shown by Eqs. (6.8a) and (6.8b) is somewhat oversimplified but sufficiently detailed to illustrate this mechanism.

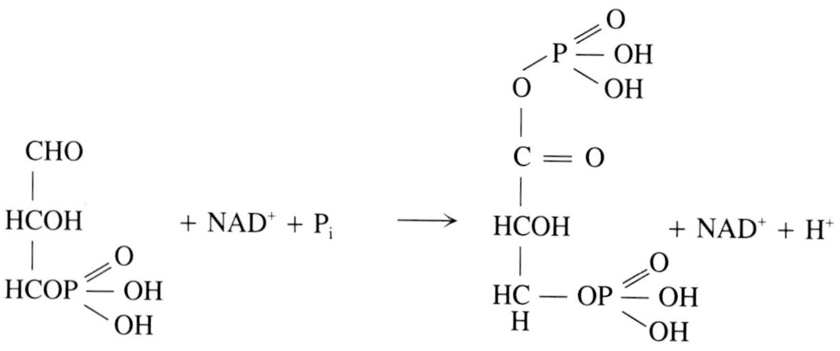

$$\begin{array}{l} \text{CHO} \\ | \\ \text{HCOH} \\ | \\ \text{HCOP(=O)(OH)OH} \end{array} + NAD^+ + P_i \longrightarrow \begin{array}{l} \text{O–P(=O)(OH)OH} \\ | \\ \text{C=O} \\ | \\ \text{HCOH} \\ | \\ \text{HC—OP(=O)(OH)OH} \\ \text{H} \end{array} + NAD^+ + H^+ \qquad (6.8a)$$

Glyceraldehyde 3–phosphate 1, 3–Phosphoglycerate

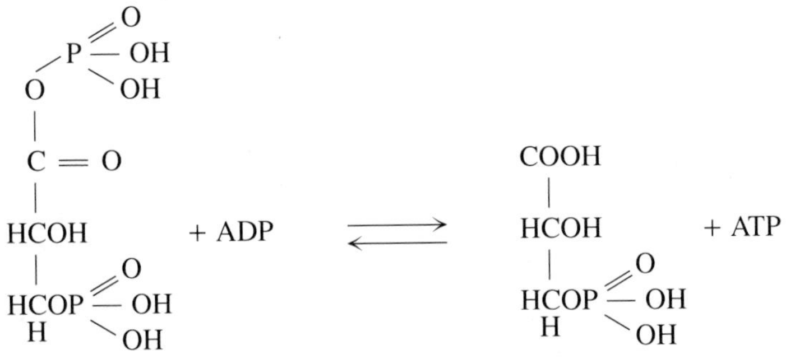

$$\begin{array}{lcclc}
\begin{array}{l} \mathrm{O{-}P({=}O)(OH)_2} \\ | \\ \mathrm{C=O} \\ | \\ \mathrm{HCOH} \\ | \\ \mathrm{HCOP(=O)(OH)_2} \\ \mathrm{H} \end{array} & + \mathrm{ADP} & \rightleftharpoons & \begin{array}{l} \mathrm{COOH} \\ | \\ \mathrm{HCOH} \\ | \\ \mathrm{HCOP(=O)(OH)_2} \\ \mathrm{H} \end{array} & + \mathrm{ATP}
\end{array} \qquad (6.8b)$$

1, 3–Phosphoglycerate 3–Phosphoglycerate

In the past, couplings similar to those represented in Eqs. (6.8a) and (6.8b) have also been postulated for the phosphorylation of ADP in the oxidative phosphorylation reactions of the cytochrome chain. The mechanism of phosphorylation involving the cytochrome chain is poorly understood; however, it does not involve a phosphorylated intermediate.

Not all the energy derived from a reaction can be used for the synthesis of ATP. Some of the energy is released as heat (Poe and Estabrook, 1969). This release of heat is of great physiological importance in mammals. The nonshivering thermogenesis that occurs in brown fat apparently results from an increase in the energy dissipated as heat in the oxidative reactions in mitochondria (Flatmark and Pedersen, 1975). This form of thermogenesis plays a fundamental role in cold adaptation and arousal from hibernation.

The idea of energy coupling, and in particular the concept that ATP can be used to power in vivo reactions, has been questioned at various times (e.g., see Banks and Vernon, 1970). Although the concepts are sometimes misunderstood and should be used with caution (e.g., see McClare, 1972), they are nevertheless valid.

III. REDOX POTENTIALS

In some reactions a reactant serves as an electron donor and another as an electron acceptor. In these cases the ability to exchange electrons can be expressed as an oxidation-reduction potential (*redox potential*). Removal of electrons from the donor and acceptance of electrons by the acceptor can be formulated as separate reactions, as shown in Eqs. (6.9a) and (6.9b). In these equations, e^- represents an electron that is being exchanged. Equation (6.9c) represents the overall reaction. "Red" and "ox" indicate, respectively, the reduced and oxidized species of the compound. Oxidation-reduction reactions must be coupled to each other in order to take place, but they can be separated out as done in Eqs. (6.9a) and (6.9b) for calculations involving the energetics of the system.

$$\mathrm{red}_1 \rightleftharpoons \mathrm{ox}_1^{n+} + ne^- \qquad (6.9a)$$

$$\mathrm{ox}_2^{n+} + ne^- \rightleftharpoons \mathrm{red}_2 \qquad (6.9b)$$

$$\mathrm{red}_1 + \mathrm{ox}_2 \rightleftharpoons \mathrm{ox}_1 + \mathrm{red}_2 \qquad (6.9c)$$

If an electrical connection is made between two containers, these reactions can actually take place separately. Each container is called a "half-cell." The system made up of two half-cells is represented in Fig. 6.2, in which the electrode of half-cell A receives an electron that serves to reduce the component of half-cell B.

$$2H^+ + 2e^- \rightleftharpoons H_2 \tag{6.10}$$

The potential of a half-cell is conventionally defined by comparison to that of an H_2 half-cell that operates as shown in Eq. (6.10). When electrons are given to the H_2 electrode, the potential is considered to be less than zero (i.e., E_h is negative). When the hydrogen half-cell is the electron donor, the potential is considered greater than zero (E_h is positive). In comparing the tendency of two reactions to take place, we are actually concerned with the difference in the redox potential between the two reactions ($E_{h1} - E_{h2}$), which is expressed as the ΔE of the two half-cells, and the ΔG of the overall reaction is represented by Eq. (6.11).

$$\Delta G = -nF\ \Delta E \tag{6.11}$$

Here ΔE is the redox potential difference (in volts), n refers to the number of electrons, and F is the Faraday constant (23 kcal/V). The use of ΔE and its interconvertibility with ΔG can best be illustrated by an example. A hypothetical redox scheme is shown in Eq. (6.12a). The two half-reactions (i.e., the portions of the reactions occurring in the two half-cells) are shown in Eqs. (6.12b) and (6.12c).

$$AH_2 + B \rightleftharpoons BH_2 + A \tag{6.12a}$$
$$AH_2 \rightleftharpoons A + 2H^+ + 2e^- \qquad E^\circ_{h_{12b}} = 0.30\ \text{V} \tag{6.12b}$$
$$B + 2H^+ + 2e^- \rightleftharpoons BH_2 \qquad E^\circ_{h_{12c}} = -0.25\ \text{V} \tag{6.12c}$$

Since reaction (6.12c) is driven by reaction (6.12b),

$$\Delta E^\circ = 0.30\ \text{V} - (-0.25\ \text{V}) = 0.55\ \text{V}$$

or

$$\Delta G^\circ = -2 \times 0.23\ \text{kcal (V mol)}^{-1} \times 0.55\ \text{V} = -25.3\ \text{kcal}$$

The reaction depicted in Eq. (6.12c) involves H^+, and therefore E will depend on the pH of the mixture (pH $= -\log[H^+]$). Usually the redox potentials have been evaluated from values determined at some standard pH (e.g., pH 7).

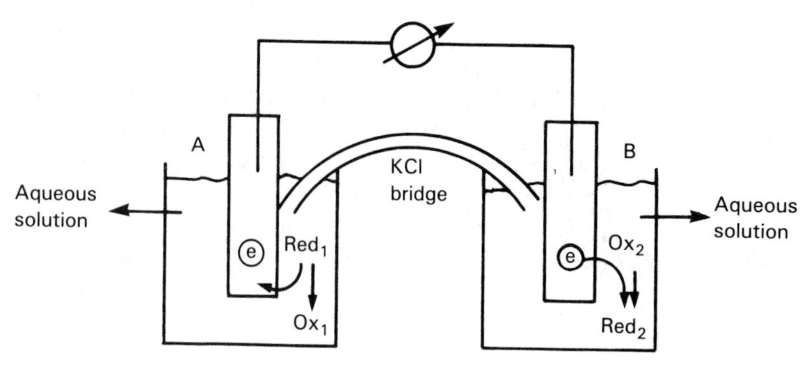

Fig. 6.2 Two half-cells.

IV. ΔG AS A FUNCTION OF THE CONCENTRATION OF REACTANTS

As already mentioned, the actual parameter that is pertinent to the energy available to perform work is ΔG rather than ΔG°, and in this section we will focus on this parameter.

The total capacity of a substance to perform useful work actually depends on its chemical potential, μ. The change of μ as a function of concentration is shown in Eq. (6.13a), which follows directly from the gas laws and is integrated in Eq. (6.13b).

$$d\mu = \frac{RT\,dC}{C} \tag{6.13a}$$

$$\mu = \mu^\circ + RT \ln C \tag{6.13b}$$

In these equations, C is the concentration of the substance in question, and μ° is the chemical potential under standard conditions. In a chemical reaction such as A→B, the ΔG will correspond to the difference in chemical potential between the two components as shown in (Eq. 6.14).

$$\begin{aligned}\Delta G = \mu_B - \mu_A &= \mu_B{}^\circ - \mu_A{}^\circ + RT \ln C_B - RT \ln C_A \\ &= \Delta G^\circ + RT \ln C_B - RT \ln C_A \\ &= \Delta G^\circ + RT \ln(C_B/C_A)\end{aligned} \tag{6.14}$$

In Eq. (6.14), $\Delta G^\circ = \mu_B - \mu_A$. Following Eq. (6.14), the ΔG for the synthesis of 1 mol of ATP would be

$$\Delta G = \Delta G^\circ + RT \ln \frac{[\mathrm{ATP}]}{[\mathrm{ADP}]\,[\mathrm{P_i}]} \tag{6.15a}$$

To convert $\log_e$ (ln) to $\log_{10}$, and to simplify the calculation, $2.3RT$ can be considered to be 1.4 kcal/mol (i.e., 2.3×2 cal mol^{-1} deg^{-1} $\times$ 300 K). Equation (6.15a) can be represented by Eq. (6.15b) if ΔG° is assumed to be 10 kcal/mol.

$$\begin{aligned}\Delta G = {} & 10\text{ kcal/mol} - 1.4\text{ kcal/mol }(\log_{10}[\mathrm{P_i}]) \\ & + 1.4\text{ kcal/mol }(\log_{10}[\mathrm{ATP}]/[\mathrm{ADP}])\end{aligned} \tag{6.15b}$$

Equations (6.15a) and (6.15b) show that the ΔG for the synthesis or, inversely, the ΔG available from the hydrolysis of ATP depend on the concentrations of the components. This fact has important consequences. For example, in muscle the ATP concentration is maintained maximally by the transfer of the phosphate from phosphocreatine. Creatine phosphokinase replenishes any ATP hydrolyzed by muscle contraction and other energy-requiring reactions. It is only when 90% of the phosphocreatine is used up that ATP begins to fall significantly, to about 10% of its resting state. The high [ATP]/[ADP] ratio permits a ΔG of -12.5 kcal/mol, a much greater magnitude than the ΔG°, which may be as low as -7.6 kcal/mol.

Generally, the hydrolysis of ATP or the reverse reaction, the synthesis of ATP, is written in a simplified form that involves only ATP, ADP, and P_i, as in (Eq. 6.3). In actuality, the reaction involves H^+, as shown in (Eq. 6.16).

$$\mathrm{ATP} \rightleftharpoons \mathrm{ADP} + \mathrm{P_i} + \mathrm{H^+} \tag{6.16}$$

The calculation of the ΔG for ATP synthesis therefore would follow Eq. (6.17a).

$$\Delta G = \Delta G^\circ + RT \ln \frac{[\mathrm{ATP}]}{[\mathrm{ADP}]\,[\mathrm{P_i}]\,[\mathrm{H^+}]} \tag{6.17a}$$

Since pH $= -\log_{10}[\mathrm{H^+}]$, Eq. (6.17a) can be represented as shown in Eq. (6.17b).

$$\Delta G = \Delta G^\circ + 1.4\frac{\text{kcal pH}}{\text{mol}} + 1.4\frac{\text{kcal}}{\text{mol}}\log_{10}\frac{[\mathrm{ATP}]}{[\mathrm{ADP}]\,[\mathrm{P_i}]} \tag{6.17b}$$

When the simpler Eq. (6.15a) is used, the ΔG° is assumed to be at a standard pH (e.g., 7.0 or 7.4). As noted previously, ΔG is significantly sensitive to pH, and a difference of one pH unit lowers the ΔG for the formation of ATP by 1.4 kcal/mol. Since the synthesis of ATP during oxidative phosphorylation (Chapter 10) or photophosphorylation (Chapter 11) is associated with membranes, the actual pH values are not known. There have been proposals that during oxidative phosphorylation or photosynthetic phosphorylation the pH of the phosphorylation sites is changed. A decrease of pH from 7 to 5, for example, would lower the ΔG as much as 2.8 kcal/mol, driving the reaction of Eq. (6.16) toward ATP synthesis.

In the present discussion, we have assumed that the ΔG° for the synthesis of ATP (i.e., the $-\Delta G^\circ$ of its hydrolysis) is in the neighborhood of 10 kcal/mol. Published estimates of ΔG° for the hydrolysis of ATP in the pH range 6 to 9 and at several Mg^{2+} concentrations (Mg^{2+} required for many biological reactions is present in the cytoplasm and binds to ATP) at 25° and 35°C range from 6.1 to 10.9 kcal/mol (see Bridger and Henderson, 1983), depending on conditions.

The energy required for the synthesis of 1 mol of ATP is frequently referred to as the *phosphate potential* and is expressed in either kilocalories or units of electrical potential (millivolts). Electrical potential units make it possible to compare the energy needed for ATP synthesis directly with the redox potentials for the cytochromes. Calculations of phosphate potentials for metabolizing isolated mitochondria have been published (Slater et al., 1973).

V. ΔG°

As mentioned in Section I, ΔG° is the ΔG under standard conditions of temperature, pressure, and concentration. ΔG° is related to the equilibrium constant (K) of the reaction. In Eq. (6.14), ΔG has been shown to correspond to $\Delta G^\circ + RT \ln(C_A/C_B)$. Since $\Delta G = 0$ at equilibrium, ΔG° can be expressed as shown in Eq. (6.18).

$$\Delta G^\circ = -RT \ln(C_B/C_A) = -RT \ln K \tag{6.18}$$

VI. ENERGY COST OF TRANSPORT

Biological systems are made up of many compartments, such as kidney tubules, cells, or subcellular organelles, enclosed by selective membranes of varying complexity. The movement of molecules from one biological compartment to another is known as transport. Transport requires an expenditure of energy whenever it occurs against an electrochemical gradient (i.e., the transport is in an uphill direction). In these cases, it is referred to as *active transport*.

For a nonelectrolyte, the chemical potential μ of a compartment in relation to a single solute is represented by Eq. (6.19a), where n represents the number of moles and μ_1 the chemical potential per mole of the solute.

$$\mu = n\mu_1 \tag{6.19a}$$

The total chemical potential is the summation of μ's for all solutes as in Eq. (6.19b). Each subscript of μ or n indicates a different solute. Here the discussion will be restricted to a single solute.

$$\mu = n_1\mu_1 + n_2\mu_2 + n_3\mu_3 + \cdots + n_i\mu_i \tag{6.19b}$$

In a system made up of two compartments separated by a biological membrane, the ΔG for the transfer of 1 mole of solute from one compartment to another will correspond to the difference in chemical potential of the solute between the two compartments. If μ_A represents the chemical potential of the solute in compartment A and μ_B the chemical potential in compartment B, the ΔG can be shown to correspond to Eq. (6.20) by subtracting the individual chemical potentials as expressed by Eq. (6.13). Since the standard chemical potential is the same in both compartments, $\mu_B{}^\circ - \mu_A{}^\circ = 0$.

$$\Delta G = \mu_B - \mu_A = \mu_B{}^\circ - \mu_A{}^\circ + RT \ln C_B - RT \ln C_A = RT \ln\left(\frac{C_B}{C_A}\right) \tag{6.20}$$

In the case of ions, the potential across the membrane and the charges of the solute molecules also have to be taken into account. We are then concerned with the electrochemical potential of the solute. The electrochemical potential (μ_ε) for a given solute corresponds to Eq. (6.21).

$$\mu_\varepsilon = \mu_1 + zF\varepsilon \tag{6.21}$$

Here ε is the electrical potential, z is the valence, and F is the Faraday constant. Consequently, the ΔG for the transfer of 1 mol of the ion is represented by Eq. (6.22), where Ψ_m is the membrane potential.

$$\Delta G = RT \ln(C_B/C_A) + zF\,\Delta\Psi_m \tag{6.22}$$

Equation (6.22) was obtained by subtracting the μ of phase B from that of phase A.

One of the compartments may contain a nondiffusible charged component such a macromolecule or a colloid. This can give rise to a special kind of equilibrium, called a Donnan equilibrium, in which the diffusible ions distribute unequally between the two phases. Ions opposite in charge to the nondiffusible component will have a higher concentration in its compartment. The reverse is true for ions with charges of the same sign as that of the nondiffusible component. The exact proportion can be calculated from Eq. (6.22). When $\Delta G = 0$, $RT \ln(C_B/C_A) = -zF\,\Delta\Psi_m$ or $C_B/C_A = [\exp(-z\Delta\Psi_m F)]/RT$. For this special case the electrical potential is known as the Donnan potential. The $\Delta\Psi_m$ is related to the charge of the nondiffusible component. A Donnan distribution does not require the presence of a membrane. The nondiffusible component may be fixed to a structure rather than restrained by a limiting membrane.

From Eq. (6.22), the energy required for the transport of ions can be calculated. For instance, consider the transport of Na^+ and K^+. In most cells the two transports are coupled, and they take place in opposite directions. The Na^+ is transported outward, whereas K^+ is transported inward. The ion concentrations used in the calculation, which are shown in Table 6.1, correspond approximately to those of the squid giant axon. Columns 2 and 3 give the internal and external concentrations of Na^+ and K^+. The potential across the membrane is shown in column 4. The inside of the axon is negative in relation to the outside. The electrical potential across the cell membrane would therefore favor the entry of K^+ and oppose the exit of Na^+ as shown in column 5, which represents the electrical potential component of ΔG. The chemical potential component of ΔG is represented in column 6, and column 7 shows the ΔG for each transfer. The energy expenditure is primarily due to the transport of the Na^+. If we assume that the transport is coupled to ATP hydrolysis, the breakdown of 1 mol of ATP would suffice for more than the simultaneous transfer of 1 mol of Na^+ and K^+, since the ΔG of hydrolysis of ATP is approximately -10 kcal/mol.

So far, the discussion has concerned the expenditure of energy necessary for active transport and the minimal energy cost (i.e., assuming 100% efficiency) has been calculated for one example. We may well want to ask the opposite question. Can the energy available from the passage of solute in the direction of the electrochemical gradient be harnessed and used for some other process? As shown by substituting the appropriate values in Eq. (6.20) or (6.22), the transfer in the direction of the gradient provides a $\Delta G < 0$, for example in the transfer of a nonelectrolyte from phase A to phase B where $C_A > C_B$. Such coupling should be feasible at least as far as the energy available for the process is concerned. The flow of one solute in the direction of its gradient can be coupled to the flow of another solute against the electrochemical gradient and in the opposite direction. This is shown, for example, in the experiment depicted in Fig. 6.3 (Rosenberg and Wilbrandt, 1958). In this experiment a suspension of human red blood cells has been equilibrated with [^{14}C]glucose. The radioactivity of the external medium is represented on the ordinate and time is shown on the abscissa. Although it has been recognized that a special mechanism is needed for the transfer of the sugars across the plasma membrane, these cells normally do not transport against a concentration gradient. However, on addition of mannose or unlabeled glucose to the medium, the cells begin to transport labeled glucose outward against the concentration gradient as shown in curves 1 and 2. Curve 3 represents a control in which only a salt solution was used; the radioactivity of the external medium corresponds to the appropriate dilution. The passage against the concentration gradient is maintained only for a limited period, and eventually the system equilibrates again. This equilibration is to be expected, since the gradient for the mannose or the unlabeled glucose is dissipated. This experiment demonstrates that the energy from the concentration gradient of the solute added to the medium can be harnessed to provide an outward flow of another solute against a concentration gradient. Outflow of one solute in response to the inflow of another has been called counterflow, and it will be discussed later.

Table 6.1 Energy Requirement for the Transport of Na^+ and K^+ in an Axon of a Marine Invertebrate[a]

(1) Ions	(2) Internal concentration (M)	(3) External concentration (M)	(4) $\Delta\Psi_m$ (V)	(5) $zF\,\Delta\Psi_m$ (kcal/mole)	(6) $2.3RT \log (C_2/C_1)$ (kcal/mole)	(7) ΔG (kcal/mole)	(8) ΔG of coupled transport (kcal/mole)
K^+	0.40	0.010	0.06	−1.4	2.2	0.8	3.5
Na^+	0.05	0.460	0.06	+1.4	1.3	2.7	

[a]2.3 RT was assumed to be 1.4 kcal/mole.

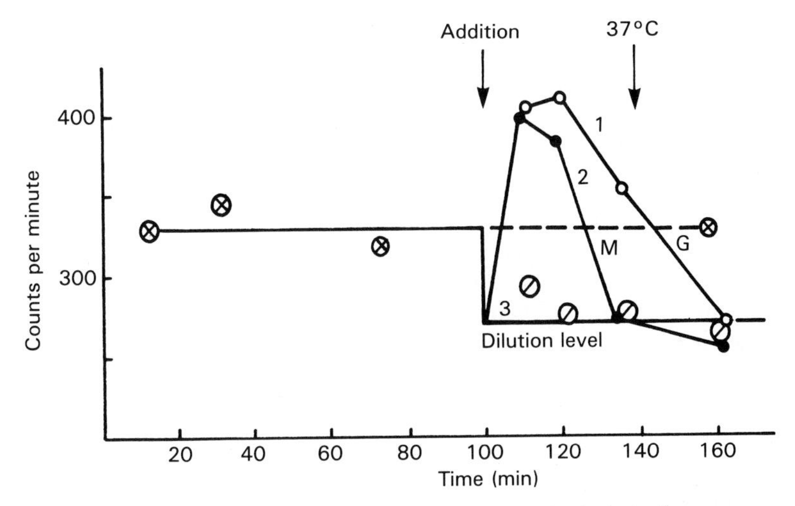

Fig. 6.3 Experiment showing uphill transport of labeled glucose across the human red cell membrane induced by counterflow of (●) mannose or (○) unlabeled glucose. Ordinate, activity in 10 µl of the external medium. Circle with × in center, activity before addition of 0.16 volume of unlabeled sugar (0.72 M in saline). Circle with bar, activity after addition of 0.16 volume of saline. Temperature 0°C until second arrow, then 37°C. (The temperature was raised to accelerate the penetration, which, however, proved to be unnecessary.) The calculated maximal concentration ratio for labeled glucose is approximately 4 (Rosenberg and Wilbrandt, 1958). Reproduced from *Journal of Gen. Physiol.,* 1958, vol. 41, pp. 289–296 by copyright permission of The Rockefeller University Press.

The transport of one solute may also proceed against its concentration gradient when it is coupled to the flow of another solute in the direction of its own concentration gradient, providing the necessary energy. This type of translocation has been called *cotransport*. A study of pigeon red cells (Vidaver, 1964) has shown inward cotransport of Na^+ and glycine, with two Na^+ transferred per glycine molecule. The dependence of the glycine transport on the Na^+ gradient has been shown in experiments where the internal concentration of ions has been varied. Red blood cells can be made leaky, exposed to a medium of the desired composition, and then resealed. Table 6.2 shows how the internal composition of human red blood cells can be varied

Table 6.2 Variation of Erythrocyte Cation Composition after Special Treatment

Concentration of cation in loading medium (mM)		Final cation content of cells (µEq/ml of cells)		
Na^+	K^+	Na^+	K^+	$Na^+ + K^+$
40	260	12	108	120
50	250	16	101	117
75	225	24	89	113
100	200	31	78	109
150	150	50	65	115
300	0	101	12	113

Source: Whittam and Ager (1965). Reprinted by permission from *Biochemistry Journal,* 97:214–227, copyright © 1965 The Biochemical Society, London.

over a wide range of concentrations. In this example, both Na^+ and K^+ levels were varied. A similar procedure was used with pigeon red cells to vary the internal concentration of Na^+. The flow of glycine in response to changes in the Na^+ gradient is shown in Table 6.3. The initial internal concentration of Na^+ is represented in column 4; that in the medium, in column 5. The initial concentration ratios of glycine (internal/external) are shown in column 6. Column 7 represents the final ratio after a period of incubation. A comparison of these ratios shows that a higher external concentration of Na^+ results in accumulation of internal glycine. When the internal concentration of Na^+ is higher than that outside, the glycine is actively transported outward (results marked with arrows). In the intact cell, since the Na^+ concentration inside the cell is low, the flow of glycine is invariably inward. In the intact cell the internal Na^+ concentration is kept low by the net outward transport of Na^+, the so-called *sodium pump,* which expends energy from the hydrolysis of ATP. Table 6.1 shows the calculation of the energy requirement for the inward transport of 1 mole of Na^+ in the squid axon.

In Chapter 14 we will examine evidence for the coupled outward transport of Na^+ and inward transport of K^+ powered by the hydrolysis of ATP, and in Chapter 15 we will examine the possible mechanism of this pump.

The converse process—transfer of ions in the direction of the electrochemical gradient coupled to the synthesis of ATP from ADP and P_i—is also possible. An experiment to test this was carried out with human red blood cells in a medium with a high external concentration of Na^+ and low external concentration of K^+ (Glynn and Lew, 1970). The synthesis of ATP from metabolic sources was inhibited with iodoacetate (which inhibits glycolysis, the major metabolic pathway in these cells). Results of the experiment are shown in Fig. 6.4. Curve 1 represents the incorporation of $^{32}P_i$ into ATP as a function of external Na^+ concentration. The concentration of Na^+ was varied without changing the osmotic pressure or the ionic strength of the medium by replacing the Na^+ with choline whenever necessary. Placing the red blood cells in a medium with a higher Na^+ concentration increases the gradient for Na^+, since the low initial internal Na^+ level remained the same. Increased external Na^+ led to increased ATP synthesis. Curve 2

Table 6.3 Glycine Accumulation and Expulsion by Intact or Lysed and Restored Cells

(1) Experiment	(2) Sample	(3) Preparation	(4) Initial cell Na^+, calculated value (mM)	(5) Na^+ in medium (mM)	(6) Initial ratio $glycine_i/glycine_o$	(7) Final ratio $glycine_i/glycine_o$
1	a	Lysed and restored	24	140	1.45	2.76
	b	Lysed and restored	24	140	1.45	3.02
2	c	Lysed and restored	24	140	1.17	2.31
	d	Intact	17.5	140	6.06	8.22
3	e	Lysed and restored	24	140	1.59	2.39
	f	Lysed and restored	24	0	1.57	→1/1.08
	g	Lysed and restored	115	0	1.43	→1/2.06
	h	Lysed and restored	126	0	1.24	→1/2.34
4	i	Lysed and restored	24	140	1.12	1.95
	j	Lysed and restored	126	0	1.07	→1/1.94
5	k	Lysed and restored	115	140	1.63	1.73
	l	Lysed and restored	84	134	1.67	1.91
	m	Lysed and restored	24	125	1.68	2.57

Source: Vidaver (1964). Reprinted with permission from *Biochemistry* 3:795–799. Copyright 1964 American Chemical Society.

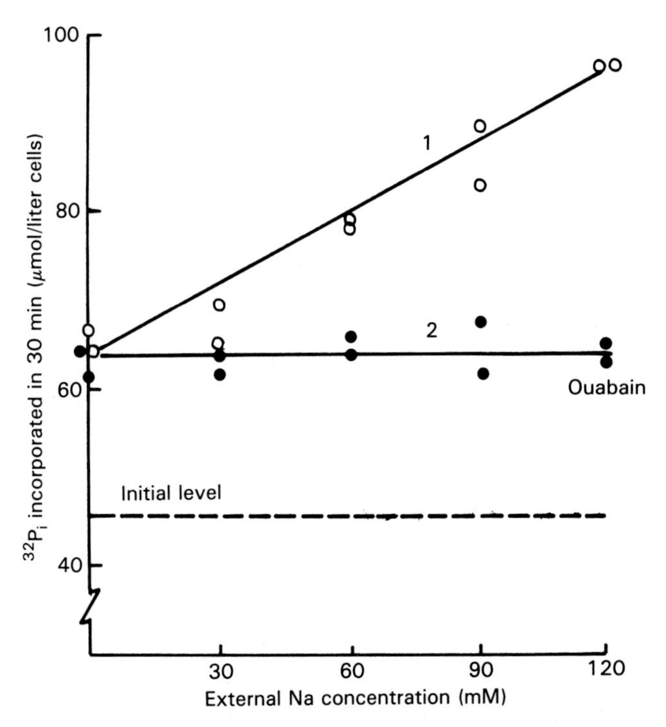

Fig. 6.4 ATP synthesis as a function of external sodium concentration. Choline was used to replace Na^+ to maintain constant osmotic pressure and ionic strength. Reprinted with permission from I. M. Glynn and V. L. Lew, *Journal of Physiology,* 207:393–402. Copyright © 1970 The Physiological Society, Oxford, England.

represents the effects of ouabain, a drug that inhibits the Na^+ pump. Since the incorporation into ATP was prevented by the inhibitor, it is most likely that the reaction does correspond to the reverse of transport against the electrochemical gradient.

The incorporation of $^{32}P_i$ into ATP can be related to the flow in the direction of the gradient by following either the penetration of Na^+ or the exit of K^+, since the transport of the two in opposite directions is coupled. In these experiments the exit of K^+ was estimated by the appearance of ^{42}K in the medium after suitable loading of the cells with the labeled ion. A correction for the hydrolysis of [$^{32}P_i$]ATP, which takes place presumably by independent processes, makes it possible to relate the K^+ exit with the ATP synthesis. The exit of 2 to 3 moles of K^+ (accompanied by the Na^+ influx) produces 1 mole of ATP.

Similar results have been obtained in experiments with other membrane-bound systems that can carry out active transport of ions. In isolated *sarcoplasmic vesicles,* the release of Ca^{2+} in the direction of the electrochemical gradient results in phosphorylation of ADP (Makinose and Hasselbach, 1971a, 1971b). The *sarcoplasmic reticulum* is a system of tubes and vesicles of striated muscle. Elements of the reticulum liberate Ca^{2+} as a signal for muscle contraction and subsequently sequester it during relaxation by an active transport. The active transport depends on the hydrolysis of ATP and is thought to resemble the Na^+ pump.

In isolated mitochondria the antibiotic valinomycin induces active transport of K^+ inward. The energy for this transport can be supplied by either hydrolysis of ATP or respiratory reactions. Conversely, the efflux of K^+ in response to the addition of valinomycin results in net synthesis of ATP from P_i and ADP (Cockrell et al., 1967). Valinomycin is one of the compounds, referred to as ionophores, that are capable of ligating ions. Because of their high solubility in the membrane lipid, they are involved in the transport of ions. The mechanism by which valinomycin affects mitochondrial transport is still a matter of debate.

These experiments show that the transport mechanism can operate in either direction. In one direction, an exergonic chemical reaction powers the transport against an electrochemical gradient. In the other direction, the passage of ions along the electrochemical gradient can be harnessed to drive the synthesis of a chemical bond.

The minimal energy expenditure necessary to transport solute is given by Eq. (6.20) or (6.22). According to these relationships, the minimal expenditure necessary to transport 1 mole of solute (i.e., the ΔG per mole transported) increases with the steepness of the gradient. Experimental examination of transport at different gradients leads to a number of conclusions that bear on the mechanism of transport.

As already discussed, it is possible to vary either the internal or external ionic composition of red blood cells. The internal composition of the red blood cell can be changed by several experimental manipulations. The result of one such procedure has been shown in Table 6.2. In other work discussed above (see Table 6.3), the transport of glycine was studied as a function of the Na^+ gradient. Other studies concerned the influx of K^+ as a function of external K^+ concentration and the efflux of Na^+ as a function of internal Na^+ concentration. The ATP hydrolyzed can be calculated from the P_i liberated by the reaction. It is necessary to correct this value by subtracting the amount of ATP produced by the red blood cell's metabolism. The latter can be calculated from the lactate produced by the glycolytic reactions that represent the major metabolic pathway of the red blood cell. In Fig. 6.5, the influx of K^+ is shown on the ordinate (Whittam and Ager, 1965). Each point represents an experimental determination at a different external K^+ concentration. The hydrolysis of ATP inhibited by ouabain is shown on the abscissa. Ouabain has little or no effect on reactions other than the active transport. The slope of the line indicates the K^+ transported per ATP hydrolyzed and remains constant (mean of 2.4 ± 0.3 K^+/ATP) over a wide range of K^+ concentrations in many independent experiments. In experiments in which the Na^+ efflux was estimated at various internal Na^+ concentrations, the Na^+ transported per ATP hydrolyzed was found to correspond to 3.3 ± 0.2. The results indicate that the energy cost of transporting one Na^+ or one K^+ by the Na^+/K^+ pump is the same regardless of the magnitude of the gradient. This suggests that the transport mechanism functions in precise

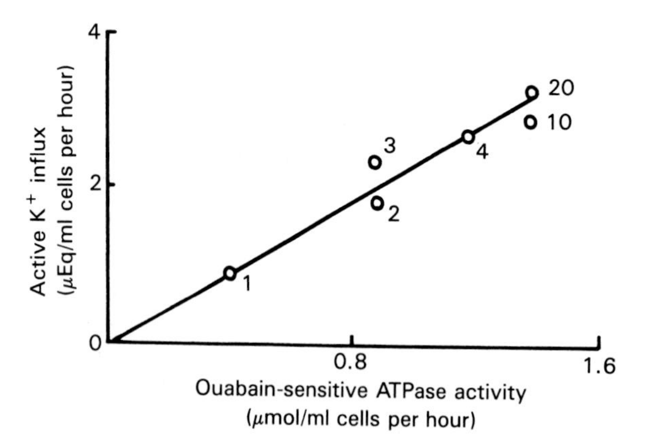

Fig. 6.5 Relationship between active K^+ influx and ouabain-sensitive ATPase activity in media with different K^+ concentrations (mM, shown by the numbers next to the data points). From Whittam and Ager (1965). Reprinted by permission from *Biochemistry Journal*, 97:214–227, copyright © 1965 The Biochemical Society, London.

Table 6.4 Calculation for ΔG of Transport Corresponding to the Hydrolysis of One ATP[a]

(1) Experiment	(2) Na_o	(3) Na_i	(4) K_o	(5) K_i	(6) Na_o/Na_i	(7) K_i/K_o	(8) $n\ 2.3\ RT \times \log_{10}\frac{Na_o}{Na_i}$ (kcal)	(9) $m\ 2.3\ RT \times \log_{10}\frac{K_i}{K_o}$ (kcal)
1	150	10	10	96	15	9.6	5.3	2.8
2	150	101	10	12	1.48	1.2	0.8	0.3

[a] $n = 3.2$, $m = 2.4$.

stoichiometry, in a manner analogous to any other biochemical reaction. The implication of this observation can best be seen by calculating transport in specific cases.

Table 6.4 shows some of the concentrations reported by Whittam and Ager (1965) and gives the calculation of ΔG for these two cases (as in Table 6.1). The external concentrations of Na^+ and K^+ are shown in columns 2 and 4 and the internal concentrations in columns 3 and 5. Columns 8 and 9 show the calculations of ΔG for the transport of Na^+ and K^+. In these calculations the membrane potential of the red blood cell has been neglected because its contribution to the energy is small. The values in columns 8 and 9 were obtained by multiplying the appropriate equation by n or m, the equivalents of Na^+ or K^+ transported per ATP hydrolyzed. The total energy required for transport under the conditions of experiment 1 is about 8.1 kcal (columns 8 and 9), and under the conditions of experiment 2 it is only 1.1 kcal. Nevertheless, as in Fig. 6.5, the ATP hydrolyzed per Na^+ or K^+ transported remains invariant. Under these conditions the ΔG for the hydrolysis of one mole of ATP is about -13 kcal, sufficient for the transport under either of the conditions. Therefore, a constant stoichiometry is maintained by varying the efficiency. When the gradient becomes smaller the efficiency drops, even if theoretically there is sufficient energy in each ATP hydrolyzed to support a much larger cation/ATP ratio.

VII. MUSCLE CONTRACTION

Biological systems expend energy in a number of processes. We examined in some detail the transport of ions. Cells also expend energy in moving by processes involved in contraction. The contraction of striated vertebrate muscle has been studied intensively. The work performed by muscle can be calculated readily, since it corresponds to the mass of the object lifted (or alternatively the tension τ exerted) times the displacement (ΔL). The energy expended that is not used to perform work is liberated as heat. The heat liberated can be calculated from the change in temperature of an insulated system whose heat capacity ($\Delta H = \mu C_H \Delta T$) is known. Note that if no work has been performed (the muscle has been held stationary in a so-called *isometric* contraction) or a contraction-relaxation cycle has been completed, all energy expended must appear as heat. The heat released by muscle is of great physiological importance in animals (even in some fish, which are considered ectothermic, or cold blooded) and in the physiological control of body temperature. In mammals the heat released by contraction-relaxation cycles is expended in the basic function of shivering.

Figure 6.6 shows the work performed (lower curve) and the heat liberated (upper curve) as a function of tension τ relative to peak tension τ_0. An apparatus kept tension constant during the contractile event. The amount of work is expressed solely by the shortening. The results show that the work performed and heat liberated undergo parallel changes. This is expressed most simply in Fig. 6.7, where the heat liberated is presented as a function of work. It follows from these results that the energy expenditure is proportional to the work performed. However, some energy is expended whether work is performed or not. In Fig. 6.7, at zero work, 2.95 mcal/g was expended; this minimal expenditure is known as the *activation energy*. The chemical events underlying contraction would, therefore, be graded with the amount of work performed. The efficiency of the system does not vary with increases in the amount of work performed. It has been shown (discussed in Chapter 17) that the amount of ATP (or phosphocreatine) hydrolyzed is proportional to the work, as we would expect from these considerations (Cain et al., 1962). Phosphocreatine is one of the energy reservoirs of muscle and can transfer its terminal phosphate to ADP to regenerate ATP.

We saw that in the transport of Na^+ and K^+, the energetic cost of transferring 1 mol of Na^+ is the same, regardless of the steepness of the gradient. Therefore, the case of ion active transport seems to be quite different in principle from that of contraction of striated muscle. The results suggest that muscle contraction corresponds to a graded process, as if contraction were the result of small finite steps. The molecular mechanisms of contraction are likely to involve such small finite steps (see Chapter 18).

As noted above, some heat is evolved by muscle whether work is performed or not. This

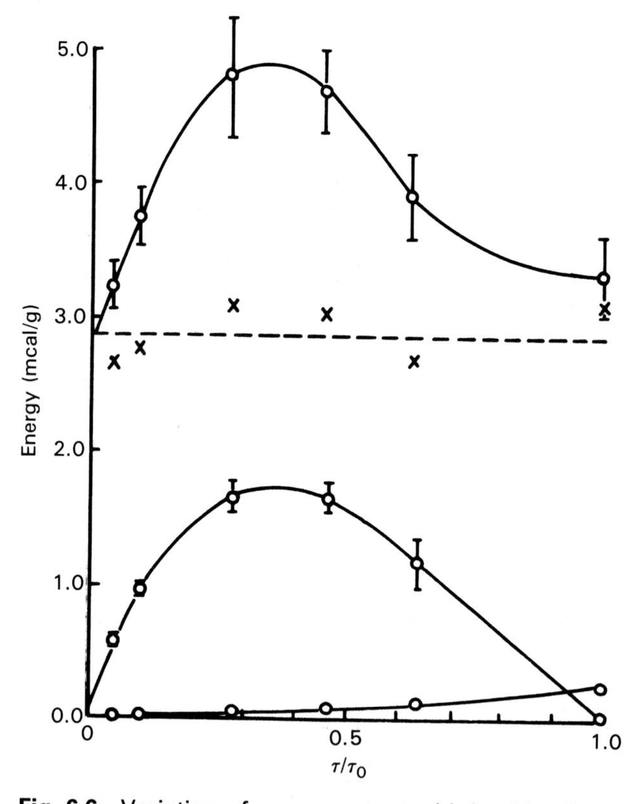

Fig. 6.6 Variation of energy output with load in afterload isotonic twitches; the load was allowed to fall during relaxation. Abscissa, load τ as a fraction of the peak isometric twitch tension τ_0. Ordinate, energy output in millicalories per gram and twitch (mean of 100 twitches, 20 by each of five muscles, $\pm$ 1 standard error plotted as a vertical bar). Upper curve, total heat; lower curves, external and internal work. Crosses, total heat minus total work. Reprinted with permission from F. D. Carlson, et al., *Journal of General Physiology,* 40:851–882. Copyright © 1963 Rockefeller University Press.

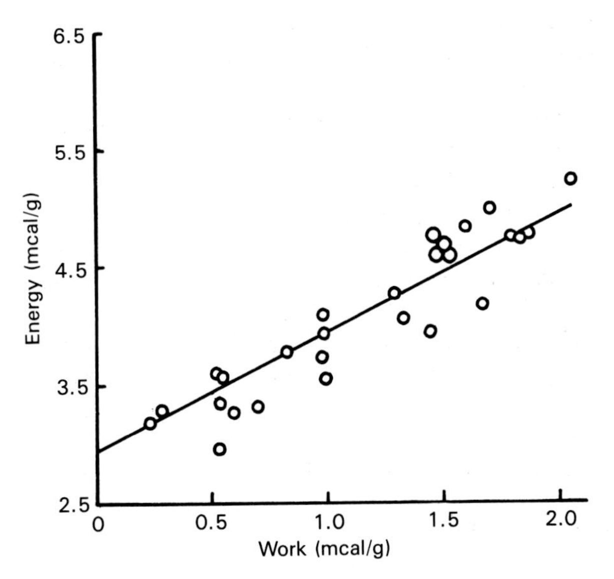

Fig. 6.7 Ordinate, total heat, millicalories per gram and twitch. Abscissa, total work, millicalories per gram and twitch. Reprinted with permission from F. D. Carlson, et al., *Journal of General Physiology,* 40:851–882. Copyright © 1963 Rockefeller University Press.

heat evolution seems inherent to the active state. When contracting muscle is slowly stretched, the total heat evolved is less than the sum of the heat of activation and the heat evolved from the work performed on the muscle by the stretching process (Hill, 1960). The difference corresponds rather closely to the work being performed on the muscle.

The results lend themselves to two possible interpretations: either the stretching has arrested the active state, or the energy input (so-called negative work) has been absorbed by reversing the contractility process. The latter alternative, which is supported by some investigators, would have rather far-reaching repercussions. The energy of contraction originates from the hydrolysis of high-energy phosphates. It has been found that the stretching of striated vertebrate muscle does not result in resynthesis of the main high-energy compound, phosphocreatine or ATP. These experiments have been interpreted as being consistent with the idea that the primary event in contraction does not directly involve the hydrolysis of ATP or phosphocreatine. However, in insect flight muscle preparations, stretching does increase the incorporation of $^{32}P_i$ into ATP (Ulbrich and Ruegg, 1971). In addition, the interpretation involving the "disappearance" of work may well be incorrect. A more likely explanation is that the events of contraction are arrested and that the heat evolved is simply a quantitative conversion of the work performed on the system into heat.

In Chapters 7 to 12 we will examine primarily the processes involved in capturing energy in a biologically utilizable form. The mechanisms involved in dissipating this energy will be taken up primarily in Chapters 13 to 18.

SUGGESTED READING

Becker, W. M. (1977) *Energy and the Living Cell.* Lippincott, Philadelphia.

Christensen, H. N. (1975) Thermodynamic aspects of transport. In *Biological Transport,* 2d ed. Benjamin, Reading, Mass.

Cramer, W. A., and Knaff, D. B. (1990) *Energy Transduction in Biological Membranes,* Chapters 1 and 2. Springer-Verlag, New York.

Dutton, P. L. (1978) Redox potentiality: determination of midpoint potential of oxidation reduction components of biological electron transfer systems. *Methods Enzymol.* 54:411–425.

Morris, J. G. (1968) *The Biologist's Physical Chemistry*. Addison-Wesley, Reading, Mass.

Woledge, R. C., Curtin, N. A., and Homsher, E. (1985) *Energetic Aspects of Muscle Contraction*, Academic Press, New York (see Chapters 1, 2, and 4).

Alternative References

Harold, F. M. (1986) *The Vital Force: A Study of Bioenergetics*, Chapters 1 to 3. W. H. Freeman, New York.

Nicholls, D. G. (1982) *Bioenergetics: An Introduction to Chemiosmotic Theory*, Chapters 1 to 4. Academic Press, New York.

REFERENCES

Banks, E. C., and Vernon, C. A. (1970) Reassessment of the role of ATP in vivo. *J. Theor. Biol.* 29:301–306.

Bridger, W. A., and Henderson, J. H. (1983) *Cell ATP*, Chapter 2. Wiley, New York.

Cain, D. F., Infante, A. A., and Davies, R. E. (1962) Chemistry of muscle contraction. Adenosine triphosphate and phosphoryl creatine as energy supplies for single contractions of working muscle. *Nature (London)* 196:214–217.

Carlson, F. D., Hardy, D. J., and Wilkie, D. R. (1963) Total energy and phosphocreatine hydrolysis in the isotonic twitch. *J. Gen. Physiol.* 46:851–882.

Cockrell, R. S., Harris, E. J., and Pressman, B. C. (1967) Synthesis of ATP driven by potassium gradient in mitochondria. *Nature (London)* 215:1487–1488.

Flatmark, T., and Pedersen, J. I. (1975) Brown adipose tissue mitochondria. *Biochim. Biophys. Acta* 416:53–103.

Glynn, I. M., and Lew, V. L. (1970) Synthesis of adenosine triphosphate at the expense of downhill cation movements in intact red cell. *J. Physiol. (London)* 207:393–402.

Hill, A. V. (1960) Production and absorption of work by muscle. *Science* 131:897–903.

McClare, C. W. F. (1972) In defense of the high energy phosphate bond. *J. Theor. Biol.* 35:233–246.

Makinose, M., and Hasselbach, W. (1971a) Calcium efflux dependent formation of ATP from ADP and orthophosphate by membranes of the sarcoplasmic vesicles. *FEBS Lett.* 12:269–270.

Makinose, M., and Hasselbach, W. (1971b) ATP synthesis by the reverse of the sarcoplasmic pump. *FEBS Lett.* 12:271–272.

Poe, M., and Estabrook, R. W. (1969) Kinetic studies of temperature changes and oxygen uptake concomitant with substrate oxidation by mitochondria: the enthalpy of succinate oxidation during ATP formation in mitochondria. *Arch. Biochem. Biophys.* 126:320–330.

Rosenberg, T., and Wilbrandt, W. (1958) Uphill transport induced by counterflow. *J. Gen. Physiol.* 41:289–296.

Rosing, J., and Slater, E. C. (1972) The value of ΔG for the hydrolysis of ATP. *Biochim. Biophys. Acta* 267:275–290.

Slater, E. C., Rosing, J., and Mol, A. (1973) The phosphorylation potential generated by respiring mitochondria. *Biochim. Biophys. Acta* 292:534–553.

Ulbrich, M., and Ruegg, J. C. (1971) Stretch induced formation of [P]ATP in glycerinated fibers of insect flight muscle. *Experientia* 27:45–46.

Vidaver, G. A. (1964) Glycine transport by hemolyzed and restored pigeon red cells. *Biochemistry* 3:795–799.

Whittam, R., and Ager, M. E. (1965) The connexion between active cation transport and metabolism in erythrocytes. *Biochem. J.* 97:214–227.

Chapter 7

Enzymes and Enzyme Complexes

The cell is a dynamic system, in continuous change and in continuous motion. The underlying events must necessarily reflect molecular changes, namely chemical reactions. Most chemical reactions occurring in living systems are catalyzed by enzymes. As a result, the study of enzyme reactions of cells can go far in explaining the basis of their functional behavior. The genetic makeup of the cells and the processes involved in the expression of the genetic information determine the kinds of enzymes or structural proteins that can be present. These enzymes and structural proteins are the ultimate functional units of the cell.

This chapter treats chemical reactions, enzyme-catalyzed reactions in particular, and then proceeds to a discussion of the properties of the enzymes, their regulation, and the organization in enzyme complexes.

I. CHEMICAL REACTIONS

In a population of molecules, the kinetic energy varies with each molecule. The distribution should be random and follow a pattern such as that represented in Fig. 7.1. In this plot, the fraction of the molecules, $(dn/dv)n_T^{-1}$, having velocities between v and $v + dv$ is represented as a function of the velocity v. A simple model describing the energy distribution in two dimensions is given by Eq. (7.1).

$$\frac{dn}{n_T\, dv} = \frac{mv}{kT} e^{-mv^2/2kT} \tag{7.1}$$

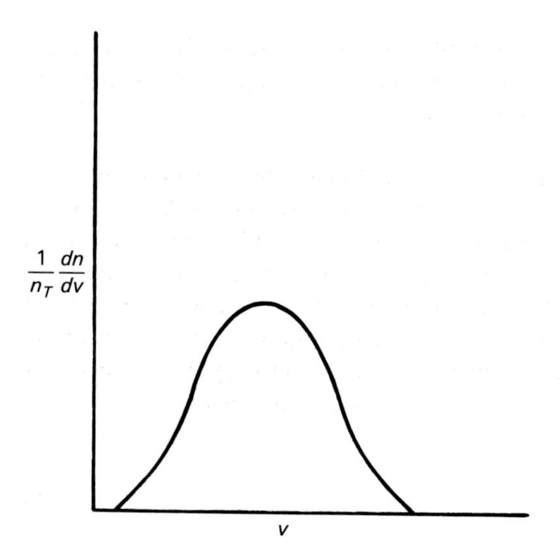

Fig. 7.1 Distribution of velocities of gas molecules or of molecules in an ideal solution. dn/n_T is the fraction of molecules with velocity in the range of v to $v + dv$. The area under the curve has a value of unity.

In this representation, m corresponds to mass, $\frac{1}{2}mv^2$ is the kinetic energy, k is the Boltzmann constant ($k = R/N$, where R is the gas constant and N is Avogadro's number), and T is temperature. Representing the kinetic energy by E and $dE = mv\,dv$, Eq. (7.1) becomes

$$\frac{dn}{n_T} = \frac{1}{kT} e^{-E/kT}\,dE \tag{7.2}$$

If we were to ask what fraction of the molecules has energy greater than a particular value of E, it becomes necessary to integrate under the curve [Eqs. (7.3) and (7.4)].

It would seem reasonable to consider that molecules need a critical minimal energy, termed the activation energy ($E^{\ddagger}$), in order to react. Equation (7.4) would then represent the proportion of molecules that surpass the activation energy at some specified temperature T and hence the proportion of molecules that can react.

$$\int_0^n \frac{dn}{n_T} = \frac{1}{kT}\int_E^{\infty} e^{-E/kT}\,dE \tag{7.3}$$

$$\frac{n}{n_T} = e^{-E/kT} \quad \text{or} \quad \ln\left(\frac{n}{n_T}\right) = \frac{-E}{kT} \tag{7.4}$$

This idea is represented in Fig. 7.2*a* and *b*. Figure 7.2*a* represents the energy of a molecule through the course of the reaction. The reaction is assumed to involve a single reactant. The molecules that can react are represented in Fig. 7.2*b* by the portion of the curve that is crosshatched. An increase in temperature would increase the kinetic energy of the system. Accordingly, more molecules would reach the critical energy level. This can be presented in a diagram in which the bell-shaped distribution is shifted to the right, as shown in Fig. 7.2*c*. As implied by Eq.(7.4), the rate of the reaction should be proportional to n/n_T. When the logarithm of the reaction rate constant (represented in this chapter as k_r) is plotted as a function of $1/T$, the slope of the line should be $-E/k$.

Generally, the relationship of Eq.(7.4) is not followed precisely. A number of other parameters are likely to come into play. For example, where more than one molecular species is involved, the two molecules must meet. Furthermore, they must meet in such a fashion that the reactive groups are specifically apposed. Therefore, we would expect the reaction rate to depend not just on temperature but on other factors as well. For example, we would expect that the shape of the molecule or the nature or location of the reactive groups would play a significant role. Accordingly, the reaction rate constant is more closely predicted by a more complex equation. Here K is a constant, mostly empirical, and h is Planck's constant.

$$k_r = e^{-E/kT} K\left(\frac{kT}{h}\right) \tag{7.5}$$

II. THE ROLE OF ENZYMES

Since the metabolic activity of the cell is to a large extent the sum of its enzymatic activities, it is very much to the point to examine some of the properties of enzymes.

Almost all enzymes studied are proteins, generally ranging between 10,000 and 500,000 in molecular weight. The reactants that interact with the enzymes are called substrates. Enzymes sometimes require small organic molecules (coenzymes) for activity. Carboxylase, for example, requires thiamine pyrophosphate, and enzymes involved in acetylation require coenzyme A. Phosphorylase, which is discussed later in this chapter, requires pyridoxal 5-phosphate. In addition, many enzymes require certain metal ions for their activity. For example, Mg^{2+} is necessary for many enzyme-catalyzed reactions.

Recently, the RNA portion of ribonuclease P (a ribonucleoprotein) has been found to have catalytic activity (Guerrier-Takada and Altman, 1984). In addition, a portion of an RNA *intron* of 395 nucleotides, released by the self-splicing of a ribosomal RNA precursor (Zaug and Cech, 1986) acts as a ribonuclease and an RNA polymerase. RNA enzymes (sometimes referred to as *ribozymes*) are thought to have a significant role in RNA processing. Introns are the noncoding

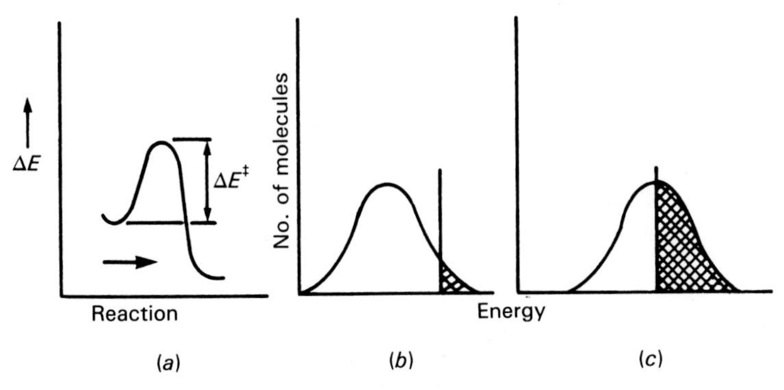

Fig. 7.2 Illustration of how an increase in temperature allows more molecules to reach the energy of activation, $\Delta E^{\ddagger}$.

sequences of either a gene or the corresponding primary RNA transcript that are excised while simultaneously the coding sequences are linked during RNA processing to form mature mRNA.

The site primarily involved in catalysis, the *active site*, complexes with substrate and may interact with cofactors such as coenzymes and metal ions. It is but a very small portion of the enzyme molecule, as suggested, for example, by the low molecular weight of many substrates. The specificity of the interaction between enzyme and substrate implies that a finite number of specific groups in the enzyme molecule are involved. The restricted nature of the active site has also been demonstrated directly. For example, low molecular weight analogs of the substrate can be covalently bonded to a single site in an enzyme molecule and thereby block its activity (Schoellmann and Shaw, 1963). The technique of x-ray crystallography has permitted the deduction of the structure of many macromolecules, including enzymes. In some cases, when the crystals of the enzyme have been exposed to the substrate, electron density maps reconstructed from the diffraction data show an increased density at a discrete small site (e.g., see Ludwig et al., 1967). The location of this site generally implicates the same groups that were previously suggested by direct chemical studies. Why a large molecule is needed for biological catalysis is still a subject of debate. Some of the conformational rearrangements thought to occur during enzyme catalysis or during the regulation of enzyme activity may require the involvement of a macromolecule.

The action of an enzyme consists of accelerating the rate of a given reaction. The tendency of a reaction to occur and its final equilibrium position are entirely expressed by the free energy ΔG of the reaction. These remain unchanged in the presence of the enzyme. The thermodynamic considerations discussed in Chapter 6 apply whether a reaction is enzyme catalyzed or not. The enzyme may be regarded as increasing the rate of a reaction entirely by lowering the energy of activation. The principle can best be illustrated by comparing Fig. 7.2 to Fig. 7.3.

Figure 7.2, as we have seen, represents a hypothetical chemical reaction. Figure 7.2*a* shows the energy as a function of the extent of the reaction, and Fig. 7.2*b* shows the proportion of molecules at any given energy. The proportion of molecules having sufficient energy to react is indicated by the crosshatched area under the curve. The energy of the molecules can be increased by heating (Fig. 7.2*c*). In contrast, introduction of an enzyme lowers the activation energy to the level represented in Fig. 7.3*a*. As a consequence, a larger proportion of the molecules now has sufficient energy to react, as indicated by the increase in the crosshatched area under the curve in Fig. 7.3*b* compared to that in Fig. 7.2*b*, without a change in their energy. Catalysis in speeding up a chemical reaction is illustrated by the data in Table 7.1 (Koshland, 1956). In this

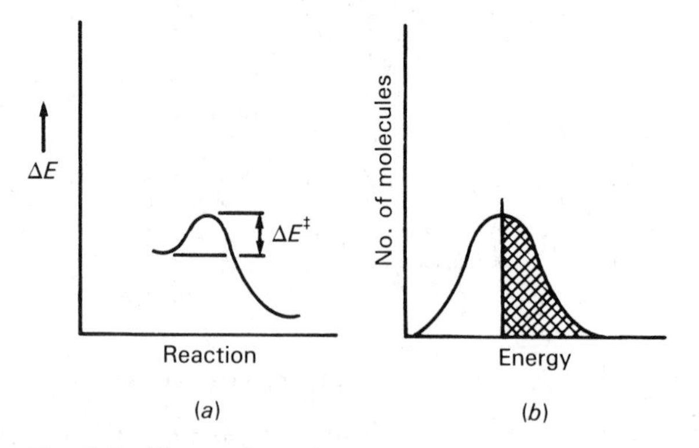

Fig. 7.3 Illustration of how an enzyme acts. The energy level of the molecules remains unchanged but the $\Delta E^{\ddagger}$ is decreased.

Table 7.1 Rate of a Chemical Reaction in the Presence (V_E) and in the Absence (V_0) of the Appropriate Enzyme

Enzyme	Substrate and concentration	$[E_T]$ (Eq/liter)[a]	Observed (mol min^{-1} liter^{-1}) V_0	V_E
Hexokinase	0.0003 M Glucose 0.002 M ATP	10^{-7}	$<1 \times 10^{-13}$	1.3×10^{-3}
Phosphorylase	0.016 M Glucose 1-phosphate 10^{-5} M Glycogen	6×10^{-8}	$<5 \times 10^{-15}$	1.6×10^{-3}
Alcohol dehydrogenase	0.0004 M NAD 0.04 M Ethanol	4×10^{-7}	$<6 \times 10^{-12}$	2.7×10^{-3}
Creatine kinase	0.024 M Creatine 0.004 M ATP	3×10^{-9}	$<3 \times 10^{-9}$	4×10^{-5}

[a]Number of moles per liter multiplied by number of active sites per molecule.

table, V_0 represents the rate or velocity of the reaction in the absence of enzyme and V_E represents the velocity in the presence of enzyme. It is apparent that enzymes can speed up reactions as much as 10^{12} times.

In a reaction involving two or more reactants, the lowering of the activation energy by the enzyme is in part the result of an increase in the number of potentially fruitful collisions between the substrate molecules. In this role, the enzyme orients the reactive groups of the substrates relative to each other in a manner most conducive for a reaction to occur (*proximity effect*) (Fig. 7.4). In contrast, the chances of the active groups free in solution coming into fruitful apposition is much lower (Fig. 7.5).

It is possible to develop some insights into the mechanisms of enzyme catalysis by studying model compounds. These compounds are simpler organic molecules that have been found to catalyze certain reactions, although with an efficiency much lower than that of enzyme catalysis.

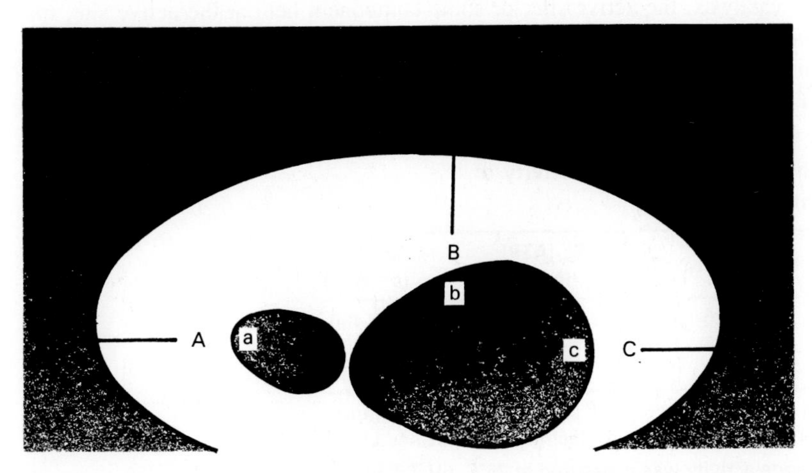

Fig. 7.4 Representation of an enzyme reaction. A, B, and C represent reactive groups of the enzyme. The egg-shaped molecules represent substrate molecules. The shaded areas in the molecules represent the reactive groups of the substrates, and a, b, and c are groups in the substrate that can interact with A, B, and C.

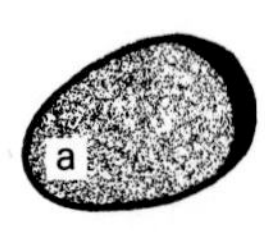

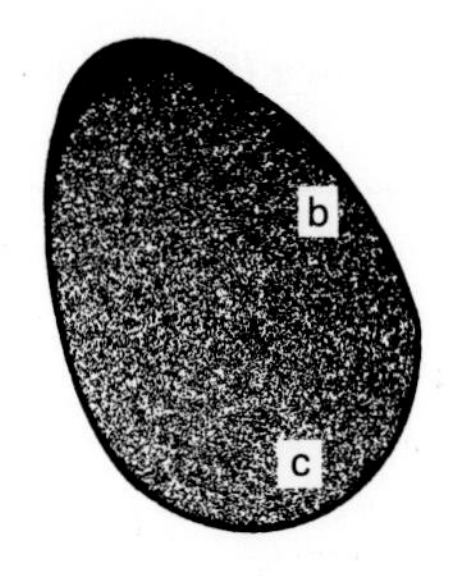

Fig. 7.5 The reaction of Fig. 7.4 without the enzyme molecule. Reactive groups A, B, and C are no longer attached. The probability of all reactive groups becoming appropriately oriented is much diminished.

Studies with simpler molecules involve fewer complications and are generally easier to interpret. The possibility of synthesizing molecules that differ from each other in specific ways permits the evaluation of various factors in the mechanisms of catalysis. In addition, from model reactions the contribution of various mechanisms that are suspected to play a role can be calculated at least approximately.

Cloning of DNA that codes specific proteins and site-directed mutagenesis (see Chapter 1) have made it possible to change systematically single amino acids in selected enzymes. An example of this approach as applied to the study of tyrosyl-tRNA synthetase (Wilkinson et al., 1984) is shown in Table 7.2. The substituted site, occupied by threonine in position 51 in the wild type, is shown in the first column, and the effects of the indicated substitutions on the kinetic constants k and K_m are shown in the rest of the table. The constant k is the rate constant of the reaction and K_m is the apparent Michaelis-Menten constant, discussed in Section III, A. The ratio of the two (k/K_m) relates the reaction rate to the concentration of free enzyme and corresponds to an index of specificity in relation to alternative or competing substrates. The study of specifically altered enzymes is a very powerful method for elucidating enzyme mechanisms. However, model compounds still provide an important alternative. Alteration of a single amino acid of an enzyme may provide results that are difficult to interpret because it may produce alterations in the overall conformation of the enzyme with dramatic effects on enzyme activity but only an indirect effect on the catalytic site.

During enzyme catalysis, the active site, or some component held at the active site, may act as a *nucleophilic* (electron-donating) or *electrophilic* (electron-attracting) agent. In this process, the enzyme binds to the substrate and forms intermediates. A model for the transfer of phosphate from ATP to creatine is shown in Fig. 7.6. The complexed Mg^{2+} holds the phosphate

Table 7.2 Pyrophosphate Exchange Activity of Tyrosyl-tRNA Synthetases[a]

Enzyme	k_{cat} (s^{-1})	K_m (ATP) (mM)	k_{cat}/K_m (s^{-1} M^{-1})
TyrTS	7.6	0.9	8,400
TyrTS (Ala 51)	8.6	0.54	15,900
TyrTS (Pro 51)	12.0	0.058	208,000

Source: Wilkinson et al. (1984). Reprinted by permission from *Nature* Vol. 307, pp. 187–188. Copyright © 1984 Macmillan Magazines Ltd.

[a] Synthetase exchange activity was measured at 25°C, pH 7.8, in 144 mM Tris-HCl, 10 mM $MgCl_2$, 0.1 mM phenylmethane sulfonyl chloride, 10 mM 2-mercaptoethanol, 2 mM pyrophosphate, 50 μM tyrosine, and 100–250 nM enzyme (assayed by active site titration).

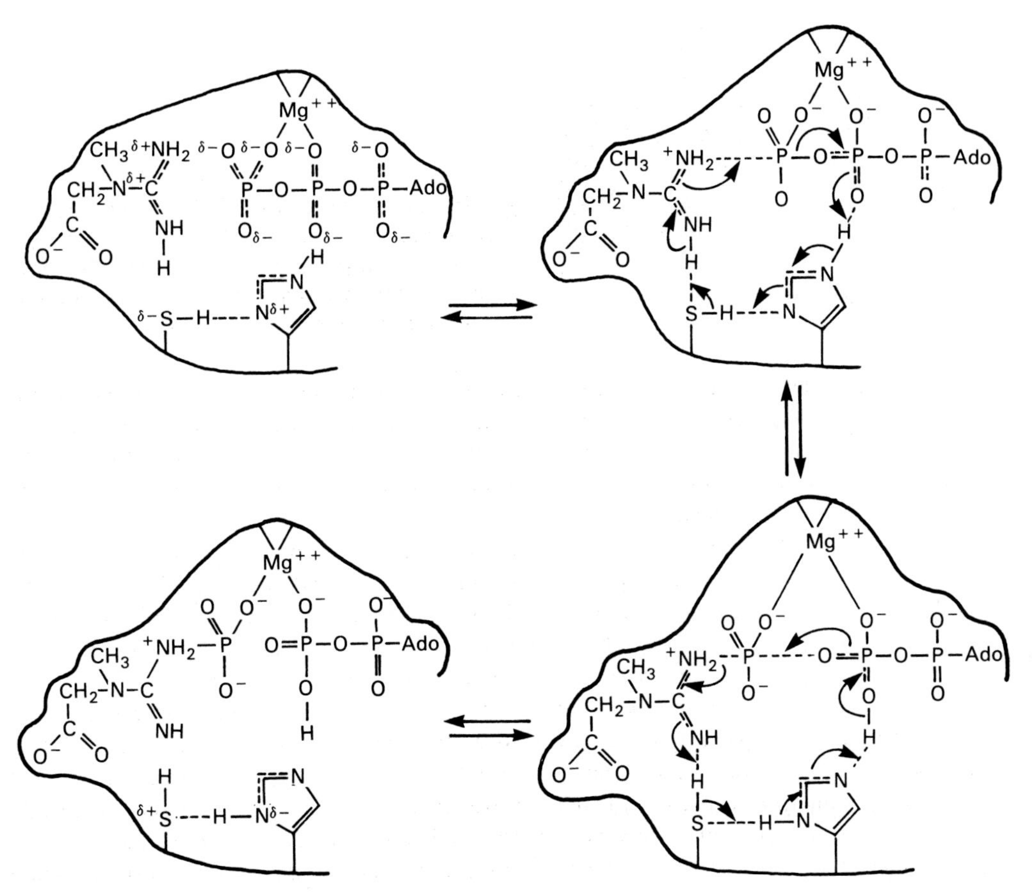

Fig. 7.6 Schematic representation of a model for the transfer of a phosphate group from ATP to creatine, catalyzed by creatine kinase. Ado represents adenosine. The first diagram (top left) shows all groups at the instant of binding by the two substrates. The last diagram (bottom left) represents the completion of the reaction before detachment. Reprinted with permission from A. White, et al., *Principles of Biochemistry*. Copyright © 1978 McGraw-Hill.

component in place. Electrophilic attack of the histidine residue of the enzyme on the phosphate, together with nucleophilic attack by the sulfhydryl group of the enzyme on the creatine, breaks the terminal phosphate bond of the ATP and simultaneously transfers the phosphate to the creatine. Detachment of the two molecules from the enzyme's active site completes the reaction.

Other mechanisms could also be involved. Residues at the active site could act as acids (i.e., H^+ donors) or bases (i.e., H^+ acceptors) in general *acid-base* catalysis. The reactivity of the substrate could be altered by a nonpolar microenvironment at the active site. By virtue of its lower dielectric constant, the hydrophobic spot would alter the ionization of the substrate. The presence of oppositely charged residues (i.e., *ion pairs*) could also have an effect on the rate of the enzyme-catalyzed reaction. However, it has been calculated that the contribution of all these factors together falls short of the rate of enzyme-catalyzed reactions. Enzyme action must, therefore, involve other mechanisms that are still poorly understood (Jenks, 1969; Koshland and Neet, 1968).

The enzyme could contribute to the orientation of the substrates much more precisely than by the proximity effects already discussed. Perhaps the interaction of the active site with the substrate could introduce a *strain* on the molecule by producing a distortion. The distortion could be produced by a straightforward interaction or could be the result of a change in the conformation of the active site which would actively behave as a rack (the *rack and strain theory*). Stryer (1988) has summarized succinctly the details of several reactions that are now reasonably well understood.

Generally, when an enzyme interacts with a substrate, the conformation of the enzyme changes (see Olson and Allgyer, 1973). In addition, changes in the shape of the enzyme molecule are frequently produced by substances involved in the regulation of enzyme activity, and these changes in conformation affect the ability of the enzyme to interact with the substrate (see Section III, C).

A number of facts that are known about enzymes can be summarized as follows.

1. The active site of an enzyme binds the substrate and interacts with the substrate to catalyze the reaction.
2. The active site is but a small portion of the total enzyme molecule.
3. The enzyme itself is regenerated after the reaction.
4. The nature of the reactive group is such that the enzyme can interact only with specific substrates that have some characteristic groupings.
5. The interaction brings about a lowering of the activation energy and thereby increases the proportion of the substrate molecules that can react at any one time. The result is an increase in the rate of the reaction.

III. KINETICS OF ENZYME REACTIONS

A. Michaelis-Menten Kinetics

We have discussed how the rate of reaction depends on the number of molecules that have reached a critical energy level, permitting the reaction to take place. The number of molecules reacting will, therefore, be directly proportional to the concentration of molecules present. Representing the reactant as A and the product as P, the reaction can be represented as in Eq. (7.6a):

$$\mathrm{A} \underset{k_2}{\overset{k_1}{\rightleftharpoons}} \mathrm{P} \qquad (7.6a)$$

where k_1 represents the rate constant of the forward reaction and k_2 that of the reverse reaction. The ratio k_1/k_2 corresponds to the equilibrium constant. At the beginning of the reaction, with only A present, the backward reaction from P to A need not be considered, as shown in Eq. (7.6b). Equation (7.6c) represents the analogous relationship when the back reaction becomes significant.

$$\frac{d(P)}{dt} = k_1(\mathrm{A}) \qquad (7.6b)$$

$$\frac{d(P)}{dt} = k_1(\mathrm{A}) - k_2(\mathrm{P}) \qquad (7.6c)$$

Where more than one reactant is involved, the same reasoning applies. In addition, the distinct kinds of molecules must meet before the reaction takes place. Again, the probability of their meeting will be proportional to their concentration. A reaction involving two reactants, A and B, can be represented by Eq. (7.7a). Accordingly, Eq. (7.7b) represents the rate of the reaction when the concentration of P is negligible.

$$A + B \underset{k_2}{\overset{k_1}{\rightleftharpoons}} P \tag{7.7a}$$

$$\frac{d(P)}{dt} = k_1(A)(B) \tag{7.7b}$$

When an enzyme is involved in the reaction, the same considerations apply, except that the enzyme itself is regenerated as a product. In a reaction involving a single substrate, either the enzyme or the substrate could represent A or B. Generally, the enzyme is present only in very low concentrations. For this reason its concentration becomes limiting in determining the rate of the reaction, and an increase in substrate concentration beyond a particular level will not increase the rate of the reaction. The equations describing the rate of an enzyme-catalyzed reaction can be modified to take this fact into consideration. An enzyme reaction involving a single substrate is represented by

$$E + S \underset{k_2}{\overset{k_1}{\rightleftharpoons}} ES \overset{k_3}{\rightharpoonup} P + E \tag{7.8a}$$

Here E and S represent enzyme and substrate, ES is the enzyme-substrate complex, and P is the product of the reaction.

As we have seen, the rate of a reaction is proportional to the concentration of reactants, and the formation of product (P) is proportional to k_3(ES). Therefore, the rate of the reaction, conventionally called the velocity of the reaction (V), can be represented as

$$V = \frac{dP}{dt} = k_3\,(ES) \tag{7.8b}$$

In most reactions that take place in a test tube or in the cell in a relatively short period of time, the total amount of enzyme (E_T) does not change (the enzyme is neither synthesized nor activated) as shown in Eq. (7.9).

$$E_T = E + ES \tag{7.9}$$

In addition, under steady-state conditions, which would be established rapidly, the concentration of ES does not change significantly, i.e., $d(ES)/dt = 0$ or

$$\frac{d(ES)}{dt} = 0 = k_1(E)(S) - k_2(ES) - k_3(ES) \tag{7.10a}$$

On rearrangement, this becomes

$$(\mathrm{E})(\mathrm{S}) = \frac{k_2 + k_3}{k_1}(\mathrm{ES}) \tag{7.10b}$$

Substituting $(k_2 + k_3)/k_1 = K_m$, the Michaelis-Menten constant, gives

$$(\mathrm{E})(\mathrm{S}) = K_m(\mathrm{ES}) \tag{7.10c}$$

Since (E) = $(\mathrm{E_T})$ − (ES) [Eq. (7.9)], Eq. (7.10c) can be rewritten in the form

$$(\mathrm{ES}) = \frac{(\mathrm{E_T})(\mathrm{S})}{K_m + (\mathrm{S})} \tag{7.10d}$$

The rate (V) of appearance of the product corresponds to k_3(ES), as shown by Eq. (7.8b). Therefore, Eq. (7.10d) can be modified to give

$$V = k_3(\mathrm{ES}) = \frac{k_3(\mathrm{E_T})(\mathrm{S})}{K_m + (\mathrm{S})} \tag{7.11}$$

$k_3(\mathrm{E_T})$ represents the maximum possible rate [when (ES) = $(\mathrm{E_T})$]; it is a constant that can be represented by V_m. Equation (7.11) now assumes the following form, known as the Michaelis-Menten equation:

$$V = \frac{V_m(\mathrm{S})}{K_m + (\mathrm{S})} \tag{7.12}$$

A close examination of Eq. (7.12) is very revealing. When the amount of substrate is low, (S) becomes negligible in the denominator and the velocity approximates $V = (V_m/K_m)(\mathrm{S})$ (see Fig. 7.7*a*). In other words, the velocity varies linearly with substrate concentration. On the other hand, when (S) is very large compared to K_m, K_m can be dropped from the equation. Consequently, (S) is a factor in both numerator and denominator and cancels out, and V now equals the constant V_m: the velocity of the reaction is independent of the substrate concentration (horizontal line in Fig. 7.7*a*). The enzyme is entirely complexed to the substrate (it is saturated); (ES) approaches $(\mathrm{E_T})$. The velocity of the enzyme reaction as a function of increasing substrate concentration, expressed in Eq. (7.12), is represented in Fig. 7.7*a* and *b*. Enzyme reactions following this pattern are said to exhibit Michaelis-Menten kinetics.

The constants V_m and K_m are characteristic of the enzyme reaction and describe the kinetic properties of the system; V_m is directly proportional to the amount of enzyme present, and K_m is characteristic of the enzyme in question. These two constants can be readily evaluated. The maximum number of substrate molecules transformed per unit time by a single enzyme molecule, V_m/(E), is the *turnover number*, a frequently used parameter of enzyme activity. The *specific activity* is the number of enzyme units (1 unit transforms 1 μmol of substrate per minute at 25°C) per milligram of protein.

The constant V_m can be estimated by evaluating the asymptote from a graphical representation such as Fig. 7.7*b*, and K_m can be readily evaluated when $V = \frac{1}{2}V_m$, where K_m corresponds to the value of (S) [see Fig. 7.7*b*; K_m =(S) where $V = \frac{1}{2}V_m$]. In practice, precise evaluation of

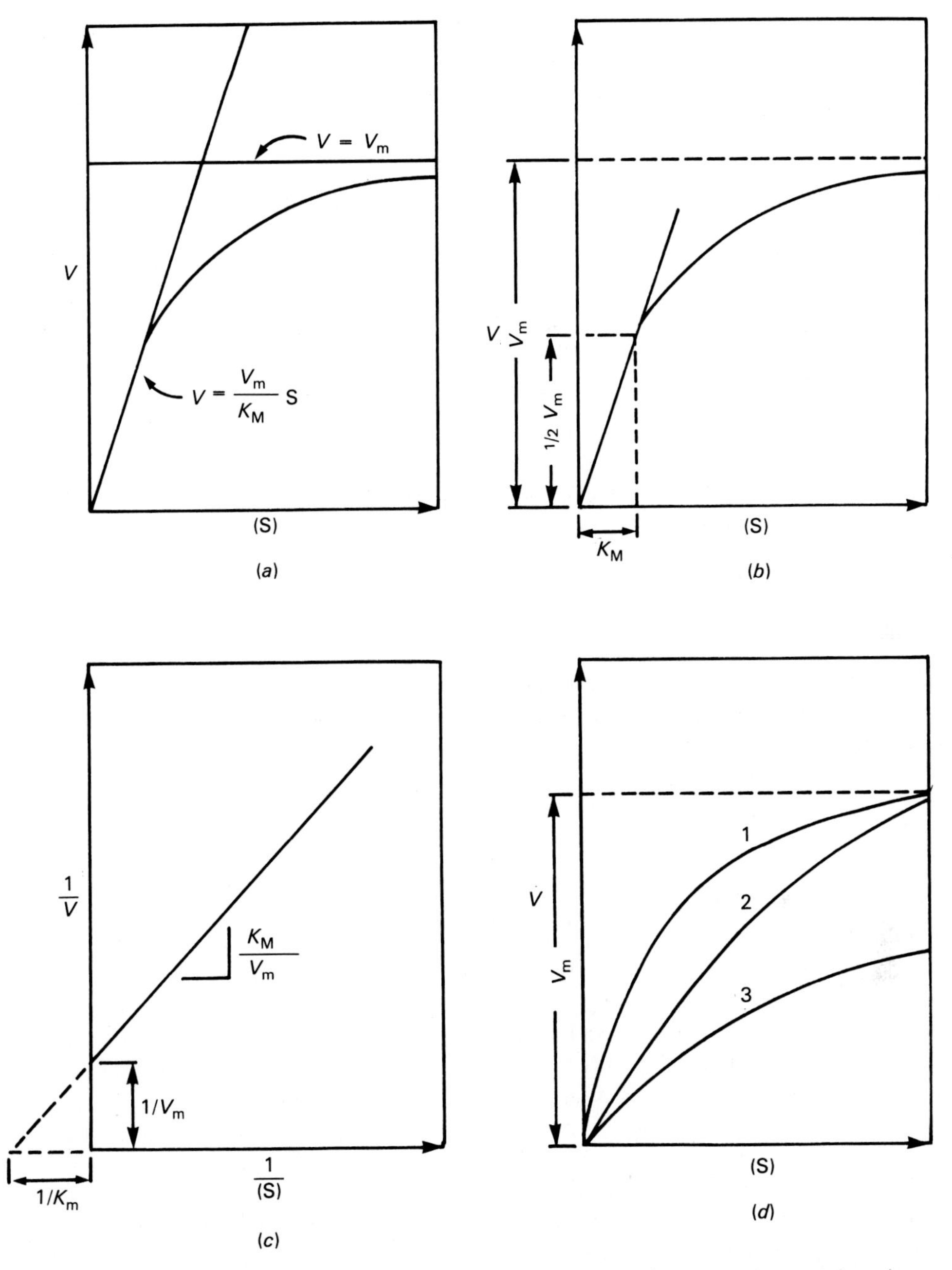

Fig. 7.7 Relationship between the various kinetic parameters in an enzyme-catalyzed reaction. (*a*) Rate of the reaction (V) as a function of substrate concentration (S). Initial slope and asymptote. (*b*) Rate of the reaction (V) as a function of substrate concentration (S). Determination of K_m from (S) when $V = V_m/2$. (*c*) Lineweaver-Burk plot. (*d*) Rate of a reaction (V) as a function of substrate concentration (S). Curve 1, control; curve 2, competitive inhibition; curve 3, noncompetitive inhibition.

V_m by such a visual method may be difficult; asymptotic levels are difficult to estimate precisely, and the range of (S) studied may be insufficient to allow for this procedure. The V_m may be more precisely evaluated from a plot of the reciprocal of V against the reciprocal of (S). The reciprocal of Eq. (7.12) takes the form

$$\frac{1}{V} = \frac{k_m}{V_m(S)} + \frac{1}{V_m} \tag{7.13}$$

In a plot of $1/V$ as a function of 1/(S) the slope is K_m/V_m. The $1/V$ intercept is $1/V_m$ and the 1/(S) intercept is $1/K_m$. This is shown in Fig. 7.7*c*, which is known as a *Lineweaver-Burk plot*. Various other possible ways of examining data from enzyme reactions are available, including curve fitting using computers.

The presence of inhibitors could alter the kinetics represented in Fig. 7.7*a*. For example, the V may be lowered (e.g., Fig. 7.7*d*; compare curves 2 and 3 to curve 1) if the inhibitor combines with the enzyme to make less enzyme available for the reaction. This would result in a decrease in rate as well as V_m (noncompetitive inhibition, curve 3). On the other hand, the inhibitor could compete directly or indirectly for the active site. In this case, the substrate at a high enough concentration should be able to compete successfully, so the V_m would remain unaffected by the presence of inhibitor (curve 2).

B. Sigmoidal Kinetics

Equations (7.12) and (7.13) and Fig. 7.7*a* represent reasonably well the kinetics of many enzyme reactions. However, many other reactions follow not these kinetics but a sigmoidal pattern, where the curve expressing rate as a function of substrate concentration is S shaped. The aspartate carbamoyltransferase reaction, which is one of these, takes place as shown in Eq. (7.14).

$$\underset{\text{Carbamoyl phosphate}}{O{=}C\begin{matrix}\diagup NH_2 \\ \diagdown O{-}P(=O)(O^-)O^-\end{matrix}} + \underset{\text{Aspartate}}{H_2N{-}CH\begin{matrix} CH_2{-}CO^-(=O) \\ CO^-(=O)\end{matrix}} \longrightarrow \underset{\text{Carbamoyl aspartate}}{O{=}C\begin{matrix}\diagup NH_2 \\ \diagdown NH{-}CH\begin{matrix} CH_2{-}CO^-(=O) \\ CO^-(=O)\end{matrix}\end{matrix}} + HOP(=O)(O^-){-}O^- \tag{7.14}$$

Carbamoyl phosphate Aspartate Carbamoyl aspartate

This reaction is the first in the pathway that converts carbamoyl phosphate and aspartate into cytidine triphosphate (CTP). The kinetics of the reaction, represented in Fig. 7.8 (Gerhart and Pardee, 1962), clearly do not follow the predictions of Eq. (7.12). In this figure, the rates of the reaction at various concentrations of aspartate are shown. The carbamoyl phosphate level is kept constant.

At low concentrations of aspartate, the rate of the reaction does not vary linearly with substrate concentration. Rather, it increases more sharply, as if the effectiveness of the enzyme were increased at the higher concentration of substrate. A number of explanations could account

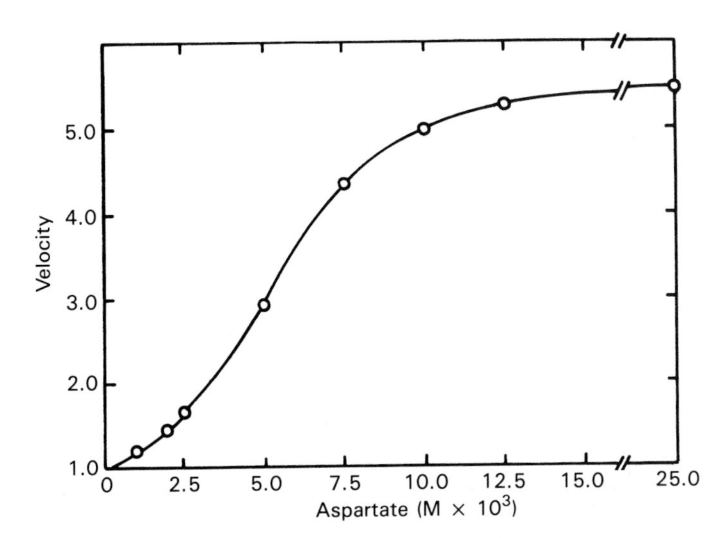

Fig. 7.8 Kinetics of aspartate carbamoyltransferase reaction. Velocity, units of activity (in this case micromoles of carbamoyl aspartate per hour) per milligram of protein $\times 10^{-3}$. Reaction mixture: 3.6×10^{-3} M carbamoyl phosphate and 9.0×10^{-3} μg of enzyme per ml. The mixture was in 0.04 M potassium phosphate buffer at pH 7.0. Reprinted with permission from J. C. Gerhart and A. B. Pardee, *Journal of Biological Chemistry,* 237:891–896. Copyright © 1962 The American Society of Biological Chemists, Inc.

for this phenomenon. The most likely one is that the reactivity of the enzyme does, in fact, change with increasing concentrations of substrate. It is difficult to visualize an effect of this kind unless we postulate that a combination of the substrate with the enzyme alters the structure of the enzyme so that it reacts more effectively with another molecule of substrate. This implies that each enzyme molecule has more than one functional unit and active site. A substrate at the active site can facilitate the reaction at another active site only if the two active sites are part of the same enzyme molecule. Many enzymes have been found to be made up of subunits that are held together by noncovalent bonds. In some enzymes these subunits are identical, whereas in others the subunits are dissimilar. Generally, but not always, the subunits alone have little or no activity. Reconstitution of the enzyme from its subunits restores the activity.

That these particular kinetics are the consequence of changes in reactivity of the carbamoyltransferase at different substrate concentrations is supported by experiments in which the enzyme is manipulated chemically. The kinetic pattern is altered when the enzyme is treated with compounds that react with sulfhydryl groups or when it is heated to 60°C for short periods. This alteration is shown in Fig. 7.9 (Gerhart and Pardee, 1962) and consists of a conversion to a Michaelis-Menten curve from the original sigmoidal curve. The lower curve in Fig. 7.9 indicates results obtained with the untreated enzyme. The upper curve indicates results obtained after treatment with $Hg(NO_3)_2$. The rates in the upper curve are increased and the V is higher, evidence that the treatment has increased the activity of the enzyme. The lack of a sigmoidal shape for the kinetics of the treated enzyme can be explained by postulating that the altered enzyme can now react only maximally and that the conformation of the enzyme has been changed by the treatment.

There is considerable evidence that the conformation of some enzymes changes when the enzymes interact with their substrates. In the case of aspartate carbamoyltransferase, the rate at which the enzyme sediments when centrifuged changes with the addition of succinate, a substrate analog (Fig. 7.10) (Gerhart and Schachman, 1968). The rate of sedimentation of macromolecules

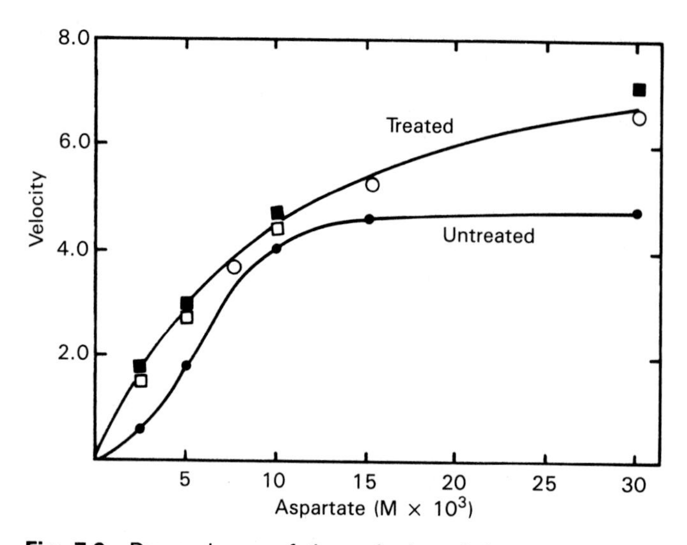

Fig. 7.9 Dependence of the velocity of the aspartate carbamoyltransferase reaction on aspartate concentration after loss of feedback inhibition. Velocity units are activity per milligram of protein × 10^{-3}. (●) Native (untreated) enzyme; (■) 10^{-6} M $Hg(NO_3)_2$ present during assay (heavy metal-treated aspartate carbamoyltransferase); (○) enzyme heated for 4 min at 60°C and cooled before assay; (□) heated enzyme assayed in presence of 2 × 10^{-1} M CTP. The reaction mixture contained 3.6 × 10^{-3} M carbamoyl phosphate; aspartate varied as indicated; 2 × 10^{-1} M CTP when used; 0.04 M potassium phosphate, pH 7.0; and 9.0 × 10^{-2} μg of enzyme protein/ml. Reprinted with permission from J. C. Gerhart and A. B. Pardee, *Journal of Biological Chemistry,* 237:891–896. Copyright © 1962 The American Society of Biological Chemists, Inc.

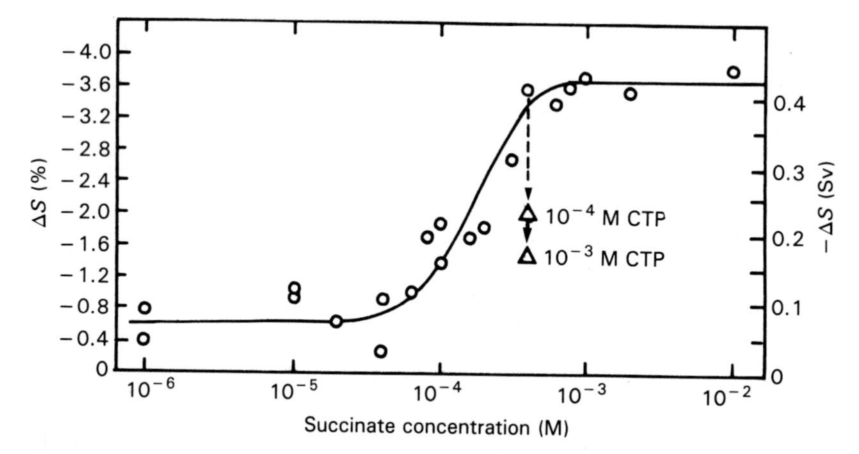

Fig. 7.10 Dependence of sedimentation coefficient S of aspartate carbamoyltransferase on the concentration of the substrate analog, succinate. All samples contained 0.04 M potassium phosphate, pH 7.0; 1.8 × 10^{-3} M dilithium carbamoyl phosphate; 4.2 mg enzyme/ml; and potassium succinate. Results are compared with simultaneous determination in which succinate is replaced by identical concentration of potassium glutarate. Preparations were centrifuged at 20°C and 60,000 rpm. The sedimentation coefficient was about 11.6 S. Reprinted with permission from Gerhart and Schachman (1968) *Biochemistry* 7:538–552, copyright 1968 by the American Chemical Society.

is a function of their size and shape (i.e., configuration). Presumably, the same changes would occur in the presence of aspartate. Note that the presence of CTP (indicated by the triangles) has the opposite effect. The metabolic significance of the regulation of aspartate carbamoyltransferase by CTP is examined in more detail in the next section (III, C). Changes produced by interactions with the substrate have also been detected by observing changes in the reactivity of some of the residues of the enzyme (e.g., —SH groups). These changes imply a change in the shielding of the residues and hence in the conformation of the enzyme. Similarly, conformational changes of some enzymes have been detected with x-ray diffraction and nuclear magnetic resonance (NMR) techniques.

C. Regulatory Sites: Allosteric Interactions

As previously mentioned, the fact that the presence of substrate facilitates the enzyme reaction argues for the presence of more than one functional unit or binding site per molecule of enzyme. In fact, treatment with mercurials (e.g., *p*-mercuribenzoate) separates the aspartate carbamoyltransferase enzyme into two types of subunits. One type is composed of two units containing the catalytic activity, whereas the other contains four units carrying the sites that react with the substances regulating the activity of the enzyme.

A great number of enzymatic reactions can be inhibited or stimulated by compounds that are chemically quite distinct from the natural substrates. These substances interact at sites distinct from the active site. Such interactions are referred to as *allosteric interactions*. Allosteric inhibitions are frequently exerted by products of metabolic pathways and may be of great physiological significance by permitting precise control of the reaction by *feedback inhibition* (end product inhibition). In this manner, overproduction of end product would lead to a slowing down of the pathway that produces it. The functional effect of such an inhibition is illustrated in Fig. 7.11*a*. In Fig. 7.11*a* and *b*, the hatched line and heavy arrow indicate stimulation and the black line and cross indicate inhibition. The pathway represented in Fig. 7.11*a* synthesizes the cell's uridine triphosphate (UTP) and CTP. Feedback inhibition by the end product (CTP) can adjust the rate of production depending on the needs of the cell. The regulation can be much more complex, particularly where the pathways are complex. The diagram of Fig. 7.11*b* represents the biosynthetic pathway of the purine nucleotides, adenosine triphosphate (ATP) and guanosine triphosphate (GTP). The pathway is branched, and its regulation includes feedback inhibition of the initial reaction before the branching point as well as inhibition of the two separate branches independently. Furthermore, accumulation of the end product of one branch of the pathway stimulates the other branch. In this pathway inosinic acid is converted into either ATP or GTP, depending on which branch of the reaction sequence is traversed. One of the early steps in the pathway, the first shown in the figure, involves transfer of an amino group to 5-phosphoribosyl-1-pyrophosphate to form 5-phosphoribosylamine (Fig. 7.11*b*). The aminotransferase catalyzing this reaction is inhibited by ATP, ADP, and AMP or, alternatively, GTP, GDP, and GMP, with the two groups of nucleotides acting at two separate regulative sites in the enzyme. The two divergent pathways are subject to separate feedback control at the point of divergence. The branch producing the G-containing nucleotides is controlled by GMP, whereas that producing the A-containing nucleotides is controlled by AMP. In addition, the end products of the two divergent pathways reciprocally facilitate each other: ATP facilitates the GTP pathway, and GTP facilitates the ATP pathway. This mode of control ensures smooth integration of the two systems synthesizing the two purine nucleotides.

The data available for the allosteric inhibition of aspartate carbamoyltransferase are shown in Fig. 7.12*a* (Gerhart and Pardee, 1962). In this experiment, the velocity of the reaction is measured at various concentrations of aspartate with the concentration of the other substrate, carbamoyl phosphate, kept constant. The upper curve represents a control in the absence of

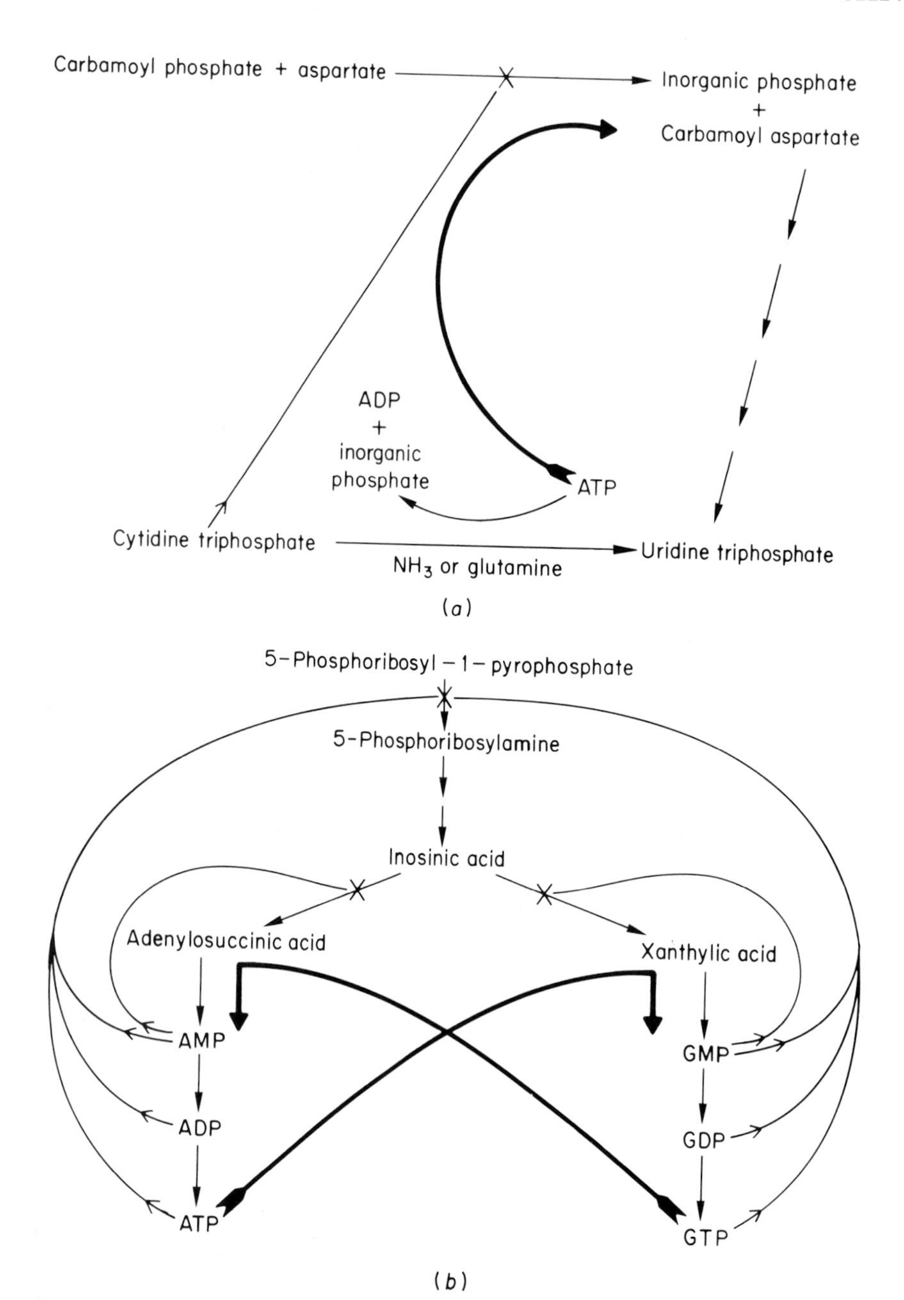

Fig. 7.11 (*a*) Feedback control of the aspartate carbamoyltransferase reaction by CTP and its stimulation by ATP. (*b*) Schematic diagram of the regulation of purine nucleotide metabolism.

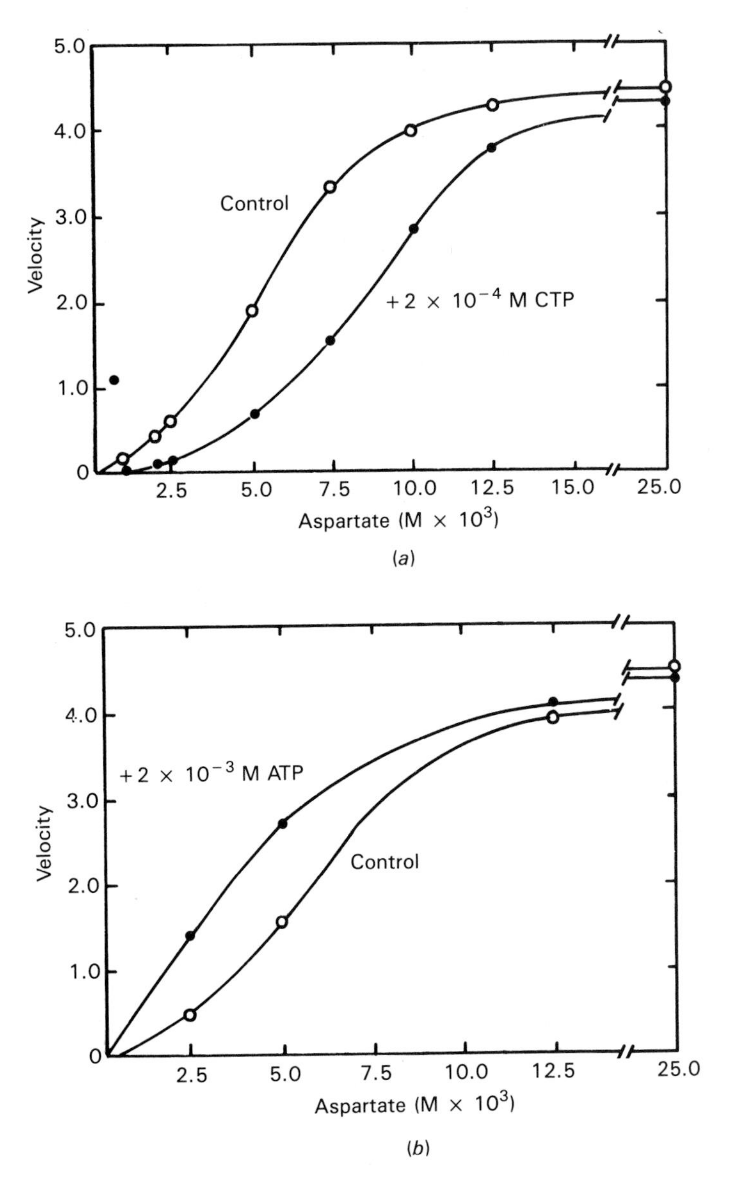

Fig. 7.12 (*a*) Reversal of CTP inhibition of aspartate carbamoyltransferase by aspartate. (*b*) Effect of ATP on reaction velocity. Velocity, units of activity per milligram protein $\times 10^{-3}$. The reaction mixture contained 3.6×10^{-3} M carbamoyl phosphate, 0.04 M potassium phosphate, pH 7.0, and 9.0×10^{-2} g of enzyme per ml. Reprinted with permission from J. C. Gerhart and A. B. Pardee, *Journal of Biological Chemistry,* 237:891–896. Copyright © 1962 The American Society of Biological Chemists, Inc.

CTP. The lower curve shows the same experiment carried out in the presence of CTP, the product of the metabolic pathway. Inhibitory compounds of this kind have been called *negative effectors* by some investigators. Other metabolites act as positive effectors; ATP acts as a positive effector in this reaction (Fig. 7.12*b*). Such a role could also be of fundamental physiological importance. Where the energy supply is plentiful (reflected in a high level of ATP), nucleic acid synthesis and consequently growth would be favored by increasing the availability of the nucleic acid precursors. On the other hand, a decrease in ATP would slow down the biosynthetic pathway.

The increase in enzymatic activity that occurs when the aspartate carbamoyltransferase is treated with mercurials (Fig. 7.9) is accompanied by loss of regulative ability. The inhibition accompanying the presence of CTP is lost completely, as shown in Table 7.3, column 1. Columns 2 and 3 correspond to the apparent V_m and K_m, respectively.

The fact that loss of regulative ability and loss of enzymatic activity do not go hand in hand speaks for two different reactive groups being responsible for the two effects. As mentioned, two different kinds of subunits can be separated out from aspartate carbamoyltransferase by treatment with *p*-mercuribenzoate. One kind of subunit involves enzymatic activity only, whereas the other, when complexed with the active portion, permits control by end product or end product analog inhibition (Fig. 7.13) or by the enhancement induced by ATP. The recoupling after removal or dilution of the mercurial compound is spontaneous.

Figure 7.13 (Gerhart and Schachman, 1965) represents a demonstration of the separation of the subunits of the enzyme by *p*-mercuribenzoate. It is possible to separate out protein molecules on the basis of their size and density differences by means of the centrifugation techniques. One such technique is centrifugation through a sucrose solution forming a density gradient. The preparation, in this case the enzyme treated with *p*-mercuribenzoate, is layered on a tube containing the sucrose gradient. Centrifugation of the tube at very high speed causes the molecules to descend through the gradient, with the heaviest and largest molecules moving most rapidly. After centrifugation, the various layers of the gradient can be separated and sampled for enzymatic activity. The assay is carried out by mixing the layer with a substrate mixture, aspartate and carbamoyl phosphate. The appearance of the product of the reaction, carbamoyl aspartate, is measured by the appropriate chemical method. In Fig. 7.13, the distribution of the enzyme in the various layers is shown by the filled circles and dashed curve. The capacity of the fractions to inhibit (after dilution to decrease the concentration of *p*-mercuribenzoate) is shown by the filled triangles and line. These were assayed as follows: one of the fractions that is unable to respond to the negative effector (fraction 6) was mixed in the presence of CTP with the fraction to be sampled for inhibitory activity. Clearly, the enzymatic

Table 7.3 Selective Destruction of Feedback Inhibition[a]

Treatment	(1) Inhibition by 2×10^{-4} M CTP (%)	(2) Maximal velocity (units/mg protein $\times 10^{-3}$)	(3) Aspartate for half-saturation (molarity $\times 10^{2}$)
1. None	70	4.5	6
2. 10^{-6} M $Hg(NO_3)_2$	0	10.0	12
3. 5×10^{-3} M *p*-hydroxymercuribenzoate	0		
4. 5×10^{-3} M mersalyl[b]	0		
5. 10^{-3} M $AgNO_3$	0		
6. Preheat 4 min at 60°C	0	9.0	12
7. 0.8 M urea	0	>4.5	

Reprinted with permission from J. C. Gerhart and A. B. Pardee, *Journal of Biological Chemistry,* 237:891–896. Copyright © 1962 The American Society of Biological Chemists, Inc.

[a] The reaction mixture contained 3.6×10^{-3} M carbamoyl phosphate, 5.0×10^{-2} M aspartate, 0.04 M potassium phosphate buffer, pH 7.0, 1.8×10^{-3} μg of enzyme protein per ml, and heavy metal compounds or urea as indicated. Sensitivity to CTP inhibition was destroyed by heating as follows. The enzyme was diluted to 1.8 μg of protein per ml in 5×10^{-3} M potassium phosphate, pH 7.0, and put in a 60°C water bath for 4 min, after which it was cooled quickly in ice water.

[b] Salicyl-(α-hydroxymercuri-β-methoxypropyl)-amide-O-acetate.

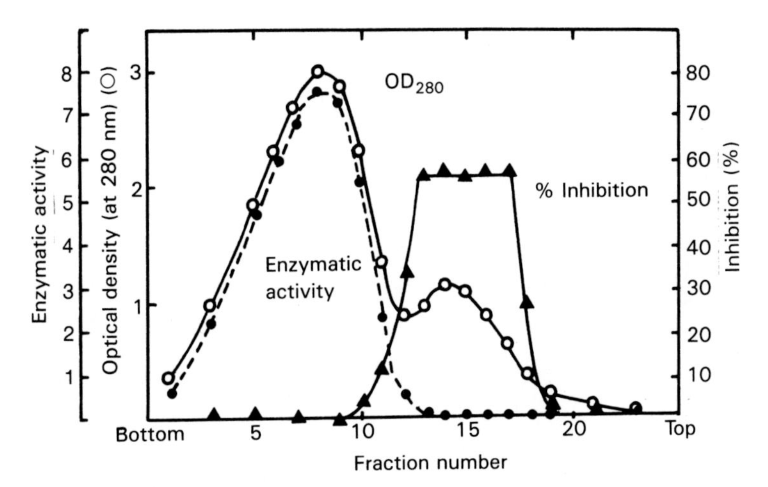

Fig. 7.13 Subunits of aspartate carbamoyltransferase separated by sucrose gradient centrifugation. The layers contained 6–25% sucrose, 0.04 M potassium phosphate, pH 7.0, and 10^{-4} M *p*-mercuribenzoate. Samples were centrifuged for 20 h at 10°C at 38,000 rpm in a W-39 rotor of a Spinco model L centrifuge. Reprinted with permission from Gerhart and Schachman (1965) *Biochemistry* 4:1054–1062, copyright 1965 by the American Chemical Society.

activity (filled circles) and the ability to respond to the inhibition (triangles) are in two separate molecules. The open circles represent the ultraviolet light absorption of each fraction (at a wavelength of 280 nm) and indicate protein concentration.

In contrast to aspartate carbamoyltransferase, many of the enzymes so far studied do not have separate subunits responsible for regulation. One of these enzymes is glutamine synthetase from *Escherichia coli*. The enzyme synthesizes glutamine from glutamate, ATP, and ammonia. Glutamine is thought to serve as an important reserve of available nitrogen and is required by several metabolic pathways. Glutamine synthetase is made up of 12 subunits, each with a molecular weight of about 50,000 (Shapiro and Stadtman, 1967). The indications so far are that the subunits are identical. The activity of this enzyme is inhibited by at least eight separate feedback inhibitors, which apparently attach to eight independent allosteric sites.

A model representing an enzyme made up of two identical subunits is shown in Fig. 7.14 (Koshland and Neet, 1968). Allosteric effectors, R_2, interact with special regulator sites of the molecule, part B, as in the case of glutamine synthetase. In some cases, such as aspartate carbamoyltransferase, part B could be a separate regulatory subunit. The shaded sites in part A of the molecule correspond to the active sites binding a substrate, S, with two residues, X_1 and X_2. The effector molecule R_1 can interact directly with the active site. The outside portion of the active site represents a flexible arm that is adjusted into position by the substrate itself and the effector R_2 (the so-called *induced fit*). There is considerable evidence that induced fit plays a significant role in enzyme catalysis.

This model depicts neither the conformational change imposed by the allosteric effector nor the conformational change induced in one subunit by the interaction of the other with the substrate. The nature of this conformational change in enzymes is still not entirely clear.

The kinetics of reactions affected by positive or negative effectors are well suited to permit substantial changes in enzymatic activity at constant substrate concentration. A major change in the rate of the enzyme reaction results from a small change in the reactivity of the enzyme. This effect is shown in Fig. 7.15 (Atkinson, 1965). The small shift in the curve depicting the

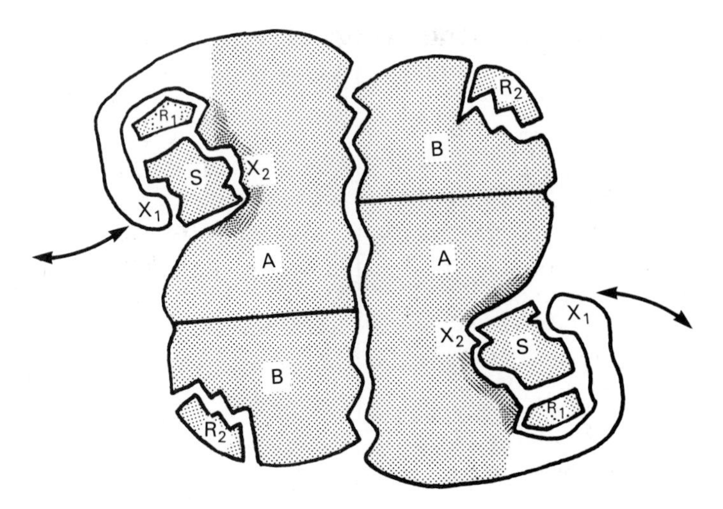

Fig. 7.14 Hypothetical structure of an enzyme containing two identical subunits. The enzyme has two residues in the active site (X_1 and X_2). The substrate is denoted by S, and R_1 and R_2 are regulator molecules, R_1 at the active site and R_2 at an allosteric site. Modified from Koshland and Neet (1968). Reproduced, with permission, from the *Annual Review of Biochemistry,* Volume 37, © 1968 by Annual Reviews Inc.

velocity of the reaction as a function of substrate concentration represents a major change in the rate of the reaction at certain fixed substrate concentrations (compare 3, the control curve, with curve 2 or 4). The velocity of the reaction at a single substrate concentration, S′, is shown by the bar diagram. At concentrations where the effect is sufficiently pronounced, the reaction may be ineffective in the presence of a negative effector or in the absence of a positive effector.

D. Transitions from Inactive to Active Forms

Some of the regulative events of the cell's biochemical reactions seem to involve allosteric interactions such as the end product inhibitions or stimulations just discussed. However, enzyme activity may also be regulated by covalent modifications in which the enzyme is converted from

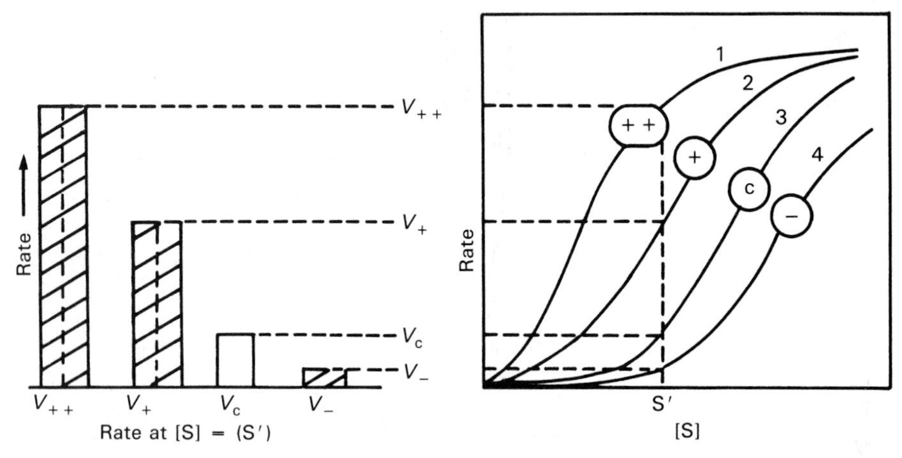

Fig. 7.15 Generalized substrate response curve for an enzyme that is regulated by positive effectors (+) or negative effectors (−). The control (no effectors) is indicated by the letter c. Reprinted with permission from D. E. Atkinson, *Science,* 150:851–857. Copyright 1965 by the AAAS.

an inactive form to an active form. Furthermore, at least in the case of phosphorylation, the modification may determine whether the enzyme can be controlled allosterically. A number of covalent modifications have been implicated, such as conversions from S — S to SH groups, acetylation or deacetylation, methylation or demethylation, adenylylation or deadenylylation. Activation of enzymes by proteolytic cleavage of the inactive form of the enzyme, the *proenzyme*, is involved in many physiological (Hall et al., 1979) and developmental processes (Hasilik and Tanner, 1978; Neurath and Walsh, 1976). At least in eukaryotic cells, many enzymes have been shown to undergo phosphorylative and dephosphorylative transitions. Table 7.4 (Krebs and Beavo, 1979) gives a relatively short list, and more are likely to be involved. Generally ATP, and in some cases GTP, is implicated in phosphorylation. The conversion from dephosphorylated to phosphorylated form proceeds as follows:

$$\text{Protein} + n\text{ATP} \underset{\text{kinase}}{\overset{\text{protein}}{\rightleftharpoons}} \text{protein-P}_n + n\text{ADP} \tag{7.15}$$

$$\text{Protein-P}_n + n\text{H}_2\text{O} \underset{\text{phosphatase}}{\overset{\text{protein}}{\rightleftharpoons}} \text{protein} + n\text{P}_i \tag{7.16}$$

In some cases the phosphorylated form is the more active form (as in the case of phosphorylase *b*), and in others it may be a less active form. Generally, the enzymes involved in a breakdown pathway are activated by phosphorylation, whereas enzymes involved in synthetic pathways are

Table 7.4 Initial Reports on Enzymes Undergoing Phosphorylation-Dephosphorylation

Enzyme	Year Reported
Glycogen phosphorylase	1955
Phosphorylase kinase	1959
Glycogen synthase	1963
Hormone-sensitive lipase	1964, 1970[a]
Fructone-l,6-biphosphatase	1966,1977[a]
Pyruvic dehydrogenase	1969
Hydroxymethylglutaryl-CoA reductase	1973
Acetyl-CoA carboxylase	1973
DNA-dependent RNA polymerase	1973
Pyruvate kinase (liver)	1974
Cholesterol ester hydrolase	1974
R subunit of type II cAMP-dependent protein kinase	1974
Reverse transcriptase	1975
Phosphofructokinase (liver)	1975
Tyrosine hydroxylase	1975
Phosphorylase phosphatase inhibitor	1975
Phenylalanine hydroxylase	1976
elF2-kinase	1977
cGMP-dependent protein kinase	1977
Tryptophan hydroxylase	1977
NAD-dependent glutamate dehydrogenase (yeast)	1978
Glycerophosphate acyltransferase	1978

Source: Krebs and Beavo (1979). Reproduced, with permission, from the *Annual Review of Biochemistry,* Volume 48, © 1979 by Annual Reviews Inc.

[a] In those instances in which a substantial period of time elapsed between initial and subsequent reports two dates are listed.

Table 7.5 Enzymes Found in the Cytoplasm of Mammalian Cells That Are Regulated by Phosphorylation

	Types of protein kinase involved[a]			
	cAMP	Ca^{2+} -calmodulin	Other	
Activation by phosphorylation				Biodegradative pathway
Glycogen phosphorylase	−	+	−	Glycogenolysis
Phosphorylase kinase	+	+[b]	−	Glycogenolysis
Myosin	−	+	−	ATP hydrolysis
Triglyceride lipase	+	−	−	Triglyceride breakdown
Cholesterol esterase	+	−	?	Cholesterol ester hydrolysis
Inactivation by phosphorylation				Biosynthetic pathway
Glycogen synthase	+	+	+	Glycogen synthesis
Acetyl-CoA carboxylase	+	−	+	Fatty acid synthesis
Glycerol phosphate acyltransferase	+	−	−	Triglyceride synthesis
HMG-CoA reductase	−	−	+	Cholesterol synthesis
Initiation factor eIF-2	−	−	+	Protein synthesis

From P. Cohen, *Molecular Aspects of Cellular Regulation,* Vol. 1. Copyright © 1980 Elsevier Science Publishers, Amsterdam.

[a] Pyruvate kinase has been omitted from this list, since it can be regarded as a biosynthetic enzyme channeling glycolytic intermediates to ATP and fatty acid synthesis or as a biodegradative enzyme involved in glucose and glycogen catabolism.

[b] Phosphorylase kinase phosphorylates itself at a low rate.

inactivated by phosphorylation. Some examples of these effects are shown in Table 7.5 (Cohen, 1980).

Equations (7.15) and (7.16) show that changes in enzymatic activity can take place through some control of the protein kinase, which catalyzes phosphorylation, or of the protein phosphatase, which catalyzes dephosphorylation. These systems are under the control of hormonal or neural signals. Cyclic AMP, cyclic GMP, Ca^{2+}-calmodulin complex, and other second messengers (see Chapter 5) have been implicated in some of these regulative events.

A general role of the phosphorylation-dephosphorylation of proteins in cellular physiology has been proposed (Greengard, 1978); the process is likely to affect the functions of chromosomes, ribosomes, microtubules, and cell or intracellular membranes.

A diverse enzyme regulatory role by ubiquitin conjugation analogous to the phosphorylative mechanism has also been proposed. Ubiquitin has been found in all cells examined (hence the name which is derived from ubiquitous). Ubiquitin is an intracellular peptide which is ligated to proteins by a series of reactions involving several enzymes (Chapter 9, II, B). A regulatory role of ubiquitin conjugation is suggested by the presence of ubiquitin hydrolases, which remove ubiquitin from conjugates. The conjugation between ubiquitin and various proteins is involved in the pathway of protein degradation that takes place in all cells (Chapter 9, II, B). However, the role of ubiquitin is likely to touch a broader spectrum of the cell's activities (Jentsch et al., 1990). Metabolically stable ubiquitin-protein conjugates are found in eukaryotes. Furthermore, a large part of chromosomal histones are conjugated to ubiquitin, suggesting a role in gene expression. Several integral membrane proteins as well as actin have also been found to be conjugated to ubiquitin. The conjugation to ribosomal enzymes enhances protein assembly. In addition, ubiquitin-conjugating enzymes appear to be involved in DNA repair and the control of the cell cycle (e.g., Goebl et al., 1988).

IV. MULTIENZYME COMPLEXES

Not all biochemical reactions occur in solution, where both enzyme and substrate are unrestrained. Some enzymes are organized in macromolecular complexes that hold together enzymes responsible for several reactions in a pathway. Some of these complexes are attached to or embedded in membranes.

The organization of functionally related enzymes in complexes concentrates catalytic activity and allows for a high steady-state concentration of intermediates so that substrates are transferred from one active site to another with a minimum of diffusional effects or dilution. Furthermore, the presence of enzymes in a complex makes possible coordinated allosteric control of several activities.

α-Ketoacid dehydrogenase and fatty acid synthetase are soluble multienzyme complexes. In this section, the pyruvate dehydrogenase complex of *E. coli* will serve as an example of a complex of several distinct enzymes. The combination of enzymes into complexes involving membranes is discussed primarily in relation to oxidative phosphorylation (Chapter 10), photosynthesis (Chapter 11), and the reactions involved in the translocation of solutes across the cell membrane (Chapters 13–15).

In *E. coli,* the pyruvate dehydrogenase reaction is carried out by enzyme aggregates made up of three types of enzyme with a total molecular weight of about 4×10^6. Similar complexes are thought to catalyze these reactions in other organisms, including mammals.

The reactions catalyzed by the pyruvate dehydrogenase complex are represented in Fig. 7.16, in which part *a* represents the overall reaction and part *b* some of the details. As shown in this figure, pyruvate in the presence of coenzyme A and NAD^+ is oxidized and decarboxylated to form acetyl coenzyme A, reduced NAD, and CO_2. This pyruvate dehydrogenation involves several separate reactions and the three enzymes represented in Fig. 7.16*b*. These enzymes are held together in the pyruvate dehydrogenase complex. In reaction 1, pyruvate reacts with pyruvate dehydrogenase (E_1). The reaction involves the coenzyme thiamine pyrophosphate (TPP). Decarboxylation of the pyruvate forms the α-hydroxyethyl-TPP-E_1 complex. Reaction

$$CH_3\overset{O}{\overset{\|}{C}}COOH + CoASH + NAD^+ \longrightarrow CH_3\overset{O}{\overset{\|}{C}}{-}SCoA + CO_2 + NADH + H^+$$

(*a*)

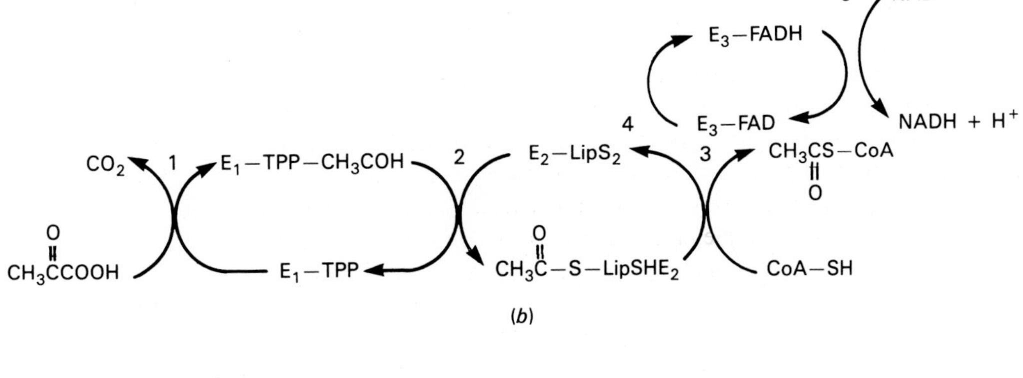

CO2 1 E1–TPP–CH3COH 2 E2–LipS2 4 E3–FAD E3–FADH 5 NAD+ NADH + H+ 3 CH3CS–CoA (C=O)
CH3CCOOH (C=O) E1–TPP CH3C–S–LipSHE2 (C=O) CoA–SH

(*b*)

Fig. 7.16 Diagrammatic representation of the pyruvate dehydrogenase reaction. E_1, Pyruvate dehydrogenase; E_2, dihydrolipoyl transacetylase; E_3, dihydrolipoyl dehydrogenase.

2 involves lipoyl reductase-transacetylase (E_2). The lipoyl moiety of E_2 oxidizes the hydroxyethyl residue that is transferred from E_1-TPP as the acetyl moiety. In reaction 3, the acetyl-reduced lipoyl-E_2 transfers the acetyl residue to coenzyme A, and in reaction 4 the lipoyl-E_2 molecule reduced in reaction 2 is oxidized by the flavoprotein dihydrolipoyl dehydrogenase (E_3-FAD). The reduced equivalents from this reaction are transferred to NAD^+ (reaction 5) and eventually oxidized by the cytochrome chain. In these reactions, the lipoyl moiety of E_2 (abbreviated Lip) interacts with both E_2-TPP (reaction 1) and E_3 (reactions 3 and 4). It serves to transfer reducing equivalents as well as acetyl groups (see Fig 7.18).

The three enzymes can be separated out from the pyruvate dehydrogenase complex by relatively gentle means, since they are held together by noncovalent bonds. The separated molecules reassemble spontaneously when mixed together at neutral pH. Alone, each kind of enzyme molecule can form an active complex made up of several subunits; these complexes have a specific morphology. Complexes can also be formed between the molecules of pyruvate dehydrogenase (E_1) or dihydrolipoyl dehydrogenase (E_3) and the transacetylase (E_2). The characteristic appearance of these complexes and that of the entire dehydrogenase complex have permitted their reconstruction. A reconstruction of the entire pyruvate dehydrogenase complex is shown in Fig. 7.17. Figure 7.17*a* shows an electron micrograph of the complex of the three enzymes after negative staining with phosphotungstate. In this procedure, the particles, suspended in a solution of phosphotungstate, are placed on a grid in a thin layer by a suitable

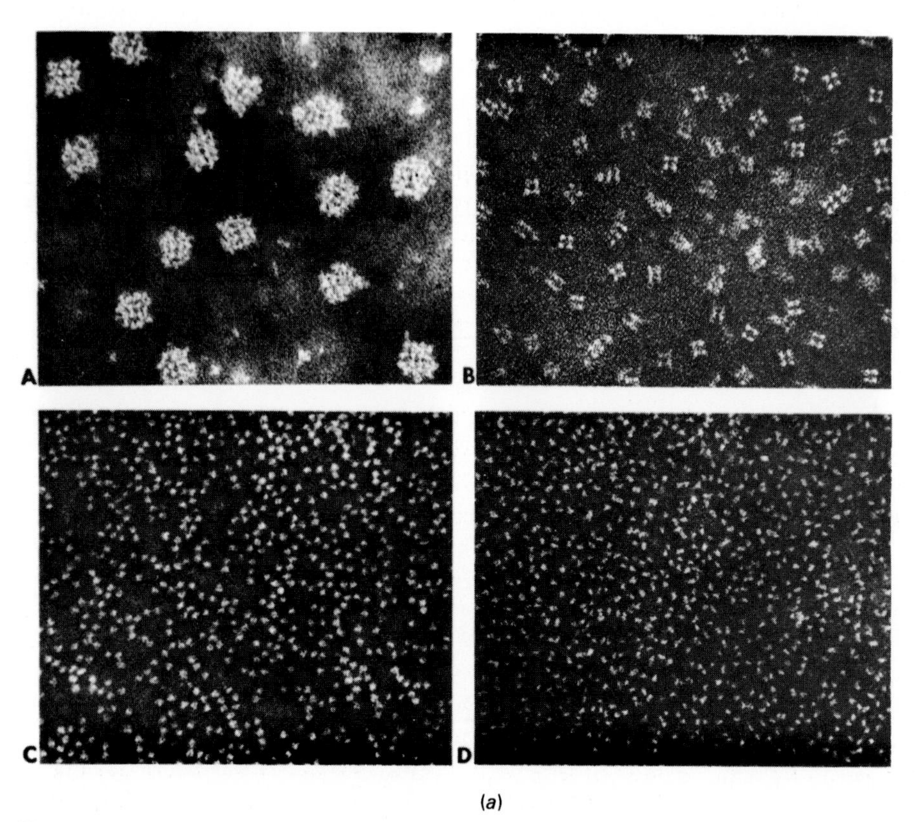

(*a*)

Fig. 7.17 (*a*) Electron micrographs of the *E. coli* pyruvate dehydrogenase complex and its component enzymes (×200,000). *A*, Pyruvate dehydrogenase complex; *B*, dihydrolipoyl transacetylase (E_2); *C*, pyruvate dehydrogenase (E_1); *D*, dihydrolipoyl dehydrogenase (E_3).

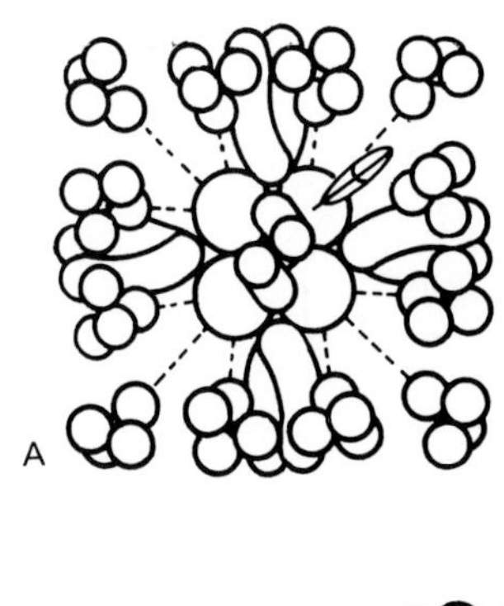

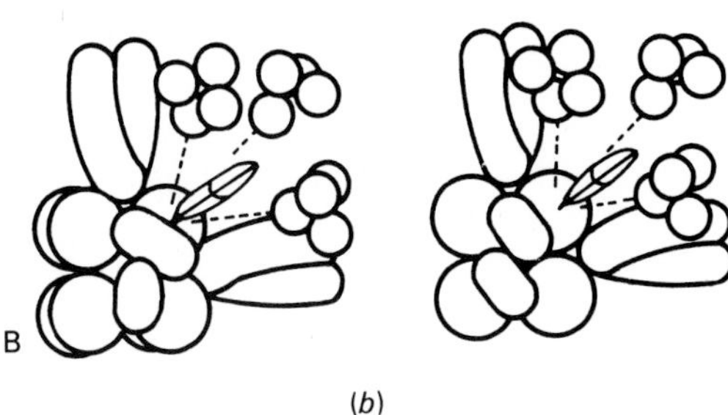

(*b*)

Fig. 7.17 (*Continued*) (*b*) Interpretative model of *E. coli* pyruvate dehydrogenase complex: *A*, Model viewed along the fourfold axis of the dihydrolipoyl transacetylase (E_2) core, illustrating the proposed architectural organization of the complex; *B*, stereoscopic drawings of a portion of the model showing the spatial relationships of the components. The eight trimers of E_2 binding domains are represented by large spheres, and the single E_2 lipoyl domain shown is represented by an ellipsoid. Pyruvate dehydrogenase (E_1) chains are represented by tetrahedra consisting of four lobes (small spheres) bound along the edges of the cubelike "inner" core of E_2 binding domains. In this projection of the model the 12 E_1 chains shown are superimposed over 12 other similarly positioned chains of E_1 (not shown). The dimers of dihydrolipoyl dehydrogenase (E_3) are located on the faces of the cubelike E_2 "inner" core and are represented by pairs of cylinders. From R. M. Oliver and L. S. Reed, Electron Microscopy, 2:1–48, 1982, with permission of Academic Press.

procedure and dried. When viewed with the electron microscope, the molecules appear white against the dense background of the phosphotungstate (Fig. 7.17*a*). A model based on various observations is shown in Fig. 7.17*b*.

The molecules of each of the three enzymes occupy fixed positions in the complex. Since the active groups correspond to a very small portion of the enzyme, it is difficult to see how an interaction can take place between the dihydrolipoyl transacetylase active site and the other two enzymes unless parts or all of the molecule move in relation to the others, first to react with one enzyme (pyruvate dehydrogenase) and then with the other (dihydrolipoyl dehydrogenase). The dihydrolipoyl transacetylase portion is composed of 24 polypeptide chains, each bearing two lipoyl moieties which function in both acetylation and redox reactions (see Fig. 7.18). A possible mechanism for this interaction is that the lipoyl-lysyl arm of dihydrolipoyl transacetylase, which is approximately 1.4 nm in length, swings between the two enzymes. However, at this time it seems much more likely that the polypeptides to which the lipoyl moiety are attached are usually mobile (Collins and Lester, 1977; Perham and Duckworth, 1981). These two models (A and B) are shown in Fig. 7.18*b*.

In the previous section we saw that enzyme complexes made up of identical subunits undergo conformational changes during enzyme-catalyzed reactions. In this section we saw how

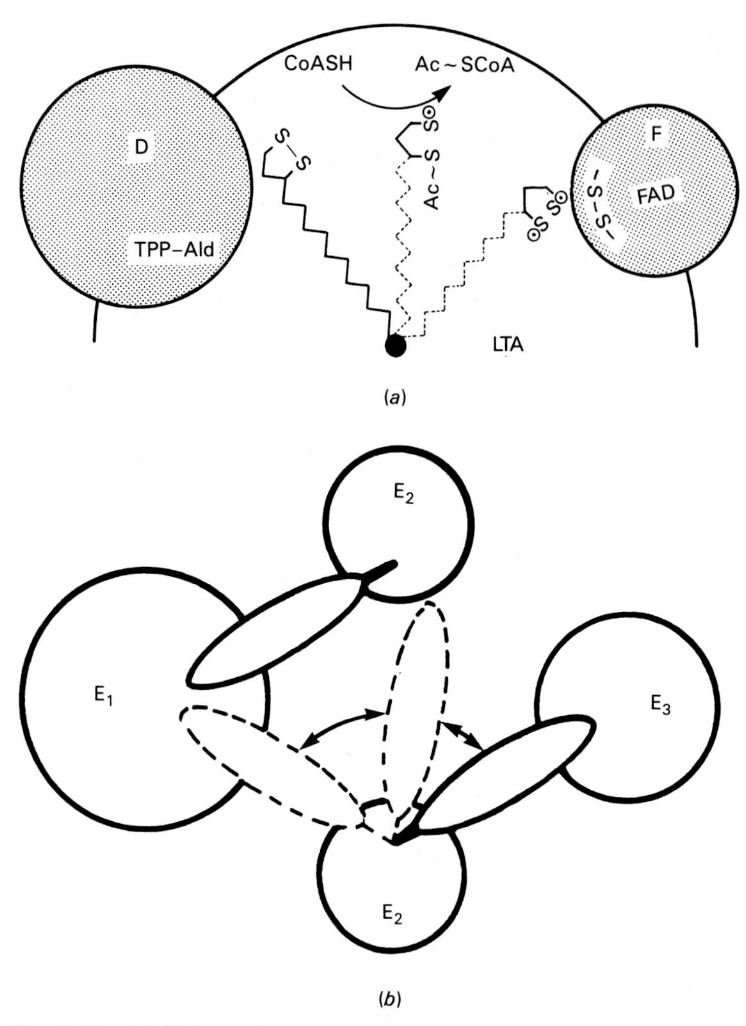

Fig. 7.18 (*a*) Schematic representation of the possible rotation of a lipoyl-lysyl moiety between α- hydroxyethylthiamine pyrophosphate (TPP-Ald) bound to pyruvate dehydrogenase (D), the site for acetyl transfer to CoASH, and the reactive disulfide of the flavo-protein (F). From Oliver and Reed (1982), with permission. (*b*) Model illustrating movement of lipoyl domains (and not simply rotation of lipoyl moieties) to span the physical gaps between catalytic sites on the complex. A dihydrolipoyl transacetylase (E_2) subunit is represented by a sphere (subunit binding domain) and its attached ellipsoid (lipoyl domain). The catalytic site for transacetylation resides on the subunit binding domain, whereas the two lipoyl moieties are on the lipoyl domain. It is visualized that movement of lipoyl domains permits their covalently attached lipoyl moieties (not shown) to service the catalytic sites on the complex. Each E_1, E_2, and E_3 subunit is serviced by at least two lipoyl moieties, which apparently reside on two separate lipoyl domains. The E_1, E_2, and E_3 subunits shown need not be adjacent to each other in the complex. From Stepp et al. (1981). Reprinted with permission from *Biochemistry* 20:4555–4560. Copyright © 1981 American Chemical Society.

the larger multienzyme complexes discussed must also involve some movement, either of the molecules of the complex in relation to each other or of a portion of one of the molecules. These movements or conformational rearrangements appear to be a fundamental property of many enzyme-catalyzed systems and may be involved in a wide range of biological mechanisms. The molecular mechanisms of muscular contraction or other events linked with cellular motility, the changes in shape of cells, and the mechanisms of solute transport across biological membranes may well involve specialized modifications of these conformational rearrangements.

The presence of organized multienzyme complexes or clusters offers significant opportunities for integrated activity and regulation, which are just beginning to be appreciated (Welch, 1977; Paul et al., 1989). There are indications, for example, that at least portions of the enzymes glyceraldehyde-3-phosphate dehydrogenase and phosphoglycerate kinase function at the inner surface of the red cell membrane to provide ATP to Na^+,K^+-ATPase and not to the rest of the cell (Mercer and Dunham, 1981). Apparently, the respiratory rate of cardiac mitochondria is under the control of creatine kinase, which catalyzes the phosphate exchange between creatine and ATP. This enzyme is present between the mitochondrial inner and outer membranes (Jacobus et al., 1983). Furthermore, the mitochondrial outer membrane has binding sites for hexokinase (Felgner et al., 1979), so the ATP synthesized can be used without delay to phosphorylate hexoses.

SUGGESTED READING

Introductory

Kraut, J. (1988) How do enzymes work? *Science* 242:533–540.

Oliver, R. M., and Reed, L. J. (1982) Multienzyme complexes. In *Electron Microscopy of Proteins*, Vol. 2 (Harris, J. R., ed.), pp. 1–48. Academic Press, New York.

Saier, M. H., Jr. (1987) *Enzymes in metabolic pathways*. Chapters 3 and 4. Harper Row, New York.

Stryer, L. (1988) Mechanisms of enzyme action. In *Biochemistry*, 3d ed. pp. 201–232. Freeman, New York.

More Advanced

Fersht, A. (1985) *Enzyme Structure and Mechanism*, 2d ed., Chapters 1–3, 8, 10 and 15. Freeman, New York.

Schachman, H. K. (1988) Can a simple model account for the allosteric transitions of aspartate transcarbamoylase? *J. Biol. Chem.* 263:18583–18586.

REFERENCES

Atkinson, D. E. (1965) Biological feedback control at the molecular level. *Science* 150:851–857.

Cohen, P. (1980) Protein phosphorylation and the coordinate control of intermediate metabolism. *Mol. Aspects Cell. Regul.* 1:255–268.

Collins, J. H., and Lester, L. J. (1977) Acyl group and electron pair relay system: a network of interacting lipoyl moieties in pyruvate and α-ketoglutarate dehydrogenase complexes from *Escherichia coli*. *Proc. Natl. Acad. Sci. U.S.A.* 74:4223–4227.

Felgner, P. L., Messer, J., and Wilson, J. E. (1979) Purification of a hexokinase-binding protein from the outer mitochondrial membrane. *J. Biol. Chem.* 254:4946–4949.

Gerhart, J. C., and Pardee, A. B. (1962) The enzymology of control by feedback inhibition. *J. Biol. Chem.* 237:891–896.

Gerhart, J. C., and Schachman, H. K. (1965) Distinct subunits for the regulation and catalytic activity of aspartate transcarbamoylase. *Biochemistry* 4:1054–1062.

Gerhart, J. C., and Schachman, H. K. (1968) Allosteric interactions in aspartate transcarbamoylase. II. Evidence for different conformational states of the protein in the presence and absence of specific ligands. *Biochemistry* 7:538–552.

Goebl, M. G., Yochem, J., Jentsch, S., McGrath, J. P., Vashavsky, A., and Byers, B. (1988) The yeast cell cycle gene CDC34 encodes a ubiquitin-conjugating enzyme. *Science* 241:1331–1335.

Greengard, P. (1978) Phosphorylated proteins as physiological effectors. *Science* 199:146–152.

Guerrier-Takada, C., and Altman, S. (1984) Catalytic activity of an RNA molecule prepared by transcription *in vitro*. *Science* 223:285–286.

Hall, E. C., McCully, V., and Gottam, G. L. (1979) Evidence for proteolytic modification of pyruvate kinase in fasted rats. *Arch. Biochem. Biophys.* 195:315–324.

Hasilik, A., and Tanner, W. (1978) Biosynthesis of the vacuolar yeast glycoprotein carboxypeptidase Y. Conversion of precursor into enzyme. *Eur. J. Biochem.* 85:599–608.

Jacobus, W. E., Moreadith, R. W., and Vandegaer, K. M. (1983) Control of heart oxidative phosphorylation by creatine kinase in mitochondrial membranes. *Ann. N. Y. Acad. Sci.* 414:73–89.

Jencks, W. P. (1969) *Catalysis in Chemistry and Enzymology*. McGraw-Hill, New York.

Jentsch, S., Seufert, W., Soomer, T., and Reins, H.-A. (1990) Ubiquitin-conjugating enzymes: novel regulators of eukaryotic cells. *Trends Biochem. Sci.* 15:195–198.

Koshland, D. E., Jr. (1956) Molecular geometry in enzyme action. *J. Cell. Comp. Physiol.* 47 (Suppl. 1):217–234.

Koshland, D. E., Jr., and Neet, K. E. (1968) The catalytic and regulatory properties of enzymes. *Annu. Rev. Biochem.* 37:349–410.

Krebs, E. G., and Beavo, J. A. (1979) Phosphorylation-dephosphorylation of enzymes. *Annu. Rev. Biochem.* 48:923–959.

Ludwig, M. L., Hartsuck, J. A., Steitz, T. A., Muirhead, H., Coppola, J. C., Reeke, G. N., and Lipscombe, W. N. (1967) The structure of carboxypeptidase A. IV. Preliminary results at 2.8 Å resolution and a substrate complex at a 6 Å resolution. *Proc. Natl. Acad. Sci. U.S.A.* 57:511–514.

Mercer, R. W., and Dunham, P. B. (1981) Membrane bound ATP fuels the Na/K pump. Studies on membrane-bound glycolytic enzymes on inside-out vesicles from human red cell membranes. *J. Gen. Physiol.* 78:567–568.

Neurath, H., and Walsh, K. A. (1976) Role of proteolytic enzymes in biological regulation. *Proc. Natl. Acad. Sci. U.S.A.* 73:3825–3832.

Oliver, R. M., and Reed, L. J. (1982) Multienzyme complexes. In *Electron Microscopy of Proteins*, Vol. 2 (Harris, J. R., ed.) pp. 1–48. Academic Press, New York.

Olson, M. S., and Allgyer, T. T. (1973) The regulation of nicotinamide adenine dinucleotide-linked substrate oxidation in mitochondria. *J. Biol. Chem.* 248:1582–1597.

Paul, R. J., Hardin, C. D., Raeymaekers, L., Wuytack, F., and Casteels, R. (1989) Preferential support of Ca^{2+} uptake in smooth muscle plasma membrane vesicles by an endogenous glycolytic cascade. *FASEB J.* 3:2298–2301.

Perham, R. N., and Duckworth, H. W. (1981) Mobility of polypeptide chain in the pyruvate dehydrogenase complex revealed by proton NMR. *Nature* (*London*) 292:474–477.

Reed, L. J., and Cox, D. J. (1966) Macromolecular organization of enzyme systems. *Annu. Rev. Biochem.* 35:57–84.

Schoellmann, G., and Shaw, E. (1963) Direct evidence for the presence of histidine in the active center of chymotrypsin. *Biochemistry* 2:252–255.

Shapiro, B. M., and Stadtman, E. R. (1967) Regulation of glutamine synthetase. IX. Reactivity of the sulfhydryl groups of the enzyme from *Escherichia coli*. *J. Biol. Chem.* 242:5069–5079.

Stryer, L. (1988) Mechanisms of enzyme action. In *Biochemistry*, 3d ed., pp. 201–232. Freeman, New York.

Welch, G. R. (1977) On the role of organized multienzyme systems in cellular metabolism: a general synthesis. *Prog. Biophys. Mol. Biol.* 32:103–191.

White, A., Handler, P., and Smith, E. L. (1978) *Principles of Biochemistry*, 6th ed. McGraw-Hill, New York.

Wilkinson, A. J., Fersht, A. R., Blow, D. M., Carter, P., and Winter, G. (1984) A large increase in enzyme substrate affinity by protein engineering. *Nature* (*London*) 307:187–188.

Zaug, A. J., and Cech, T. R. (1986) The intervening sequence RNA of *Tetrahymena* is an enzyme. *Science* 231:470–475.

Chapter 8

Regulation of Metabolism

As we have discussed in Chapter 7, the regulation of metabolism may be through control of the activity of the enzymes. We have seen how positive and negative effectors can change the rate of an enzyme reaction or how a whole pathway can be activated or inhibited by converting an enzyme into a more active or less active form.

In addition, the regulation of metabolism may involve either the production or the degradation of enzyme molecules. The control of the synthesis of the enzyme may be transcriptional (through control of gene expression) or translational (through any mechanism affecting the production of the enzyme). These controls will be discussed in Chapter 9.

It follows that the overall control of metabolism will be the result of an interplay of the various regulatory effects. The present chapter will concentrate on the control of enzyme activity in the metabolic reactions of the cell, in particular the energy-yielding reactions. These mechanisms are most likely to have a significant effect on moment-to-moment adaptive changes.

The metabolism of cells and its regulation are so complex that it is difficult to arrive at a quantitative evaluation of the effect of control mechanisms on the overall economy of the cell. A good deal is known about the properties of the enzymes involved in the cell's metabolic pathways. The consequences of altering metabolic conditions have been studied in some detail in intact cells. Nevertheless, it can be extremely difficult to correlate events occurring in the cell with the known properties of the purified enzymes. The cellular environment of the enzymes and the diffusion barriers within the cell are understood only in outline. The internal components of intact cells can seldom be manipulated experimentally without gross disruptions frequently unrelated to the effect under study. As we shall see in the discussion that follows, some inferences can be drawn from the properties of the enzymes involved in particular pathways. It should be born in mind, however, that although the general patterns and principles remain the same, various organisms and tissues of the same organism may differ in detail.

Enzyme regulation can take place by different mechanisms. Regulation may occur by the phosphorylation-dephosphorylation of enzymes by specific kinases (Chapter 7, III, D). The kinases respond to the level of a metabolite or, when responding to hormonal signals, to second messengers. This kind of regulation generally affects key entry reactions. For example, in this chapter we will discuss the activation of glycogen phosphorylase (I, A) which is activated by phosphorylation and that of pyruvate dehydrogenase (III, A) which is inhibited by phosphorylation. These two enzymes control the entry into the glycolytic pathway (when glycogen is the substrate) and the tricarboxylic acid (TCA) cycle respectively.

As discussed in more detail below, particularly in relation to glycolysis, the control can also be through allosteric regulation, the elaboration of regulatory effectors by side reactions of the pathway and the presence of substrate cycles. Regulation of the TCA cycle involves simpler principles, since it is controlled primarily through substrate availability and feedback inhibition. However, the wealth of detail offers much complexity and the precise role of each regulatory reaction on the whole is not always easy to assess.

I. SOME PRINCIPLES OF REGULATION

The regulation of metabolic pathways through the control of the activity of enzymes is complex. Fortunately, some principles which merit a separate discussion have emerged.

A. Cascades Under Hormonal Control Amplify Responses

As already discussed, some enzymes may be present in two forms, one active and one inactive or less active. Conversion of the inactive species into active molecules may be controlled by a complex mechanism in which a hormonal signal is amplified by a series of biochemical reactions (a *cascade* mechanism). Glycogen phosphorylase of skeletal muscle is controlled in this fashion, as shown in Fig. 8.1 (Fisher et al., 1970). The less active form of phosphorylase, phosphorylase *b*, is converted to the active phosphorylated species, phosphorylase *a* (reaction 5 of Fig. 8.1). In turn, phosphorylase *a* can be inactivated by the appropriate phosphatase. Phosphorylase *a* releases glucose-1-phosphate from glycogen (reaction 6). Glucose-1-phosphate can then be metabolized by the muscle cell's glycolytic machinery.

Phosphorylase is therefore very significant in the mobilization of the carbohydrate stores of muscle. The conversion of phosphorylase *b* to phosphorylase *a* is indirectly regulated by the hormone *epinephrine* secreted by the adrenal medulla, a secretion under the control of the nervous system. In muscle, epinephrine activates adenyl cyclase (reaction 1 in Fig. 8.1), which catalyzes the conversion of ATP to cyclic AMP (reaction 2). Cyclic AMP in turn activates (reaction 3) the enzyme that phosphorylates phosphorylase kinase (i.e., a kinase kinase) in the presence of Ca^{2+} (reaction 4). Then the phosphorylated active form of phosphorylase kinase and ATP phosphorylate the phosphorylase *b*, thereby producing phosphorylase *a* (reaction 5). The binding of a small amount of epinephrine has triggered the production of glucose-1-phosphate and thereby stimulated the major metabolic pathway of muscle through a cascade of reactions. The activity of phosphorylase is also under the control of several effectors such as ATP, glucose-6-phosphate, glucose, and uridine diphosphoglucose.

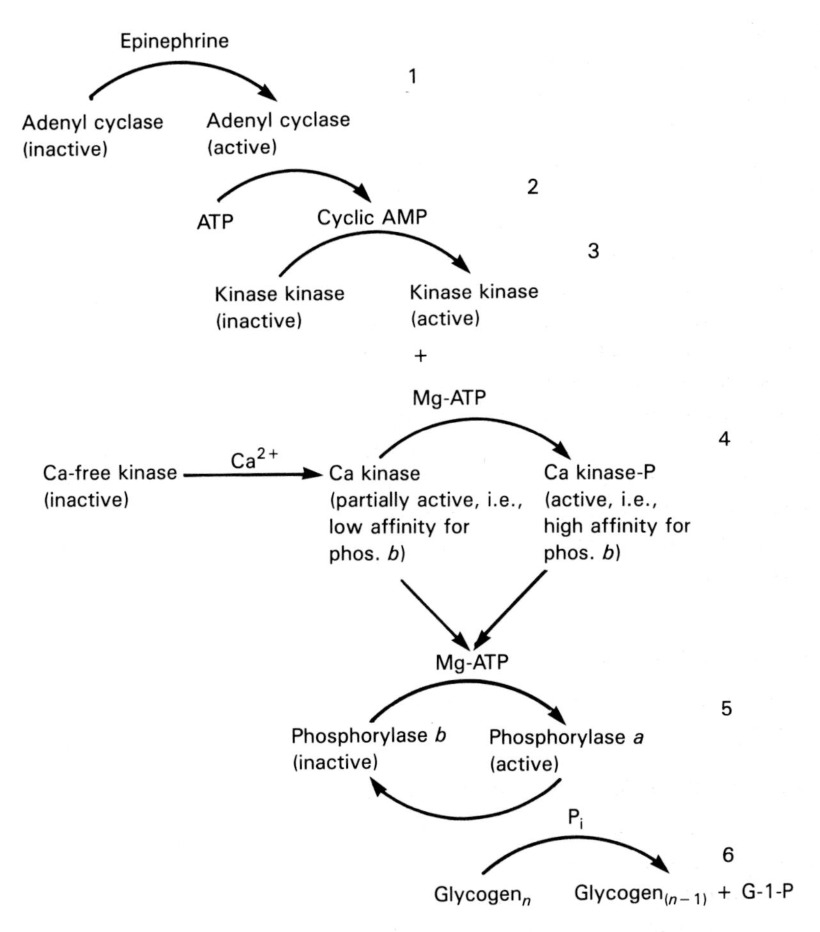

Fig. 8.1 Control of glycogen phosphorylase (Fisher et al., 1970). Reprinted by permission from *Essays in Biochemistry* 6:23–68, copyright © 1970 The Biochemical Society, London.

The formula for cyclic AMP (cAMP; cyclic adenosine-3′,5′-phosphate) is shown in Fig. 8.2. Cyclic AMP appears to mediate the effects of many hormones. It fulfills the role in intracellular signaling of a *second messenger,* discussed in Chapter 5.

The control of the release of glucose by epinephrine in muscle is but one of the reactions involved in mobilizing glucose in mammals (Exton et al., 1971). The mechanism involving muscle phosphorylase increases the glucose available for glycolysis in muscle. In liver, phosphorylase is also involved in the regulation of blood glucose and hence in the metabolism of the organism as a whole. In addition to an effect on phosphorylase, cAMP mediates the transformation of glycogen synthetase to a less active form, thereby blocking the storage of glucose. Similarly, cAMP inhibits phosphorylase phosphatase, thereby prolonging the presence of the enzyme in its active form.

Epinephrine is not the only hormone involved in the release of cAMP. *Glucagon* also tends to increase the concentration of cAMP, and *insulin* tends to decrease the concentration. The interactions between the effects of the three hormones are responsible for the final effect on glucose metabolism. Both glucagon and insulin are pancreatic endocrine hormones. A number of key enzymes controlled by a variety of signals are also regulated through a phosphorylation-dephosphorylation mechanism; an example is pyruvate dehydrogenase. However, in contrast to the phosphorylase, phosphorylated pyruvate dehydrogenase corresponds to the inactive form (see Chapter 7).

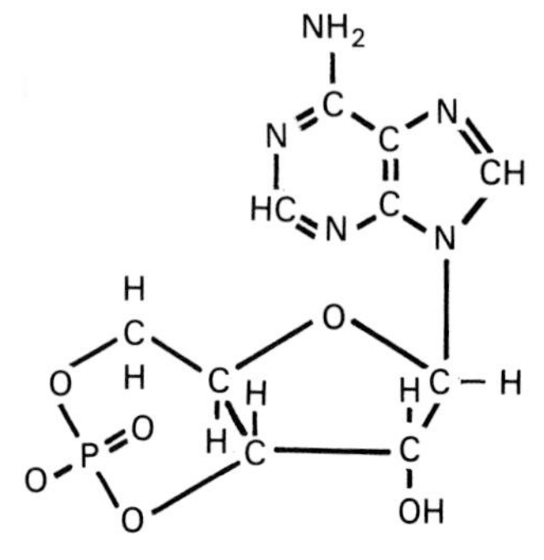

Fig. 8.2 Structure of cyclic AMP.

B. Regulated Steps are Exergonic

Most of the component reactions in a pathway proceed rapidly and are close to equilibrium (i.e., $\Delta G \approx 0$). Therefore they cannot be subject to regulation. The metabolic pathways themselves are highly exergonic; that is, they are virtually irreversible, the result of at least one virtually irreversible highly exergonic step. This step, the so-called first committed step, is frequently at the beginning of the metabolic pathway. Since it is far from equilibrium, it can be regulated. The regulation through the control of a rate-limiting exergonic step is analogous to the way in which a dam controls the flow of a whole river by adjusting the flow of the water passing through the dam itself. However, this is not the only necessary condition for regulation. Obviously the enzyme must be capable of responding to the appropriate regulator. Furthermore, the substrate concentration should be in a range allowing regulation; that is, the enzyme should not be saturated in the physiological range of concentrations.

In the regulated steps, generally forward and backward reactions are catalyzed by separate enzymes. This feature allows irreversible steps in either direction and the independent regulation of the forward and backward pathways. The latter is important since obviously the forward and backward reactions have opposite functional roles. In addition, it allows for an important regulatory device dependent on substrate cycling. The increase in rate of a limiting step by an allosteric mechanism is at most tenfold (see Fig. 7.15, previous chapter). The increase in fluxes through a metabolic pathway surpasses this value and can be as high as one hundredfold. This is possible because of the presence of two separate enzymes, which are separately regulated.

C. Substrate Cycling: An Effective Mechanism

The existence of a metabolic step in which two separate enzymes catalyze the forward and the backward reactions results in the continuous formation and degradation of a metabolite. This has been called *substrate cycling*. Since its purpose was not originally understood and its results undoubtedly wasteful in terms of energy, such a combination of reactions was originally called *futile cycling*.

Let us examine this process in more detail and assess its function (Newholme et al., 1984). The flow of metabolites through a pathway, that is, the flux (J_{net}) through a rate-limiting step, can be represented as the difference between the forward (J_f) and the backward flux (J_b), as we saw catalyzed predominantly by two different enzymes. This can be represented as:

$$J_{net} = J_f - J_b \tag{8.1}$$

Selecting reasonable values in arbitrary units [based on the regulation of the phosphorylation of fructose-6-phosphate (F6P) to form fructose-1,6-bisphosphate (F1,6P)], J_f may be 10 and J_b, 9.

J_{net} will be 1. If J_f is increased tenfold and J_b decreased tenfold, J_{net} will be 99, i.e., approximately a one hundredfold increase. In the case of this metabolic step, the forward reaction is catalyzed by phosphofructokinase (PFK), whereas the reverse reaction is catalyzed by fructose-1,6-bisphosphate phosphatase (FBPase). The two enzymes are regulated by AMP which stimulates the kinase and inhibits the phosphatase. The stimulation of PFK by AMP actually results from blocking an inhibition by ATP (Mansour and Ahlfors, 1968).

This metabolic step is also sensitive to a variety of other metabolites, illustrating how complex the regulative pattern can be. One of the allosteric activators of PFK is cAMP, a second messenger under hormonal control.

The fact that in the case discussed, the regulating metabolite is primarily AMP, leads us to another basic principle of regulation. The effector must be a chemical which changes in concentration significantly with physiological conditions. This question will be discussed in the next section.

D. An Effector Must Be Very Sensitive to Small Changes

Effectors are produced in two distinct ways. The nature of the pathway itself may be such as to provide a metabolite that can serve as a sensitive indicator of metabolic conditions. Basically this mechanism corresponds to feedback inhibition discussed in Chapter 7. The first reaction of the glycolysis, the phosphorylation of glucose to form glucose-6-phosphate (G-6-P), is catalyzed by hexokinase (Hk). Hk is inhibited allosterically by G-6-P (McDonald et al., 1979). In some cells, this effect is very important, and it can regulate the whole pathway. Alternatively, the enzymatic machinery of the cell can produce an effector which is the result of side reactions, sometimes under the control of a hormone. The concentration of AMP, which we discussed above, is such an indicator. Fructose-2,6-bisphosphate (F2,6P), one of the most powerful allosteric regulators now known, is also generated in a step that is not in the main metabolic pathway. The regulative role of F2,6P differs with the function of the tissue. Glucose-1,6-phosphate (G1,6P), which has an important role as a cofactor, is also a significant effector in glycolysis. The regulation by AMP, F2,6P , and G1,6P are discussed in more detail below.

AMP as a Metabolic Effector As we saw from the example of PFK, it is AMP, not ATP, that is the major controlling effector. Why not ATP which is directly involved in supplying the power to cells? Many enzymes (including PFK) are in fact regulated by ATP. However, in this case, and particularly in heart muscle, the effect of AMP is preeminent. The answer lies in the fact that the ATP and ADP concentrations remain relatively constant (Helmreich and Cori, 1965). In vertebrate muscle and nerve it is rapidly regenerated by two reactions: one, the creatine kinase reaction where ATP is formed from ADP and phosphocreatine, as shown in Eq. (8.2); the other, the adenylate kinase reaction regenerates ADP to produce AMP as shown in Eq. (8.3).

$$\text{creatine}{\sim}\text{P} + \text{ADP} \rightleftarrows \text{ATP} + \text{creatine} \tag{8.2}$$

$$2\ \text{ADP} \rightleftarrows \text{ATP} + \text{AMP} \tag{8.3}$$

The adenylate kinase reaction rate is rapid and can be considered at equilibrium. Because of its regulation of AMP concentration, it plays a very important physiological role. The equilibrium constant K_{eq} for the reaction depicted in Eq. 8.4

$$K_{eq} = (\text{ATP})\,(\text{AMP})\ /\ (\text{ADP})^2 \tag{8.4}$$

is close to unity, and the concentrations of ADP and ATP are rather large compared to AMP. A small increase in ATP will produce a large effect (e.g., as seen qualitatively from Eq. 8.3).

Actual calculations assuming a decrease in ATP of 10% will produce a large decrease in AMP (approximately 4-fold), so that AMP can serve as a sensitive indicator of the energy stores of the cell.

F2,6P F2,6P is an important regulator of glycolysis in several mammalian tissues (Hers et al., 1982). In liver it acts to block glycolysis, thereby favoring the release of glucose produced by the concurrent degradation of glycogen. This is in line with one of the roles of liver which releases or removes glucose to maintain the level in the blood constant. In heart muscle, which degrades glycogen for its own glycolysis, F2,6P stimulates glycolysis.

The level of F2,6P depends on the balance between the enzyme activity responsible for the phosphorylation of F6P (phosphofructokinase 2, PFK-2) and the enzyme activity responsible for its hydrolysis (fructose-2,6-bisphosphatase, FBPase-2). The same protein is responsible for the two activities. In liver, the cAMP-dependent phosphorylation of the enzyme by a protein kinase inhibits PFK-2 and activates the phosphatase. In heart, the opposite is true—phosphorylation activates PFK-2 and blocks FBPase-2.

G1,6P G1,6P, a cofactor of phosphoglucomutase reaction which converts glucose-1-phosphate (G1P) to glucose-6-phosphate (G6P), is formed by the phosphorylation of G1P and degraded by a specific phosphatase. G1,6P is an activator of PFK, pyruvate kinase (PK) and is an inhibitor of HK and FBPase. It also releases phosphoglucomutase from ATP and citrate inhibition. Since its level is subject to a variety of physiological controls, including the presence of several second messengers (summarized in Table 8.1), it must act as a significant physiological signal (Beitner, 1984).

II. CONTROL OF GLYCOLYSIS

The regulation of glycolysis is fairly complex, involving most of the mechanisms discussed. Many of these regulatory effects are summarized in Fig. 8.3, part A.

A. ΔG and Regulation

Insights into the regulation of a metabolic pathway require information that is not always available. As discussed above, regulated steps must be exergonic. Therefore, the ΔG of the various reactions can predict in what steps the regulation may occur inside cells. However, ΔG alone merely indicates feasibility of regulation. As already mentioned, the enzyme catalyzing the step must be capable of responding to the appropriate effectors. As we saw in Chapter 6, calculation of ΔG requires knowing not only the $\Delta G°$ of the reactions (and this can be calculated, see Chapter 6), but also the actual substrate concentrations. Fortunately, data is available for some tissues, for example, heart muscle cells (Newsholme and Start, 1973). The calculated ΔGs show that regulation is likely to occur only at three glycolytic steps: the HK ($\Delta G = -6.5$ kcal), PFK ($\Delta G = -6.2$ kcal), and PK ($\Delta G = -3.3$kcal) reactions. The major negative effectors of these enzymes are G6P for HK and ATP for PK. ATP and citrate are inhibitors for PFK. Only PFK has positive effectors, among which is AMP. As we saw in heart muscle, AMP is very sensitive to the energy balance of the cell. Furthermore, the regulatory effect is enhanced by the mechanism of substrate cycling, permitting as much as a one hundredfold activation. For these reasons, in most mammalian tissues PFK is thought to be the most important glycolytic control site (e.g., see Boscá and Corredor, 1984). These regulatory steps are indicated in Fig. 8.3, part A.

Table 8.1 Factors and Conditions That Control Cellular G1,6biP Levels

Effectors	Tissues	Changes in G1,6biP levels[a]
Cyclic AMP	Diaphragm	↑
	Cultured muscle	↑
Cyclic GMP	Diaphragm	↓
	Cultured muscle	↓
Ca^{2+}	Diaphragm	↓
	Cultured muscle	↓
Hormones		
Epinephrine	Diaphragm	↑
	Cultured muscle	↑
Epinephrine + propranolol	Cultured muscle	↓
Vasopressin	Cultured muscle	↓
Serotonin	Skeletal muscle	↓
	Skin	↓
Bradykinin	Skeletal muscle	↓
	Skin	↓
Phospholipase A_2	Diaphragm	↓
Lysolecithin	Diaphragm	↓
Muscular dystrophy	Skeletal muscle	↓
Fasting	Skeletal muscle	
	Normal	↓
	Dystrophic	↓
Refeeding	Skeletal muscle	
	Normal	↑
	Dystrophic	↑
Anoxia	Diaphragm	↓
Ischemia	Brain	↓
Aerobiosis	Diaphragm	↑
Differentiation (fusion)	Cultured muscle	↑
Growth	Skeletal muscle	↑
	Heart	↔
	Brain	↓
	Skin	↓
Old age	Skeletal muscle	↓
Glucose	Pancreatic islets	↑
Pharmacologic agents		
Local anesthetics	Diaphragm	↓
Lithium	Skeletal muscle	
	Normal	↓
	Dystrophic	↓
	Diaphragm	↓
	Brain	↓
	Liver	↔
Trifluoperazine	Skeletal muscle	↑

Source: Beitner (1984). Reprinted with permission from *International Journal of Biochemistry* 16. Beitner, R., Control levels of glucose 1,6-biphosphate, copyright 1984 Pergamon Journals Ltd.

[a] ↑ = increase; ↓ = decrease; ↔ = no change.

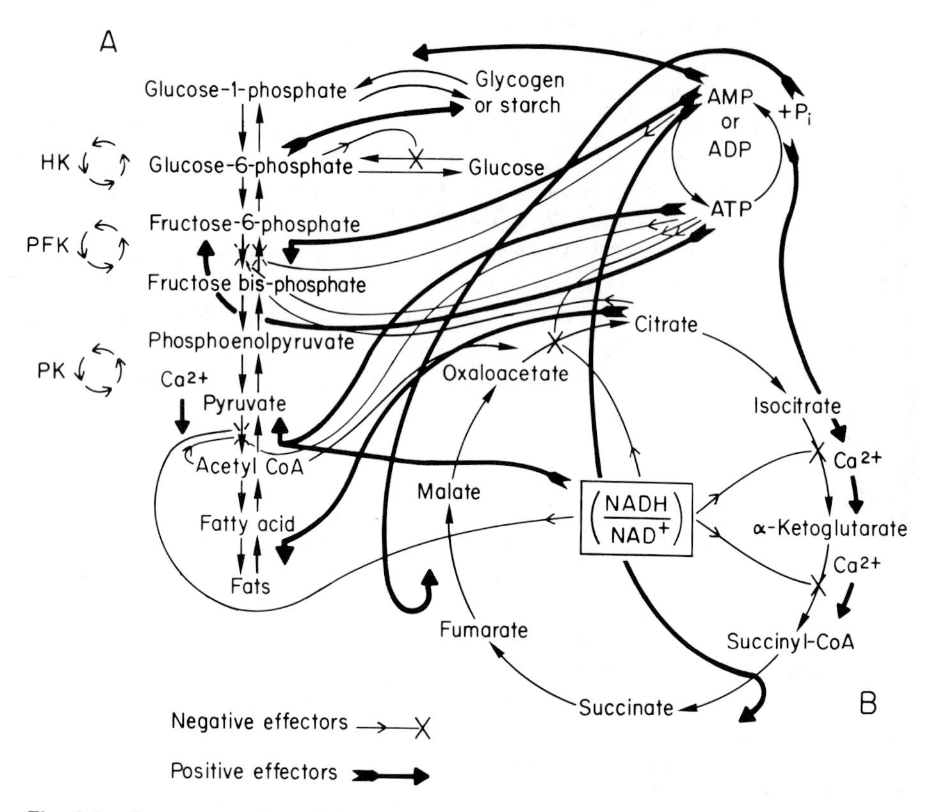

Fig. 8.3 Representation of the control of metabolism. The phosphorylative steps were left out to simplify the diagram.

B. Substrate Cycling

Substrate cycling occurs not only in the F6P to F1,6P step (e.g., Clark, D.G. et al., 1973; Clark, M.G. et al., 1973; Rognstad and Katz, 1980; Challis et al., 1984), but also in glucose to G6P (Surholt and Newshome, 1983) and the PK step (Freidmann et al., 1971), suggesting that in at least some of the systems studied, these steps can be regulated through this mechanism. The substrate cycling steps are indicated in Fig. 8.3 on the left of part A of the diagram.

III. CONTROL OF THE TRICARBOXYLIC ACID CYCLE

The TCA cycle acts in concert with the NADH oxidation and ATP synthesis. It is therefore inexorably linked to glycolysis and other sources of acetylCoA on the one hand and the electron transport system which oxidizes the NADH on the other. In many ways it makes more sense to examine the regulation of the tricarboxylic acid cycle in relation to the whole metabolic system as done in the next section (IV). However, some of the details facilitate our understanding of the whole.

Regulation of the TCA cycle primarily involves the regulation of the availability of the entry substrate and feedback inhibition (see Hansford, 1980) (see Fig. 8.3, part B). Since the entry substrates acetylCoA and oxaloacetate are present in less than saturation level, their influx into the system can control the flow through the cycle. A major product of the cycle is NADH, and as might be expected, it acts as a feedback inhibitor.

A. Substrate Availability

Pyruvate dehydrogenase (PDH), a multienzyme complex, is responsible for the production of one molecule of acetylCoA and one of NADH from the oxidation of one pyruvate. This reaction controls the entry of substrate from the glycolytic pathway (see Randle et al., 1978). PDH is inactivated by phosphorylation catalyzed by a specific kinase. The kinase in turn is activated by NADH and acetylCoA (see Reed et al., 1985). NADH and acetylCoA also act directly on PDH as feedback inhibitors. Both mechanisms then inhibit the production of acetylCoA in response to excess products. PDH is also regulated by nucleotides (GTP inhibiting and AMP activating), and its phosphorylation is enhanced by high ATP/ADP ratios.

Oxaloacetate is produced by malate dehydrogenase. The reaction is in equilibrium so that

$$K_{eq} = (\text{oxaloacetate})\ (\text{NADH})\ /\ (\text{NAD}^+)\ (\text{malate}). \tag{8.5}$$

As shown in Eq. (8.5), the oxaloacetate concentration varies inversely with that of NADH. Conversely, the increase in NAD^+ (and consequent decrease in NADH) will increase the level of oxaloacetate. This reaction serves as a delicate sensor of redox state of NAD^+ and responds by adjusting the amount of oxaloacetate fed into the TCA cycle.

B. Inhibition by Products and Intermediates

As already discussed, regulation is likely to occur at exergonic steps in the tricarboxylic acid cycle citrate synthase, isocitrate dehydrogenase, and α-ketoglutarate dehydrogenase. These in fact appear to be control steps (see Fig. 8.3, part B). NADH produced by the cycle inhibits all three. In addition, citrate inhibits citrate synthase, and succinylCoA inhibits α-ketoglutarate dehydrogenase and citrate synthase. ADP stimulates isocitrate dehydrogenase, whereas ATP inhibits it.

An interesting role is played by Ca^+, a second messenger (see Chapter 5), which stimulates pyruvate, isocitrate, and α-ketoglutarate dehydrogenase, thereby tending to increase the metabolic rate.

IV. CONTROL OF OXIDATIVE PHOSPHORYLATION AND GLYCOLYSIS IN TANDEM

The interactions of the various components of the metabolic reactions cannot be appreciated without examining them together. As we saw in the discussion of glycolysis and TCA cycle, the availability of substrates and regulation by indicators of energy metabolism are important components. This section will discuss these factors in some detail.

A. Regeneration of NAD^+

One of the products of glycolysis is NADH, which is formed from NAD^+ by the glyceraldehyde-3-phosphate dehydrogenase reaction as shown schematically in Fig. 8.4. The total amount of NAD, reduced or oxidized, is finite, and the NAD^+ has to be regenerated if glycolysis is to proceed. During oxidative metabolism, NADH can be oxidized by the electron transport chain (see Chapter 10 and discussion below). However, when the oxidative metabolism cannot keep up with the recycling of NADH, NAD^+ is regenerated by an anaerobic reaction. In mammalian

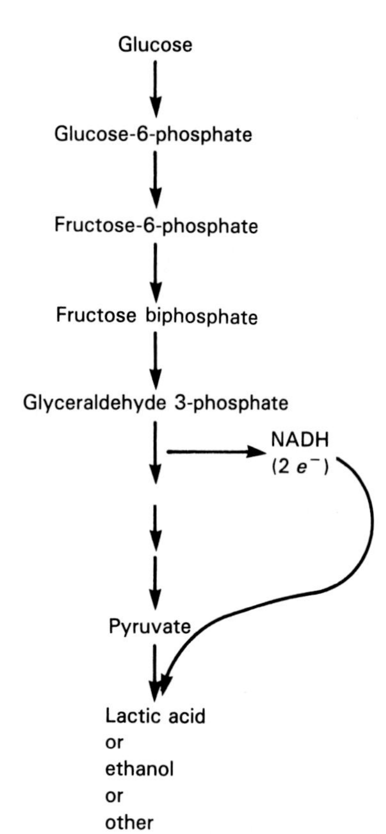

Fig. 8.4 Involvement of NADH in anaerobic glycolysis.

tissues, lactate is produced from pyruvate by lactate dehydrogenase, as shown in Fig. 8.4 and Eq. (8.6). In yeast, the decarboxylation of pyruvate by the pyruvate decarboxylase (Eq. 8.7) is followed by the reduction of acetaldehyde to produce ethanol (Eq. 8.8).

$$\begin{matrix} COO^- \\ | \\ C=O \\ | \\ CH_3 \end{matrix} + NADH + H^+ \longrightarrow \begin{matrix} COO^- \\ | \\ HCOH \\ | \\ CH_3 \end{matrix} + NAD^+ \tag{8.6}$$

$$\begin{matrix} COO^- \\ | \\ C=O \\ | \\ CH_3 \end{matrix} \longrightarrow \begin{matrix} CH_3 \\ | \\ C-H \\ \| \\ O \end{matrix} + CO_2 \tag{8.7}$$

$$\begin{matrix} CH_3 \\ | \\ C=O \\ | \\ H \end{matrix} + NADH + H^+ \longrightarrow CH_3CH_2OH + NAD^+ \tag{8.8}$$

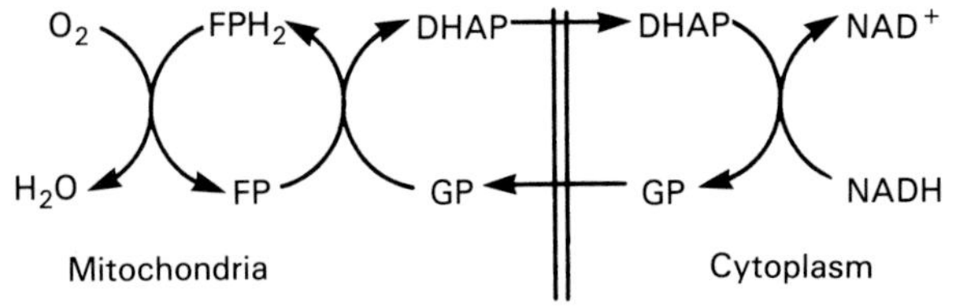

Fig. 8.5 Representation of shuttle carrying reducing equivalents from cytoplasmic NADH to intramitochondrial NAD^+.

Mitochondrial NADH can be oxidized readily by the electron transport system (Chapter 10). However, glycolysis occurs in the cytoplasm. How does the NADH reach the NADH-coenzyme Q reductase (complex I) of mitochondria? This multienzyme complex reacts with the NADH only on the matrix side of the mitochondrial inner membrane. Apparently the oxidation of the cytoplasmic NADH is mediated by shuttles. Two possible shuttles are likely to take place. Dihydroacetone phosphate (DHAP), one of the intermediates of glycolysis, can be reduced to form glycerophosphate (GP), a reaction catalyzed by glycerol phosphate dehydrogenase (Eq. 8.9).

$$\begin{array}{l} CH_2OH \\ | \\ C=O \\ | \\ CH_2OPO_3H_2 \end{array} + NADH + H^+ \longrightarrow \begin{array}{l} CH_2OH \\ | \\ CHOH \\ | \\ CH_2OPO_3H_2 \end{array} + NAD^+ \qquad (8.9)$$

The mitochondria from muscle can oxidize GP to regenerate DHAP which can then be transferred to the cytoplasm (Fig. 8.5). A similar shuttle may occur in liver where β-hydroxybutyrate can be formed from acetoacetate in the cytoplasm which can be oxidized in the mitochondria regenerating the β-hydroxybutyrate.

Shuttles such as these may play an important role in the regulation of metabolism. For example, inhibition of GP shuttle would favor the formation of lactic acid. Lactic acid is in fact favored in some cells, for example, tumor cells. This effect has been referred to as the *Crabtree effect*. The actual mechanism for the Crabtree effect is not known, and a number of other possibilities have been proposed. In other systems under aerobic conditions, no lactic acid is accumulated, a phenomenon referred to as the *Pasteur effect*. Again the mechanism is not well understood, and it has been suggested that a competition between the glycolytic and the electron transport systems for NADH, ADP, and P_i may be responsible. P_i may be involved in the expression of both the Crabtree and the Pasteur effect in some cells (Koobs, 1972). In some tissues the oxidation of NADH by peroxisomes may have an effect on the regulation of metabolism.

B. Energy Balance

In Chapter 7, we have seen the importance of the adenosine phosphates acting as allosteric regulators in directing metabolites in the purine nucleotide biosynthetic pathway (e.g., Fig. 7.11), where ATP is a positive effector of reactions favoring the synthesis of macromolecules. The role of ATP, ADP, and AMP in metabolic regulation was discussed above (sections I,D; II,A; and III,B) from the limited perspective of the individual pathway, for example, glycolysis and the TCA cycle. These compounds, by indicating the energy balance of the cell, could serve as ideal general regulators of metabolism. Therefore, a more global perspective can be very useful. The role of these compounds in regulating metabolism has long been recognized and referred to as the *energy charge* (Atkinson, 1977). A high energy charge (high ATP concentration) would inhibit ATP-generating reactions but facilitate energy-yielding reactions. Generally, ATP acts in the opposite direction than AMP or ADP, although different reactions may be the target of the individual adenine nucleotide.

The examination of the regulation of the individual steps clearly indicates such a role for these compounds. For example, in the PK reaction, ATP present in high concentrations favors the formation of the inactive form of the enzyme, whereas ADP has the opposite effect. This, in effect, can regulate the amount of pyruvate that can enter the TCA cycle.

AMP or ADP (depending on the organism) favors the hydrolysis of polysaccharide by the polysaccharide phosphorylase reaction, thereby favoring the mobilization of glucose. They speed up the machinery of glycolysis through their activation of PFK and inhibit the formation of polysaccharide stores by inhibiting FBPase. Through their activation of isocitrate dehydrogenase, they stimulate the turnover of the TCA cycle. In the same way, P_i stimulates succinate dehydrogenase and fumarase. ATP, in contrast, has the opposite effect; by activating FBPase, it favors the synthesis of polysaccharide. Likewise, ATP inhibits isocitrate dehydrogenase and tends to slow down the TCA cycle.

We have already seen that the ATP concentration does not vary significantly in muscle since it can be regenerated from creatine phosphate, AMP being a much more sensitive indicator of the energy level. However, phosphocreatine (PC), which decreases during muscular activity, can also act as a regulative signal. The concentration of PC activates FDPase (Fu and Kemp, 1973) and inhibits PFK (Uyeda and Racker, 1965) and PK (Moyed, 1961). Molecules which are interconvertible with ATP can also have an effect; for example, GTP inhibits α-ketoglutarate dehydrogenase (Olson and Algyer, 1973).

There are other controls superimposed on this basic pattern. G6P is a positive effector for the polysaccharide synthase reaction (favoring the laying down of carbohydrate stores) while discouraging the glycolysis through its inhibitory effect on hexokinase. Citrate, which inhibits citrate synthase, stimulates fatty acid synthesis from acetylCoA, thereby favoring the laying down of fat deposits. Succinate dehydrogenase is inhibited by oxaloacetate (Pardee and Potter, 1948) and activated by succinate, ATP, reduced CoQ, and succinylCoA (Pardee and Potter, 1948; Kearney et al., 1972). Some of these reactions, other reactions previously discussed, and their regulation are shown in Fig. 8.3. The heavy arrows indicate positive effects, whereas lighter lines with crosses indicate negative effects. For simplicity, the figure leaves out the regulation by F2,6P and G1,P.

C. Control of Oxidative Phosphorylation in Mitochondria

In eukaryotic cells, mitochondrial oxidative phosphorylation provides 95% of the total ATP required. It follows that understanding how the mitochondrial system is regulated is of paramount importance in understanding the regulation of metabolism. The mitochondrial electron transport chain is responsible for accepting reducing equivalents from NADH and reduced flavoprotein (FPH_2) and using O_2, oxidizing them to produce water and ATP. The chain is a multienzyme system constituted of four complexes embedded in the mitochondrial inner membrane (see Chapter 10). The appropriate dehydrogenases and the ATP synthases are located on the inner face of the mitochondrial inner membrane. The transfer of ATP to the cytoplasm, where it is mostly used, requires the adenine nucleotide translocator, an integral protein of the inner membrane which mediates the exchange between the mitochondrial ATP^{4-} and the cytoplasmic ADP^{3-}.

Two major hypotheses have been advanced to explain the regulation of mitochondrial oxidative phosphorylation. One such hypothesis proposes that the first two sites of oxidative phosphorylation (in complexes I and III) are at near equilibrium. The regulation of the system would result from the virtually irreversible reactions in which reduced cytochrome c (2 cyt c^{2+}) is oxidized by O_2 in complex IV. This proposal has been referred to as the *near equilibrium hypothesis* (see Erecińska and Wilson, 1982).

An alternative hypothesis proposes a key role of the adenine nucleotide translocase. This latter proposal has been referred to as the *translocase hypothesis* (Klingenberg, 1980). Basically, it proposes that the transport mediated by the adenine nucleotide translocase (the ADP/ATP transporter, see Chapter 2, p. 62) is rate limiting and is responsible for setting the metabolic rate.

The concept of the near equilibrium hypothesis can be illustrated most simply by representing the events of the first two phosphorylative sites as follows:

$$\text{NADH} + 2\text{ cyt c}^{3+} + 2\text{ ADP} + \text{P}_i \rightleftarrows \text{NAD}^+ + 2\text{ cyt c}^{2+} + 2\text{ ATP}. \quad (8.10)$$

In this representation cyt c^{3+} and cyt c^{2+} represent the oxidized and the reduced species of cytochrome c. If we now assume that the reactions are at equilibrium:

$$K_{eq} = [(\text{NAD+})/(\text{NADH})]^{1/2}\,(\text{ATP})/(\text{ADP})(\text{P}_i)\,(\text{cyt c}^{2+})/(\text{cyt}^{3+}) \quad (8.11)$$

so that at equilibrium,

$$(\text{cyt c}^{2+})/(\text{cyt c}^{3+}) = (\text{NADH})^{1/2}/(\text{NAD}^+)^{1/2}\,(\text{ADP})\,(\text{P}_i)/(\text{ATP})\,(1/K_{eq}). \quad (8.12)$$

As shown in Eq. (8.12), any increase in the NADH/NAD^+ or (ADP) (P_i)/ATP ratios will be reflected in an increase in cyt c^{2+}. This increase in cyt c^{2+} will therefore lead to an increase in oxidation since the cyt c-oxidase is not saturated and the reaction is irreversible.

The key and contrasting points of the two models have to do with the equilibrium position set by the two models. The near equilibrium hypothesis assumes equilibrium for the first two phosphorylating steps which include adenylate translocase. The translocase hypothesis assumes that the adenylate translocase is not at equilibrium.

Can we support one model over the other at this time? Examination of the available data shown in Table 8.2 indicates that at steady state in many biological systems these reactions appear to be close to equilibrium (Erećinska and Wilson, 1978). The table calculates the redox potential for the reactions between NADH and cyt c by subtraction. The values were calculated from the experimentally determined concentrations for the system listed in the first column. The corresponding ΔG values are listed in the fourth column. When the ΔG for the synthesis of ATP, also calculated from the experimentally determined concentration (column 5), is subtracted

Table 8.2 Free-Energy Relationships Between the Oxidation-Reduction Reactions of the Respiratory Chain and ATP Synthesis (Experimental)[a]

Type of material	E_h cytochrome *c*	E_h NAD (V)	ΔE (V)	$\Delta G_{ox\text{-}red}$ (kcal/2 e^-)	ΔG_{ATP} (kcal/2 ATP)	$\Delta\Delta G$ (kcal)
Liver cells (no substrate)	0.272	−0.242	0.514	23.7	23.8	−0.1
Liver cells (lactate + ethanol)	0.269	−0.260	0.529	24.4	24.2	0.2
Perfused liver (no substrate)	0.253	−0.263	0.516	23.8	22.6	1.2
Ascites tumor cells	0.260	−0.270	0.530	24.4	23.6	0.8
Cultured kidney cells	0.271	−0.252	0.523	24.1	24.2	−0.1
Tetrahymena pyriformis cells	0.251	−0.236	0.487	22.5	22.4	0.1
Paracoccus denitrificans cells	0.276	−0.244	0.520	24.0	24.1	−0.1
Perfused heart (80 cm H_2O)	0.253	−0.313	0.566	26.1	26.3	−0.2
Pigeon heart mitochondria (succinate)	0.270	−0.343	0.613	28.4	29.8	−1.4

Reprinted with permission from M. Erećinska and D. F. Wilson, Trends in Biochemical Sciences, 3:219–223.

[a] $\Delta E = E_{h\,NAD} - E_{h\,cyt\,c}$ where $E_h = E_m + 2.3RT/nF \log[\text{ox}]/[\text{red}]$; $\Delta G_{ox\text{-}red} = -nF\,\Delta E$; $\Delta G_{ATP} = \Delta G_0' + 1.36 \log[\text{ADP}][\text{P}_i]/[\text{ATP}]$; $\Delta\Delta G = \Delta G_{ox\text{-}red} - \Delta G_{ATP}$.

from this value (displayed in column 6 and represented as $\Delta\Delta G$), the results show little deviations from the equilibrium (i.e., $\Delta G = 0$). ΔG (expressed as $\Delta\Delta G$) for the whole system is close to zero.

Unfortunately, arguments for or against either model crucially depend on the accuracy of the rather complex measurements needed to calculate concentrations. Furthermore, some of the components may be bound and not free in solution. Therefore, at this time it is difficult to reach a firm conclusion (see Erecińska and Wilson, 1982 for a discussion).

V. CONTROL OF CARBOHYDRATE METABOLISM IN PLANTS

In plants the energy captured from light is trapped and converted into ATP (synthesized from ADP and P_i) and NADPH (reduced from $NADP^+$) (see Chapter 11). Eventually, after a series of reactions that fix CO_2, the energy is stored in the synthesis of starch or sucrose. Starch is formed in the chloroplasts, whereas sucrose is formed in the cytoplasm. The biochemical pathway generates dihydroxyacetone phosphate (DHAP) or triose phosphates derived from DHAP. In the production of starch, DHAP is converted first to hexose phosphate. In the production of sucrose, triose phosphate is first transported across the chloroplast envelope to the cytoplasm. In the absence of light, the plant cells can draw on energy reserves by breaking down starch to form triose phosphates, which can be metabolized further. These two alternatives are summarized in Fig. 8.6 (Buchanan, 1984). As shown, the enzymes that synthesize starch and those that catalyze its breakdown coexist in the chloroplast. For this reason the system is tightly regulated. The enzymes involved in synthesis are activated by light (when energy is available from light) and the degradative enzymes are deactivated by light, so that in plants carbon assimilation predominates during the day and carbohydrate degradation during the night. Light regulates the enzymes indirectly, through the normal functioning of the photosynthetic machinery of the chloroplast. Ferredoxin, a component of the electron transport chain of the thylakoid vesicles, is reduced during photosynthesis (see Chapter 11). Light controls the enzymes through the redox changes of ferredoxin.

Part of the control involves redox changes in *thioredoxin* (Td) (Holmgren, 1989). Thiore-

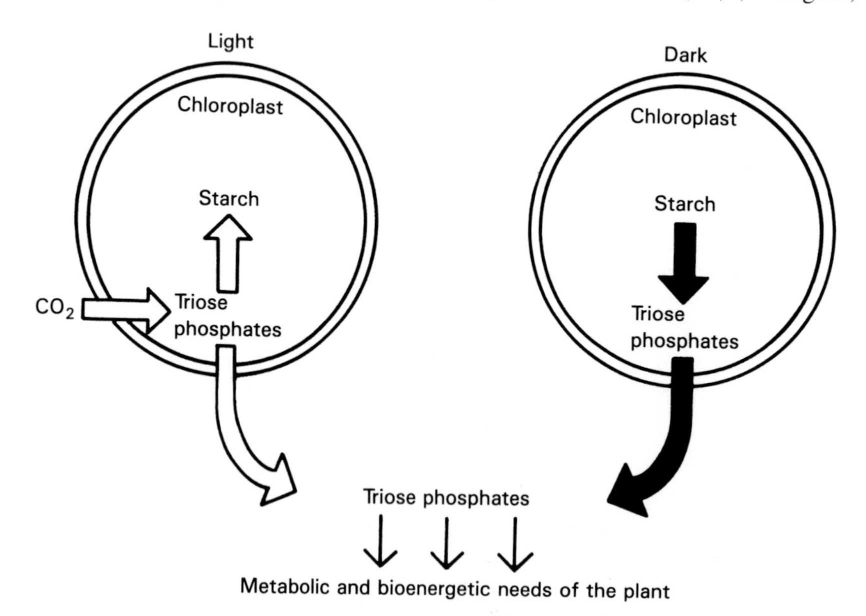

Fig. 8.6 Role of chloroplasts in producing and storing energy-rich compounds in the plant. From B. B. Buchanan, BioScience, 34:378–383. Copyright 1984 by the American Institute of Biological Sciences.

doxin corresponds to a group of small proteins about 12 kDa in size and is probably present in all organisms. The thiol groups of Td undergo redox changes so that $2SH \rightleftharpoons S—S$. Thioredoxin is reduced by ferredoxin (Fd) in a reaction catalyzed by ferredoxin-thioredoxin reductase, an iron-sulfur protein. The functioning of the ferredoxin-thioredoxin system is summarized in Fig. 8.7. Reduced thioredoxin activates the enzymes of carbohydrate biosynthesis and those of the reductive pentose phosphate cycle and deactivates glucose-6-phosphate dehydrogenase, which is needed for degradation through the pentose phosphate cycle. The reductive pentose phosphate and the regulation by light are shown in Fig. 8.8 (Buchanan, 1984). A summary is given in Table 8.3 (Anderson et al., 1982). Thioredoxin acts by reducing or oxidizing the enzymes, which can also undergo thiol-disulfide transitions. The action of two different varieties of Td is

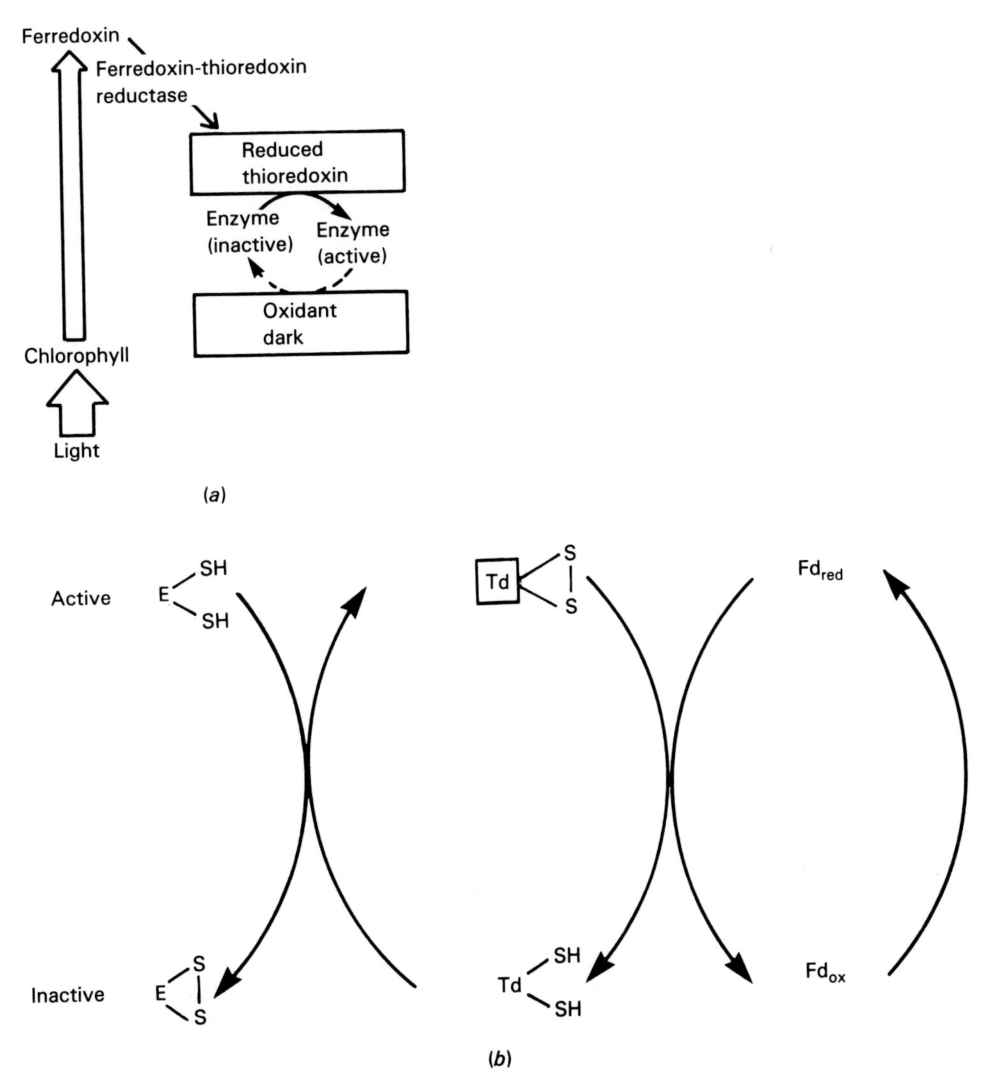

Fig. 8.7 (*a*) Ferredoxin-thioredoxin system of enzyme photoregulation (light activation/dark deactivation). From B. B. Buchanan, BioScience, 34:378–383. Copyright 1984 by the American Institute of Biological Sciences. (*b*) Outline of the mechanism of activation and inactivation of NADP-MDH. E, Enzyme; Td, thioredoxin; Fd, ferredoxin.

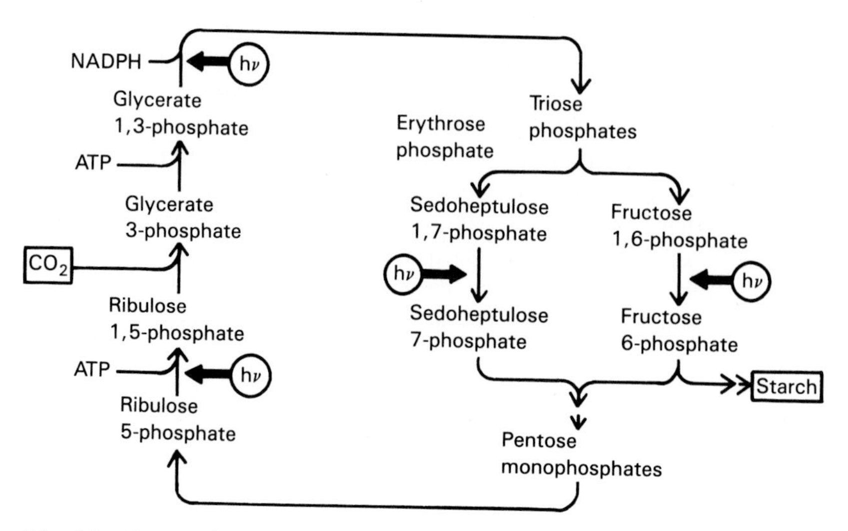

Fig. 8.8 Steps of the reductive pentose phosphate cycle regulated by light through the Td system. The light dependent reactions are indicated by hu and the heavy arrows. From B. B. Buchanan, BioScience, 34:378–383. Copyright 1984 by the American Institute of Biological Sciences.

summarized in Fig. 8.9. As shown, the ferredoxin-thioredoxin system also activates NADP-malate dehydrogenase (NADP-MDH) and ATP synthase (CF_1 ATPase). In nonphotosynthesizing tissues, Td can be reduced by NADPH.

NADPH-MDH is an example of the evidence available for the ferredoxin-thioredoxin control system in plants. NADP-MDH catalyzes the reduction of oxalate by NADPH:

Table 8.3 Light Modulation in Different Plants

	Prokaryote, cyanobacterium *Anacystis nidulans*	Eukaryotes						
		Algae		C_3 plants		C_4 plants		CAM
		Green	Brown	*Pisum*	*Spinacia*	*Zea*	*Tidestromia*	*Kalanchoe*
Light-activated enzymes				Light stimulation *x*-fold				
NADP-linked malic dehydrogenase				14	∞	50	3.3	1.7
NADP-linked glyceraldehyde-3-P dehydrogenase	Nil	~2	Nil	2.4	5	2	3.1	2
Ribulose-5-P kinase	2			7.7	3.2	1.6	4	4.4
Fructose-1,6-P_2 phosphatase	Nil			1.7	2.2		Nil	Nil
Sedoheptulose-1,7-P_2 phosphatase	Nil			1.8	1.7			1.7
Pyruvate, orthophosphate dikinase						~12		
NAD-linked malic dehydrogenase				1.7				
Dark-activated enzymes				Dark stimulation *x*-fold				
Glucose-6-P dehydrogenase	4.8			2	3			Nil
				1.4				
Phosphofructokinase				2				3
				11				
Phosphorylase				1.6				
Phosphoglucomutase				3.7				
Phosphoglucoisomerase				3.6				

From L. E. Anderson, et al., BioScience, 32(2):104. Copyright 1982 by the American Institute of Biological Sciences, reprinted with permission.

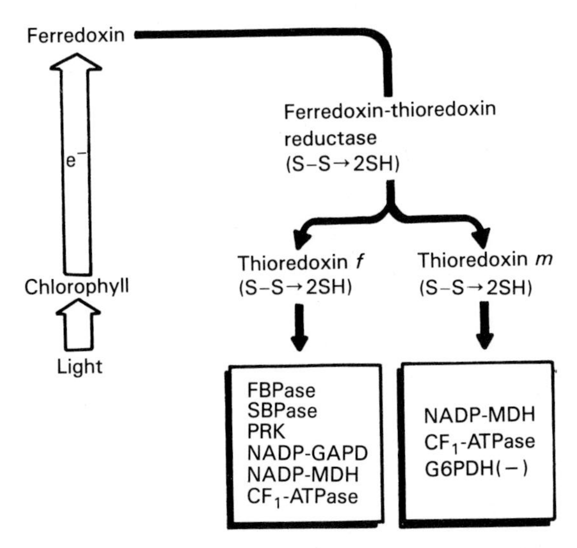

Fig. 8.9 Enzymes regulated by the ferredoxin-thioredoxin system. From Cseke and Buchanan, Biochimica et Biophysica Acta, 853:43–63. Copyright © 1986 Elsevier Science Publishers, Amsterdam.

$$\text{Oxaloacetate} + \text{NADP} + \text{H}^+ \rightleftharpoons \text{malate} + \text{NADP}^+ \tag{8.13}$$

The enzyme is rapidly inactivated when leaves are kept in the dark and is reactivated by light (Johnson, 1971; Johnson and Hatch, 1970; Scheibe and Anderson, 1981). Thiols keep the enzyme active in the presence of oxygen and reactivate the enzyme when it has been inactivated. The light activation is blocked by 3-(3,4-dichlorophenyl)-1,1-dimethylurea (DCMU) (Hatch, 1977), which blocks the electron transport chain of the chloroplast. Activation of the inactive enzyme requires the presence of a low molecular weight protein (Kagawa and Hatch, 1977), thioredoxin, which also undergoes dithiol-disulfide transitions. The available information argues for the scheme indicated in Fig. 8.7, already discussed.

Enzyme modulation by reduction through the involvement of thioredoxin or similar small proteins such as glutaredoxin has a role in *E. coli* in the reduction of ribonucleotides required for DNA synthesis. Thioredoxin is thought to play a regulatory role in mammalian cells, including the control of growth (Holmgren, 1989). Sulfides and sulfhydril groups are pervasive in enzymes. Similarly, there are a variety of enzymes which catalyze transitions between the two and in addition many cells are rich in gluthione. These observations suggest that the redox state of cells may play an important role (Ziegler, 1985), apart from the ferredoxin-thioredoxin system mechanism of plants.

VI. INTERACTIONS BETWEEN THE VARIOUS MECHANISMS

The control of biochemical reactions could undoubtedly be the result of control of the activities of specific enzymes. In addition, repression and induction could turn on or off the production of the enzymes themselves. It is probable that both mechanisms are significant.

Where the production of enzymes and the modulation of enzyme activity both play a role, it is likely that the first process serves as a gross on-and-off switch and the second as a finer and more rapid control. That this is probably correct has been shown by experiments in which the growth rate of bacterial mutants lacking the finer feedback control is compared to that of the original or wild-type strain.

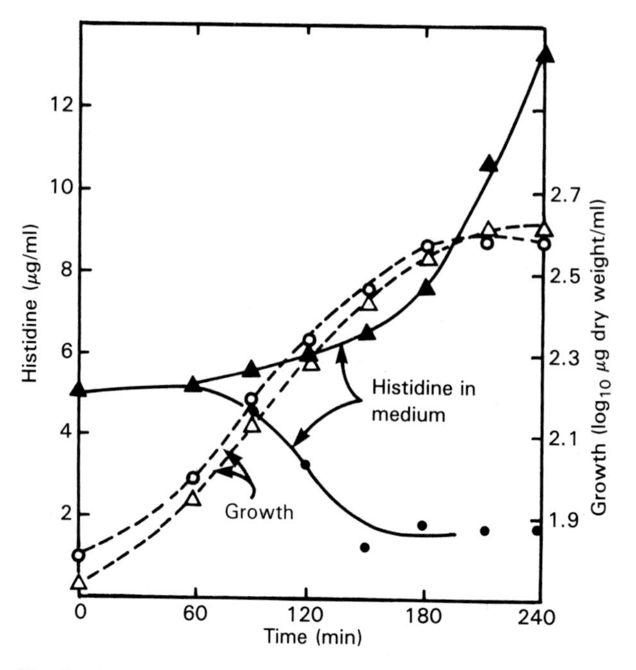

Fig. 8.10 Histidine utilization and excretion during growth. The minimum medium in this experiment was supplemented with histidine (5 mg/liter). (△, ▲) Mutant; (○,●) original strain. Reprinted with permission from H. S. Moyed, Cold Spring Harbor Symposia on Quantitative Biology, 26:323–329. Copyright © 1961 Cold Spring Harbor Press.

In *E. coli,* the histidine-synthesizing pathway is controlled by both repression regulating the synthesis of the enzyme and feedback inhibition by histidine, which modulates enzyme activity. Experiments have been carried out (Moyed, 1961) in which bacterial growth and the level of histidine in the medium were monitored for the wild strain and for a mutant lacking the feedback control. During growth, the mutant culture was unable to cut down its production of histidine, which spilled out into the medium (Fig. 8.10, black line and triangles). The wild type, however, could remove histidine from the medium and must also have inhibited its synthesis (Fig. 8.10, lower line and filled circles). Although the mutant must be less efficient, the growth of both cultures is about the same (Fig. 8.10, dashed lines). However, synthesis of unneeded histidine must be wasteful, since it requires both nutrients and energy. The feedback control of histidine synthesis may have a much more significant role when the nutrient supply of the culture is limited. Furthermore, the cells undoubtedly possess many feedback controls on their metabolic reactions. Although the effect of each one might be small, presumably their effect on the economy of the cell will be additive.

The possibility of regulatory systems yet to be recognized should also be considered. There are suggestions that thiol-group redox changes, well established in plant carbohydrate metabolism (see Section II) may play an important role in other systems as well.

SUGGESTED READING

Beitner, R. (1984) Control of levels of glucose 1,6-bisphosphate. *Int. J. Biochem.* 16:579–585.

Cséke, C., and Buchanan, B. B. (1986) Regulation of the formation and utilization of photosynthate in leaves. *Biochim. Biophys. Acta* 853:43–63.

Erećinska, M., and Wilson, D. F. (1982) Regulation of cellular energy metabolism. *J. Membr. Biol.* 70:1–14.

Goodwin, T. W., and Mercer, E. I. (1983) *Introduction to Plant Biochemistry*, 2nd ed. Pergamon, New York. See pp. 139–159.
Klingenberg, M. (1979) ATP shuttle of the mitochondrion. *Trends in Biochem. Sci.* 4:249–252.
Newsholme, E. A., Challis, R. A., and Crabtree, B. (1984) Substrate cycles: their role in improving sensitivity in metabolic control. *Trends in Biochem. Sci.* 9:227–280.
Saier, M. H., Jr. (1987) *Enzymes in Metabolic Pathways*. Chapters 5 and 6. Harper and Row, New York.

REFERENCES

Anderson, L. E., Ashton, A. R., Mohamed, A. H., and Sheibe, R. (1982) Light/dark modulation of enzyme activity in photosynthesis. *BioScience* 32:103–107.
Atkinson, D. E. (1977) *Cellular energy metabolism and its regulation*. Academic Press, New York.
Beitner, R. (1984) Control of levels of glucose 1,6-bisphosphate. *Int. J. Biochem.* 16:579–585.
Boscá, L., and Corredor, C. (1984) Is phosphofructokinase the rate-limiting step of glycolysis? *Trends in Biochem. Sci.* 9:372–373.
Buchanan, B. B. (1984) The ferredoxin/thioredoxin system: a key element in the regulatory function of light in photosynthesis. *BioScience* 34:378–383.
Challiss, R. A. J., Arch, J. R. S., and Newsholme, E. A. (1984) The rate of substrate cycling between fructose 6-phosphate and fructose-1,6-bisphosphate in skeletal muscle. *Biochem. J.* 221:153–161.
Clark, D. G., Rognstad, R., and Katz, J. (1973) Isotopic evidence for futile cycles in liver cells. *Biochem. Biophys. Res. Comm.* 54:1141–1148.
Clark, M. G., Bloxham, D. P., Holland, P. C., and Lardy, H. A. (1973) Estimation of fructose diphosphatase-phosphokinase substrate cycle in the flight muscle of *Bombus affinis*, *Biochem. J.* 134:589–597.
Cséke, C., and Buchanan, B. B. (1986) Regulation of the formation and utilization of photosynthate in leaves. *Biochim. Biophys. Acta* 853:43–63.
Erećinska, M., and Wilson, D. F. (1978) Homeostatic regulation of cellular energy metabolism. *Trends Biochem. Sci.* 3:219–223.
Erećinska, M., and Wilson, D. F. (1982) Regulation of cellular energy metabolism. *J. Membr. Biol.* 70:1–14.
Exton, J. H., Lewis, S. B., Ho, R. J., Robinson, G. A., and Park, C. R. (1971) The role of cyclic AMP in the interaction of glucagon and insulin in the control of the liver. *Ann. N.Y. Acad. Sci.* 185:85–100.
Fisher, E. H., Packer, A., and Saari, J. C. (1970) The structure, function and control of glycogen phosphorylase. *Essays Biochem.* 6:23–68.
Freidmann, B., Goodman, E. H. Jr., Saunders, H. L., Kostos, V., and Weinhouse, S. (1971) An estimation of pyruvate recycling during gluconeogenesis in the perfused rat liver. *Arch. Biochem. Biophys.* 143:566–578.
Fu, J. Y., and Kemp, R. G. (1973) Activation of muscle fructose-1,6-diphosphatase by creatine phosphate and citrate. *J. Biol. Chem.* 248:1124–1125.
Hansford, R. G. (1980) Control of mitochondrial substrate oxidation. *Curr. Topics Bioenerg.* 10:217–278.
Hatch, M. D. (1977) Light/dark mediated activation and inactivation of NADP-malate dehydrogenase in isolated chloroplasts from *Zea mays*. In *Photosynthetic Organelles* (Miyachi, S., Katoh, S., Fujita, Y., and Shibata, J., eds.), pp. 311–314. *Plant Cell Physiology*, special issue.
Helmreich, E., and Cori, C. F. (1965) Regulation of glycolysis in muscle. *Adv. Enzyme Regul.* 3:91–107.
Hers, H.-G., Hue, L., and Van Scaftingen, E. (1982) Fructose 2,6 bisphosphate. *Trends in Biochem. Sci.* 7:329–333 (1982).
Hofmann, E. (1978) Phosphofructokinase—a favourite of enzymologists and of students of metabolic regulation. *Trends Biochem. Sci.* 3:145–147.
Holmgren, A. (1989) Thioredoxin and glutaredoxin systems. Minireview. *J. Biol. Chem.* 264:13963–13966.
Johnson, H. S. (1971) NADP-malate dehydrogenase: photoactivation in leaves of plants with Calvin cycle photosynthesis. *Biochem. Biophys. Res. Commun.* 43:703–709.

Johnson, H. S., and Hatch, M. D. (1970) Properties and regulation of leaf nicotinamide-adenine dinucleotide phosphate-malate dehydrogenase and 'malic' enzyme in plants with C_4-dicarboxylic acid pathway in photosynthesis. *Biochem. J.* 119:273–280.

Kagawa, T., and Hatch, M. D. (1977) Regulation of C_4 photosynthesis: characterization of a protein factor mediating the activation and inactivation of NADP-malate dehydrogenase. *Arch. Biochem. Biophys.* 84:290–297.

Kearney, E. B., Mayer, M., and Singer, T. P. (1972) Regulatory properties of succinate dehydrogenase: activation by succinyl CoA, pH and anions. *Biochem. Biophys. Res. Commun.* 46:531–537.

Kemp, R. G. (1973) Inhibition of muscle pyruvate kinase by creatine phosphate. *J. Biol. Chem.* 248:3963–3967.

Klingenberg, M. (1980) The ADP-ATP translocation in mitochondria, a membrane potential controlled transport. *J. Membr. Biol.* 56:97–105.

Koobs, D. H. (1972) Phosphate mediation of the Crabtree and Pasteur effects. *Science* 178:127–133.

Krebs, H. A. (1972) The Pasteur effect and the relation between respiration and fermentation. *Essays Biochem.* 8:1–34.

Mansour, T. E., and Ahlfors, C. E. (1968) Studies in heart phosphofructokinase. Some kinetic and physical properties of the crystalline enzyme. *J. Biol. Chem.* 243:2523–2533.

McDonald, R. C., Steitz, T. A., and Engleman, D. M. (1979) Yeast hexokinase in solution exhibits a large conformational change upon binding glucose or glucose 6-phosphate. *Biochemistry* 18:338–342.

Moyed, H. S. (1961) Interference with feedback control of enzyme activity. *Cold Spring Harbor Symp. Quant. Biol.* 26:323–329.

Newsholme, E. A., Challis, R. A., and Crabtree, B. (1984) Substrate cycles: their role in improving sensitivity in metabolic control. *Trends Biochem. Sci.* 9:227–280.

Newsholme, E. A., and Start, C. (1973) *Regulation in metabolism,* p. 97. Wiley, New York.

Olson, M. S., and Allgyer, T. T. (1973) The regulation of nicotinamide adenine dinucleotide-linked substrate oxidation in mitochondria. *J. Biol. Chem.* 248:1582–1597.

Pardee, A. B., and Potter, V. R. (1948) Inhibition of succinic dehydrogenase by oxaloacetate, *J. Biol. Chem.* 176:1085–1094.

Randle, P. J. (1978) Pyruvate dehydrogenase complex—meticulous regulator of glucose disposal in animals. *Trends Biochem. Sci.* 31:2217–2219.

Reed, L. J., Damuni, Z., and Merryfield, M. L. (1985) Regulation of mammalian pyruvate and branched-chain α-keto-acid dehydrogenase complexes by phosphorylation and dephosphorylation. *Curr. Topics Cell Reg.* 27:41–49.

Rognstad, R., and Katz, J. (1980) Control of glycolysis in the liver by glucagon at the phosphofructokinase-fructose 1,6-diphosphatase site. *Arch. Biochem. Biophys.* 203:642–646.

Scheibe, R., and Anderson, L. E. (1981) Dark modulation of NADP-dependent malate dehydrogenase and glucose-6-phosphate dehydrogenase in chloroplast. *Biochim. Biophys. Acta* 636:58–64.

Surholt, B., and Newsholme, E. A. (1983) The rate of substrate cycling between glucose and glucose 6-phosphate in muscle and fat body of the hawk moth (*Acherontia atropos*) at rest and during flight. *Biochem. J.* 210:49–54.

Uyeda, K., and Racker, E. (1965) Regulatory mechanisms in carbohydrate metabolism. VII. Hexokinase and phosphofructokinase. *J. Biol. Chem.* 240:4682–4688.

Ziegler, D. M. (1985) Role of oxidation-reduction of enzyme thiols-disulfides in metabolic regulation. *Annu. Rev. Biochem.* 54:305–329.

Chapter 9

Regulation by Degradation and Synthesis of Macromolecules

As discussed in Chapters 7 and 8, metabolic activity can be regulated by adjusting the rates of the cell's biochemical reactions either by activating previously inactive enzymes or, more subtly, by adjusting the activity of the appropriate enzymes. Alternatively, cells may alter the concentration of enzymes and hence the rate of certain metabolic reactions.

In eukaryotes, the concentration of any one enzyme is controlled by the balance between the ability of cells to synthesize the enzyme and their ability to degrade it. In bacteria, the enzymes are generally stable (Ballard, 1977). However, because of their much shorter generation time, in exponentially growing cells unneeded enzymes would be rapidly diluted by cell division.

The synthesis of specific enzymes can be regulated at various biochemical levels, from transcription of the gene to the actual synthesis of the enzyme. The present chapter explores regulation of these events, first in bacteria and then in eukaryotes.

I. REGULATION IN BACTERIA

In bacteria, a number of enzymes are controlled by the concentrations of some metabolites. All indications suggest that control of transcription and control of translation are the most significant regulatory mechanisms. Selective degradation of mRNA has been observed (Blundell et al., 1972), but its physiological role is far from clear.

A. Induction and Repression

Operons and Their Regulation Enzymes that belong to a metabolic pathway are likely to be required in fixed stoichiometric proportions. It would therefore be an efficient way to regulate the expression of the corresponding genes if they were part of the same transcriptional unit, and they would be transcribed to form *polycistronic* mRNA, i.e., a single mRNA transcript coded for many polypeptides. This is the predominant pattern in bacteria. In *Escherichia coli*, most of the structural genes are organized into *operons* consisting of two to eight closely linked *cistrons* (Bachman and Low, 1980). A major question has been how the transcription and translation of operons are adjusted by specific protein regulators—activators and repressors.

Genetic evidence suggests that the transcription of some operons is blocked by binding of a *repressor* to a specific site on the DNA. This could be in response to the presence of a metabolite, which could bind to an *aporepressor* molecule to form the active *repressor*. Similarly, transcription could be *induced* by inactivation of a repressor when it binds a metabolite. The mechanisms resulting from repressors are *negative* controls. Regulation can also take place through a *positive* mechanism by means of *activators* that facilitate transcription. An activator could interact with the operon indirectly, for instance, by some effect on the RNA polymerase. These two mechanisms are represented in Figs. 9.1 and 9.2. Either mechanism can function to induce a pathway in response to the presence of substrate or to inhibit the response to a feedback signal. Examples of negative and positive controls are listed in Table 9.1.

The induction of β-galactosidase in the presence of an inducer is shown in Fig. 9.3 (Cohn, 1957). This enzyme is part of the *lac* operon discussed below. In Fig. 9.3, the amount of β-galactosidase present in an *E. coli* culture is expressed as a function of the total bacterial protein. Addition of inducer (lower arrow) initiates the production of the enzymes from a negligible level. The amount of galactosidase keeps pace with the multiplication of the cells as 6.6% of the cells' protein, as indicated by the slope of the line. Removal of the inducer (upper arrow) results in an abrupt interruption of the production of galactosidase, which is now diluted by the continued growth of the culture. The amount of enzyme shown in Fig. 9.3 was calculated from

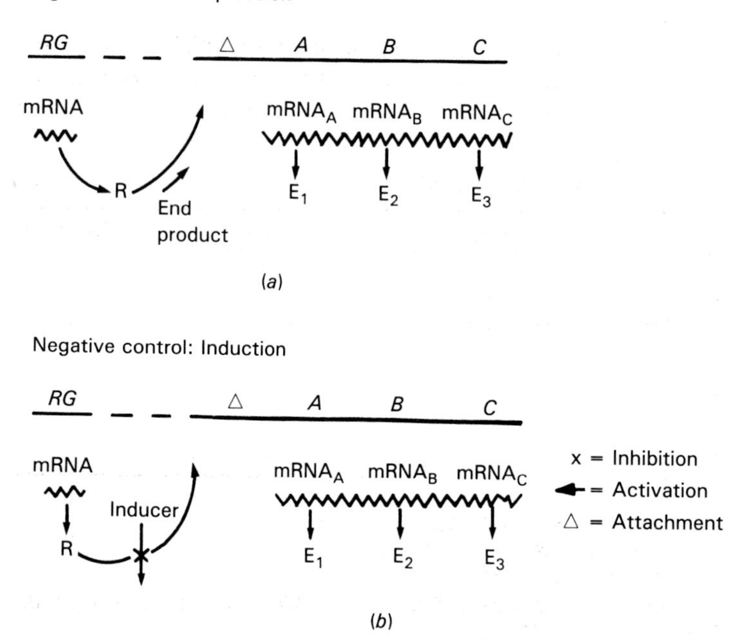

Fig. 9.1 Models of transcriptional control using a repressor. (*a*) The presence of an end product facilitates the repressor's ability to block transcription. (*b*) Inhibition of the repressor by an inducer.

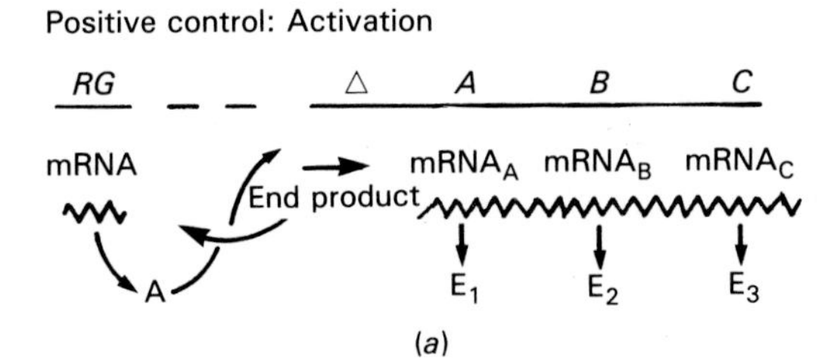

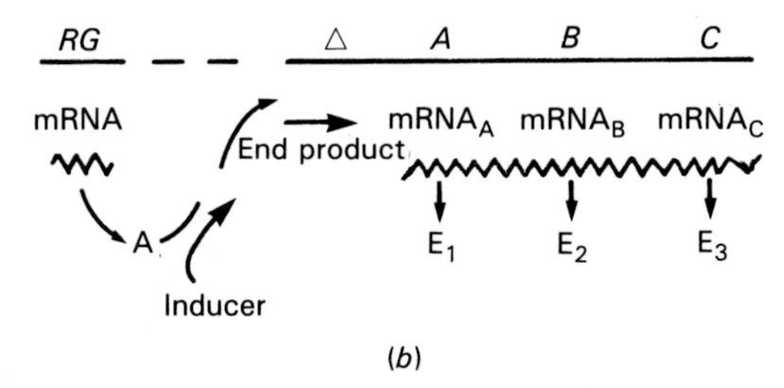

Fig. 9.2 Models of transcriptional control using an activator. (*a*) An end product inhibits the activator so that transcription is inhibited. (*b*) An inducer allows the activator to facilitate transcription.

knowledge of the activity of purified enzyme preparations and the activity of the enzyme measured in this experiment. The enzyme is synthesized de novo from free amino acids (Hogness et al., 1955; Rotman and Spiegelman, 1954).

β-Galactosidase is the hydrolytic transglucosylase responsible for the initial reaction in the metabolism of β-galactosides, such as in lactose *E.coli*. Other enzymes are associated with this metabolic pathway, among them a transporter molecule, *a permease*, and an enzyme catalyzing

Table 9.1 Some Metabolic Control Systems in Bacteria

Probable control	Operon or enzyme	Status[a]	Reference
Negative control	*lac*	+	*b*
	Tryptophan synthetase pathway	−	*c*
	Arginine synthetase pathway	−	*d*
			e
Positive control	*gal*	+	*f*
	Arabinose	+	*g*
	Maltose	+	*h*
	Alkaline phosphatase	−	*i*

[a]+, induction; −, repression.
[b]Jacob, F., and Monod, J. *J. Mol. Biol.* 3:318–356 (1961).
[c]Cohen, G., and Jacob, F. *C. R. Acad. Sci.* 248:3490–3492 (1959).
[d]Vogel, F. *Proc. Natl. Acad. Sci. U.S.A.* 43:491 (1957).
[e]Gorini, L., and Maas, W. K. In *The Chemical Basis of Development* (McElroy, W.D., and Glass, B., eds.), p. 469. Johns Hopkins Press, Baltimore (1958).
[f]Markovitz, A. *Proc. Natl. Acad. Sci. U.S.A.* 51:239–249 (1964).
[g]Englesberg, E., Irr, J., Power, J., and Lee, N. *J. Bacteriol.* 90:946–957 (1965).
[h]Hatfield, D., Hoffnung, M., and Schwartz, M. *J. Bacteriol.* 100:1311–1315 (1969).
[i]Garen, R., and Echols, H. *Proc. Natl. Acad. Sci. U.S.A.* 48:1398–1402 (1962).

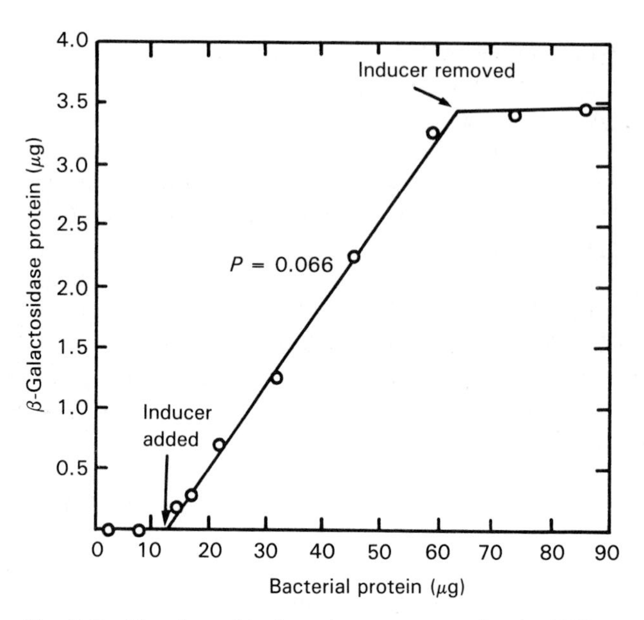

Fig. 9.3 Kinetics of induced enzyme synthesis. Differential plot expressing accumulation of β-galactosidase as a function of increase of mass of cells in a growing culture of *E. coli.* Since abscissa and ordinates are expressed in the same units (micrograms of protein), the slope of the straight line gives galactosidase as the fraction (P) of total protein synthesized in the presence of inducer. Reprinted with permission from M. Cohn, *Bacteriological Review,* 21:140–168. Copyright © 1957 American Society for Microbiology.

the transacetylation from acetyl-CoA to thiogalactoside. The three enzymes, which are coded by the *lac* operon, are coordinately induced by a mechanism involving the binding of a repressor by an inducer. A site (i) away from the operon has been found to be responsible for the induction of the *lac* operon in accordance with the model represented in Fig. 9.1*b*. In the diagram, i corresponds to RG.

The regulation of the enzymes associated with the breakdown of arabinose is somewhat more complex (Englesberg et al., 1965), as shown in Fig. 9.4. In this figure, the enzymes coded for by the arabinose operon are shown in the first line (item 1). The corresponding genes in the order in which they occur are shown in the line representing the chromosome (item 3); D, A, B, and C code for the enzymes of the operon. Item 2 represents the mRNAs for the operon. Gene C (see items 3 and 4) codes for a repressor (P_1) that can be converted into an activator (P_2). Arabinose controls the reversible conversion favoring the formation of activator. The repressor and the activator act as specific sites in the chromosome (O and I, respectively) to control the transcription of mRNA for genes D, A, and B.

So far, the repressor or activator molecules discussed have been proteins coded by the regulator gene. However, the regulating proteins may be part of the operon itself and so can regulate the transcription of their own mRNA (autoregulation; see Brawerman, 1981), and in fact the aporepressor may be an enzyme coded by the operon. In the histidine biosynthetic pathway of *Salmonella typhimurium*, the aporepressor is the first enzyme of the pathway (pyrophosphorylase), which is coded by the first gene in the operon. The repressor is this enzyme attached to histidine tRNA. In this case the production of histidine is closely regulated by the need of the protein-synthesizing machinery. Any slowdown in the manufacture of protein would immediately lead to accumulation of the histidine tRNA and hence a shutdown of the enzyme synthesis (Brawerman, 1981). Autoregulation is well suited to fine and continuous regulation of a pathway.

In summary, the data from genetic experiments are consistent with the interpretation that the *lac* operon is controlled by means of a repressor. By combining with a specific site in the operon's DNA, the repressor prevents the transcription of the *lac* mRNA. An inducer releases the inhibition by combining with the repressor so that, as the consequence of some allosteric effect, the repressor can no longer bind to the DNA. The level of repressor remains the same during induction. The arabinose operon has some of the same features; however, the repressor may also function as an activator.

The Regulatory Proteins Isolation of the *lac* repressor from *E. coli* required an assay for recognition of its presence, which was provided by binding to the radioactively labeled artificial inducer isopropyl thiogalactoside (IPTG). The degree of binding and the amount of repressor can be increased by using a mutant strain, i^t, that can bind the inducer more tightly than the normal repressor, as well as diploid cells that have double the amount of repressor present in the wild type.

Curve 1 of Fig. 9.5 (Gilbert and Müller-Hill, 1966) shows the results obtained with i^t repressor, and curve 2 shows the results obtained with the same protein fraction from i^s strains lacking the repressor. The percentage of inducer bound per 10 mg/ml of repressor fraction is plotted against the concentration of IPTG. Only the i^t extract binds the inducer.

Purification of the repressor by ultracentrifugation through a glycerol gradient reveals a molecular weight between 150,000 and 200,000 (corresponding to 7.6–8S). The repressor is a protein, since it is heat labile, insensitive to DNase and RNase, and inactivated when treated with pronase, an enzyme that hydrolyzes proteins. As might be expected from the model of repressor control, the concentration of repressor is not affected by induction.

Another part of the concept derived from genetic data summarized by the models of Fig.

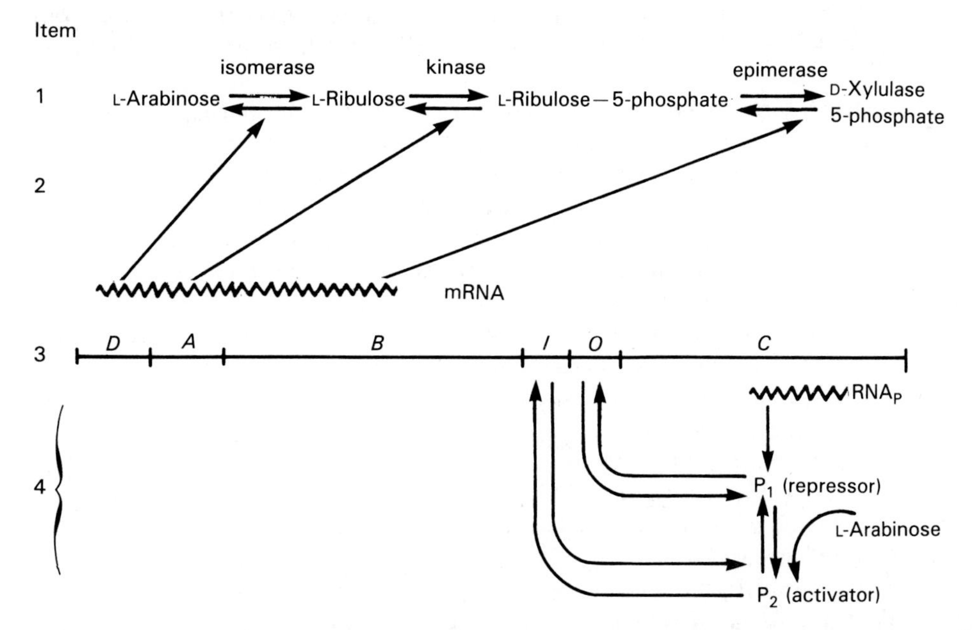

Fig. 9.4 Regulation of the enzymes associated with the catabolism of arabinose. (1) Enzymes involved in the catabolism of arabinose. (2) The mRNA for the operon. (3) Structural genes controlling the enzymes of the arabinose pathway. (4) Scheme by which the repressor can be converted into an activator.

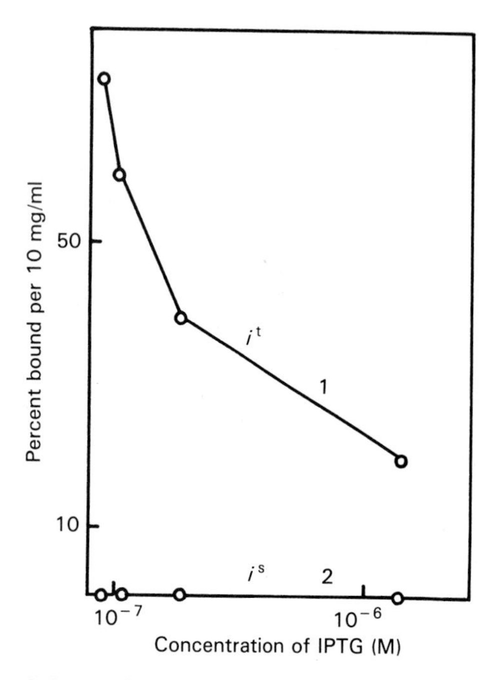

Fig. 9.5 Binding ability of an i^s and an i^t strain. The repressor fraction was isolated in parallel from 50-g lots of isogenic F' i^s/i^s and F' i^t/i^t cells. Reprinted with permission from W. Gilbert and B. Müller-Hill, *Proceedings of the National Academy of Sciences,* 56:1891–1898, 1966.

9.1*a* can be tested. Does the repressor bind to the DNA corresponding to the *lac* operon? In centrifugal sedimentation experiments, the small portion of the labeled protein corresponding to the repressor sediments with the DNA containing the *lac* operon (Figs. 9.6*a* and 9.7, curve 2) rather than with the bulk of the protein (curve 1 in both sets of figures) and this binding is blocked by IPTG (curve 2, Fig. 9.6*b*). In addition, DNA lacking the *lac* operon fails to bind, as shown in Fig. 9.7, curve 3, and Fig. 9.8 (Gilbert and Müller-Hill, 1967).

A similar procedure has resulted in isolation of the repressor for the galactose operon (Parks et al., 1971), which controls several enzymes of galactose metabolism (Jordan et al., 1962). In this study, the titer of repressor was increased substantially by first infecting the *E. coli* cells with the phage linked to the *gal* regulator gene. Production of the phage increases the titer of repressor. The molecules of repressor were isolated by eluting the extract of the cells in a column in which the artificial inducer molecule *p*-aminophenyl-β-D-thiogalactoside was covalently bound to agarose beads. The inducer, by specifically binding the repressor, slowed its passage through the column.

Activators are generally present in very small amounts and their isolation has been difficult. However, several have been purified (Raibaud and Schwartz, 1984). They generally are accessory factors for RNA polymerase or actually displace one of the subunits of the polymerase.

Several other features of the operon have been recognized, primarily from genetic experiments, and are summarized in Fig. 9.9 on p. 247 (Yanofsky, 1981), which corresponds to a generalized genetic map of an operon. The *operator* site is the binding site of modifiers of transcriptional activity, i.e., repressors and activators. The *promoter* is the site of attachment of the RNA polymerase, which begins transcribing at the *transcription site,* generally adjacent to the operator. Genes *A, B, C, D,* and *E* code for enzymes. Other regulatory sites in the diagram function in mechanisms that are discussed in subsections B and C below.

B. Attenuation

Repressors and activators regulate the initiation of transcription. *Attenuation* affects the transcription of the structural genes by terminating the transcription in the *leader region* (TrpL in Fig. 9.9 on p. 247), a site preceding the first structural gene.

In the tryptophan (*trp*) operon of *E. coli,* attenuation is quantitatively less significant than repression and is probably responsible only for fine tuning. However, attenuation is the sole regulatory mechanism for operons controlling the synthesis of several other amino acids.

The *trp* operon consists of a transcription regulatory region and five structural genes coding for the enzymes involved in the production of tryptophan. The initiation of transcription is regulated by a tryptophan-activated repressor protein that controls the availability of the operon to RNA polymerase (Oppenheim et al., 1980). In contrast, attenuation involves the translation of *trp* codons in the leader region of the operon. When the tryptophan concentration is low, the availability of tRNA is also low. Therefore, the translational machinery, the ribosome-nascent mRNA complex, would be unable to get past these codons. Apparently, when the ribosomes are stalled, the secondary structure of the leader transcript is changed so that both translation and transcription are blocked. The various features of the control of the *trp* operon have been incorporated in Fig. 9.9.

Although attenuation controls other amino acids operons, these have been less well studied.

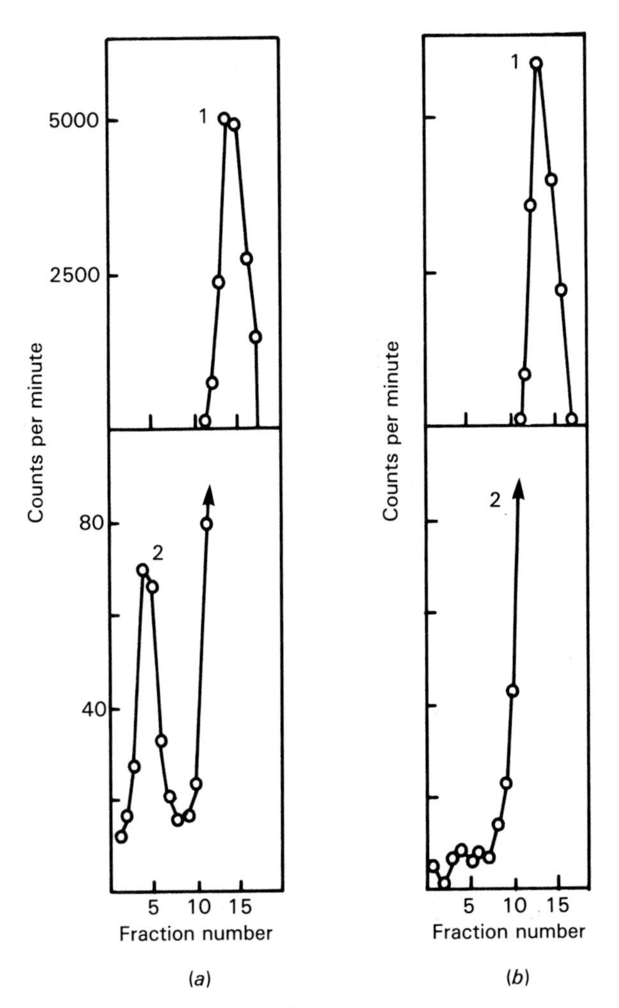

Fig. 9.6 Binding of the *lac* repressor to d*lac* phage DNA (*a*) and its release by inducer (*b*). The radioactivity of the bulk of the proteins is shown in the upper curves (curve 1, *a* or *b*). Reprinted with permission from W. Gilbert and B. Müller-Hill, *Proceedings of the National Academy of Sciences,* 58:2415–2421, 1967.

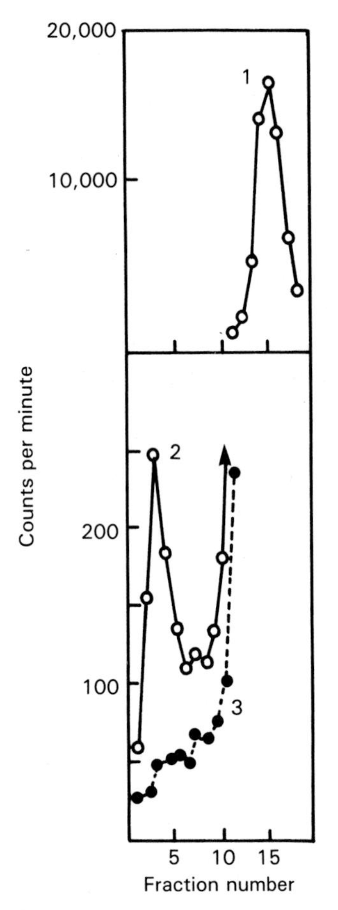

Fig. 9.7 Specificity of the binding. In parallel gradients the same repressor preparation was run with two different DNAs. One reaction mixture contained pure d*lac* phage DNA (curve 2); the others contained parental φ80-λ hybrid (curve 3). The radioactivity of the bulk of the proteins is shown in curve 1. Reprinted with permission from W. Gilbert and B. Müller-Hill, *Proceedings of the National Academy of Sciences,* 58:2415–2421, 1967.

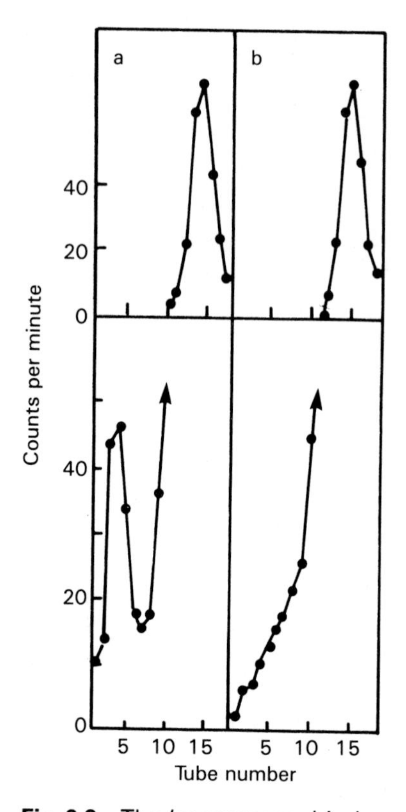

Fig. 9.8 The *lac* repressor binds specifically to the *lac* operator. In parallel gradients the same repressor preparation was run with different purified d*lac* phage DNAs. (*a*) Control; (*b*) O^c mutant DNA. Reprinted with permission from W. Gilbert and B. Müller-Hill, *Proceedings of the National Academy of Sciences,* 58:2415–2421, 1967.

All indications are that the mechanism is similar, since the leader sequences also contain codons for the amino acids controlled by the operon.

C. Translational Control

In bacteria, coordinate regulation of transcription and translation can take place readily since the two occur almost simultaneously; the nascent mRNA is used for translation. We have seen how this operates in the events of attenuation.

Evidence of translational control independent of transcription exists for a number of cases, epitomized by the control of ribosomal protein. Fifty-two ribosomal proteins (*rproteins*) are coordinately produced. One ribosome contains one copy of each protein except L7/L2, of which there are four copies (Hardy, 1975). Free rproteins and rRNAs are normally present in very small amounts (Gausing, 1974).

The evidence indicates that certain rproteins inhibit translation of the mRNA coding for proteins in feedback manner (Engelsberg et al.; 1965; Yates and Nomura, 1980) by attaching to a small region on the polycistronic mRNA near or at the initiation site of the first cistron of the regulatory unit.

Transcriptional regulation is not likely to be a major factor in the control of the ribosomal operon, since increasing the amount of ribosomal mRNA by introducing many ribosomal genes in the same cells does not increase the number of ribosomes.

Ribosomal concentration is regulated in a similar way during growth. The ribosomal level is regulated by the ribosomes available by repressing the rRNA genes by a feedback mechanism (Dean et al., 1981).

Translational control also occurs in the osmoregulation of the major outer membrane proteins of *E. coli,* OmpF and OmpC (Mizuno et al., 1984). The *ompC* locus is transcribed bidirectionally at high osmolarity. A 174-base transcript encoded upstream (mRNA interfering complementary RNA, *micRNA*) inhibits OmpF production by combining with *ompF* mRNA. The translation is blocked by premature termination of the *ompF* mRNA, presumably by hybridization.

D. Regulatory Switches

Under some physiological conditions several operons may be regulated together, even when they are scattered physically throughout the genome and may represent functions that are not clearly related. The operons of sugar metabolism respond to the presence or absence of glucose in the medium (Magasanik, 1962). The response depends on the cellular concentration of cAMP, as shown by experiments carried out with mutants of genes that are involved in either synthesizing cAMP or sensing its concentration. Apparently, a catabolite repression protein acts in the presence of cAMP as a positive regulator of the transcription of the lactose, maltose, and arabinose operons.

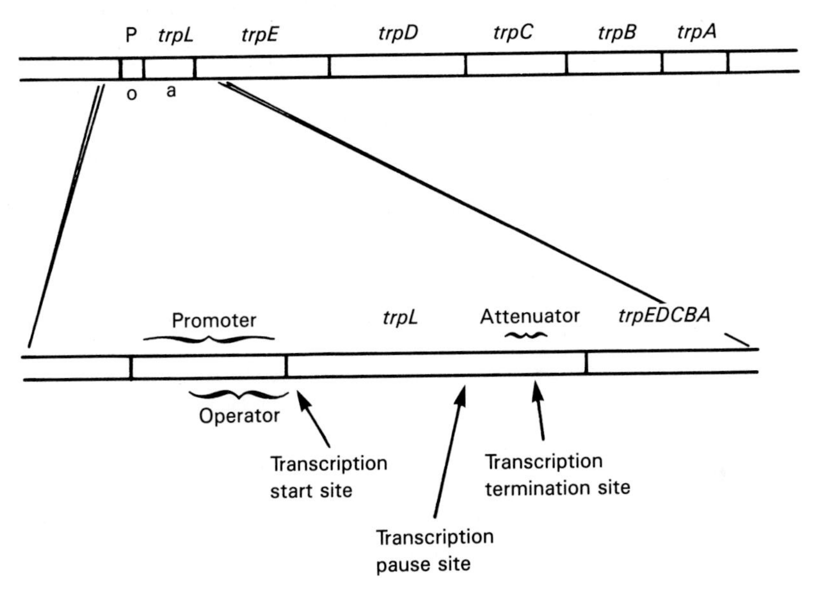

Fig. 9.9 Regulatory and structural gene regions of the *trp* operon of *E. coli.* Reprinted by permission from *Nature,* Vol. 289:451–457. Copyright © 1981 Macmillan Magazines Ltd.

Considerable evidence indicates that there are many complex and interlocking regulatory networks, each involving more than one operon. Each component of a network responds to a common regulatory signal released inside the cells with changes in environment or sudden stress. A unit capable of being regulated by a single repressor, whether it consists of one or more operons, has been called a *regulon* (Maas and Clark, 1964). The signal molecules implicated so far are *cAMP* and guanosine tetraphosphate (ppGpp). Others are likely to exist as well. As in the case of the operons of sugar catabolism, the signal acts by modifying a protein regulator molecule. The regulatory protein may act directly with the signal molecule and the regulated operons.

Several regulons have been studied, in particular a high-temperature regulon and a regulon (known as SOS) involved in DNA repair. Other regulons respond to nutrient limitation, limitations in inorganic phosphate, and limitations in fixed nitrogen.

II. REGULATION IN EUKARYOTES

In bacteria, the adjustment of the enzymatic composition to face an environmental challenge, such as the presence of a new nutrient in the medium, is a major function of the regulatory mechanisms. As we saw, regulation is primarily at the transcriptional and translational levels.

Lower eukaryotes, such as yeast, may also be subject to the same nutritional challenges. The problem is different for multicellular organisms. Since their tissues are specialized, only some (for example, the liver) have to adjust to nutritional changes. The environment of other tissues is maintained relatively constant in nutrient composition. However, protein and enzyme synthesis is regulated by hormones or growth factors. Furthermore, the protein composition of a cell depends on its developmental stage. Development reflects differential gene expression that results in differentiation. In the differentiated state, only some of the proteins encoded by the genome are fully expressed. In multicellular eukaryotes the control of protein or enzyme levels is very complex. As in prokaryotes, protein and enzyme synthesis is controlled by transcriptional and translational mechanisms. However, specific enzymes or their corresponding mRNAs are also differentially degraded, and the transfer of mRNA from the nucleus may be regulated.

A. Production and Degradation of Specific Enzymes

The balance between production and breakdown of enzymes plays an important role in the regulation of the cell's metabolic machinery in eukaryotes. Alteration of either the synthesis or breakdown of an enzyme will have an appropriate effect on its concentration.

The concentration of an enzyme can be increased in response to the presence of its substrate, to the secretion of a hormone, or to general nutritional conditions. In rat liver, the level of the enzyme tryptophan pyrrolase (also called tryptophan 2,3-dioxygenase) is increased sharply by administering hydrocortisone or a large concentration of tryptophan. The enzyme catalyzes the reaction shown in Eq. (9.1).

$$\text{Tryptophan (indole}-CH_2-CH(NH_2)-COOH) + O_2 \longrightarrow \text{N-Formylkynurenine (C}_6H_4(-C(=O)-CH_2-CH(NH_2)-COOH)(-NH-C(=O)-H)) \quad (9.1)$$

Tryptophan N–Formylkynurenine

This reaction is the first in one of the two major biochemical pathways responsible for the breakdown of tryptophan. The formation of the product of the reaction, *N*-formylkynurenine, can be followed with time spectrophotometrically.

The adrenal cortex secretes a number of steroid hormones, some of which increase the breakdown of proteins and the formation of glucose (gluconeogenesis) in the intact animal. Steroid induction of enzymes such as tryptophan pyrrolase, tyrosine transaminase, serine dehydratase, serine deaminase, and phosphoenolpyruvate carboxykinase is likely to be part of these two functions. The hormonal control of tyrosine transaminase and serine dehydratase is discussed in more detail later in this chapter.

In the experiment of Table 9.2 (Schimke et al., 1965), adrenalectomized rats (i.e., rats whose adrenal gland was surgically removed) were used to avoid the complication of endogenous release of the hormone by the adrenal cortex. The tryptophan pyrrolase activity in a liver extract increased with the addition of tryptophan and hydrocortisone (column 3 of Table 9.2). Other enzyme activities also increased (columns 4 and 5), although more moderately. Compared to an untreated control (in which a NaCl solution was injected), both hydrocortisone- and tryptophan-treated animals had large increases in the activity of tryptophan pyrrolase. The two treatments administered together had an even greater effect.

Since a latent form of tryptophan pyrrolase is known, the increase in enzyme activity could result from activation of the enzyme rather than de novo synthesis. An immunological assay can resolve which of the two possible mechanisms is responsible for the effect.

Immunological techniques depend on the interaction between antibodies and their corresponding antigens. Serum antibodies specific for an enzyme can be produced by injecting the pure protein into animals, generally rabbits. The amount of antibody-antigen complex formed after the addition of excess antibody to an extract quantitatively measures the amount of enzyme present. In the *radioimmunoassay* technique, which is commonly used, radioactive iodine is introduced into the antibody molecule and the antibody-antigen complex is assayed in the precipitate by simply measuring its radioactivity. Similarly, if the proteins of an animal are made radioactive by injection of radiolabeled amino acids, an enzyme can be recognized with an antibody in the same way. Experiments using this latter approach will be discussed later.

In the experiment of Fig. 9.10 (Schimke et al., 1965), aliquots of liver homogenate were added sequentially in increasing amounts to a fixed amount of antibody. The supernatant was sampled for the enzyme activity that remains after precipitation of the antibody-enzyme complex. This manipulation is analogous to an analytical procedure in which aliquots of a reagent are added to a sample until an end point is reached—in this case, the appearance of the enzymatic activity in the supernatant. The enzyme activity of the aliquots, before addition to the antibody,

Table 9.2 Effect of Repeated Doses of Tryptophan and Hydrocortisone on Liver Weight and Tryptophan Pyrrolase, Formylkynurenine Formylase, and Arginase of Rat Liver

(1)	(2) Total liver weight[a] (g)	(3) Enzyme activity (μmol product/h per g liver weight)	(4)	(5)
Treatment		Tryptophan pyrrolase	Formylase	Arginase
NaCl	25.4	3.0	1800	13,800
Hydrocortisone	27.4	38	1810	19,740
Tryptophan	28.7	28	2060	15,200
Hydrocortisone + tryptophan	25.0	116	2180	20,160

Reprinted with permission from R. T. Schimke, et al., *Journal of Biological Chemistry,* 240:322–331.

[a]Combined weight of four animals.

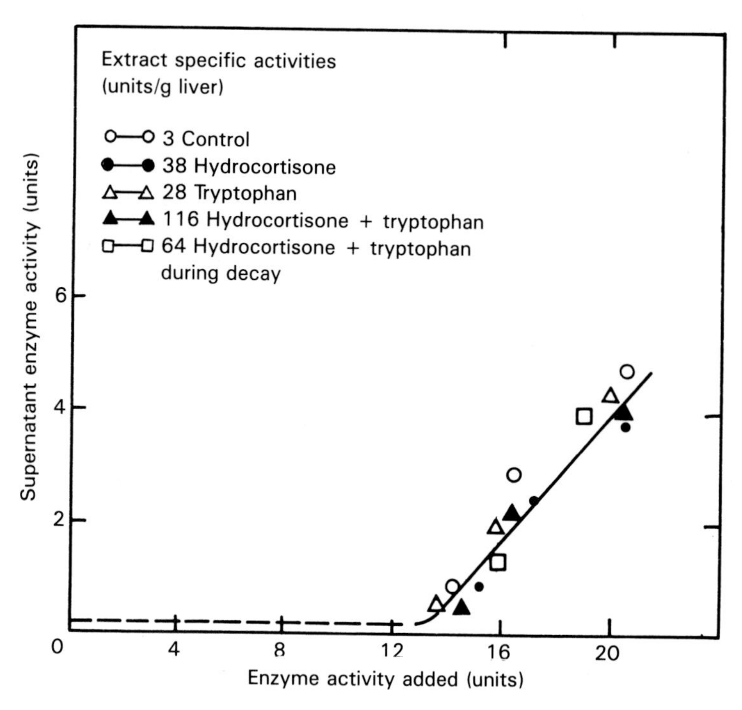

Fig. 9.10 Immunological analysis of tryptophan pyrrolase activity of rat liver extracts. Adrenalectomized rats weighing 150–170 g were given repeated administrations of 0.85% NaCl (control) or hydrocortisone 21-phosphate, L-tryptophan, or both. At the end of 16 h, the livers were removed. Two rats that had received both hydrocortisone and tryptophan were killed 3 h later, at a time when the enzyme level was falling. The dashed line indicates that no enzyme activity was detectable in the supernatant fluid. Reprinted with permission from R. T. Schimke, et al., *Journal of Biological Chemistry,* 240:322–331. Copyright © 1965 American Society of Biological Chemists, Inc.

is shown on the abscissa and the enzyme activity remaining in solution after precipitation is shown on the ordinate. As soon as the added enzyme is in excess of the antibody present, the activity appears in the supernatant. Presumably, the enzyme would interact with the antibody whether it was active or not. Inactive enzyme should remove antibody but would not be detected in the assay of the activity of the homogenate. Any masked enzyme would therefore show up in a graph such as that of Fig. 9.10 as an earlier appearance of activity (at a lower homogenate activity) than if all the enzyme were active, i.e., at a point closer to the origin. In this experiment, regardless of whether the animal received only weak salt solution (the control) or was treated with hydrocortisone, tryptophan, or both, the results are the same. The appearance of enzymatic activity in the supernatant occurs at precisely the same level of added enzyme activity in all experimental preparations, indicating that the active enzyme and the antigen bound by the antibody are precisely equivalent. Other criteria (see below) are in agreement with the conclusions of this study.

Since the amount of enzyme has increased, either more enzyme is being made or less broken down. After ^{14}C-labeled amino acid (in these experiments [^{14}C]leucine) is injected into rats, its immunological precipitation in homogenates serves as a measure of synthesis of the enzyme, assuming that [^{14}C]leucine inside the cells does not change under the different experimental conditions. The results of an experiment depicted in Fig. 9.11 (Schimke et al., 1965) show that this assumption is correct. Curve 2 in the figure represents the amount of radioactive trichloroace-

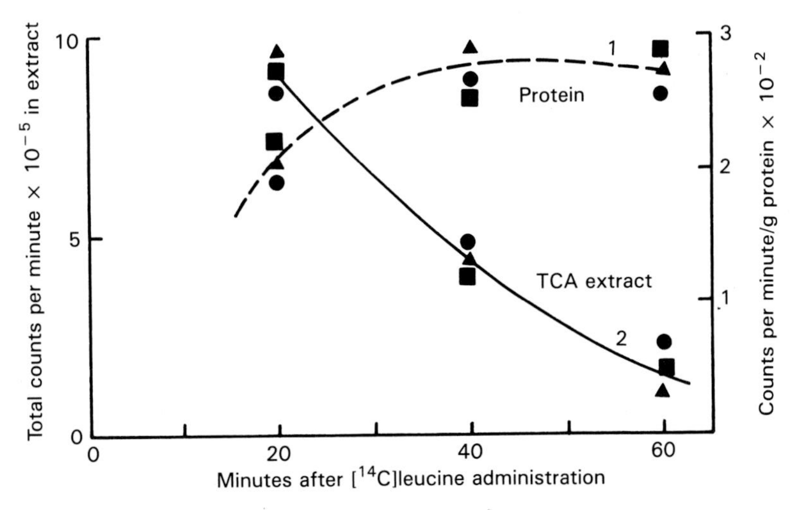

Fig. 9.11 Effect of hydrocortisone 21-phosphate and tryptophan administration on free pool of administered L-[^{14}C]leucine and its incorporation into liver protein. Reprinted with permission from R. T. Schimke, et al., *Journal of Biological Chemistry,* 240:322–331. Copyright © 1965 American Society of Biological Chemists, Inc.

tic acid-soluble material (TCA extract) which corresponds primarily to free amino acid. The radioactivity in the protein corresponds to the material precipitated with TCA (curve 1). Each point corresponds to a different group of rats treated in the same way. In Fig. 9.11 the squares and the triangles represent the radioactivity present in liver extracts of hydrocortisone- and tryptophan-treated animals, respectively, and the circles represent controls. The radioactivity is equivalent in all three cases. Therefore the concentration of [^{14}C]leucine incorporated represents the same degree of protein synthesis in all three cases.

The incorporation of a label into immunologically precipitated tryptophan pyrrolase is shown in Table 9.3. The label incorporated in the enzymes of the saline-treated control animals corresponds to about 1400 counts/min (column 4). The incorporation in the hydrocortisone-treated animals (column 4) is considerably higher, about 9500 counts/minute. Since this corre-

Table 9.3 Immunological Precipitation of Tryptophan Pyrrolase from Liver Extracts of Rats that Received [^{14}C]L-lysine[a]

	Total enzyme activity in (units/hour)			
	(1)	(2)	(3)	(4)
Treatment	Homogenate	Supernatant	DEAE-cellulose eluate	Corrected incorporation (total counts/min)
NaCl	41.1	26.0	23.1	1406
Hydrocortisone	189	118	107	9466
Tryptophan	80.1	56.7	51.3	1954

Reprinted with permission from R. T. Schimke, et al., *Journal of Biological Chemistry,* 240:322–331. Copyright © 1965 American Society of Biological Chemists, Inc.

[a]Adrenalectomized rats weighing 150 to 160 g each were given single intraperitoneal injections of 10 ml of 0.85% NaCl, 8.0 mg of hydrocortizone 21-phosphate, or 150 mg of L-tryptophan in 10 ml of 0.85% NaCl. After 3 h and 20 min each animal was given an intraperitoneal injection of 20 μCi of L-[^{14}C]lysine (specific activity, 80 mCi/mmol) in 1 ml of 0.85% NaCl. After 40 min the animals were killed, and extracts were prepared for immunological precipitation.

sponds approximately to a sevenfold increase in incorporation, the system behaves as if there were an increase in the synthesis of the enzyme in response to the presence of the hormone.

In contrast, the incorporation in the tryptophan-treated animals (column 4) is very small, about 40% more than the control value, despite the fact that the activity of the enzyme doubled (compare tryptophan and control values in columns 1, 2, and 3). Does the increase in available enzyme result from a decrease in its breakdown? This question can be examined experimentally by observing the temporal disappearance of the label from the previously labeled tryptophan pyrrolase molecule.

The results of these experiments are shown in Fig. 9.12. After a single injection of [^{14}C]leucine, some of the animals were sacrificed at the times shown on the abscissa. The radioactivity of the total unfractionated protein remains the same in the controls (Fig. 9.12a, squares) and the tryptophan-treated animals (Fig. 9.12b). However, the decrease in radioactivity of the enzyme (Fig. 9.12a, triangles) with time is rapid in the controls while the enzymatic activity remains constant (Fig. 9.12a, circles). The rate of synthesis must precisely correspond to the rate of breakdown. In contrast, in the tryptophan-treated animals the radioactivity does not decrease with time (Fig. 9.12b, triangles) while enzyme activity increases (Fig. 9.12b, circles). The increase in activity must result from inhibition of the degradation of the pyrrolase.

Rat liver serine dehydratase exhibits a similar response to a dietary intake and the presence of the hormone (Jost et al., 1968). This enzyme catalyzes the dehydration and deamination of serine to produce pyruvate. The pyruvate subsequently can be oxidized or used in the production of glucose. Administration of a mixture of amino acids to rats results in introduction of the enzyme unaccompanied by changes in the total protein content of the liver. Glucagon, a polypeptide hormone secreted by the pancreas that favors gluconeogenesis, has a similar effect. Administration of glucagon or the amino acids produces an increase in the incorporation of

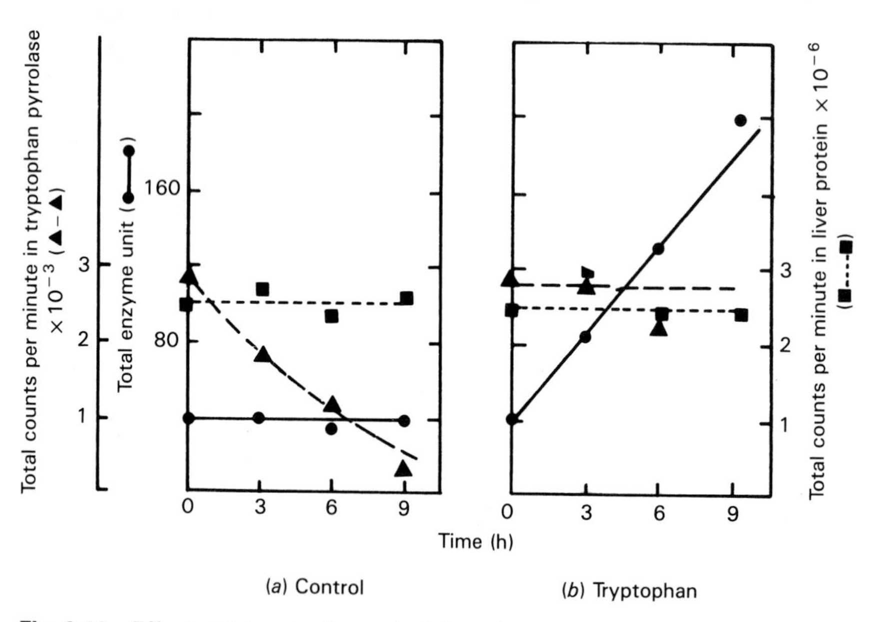

Fig. 9.12 Effect of L-tryptophan administration on loss of tryptophan pyrrolase prelabeled with L-[^{14}C]leucine. The values represent total enzyme activity present in the combined extracts from two animals (●). Counts per minute represent total radioactivity present in protein precipitated by the tryptophan pyrrolase antiserum (▲) or in total cellular protein (■) from two livers. Reprinted with permission from R. T. Schimke, et al., *Journal of Biological Chemistry,* 240:322–331. Copyright © 1965 American Society of Biological Chemists, Inc.

[^{14}C]valine into the enzyme molecule, as shown by an immunological precipitation similar to that in the experiment of Table 9.3. In contrast, glucose inhibits production of the enzyme and increases its rate of breakdown (shown by experiments analogous to those of Table 9.3 and Fig. 9.12).

B. The Degradative Pathways

The degradation of proteins contributes to the turnover of the cellular components. Under basal conditions, i.e., at rest and with a sufficient food supply, protein turnover has been estimated to account for 15% of the energy expenditure of humans. In addition to its role in the regulation of enzyme concentration, proteolysis is also thought to function in removing abnormal proteins produced by mistakes in translation (Goldberg, 1972). Such proteins would fulfill no function, and their breakdown allows their use as a metabolic fuel.

We have seen how degradation plays a role in the regulation of enzyme activity. In yeast, changes in the supply of nutrients, such as addition of glucose or removal of carbon sources (carbon starvation) (Blundell et al., 1972), irreversibly inactivate certain enzymes, and at least glutamine synthetase (Ferguson and Sims, 1974) is degraded. The enzyme activity returns after a return to the original conditions but only if protein synthesis is not blocked by inhibitors such as cycloheximide.

Degradation as a General Regulatory Mechanism In eukaryotes, the rate of proteolysis is as significant as the rate of synthesis. This follows from the demonstration that most enzymes have a characteristic lifetime.

At steady state, the balance between synthesis and degradation is generally expressed as a half-time ($t_{1/2}$). The $t_{1/2}$ can be estimated from data such as those shown in Fig. 9.12a representing the decay of the radioactivity in the enzyme molecule (E). This rate, dE/dt, can be expressed as a function of the rate constant for synthesis k_s (which is a constant if short synthetic periods are considered) and a degradation rate constant k_d, as shown in Eq. (9.2).

$$\frac{dE}{dt} = k_s - k_d E \tag{9.2}$$

If the original radioactivity has been rapidly diluted by adding unlabeled amino acid (i.e., by a chase) $k_s = 0$. Equation (9.2) can be integrated to the form

$$\ln \frac{E}{E_0} = -k_d t \tag{9.3}$$

where E_0 represents the total amount of radioactivity present at time zero. When half of the enzyme has decayed, the half-time can be expressed as shown by Eq. (9.4); k_d can be readily calculated from a logarithmic plot of E/E_0 as a function of time, where it would correspond to the slope.

$$t_{1/2} = \frac{\ln 2}{k_d} \tag{9.4}$$

The regulation of an enzyme through its turnover rate must be reflected in a characteristic $t_{1/2}$ for the enzyme. Table 9.4 (Ballard, 1977) shows that enzymes indeed have characteristic half-times, in agreement with this idea. Therefore proteolytic degradation seems to choose

Table 9.4 Characteristic Half-Times of Enzymes

	$t_{1/2}$ in vivo (days)
Lactate dehydrogenase	6.0
Fructose bisphosphate aldolase	4.9
Glucose 6-phosphate dehydrogenase	1
Glucokinase	1
Phosphoenolpyruvate carboxykinase (GTP)	0.3
Thymidine kinase	0.1
Ornithine decarboxylase	0.008
RNA polymerase I	0.05
Tyrosine aminotransferase	0.06
Tryptophan oxygenase	0.08
Phosphoenolpyruvate carboxykinase	0.25
Acetyl-CoA carboxylase	2
Glyceraldehyde phosphate dehydrogenase	3.4
Arginase	4.5

Source: Ballard (1977) and (1978). Reprinted with permission from *Essays in Biochemistry* 13:1–37, copyright © 1977 The Biochemical Society, London.

specific targets. What mechanism or mechanisms could provide such precise specificity? It might be possible to clarify these processes by examining the properties of known cellular proteinases (Barrett, 1980; Matern and Holzer, 1979). These enzymes differ significantly from each other, but many have a common characteristic: they must be activated or they need special conditions for their action. However, aside from the fact that they are frequently located at specific sites (e.g., microsomal or mitochondrial membranes, the Z band of the sarcomere), nothing suggests that they can act on certain proteins and not on others. An initial inactivation has frequently been found to precede the actual hydrolysis. There are indications that some enzymes are marked for proteolysis by phosphorylation. In other cases, other forms of modification, such as the inactivation by oxidation, may mark an enzyme for proteolysis (e.g., Rivett, 1985).

Biochemical Pathways of Proteolysis Although the mechanism for the specificity of proteolysis is not clear, a good deal is known about the biochemical pathways of the breakdown. Present evidence indicates that there are two distinct processes.

Perfusion of rat liver or incubation of cell cultures with a medium devoid of amino acids induces an increase in the breakdown of proteins. This enhancement of proteolysis can be blocked by supplementing the medium with amino acids (Woodside and Mortimore, 1972) or by introducing insulin (Mortimore and Mondon, 1970) or some other growth factor.

The breakdown of a number of components has been shown to occur in lysosomes after endocytosis. Therefore, a general role of lysosomes in proteolysis would not be surprising. In fact, proteolysis, which accompanies amino acid deprivation, correlates well with the enlargement of lysosomes and the presence of *autophagic vacuoles* (Mortimore and Schworer, 1977), i.e., vacuoles that enclose cellular structures and function in their digestion. Conversely, these events are prevented by the presence of amino acid supplements.

How can this problem be studied in more detail? Since many cell organelles continue to function after isolation, it would seem possible to examine lysosomal activity after isolation of the vesicle. Under a variety of conditions, including amino acid deprivation and amino acid supplementation (Mortimore and Ward, 1981), lysosomes isolated from amino acid-deprived animals release amino acids or acid-soluble material at a rate compatible with the overall protein breakdown (Mortimore et al., 1973). Despite this observation, it is difficult to imagine that lysosomes are responsible for the very specific breakdown of the rapidly turning over enzymes we discussed. In contrast, it would be easy to imagine the involvement of lysosomes in the nonspecific hydrolysis of protein at a lower rate, in response to a generalized signal (however, see below).

The fate of these two kinds of proteins could be followed after a short pulse of a radioactive amino acid. Such a pulse should predominantly label the rapidly turning over proteins. A more prolonged pulse would label both kinds of protein. The view that there are two different processes is supported by the finding that inhibitors of lysosomal proteases, which have little or no effect on the proteolysis of short-lived or abnormal proteins (Neff et al., 1979), do block the proteolysis of the longer-lived proteins. Furthermore, a temperature-sensitive mutant has been found for one of the enzymes of the ubiquitin conjugation system (Finley et al., 1984). The ubiquitin system corresponds to a second proteolytic pathway and, as we shall see later in this section, is responsible for the rapid turnover of proteins. In the mutant, at a permissive temperature, 70% of the rapidly turning over proteins are degraded in 4 h (Ciechanover et al., 1984). However, only 15% of this fraction is degraded at the nonpermissive temperature.

The effect of protease inhibitors, which inhibit lysosomal proteolysis, is shown in Table 9.5 (Neff et al., 1979). The experiment was carried out with cultured hepatocytes. Part A of the table shows the appearance of ^{3}H-labeled material in the medium of cells that have either been labeled for long periods (column 1), pulse-labeled with [^{3}H]leucine (column 2), or pulse-labeled with an analog that will be incorporated to form an abnormal protein (column 3). The inhibitors have no effect on the pulse-labeled proteins or abnormal proteins but inhibit the proteins that have been labeled for a long period. Although the effect is small, it is significant. Table 9.5B

Table 9.5A Effects of Protease Inhibitors on Degradation of Different Classes of Cellular Proteins

	Normal		Analog containing
Inhibitor	(1) Chronically labeled	(2) Pulse labeled	(3) Pulse labeled
	Protein degraded/h (% of total)		
None	1.65±0.13	21.2±0.7	26.7±2.6
	Percent inhibition		
Leupeptin	21	0	0
Chymostatin	27	0	0
Antipain	18	0	0
Pepstatin	0	—	—

Table 9.5B Lack of Additivity of Inhibitors on Protein Degradation

Inhibitor	Protein degraded/h (% of total)	Inhibition (%)
None	3.32±0.11	—
Leupeptin	2.62±0.17	21
Chymostatin	1.72±0.09	48
Antipain	2.97±0.16	11
Leupeptin + chymostatin	1.54±0.08	54
Leupeptin + chymostatin + antipain	1.72±0.05	48

From N. Neff, et al., *Journal of Cellular Physiology,* 101:439–458.

shows a much more marked effect for some of the inhibitors—as much as 54% inhibition. In addition, the effect is not additive, suggesting that the inhibitors are acting on the same mechanism. The amount of inhibition is not complete, possibly because of permeability barriers to the inhibitors. More significantly, the inhibitors failed to inhibit the breakdown of defective protein, even when its concentration was sharply increased by prolonged incubation. The results are therefore consistent with the presence of two independent pathways: one for the slowly turning over proteins broken down by lysosomes and the other for the breakdown of rapidly turning over proteins including abnormal proteins (however, see Gronostajski et al., 1985).

Three pathways of lysosomal proteolysis have been recognized. Microautophagy takes place in well-nourished cells and is non-selective. The lysosomal membrane invaginates at multiple locations (Marzella and Glaumann, 1987) and internalizes proteins which are then digested by lysosomal proteases. Macroautophagy (Seglen et al., 1990) which is induced in cell cultures, for example, by removal of growth factors, is similarly nonspecific. Autophagic vacuoles are formed in the cytoplasm in which newly formed membranes sequester cytoplasmic components. These are then exposed to the lysosomal hydrolytic enzymes. An additional pathway induced in confluent cultured cells (in which growth is arrested) by serum deprivation is selective. In this mechanism, proteins with a specific sequence of five amino acids are imported into the lysosomes (Dice et al., 1990) and subsequently digested.

The nonlysosomal pathway has been partially elucidated by use of a special strategy (Gronostajski et al., 1985). All protein breakdown in the cell requires energy. Therefore, proteolysis can be arrested in extracts by the simple expedient of leaving ATP out of the incubation medium. This would allow the fractionation and subsequent reconstitution of the system, which can then be activated at will by the addition of ATP. Reticulocyte extracts have proved particularly useful since they synthesize almost exclusively hemoglobin and also possess a very active proteolytic system to dispose of abnormal globin.

This approach can best be illustrated by showing the results of the original experiments, which defined the ubiquitin proteolytic pathway and were carried out with the reticulocyte extract (Table 9.6) (Ciechanover et al., 1978). The degradation of [^{3}H]globin, previously labeled by incubation of the intact cells with ^{3}H-labeled amino acids, provided a measure of proteolytic activity. As shown in line 1, ATP is required for proteolysis. The two fractions derived from the extract, fractions I and II, have little activity when alone, as shown by lines 2 and 3. However, the two together reconstitute the proteolytic system as long as ATP is present, as shown in line 4. Further extraction of fraction I demonstrated the presence of a heat-stable factor with a molecular weight of 10,000 (Ciechanover et al., 1980a) and containing 76 amino acids. In the complete system, this low molecular weight component was found to bind to protein when the system was incubated in the presence of ATP and fraction II (Ciechanover et al., 1980b). The binding was covalent, since the association was resistant to severe treatment such

Table 9.6 Resolution of the ATP-Dependent Cell-Free Proteolytic System into Complementing Activities

	Degradation of [^{3}H]globin (%/h)	
Enzyme fraction	**−ATP**	**+ATP**
1. Extract	1.5	10.0
2. Fraction I	0	0
3. Fraction II	1.5	2.7
4. Fractions I and II	1.6	10.6

Source: Ciechanover et al. (1978).

as heat denaturation, exposure to acid and alkali, or exposure to a chemical reducing agent. The heat-resistant protein corresponds to a form of ubiquitin containing two extra glycine residues. Several molecules of this factor bind a molecule of higher molecular weight protein. The ubiquitin attaches through its carboxyl terminal to the ε amino groups of the lysines of the higher molecular weight proteins.

When ATP is removed from the extract, covalent binding of diglycine ubiquitin to proteins ceases. However, the extract continues to break down the protein portion of the protein-ubiquitin complex (Hershko et al., 1980). In this fashion ubiquitin is regenerated and in the presence of ATP can be used again (Ciechanover et al., 1984). A recently formulated version of this cycle is illustrated in Fig. 9.13 (Hershko, 1988) where E_1, E_2, and E_3 correspond to a family of enzymes required for ubiquitin ligation and protein breakdown.

Conjugation of proteins to ubiquitin has been observed in many mammalian tissues (e.g., Ciechanover et al., 1984), so the ubiquitin-proteolytic system seems to be a general mechanism not limited to reticulocytes. Furthermore, mutations in two of the three yeast genes coding for E_2-proteins greatly reduce protein degradation (Seufert and Jentsch, 1990). Deletion of all three is lethal (Seufert et al., 1990). It follows that the ubiquitin-linked proteolysis system plays a vital role.

The conjugation with ubiquitin marks a protein for degradation. But what determines what protein is bound to ubiquitin? There are several possibilities. The conjugation of several ubiquitin molecules per protein provides room for regulating degradation by altering one of these reactions. The binding of the proteins to E_3 (Fig. 9.13) has a good deal of specificity for certain amino acid residues at the amino termini of the proteins (Hershko, 1988; Bartel et al., 1990). In fact, changing the amino-terminal residue of proteins expressed in yeast dramatically accelerates degradation (Bachmair et al., 1986). Similarly, tRNA-mediated transfer of arginine to the amino terminus converts proteins with acidic residues into good substrates for the ubiquitin system (Ciechanover and Schwartz, 1989). However, the ubiquitin system can degrade proteins where the amino terminal is acetylated (Mayer et al., 1989), and therefore other factors are also likely to be involved as well. Furthermore, the various combinations of possible amino acids in this position are unlikely to suffice for the variety of half-lives known to occur. Oxidation of

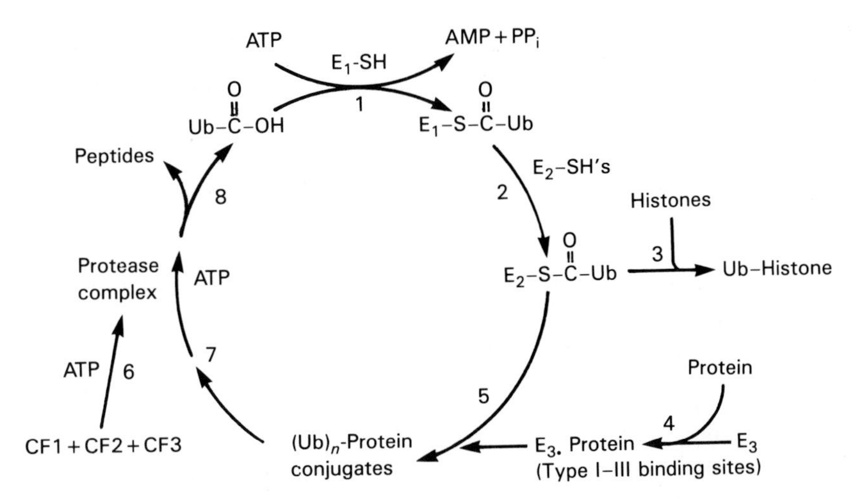

Fig. 9.13 Proposed sequence of events in ubiquitin-mediated protein breakdown. Ub, Ubiquitin; CF1, CF2, and CF3, conjugate-degrading factors 1–3, respectively; E_1, E_2, and E_3 are enzymes. Reprinted with permission from A. Hershko, *Journal of Biological Chemistry*, 263:15237–15240. Copyright © 1988 American Society of Biological Chemists, Inc.

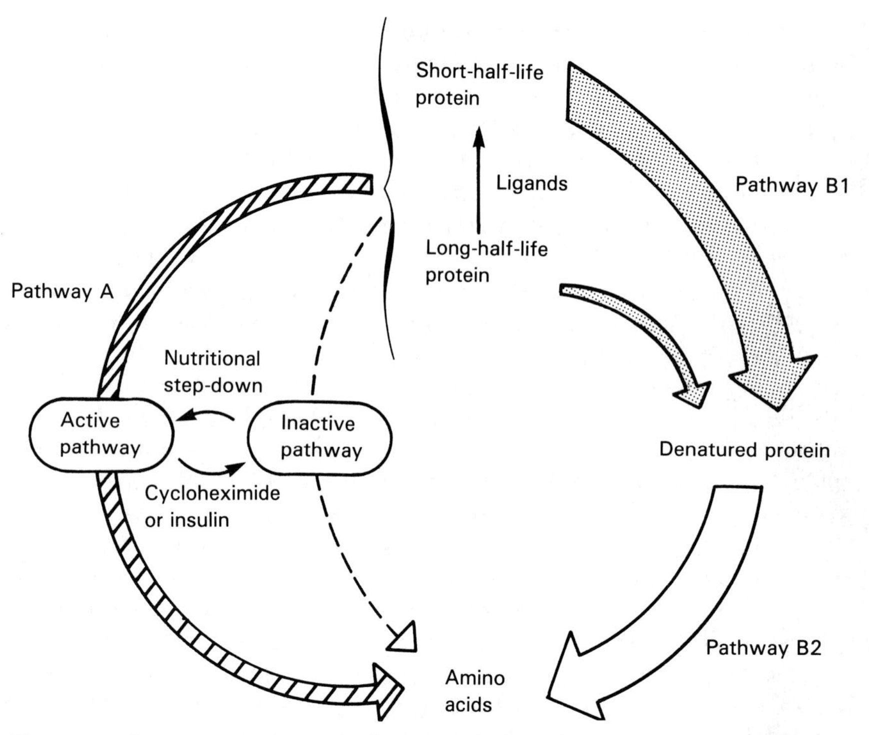

Fig. 9.14 Scheme for protein degradation. The various pathways and regulatory controls are discussed in text (Knowles and Ballard, 1976). A corresponds to the nonspecific pathway and B2 to the specific pathway. Reprinted with permission from *Biochemistry Journal* 156:609–617, copyright © 1976 by The Biochemical Society, London.

methionine residues and histidine (Fucci et al., 1983) and the presence of enrichment in four amino acids (Roger et al., 1986; Rechsteiner, 1990) have been implicated in degradation, although not necessarily by the ubiquitin pathway. Furthermore, a protease inhibitor has been identified (Speiser and Etlinger, 1983). Perhaps some interaction with the inhibitor could have a regulative effect. In addition, several ubiquitin-protein isopeptidases have been identified in animal cells (Mortimore and Mondon, 1970). These peptidases release protein and ubiquitin intact, providing other possible regulative routes.

Despite the important advances in our understanding of the proteolytic degradation pathways, the precise role of the ubiquitin system and the mechanism that imparts specificity is still not clear, and E_3s generally required for protein degradation are not always involved (Haas et al., 1990).

What protease system degrades proteins ligated to ubiquitin? One of the ATP-dependent proteases, a 1,300 kDa complex, the so-called 26S protease, degrades these proteins (Hough et al., 1987). A 20S multicatalytic protease complex that had been studied independently, the proteasome, is part of the 26S protease (Eytan et al., 1989; Driscoll and Goldberg, 1990). In yeast, mutational alteration of the proteasome caused accumulation of ubiquitin-protein conjugates (Heinemeyer et al., 1991). Deletion of the corresponding gene is lethal (Fujiwara et al., 1990; Heinemeyer et al., 1991), confirming its important role in the ubiquitin-protease pathway.

In summary, it seems that two different pathways operate in proteolysis. A generalized pathway involving all proteins operates through the action of lysosomes (A in Fig. 9.14; Knowles and Ballard, 1976). This pathway, which is continuously present, can be stimulated by major switches such as nutritional step-down or insulin. The proteins broken down in this pathway are those with a long half-life. Proteins with a short half-life, including defective proteins, are broken down in an alternative pathway that involves primarily covalent binding to ubiquitin (B1 and B2 in Fig. 9.14). There may be additional ubiquitin-independent pathways of degradation (McGuire et al., 1988). The mechanism conferring specificity to the degradation has still not been elucidated.

The ubiquitin-conjugating system has actually been implicated in many important and diverse regulatory functions other than protein degradation (see Chapter 7, III, D), ranging from a possible role in gene expression to a role in the regulation of enzyme activity, the assembly of ribosomal proteins, DNA repair, yeast sporulation, and regulation of the cell cycle (see Jentsch et al., 1990).

C. Regulation of Enzyme Synthesis

In eukaryotes, the control of the synthesis of specific proteins may depend on a variety of regulatory mechanisms much more complex than those occurring in bacteria. Animal cells are thought to contain at least 1000 times more genetic information than bacteria. This genetic information is contained in the chromosomes, where DNA is combined with proteins such as histones that may have a role in controlling transcription. The structure of the chromosome itself is likely to have an effect on its availability for transcription. The nuclear envelope may play a role in the release of mRNA and its modification after transcription. Furthermore, the mRNA used in transcription is constantly being broken down, although this process takes place much more slowly in animal cells than in prokaryotes. The possible mechanisms of control in eukaryotic cells are depicted in Fig. 9.15 (Walker, 1977).

Transcriptional Control The control of the production of an enzyme could be the result of a mechanism similar to that controlling the operons of *E. coli*. Transcription of the appropriate genes could be repressed or activated, depending on the signal received. There are many indications that the level of certain enzymes such as xanthine oxidase is regulated by a transcriptional mechanism. Xanthine oxidase is an enzyme with a molecular weight of about 300,000 and contains two flavin nucleotides. The enzyme acts in the pathway responsible for the degradation of purine rings and catalyzes the oxidation of xanthine to form uric acid as shown in Eq. (9.5).

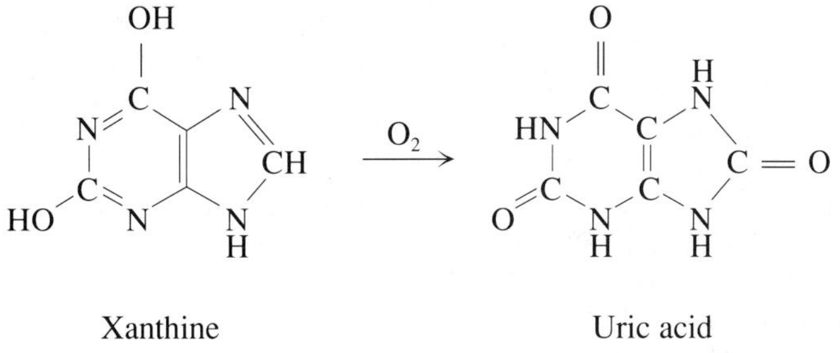

$$\text{Xanthine} \xrightarrow{O_2} \text{Uric acid} \qquad (9.5)$$

Xanthine   Uric acid

The uric acid produced can be followed spectrophotometrically by measuring the light absorbed at 292 nm, as done in the experiments represented in Fig. 9.16, Table 9.7, and Fig. 9.17. The optical density represented in the figures can be considered to be directly proportional to the

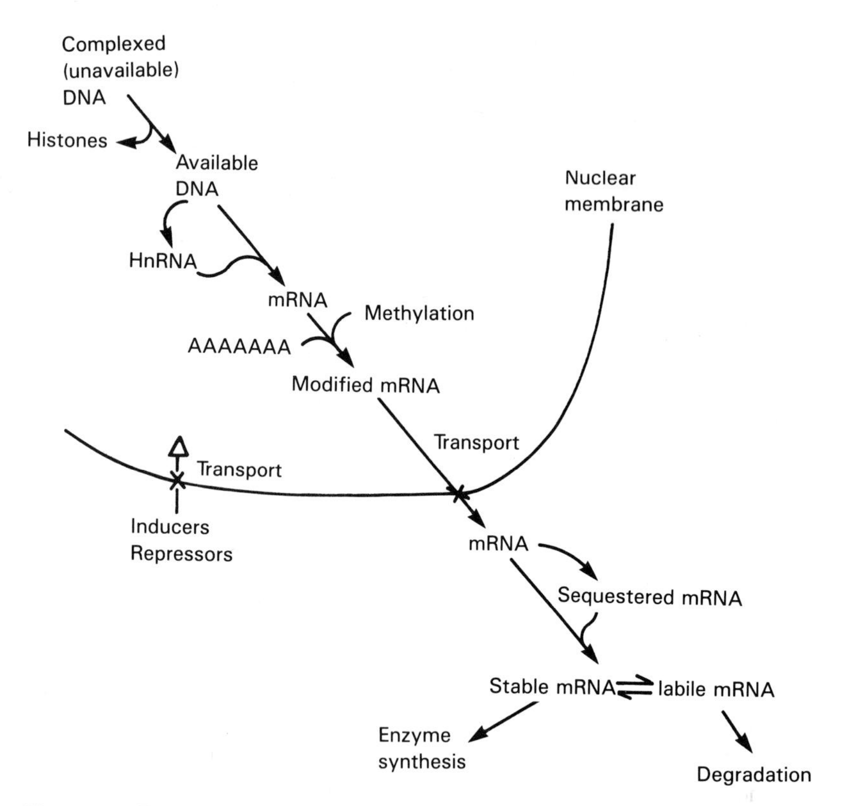

Fig. 9.15 Transcriptional and posttranscriptional stages in the production of translatable mRNA in animal cells (Walker, 1977). Reprinted with permission from *Essays in Biochemistry* 13:39–69, copyright © 1977 by The Biochemical Society, London.

amount of uric acid produced and is expressed per milligram of liver protein. In contrast, the results of Table 9.7 (Rowe and Wyngaarden, 1966) are expressed as specific activity of the enzyme, i.e., activity per milligram of purified enzyme.

In some species, uric acid is the final breakdown product of the pathway, and this product is excreted directly. In the rat, as in most other mammals, uric acid is further degraded to produce urea. In the rat, the xanthine oxidase activity of the liver is increased by a high-protein diet (e.g., 23% protein) and decreased by a low-protein diet (e.g., 8%). The effect of diet is shown in Fig. 9.16. Rats normally fed a 23% protein diet are switched at time zero to an 8% diet (first arrow). After 14 days on the low-protein diet, the animals are again placed on a high-protein diet (second arrow), and a subsequent increase in xanthine oxidase is observed. In Figs. 9.16 and 9.17, the xanthine oxidase activity is expressed in terms of optical density change (in relative units). The values represent the amount of uric acid produced in the test tube assay in relative units. As shown in Fig. 9.16 (first part of the curve), switching to a low-protein diet causes a 10-fold drop in activity. The return to a high-protein diet after a short lag dramatically increases the level of the enzyme. The enzyme activity after reintroduction of the high-protein diet is also shown in Fig. 9.17 (Rowe and Wyngaarden, 1966), curve 1.

The results of Table 9.7 (Rowe and Wyngaarden, 1966) show that the increase in enzyme activity does not correspond to activation of preexisting molecules. Rats were injected with [^{14}C]leucine 6 h before being sacrificed. Items 1 and 2 represent controls that were maintained on 23% protein for 14 days, and items 3 and 4 represent results obtained with rats on an 8%

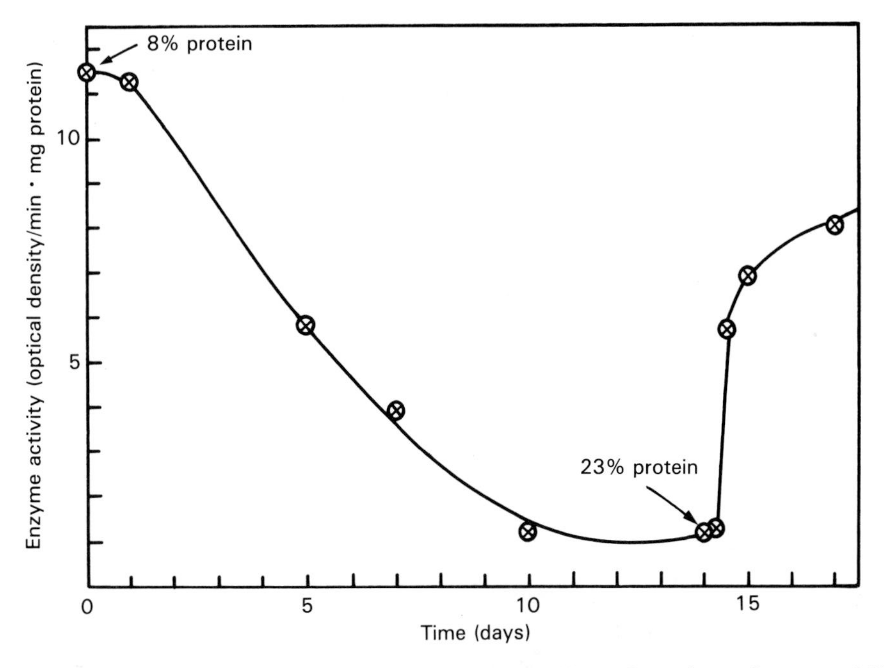

Fig. 9.16 Induction of xanthine oxidase in rat liver. Data from Rowe and Wyngaarden (1966), expressed per milligram of liver protein.

protein diet. Items 5 and 6 represent animals treated as in items 3 and 4 but subsequently fed the high-protein diet for 12 h. The activity per milligram of protein of the purified enzyme (specific activity of the enzyme) (column a) is the same regardless of physiological condition. Activation would be expected to increase the enzyme activity per milligram of enzyme. The incorporation of label into the enzyme molecule (column b) is the same for the control animals and the animals maintained on an 8% protein diet and with low xanthine oxidase activity. Hence

Table 9.7 Effect of Diet on Specific Enzyme Activity and ^{14}C-Labeled Amino Acid Incorporation of Xanthine Oxidase

Group	(a) Specific activity (optical density/min per mg enzyme)	(b) Enzyme specific radioactivity (counts/min per mg protein)
Control (23% protein diet)		
1	2000	170
2	3260	195
8% Protein diet		
3	2000	158
4	3180	200
8% Protein diet for 14 days—shifted to 23% protein diet for 12h		
5	2100	778
6	3245	900

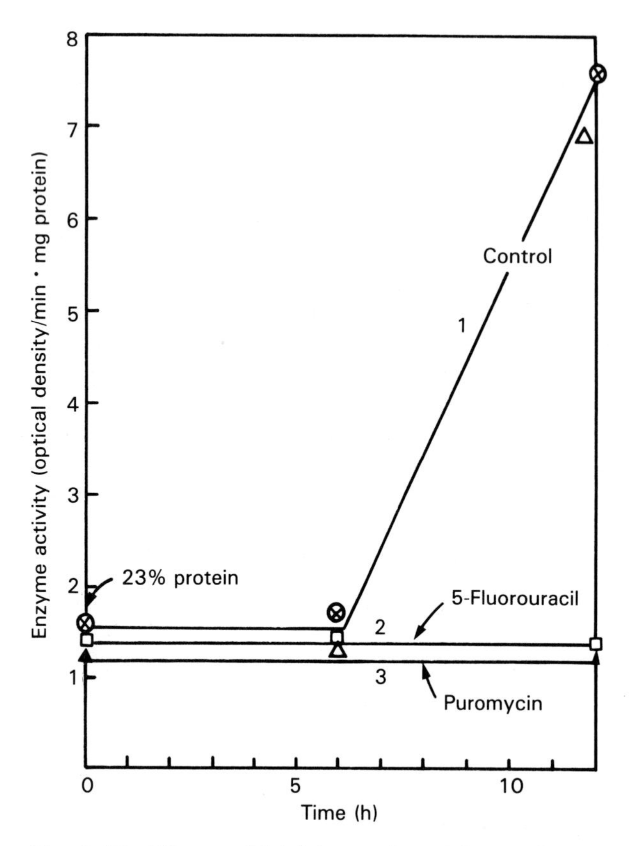

Fig. 9.17 Effects of inhibitors of protein synthesis or RNA in the induction of xanthine oxidase. Data from Rowe and Wyngaarden (1966), expressed per milligram of liver protein.

the decrease in the activity of xanthine oxidase probably corresponds to a higher rate of breakdown. In animals for which the low-protein diet was replaced after 14 days by 23% protein, there is an increase in radioactivity that at least roughly corresponds to the fivefold increase in enzymatic activity shown in Fig. 9.16. Therefore, the increase in enzyme activity appears to be the result of increased synthesis.

This interpretation is supported by the results shown in Fig. 9.17. The controls, on a 23% protein diet, exhibit an increase in xanthine oxidase activity (curve 1). Puromycin (curve 3), 5-fluorouracyl (curve 2), or actinomycin D (not shown) blocks the increase. Puromycin blocks protein synthesis, whereas 5-fluorouracyl and actinomycin D block RNA synthesis. The control is therefore transcriptional. Similar results were obtained for the induction of rat liver serine dehydratase (Jost et al., 1968).

Many other types of control are likely to involve a mechanism at the level of transcription. For example, the synthesis by guinea pig peritoneal cells of one of the serum proteins associated with the inflammatory response (protein C4) seems to depend on a factor that switches on the production of C4. The response can be inhibited by actinomycin D (Cahoun and Hatfield, 1975). Similarly, purine biosynthesis in hepatoma cells in tissue culture is repressed by adenine (Martin and Owen, 1972). The derepression is sensitive to both actinomycin D and cycloheximide. A role of transcription in the induction of tryptophan pyrrolase has been shown by the demonstration that the amount of mRNA coding for the enzyme parallels the amount of enzyme induced (Schutz et al., 1975). A number of hormonal controls are thought to be exerted at the transcriptional level, and some examples are shown in Table 9.8 (e.g., Neff et al., 1979). Enzyme production is not always affected by the mechanisms controlling transcription. In some cases the control is probably at a posttranscriptional level.

An increase in specific mRNA in response to a hormone has been clearly demonstrated in the induction of yolk proteins by estrogen in rooster liver. DNA synthesized by reverse transcriptase (cDNA) from purified specific mRNA can recognize the mRNA in *hybridization* tests, where nucleic acid strands combine by virtue of their complementariness (see Chapter 1). The mRNA used as a template is not difficult to isolate by standard procedures provided that it is present in the tissue in sufficient amounts. In vitro translation systems can confirm the nature of the mRNA isolated. After production of the cDNA, the template RNA can be removed by degradation in alkali, and DNA polymerase can be used to replicate the missing complementary strand of the DNA duplex. The duplex inserted into a vector, either a bacterial plasmid or a phage, can reproduce multiple copies, or *clones*, in a bacterial host. Hybridization of the cDNA bound to a filter can then be used to recognize the appropriate mRNA.

Administration of estrogen induces the production of egg yolk protein. Figure 9.18 (Wiskocil et al., 1980) shows the various specific mRNAs as a function of time in the liver of roosters after a single injection of estrogen. The mRNA for albumin remains relatively unchanged (curve 1). However, the mRNAs for the yolk proteins apoVLDLII (curve 2) and vitellogenin (curve 3) are induced. Interestingly, after the initial synthesis they are also rapidly degraded, suggesting that both transcription and degradation of the mRNA play a regulatory role.

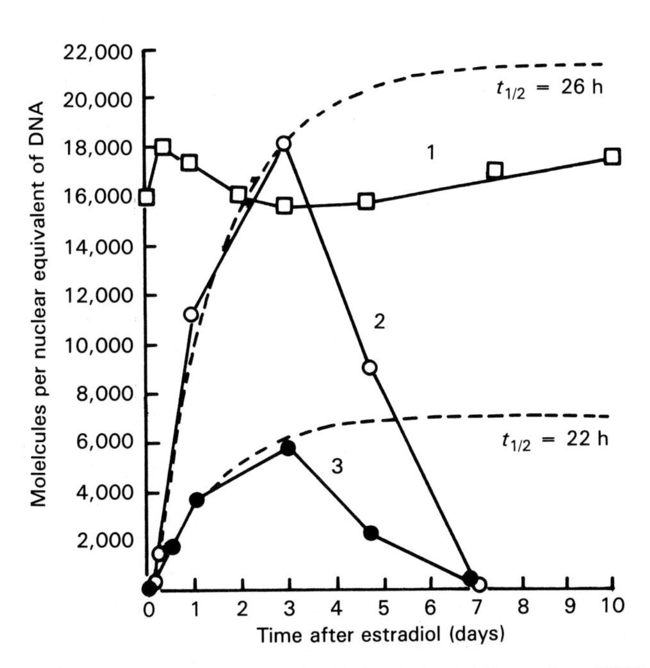

Fig. 9.18 Levels of apoVLDLII mRNA, vitellogenin mRNA, and serum albumin mRNA during 10 days after primary stimulation with 17β-estradiol. Absolute levels of serum albumin mRNA (□), apoVLDLII mRNA (○), and vitellogenin mRNA (●) after injection of hormone. The data are expressed as molecules of mRNA per nuclear equivalent of DNA. The theoretical accumulation curves (---) were calculated from data of Wiskocil et al., with permission, *Proceedings of the National Academy of Sciences,* 77:4474–4478, 1980.

Table 9.8 Hormonal Control of Macromolecular Synthesis[a]

Hormone	Experimental system or target organ	Evidence for level of control: Transcriptional	Evidence for level of control: Translational	References
Protein hormones				
Growth	Rat tissues (liver, muscle and others)	RNA ↑ (including mRNA), RNA polymerase ↑	In vitro stimulation of protein synthesis	*b*
Insulin	Muscle	RNA ↑, RNA polymerase ↑	Actinomycin D inability to block some actions, RNA and ribosome interaction aided	*b*
Peptide hormones				
ACTH	Adrenals	Not clear		*b*
TSH	Thyroid	RNA ↑		*b*
LH	Ovaries	Not clear		*b*
Thyroxine	Rat and amphibian liver	RNA ↑, including mRNA	Amino acid incorporation before increased transcription and with administration to cell-free system	*b*
Steroid				
Androgen	Accessory sex tissue	RNA ↑, RNA polymerase ↑, ribosomes ↑		*b*
Estrogen	Uterus	RNA ↑, actinomycin D inhibition, increased template activity of chromatin, appropriate RNA mimics the effect of hormone, RNA polymerase ↑		*b*
	Chick oviduct	New class of RNA (hybridization) RNA polymerase ↑		*b*
Corticosteroids	Embryonic chick retina	Inhibitors of RNA or protein synthesis block effects, new mRNA appears for one induced enzyme	New mRNA appears for one induced enzyme	*c, d, e*
	Rat kidney cortex	Appearance of new mRNA		*f*
	Kidney cortex	RNA ↑		*f*
	Chick embryo	Specific RNA ↑		*b*
	Retina, liver	Increased template activity,		*b*
	Thymus nuclei	hormone blocks mRNA synthesis		*g*
	Erythroid cells	Actinomycin D inhibition of hemoglobin synthesis	In vitro stimulation of cytoplasmic extract	
Ecdysone	*Calliphora* larva	RNA ↑ in chromosomal loci		*b*
Progesterone	Chick oviduct	Actinomycin D inhibition of avidin synthesis, RNA polymerase ↑, new RNA made (hybridization)		*b*

[a]Upward arrows indicate increases.
[b]O'Malley, B. W. *Trans. N. Y. Acad. Sci. [2]* 31:478–503 (1969).
[c]Moscona, A. A., Moscona, M. H., and Saenz, N. *Proc. Natl. Acad. Sci. U.S.A.* 61:160–167 (1968).
[d]Reif-Lehrer, L., and Amos, H. *Biochem. J.* 106:425–430 (1968).
[e]Schwartz, R. J. *Nature (London) New Biol.* 237:121–125 (1972).
[f]Congote, L. F., and Trachewsky, D. *Biochem. Biophys. Res. Commun.* 46:957–971 (1972).
[g]Abraham, A. D., and Sekeris, C. E. *Biochem. Biophys. Res. Commun.* 247:562–(1971).

Posttranscriptional Control Transcriptional control mechanisms seem to play a prominent role in the regulation of enzyme synthesis. As outlined in Fig. 9.13, a number of regulative events could occur after the transcription. Several experiments indicate that the specific degradation of mRNA is significantly involved in the regulation (Styles et al., 1976). Conceivably, the rate of synthesis of an enzyme could also be regulated by changing the rate at which the translational events take place. This section addresses the various possible posttranscriptional control systems.

The induction of tyrosine transaminase in rat liver or rat liver tumor cells (HTC cells) by adrenal steroid hormones or their analogs (e.g., dexamethasone phosphate) suggests that the control is exerted after the formation of the appropriate mRNA.

$$\left.\begin{array}{c} HO-C_6H_4-CH_2CH(NH_2)COOH \\ \text{Tyrosine} \\ + \\ HOOC-(CH_2)_2-CO-COOH \\ \alpha\text{-Ketoglutaric acid} \end{array}\right\} \longrightarrow \left\{\begin{array}{c} HO-C_6H_4-CH_2C(=O)COOH \\ \beta\text{-Hydroxyphenyl pyruvate} \\ \\ HOOC-CH(NH_2)-CH_2-COOH \\ \text{Glutamic acid} \end{array}\right. \quad (9.6)$$

The transamination of tyrosine is represented in Eq. (9.6). In this reaction the α amino group of the amino acid is transferred to α-ketoglutaric acid with the formation of β-hydroxyphenyl pyruvate and glutamic acid. The β-hydroxyphenyl pyruvate can be broken down by several oxidative steps that eventually lead to the formation of acetoacetate, which can be metabolized by the mitochondria after the appropriate activation reaction. The regulation of tyrosine transaminase may shed some light on the physiological mode of action of the adrenal steroid hormones, as mentioned for tryptophan pyrrolase. The experiment illustrated in Fig. 9.19 (Thompson et al., 1966) shows that induction of the enzyme in HTC cells is triggered by the presence of dexamethasone phosphate. The inducer brings about a 10-fold increase in enzyme activity after a lag period of 1½ to 2 h. The induction occurs in the presence of actinomycin D (Tomkins et al., 1969). Immunological precipitation (as described in Fig. 9.10 for the tryptophan pyrrolase assay) demonstrates that the induction corresponds to actual formation of the enzyme and not to activation of a preexisting molecule (Granner et al., 1968). The degradation of the enzyme, which normally takes 3 to 7 h, seems to be unchanged by administration of the drug (Auricchio et al., 1969; Martin et al., 1969).

The data presently available permit the formulation of a model, shown in Fig. 9.20, which has some similarities to that presented for the *lac* operon of *E. coli*. However, it also has a number of significant differences. The enzyme is coded by gene G^S and the repressor by G^R

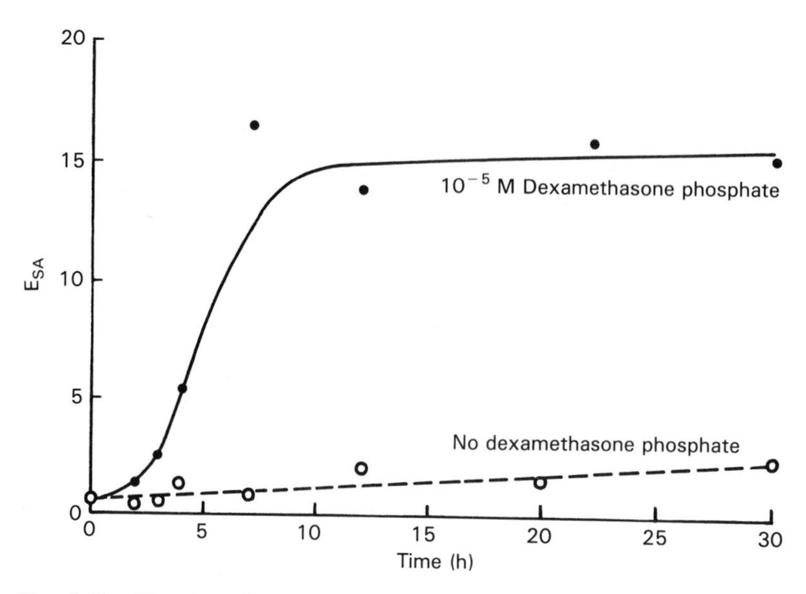

Fig. 9.19 Kinetics of induction of tyrosine transaminase activity in HTC cells at 37°C. E_{SA} refers to tyrosine transaminase specific activity. Reprinted with permission from E. B. Thompson, et al., *Proceedings of the National Academy of Sciences,* 56:296–303, 1966.

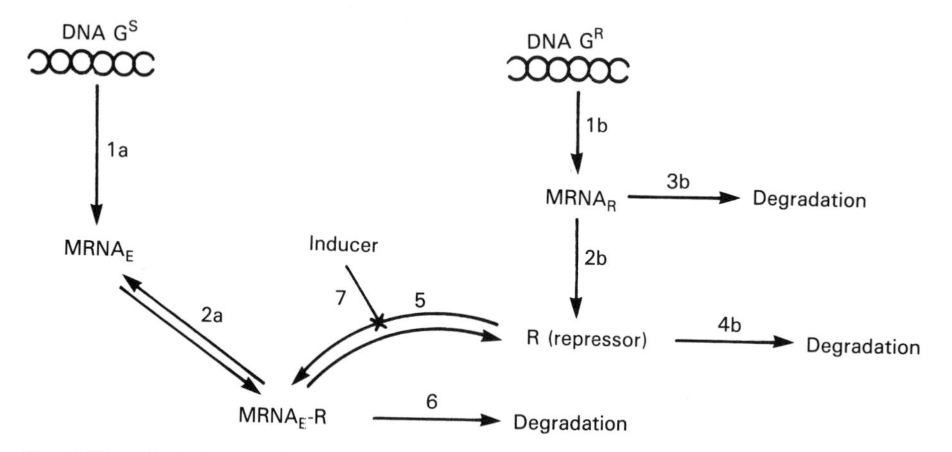

Fig. 9.20 Model to explain the observed results in the regulation of tyrosine transaminase production (see text). Reprinted with permission from E. B. Thompson, et al., *Proceedings of the National Academy of Sciences,* 56:296–303, 1966.

(Fig. 9.20). Both genes are transcribed to produce the corresponding mRNAs ($mRNA_E$ and $mRNA_R$, respectively) in steps 1a and 1b of Fig. 9.20. The two mRNAs can be translated to produce enzyme (E) or repressor (R) (steps 2a and 2b). The repressor acts by binding to the mRNA (reaction 5), thereby making it unavailable for production of the enzyme. In this model, the inducer (adrenal steroid) can complex with a protein repressor, thus removing its inhibitory ability to bind to the mRNA coding for the tyrosine transaminase (step 7). Normally, the mRNA-repressor combination is degraded more rapidly than the free mRNA. The model assumes that the repressor and the mRNA corresponding to the repressor are labile (reactions 3b and 4b in Fig. 9.20).

These features have been proposed on the basis of several experiments. The addition of cycloheximide to the HTC cells inhibits protein synthesis by about 97%. Naturally, this technique blocks induction, since the enzyme cannot be formed (Permutt and Kipnis, 1972). However, on washing the cells and adding the inducer, the system is again capable of responding. If the inducer is added after washing without pretreatment with cycloheximide, induction occurs after a lag period of 1.5 to 2 h (Fig. 9.21a, curve 2) (Peterkovsky and Tomkins, 1968). The washing itself has no effect on the induction, as shown in curve 1, where no inducer has been added. After the cycloheximide is washed off in the presence of inducer, the cells synthesize the enzyme (Fig. 9.21b). When the inducer is added after washing, the customary lag period occurs (curve 2). However, when the inducer and cycloheximide are present from the beginning, the synthesis

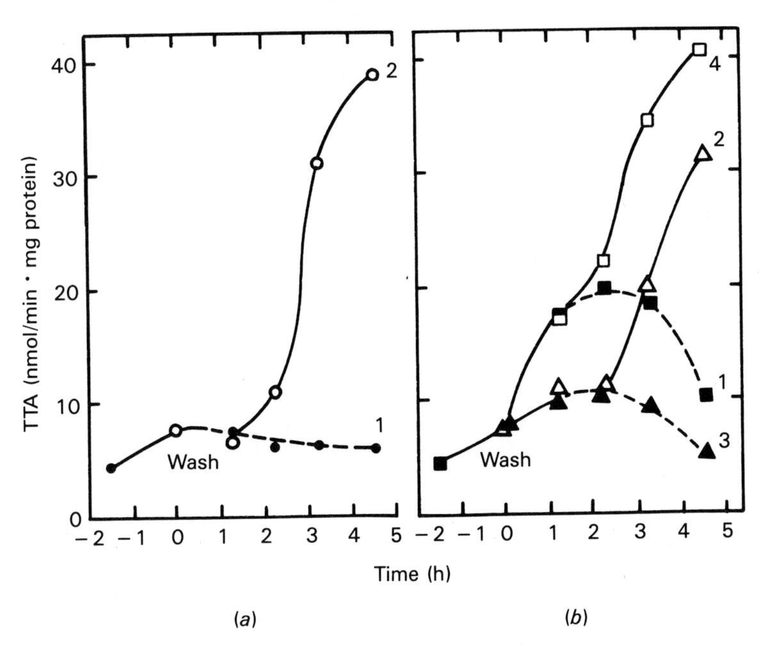

Fig. 9.21 Level of tyrosine transaminase (TTA) activity after preincubation of HTC cells with or without various additions. Cells were preincubated for 1.5 h, washed with medium, and reincubated. (*a*) Curve 1 (●), no preincubation addition, no reincubation addition; curve 2 (○), no preincubation addition, Dex addition at reincubation. (*b*) Curve 1 (■), preincubation CH + Dex addition, no reincubation addition; curve 2 (△), preincubation CH addition, Dex addition reincubation; curve 3 (▲), preincubation CH addition, no reincubation addition, curve 4 (□), preincubation CH + Dex addition, Dex addition reincubation. Dex, dexamethasone phosphate; CH, cycloheximide. Reprinted with permission from B. Peterkofsky and G. M. Tomkins, *Proceedings of the National Academy of Sciences,* 60:222–228, 1968.

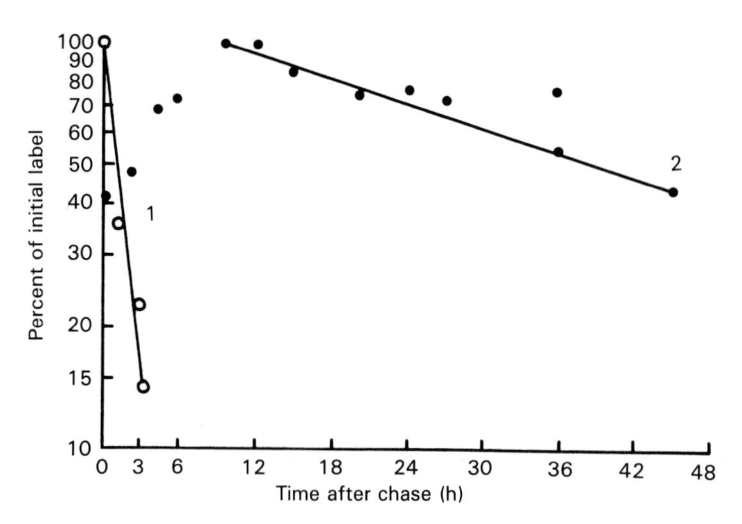

Fig. 9.22 Degradation of casein mRNA. Both control (curve 1, ○) and experimental (curve 2, ●) runs were in the presence of insulin and hydrocortisone. In the absence of prolactin the half-life corresponds to 1.1 h and in its presence to 28.5 h. Reprinted with permission from Guyette et al. (1979); copyright 1979 by Cell Press.

of the enzyme no longer involves a lag period; it begins without delay (curve 4). In fact, synthesis occurs when the inducer, present in the original incubation medium together with cycloheximide, is washed away and is not reintroduced (curve 1). The cycloheximide alone has no significant effect on induction (curve 3). The accumulation of the capacity to produce the enzyme in the absence of protein synthesis suggests that it is the result of the accumulation of mRNA.

A role of mRNA degradation in posttranscriptional regulation has been shown directly in a number of cases, e.g., for the casein mRNA of breast cells.

Casein is the major protein in milk secreted by breast cells. The secretion is hormonally controlled in part by prolactin. Cultured breast cells respond to the presence of prolactin by increasing the casein mRNA 100-fold. However, the nuclei of breast cells can increase the synthesis of casein mRNA only threefold in response to prolactin—apparently, the major effect is reduction of the degradation of mRNA. Results of an experiment showing this are presented in Fig. 9.22 (Guyette et al., 1979).

In this experiment cultured breast cells were pulse-chased with ^{3}H-labeled uridine in the presence (curve 1) and absence of prolactin (curve 2). The casein mRNA (ordinate) was estimated from the radioactivity hybridized to the corresponding cDNA probes (see Chapter 1). The points correspond to cell samples withdrawn at different times. The results show that prolactin blocks the degradation of casein mRNA.

Various hormones alter the half-life of specific mRNAs, indicating that this control mechanism is common. Some of these hormones and the corresponding mRNA half-lives are shown in Table 9.9 (Shapiro and Brock, 1985).

In some cases, the mechanism of posttranscriptional regulation involves the synthesis of protein. There are indications that such controls can take place in other systems. The regulatory effect of the hormone insulin depends in part on the synthesis of new protein. This effect, however, is not inhibited by actinomycin D (Eboue-Bonis et al., 1963; Tomkins et al., 1969). When muscle ribosome subunits of insulin-deficient rats are reassociated, they synthesize polypeptides that use polyuridylic RNA less efficiently than subunits isolated from normal rats (Martin and Wool, 1968). A similar translational control probably also occurs with thyroxine.

Table 9.9 Biological Systems That Exhibit Regulation of mRNA Stability

mRNA	Tissue	Regulatory signal	mRNA Half-life[a] + Effector	mRNA Half-life[a] − Effector
Vitellogenin	*Xenopus* liver	Estrogen	+E; 500 h	−E; 16 h
Vitellogenin	Rooster liver	Estrogen	+E; ~24 h	−E; $<$3 h
Ovalbumin, conalbumin	Hen oviduct	Estrogen, progesterone	+E; ~24 h	−E; 2–5 h
Casein	Rat mammary gland	Prolactin	+Pro; 92 h	−Pro; 5 h
Prostatic steriod-binding protein	Rat ventral prostate	Androgen	$\Delta t_{1/2} \sim 30\times$	
Lactate dehydrogenase A subunit	Rat C6 glioma cells	cAMP, isoproterenol, dibutyl cAMP	+Ipt; 2.5 h	−Ipt; 45 min
β-Interferon	Human fibroblasts	Poly(I·C) vs. poly (I·C) + cycloheximide or Newcastle's virus	(I·C) +ChX or (I·C) + N.V., $t_{1/2} > 12$ h	(I·C) $t_{1/2} < 30$ min
Histones	HeLa cells	DNA replication	During replication 11 h after 13 min	
Histones	Yeast	DNA replication	During replication ~15 min after ~5 min	
Adenovirus 1A (9S), 1B (14S)	HeLa cells	Early/late infection	Late, 60–100 min; early, 6–10 min	
L3 (ribosomal protein)	Yeast	mRNA overproduction	$\Delta t_{1/2} \sim 2\times$	
γ integrase (int)	γ infected *E. coli*	Early/late infection (terminator read-through)	$\Delta t_{1/2} > 10\times$	

Source: D. J. Shapiro and M. L. Brock, *Biochem. Action Hormones,* 12:139–172, 1985.

[a]Abbreviations: E, estrogen; pro, prolactin; Ipt, isoproterenol; IC, poly(IC); ChX, cycloheximiole; NV, Newcastle virus.

This hormone can stimulate reticulocyte extracts to incorporate amino acids, and this effect for some reason requires the presence of mitochondria (Krause and Sokologg, 1967).

Molecular Events in Transcriptional and Posttranscriptional Regulation Knowledge of the molecular details underlying regulation of the production of macromolecules is still incomplete. We have seen that the control of transcription is of major significance. Analogy with the prokaryotic systems suggests that interactions involving the binding of specific protein molecules to DNA sites play an important role. Although this picture is correct at least in outline for eukaryotes as well, the study of the regulation of specific genes has revealed a far more complex system, in part due to the complexity of the transcriptional system itself (Dynan and Tjian, 1985).

The transcriptional unit is much larger than the final mRNA molecule, and several steps are necessary to form mature mRNA. Sequences that are not part of the message (*introns*) are removed and the RNA pieces that make up the message (the *exons*) are spliced (Gilbert, 1978). Before this happens, however, a methylated guanylate residue (*cap*) is added to the 5′ end of the transcript (Shatkin, 1976) and adenylate residues are added to a specific site at the 3′ end (Brawerman, 1981). The 5′ cap is thought to prevent degradation and to be involved in the initiation of translation. In some cases the original transcript contains more than one kind of mRNA and some process must occur to separate them.

Transcription requires the complex interaction of specific sites of the DNA, protein molecules (Weisbrod, 1982), and RNA polymerase. In eukaryotes the transcription of genes encoding proteins is carried out by polymerase II, a different enzyme from the two polymerases involved in the transcription of the precursor of ribosomal RNA (RNA polymerase I) or of small RNAs such as the tRNAs and 5S ribosomal RNA (RNA polymerase III). Apparently, initiation of transcription involves specific regions on the DNA 100 to 300 base pairs (bp) from the transcribed region. These most frequently include the 25- to 30-bp consensus sequence TATAAA (the TATA box). Further upstream from the TATA box other *promoter* regions are present. A separate *enhancer* sequence is present in the same DNA molecule but as far as 1000 bp from the transcribed region (Dynan and Tjian, 1985).

Interactions involving these elements and polymerase II, i.e., the initiation events, are thought to be most significant in the control of transcription. A DNA region responsible for the termination of transcription (Hofer et al., 1982) may also be involved in regulation. RNA polymerase II does not recognize promoter regions in vitro unless additional *regulator proteins* are present. These proteins may be common for all promoters, and others may be promoter specific. Many of the promoter-specific proteins have been shown to bind to the promoter DNA sequences.

In addition to regulator proteins, the conformation of the DNA-histone complexes and the state of the DNA itself are thought to play a role in the regulation of transcription. Both of these ideas have suggested experiments that have provided a good deal of insight on the mechanism of transcriptional regulation.

Regulator Proteins Two examples of regulation by regulator proteins are the transcriptional control of galactose metabolism in yeast and the control of a gene by a glucocorticoid receptor.

In yeast, galactose metabolism requires three enzymes that convert galactose to glucose-6-phosphate [Eq. (9.7)]. These enzymes are coded in genes present in chromosome II. The galactose transporter gene is in chromosome XII, and a gene responsible for the enzyme involved in the conversion of melibiose into galactose is in chromosome U_1. Production of all these enzymes is induced by a metabolic product of galactose. The induction requires the presence of gene *gal* 4 in chromosome X. The *gal* 4 gene has been cloned and the protein coded by it, Gal 4, has been produced in quantity (see Chapter 1). Gal 4 binds 150 to 250 bp upstream from the initiation site for the transcription of each of the galactose metabolism genes. Gal 4 is therefore a positive effector that favors transcription and acts at several sites. Negative effectors are also present, including one that regulates the transcription of *gal* 4 (Broach, 1979; Hopper et al., 1978).

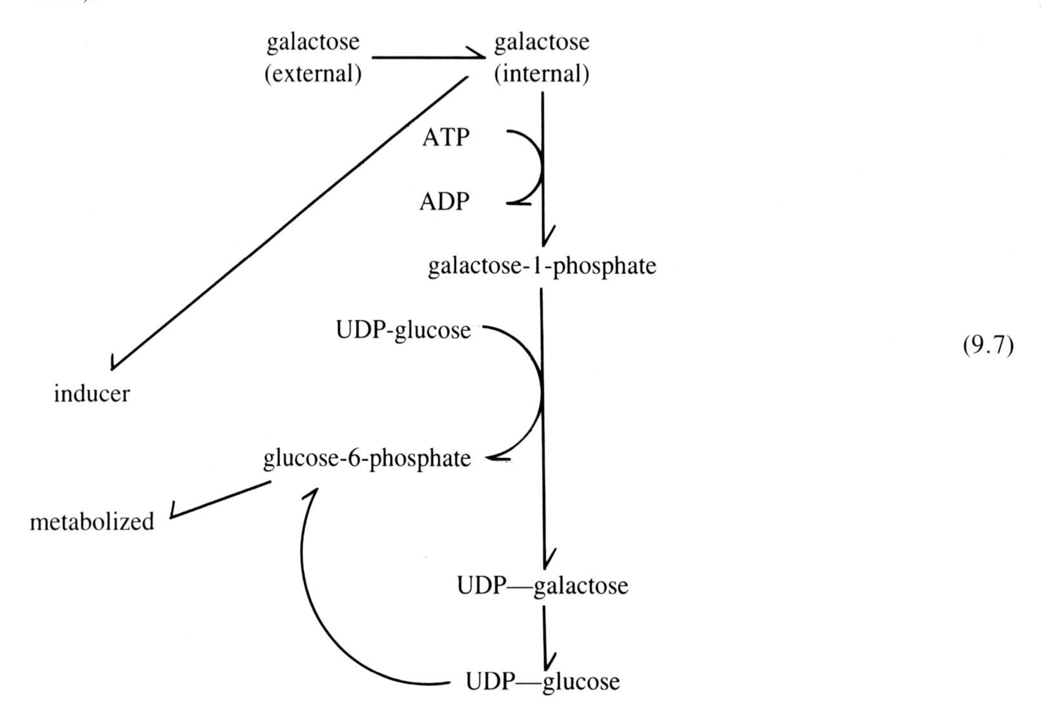

(9.7)

Steroids act on responsive cells by binding to receptor proteins (Gardner and Tomkins, 1969). The receptors for various growth factors and hormones discussed in Chapter 6 are integral membrane proteins with binding groups at the cell surface. In contrast, steroid receptors are present in the cytoplasm, since the lipid-soluble steroid hormones can enter the cell rapidly. Labeling of hormones with radioactive isotopes has shown that the hormone-receptor complex is transferred to the nucleus. Mouse mammary tumor cells are responsive to glucocorticoid. A piece of the DNA of *mouse mammary tumor virus* (MMTV) corresponds to the promoter region of the steroid-activated gene. This piece binds to the glucocorticoid-receptor complex specifically, as shown by electron micrographs of specific fragments of the MMTV DNA and the protective effects of the binding of the receptor from digestion by DNase (Payvar et al., 1983).

Chromosomal Structure and Transcription Changes in the arrangement of the components of the chromosomes, DNA and histones, have been implicated in the control of transcription. However, the precise nature of the changes and their effects in transcription are not always clear.

Active genes are more sensitive to DNase than inactive genes. This observation suggests that the inactive region may bind to histones that shield them from the digestion (Stalder et al., 1980).

Inactive genes have also been found to be more methylated than active genes (Dynan and Tjian, 1985). The C present in 5′CG3′ regions is frequently methylated immediately after DNA replication. The presence of methylated sequences can be recognized experimentally by using restriction enzymes, some of which are blocked by the methylation (Weintraub et al., 1981). The opposite is also true: genes are activated in cells grown in the presence of 5′-azacytidine, an analog of cytidine that cannot be methylated.

Multiplicity of Controls Regulative effects need not occur through a single mechanism. For example, a hormone such as insulin can increase the permeability of muscle to glucose or amino acids. In addition to its effect on translation mentioned before, the hormone induces an increase in RNA synthesis, which indicates a possible effect on the transcriptional mechanism.

The control can also be exerted in a much more complex manner, as it is with the regulon of bacteria (see Section I,D). For example, the target organ of a hormone may be specific. The production of one macromolecule rather than another will be affected because the cells in question manufacture only a limited number of proteins. The effect may be indirect and nonspecific (e.g., a general increase in metabolism); nevertheless, the end result can be an increase in the synthesis of a single protein by the specialized cells. In addition, a control mechanism triggered by a hormone may induce the multiplication of cells. Although the hormone may not have a direct effect on protein synthesis per se, an increase in the rate of mitosis will eventually lead to the production of more protein molecules as more cells are involved in the synthesis.

Interestingly, the production of some hormones may in turn be under the control of a transcriptional or translational mechanism, responding to the presence of a simple metabolite whose metabolism is controlled by the hormone in question. For example, the biosynthesis of insulin is increased by an excess of glucose and this increase is blocked by actinomycin D (Permutt and Kipnis, 1972).

SUGGESTED READING

Protein and RNA Degradation

Ciechanover, A., and Gonen, H. (1990) The ubiquitin-mediated proteolytic pathway: enzymology and mechanisms of recognition of the proteolytic substrates. *Seminars in Cell Biol.* 1:415–422.

Finley, D., and Chau, V. (1991) Ubiquitination. *Annu. Rev. Cell Biol.* 7:25–69.

Hargrove, J. L., and Schmidt, F. H. (1989) The role of mRNA and protein stability in gene expression. *FASEB J.* 3:2360–2370.

Hershko, A. (1988) Ubiquitin-mediated protein degradation. *J. Biol. Chem.* 263:15237–15240.

Olson, T. C., and Dice, J. F. (1989) Regulation of protein degradation rates in eukaryotes. *Curr. Opin. Cell Biol.* 1:1194–1200.

Prokaryotic Transcriptional Controls

Adhya, S., and Garges, S. (1990) Positive control. Minireview. *J. Biol. Chem.* 265:10797–10800.

Yanofsky, C. (1981) Attenuation in the control of expression of bacterial operons. *Nature (London)* 289:751–757.

Yanofsky, C. (1988) Transcription attenuation. Minireview. *J. Biol. Chem.* 263:609–612.

Eukaryotic Transcriptional Controls

Darnell, J. E., Jr. (1982) Variety in the level of gene control in eukaryotic cells. *Nature (London)* 297:365–371.

Dynan, W. S., and Tjian, R. (1985) Control of eukaryotic messenger RNA synthesis by sequence-specific DNA-binding proteins. *Nature (London)* 316:774–778.

Evans, R. M. (1988) The steroid and thyroid hormone receptor superfamily. *Science* 240:889–894.

Guarente, L. (1988) UASs and enhancers: common mechanism of transcriptional activation in yeast and mammals. *Cell* 52:303–305.

Jones, N. C., Rigby, P. W. J., and Ziff, E. B. (1988) *Trans*-acting protein factors and the regulation of eukaryotic transcription: lessons from studies on DNA tumor viruses. *Genes Dev.* 2:267–281.

Sollner-Webb, B. (1988) Surprises in polymerase III transcription. *Cell* 52:153–154.

Watson, J. D., Gilman, M., Witkowski, J., and Zoller, M. (1992) Chapter 4 of *Recombinant DNA*, Second Edition, Scientific American Books (distributed by W. H. Freeman and Co., New York).

REFERENCES

Auricchio, F., Martin, D., Jr., and Tomkins, G. (1969) Control of degradation and synthesis of induced tyrosine aminotransferase studied in hepatoma cells in culture. *Nature (London)* 224:806–808.

Bachman, B. J., and Low, K. B. (1980) Linkage map of *Escherichia coli* K-12 edition 6. *Microbiol. Rev.* 44:1–56.

Bachmair, A., Finley, D., and Varshavsky, A. (1986) In vivo half-life of a protein is a function of its amino-terminal residue. *Science* 234:179–186.

Ballard, F. J. (1977) Intracellular protein degradation. *Essays Biochem.* 13:1–37.

Barrett, A. J. (1980) The many forms and functions of cellular proteinases. *Fed. Proc.* 39:9–14.

Bartel, B., Wunning, I., and Varshavsky, A. (1990) The recognition component of the N-end rule pathway. *EMBO J.* 9:3179–3189.

Blundell, M., Craig, E., and Kennell, D. (1972) Decay rates of different mRNA's in *E. coli* and models of decay. *Nature (London) New Biol* 238:46–49.

Brawerman, G. (1981) The role of the poly(A) sequence in messenger RNA. *CRC Crit. Rev. Biochem.* 10:1–38.

Broach, J. R. (1979) Galactose regulation in *Saccharomyces cerevisiae*. The enzymes encoded by the Gal7, 10, 1 cluster are co-ordinately controlled and separately translated. *J. Mol. Biol.* 131:41–53.

Cahoun, D. H., and Hatfield, G. W. (1975) Autoregulation of gene expression. *Annu. Rev. Microbiol.* 29:275–299.

Ciechanover, A., Hod, Y., and Hershko, A. (1978) A heat-stable polypeptide component of an ATP-dependent proteolytic system from reticulocytes. *Biochem. Biophys. Res. Commun.* 81:1100–1105.

Ciechanover, A., Elias, S., Heller, H., Ferber, S., and Hershko, A. (1980a) Characterization of the heat-stable polypeptide of the ATP-dependent proteolytic system from reticulocytes. *J. Biol. Chem.* 255:7525–7528.

Ciechanover, A., Heller, H., Elias, S., Haas, A. L., and Hershko, A. (1980b) ATP-dependent conjugation of reticulocyte proteins with the polypeptide required for protein degradation. *Proc. Natl. Acad. Sci. U.S.A.* 77:1365–1368.

Ciechanover, A., Finley, D., and Varshavsky, A. (1984) Ubiquitin dependence of selective protein degradation demonstrated in the mammalian cell cycle mutant ts85. *Cell* 37:57–66.

Ciechanover, A., and Schwartz, A. L. (1989) How are substrates recognized by the ubiquitin-mediated proteolytic system? *Trends Biochem. Sci.* 14:483–488.

Cohn, M. (1957) Contributions of studies on the β-galactosidase of *Escherichia coli* to our understanding of enzyme synthesis. *Bacteriol. Rev.* 21:140–168.

Dean, D., Yates, J. C., and Nomura, M. (1981) Identification of ribosomal protein S7 as a repressor of translation within the *str* operon of *E.coli*. *Cell* 24:413–419.

Dice, J. F., Terlecky, S. R., Chiang, H.-L., Olson, T. S., Isenman, L. D., Short-Russell, S. R., Freundlieb, S., and Terlecky, L. J. (1990) A selective pathway for degradation of cytosolic proteins by lysosomes. *Seminars in Cell Biol.* 1:449–455.

Driscoll, J., and Goldberg, A. L. (1990) The proteasome multicatalytic protease is a component of the 1500-kDa proteolytic complex which degrades ubiquitin-conjugated proteins. *J. Biol. Chem.* 265:4789–4792.

Dynan, W. S., and Tjian, R. (1985) Control of eukaryotic messenger RNA synthesis by sequence-specific DNA-binding proteins. *Nature (London)* 316:774–778.

Eboue-Bonis, D., Chambaut, A. M., Volfin, P., and Clauser, H. (1963) Action of insulin on the isolated rat diaphragm in the presence of actinomycin D and puromycin. *Nature (London)* 199:1183–1184.

Engelsberg, E., Irr, J., Power, J., and Lee, N. (1965) Positive control of enzyme synthesis by gene *C* in the L-arabinose system. *J. Bacteriol.* 90:946–957.

Eytan, E., Ganoth, D., Armon, T., and Hershko, A. (1989) ATP-dependent incorporation of 20 S protease into 26 S complex that degrades proteins conjugated to ubiquitin. *Proc. Natl. Acad. Sci.* 86:7751–7755.

Ferguson, A. R., and Sims, A. P. (1974) The regulation of glutamine metabolism in *Candida utilis:* the inactivation of glutamine synthetase. *J. Gen. Microbiol.* 80:173–185.

Finley, D., Ciechanover, A., and Varshavsky, A. (1984) Thermolability of ubiquitin-activating enzyme from the mammalian cell cycle mutant ts85. *Cell* 37:43.

Fucci, L., Oliver, C. N., Coon, N. J., and Stadtman, E. R. (1983) Inactivation of key metabolic enzymes by mixed-function oxidation reactions: possible implications in protein turnover and aging. *Proc. Natl. Acad. Sci. U.S.A.* 80:1521–1528.

Fujiwara, T., Tanaka, K., Orino, E., Yoshimura, T., Kumatori, A., Tamura, T., Hung, C. H., Nakai, T., Yamaguchi, K., Sin, S., Kakizuka, A., Nakanishi, A., and Ichihara, A. (1990) Proteasomes are essential for yeast proliferation. cDNA cloning and gene disruption of two major subunits. *J. Biol. Chem.* 265:16604–16613.

Gardner, R. S., and Tomkins, G. M. (1969) Steroid hormone binding to a macromolecule from hepatoma tissue culture cells. *J. Biol. Chem.* 244:4761–4767.

Gausing, K. (1974) Ribosomal protein in *E. coli:* rate of synthesis and pool size at different growth rates. *Mol. Gen. Genet.* 129:61–75.

Gilbert, W. (1978) Why genes in pieces? *Nature (London)* 271:501.

Gilbert, W., and Müller-Hill, B. (1966) Isolation of the *lac* repressor. *Proc Natl. Acad. Sci. U.S.A.* 56:1891–1898.

Gilbert, W., and Müller-Hill, B. (1967) The *lac* operator is DNA. *Proc. Natl. Acad. Sci. U.S.A.* 58:2415–2421.

Goldberg, A. L. (1972) Degradation of abnormal proteins in *Escherichia coli. Proc. Natl. Acad. Sci. U.S.A.* 69:422–426.

Granner, D. K., Hayashi, S.-I., Thompson, E. B., and Tomkins, G. M. (1968) Stimulation of tyrosine aminotransferase synthesis by dexamethasone phosphate in cell culture. *J. Mol. Biol.* 35:291–301.

Gronostajski, R. M., Pardee, A. B., and Goldberg, A. L. (1985) The ATP dependence of the degradation of short- and long-lived proteins in growing fibroblasts. *J. Biol. Chem.* 260:3344–3349.

Guyette, W. A., Matusik, R. J., and Rosen, J. M. (1979) Prolactin-mediated transcriptional and post-transcriptional control of casein gene expression. *Cell* 17:1013–1023.

Haas, A., Reback, M., Pratt, G., and Rechsteiner, M. (1990) Ubiquitin-mediated degradation of histone H3 does not require the substrate-binding ubiquitin protein ligase, E3, or attachment of polyubiquitin chains. *J. Biol. Chem.* 265:21664–21669.

Hardy, S. J. S. (1975) The stoichiometry of the ribosomal proteins in *Escherichia coli. Mol. Gen. Genet.* 140:253–274.

Heinemeyer, W., Kleinschmidt, J. A., Saidowsly, C. E., and Wolf, D. H. (1991) Proteinase YScE, the yeast proteasome/multicatalytic-multifunctional proteinase: mutants unravel its function in stress induced proteolysis and uncover its necessity for cell survival. *EMBO J.* 10:555–562.

Hershko, A. (1988) Ubiquitin-mediated protein degradation. *J. Biol. Chem.* 263:15237–15240.

Hershko, A., Ciechanover, A., Heller, H., Haas, A. L., and Rose, I. A. (1980) Proposed role of ATP in protein breakdown: conjugation of proteins with multiple chains of the polypeptide of ATP- dependent proteolysis. *Proc. Natl. Acad. Sci. U.S.A.* 77:1783–1786.

Hershko, A., Heller, H., Eytan, E., Reis, Y. (1986) The protein substrate binding site of the ubiquitin-protein ligase system. *J. Biol. Chem.* 261:11992–11999.

Hofer, E., Hofer-Warbinek, R., and Darnell, R. E., Jr. (1982) Globin RNA transcription: a possible termination site and demonstration of transcriptional còntrol correlated with altered chromatin structure. *Cell* 29:887–893.

Hogness, D. S., Cohn, M., and Monod, J. (1955) Studies on the induced synthesis of β-galactosidase in *Escherichia coli:* the kinetics and mechanisms of sulfur incorporation. *Biochim. Biophys. Acta* 16:99–116.

Hopper, J. E., Broach, J. R., and Rowe, L. B. (1978) Regulation of the galactose pathway in *Saccharomyces cerevisiae:* induction of uridyl transferase mRNA and dependency of GAL4 gene function. *Proc. Natl. Acad. Sci. U.S.A.* 6:2878–2882.

Hough, R., Pratt, G., and Rechsteiner, M. (1987) Purification of two high molecular weight proteases from rabbit reticulocyte lysate. *J. Biol. Chem.* 262:8303–8313.

Jentsch, S., Seufert, W., Soomer, T., and Reins, H.-A. (1990) Ubiquitin-conjugating enzymes: novel regulators of eukaryotic cells. *Trends in Biochem. Sci.* 15:195–198.

Jordan, E., Yarmolinsky, M. B., and Kalckar, H. M. (1962) Control of inducibility of enzymes of the galactose sequence in *Escherichia coli. Proc. Natl. Acad. Sci. U.S.A.* 48:32–40.

Jost, J., Khairallah, E. A., and Pitot, H. C. (1968) Studies on the induction and repression of enzymes in rat liver. V. Regulation of the rate of synthesis and degradation of serine dehydratase by dietary amino acids and glucose. *J. Biol. Chem.* 243:3057–3066.

Kay, J. (1978) Intracellular protein degradation. *Trans. Biochem. Soc.* 6:789–797.

Knowles, S. E., and Ballard, F. J. (1976) Selective control of degradation of normal and abberrant proteins in Reuber H35 hepatoma cells. *Biochem. J.* 156:609–617.

Krause, R. L., and Sokologg, L. (1967) Effects of thyroxine on initiation and completion of protein chains of hemoglobin *in vitro. J. Biol. Chem.* 242:1431–1438.

Maas, W. K., and Clark, A. J. (1964) Studies in the mechanism of repression of arginine biosynthesis in *Escherichia coli.* II. Dominance of repressibility in diploids. *J. Mol. Biol.* 8:365–370.

Magasanik, B. (1962) Catabolite repression. *Cold Spring Harbor Symp. Quant. Biol.* 26:249–256.

Martin, D. W., Jr., and Owen, N. T. (1972) Repression and derepression of purine biosynthesis in mammalian hepatoma cells in culture. *J. Biol. Chem.* 247:5477–5485.

Martin, D. W., Jr., Tomkins, G. M., and Bresler, M. A. (1969) Control of specific gene expression examined in synchronized mammalian cells. *Proc. Natl. Acad. Sci. U.S.A.* 63:842–849.

Martin, T. E., and Wool, I. G. (1968) Formation of active hybrids from subunits of muscle ribosomes from normal and diabetic rats. *Proc. Natl. Acad. Sci. U.S.A.* 60:569–574.

Marzella, L., and Glaumann, H. (1987) Autophagy, microautophagy, and crinophagy as mechanisms for protein degradation. In *Lysosomes: their role in protein breakdown* (Glaumann, H., and Ballard, F. J., eds.) pp. 319–367. Academic Press, New York.

Matern, H., and Holzer, H. (1979) Endogenous proteolytic modulation of yeast enzymes. In *Modulation of Protein Function* (Atkinson, D. E., and Fox, C. F., eds.) pp. 81–92. Academic Press, New York.

Mayer, A., Siegel, N. R., Schwartz, A. L., and Ciechanover, A. (1989) Degradation of proteins with acetylated amino-termini by the ubiquitin system. *Science* 244:1480–1483.

McGuire, M. J., Croall, D. E., and DeMartino, G. N. (1988) ATP-stimulated proteolysis in soluble extracts on BHK 21/CB cells. Evidence for multiple pathways and a role of an enzyme related to the high-molecular weight protease macropain. *Arch. Biochem. Biophys.* 262:273–285.

Mizuno, T., Chou, M.-Y., and Inouye, M. (1984) A unique mechanism regulating gene expression: translational inhibition by a complementary RNA transcript (micRNA). *Proc Natl. Acad. Sci. U.S.A.* 81:1966–1970.

Mortimore, G. E., and Mondon, C. E. (1970) Inhibition by insulin of valine turnover in liver. *J. Biol. Chem.* 245:2375–2383.

Mortimore, G. E., and Schworer, C. M. (1977) Induction of autophagy by amino-acid deprivation in perfused rat liver. *Nature (London)* 270:174–176.

Mortimore, G. E., and Ward, W. F. (1981) Internalization of cytoplastic protein by hepatic lysosomes in basal and deprivation-induced proteolytic states. *J. Biol. Chem.* 256:7659–7665.

Mortimore, G. E., Neely, A. N., Cox, J. R., and Guinivan, R. A. (1973) Proteolysis in homogenates of perfused rat liver: responses to insulin, glucagon and amino acids. *Biochem. Biophys. Res. Commun.* 54:89–95.

Neff, N., DeMartino, G. N., and Goldberg, A. L. (1979) The effect of protease inhibitors and decreased temperature on the degradation of different classes of proteins in cultured hepatocytes. *J. Cell Physiol.* 101:439–458.

Oppenheim, D. S., Bennett, G. N., and Yanofsky, C. (1980) *Escherichia coli* RNA polymerase and *trp* repressor interaction with the promoter-operator reglon of the tryptophan operon of *Salmonella typhimurium*. *J. Mol. Biol.* 144:133–142.

Parks, J. S., Gottesman, M., Shimada, K., Weisberg, R. A., Perlman, R. L., and Pastan, I. (1971) Isolation of the *gal* repressor. *Proc. Natl. Acad. Sci. U.S.A.* 68:1891–1895.

Payvar, F., DeFranco, D., Firestone, G. L., Edgar, B., Wrange, O., Okret, S., Gustafsson, J.-A., and Yamamoto, K. R. (1983) Sequence-specific binding of glucocorticoid receptor to MTV DNA at sites within and upstream of the transcribed region. *Cell* 35:381–392.

Permutt, M. A., and Kipnis, D. M. (1972) Insulin biosynthesis. 1. On the mechanism of glucose stimulation. *J. Biol. Chem.* 247:1194–1199.

Peterkofsky, B., and Tomkins, G. M. (1968) Evidence for the steroid induced accumulation of tyrosine aminotransferase messenger RNA in the absence of protein synthesis. *Proc. Natl. Acad. Sci. U.S.A.* 60:222–228.

Raibaud, O., and Schwartz, M. (1984) Positive control of transcription initiation in bacteria. *Annu. Rev. Genet.* 18:173–206; see pp. 180–181.

Rechsteiner, M. (1990) PEST sequences are signals for rapid intracellular proteolysis. *Seminars in Cell Biol.* 1:433–440.

Rivett, A. J. (1985) Preferential degradation of the oxidatively modified form of glutamine synthetase by intracellular mammalian proteases. *J. Biol. Chem.* 260:300–305.

Roger, S., Wells, R., and Rechsteiner, M. (1986) Amino acid sequences common to rapidly degraded proteins: the PEST hypothesis. *Science* 234:364–368.

Rotman, B., and Spiegelman, S. (1954) On the origin of the carbon induced synthesis β-galactosidase in *Escherichia coli*. *J. Bacteriol.* 68:419–429.

Rowe, P. B., and Wyngaarden, J. B. (1966) The mechanism of dietary alterations in rat hepatic xanthine oxidase levels. *J. Biol. Chem.* 241:5571–5576.

Schimke, R. T., Sweeney, E. W., and Berlin, C. M. (1965) The roles of synthesis and degradation in the control of rat liver tryptophan pyrrolase. *J. Biol. Chem.* 240:322–331.

Schutz, G., Killewich, L., Chen, G., and Feigelson, P. (1975) Control of the mRNA or hepatic tryptophan oxygenase during hormonal and substrate induction. *Proc. Natl. Acad. Sci. U.S.A.* 72:1017–1020.

Seglen, P. O., Gordon, P. B., and Holen, I. (1990) Non-selective autophagy. *Seminars in Cell Biol.* 1:441–448.

Seufert, W., and Jentsch, S. (1990) Ubiquitin-conjugating enzymes UBC4 and UBC5 mediate selective degradation of short-lived and abnormal proteins. *EMBO J.* 9:543–550.

Seufert, W., McGrath, J. P., and Jentsch, S. (1990) UBC1 encodes a novel member of an essential subfamily of yeast ubiquitin-conjugating enzymes involved in protein degradation. *EMBO J.* 9:4535–4541.

Shapiro, D. J., and Brock, M. L. (1985) Messenger RNA stabilization and gene transcription in the estrogen induction of vitellogenin mRNA. *Biochem. Action Hormones* 12:139–172.

Shatkin, A. J. (1976) Capping of eukaryotic mRNA. *Cell* 9:645–653.

Speiser, S., and Etlinger, J. D. (1983) ATP stimulates proteolysis in reticulocyte extracts by repressing endogenous protease inhibitor. *Proc. Natl. Acad. Sci. U.S.A.* 80:3577–3580.

Stalder, J., Groudine, M., Dodgson, J. B., Engel, J. D., and Weintraub, H. (1980) Hb switching in chickens. *Cell* 19:973–980.

Styles, C. D., Lee, K. L., and Kenney, F. T. (1976) Differential degradation of messenger RNAs in mammalian cells. *Proc. Natl. Acad. Sci. U.S.A.* 73:2634–2638.

Thompson, E., Tomkins, G. M., and Curran, F. (1966) Induction of tyrosine α-keto-glutarate transaminase by steroid hormones in a newly established tissue culture cell line. *Proc. Natl. Acad. Sci. U.S.A.* 56:296–303.

Tomkins, G. M., Gelehrter, T. D., Granner, D., Martin, D., Jr., Samuels, H., and Thompson, E. B. (1969) Control of specific gene expression in higher organisms. *Science* 166:1474–1480.

Walker, P. R. (1977) The regulation of enzyme synthesis in animal cells. *Essays Biochem.* 13:39–69.

Weintraub, H., Larsen, A., and Groudine, M. (1981) α-Globin-gene switching during the development of chicken embryos: expression and chromosome structure. *Cell* 24:333–344.

Weisbrod, S. (1982) Active chromatin. *Nature (London)* 297:289–295.

Wiskocil, R., Bensky, P., Dower, W., Goldberger, R. F., Gordon, J. I., and Deely, R. G. (1980) Coordinate regulation of two estrogen dependent genes in avian liver. *Proc. Natl. Acad. Sci. U.S.A.* 77:4474–4478.

Woodside, K. H., and Mortimore, G. E. (1972) Suppression of protein turnover by amino acids in the perfused rat liver. *J. Biol. Chem.* 247:6474–6481.

Yates, J. L., and Nomura, M. (1980) *E. coli* ribosomal protein L4 is a feedback regulatory protein. *Cell* 21:517–522.

Yanofsky, C. (1981) Attenuation in the control of expression of bacterial operons. *Nature (London)* 289:751–757.

Chapter 10

Oxidative Phosphorylation and Mitochondrial Organization

I. GENERAL CONSIDERATIONS

A. Photosynthesis and Oxidative Phosphorylation

In green plants and in photosynthetic bacteria, energy is generated by photosynthetic reactions. In most other cells, the energy is generated primarily by oxidative phosphorylation. The products of photosynthesis in plants fuel oxidative phosphorylation in animals. The role of these two groups of reactions is so fundamental and their mechanisms are so intriguing that a good deal of effort has gone into their study. Although photosynthesis and oxidative phosphorylation are formally very different, they share many features. Differences and similarities are summarized in Table 10.1. Another pattern that occurs in some bacteria is *chemolithotrophy,* i.e., the oxidation of reduced inorganic substrates, which occupies an important biological niche (see Wood, 1988; Ehrlich, 1990) but is not discussed further in this book.

In oxidative phosphorylation, the oxidation of substrates takes place by a process significantly different from that of nonbiological systems. The latter involves oxygen or other oxidants directly and results in the release of energy in the form of heat. In contrast, the oxidative reactions of the cell proceed in steps, each progressively dissipating part of the energy as heat and part to phosphorylate ADP. Oxygen is involved only in the terminal reaction. The discrete steps of oxidative phosphorylation may be regarded as the passage of electrons through the *electron transport* chain until finally oxygen itself is used and water is formed. In photosynthesis, light is the source of energy. The photosynthetic pigments, chlorophyll and the accessory pigments, absorb the radiant energy. The excited chlorophyll of the reaction centers transfers an electron to the primary acceptor and then to the electron transport system. In both oxidative phosphorylation and photosynthetic reactions, the reactions involved in the passage of elec-

Table 10.1 Energy Capturing Systems of Cells

(1) Characteristics	(2) Photosynthesizing systems	(3) Oxidative phosphorylation systems
Source of energy	Entirely or in part energy from sunlight	Oxidizable substrates
Location in cells	Membrane structure: chloroplasts or chromatophores	Membrane structure: mitochondria or mesosomes
Biochemical organization	Electron transport systems involving cytochromes	Electron transport systems involving cytochromes
Primary form of chemical energy trapping	Synthesis of ATP, reduction of NADP	Synthesis of ATP
Overall results	a. Release of O_2 or other oxidized component (e.g., S or H_2SO_4) b. Fixation of CO_2 or some other C source	O_2 uptake CO_2 release
General pattern of system	Light → photosynthetic reaction involving chlorophyll ↓ e^- ↑ electron transport steps e^- O_2 ← H_2O photosynthetic reaction involving chlorophyll	Substrate → oxidized substrate ↓ e^- electron transport steps ↓ e^- $O_2 \rightarrow H_2O$

trons—i.e., the reactions of the electron transport chain—are associated with the internal membranes of specialized organelles or structures. Oxidative phosphorylation takes place in *mitochondria* or in bacterial membranes. Similarly, photosynthesis in green plants occurs in the *chloroplasts* and in bacterial membranes.

The idea that electron transport and possibly ATP synthesis could occur in the plasma membrane of eukaryotic cells is intriguing (Goldenberg, 1982; Sun et al., 1984); however, the information available at this time remains too sketchy for critical judgments.

B. Oxidation without Phosphorylation

In brown fat, a number of substrates can be oxidized in mitochondria without phosphorylation (Nedergard and Cannon, 1984). In nonshivering thermogenesis, the mitochondria of brown adipose tissue oxidize substrates but the energy is dissipated as heat to warm the animal. This takes place by the uncoupling of oxidation from phosphorylation by a special mechanism (Flatmark and Pederson, 1975). More commonly, oxidative reactions that do not produce ATP take place elsewhere in the cell.

A group of hemoproteins, the P450 cytochromes (see Guengerich, 1992; Coon et al., 1992), oxidizes a variety of important chemicals including drugs, carcinogens, steroids, pesticides, hydrocarbon and endogenous compounds such as fatty acids, steroids, acetone, vitamins, and many others. Accordingly, they are significant in understanding pharmacology and human disease. In mammals, the enzymes are present in the endoplasmic reticulum membranes and to a lesser extent in mitochondria. Their ubiquitous presence in other organisms, including prokaryotes such as bacteria, is not only of general interest but has facilitated research on characterizing the system. The P450 enzymes are coded by a gene superfamily and 150 of these have so far been identified. Each isoform may be fairly specific, but others, such as those in the hepatic endoplasmic reticulum, have been estimated to catalyze about 250,000 different reactions involving foreign substances. Newly created chemicals synthesized in the laboratory are likely to become substrates.

P450 are generally but not always monoxygenases; that is, the steps involved in the oxidation of a substrate (e.g., RH) involve one oxygen atom (to form ROH).

Studies of the P450 system should make it possible to screen drugs with liver microsomal preparations or reconstituted P450 systems which would permit predictions of metabolic stability and toxicity. Similarly, it may be possible to place the P450 systems in the appropriate microorganisms using recombinant DNA techniques (see Chapter 1, II, A) to dispose of toxic compounds in the environment. The desaturation of fatty acids, which requires O_2 and NADPH, and their ω-oxidation (i.e., the oxidation at the ω-carbon atom) is carried out in the endoplasmic reticulum by the P450 system. The function of the ω-oxidation is unknown. Production of the endoplasmic system and P450 can be induced by exposure of the animal to foreign substances.

The *peroxisomal* oxidative system does not involve cytochromes (Huang et al., 1983). The enzymes of this system are contained in small vesicular cellular elements, the *peroxisomes*, which are enclosed by a single membrane. The peroxisomal matrix is granular and, at least in rat liver, contains a crystalline or polytubular structure made of urate oxidase. Peroxisomes contain enzymes capable of carrying out the β-oxidation of fatty acids (see below). The peroxisome system can oxidize a variety of substances in two steps. In one step, H_2O_2 is formed from oxygen using reducing equivalents from the substrate, and substances such as urate, D- and L-amino acids, L-α-hydroxy acids, and glyoxylate are oxidized. In the second step, the H_2O_2 is used as an oxidant; phenols, nitrites, ethanol, methanol, and formate can be oxidized. The system can also operate to oxidize NADH indirectly. The glycolate oxidized by the peroxisomes to form glyoxylate can be regenerated in a reaction that uses NADH as a source of reducing equivalents. Catalytic amounts of these two metabolites operating in oxidation-reduction cycles can oxidize NADH indefinitely. In plants, a similar cycle is thought to be responsible for *photorespiration* (see below).

The purpose of these oxidative reactions is not entirely clear. In mammalian liver or intestine, where peroxisomes are also present, they may be important in the regulation of the concentration of metabolites. The oxidation of NADH by the peroxisomes may be significant in the regulation of glycolysis that requires the regeneration of NAD^+ for its continuous operation. NADH oxidation may also function in avoiding high levels of O_2 that might be damaging.

Production of the peroxisomes can be induced by certain drugs or a high-fat diet. The enzyme system capable of β-oxidation is likely to function in the breakdown of the fatty acids, which are poorly processed by mitochondria, and may play a significant role when the need for fatty acid breakdown is high (Hashimoto, 1982; Mannaerts and Debeer, 1982; Osmundsen, 1982). However, the peroxisomes are most likely to act in concert with mitochondria, since they are not capable of processing short-chain fatty acids.

In higher plants, peroxisomes are found in the oil-rich tissues of seeds, where they also known as *glyoxysomes* (Huang et al., 1983). In these tissues they are exclusively responsible for β-oxidation of fatty acids and they contain glyoxylate cycle enzymes. The glyoxylate cycle can convert 2 mol of acetyl-CoA to 1 mol of succinate, thereby permitting the synthesis of sugars in conjunction with reactions occurring in the cytoplasm and in mitochondria.

Leaf peroxisomes (Huang et al., 1983) are involved in photorespiration, in which they oxidize reducing equivalents formed by photosynthesis in conjunction with reactions taking place in the cytoplasm and the mitochondria. The complex reactions can function in the carbon reduction pathway of the chloroplasts or in the oxidative pathway, depending on the competition between oxygen and CO_2. Oxygen favors the oxidation of ribulose-1,5-biphosphate in the oxidative pathway, whereas CO_2 favors the carboxylation of ribulose-1,5-biphosphate with the formation of phosphoglycerate.

The precise physiological role of photorespiration is not entirely clear, but the process appears to be essential. Mutants lacking the system (Sommerville and Ogren, 1980) do not survive in the presence of oxygen. Without the photorespiratory system, oxygen and light cause a loss of photosynthetic capacity, which suggests that photorespiration is involved in maintaining a low oxygen concentration. Photorespiration may also be required to consume excess photosynthetic assimilatory power.

This chapter focuses primarily on the electron transport system of mitochondria. Sections II and III describe the structural and biochemical organization of mitochondria. Chapter 12 presents in more detail electron transport and its coupling to ATP synthesis and ion transport.

II. STRUCTURAL ORGANIZATION OF MITOCHONDRIA

The macromolecules constituting the machinery of oxidative phosphorylation are organized in the mitochondria or, in bacteria, in infoldings of the plasma membrane. Mitochondria, seen with the electron microscope, are distinctive. Their appearance varies with the preparative technique. Transmission electron microscopy of thin sections of tissues or mitochondrial preparations has led us to most of the current concepts of mitochondrial structure. The mitochondria appear either spherical or tubular, depending on the tissue. Figure 10.1*a* is a reconstruction of a tubular mitochondrion and its system of membranes, based on electron micrographs of thin sections fixed with either OsO_4 or permanganate. An external membrane encloses the whole organelle. Another uninterrupted convoluted membrane lines the mitochondrial lumen, and its folds form deep lamellar partitions of the lumen, called the *cristae*. Figure 10.1*a* shows both membranes as single lines, as they appeared in many of the early electron micrographs. Each

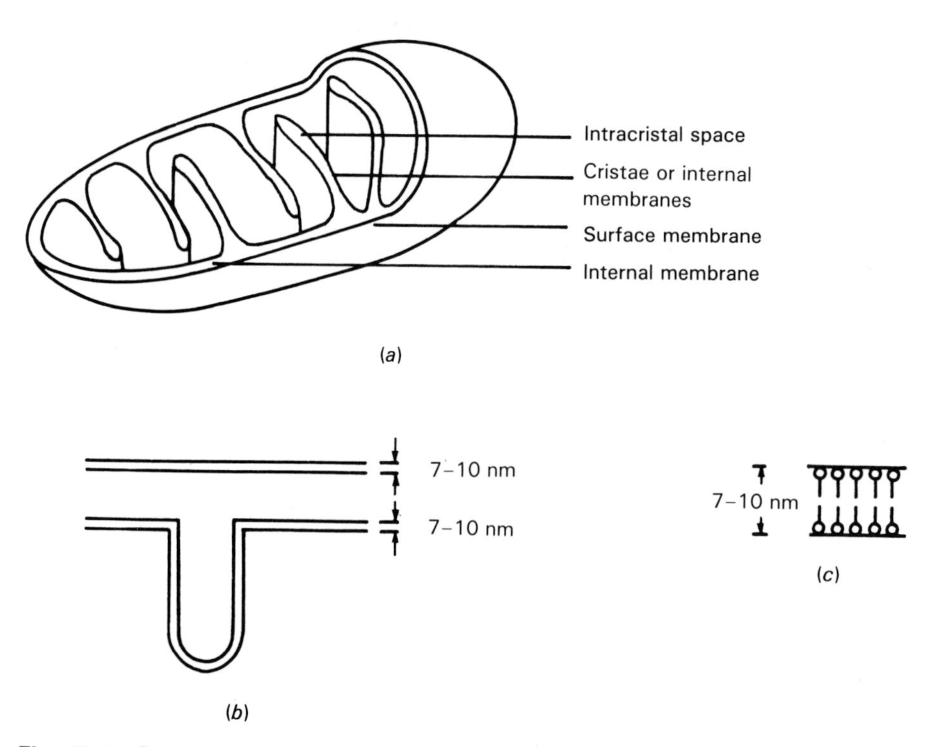

Fig. 10.1 Schematic representation of mitochondrial structure based on electron microscopy with thin sections. (*a*) Model of mitochondrial structural. (*b*) Model of a cross section of the mitochondrial membranes at the site of a crista. (*c*) Model of a bimolecular lipid layer or unit membrane in cross section.

single line has now been resolved into two distinct, evenly spaced lines. Figure 10.1*b* shows a cross section of the membranes to illustrate this point. Each double line is about 7 to 10 nm thick. These lamellae may assume a granular appearance with certain methods of fixation (Malhorta, 1966).

Artificially produced lipid or phospholipid leaflets are thought to represent bimolecular lipid layers when prepared for electron microscopy. These leaflets also exhibit a regular double array of lines but of smaller dimension. A bimolecular lipid layer (*bilayer*) model of a membrane is represented in Fig. 10.1*c*. Following convention, the circles in the figure represent the polar, hydrophilic heads of the lipid molecules, which are dipped in the water phase, and the rods correspond to the hydrophobic chains of the molecules. The interaction of the polar ends with water, the lateral interactions between the hydrophobic chains, and more importantly the preference of water to bind to other water molecules impart a high degree of stability to the structure. Membranes that can be interpreted by such a simple model have been called *unit membranes*. The dark lines seen with the electron microscope in osmium- or permanganate-fixed sections probably correspond to deposits of fixation products at the position of the hydrophilic groups.

Although these models have proved useful, they are too simple to explain some of the available information. Bilayers prepared from mitochondrial lipids are 5 to 6 nm thick (Gulk-Krzywicki et al., 1967), not the 7 to 10 nm shown by conventional transmission electron microscopy or the 11.5 to 12 nm shown by x-ray diffraction of packed mitochondrial membranes (Thompson et al., 1968). Perhaps considerations of the protein composition of these membranes may lead us to more realistic models.

The inner mitochondrial membrane contains as much as 75% proteins, only half of which are integral proteins (i.e., proteins embedded in the lipid framework) (Hatefi, 1985). The various proteins and protein complexes known to be associated with the mitochondrial inner membrane are discussed later. A reconstruction of their size and orientation based in part on x-ray diffraction and electron microscopy (e.g., Capaldi et al., 1987) (see Fig. 10.9) indicates that the protein complexes are likely to project into the water phase, which explains the dimensions in excess of those of a simple bilayer. Furthermore, virtually all the lipid can be extracted without changing the dimensions of the double layer (Fleischer et al., 1967). Phospholipids and the protein components held together by hydrophobic bonds (Green and Fleischer, 1963; Green et al., 1967) form complexes that can be fractionated from intact mitochondria (see Section III,B) and then reconstituted to reform functional membranes (Green and Fleischer, 1963; Green et al., 1967; Tzagaloff et al., 1967).

The nature of the space between the internal and external mitochondrial membranes or in the folds of the cristae shown in Fig. 10.1*a* and *b* is not clear. Some studies have demonstrated stainable material between the folds of the cristae (Hall and Crane, 1970). Variations in the volume of the internal mitochondrial lumen would increase or decrease the separation between the internal and external membranes. There is evidence that these structural alterations actually take place during changes in metabolic state (Hackenbrock, 1966), and it has been suggested that these changes in the volume of the lumen are osmotic in nature (Anagnosti and Tedeschi, 1970; Izzard and Tedeschi, 1970).

In negatively stained preparations (Fig. 10.2) (Parson, 1963), the mitochondrial membranes have the appearance of tubes. The inner surfaces of the internal membranes are studded with particles 8 nm in diameter, which are attached to the membranes by a stalk. At times, this arrangement has been observed with conventional electron microscopy (Ashhurst, 1965; Schneider et al., 1972). It has been suggested that these particles are not always visible with the usual preparatory procedures because of their lability, since they do not appear with negative staining after conventional fixation. In submitochondrial preparations or reconstituted systems, these granular structures are correlated with the presence of F_1, the ATP synthase.

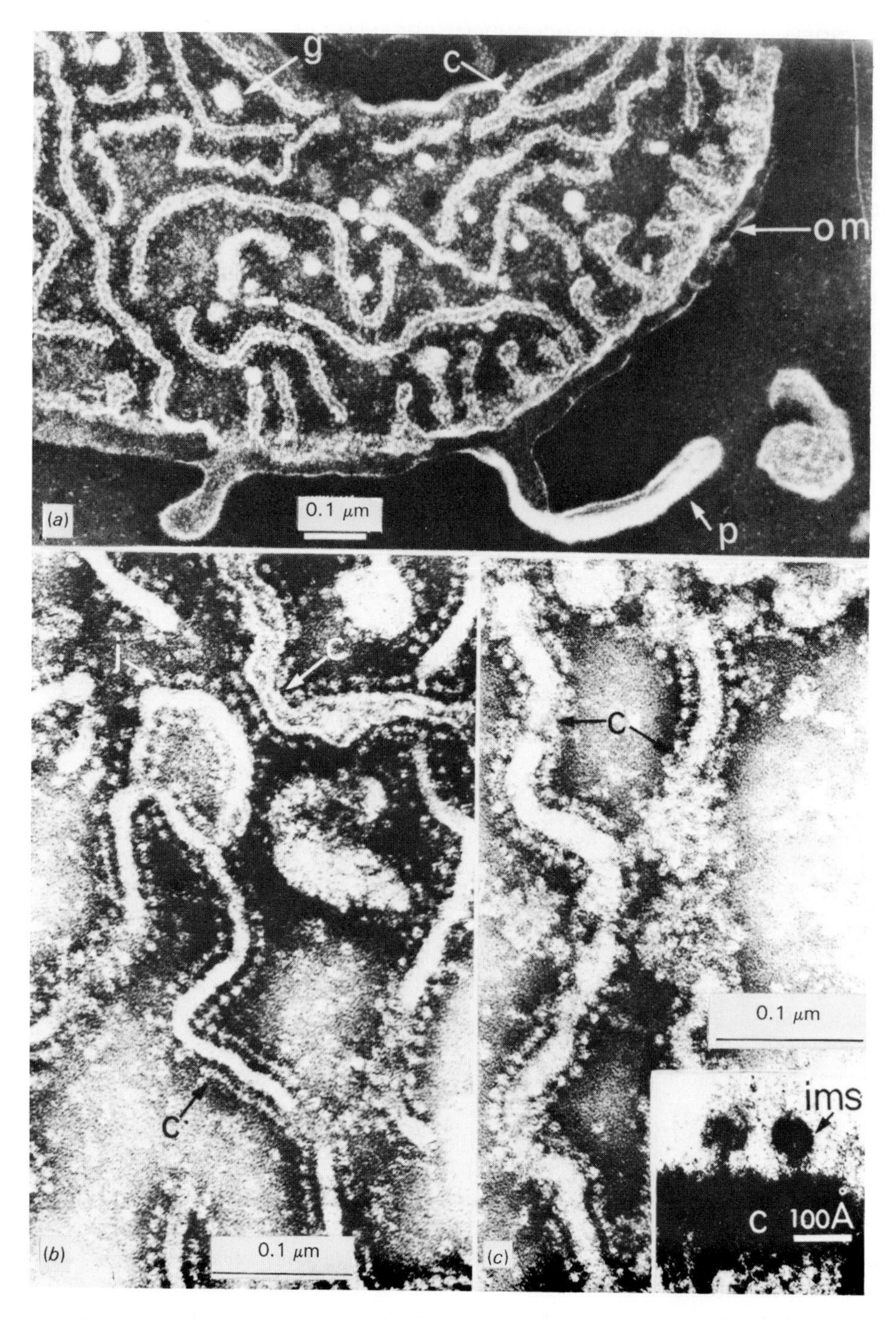

Fig. 10.2 (*a*) Negatively stained mitochondria spread on potassium phosphotungstate. (*b* and *c*) Subunits associated with the inner membrane or cristae. The particles are shown in more detail in the inset (in which the contrast is reversed). Abbreviations: om, outer membrane; c, a crista; g, an intramitochondrial granule; ims, a particle; p, projection of outer membrane; j, branching of crista. Reprinted with permission from Parson (1963), copyright 1963 by the AAAS.

Freeze-fracture, also used in the study of membranes, is entirely different from the two electron microscopic methods already discussed. In this method, a very small piece of tissue is rapidly frozen at very low temperature. The preparation is then fractured by using a chilled microhammer. The fractured preparation is etched by sublimating the water to a depth of a few tens of nanometers, and the exposed surface is replicated by condensing on it a carbon or platinum and carbon layer. The specimen is removed and the replica is viewed with the electron microscope. The fracture is thought to expose the inner hydrophobic central surface of the unit membrane. Observations suggest that the inner hydrophobic face of the inner membrane has many particles ranging in diameter between 10 and 15 nm that probably correspond to intrinsic proteins, i.e., proteins embedded in the lipid framework (Wrigglesworth et al., 1970).

III. BIOCHEMICAL ORGANIZATION OF MITOCHONDRIA

Substrate oxidation coupled to phosphorylation involves three kinds of reaction sequences: reactions in which the substrates themselves are oxidized (e.g., in the tricarboxylic acid cycle), reactions in which the electrons are transported through the cytochrome chain, and reactions involved in the coupling of the electron transport to ATP synthesis or ion transport.

The three kinds of reactions can actually be dissociated by isolating functional complexes or by lysing the mitochondria. Some of the enzymes concerned directly with the oxidation of substrates can be isolated by relatively simple procedures. They are thought to be either loosely bound to the membrane or dissolved in the mitochondrial lumen. The isolation of succinate dehydrogenase, α-glycerophosphate dehydrogenase, and the complexes involved in the electron transport requires disruption of the membranes, since these enzymes are either firmly attached or part of the lipoprotein framework of the inner mitochondrial membrane. Part of the ATP synthase complex (the F_0 portion) requires disruption of the lipoprotein framework, and part (the F_1 portion) can be removed by simpler disruptive procedures.

Interactions between the matrix enzymes suggest that they may be associated in physiologically significant complexes. The measured rates in a pathway are in excess of the rate calculated from the known intramitochondrial concentrations of substrates and enzymes assuming that they are free in solution (e.g., Halpen and Srere, 1977). Furthermore, enzymes that catalyze consecutive reactions in a metabolic pathway tend to associate with each other (Beckman and Kanarek, 1981), and at least some of them are capable of being bound to the inner mitochondrial membrane (Sumegi and Srere, 1984). This idea, however, is still highly controversial.

In intact mitochondria or submitochondrial vesicles (i.e., vesicles derived from mitochondria), electron transport may be coupled with or uncoupled from phosphorylation. Except for the evolution of heat in mitochondria of brown fat cells, mitochondria are involved in oxidative phosphorylation. Mitochondria capable of phosphorylation may be uncoupled by the addition of chemicals, so-called *uncouplers,* such as 2,4-dinitrophenol. In this condition they continue to oxidize substrates, frequently at an increased rate, but without producing ATP. Other chemicals have other distinct effects. Some of them block the flow of electrons at specific sites without substantially affecting the coupling. In contrast, oligomycin and aurovertin block ATP synthesis and hydrolysis without directly interfering with coupling or electron transport. Uncouplers and inhibitors have been used extensively in studies of the mechanisms of electron transport and phosphorylation. The points at which some of the inhibitors block the respiratory chain are discussed later.

A. The Electron Transport Chain

The metabolic breakdown of carbohydrates and fatty acids eventually results in the formation of reduced equivalents and acetyl-CoA. The reactions that take place in the mitochondria are concerned with the formation of acetyl-CoA, further oxidation of the acetyl moiety through the enzymes of the tricarboxylic acid cycle, and the oxidation of reducing equivalents through the electron transport chain. Some of the amino acids are also metabolized in reactions in which they form acetyl-CoA, although the carbon skeleton of other amino acids may enter the catabolic pathway through the tricarboxylic acid cycle or the glycolytic pathway (see Fig. 10.3). This chapter treats primarily the reactions involving the electron transport chain, which corresponds to a series of oxidative-reductive steps in which each reduced component is in turn oxidized by the one that follows. Finally, in the last oxidative step, the reducing equivalents are directly oxidized by oxygen. In essence, the electrons are passed from one component of the electron transport chain to another that follows. The reactions can be represented as shown below.

$$\mathrm{NADH} + \mathrm{FP} + \mathrm{H}^{+} \rightarrow \mathrm{NAD}^{+} + \mathrm{FPH}_2 \tag{10.1}$$

$$\mathrm{FPH}_2 + 2\,\mathrm{cyt}\,b^{3+} \rightarrow \mathrm{FP} + 2\mathrm{H}^{+} + 2\,\mathrm{cyt}\,b^{2+} \tag{10.2}$$

$$2\,\mathrm{cyt}\,b^{2+} + 2\,\mathrm{cyt}\,c^{3+} \rightarrow 2\,\mathrm{cyt}\,c^{2+} + 2\,\mathrm{cyt}\,b^{3+} \tag{10.3}$$

Here FP is flavoprotein and cyt a cytochrome. The final reaction of the series involves oxygen:

$$2\,\mathrm{cyt}\,a_3^{2+} + \tfrac{1}{2}\mathrm{O}_2 + 2\mathrm{H}^{+} \rightarrow 2\,\mathrm{cyt}\,a_3^{3+} + \mathrm{H_2O} \tag{10.4}$$

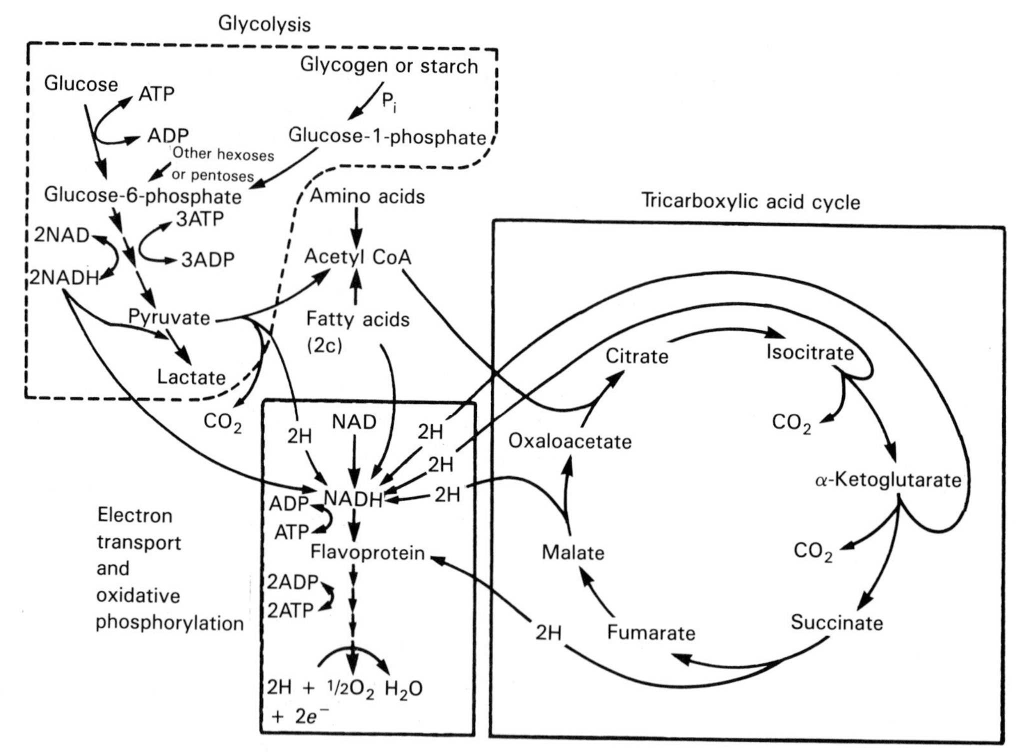

Fig. 10.3 Diagram of metabolic reactions show the role of reactions taking place in mitochondria, shown in the solid line boxes.

More schematically, the reactions of Eqs. (10.1) to (10.4) can be represented as a single equation:

$$\text{NAD} \rightarrow \text{FP} \rightarrow 2\ \text{cyt}\ b \rightarrow 2\ \text{cyt}\ c \rightarrow \cdots \rightarrow \text{cyt}\ a_3 \rightarrow \text{O}_2 \tag{10.5}$$

In this representation the arrows indicate the direction of the electron flow.

Many of the components of the electron transport chain have been known and intensely studied for a number of years. To evaluate the role of each of these components in oxidative phosphorylation, it is necessary first to know whether they are actually functioning in the redox reaction chain and then to determine their order in the chain. The electron transport chain includes NAD, the cytochromes, flavoproteins, coenzyme Q (CoQ), and nonheme iron. Copper is involved in oxidase activity (e.g., via cytochrome-*c* oxidase) in a way that is still unclear. There is some evidence that one of the coppers in cytochrome-*c* oxidase is an electron carrier and the other is involved in ligand binding. The flavoproteins are specific proteins containing flavin groups (riboflavin-5′-phosphate or flavin-adenine dinucleotide). The cytochromes contain as a redox component Fe^{3+} (or Fe^{2+}) in the form of iron porphyrins. CoQ (ubiquinone) is a lipid-soluble quinone. Some of the known components of the cytochrome chain and their probable position in the electron transport chain are discussed in more detail below (see Figs. 10.6, 10.7, and 10.10).

Oxidation and reduction of the cytochrome system are reflected in changes in the light absorbed, as shown in Fig. 10.4 (Chance and Williams, 1956). The solid line in Fig. 10.4 represents the difference in absorption spectra between mitochondria in the presence of substrate under anaerobic conditions (more reduced) and under aerobic conditions (more oxidized). The various peaks correspond to the components indicated. Curve 1 corresponds to the present discussion; curve 2 is discussed later.

In addition to their early study by light spectroscopy, the redox changes of some of the components of the respiratory chain have been successfully studied with *electron paramagnetic resonance* (EPR), also called electron spin resonance (ESR). This technique has been very useful in the study of the electron transport chain, particularly at low temperatures (e.g., 13°K). Electrons usually occur in pairs, and the paired electrons have opposed spins. Unpaired electrons can be present either in free radicals or in compounds such as the electron transport components. The unpaired electrons respond to changes in the magnetic field, and the interaction can be detected by changes in the absorption of microwaves. In EPR, generally the magnetic field is varied and the transmitted radiation is amplified, recorded, and displayed, most commonly as the derivative of the absorbed radiation in relative units. The relationship between the magnetic field and the absorption is dependent on the constant g or g factor, following the relationship of Eq. (10.6). In this equation h is Planck's constant; β is the Bohr magneton, also a constant; and H and ν are the strength of the magnetic field and the frequency of the microwave radiation, respectively. For an electron without any disturbances $g = 2.00232$. However, other charged groups may disturb this pattern so that the g value is different, the usual case in biological samples.

$$h\nu = g\beta H \tag{10.6}$$

Since the frequency is generally maintained constant, the position of resonance in the spectrum (i.e., the absorption of the radiation) can be indicated by the magnitude of either H or g. In biological studies it is frequently indicated by the g value. Figure 10.5 shows the parameters of the EPR spectrum of complex I from beef heart mitochondria (see Section III,B) at 11 K (obtained with cold helium gas).

The proportion of the components reduced under various conditions can provide information

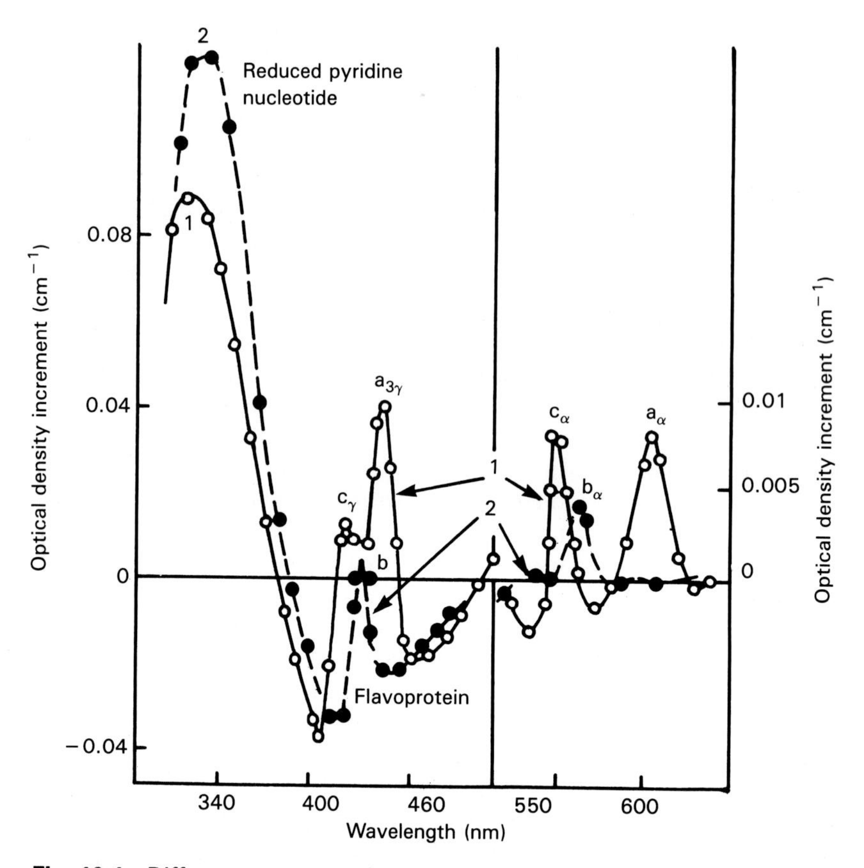

Fig. 10.4 Difference spectra of the respiratory carriers in rat liver mitochondria. The solid line represents absorbancy changes brought about by the presence of substrate under anaerobic conditions. The dashed line represents the change produced by substrate in the presence of antimycin A. The results are measured as differences from the absorption under aerobic conditions. From B. Chance and G. R. Williams (1956), *Advances in Enzymology,* vol. 17, p. 74, copyright © 1956 by John Wiley & Sons, Inc. Data from *The Journal of Biological Chemistry.*

about the organization of the electron transport chain. If the complete system (mitochondria, substrate, ADP, and O_2) is smoothly integrated and the rate of passage of electrons is the same throughout the chain, the steady-state reduced level of each electron carrier should reflect its position in the chain.

The order in which the cytochromes function should also be deducible from the order in which they are oxidized when O_2 is suddenly introduced into the system. If the mitochondria are first exposed to a nitrogen atmosphere, the cytochromes become entirely reduced; the electrons originating from the substrate no longer have a terminal acceptor. When O_2 is introduced, the components closer to O_2 should be oxidized first. The results indicate that cytochrome *a* is oxidized first, followed by cyt *c*, cyt *b*, and flavoprotein.

Selective blocks can be introduced at various points in the chain. The components on the NADH side of the block, no longer connected to an electron acceptor, will become more reduced. The components on the oxygen side of the block will be unable to receive electrons and will become more oxidized. The location of a block of this kind, where the redox state of two neighboring components sharply differs, is known as a *crossover point*. Specific reactions in the sequence can be blocked by means of inhibitors. Addition of the inhibitor antimycin to a metabolizing suspension blocks the oxidation. Changes in the absorption spectrum induced by the presence of antimycin are shown by the dashed line (curve 2) of Fig. 10.4, where it is clear that cytochromes *c* and *a* are fully oxidized whereas cytochrome *b*, flavoprotein, and pyridine nucleotides remain reduced. Thus cytochromes *c* and *a* are situated on the oxygen side of the block. Similar experiments can be carried out with other inhibitors such as amytal, which blocks between the pyridine nucleotide-linked flavoproteins and CoQ. A number of others, such as British antilewisite (BAL), leave only cytochrome *b*, flavoprotein, and pyridine nucleotide in reduced forms. In contrast, in the presence of substrate and cyanide, all carriers remain reduced. Cyanide is known to react with cytochrome a_3; thus cytochrome a_3 can be considered the terminal component of the chain, in line with its well-known reactivity with CO and the competition for binding between CO and O_2. We can conclude that, as far as this experimental approach can be carried, the findings are entirely in agreement with the order shown in Figs. 10.6 and 10.7. Figure 10.6 summarizes the steps in the cytochrome chain including the various substrates. Figure 10.7 incorporates the iron sulfur protein detected with EPR techniques (Ohnishi and Salerno, 1982).

Any functional component of the electron transport chain must respond sufficiently rapidly to account for the rates observed under limiting conditions, such as sudden admission of O_2 or ADP, withdrawal of ADP, or inhibition of electron transport with an appropriate inhibitor (e.g., see Klingenberg and Kroger, 1967).

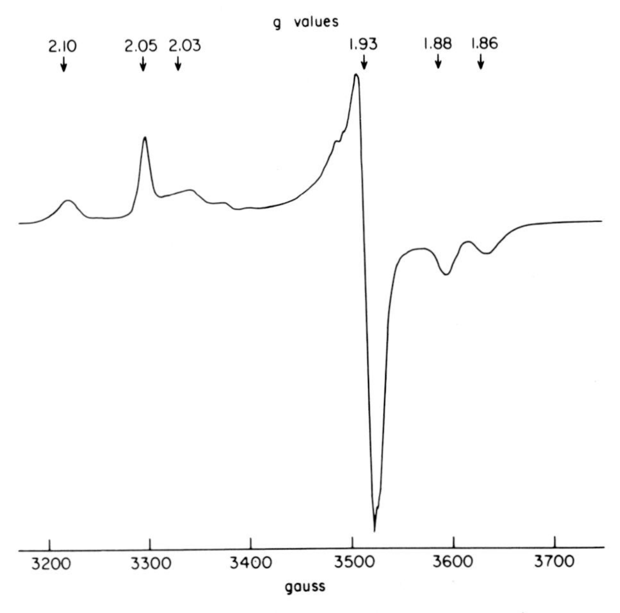

Fig. 10.5 Parameters of the electron paramagnetic resonance spectrum of beef heart complex I at 12 mW, 9.49 MHz, at 11°K. (Courtesy of John Salerno.) The centers, *g* values, and redox potentials are as follows:

	Center	*g*	*g*	*g*	E_h (mV)
(2Fe, 2S)	N-Ia	2.03	1.94	1.91	−400
	b	2.03	1.94	1.91	−250
(4Fe, 4S)	N-2	2.05	1.93	1.93	−20
	N-3	2.04	1.93	1.86	−250
	N-4	2.10	1.93	1.88	−250

B. The Complexes

From the very first studies of mitochondria, biochemists have attempted to break down the system into simpler fractions more amenable to elucidating the mechanisms of the reactions. The close association of the electron transport components with the mitochondrial membrane structure has made this type of analysis very difficult to carry out.

It is possible to separate the enzymes of the citric acid cycle and of fatty acid metabolism from the electron transport elements (MacLennan, 1970). The citric acid cycle enzymes (e.g., isocitrate dehydrogenase, malate dehydrogenase, fumarase, condensing enzyme, and aconitase) and the enzymes for fatty acid oxidations (fatty acid synthase and thiokinases) can be brought into solution by relatively mild treatments of rat liver or beef heart mitochondria. For this reason they are thought to be in solution in the matrix or loosely attached to the inner mitochondrial membrane.

The so-called electron transport particles (ETPs), vesicles prepared from mitochondria by sonication or mechanical disruption, can carry out all electron transport reactions from either NADH or succinate. Some lipoprotein complexes extracted from mitochondria or ETPs can carry out electron transport, but they are unable to phosphorylate. Four such complexes corresponding to portions of the electron transport chain have been isolated (see Hatefi, 1985). The reconstitution of such a system is represented in Fig. 10.8. The complexes do not function together without the addition of CoQ and cyt *c*, which are therefore presumed to have been lost during the preparation and to operate normally in the spans shown in the figure. Further fractionation of this preparation is possible with a number of disruptive procedures.

As seen previously, the inner mitochondrial membranes are encrusted with particles attached to the membranes by stalks. Stalks, particles, and the corresponding piece of membrane (the so-called base piece or F_0) constitute the basic unit of phosphorylation. The particles (which have been called F_1) have ATPase activity and by all indications normally function as an ATP

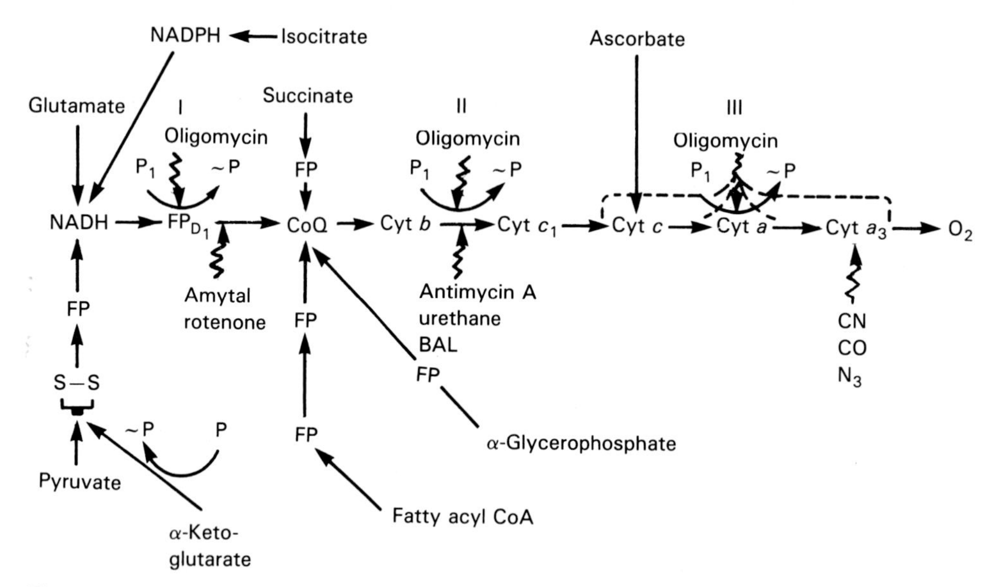

Fig. 10.6 Diagrammatic presentation of the respiratory chain. Roman numerals indicate probable sites of phosphorylation. Wavy arrows indicate probable sites of action of inhibitors.

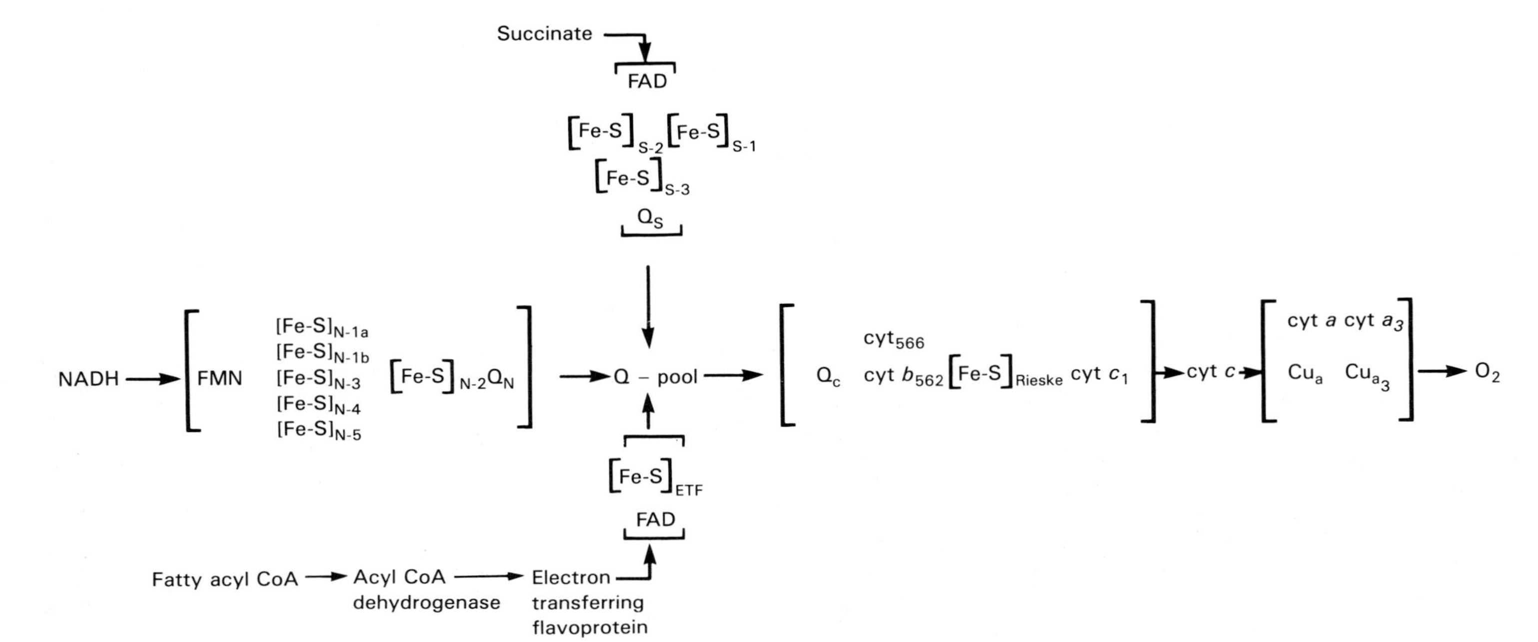

Fig. 10.7 Respiratory chain redox components present in the inner mitochondrial membrane. Fe-S clusters associated with NADH-UQ and succinate-UQ reductase segments are designated with suffixes N-x and S-x, respectively. Q_S, Q_N, and Q_C are protein-associated pools of ubiquinone in succinate-UQ NADH-UQ, and ubiquinol-cytochrome *c* reductase segments, respectively, which can be distinguished from the bulk ubiquinone pool. From Ohnishi and Salerno (1982), Iron-sulfur clusters in the mitochondria electron-transport chain, in *Iron Sulfur Proteins*, ed. T. G. Spiro, p. 288, copyright © 1982 by John Wiley & Sons, Inc.

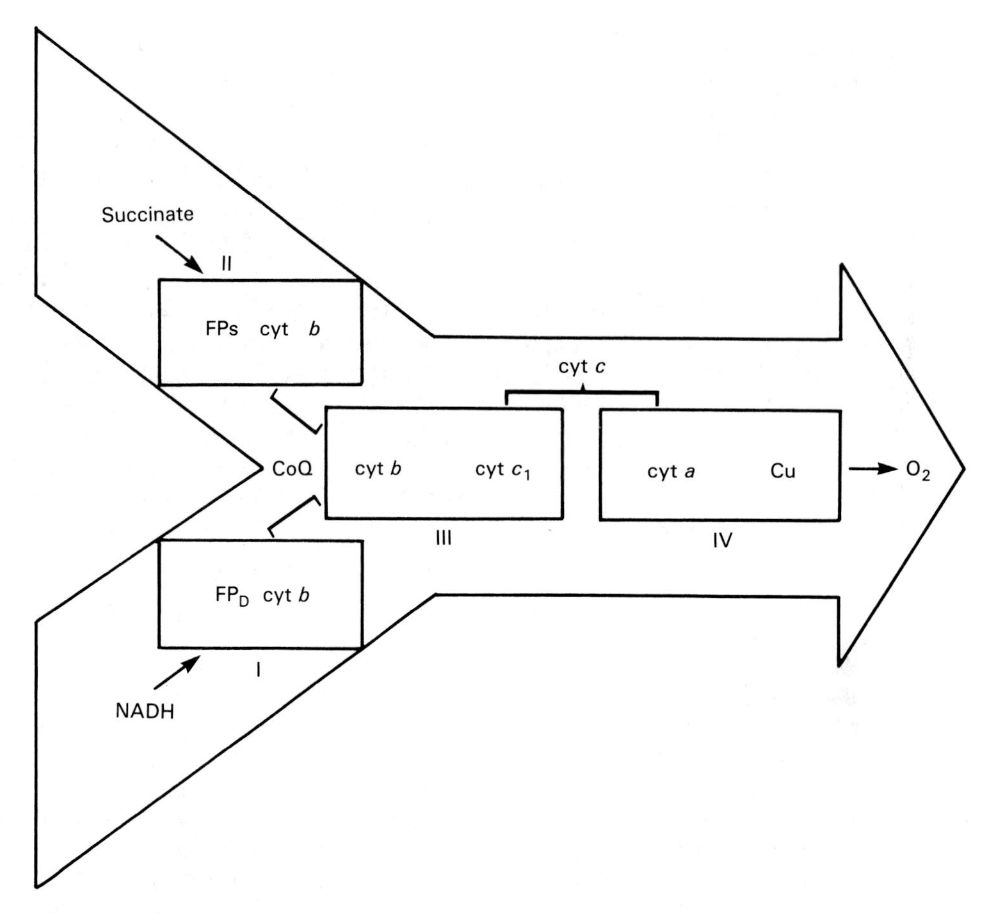

Fig. 10.8 Diagrammatic representation of the sections of the electron transport chain that can be isolated.

synthase. The stalk piece is necessary for the attachment to the membrane (MacLennan and Asai, 1968).

The presence of enzymes and enzyme complexes fixed to a membrane may also have other implications. Since the position of each enzyme molecule is critical, components with a structural role may be significant in the functioning of the assembly. This may be one of the reasons why the functioning of the cytochrome chain depends on the presence of lipid components.

This dependence is illustrated by the experiment represented in Table 10.2 (Fleischer et al., 1967). As shown by sample a in the table, mitochondria can readily oxidize succinate with the concomitant reduction of cytochrome *c* (columns 3–5). These reactions are essentially abolished by the removal of lipids or CoQ (ubiquinone) from the system by acetone extractions (column 3, samples b–d). The reactions are readily reestablished by adding the missing components (samples b, c, and d, column 5; or sample d, column 4). This reactivation can be accounted for by the addition of CoQ to the systems containing some phospholipid (samples b and d), but where the amount of phospholipid is low, addition of CoQ by itself is insufficient to reestablish activity (samples c and e, column 4). Thus, phospholipid components are essential for the maintenance of the enzymatic activity. Apparently, phospholipid is required by the electron transport chain in all three of the spans tested (from succinate to CoQ, from reduced CoQ to cytochrome *c*, and from reduced cytochrome *c* to oxygen) (Green and Fleischer, 1963). Although the lipid components of the membrane may play a role by holding the various enzymes and electron carriers in appropriate positions, it is also conceivable that they help to maintain them in their appropriate conformation.

Results of a variety of studies indicate that some specific components of the mitochondria are located on different sides of the mitochondrial inner membrane. Intact mitochondria and submitochondrial preparations derived from mitochondria after ultrasonic irradiation have been useful for these studies. The submitochondrial preparations most likely correspond to vesicles formed by pinching off of the folds of the inner mitochondrial membrane, i.e., the folds of the cristae. As a consequence, the outside of the vesicles corresponds to the part of the membrane that was originally internal, facing the mitochondrial lumen. In agreement with this interpretation, the two preparations generally have opposite polarity (sidedness). Several techniques have been used to determine the location of the components. In one method, the capacity of ferricyanide to act as an electron acceptor for a component of the chain serves to determine the component's location in relation to the membrane (Klingenberg and von Jagow, 1970; Lee, 1970). Ferricyanide does not penetrate either mitochondria or submitochondrial vesicles. Hence, any component reacting with ferricyanide must be external to the vesicles or the mitochondria. Antibodies against the purified mitochondrial components have also been found useful (Racker et al., 1970). Since the antibodies do not penetrate significantly into either preparation, the inactivation of the corresponding system indicates a location external to the space enclosed by the membrane. Similarly, [^{35}S]diazobenzene sulfonate and polylysine do not pass through the membranes. However, they can be cross-linked to the mitochondrial components. These modified components are subsequently isolated and examined to see whether they have become attached to the probe molecules (Schneider et al., 1972), which determines their location in relation to the membrane. The localization summarized in Table 10.3 is consistent with results of these studies.

Figure 10.9 (Capaldi et al., 1987; Ohnishi, 1987; Ragan, 1987; Weiss, 1987) summarizes our present understanding of the organization of the mitochondrial electron transport complexes in relation to the inner mitochondrial membrane. The information comes from a variety of experimental approaches involving the sidedness of reactive sites discussed in relation to Table 10.3 and several other approaches, including electron microscopy of two-dimensional crystals

Table 10.2 Phospholipid Content and Enzymic Activity of Beef Heart Mitochondria

			Succinate cytochrome reductase[a] activity after addition of		
	(1)	(2)	(3)	(4)	(5)
Sample	Treatment	P (μg/mg protein)	Nothing	CoQ	CoQ + MPL
a. Mitochondria	None	16.8	0.59	0.52	0.53
b. "Neutral lipid-depleted" mitochondria	4% water in acetone	14.5	0.02	0.84	1.00
c. "Lipid-deficient" mitochondria	10% water in acetone	3.7	0.01	0.08	0.97
d. c + MPL		10.5	0.07	0.77	1.03
e. "Lipid-free" mitochondria	10% water in acetone + NH_3	2.2	0.01	0.01	0.39

Source: Fleischer et al. (1967). Reproduced from *The Journal of General Physiology,* 1967, vol. 32, pp. 193–208, by copyright permission of the Rockefeller University Press.

[a]Micromoles cytochrome reduced per minute per milligram of protein at 30°C; CoQ, co-enzyme Q; MPL, mitochondrial phospholipid.

(see Chapter 1); the use of photoactive hydrophobic probes to label the portions in contact with the lipid components (see Chapter 1 for photoactive reagent techniques); chemical cross-linking, which provides information about the spatial relationships between the polypeptides of a complex; and the predictions of hydropathy profiles (Chapter 2).

In Chapter 7 we saw that in enzyme-catalyzed reactions, the presence of different enzymes of one pathway in a multienzyme complex may require movement of part of an enzyme or of the enzyme molecules in relation to each other. The reactions of the cytochrome chain have an analogous requirement. Several cytochromes and other electron carriers are attached to the mitochondrial membrane. The chain can be regarded as an electron relay system in which an electron pair is passed from one cytochrome to the next one in the sequence.

With the exception of electron tunneling (see page 294), the relay of electrons through a number of fixed components cannot be explained conventionally without assuming some sort of movement, since the transfer of electrons involves active groups that are probably spatially separated. The transfer could occur if the protein containing the active group actually rotated or moved in some manner. However, there is evidence that transfer occurs by the shuttling of electrons between four complexes by simpler components. The major complexes that make up the electron transport chain and the ATP synthase complex can be separated (Hatefi, 1985) by appropriate techniques (see Figs. 10.8 and 10.10) and can be recombined with reconstitution of activity. These complexes are thought to represent enzyme assemblies present as independent units in the native membrane. Studies of the mobility of the four complexes, cytochrome *c*, and ubiquinone by a photobleaching and recovery technique similar to the one described in Chapter 2 suggest that all these components are free to move in the inner mitochondrial membrane (Hackenbrock et al., 1986). The kinetics of the appropriate redox reactions are consistent with a diffusional control of electron transport (Hackenbrock et al., 1986). We would expect ubiquinone (lipid soluble) and cytochrome *c* (water soluble), which have high diffusion coefficients and link the electron transport between the various complexes, to play an important role as shuttles in these exchanges. In Chapter 11 we will see that plastoquinone (lipid soluble) and plastocyanin (water soluble) play a similar role in chloroplast electron transport.

The idea that the components of the electron transport chain are free to move in the plane

Table 10.3 Presumed Location of Mitochondrial Components in Relation to the Mitochondrial Inner Membrane[a]

Component	Probable location	Evidence
Cytochromes aa_3	Possibly across	Diazobenzene sulfonate labeling
Cytochrome a	Outside	Polylysine cross-linking
Cytochrome a_3	Inside	Polylysine cross-linking
Cytochrome c	Outside	Reaction with ferricyanide, antibodies react with intact mitochondria
Succinate dehydrogenase	Inside	Reaction with ferricyanide, depends on penetration of succinate
NADH dehydrogenase	Inside	Reaction with ferricyanide, depends on penetration of NADH
Glycerophosphate dehydrogenase	Outside	Substrate is metabolized but cannot penetrate, reaction with ferricyanide
F_1	Inside	Visualization with the electron microscope, diazobenzene sulfonate labeling

[a]See text for explanation of criteria used.

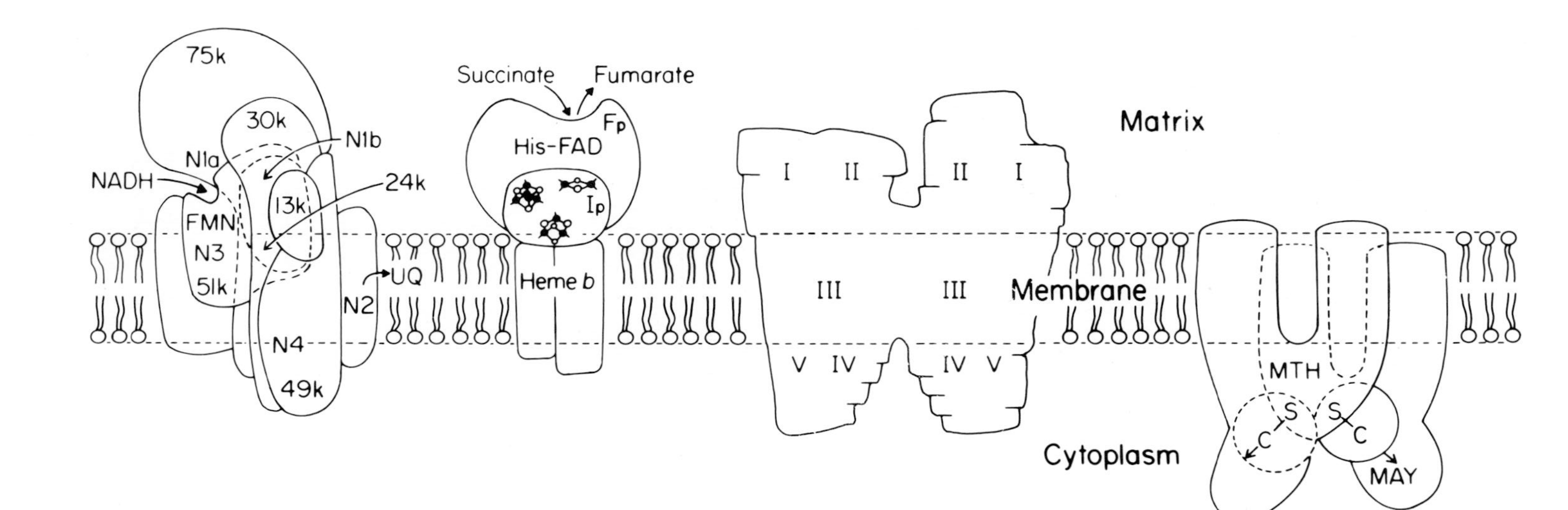

Fig. 10.9 Representation of the structure of the various electron transport complexes, their components, and their arrangement in relation to the mitochondrial inner membrane. From left to right: NADH-ubiquinone reductase (complex I) (Ragan, 1987), succinate-ubiquinone oxidoreductase (complex II) (Ohnishi, 1987), ubiquinone-cytochrome-*c* reductase (complex III) (Weiss et al., 1987), and cytochrome-*c* oxidase (complex IV) (Capaldi et al., 1987).

of the membrane is supported by other observations. For example, when some of the electron transport chains are partially blocked with CO, many of them are nevertheless oxidized but at a lower rate (Chance et al., 1970). From this observation it may be postulated that the electron transport chains are interconnected in some way. However, since the interaction is relatively slow, the possibility of diffusional movement in the plane of the membrane is attractive. A summary of the complexes, associated components, and their stoichiometry is represented in Fig. 10.10 (Hatefi, 1985).

The transfer of electrons within a complex may involve adjacent amino acids, such as tryptophan, that are capable of transferring electrons, or it may take place through a distance by the phenomenon of electron tunneling. Electron tunneling is thought to be involved in a variety of electron transfers in mitochondria. In part, this is suggested by the evidence for short distances between the active centers (Salerno and Ohnishi, 1979). In part, its likelihood is apparent from analogy to photosynthetic electron transport. Cytochrome oxidation in photosynthetic bacteria is independent of temperature at very low temperatures (4–100 K), and this is also the case for the transfer of electrons from the primary electron acceptor back to the oxidized reaction center (DeVault, 1979).

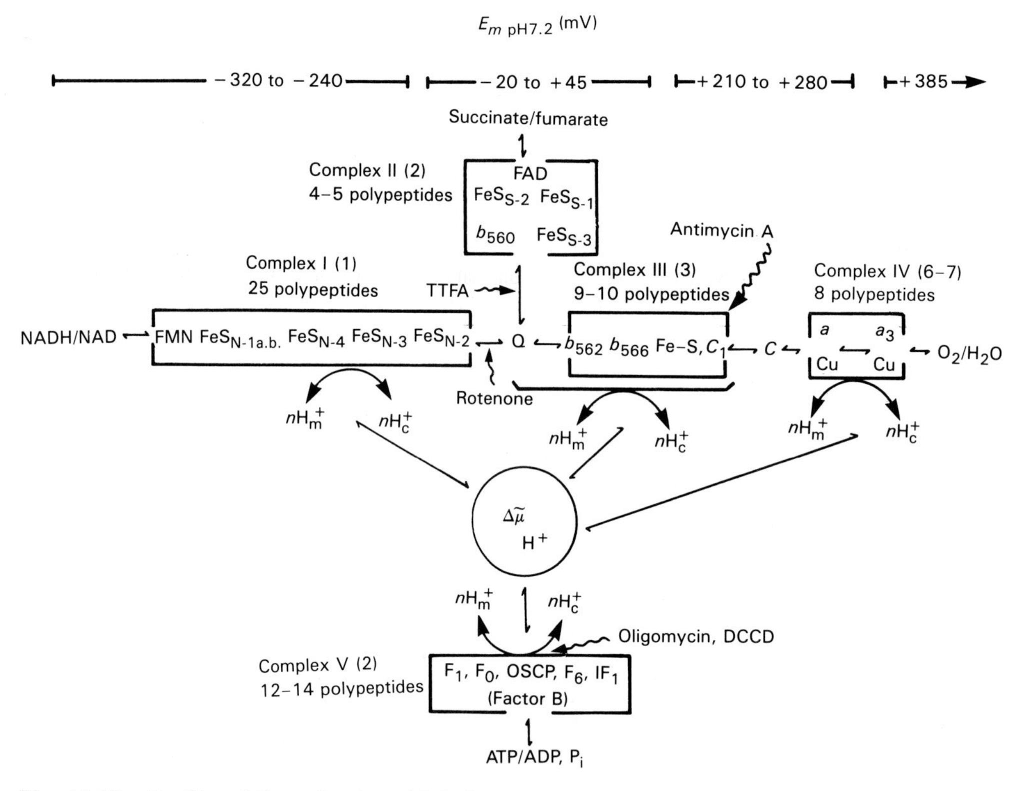

Fig. 10.10 Profile of the mitochondrial electron transport–oxidative phosphorylation system showing the well-characterized components of complexes I, II, III, IV, and V. DCCD stands for N,N′-dicyclohexylcarbodiimidole and TTFA for thenoyltrifluoroacetone. The abbreviations in complex V represent various peptides or peptide assemblies (Hatefi, 1985). Reproduced, with permission, from the *Annual Review of Biochemistry,* vol. 54, © 1985 by Annual Reviews Inc.

A summary of our present understanding of the biochemical organization of the mitochondrial complexes is shown in Fig. 10.10 (Hatefi, 1985). In this diagram the top line represents the redox potentials of the components, and below this line the mitochondrial electron transport complexes are indicated. The numbers in parentheses indicate the stoichiometry of the complexes in the intact mitochondrion compared to complex I (which is taken as unity). The wavy lines indicate the site at which the listed inhibitors act. As shown, the electron transport reactions of each complex are thought to produce a translocation of protons from the mitochondrial matrix (H^+_m) to the cytoplasmic phase (H^+_c) to produce an electrochemical proton gradient (indicated by μ^+_H). The translocation of protons in complex V (the F_0F_1 complex) in the opposite direction generates ATP from ADP and P_i. This process is discussed in more detail in Chapter 12.

Presumably, the main function of mitochondrial electron transport is the production of ATP from ADP and P_i. Analogously, one of the functions of the light-requiring reactions of photosynthesis, which also involve electron transport, is the synthesis of ATP. The similarity between the two systems is considerable, so the oxidative and photophosphorylative reactions are best discussed together. Chapter 11 concerns photosynthesis. The phosphorylative reactions of the mitochondrial and chloroplast electron transport chains are treated together in Chapter 12.

SUGGESTED READING

Cramer, A. W., and Knaff, D. B. (1990) *Energy Transduction in Biological Membranes,* Chapters 3, 4, and 5. Springer-Verlag, New York.

Nicholls, D. (1984) Mechanisms of energy transduction. In *Bioenergetics* (Ernster, L., ed.), pp. 29–48. Elsevier, New York.

Tzagoloff, A. (1982) *Mitochondria,* Chapters 2–5. Plenum, New York.

Wikström, M., and Saraste, M. (1984) The mitochondrial respiratory chain. In *Bioenergetics* (Ernster, L., ed.), pp. 49–94. Elsevier, New York.

Alternative References

Harold, F. M. (1986) *The Vital Force: A Study of Bioenergetics,* Chapter 7, pp. 197–250. W. H. Freeman, New York.

Nicholls, D. G. (1982) *Bioenergetics: An Introduction to Chemiosmotic Theory,* Chapters 5 and 7. Academic Press, New York.

Special Aspects

Capaldi, R. A., Takamiya, S., Zhang, Y. -Z., Gonzalez- Halphen, D., and Yanamura, W. (1987) Structure of cytochrome-*c* oxidase. *Curr. Top. Bioenerg.* 15:91–108.

Ehrlich, H. (1990) *Geomicrobiology,* 2d ed., Chapter 6. Dekker, New York.

Hackenbrock, C. R., Chazotte, B., and Gupte, S. S. (1986) The random collision model and a critical assessment of diffusion and collision in mitochondrial electron transport. *J. Bioenerg. Biomembr.* 18:331–368.

Hatefi, Y., Ragan, I., and Galante, Y. M. (1985) The enzymes and the enzyme complexes of the mitochondrial oxidative phosphorylation system. In *The Enzymes of Biological Membranes,* 2d ed., Vol. 4, pp. 1–70 (Martonosi, A. N., ed.). Plenum, New York.

Jones, C. W. (1988) Membrane associated energy conservation in bacteria: a general introduction. In *Bacterial Energy Transduction* (Anthony, C., ed.), pp. 1–82. Academic Press, New York.

Kadenbach, B., Kuhn-Nentwig, I., and Buge, U. (1987) Evolution of a regulatory enzyme: cytochrome-*c* oxidase (complex IV). *Curr. Top. Bioenerg.* 15:114–151.

Mitchell, P. (1987) Respiratory chain systems in theory and practice. In *Advances in Membrane Biochemistry and Bioenergetics* (Kim, C. H., Tedeschi, H., Diwan, J. J., and Salerno, J. C., eds.), pp. 25-52. Plenum, New York.

Ohnishi, T. (1987) Structure of the succinate-ubiquinone oxido-reductase (complex II). *Curr. Top. Bioenerg.* 15:37–66.

Ragan, C. I. (1987) Structure of NADH-ubiquinone reductase (complex I). *Curr. Top. Bioenerg.* 15:11–36.

Senior, A. E. (1988) ATP synthesis by oxidative phosphorylation. *Physiol. Rev.* 68:177–231.

Senior, A. E. (1990) The proton translocating ATPase of *Escherichia coli*. *Annu. Rev. Biophys. Chem.* 19:7–41.

Tolbert, N. E. (1981) Metabolic pathways in peroxisomes and glyoxysomes. *Annu. Rev. Biochem.* 50:133–158.

Weiss, H. (1987) Structure of mitochondrial ubiquinol-cytochrome *c* reductase (complex III). *Curr. Top. Bioenerg.* 15:67–90.

REFERENCES

Anagnosti, E., and Tedeschi, H. (1970) Mechanism of low-amplitude orthophosphate-induced swelling in isolated mitochondria. *J. Cell Biol.* 47:520–525.

Ashhurst, D. E. (1965) Mitochondrial particles seen in sections. *J. Cell Biol.* 244:497–499.

Beckman, S., and Kanarek, L. (1981) Demonstration of physical interaction between consecutive enzymes of the citric acid cycle and the aspartate-malate shuttle. *Eur. J. Biochem.* 117:527–535.

Capaldi, R. A., Takamiya, S., Zhang, Y. -Z., Gonzalez-Halphen, D., and Yanamura, W. (1987) Structure of cytochrome-*c* oxidase. *Curr. Top. Bioenerg.* 15:91–113.

Chance, B., and Williams, G. R. (1956) The respiratory chain and oxidative phosphorylation. *Adv. Enzymol.* 17:65–134.

Chance, B., Erećinska, M., and Wagner, M. (1970) Mitochondrial responses to carbon monoxide toxicity. *Ann. N.Y. Acad. Sci.* 174:193–204.

Coon, M. J., Ding, X., Pernecky, S. J., and Vaz, A. D. N. (1992) Cytochrome P450: progress and predictions. *FASEB J.* 6:669–673.

DeVault, D. (1979) Introduction to biological aspects. In *Tunneling in Biological Systems* (Chance, B., DeVault, D. C., Fraunfelder, H., Marcus, R. A., Schrieffer, J. R., and Sutin, N., eds.), pp. 303–316. Academic Press, New York.

Ehrlich, H. (1990) *Geomicrobiology*, 2d ed., Chapter 6. Decker, New York.

Flatmark, T., and Pederson, J. I. (1975) Brown adipose tissue mitochondria. *Biochim. Biophys. Acta* 416:53–103.

Fleischer, S., Fleischer, B., and Stoekenius, W. (1967) Fine structure of lipid depleted mitochondria. *J. Cell Biol.* 32:193–208.

Goldenberg, H. (1982) Plasma membrane redox activities. *Biochim. Biophys. Acta* 694:203–223.

Green, D. E., and Fleischer, S. (1963) The role of lipids in mitochondrial electron transfer and oxidative phosphorylation. *Biochim. Biophys. Acta* 70:554–581.

Green, D. E., Allman, D. W., Bachmann, E., Baum, H., Kopaczyk, K., Korman, E. F., Lipton, S., MacLennan, D. H., McConnell, D. G., Perdue, J. F., Rieske, J. S., and Tzagoloff, A. (1967) Formation of membranes by repeating units. *Arch. Biochem. Biophys.* 119:312–325.

Guenngerich, F. P. (1992) Cytochrome P450: advances and prospects. *FASEB J.* 6, 667–668.

Gulk-Krzywicki, T. E., Rivas, E., and Luzzati, V. (1967) Structure et polymorphysme des lipides: Étude par diffraction des rayon X du système forme de lipides des mitochondries de coeur de boeuf et d'eau. *J. Mol. Biol.* 27:303–322.

Hackenbrock, C. R. (1966) Ultrastructural bases for metabolically linked mechanical activity in mitochondria. I. Reversible ultrastructural changes with change in metabolic steady state in isolated liver mitochondria. *J. Cell Biol.* 30:269–297.

Hackenbrock, C. R., Chazotte, B., and Gupte, S. S. (1986) The random collision model and a critical assessment of diffusion and collision in mitochondrial electron transport. *J. Bioenerg. Biomembr.* 18:331–368.

Hall, J. D., and Crane, F. L. (1970) An intracristal structure in beef heart mitochondria. *Exp. Cell Res.* 62:480–483.

Halper, L. A., and Srere, P. A. (1977) Interaction of citrate synthase and mitochondrial malate dehydrogenase in the presence of polyethelene glycol. *J. Biol. Chem.* 184:529–534.

Hashimoto, T. (1982) Individual peroxisomal β-oxidation enzymes. *Ann. N.Y. Acad. Sci.* 386:5–12.

Hatefi, Y. (1985) The mitochondrial electron transport and oxidative phosphorylation system. *Annu. Rev. Biochem.* 54:1015–1069.

Huang, A. H. C., Trelease, R. N., and Moore, T. S., Jr. (1983) *Plant Peroxisomes,* Chapter 4, pp. 87–155. Academic Press, New York.

Izzard, S., and Tedeschi, H. (1970) Ion transport underlying metabolically controlled volume changes of isolated mitochondria. *Proc. Natl. Acad. Sci. U.S.A.* 67:702–709.

Klingenberg, M., and Kroger, A. (1967) On the role of ubiquinone in the respiratory chain. In *Biochemistry of Mitochondria* (Slater, E. C., Kaniuga, Z., and Wojtczak, L., eds.), pp. 11–27. Academic Press, New York.

Klingenberg, M., and von Jagow, G. (1970) Topochemistry of the respiratory chain in the mitochondrial (cristae) membrane. In *Electron Transport and Energy Conservation* (Tager, J. M., Papa, S., Quagliariello, E., and Slater, E. C., eds.), pp. 281–290. Adriatica Editrice, Bari, Italy.

Lee, C. P. (1970) Orientation of the respiratory chain in the mitochondrial inner membrane. In *Electron Transport and Energy Conservation* (Tager, J. M., Papa, S., Quagliariello, E., and Slater, E. C., eds.), pp. 291–300. Adriatica Editrice, Bari, Italy.

MacLennan, D. H. (1970) Molecular architecture of the mitochondrion. *Curr. Top. Membr. Transport* 1:177–232.

MacLennan, D. H., and Asai, J. (1968) Studies on the mitochondrial adenosine triphosphatase system. V. Localization of the oligomycin-sensitivity conferring protein. *Biochem. Biophys. Res. Commun.* 33:441–447.

Malhorta, S. K. (1966) A study of structure of the mitochondrial membrane system. *J. Ultrastruct. Res.* 15:14–37.

Mannaerts, G. P., and Debeer, L. J. (1982) Mitochondrial and peroxisomal β-oxidation of fatty acids in rat liver. *Ann. N.Y. Acad. Sci.* 386:30–38.

Nedergard, J., and Cannon, B. (1984) Thermogenic mitochondria. In *Bioenergetics* (Ernster, L., ed.), pp. 291–314. Elsevier, New York.

Ohnishi, T. (1987) Structure of the succinate-ubiquinone oxidoreductase (complex II). *Curr. Top. Bioenerg.* 15:37–66.

Ohnishi, T., and Salerno, J. C. (1982) Iron-sulfur clusters in the mitochondrial electron-transport chain. In *Iron Sulfur Proteins* (Spiro, T. G., ed.), pp. 285–327. Wiley, New York.

Osmundsen, H. (1982) Peroxysomal β-oxidation of long fatty acids: effects of high fat diet. *Ann. N.Y. Acad. Sci.* 386:12–27.

Parson, D. F. (1963) Mitochondrial structure: two types of subunits on negatively stained mitochondrial membranes. *Science* 140:985–988.

Racker, E., Brunestein, C., Loyter, A., and Christiansen, R. O. (1970) The sidedness of the inner mitochondrial membrane. In *Electron Transport and Energy Conservation* (Tager, J. M., Papa, S., Quagliariello, E., and Slater, E. C., eds.), pp. 235–252. Adriatica Editrice, Bari, Italy.

Ragan, C. I. (1987) Structure of NADH-ubiquinone reductase (complex I). *Curr. Top. Bioenerg.* 15:1–36.

Salerno, J. C., and Ohnishi, T. (1979) Electron transport in the succinate ubiquinone segment of the respiratory chain. In *Tunneling in Biological Systems* (Chance, B., DeVault, D. C., Fraunfelder, H., Marcus, R. A., Schrieffer, J. R., and Sutin, N., eds.), pp. 473–482. Academic Press, New York.

Schneider, D. L., Kagawa, Y., and Racker, E. (1972) Chemical modification of the inner mitochondrial membrane. *J. Biol. Chem.* 247:4074–4079.

Sommerville, C. R., and Ogren, W. L. (1980) Photorespiration mutants in *Arabidopsis thaliana* deficient in serine-glyoxylate aminotransferase. *Proc. Natl. Acad. Sci. U.S.A.* 77:2684–2687.

Sumegi, B., and Srere, A. (1984) Complex I binds several mitochondrial NAD-coupled dehydrogenases. *J. Biol. Chem.* 259:15040–15045.

Sun, I. L., Crane, F. L., Grebing, C., and Low, H. (1984) Properties of a transplasma membrane electron transport system in HeLa cells.' *J. Bioenerg. Biomembr.* 16:583–595.

Thompson, J. E., Coleman, R., and Finean, J. B. (1968) Comparative x-ray diffraction and electron microscope studies of isolated mitochondrial membranes. *Biochim. Biophys. Acta* 150:405–414.

Tzagaloff, A., MacLennan, D. H., McConnell, D. G., and Green, D. E. (1967) Studies on the electron transfer system. LXVIII. Formation of membranes as basis of the reconstitution of the mitochondrial electron transfer system. *J. Biol. Chem.* 242:2051–2061.

Weiss, H. (1987) Structure of ubiquinol-cytochrome *c* reductase (complex III). *Curr. Top. Bioenerg.* 15:67–90.

Wood, P. M. (1988) Chemolithotrophy. In *Bacterial Energy Transduction* (Anthony, C., ed.), pp. 183–230. Academic Press, New York.

Wrigglesworth, J. M., Packer, L. and Branton, D. (1970) Organization of mitochondrial structure as revealed by freeze-etching. *Biochim. Biophys. Acta* 205:125–135.

Chapter 11

Photosynthesis

With few exceptions, such as the chemolithotrophic bacteria, the ultimate source of most of the energy used by biological systems is sunlight. Consequently, almost all organisms depend directly or indirectly on the conversion of radiant energy to chemical free energy. The *heterotrophs,* which do not carry out photosynthesis, degrade complex molecules provided by other organisms.

The overall process of photosynthesis is frequently represented as shown in Eq. (11.1).

$$CO_2 + 2H_2A + \text{light} \rightarrow [CHOH] + H_2O + 2A \quad (11.1)$$

In green plants, A corresponds to oxygen, which originates from water as shown by the equation. This has been demonstrated with ^{18}O-labeled water (Stemler and Radmer, 1975). In some bacteria A corresponds to S [Eq. (11.2)] and even S and H_2O [Eq. (11.3)].

$$CO_2 + 2H_2S + \text{light} \rightarrow (CHOH) + H_2O + 2S \quad (11.2)$$

$$3CO_2 + 2S + 5H_2O + \text{light} \rightarrow 3(CHOH) + 2H_2SO_4 \quad (11.3)$$

In bacteria the source of carbon may be not CO_2 but rather an organic molecule, and the final product may be not a carbohydrate but some other reduced organic compound. Several of the known schemes are summarized in Table 11.1 (Stanier, 1961).

The absorption of light quanta and the subsequent reactions that convert radiant energy to chemical energy are included in what we call photosynthesis. The *chlorophylls* (sometimes

Table 11.1 Characteristics of Photosynthetic Organisms[a]

	Green plants	Cyanobacteria	Green sulfur bacteria	Purple bacteria
Source of reducing power	H_2O	H_2O	H_2S, reduced inorganic compounds	H_2S, reduced inorganic and organic compounds
Photosynthetic oxygen evolution	Yes	Yes	No	No
Principal source of carbon	CO_2	CO_2	CO_2	CO_2 or organic compounds
Relation to oxygen	Aerobic	Aerobic	Strictly anaerobic	Strictly anaerobic or facultatively anaerobic
Site	Chloroplast	Membrane-like	Chlorosomes	Internal membranes (chromatophores)

[a]Modified from K. Y. Stanier, *Bacterial Review,* 25:1–17. Copyright © 1961 American Society for Microbiology.

abbreviated in this discussion as Chl) are the most abundant pigments, but many others (such as carotenoids and, in some algae, phycobilins) are also present. Many of these pigment molecules absorb light and funnel the energy to the reaction centers, which are involved more directly in energy transduction. The characteristics of some of the pigments are shown in Fig. 11.1. Chlorophylls *a* and *b* are the predominant types of Chl in plants and cyanobacteria (Fig. 11.1*a*). Purple photosynthetic bacteria contain bacteriochlorophyll (BChl), either *a* or *b*, whereas green bacteria have a third type of BChl. Chlorophyll and pheophytins, shown in Fig. 11.1*a*, resemble protoporphyrin IX, the prosthetic group of hemoglobin and the *c*-type cytochromes. In contrast to those proteins, in Chl Mg^{2+} and in pheophytins $2H^+$ replace Fe. Figure 11.1*b* shows β-carotene, a major carotene in many plants, and spheroidine, a bacterial carotenoid. The phycobilins, many of which are poorly characterized, are not shown; they are proteins containing tetrapyrroles similar to the bile pigments in animals. They are present in cyanobacteria (blue-green algae) and red algae in large complexes, the *phycobilisomes*, which absorb light over a wide spectral region.

All the reactions involved in light absorption, transfer of energy, electron transport, and phosphorylation are associated with membranes. Not surprisingly, the latter two resemble the processes in oxidative phosphorylation discussed in Chapter 10.

I. THE PHOTOSYNTHETIC MEMBRANES

The chloroplasts present in eukaryotes are discrete structures enclosed by membranes. They may be spiral-shaped, cup-shaped, star-shaped, or irregular, and in higher plants and bryophytes they are generally shaped like saucers and are approximately 5 to 10 μm in diameter. The light-driven reactions as well as the biochemical pathways involved in the fixation of CO_2 occur in chloroplasts. Figure 11.2 (Staehelin, 1986) shows a chloroplast of a red alga. A thin section of chloroplast of a higher plant as seen with the electron microscope is shown in Fig. 11.3 (p. 303) (Park, 1966). Figure 11.4 (p. 304) illustrates the probable arrangement of a chloroplast similar to that of Fig. 11.3. The membrane enclosing the organelles is double. The internal membrane system of flattened vesicles or sacs, the *thylakoids,* is embedded in the chloroplast interior, the *stroma*. These vesicles may be closely apposed in a combination of shorter and longer lamellae. These composite structures form stacks, the *grana*. The grana stacks are interconnected by lamellae (the *stroma lamellae*), so at least some of the thylakoid inner spaces are continuous

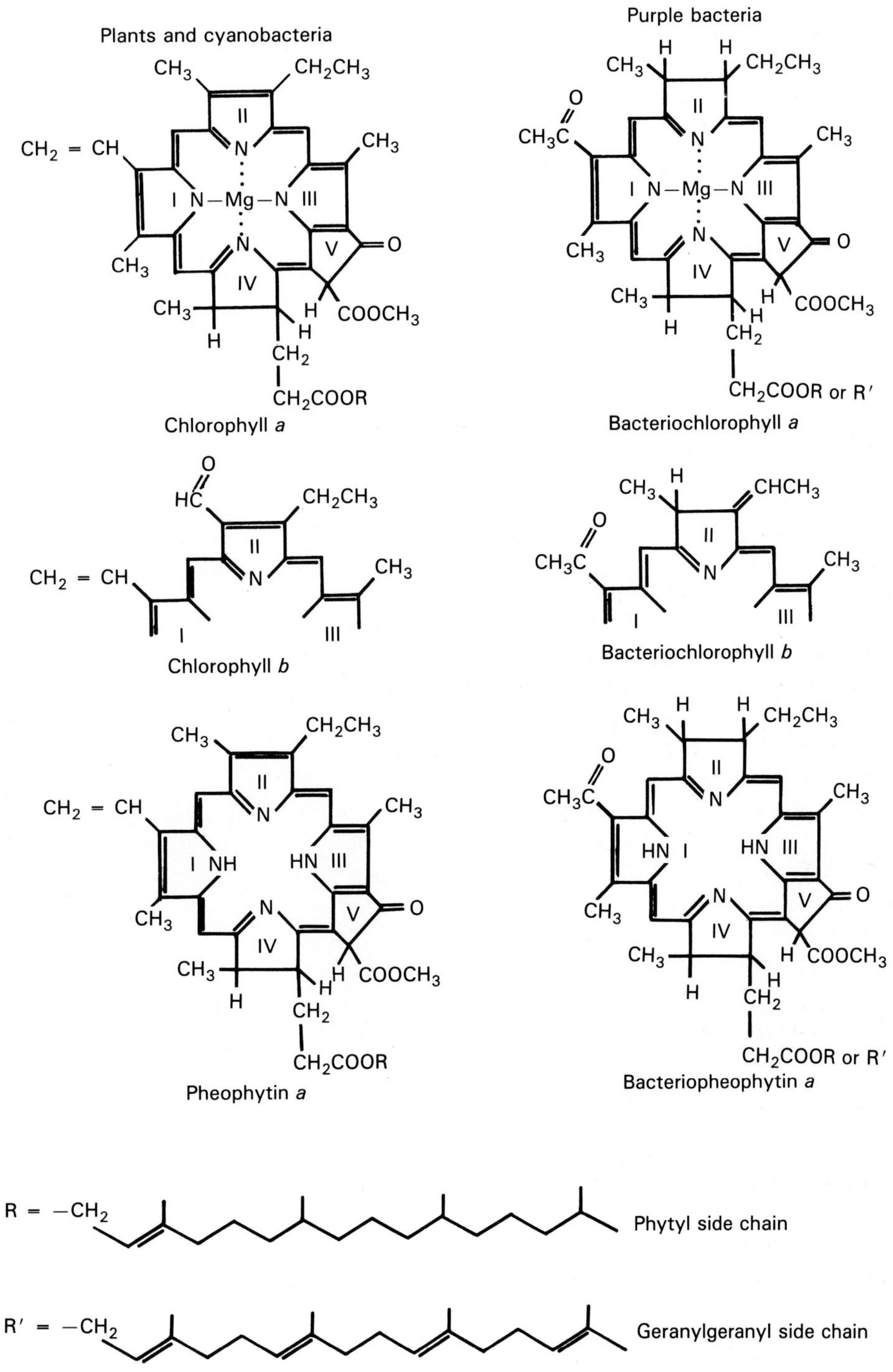

Plants and cyanobacteria

Chlorophyll *a*

Purple bacteria

Bacteriochlorophyll *a*

Chlorophyll *b*

Bacteriochlorophyll *b*

Pheophytin *a*

Bacteriopheophytin *a*

R = $-CH_2$ — Phytyl side chain

R′ = $-CH_2$ — Geranylgeranyl side chain

(*a*)

Fig. 11.1 Structures of photosynthetic pigments. (*a*) Chlorophylls & pheophytins.

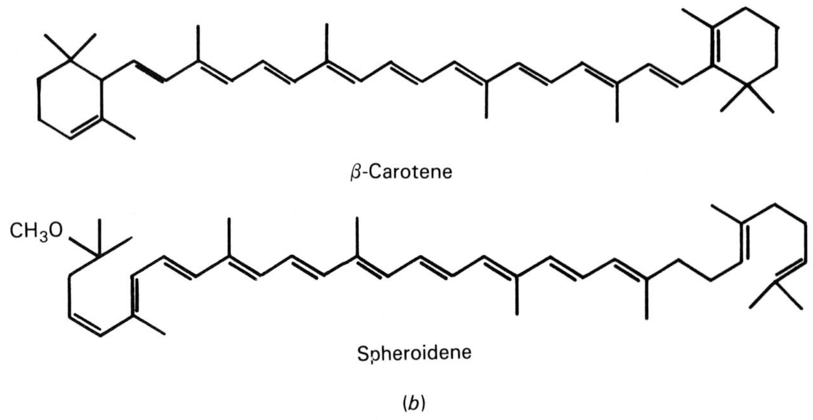

β-Carotene

CH_3O

Spheroidene

(*b*)

Fig. 11.1 Structures of photosynthetic pigments (*Continued*). (*b*) Carotenoids.

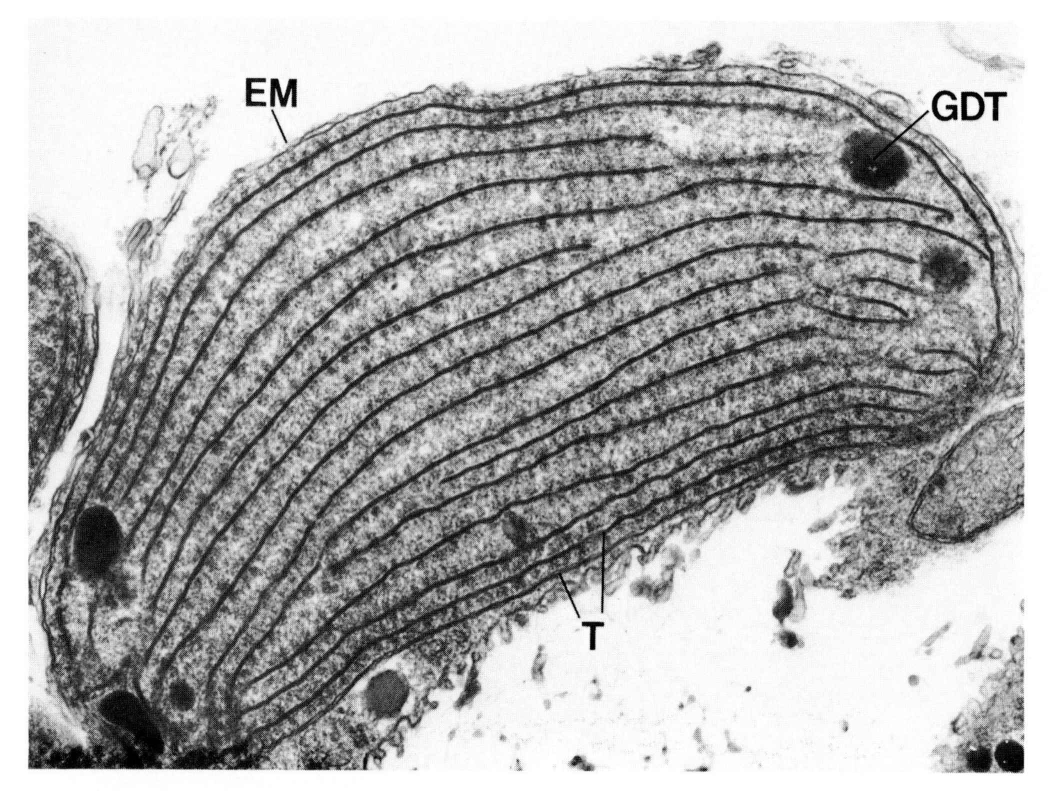

Fig. 11.2 Section through a chloroplast of the red alga *Spermothamnion tuneri*. The phycobilosomes are attached to the thylakoids (T). EM, Envelope membrane; GDT, girdle thylakoid. Reprinted by permission from Staehelin (1986).

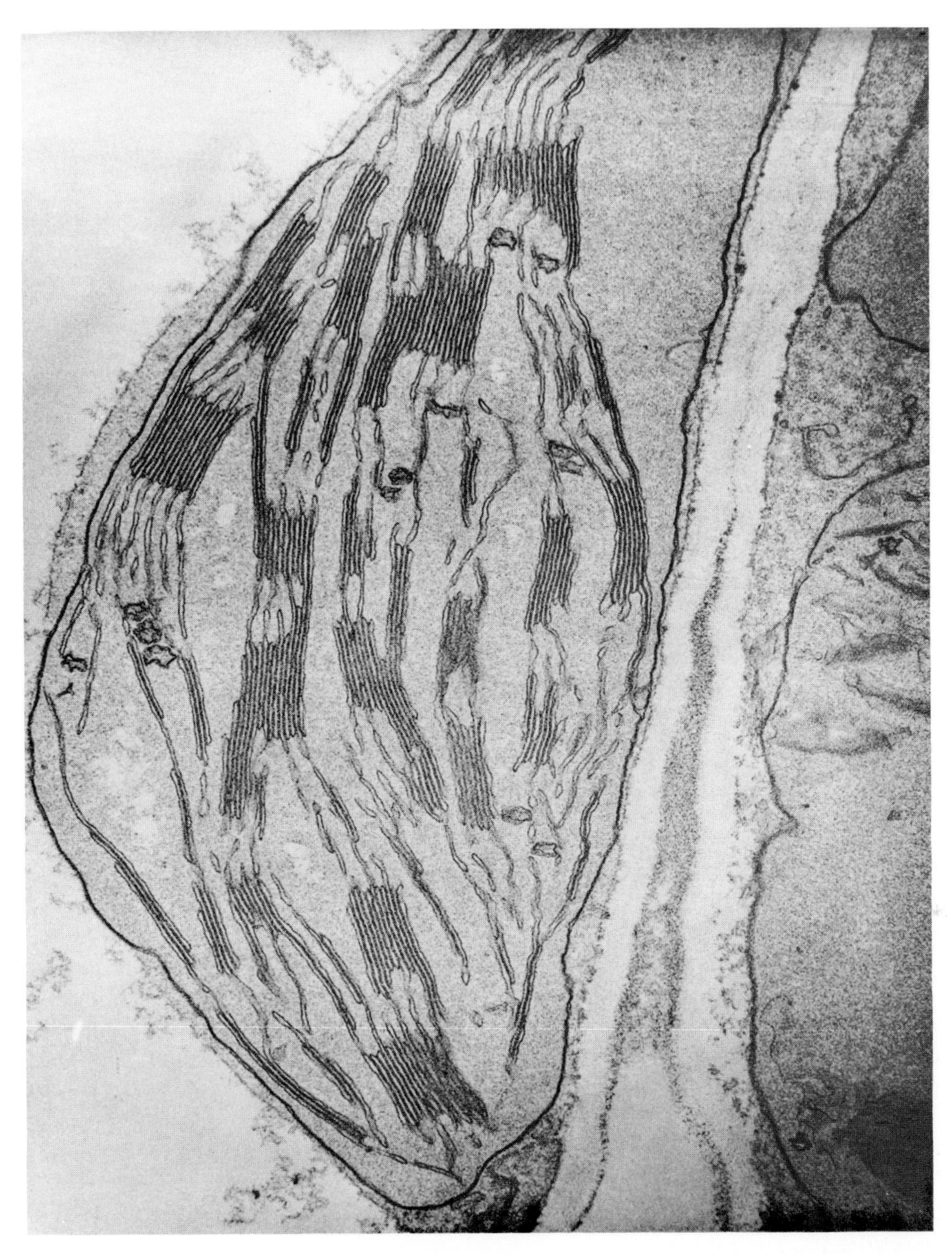

Fig. 11.3 Thin section of $KMnO_4$-fixed *Spinacea oleracea* chloroplast, ×19,500. Reprinted by permission from Park (1966).

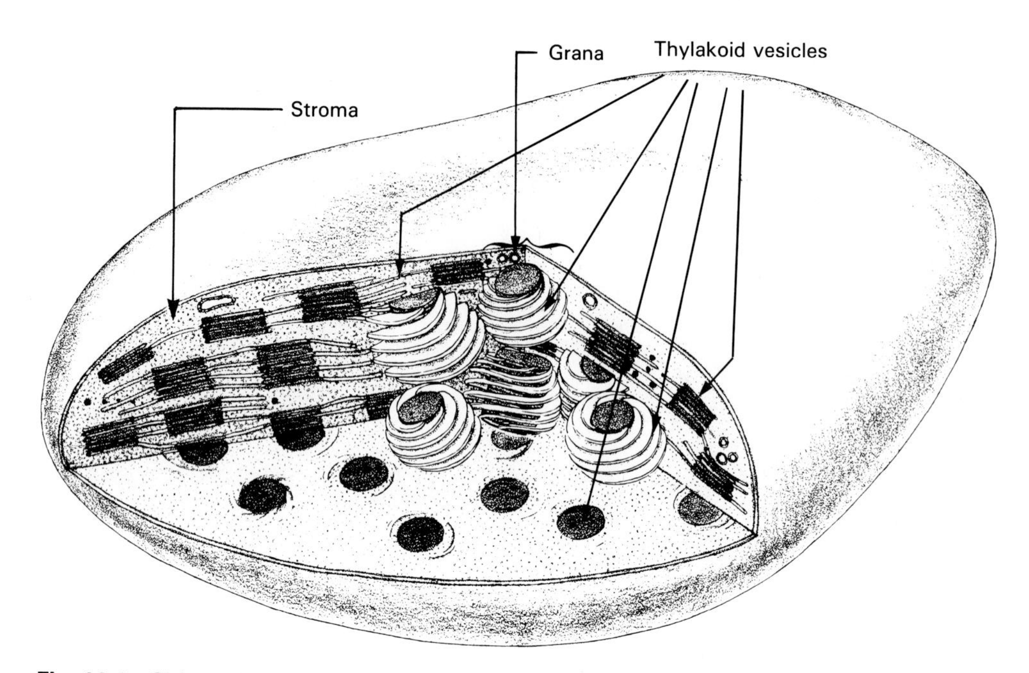

Fig. 11.4 Chloroplast thylakoid membrane architecture viewed in two and in three dimensions. This drawing of a sectioned chloroplast is intended to aid in visualizing the relationship between the appearance of thylakoid membranes in two dimensions with that in three dimensions. From D. R. Ort (1986), *Encyclopedia of Plant Physiology,* with permission. Copyright © Springer-Verlag, West Germany.

with those in other stacks. Parts of the thylakoid surface are apposed to other thylakoid surfaces and other parts, such as the stroma lamellae, are exposed to the chloroplast interior. Although the general features remain the same, the details may vary considerably, even within the same cell. The thylakoids have been shown to contain the chlorophyll as well as the light-harvesting complexes (LHCs), the reaction centers (RCs), and the components of the electron transport chain, which will be discussed later. The soluble material seems to contain the enzymes responsible for carbon dioxide fixation and the biochemical synthetic pathways. The significance of structural details such as the stacking is not clear, since mutants lacking grana are functional (Goodenough et al., 1969; Goodenough and Staehelin, 1971).

In bacteria (Fig. 11.5) (Sprague and Varga, 1986), the morphology of the photosynthetic membranes varies. The cell membrane may be directly involved, or complex invaginations of the cell membrane or special vesicles such as the chlorosomes may be present. Vesicles prepared from photosynthetic bacteria, the *chromatophores,* have been found to be active in photosynthesis.

II. THE EVENTS OF PHOTOSYNTHESIS: LIGHT AND DARK REACTIONS

In photosynthesis, pigments present in specialized complexes, the *antennae,* or light harvesting complexes (LHCs), absorb most of the light. Absorption of photons of incident visible or ultraviolet light produces a change in the electron orbitals of the absorbing molecules (Sauer, 1986). The electrons generally occupy molecular orbitals in the lowest possible energy state, where they are present in one pair per orbital. In the excited state produced by the photon

GREEN BACTERIA (CHLOROSOMES)

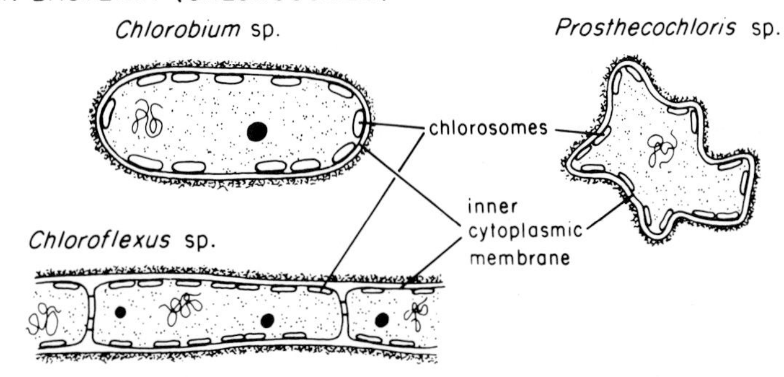

SIMPLE PHOTOSYNTHETIC BACTERIA
(NO CHLOROSOMES OR INTERNAL MEMBRANES)

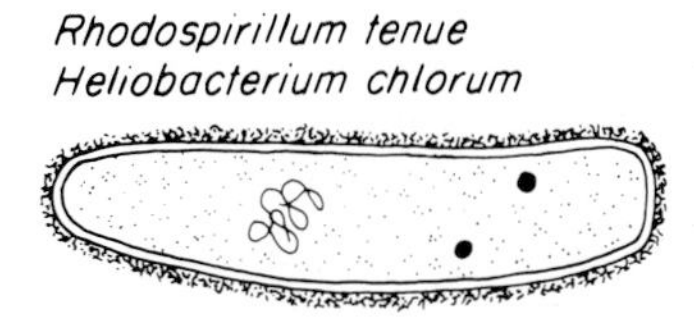

PURPLE BACTERIA (INTRACYTOPLASMIC MEMBRANES)

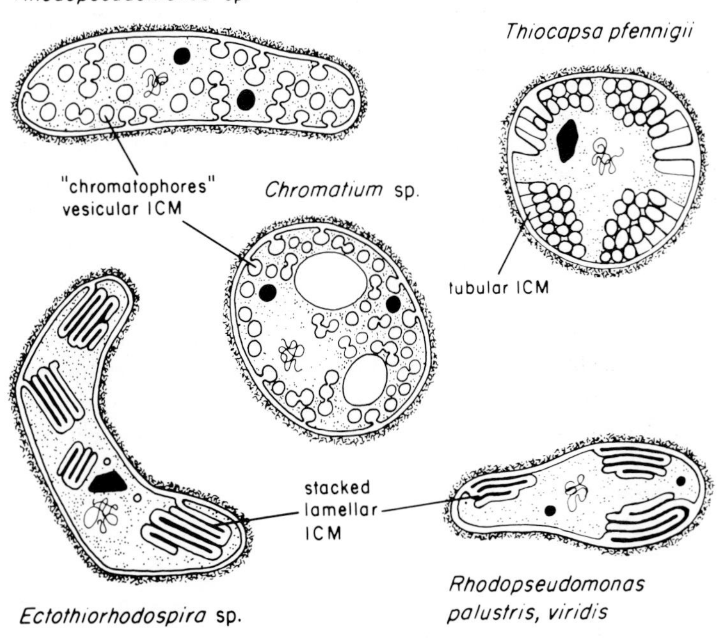

Fig. 11.5 Morphologies observed in the anoxygenic photosynthetic bacteria for housing the photosynthetic apparatus. The green bacteria contain chlorosomes attached to the inner side of the cytoplasmic membrane (CM). Most purple bacteria elaborate an extensive membrane system within the cytoplasm for the photosynthetic components. A few "simple" bacteria contain only a single CM, which contains the pigments and proteins necessary for photosynthetic growth. From S. G. Sprague and A. R. Varga (1986), *Encyclopedia of Plant Physiology,* with permission. Copyright © Springer-Verlag, West Germany.

absorption, one of the electrons is moved to an orbital other than the ground state to produce a singlet excited state. The energy of the photon is converted into an orbital electronic energy of the excited state. The excitation is then transferred from pigment to pigment by resonance energy transfer, over distances as long as 8 to 10 nm. The energy transferred to the RC eventually results in the oxidation of a pigment, for example, P680 and P700 in green plants and P870 in some bacteria (see below)—i.e., the loss of one electron that initiates the events of electron transport (Sauer, 1986). Energy can also be transferred from PSII to PSI.

Conceivably, photosynthesis could involve separate processes: one set of processes involving the trapping of radiant energy, electron transport, and the production of reducing equivalents, and the other set involving the fixation of CO_2. Such a dichotomy would be similar to the events of oxidative phosphorylation, where the enzymes of the tricarboxylic acid cycle can be considered separate from those of electron transport and oxidative phosphorylation. The information presently available is in agreement with this view.

The reactions responsible for the fixation of CO_2 or some other carbon donor clearly belong to a separate system from those involved in the capture of light quanta. These *dark reactions* do not require light. They can take place if reduced NADPH and ATP are present or if the system containing oxidized $NADP^+$ and ADP is exposed to the light. The simplest way to test this proposition is to incubate the photosynthetic system in the dark in the presence of $^{14}CO_2$ (or, for bacteria, some other ^{14}C compound) and other required components after illumination. The data from this kind of experiment are shown in Table 11.2 (Trebst et al., 1958). In sample 1, the mixture containing chloroplast membranes (presumably thylakoids), $NADP^+$, and ADP was exposed to light. After removal of the chloroplast membranes, the $^{14}CO_2$ was introduced into the system in the dark. The counts shown represent CO_2 assimilated. For sample 2 the procedure was identical except that ADP and $NADP^+$ were left out. The amount of CO_2 fixed is very low and probably reflects the presence of a small residual amount of ATP and NADPH. Sample 3 represents the complete system exposed to light with both $^{14}CO_2$ and chloroplast membranes present. Sample 4 corresponds to the complete system maintained in the dark. It is clear that considerable $^{14}CO_2$ fixation occurs in the dark as long as energy has been supplied to the system in the form of ATP and reducing equivalents. Evidently, once the energy is stored in a biochemically utilizable form, neither light nor the photosynthetic machinery of the chloroplast is needed.

Similar experiments have been carried out with bacterial systems, for instance, with extracts of the purple sulfur bacteria *Chromatium*, which incorporate [^{14}C]acetate (Table 11.3) (Losada et al., 1960). As long as ATP (item 2) or light (item 5) is present, the preparation incorporates [^{14}C]acetate. Hexokinase and glucose together (items 4 and 7) largely eliminate the incorporation, probably by decreasing the amount of ATP available. In this case ATP alone suffices, since the

Table 11.2 CO_2 Fixation by a Chlorophyll-Free Extract and a Complete Chloroplast System from Spinach[a]

Treatment	$^{14}CO_2$ fixed (counts/min × 10^3)
1. Chlorophyll-free extract, dark (NADP and ADP present)	134
2. Chlorophyll-free extract, dark (no NADP and no ADP)	9
3. Complete chloroplast system, light	200
4. Complete chloroplast system, dark	20

Source: Trebst et al. (1958). Reprinted by permission from *Nature* 182, 351–355, copyright © 1958 by Macmillan Magazines Limited.

[a] All samples contained NAD and ADP before illumination, except for sample 2.

Table 11.3 Equivalence of ATP and Light in the Assimilation of [^{14}C]Acetate by Cell-Free Preparations of Chromatium

Treatment	^{14}C fixed in soluble compounds (counts/min × 10^3)
1. Dark, control	27
2. Dark, ATP	180
3. Dark, ATP, hexokinase	186
4. Dark, ATP, hexokinase, glucose	6
5. Light, control	414
6. Light, hexokinase	348
7. Light, hexokinase, glucose	20

Source: Losada et al. (1960). Reprinted by permission from *Nature,* 186, 753–760, copyright © 1960 by Macmillan Magazines Limited.

reducing equivalents can be produced from ATP hydrolysis by running electron transport in reverse (see Section III,C).

Conceivably, the biochemical pathway followed may depend on whether light is present. However, the radioautographic paper chromatographs of Fig. 11.6*a* and *b* (Trebst et al., 1958), obtained with the reaction products of experiments such as those of Table 11.2, show that this is not the case. The results show that the intermediates accumulating under the two sets of conditions (i.e., in the dark, after illumination or under continuous illumination) are the same. The principles of paper chromatography used to separate these molecules are discussed in the legend of the figure.

The results just discussed show that the light-requiring reactions per se are not involved in carbon fixation, except to supply chemical free energy in the form of NADPH and ATP. Seen from this point of view, photosynthesis is concerned directly only with trapping radiant energy and converting it into biologically utilizable forms: reducing equivalents and a phosphate of high group-transfer potential.

In different organisms photosynthesis differs in detail. However, the light-requiring processes correspond to the production of ATP and reducing equivalents. The subsequent biochemical reactions can take place in the dark.

III. PHOTOOXIDATION

The sections that follow are concerned with the events that result in the production of ATP and reducing equivalents during photosynthesis.

After the primary photochemical event—the light-energized oxidation of Chl in the reaction centers—the return of the excited electrons through electron transport carriers could provide enough energy for the synthesis of ATP from ADP and P_i by a process analogous to that taking place in mitochondria. Similarly, these electrons could be used to reduce electron carriers with the eventual production of NADPH.

Let us begin with the photooxidation of chlorophyll. When Chl or BChl are oxidized one electron is removed, not from the metal (in this case Mg) as in the case of cytochromes, but from the aromatic π-electron system of the molecule. The spin of the remaining electron is delocalized over the whole π-electron system.

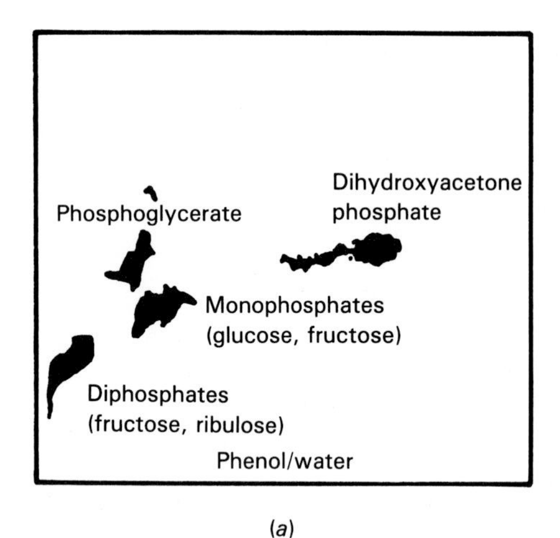

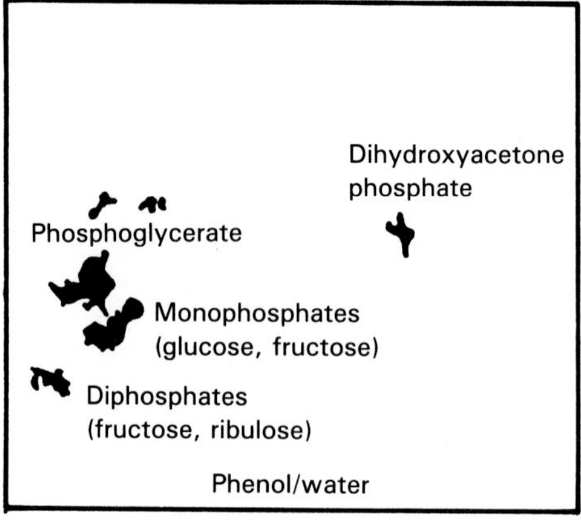

(*a*) (*b*)

Fig. 11.6 Tracings of radioautographs of chromatograms showing (*a*) products of photosynthetic $^{14}CO_2$ assimilation by illuminated spinach chloroplasts and (*b*) products of dark $^{14}CO_2$ assimilation by chlorophyll-free extract of chloroplasts. In paper chromatography, a mixture is placed at one end of the paper. The solution moves by capillary action, and it is submitted to minute partition steps between the solvent and the wet cellulose fibers of the paper. The migration of various compounds on the paper and in a particular medium will depend on their individual solubility in the solvent and the location can be seen by means of reactions generating colored compounds or, in the case of radioactive compounds, by autoradiography. Ideally, each spot is unique and corresponds to a single compound. In the chromatographs, the compounds have migrated in one direction in one particular solvent system (phenol:water) and in the other direction, at right angles from the first, in another solvent system (butanol:acetic acid). Separation with such a technique depends on the solubility of the compounds in the different solvents, resulting in increased separation. The position of the compounds was recorded by placing the paper on photographic paper. From Trebst, et al. (1958). Reprinted by permission from *Nature*, 182:351–355, copyright © 1958 Macmillan Magazines Limited.

The photooxidation can be studied and detected by electron paramagnetic resonance (EPR) and also by light absorption measurements. For the latter, the experimental design and the apparatus used for spectrophotometric measurements in photosynthetic systems are illustrated in Fig. 11.7 (Clayton, 1980). In this representation, the monitoring beam provides the light needed to measure changes in absorption. The beam is weak and ideally has no effect on the pigments and therefore none on the absorption spectrum. The detector, which measures the light transmitted through the sample, is shielded in some way from stimulation by the excitatory beam. Since excitatory and monitoring beams are generally at different wavelengths, this can be done with light filters.

Various experiments have demonstrated that the primary photosynthetic event is the oxidation of chlorophyll by light. A mutant of the purple photosynthetic bacteria *Rhodobacterium sphaeroides* that lacks carotenoids was used in the experiment of Fig. 11.8 (Clayton, 1980). Figure 11.8*a* represents the absorption spectrum of the isolated RCs. The pigments responsible for the peaks are indicated. Figure 11.8*b* represents the change in optical density brought about by illumination, which occurs primarily in the region of absorption of BChl. The effect is an oxidation, as shown by the inhibition of the effect with a reducing compound (Fig. 11.8*c*).

The chlorophylls that are photooxidized have been named after the wavelength, in nanometers, of the major absorbance decrease produced by the photooxidation, in this case P870 (P stands for pigment). In chloroplasts, two such reactive complexes have been detected spectrophotometrically: P700 and P680, corresponding to the two photosystems (I and II) discussed below. In bacteria, two BChl *a* or *b* molecules, depending on the bacteria, appear to be involved in the photooxidation, since the remaining unpaired electron is delocalized over two BChl molecules. It has been suggested that two Chl *a* molecules are also involved in the chloroplasts, but at this time the evidence is contradictory.

A. Organization of the Reaction Centers

The chlorophylls that undergo photooxidation are complexed to protein forming the reaction center. The reaction centers of several bacteria have been purified and crystallized (Parson, 1987; Pierson and Olson, 1987). In this section we consider only the RC-2 bacteria (filamentous and purple bacteria). The RC-1 bacteria (green sulfur and gram positive) differ in having BChl as the primary acceptor and an Fe-S center as the secondary acceptor (Malkin et al., 1981). Each center of RC-2 bacteria contains four molecules of BChl, two of bacterial pheophytin (BPh), one or two quinones, and one nonheme Fe, which can be replaced by Mn or Zn. The BPh corresponds to BChl *a* without Mg. The reaction centers generally also contain a cytochrome with four *c*-type hemes. Three polypeptides, generally referred to as L, M, and H, are present in 1:1:1 stoichiometry in most bacteria studied, and their molecular masses are 32, 34, and 29 kDa. The H polypeptide is not present in all bacteria and, when present, can be removed without affecting function. Two of the BChl molecules are packed close together, and their optical absorption changes with illumination identify them as P870.

Figure 11.9 shows the arrangements of the prosthetic groups of the center as identified by the three-dimensional reconstruction of x-ray crystallography (Deisenhofer et al., 1984, 1985). The four hemes are in the cytochrome subunit, and the other components shown are in the L-M

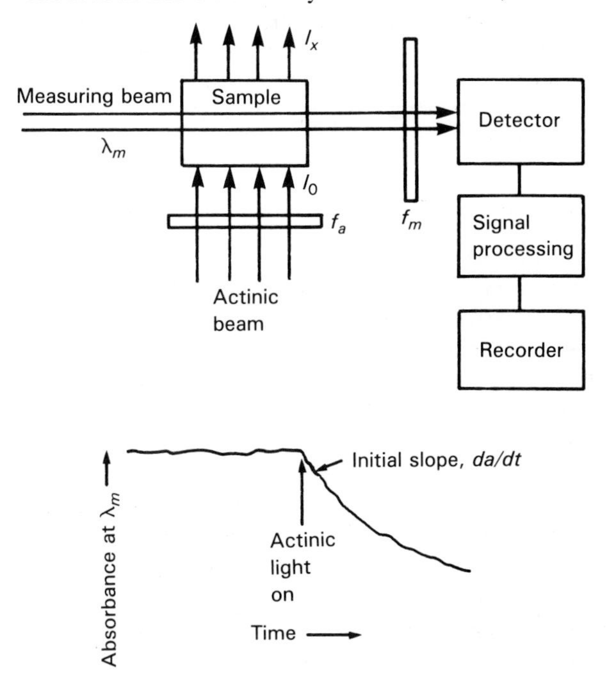

Fig. 11.7 System for measuring light-induced absorbance changes and the quantum efficiency of a photochemical process. From Clayton (1980). Copyright © 1980 by Cambridge University Press, reprinted with the permission of Cambridge University Press.

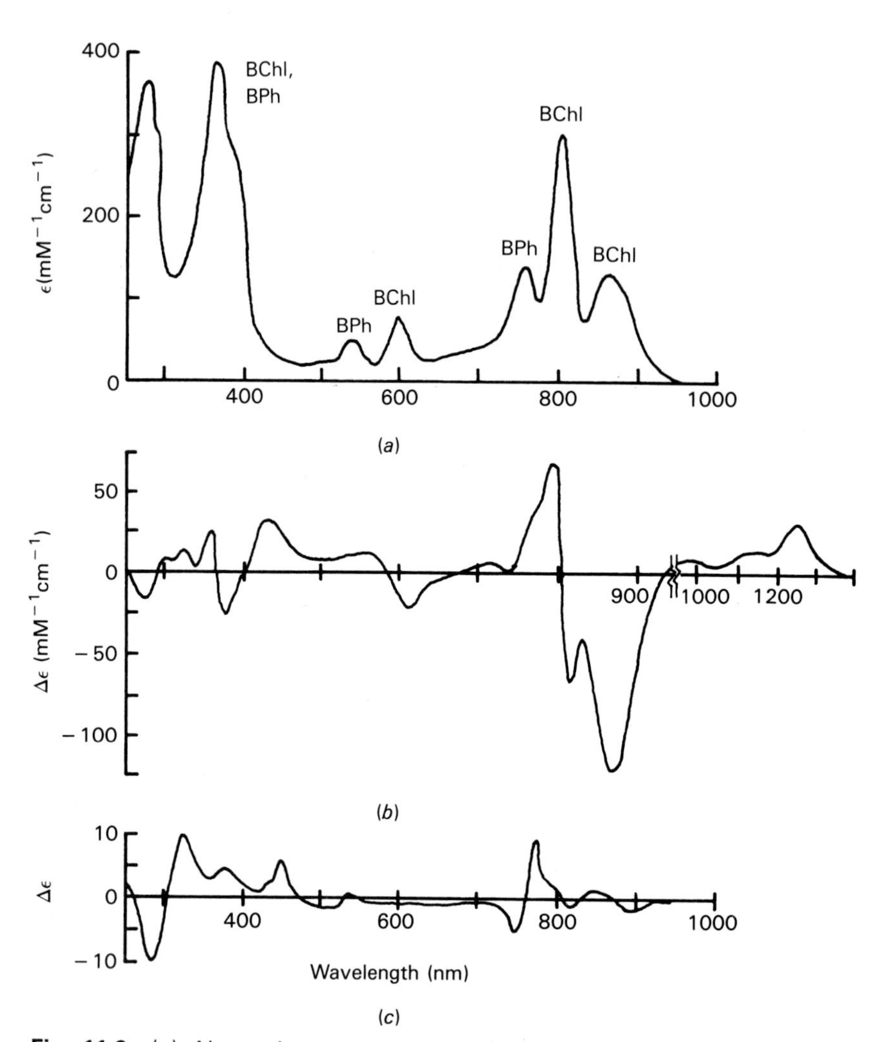

Fig. 11.8 (*a*) Absorption spectrum and (*b* and *c*) spectra of light-induced absorbance changes of reaction centers isolated from carotenoidless mutant *R. sphaeroides*. (*b*) Reaction centers alone. (*c*) Reaction centers plus an electron donor to prevent the accumulation of oxidized bacteriochlorophyll during illumination. Changes due to oxidation of the donor have been discounted. From Clayton (1980). Copyright © 1980 by Cambridge University Press, reprinted with the permission of Cambridge University Press.

complex. The L and M proteins traverse the membrane, with the periplasmic side represented at the top and the cytoplasmic side at the bottom. The orientation of the quinone in this diagram is arbitrary.

The reaction centers of the chloroplasts corresponding to P700 and P680 have not been purified, as discussed in Section VI.

B. Cyclic Photophosphorylation of Bacteria

When RCs of isolated bacteria are excited with a very short flash, a transient P870 excited single state is formed and decays in picoseconds, forming a $P870^{+}BPh^{-}$ radical pair. Accordingly, the absorption bands of P870 and BPh disappear, and new bands that have been attributed to the radicals are formed. This is followed by the reduction of one quinone to produce a semiquinone radical, Q_A^-. The reactions occur extremely rapidly, essentially with a quantum efficiency of 1 and independent of temperature. This supports the idea that these components are held tightly together with little motion (Dutton, 1986).

The electron of Q_A^- is transferred to a second quinone, Q_B. This electron displacement is a much slower process and is temperature dependent. These reactions and the redox potentials are outlined in Fig. 11.10.

The electron displaced from P870 is replaced from the *c*-type cytochrome, so the RC can respond again to light. After a second electron is removed from P870, the Q_B has become fully reduced and Q_B^{2-} picks up two H^+ from the cytoplasmic side of the membrane. The cytochrome bc_1 complex oxidizes Q_BH_2. These reactions are thought to proceed with a net proton efflux of as many as four H^+ per two electrons (Dutton, 1986). The passage of these protons through a channel in the F_0 portion of the ATP synthase is thought to be responsible for ATP synthesis (see Chapter 12). These details are summarized in Fig. 11.11.

The series of reactions just discussed is entirely cyclic. The electron displaced from the photooxidation of P870 is replaced by the cytochrome bc_1 complex, which in turn recovers it from the oxidation of quinone. Thus, the energy from the light absorption is converted into ATP

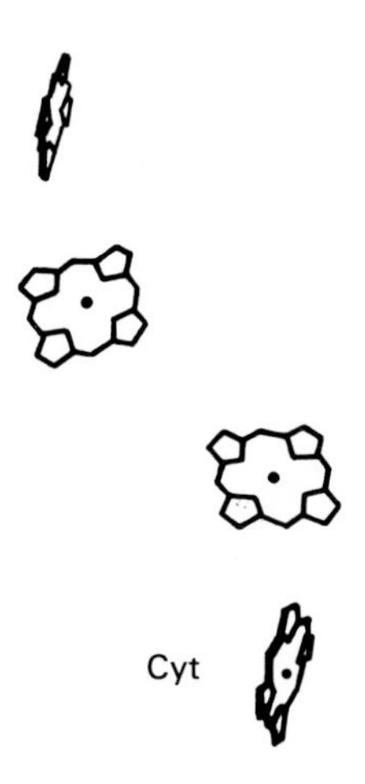

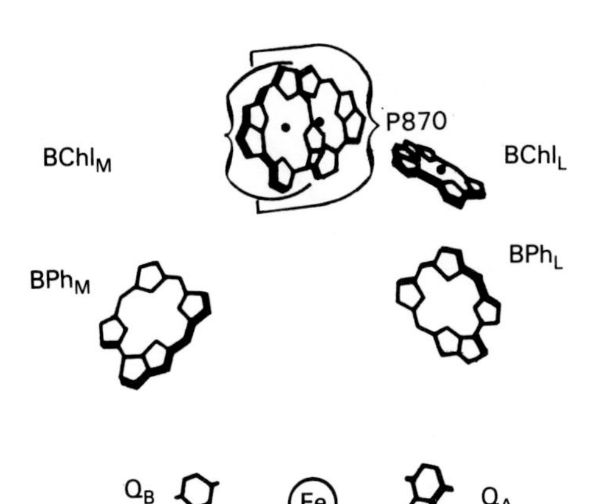

Fig. 11.9 Arrangement of the prosthetic groups in the *Rhodopseudomonas viridis* reaction center. Q_B is shown at the site identified by Deisenhofer et al. (1984) but the orientation of the quinone in this site is drawn arbitrarily; the exact orientation of Q_B in the crystal structure has not been described. The four hemes at the top are in the cytochrome subunit; the other components are in the L-M complex as indicated by the subscripts. The plane normal to the chromatophore membrane is approximately vertical and the periplasmic side of the complex is at the top. Reprinted by permission from W. W. Parson, *Photosynthesis.* Copyright © 1987 Elsevier Science Publishers, Amsterdam.

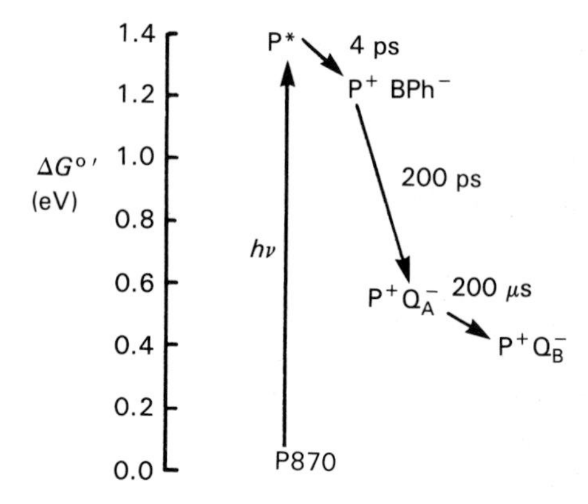

Fig. 11.10 Kinetics and standard free energy changes of electron transfer steps in reaction centers isolated from *Rb. sphaeroides*. The rates of the transfers are indicated next to the arrows.

with no net change in electrons. Most photosynthetic eubacteria carry out cyclic electron transport (Pierson and Olson, 1987).

C. Production of Reducing Equivalents

What accounts for the production of reducing equivalents? In the RC-2 bacteria, NAD^+ is probably reduced by reverse (energy-requiring) electron transport from the quinone pool. In

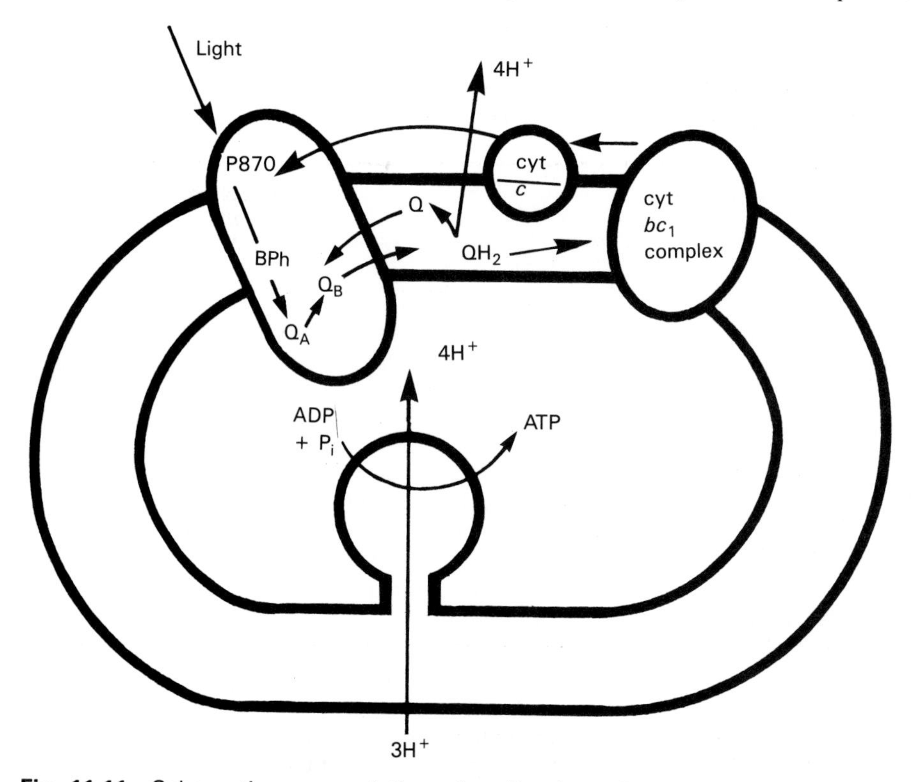

Fig. 11.11 Schematic representation of cyclic photophosphorylation in purple bacteria. Reducing equivalents are transferred from the P870 complex to the cytochrome bc_1 complex by quinone (dissolved in the membrane), which is then returned to the P870 complex by cytochrome *c* (water soluble). The protons generated by the electron transport return through the ATP synthase, providing the energy to synthesize ATP from ADP and P_i.

contrast, green sulfur bacteria reduce ferredoxin directly from the secondary acceptor. Both cases require an external reductant such as succinate or H_2S.

Experiments with chromatophores isolated from *Rhodospirillum rubrum* and other nonsulfur purple bacteria show that electrons can be transferred from succinate to NAD^+ in the dark, provided that ATP or pyrophosphate hydrolysis (presumably through the action of ATP synthase, proceeding in reverse) is available to supply energy (Jones and Vernon, 1969). In the absence of ATP, the reduction required cyclic electron flow to pump electrons uphill from organic reductants such as succinate. A block of the cyclic electron flow blocked NAD^+ photoreduction but not the ATP-driven reduction. In addition, uncouplers of oxidative phosphorylation or photophosphorylation blocked the photoreduction and the ATP-dependent reduction (Hauska et al., 1983; Malkin, 1987). Uncouplers are thought to collapse the energy-dependent H^+ electrochemical gradient, i.e., the proton motive force. Therefore, these experiments suggest that the energy is provided indirectly through the proton motive gradient. The reduction also requires the presence of a functional NADH dehydrogenase. The mechanism suggested by these experiments is summarized in Fig. 11.12*a* (Knaff and Kampf, 1987).

In green sulfur bacteria, however, the energy is thought to be directly available, since there is no inhibition by uncouplers (Knaff and Kampf, 1987). The scheme consistent with the information available for these bacteria is summarized in Fig. 11.12*b*.

IV. THE LIGHT-HARVESTING SYSTEM

The number of Chl molecules per RC is very large. It has been estimated to be from 25 to several hundred, depending on the bacteria, to about 300 in the chloroplast. The function of

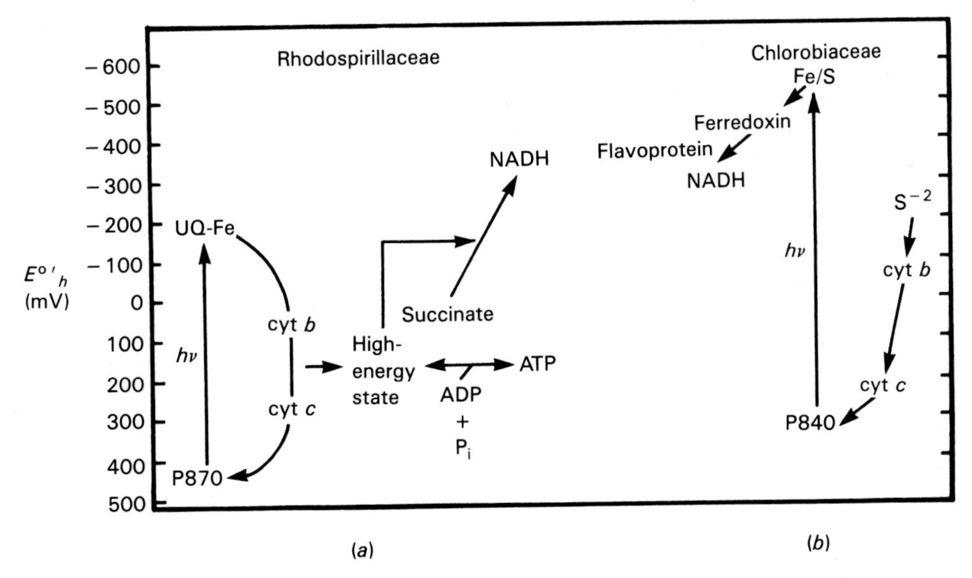

Fig. 11.12 Mechanism of NAD^+ photoreduction in (*a*) purple bacteria and (*b*) green sulfur bacteria. UQ-Fe represents the Fe^{2+}-quinone complex present at the primary quinone site of purple bacteria, although in some species menaquinone replaces ubiquinone. Fe/S represents the iron-sulfur center that functions as an early acceptor in green sulfur bacteria. The earliest electron acceptors have been omitted for both green and purple bacteria. The involvement of cyt *b* in S^{2-} oxidation in green bacteria is speculative but is based on inhibition by antimycin A. Reprinted by permission from D. B. Knaff and C. Kampf, *Photosynthesis.* Copyright © 1987 Elsevier Science Publishers, Amsterdam.

these molecules is to act as an antenna, i.e., absorb light and transfer the energy to neighboring molecules by resonance transfer until it is trapped in the reactions of the RC. Part of the evidence for this mechanism stems from the fact that in oxygen-producing systems the amount of oxygen evolved increases with intensity of the flash until a maximal saturation rate is obtained. The amount of O_2 produced is very small in relation to the Chl content because the RCs are finite in number and must be reduced again before participating in another photooxidative event. About 2500 Chl molecules are present per O_2 evolved under saturating conditions or 300 per RC (8 photons are required per O_2, see Fig. 11.20).

The light-harvesting complexes (LHCs) contain Chl or BChl and a variety of other pigments such as carotenoids (Zuber et al., 1987). Thus they can absorb light from a very broad region of the spectrum. The pigment molecules are noncovalently attached to integral proteins. A few pigment molecules are associated with each protein. The energy is transferred from short-wavelength-absorbing to longer-wavelength-absorbing pigments by inductive resonance and the formation of exciton states.

V. THE TWO PHOTOCHEMICAL SYSTEMS OF CHLOROPLASTS

As already discussed, chloroplasts have two distinct chlorophyll complexes in the RCs: P700 and P680. With their respective electron acceptors and donors, these complexes constitute two distinct assemblies, which have been called photosystem I (PSI) and photosystem II (PSII), respectively.

A summary of our present understanding of their organization is shown in Fig. 11.13 (Blakenship and Prince, 1986). The scale on the left represents the redox potentials of the electron carriers. The large arrows represent the excitation of PSI or PSII by light. As in the figures in Chapter 10, the small arrows represent electron transport steps. As shown in Fig. 11.13, PSI and PSII are connected in series; i.e., the electron transport chain that receives electrons from PSII replaces the electrons displaced by the light reaction in PSI. This arrangement also imposes two distinct features. One corresponds to a mechanism that replaces the electron lost by the photooxidation of PSII, carried out by the oxygen-evolving complex (OEC) (Dunahay et al., 1984), and the other corresponds to a series of reactions that accept the electrons from PSI to eventually reduce $NADP^+$.

Biochemical fractionation of the chloroplast photosynthetic system has shown that the complete system consists of three separate transmembrane complexes: PSI, PSII, and cytochrome b_6f complex. The electron transport between PSII and the cytochrome b_6f complex is mediated by plastoquinone (PQ). Plastocyanin (PC) mediates the electron transport between the cytochrome b_6f complex and PSI. The three complexes are represented in Fig. 11.14 in diagrammatic form (Cramer et al., 1985).

Plastoquinone closely resemble ubiquinone (CoQ), a component of the electron transport chain of mitochondria. *Plastocyanin* is a water-soluble 11-kDa protein containing copper ion coordinated to the side chains of cystein, methionine, and two histidines. Reduced PC and oxidized PC contain Cu^+ and Cu^{2+}, respectively.

The arrangement of Fig. 11.14 is reminiscent of the four complexes of mitochondria in which electron transport is carried out between complexes by either CoQ or cytochrome *c* (Chapter 10, Figs. 10.9 and 10.10).

The stoichiometry of the photosystems and the Chl distribution in higher plant chloroplasts are summarized in Table 11.4 (Glazer and Melis, 1987; Whitmarsh and Ort, 1984).

The complexes are not randomly distributed in the membrane; rather, PSII is primarily located in the appressed grana and PSI primarily in the stroma lamellae (Anderson and Anderson,

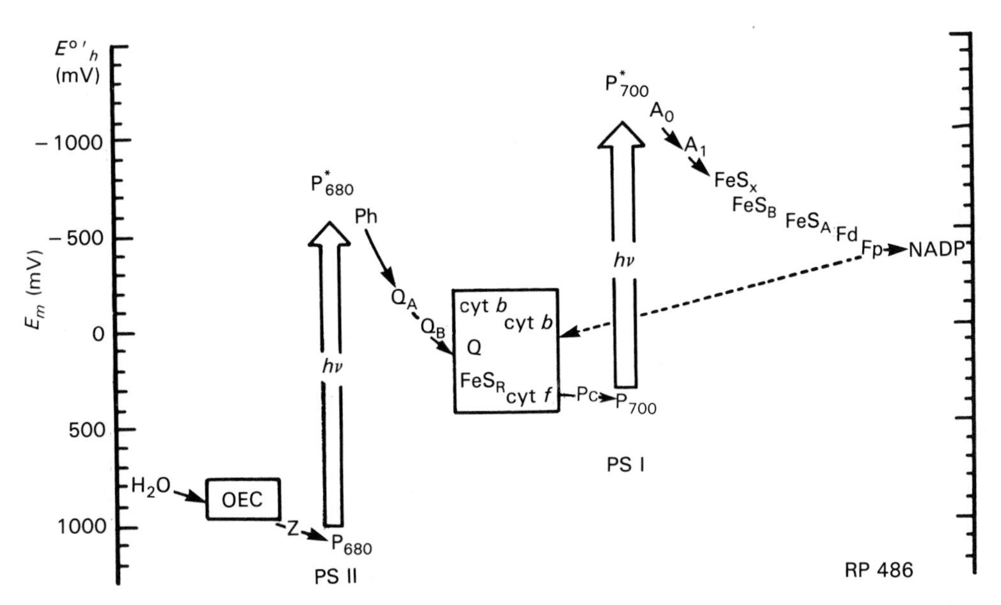

Fig. 11.13 The Z scheme for oxygenic photosynthesis constructed using excited-state redox potentials. The carriers are placed at the accepted midpoint redox potentials (at pH 7) where they have been measured directly; other potentials are estimated. OEC, oxygen-evolving complex; Z, donor to photosystem II (PSII); P_{680}, reaction center chlorophyll of PSII; Ph, pheophytin acceptor of PSII; Q, quinone; cyt, cytochrome; FeS_R, Rieske iron-sulfur protein; Pc, plastocyanin; P_{700}, reaction center chlorophyll of PSI; A_0 and A_1, early acceptors of PSI (possibly chlorophyll and quinone species); FeS_x, FeS_B, and FeS_A, bound iron-sulfur protein acceptors of PSI; Fd, soluble ferredoxin; Fp, flavoprotein (ferredoxin-NADP reductase). FeS_B and FeS_A may operate in parallel. The dashed line indicates cyclic electron flow around PSI. The pathway of electron flow through the cyctochrome b_6f complex is outlined by a box. Reprinted by permission from R. E. Blankenship and R. C. Prince, *Trends in Biochemical Sciences,* 10:382–383. Copyright © 1985 Elsevier Science Publishers, England.

1982). The interaction between the three complexes is thought to occur by the diffusion of plastoquinone in the membrane from PSII to the cytochrome b_6f complex and by the location of the cytochrome complex close to PSI. The low molecular weight and high lipid solubility of plastoquinone would permit fast diffusion in the plane of the membrane, sufficient to account for the electron transport rate (Haehnel, 1984). This arrangement is shown in Fig. 11.15 (Haehnel, 1984).

What is the evidence for two photosynthetic events? Part of the story includes results of experiments on the effectiveness of various wavelengths. The effectiveness of light in photosynthesis can be expressed as the yield of O_2 evolved per einstein of light. The einstein, the energy equivalent of photons, corresponds to $Nh\nu$, where N is Avogadro's number (6×10^{23}), ν is the frequency of the light (vibrations per second), and h is Planck's constant (1.6×10^{-34}). Results of an experiment measuring the yield as a function of wavelength for the grana alga, *Chlorella,* are shown in Fig. 11.16 (Emerson et al., 1957). Curve 1 represents the yield with a single beam of light. Curve 2 represents the yield with two beams of light, the second beam being of lower intensity maintained constant at a shorter wavelength. The yield is greater for curve 2, particularly at longer wavelengths. The effect must involve separate light-activated units, since it occurs even when the two beams are alternated (Fig. 11.17) (Myers and French, 1960).

The excitation of PSI and PSII is necessary for optimal functioning of chloroplast photosynthesis, but the two also have separate functional roles. The photochemical O_2-generating system

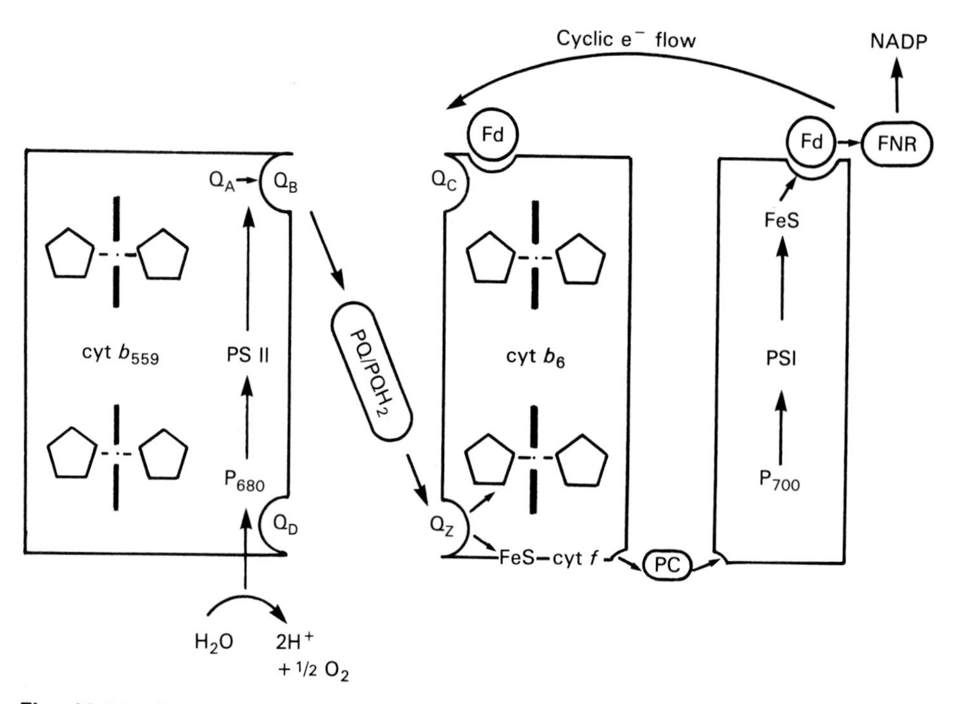

Fig. 11.14 Arrangement of the three major transmembrane protein electron transport complexes in the thylakoid membrane. The photosystem II complex is involved in water splitting, in electron transfer to noncyclic electron transport chain and proton deposition to the lumen. It includes the 51- and 44-k Da chlorophyll-binding polypeptides that also bind the Q_A acceptor and (perhaps) the Q_D donor plastoquinone; two 32-kDa putative plastoquinone-binding proteins to which herbicides and probably the Q_B plastoquinone also bind; and cytochrome b_{559} containing two hemes per P680 reaction center that are hypothesized to span the membrane as shown. Electrons from the PSII complex are transferred by a plastoquinone pool to the cytochrome b_6f complex through a plastoquinone binding site, Q_Z. The *bf* complex consists of four subunits: 34-kDa cytochrome *f*, 23-kDa cytochrome b_6 (with two heme groups), a 20-kDa polypeptide FeS redox center, and a fourth 17-kDa polypeptide (not shown). The cytochrome *bf* complex is connected to photosystem I by plastocyanin (PC). The transmembrane PSI P700 reaction center complex possibly contains two large (M_r 70,000) P700 reaction center chlorophyll polypeptides and at least three smaller peptides that may contain the FeS electron receptor complex used to reduce ferredoxin (Fd) and NADP. The involvement of Fd in PSI cyclic phosphorylation, possibly through a plastoquinone binding site, Q_C, is shown. The fourth membrane-spanning ATP synthase complex is not shown. Reprinted by permission from W. A. Cramer, et al., *Trends in Biochemical Sciences,* 10:125. Copyright © 1985 Elsevier Science Publishers, England.

Table 11.4 Photosystem Stoichiometry and Chlorophyll Distribution in Wild-Type Higher Plant Chloroplasts

	PSII[a]	PSI[a]	Cyt *b*/*f*[b]
Stoichiometry	1.7	1.0	1.10
Antenna size	330	200	
Chl *a*	235	180	
Chl *b*	95	20	
Total Chl, %	63	37	

[a]Data from Glazer and Melis (1987). Reproduced, with permission, from the Annual Review of Plant Physiology, vol. 38, © 1987 by Annual Reviews Inc.

[b]Data from Whitmarsh and Ort (1984).

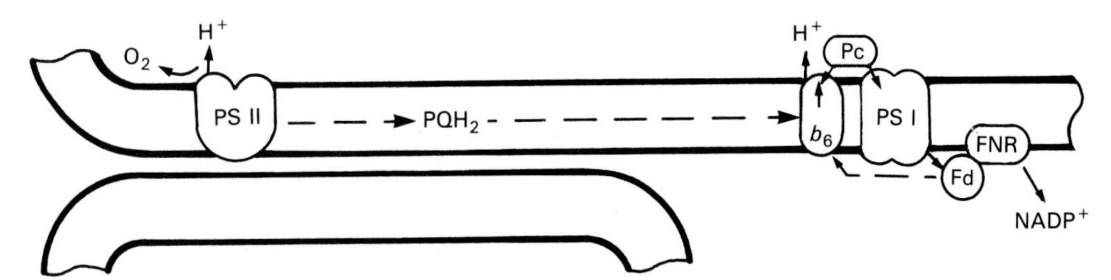

Fig. 11.15 Schematic representation of photosynthetic electron transport distributions of the integral complexes in chloroplasts with grana stacks. PQH_2, Plastoquinol; b_6f, cytochrome b_6f complex; Pc, plastocyanin; Fd, ferredoxin; and FNR, ferredoxin-$NADP^+$ reductase. Not shown is the H^+ uptake from outside. The long dashed arrows indicate assumed long-range diffusion of mobile electron carriers. From Haehnel (1984). Reproduced, with permission, from the *Annual Review of Plant Physiology,* volume 35, © 1984 by Annual Reviews Inc.

can be dissociated from other events by providing the system with an electron acceptor (e.g., indophenol blue). Such a reaction could be visualized as follows:

$$2OH^- \rightarrow 2e^- + 2H^+ + \tfrac{1}{2}O_2$$

The electrons would be picked up by the electron acceptor (not shown in the equation). On the other hand, the production of reducing equivalents in the form of NADPH would require an electron donor (e.g., ascorbate):

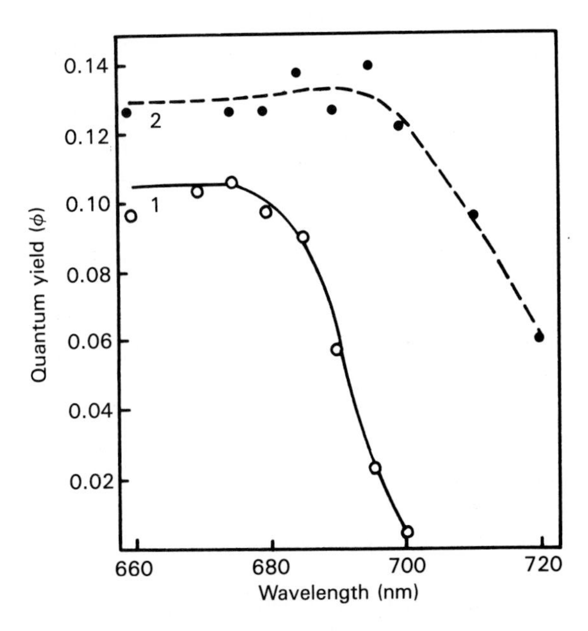

Fig. 11.16 Effect of supplementary light on quantum yield. Curve 1, control; curve 2, same conditions as used for curve 1 but with supplementary light. From R. Emerson, *Science,* 127:1059–1060. Copyright 1958 by the AAAS.

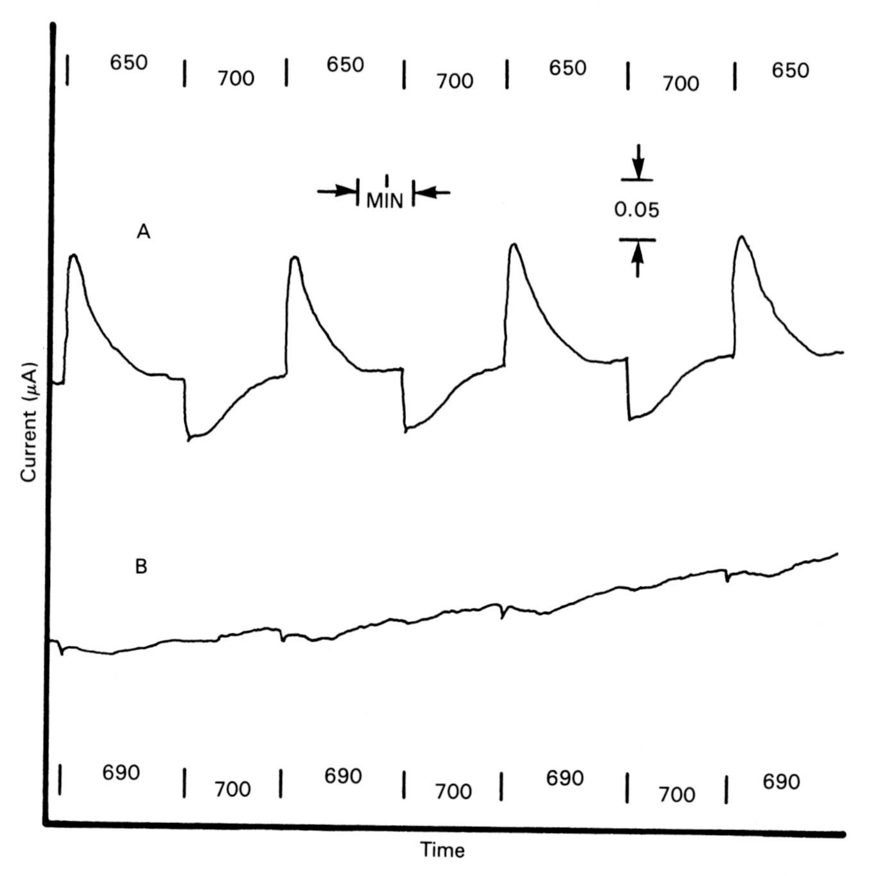

Fig. 11.17 Chromatic transients observed by alternating two light beams of intensities adjusted to sustain equal steady rates of photosynthesis. The current recorded from a platinum electrode is proportional to the oxygen concentration. Record A, alternating beams of 650 and 700 nm. Record B, alternating beams of 690 and 700 nm. From Myers and French (1960). Reproduced from *The Journal of General Physiology*, 1960, 43:723–736, by copyright permission of the Rockefeller University Press.

$$NADP^+ + H^+ + 2e^- \rightarrow NADPH$$

Figure 11.18 (Arnon, 1961) shows the wavelengths most effective for oxygen production or NADP reduction over the narrow wavelength range where the absorption spectra of the two chlorophylls do not overlap. Clearly, the evolution of O_2 is most effectively produced by a lower wavelength than the reduction of $NADP^+$, testifying to the involvement of PSII. On the other hand, $NADP^+$ reduction has action peaks at much longer wavelengths, revealing an association primarily with PSI. Interestingly, ATP production seems to be associated with PSII.

As represented in Fig. 11.13 or 11.14, PSI and PSII are connected in series by the cytochrome b_6f complex. The evidence for such an arrangement is the effect of illumination of either PSI or PSII on the redox state of the cytochrome complex. Stimulation of PSII reduces the cytochrome complex, whereas illumination of PSI oxidizes it. The experiment represented in Fig. 11.19 (Duysens et al., 1961), carried out with the red alga *Porphyridium*, shows the absorption of cytochromes. In this record, the upward deflections reflect oxidation and the

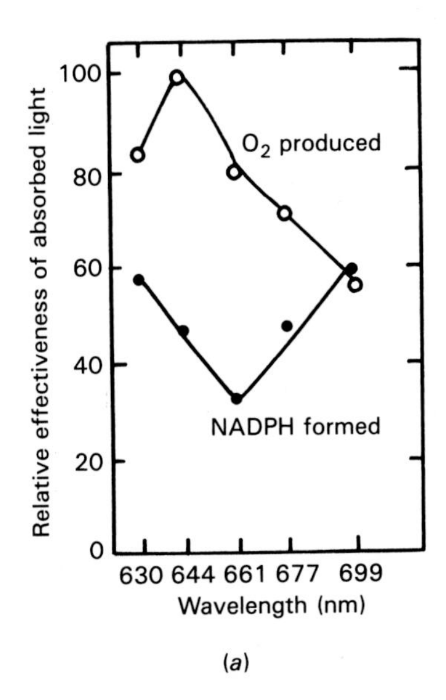

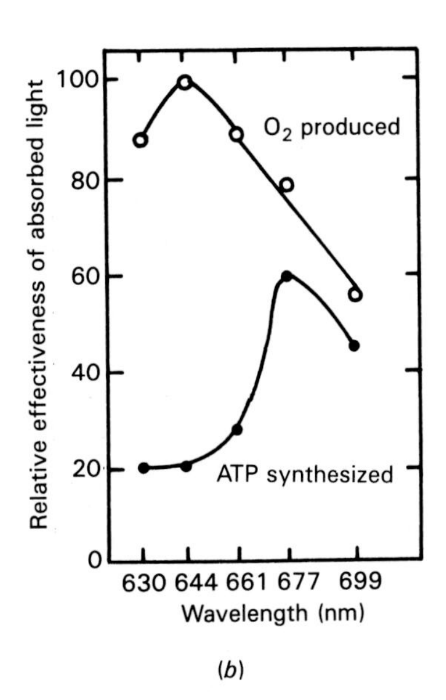

Fig. 11.18 (*a*) Effectiveness of monochromatic light, in the red region of the spectrum, in oxygen evolution and NADP reduction by isolated chloroplasts. Ascorbate was used as an electron donor for the photoreduction of NADP. (*b*) Effectiveness of monochromatic light, in the red region of the spectrum, in oxygen evolution and cyclic photophosphorylation by isolated chloroplasts. The 100 on the ordinate scale is equivalent to 0.16 μatom oxygen evolved per micromole of light quanta absorbed; 60 on the ordinate scale is equivalent to 0.10 μmol ATP formed per micromole of light quanta absorbed. From D. Arnon, *Bulletin of Torrey Botany Club,* 88:215–259, with permission of Johns Hopkins Press, 1961.

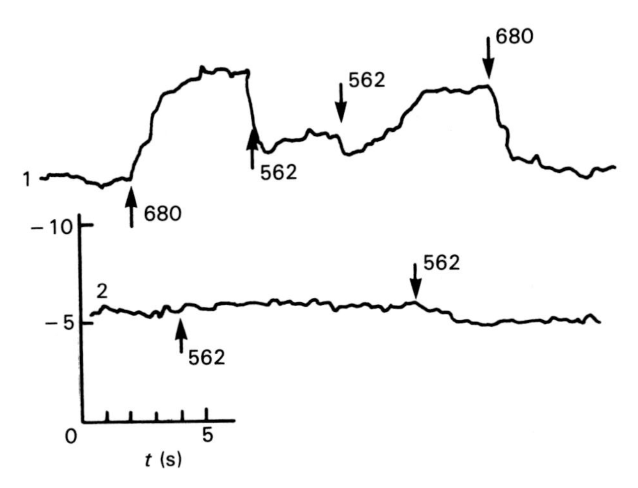

Fig. 11.19 Time course of cytochrome oxidation in *Porphyridium* in light of 680 and 562 nm of intensities 4.4×10^{-10} and 5.3×10^{-10} einstein cm^{-2} s^{-1}. Upward and downward arrows indicate that the light is switched on and off, respectively. Tracing 1 shows that cytochrome is oxidized by light of 680 nm but reduced by light of 562 nm. Curve 2 shows no effect by light of 562 nm. From Duysens et al. (1961). Reprinted by permission from *Nature,* 190:510–514, copyright © 1961 Macmillan Magazines Limited.

downward arrows the turning off of light. Illumination at 680 nm corresponds to excitation of PSI, and that at 562 nm corresponds to excitation of PSII. As predicted by the model in which the two systems are in series, light excitation of PSI oxidizes the cytochromes and excitation of PSII reduces them.

VI. COMPONENTS OF THE CHLOROPLAST PHOTOSYNTHETIC SYSTEMS

Knowledge of the topology and function of the various components of the thylakoid membrane has advanced rapidly thanks to the application of spectroscopic techniques, fractionation procedures, and conventional reagents or antibodies. More recently, sequencing of the appropriate cDNA and computer predictions of structure based on the hydrophobicity of the amino acid side chains of the polypeptides have provided new insights. This section presents some of the information that has been gained by the use of several of these techniques.

A. PSII

The LHC complex of PSII, referred to as LHC II, contains Chl *a* and *b* in approximately equal amounts. Part of the complex is tightly bound to PSII. A mobile or peripheral portion not closely associated with PSII accounts for 120 Chl molecules. The total corresponds to approximately 250 Chl molecules (Staehelin, 1986).

The reaction center of PSII resembles that of purple photosynthetic bacteria, which has been

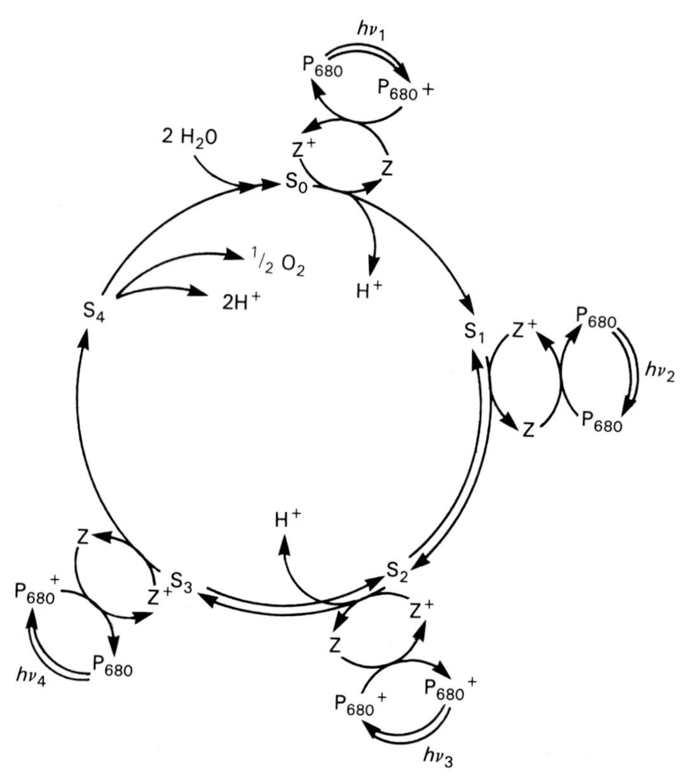

Fig. 11.20 Scheme in which one electron is removed in each transition from S_0 to S_3. State S_4 decays spontaneously, releasing O_2.

isolated and studied by x-ray crystallography (Deisenhofer et al., 1984). As discussed, the bacterial photochemical activity depends on two polypeptides, L and M, which span the membrane. The PSII reaction center contains a 32-kDa polypeptide, D_1, and a 34-kDa polypeptide, D_2. These two polypeptides have important homologies to L and M (Trebst, 1986) and are therefore thought to correspond in function to the bacterial polypeptides. In addition to D_1 and D_2, the isolated RC of PSII (Namba and Satoh, 1987) contains one molecule of cytochrome *b*-559, five Chl, pheophytin, and two β-carotenes.

P680 is photooxidized on illumination (Doring et al., 1967). The primary electron acceptor is pheophytin (Klimov et al., 1980), a porphyrin identical to Chl *a* but lacking Mg. A specialized plastoquinone, Q_A, is a secondary acceptor. These transfers produce an endergonic charge separation between P680 and pheophytin, followed by electron transfer to Q_A.

An electron from Z replaces the electron removed by the photooxidation of P680 *b*. A tyrosine residue of D_1 is thought to be the electron donor Z (Debus et al., 1988).

Both Z^+ and $P680^+$ can be detected with EPR, and the kinetics of the ESR changes implicate Z as the immediate donor (Boska et al., 1983). The electron lost by Z is replaced by the oxidation of water and the evolution of O_2, carried out by the oxygen-evolving complex (see Murata and Miyao, 1985), which is tightly bound to PSII. Each individual OEC undergoes a succession of increasing oxidation states from S_0 to S_4. Oxygen is liberated at S_4 (Kok et al., 1970). The reactions of PSII are represented by Eq. (11.4), and the S scheme is summarized by Fig. 11.20.

$$
\begin{array}{l}
2H_2O \searrow \ \text{OEC} \rightarrow Z \rightarrow P680 \cdot Ph \cdot Q_A \xrightarrow[\text{energy}]{\text{light}} P680^* \cdot Ph \cdot Q_A \ \swarrow O_2 \\
P680^* \cdot Ph \cdot Q_A \xrightarrow{<10\ \text{ps}} P680^+ \cdot Ph^- \cdot Q_A \\
P680^+ \cdot Ph^- \cdot Q_A \xrightarrow[\text{ps}]{\sim 100} P680^+ \cdot Ph \cdot Q_A^- \\
P680^+ \cdot Ph \cdot Q_A^- \xrightarrow[\text{ns}]{50} P680 \cdot Ph \cdot Q_A^- \\
P680 \cdot Ph \cdot Q_A^- \xrightarrow[\mu s]{200} P680 \cdot Ph \cdot Q_A
\end{array}
\tag{11.4}
$$

Incorporation of the S scheme into our original representation of the two-photosynthetic-center Z scheme produces a more realistic model, shown in Fig. 11.21 (Ort, 1986).

B. PSI

At least six membrane components are involved in PSI (Malkin, 1987). As already discussed, P700 is the electron donor. The primary electron acceptor is A_0 and the intermediate acceptor is A_1. Three iron-sulfur centers, F_X, F_B, and F_A, have been identified. The sequence and kinetic constants are shown in Eq. (11.5).

$$
\begin{array}{l}
P700\ A_0\ A_1\ XBA \xrightarrow[\text{energy}]{\text{light}} P700^*\ A_0\ A_1\ XBA \xrightarrow{<10\ \text{ps}} P700^+\ A_0^-\ A_1\ XBA \\
P700^+\ A_0^-\ A_1\ XBA \xrightarrow[\text{ps}]{\sim 100} P700^+\ A_0\ A_1^-\ XBA \longrightarrow P700^+\ A_0\ A_1\ X^-BA
\end{array}
\tag{11.5}
$$

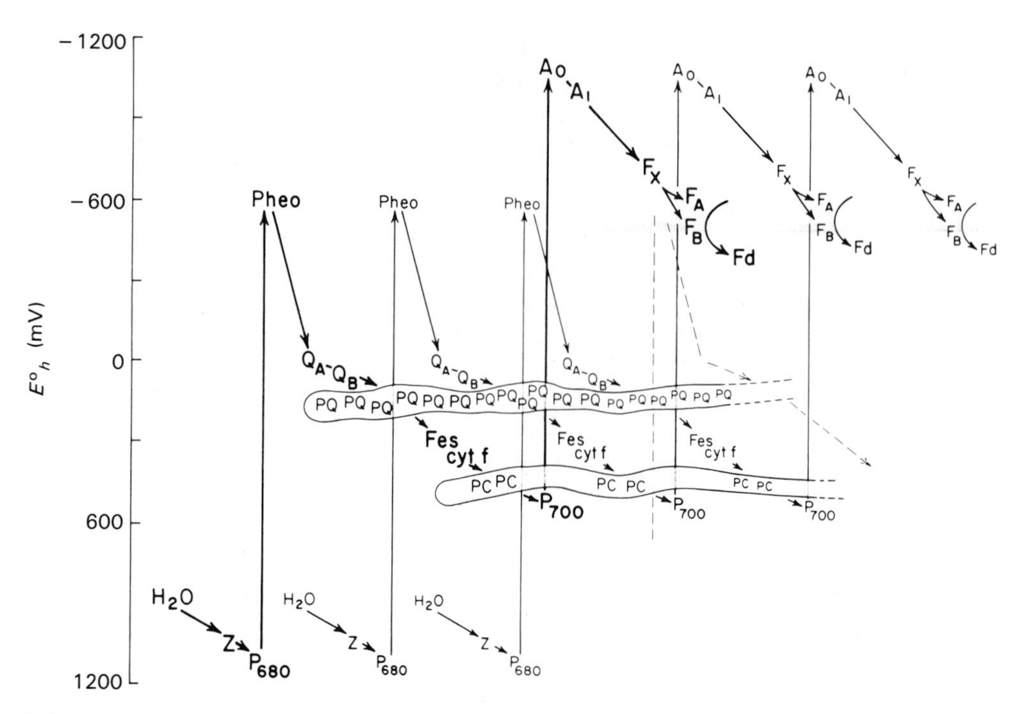

Fig. 11.21 A contemporary Z scheme depicting the sequence and oxidation-reduction midpoint potential ($E_{m,7}$) of the electron carriers in oxygenic photosynthesis. This differs from the familiar Z scheme format in order to convey the concept that the photosynthetic electron transfer from H_2O to ferredoxin (Fd) is accomplished by three electron transport complexes operating in series interconnected by two pools of mobile carriers. From D. R. Ort (1986), *Encyclopedia of Plant Physiology,* with permission. Copyright © Springer-Verlag, West Germany.

P700 is a Chl molecule, although its nature and structure (e.g., whether it is a dimer or a monomer) are still unknown. A_0 is also a Chl or a chlorophyll-like molecule, and A_1 is phylloquinone (vitamin K_1) present at a stoichiometry of two per P700.

P700 is associated with a core antenna consisting of a 70-kDa tetramer (Vierling and Alberte, 1983), which also binds 130 Chl *a* and 16 carotenoid molecules. The purified PSI also contains other polypeptides (Bengis and Nelson, 1977).

The accessory light-harvesting system of PSI, LHC I, contains an additional 60 to 80 Chl *a* and *b* molecules with approximately three times more Chl *a*. The LHC I polypeptides are in the range of 20 to 25 kDa and are immunologically distinct from those of PSII (Lam et al., 1984). Since LHC I can be separated into two fractions, each could have a distinct function as in the case of LHC II, where one is considered laterally mobile and the other is closely associated with PSII.

C. The Cytochrome b_6f Complex

The cytochrome b_6f complex (Cramer et al., 1987) contains plastoquinone, and plastocyanin oxidoreductase has been isolated (Hauska et al., 1983). In addition, the complex was found to include the 34-kDa polypeptide of cytochrome *f*, a 23-kDa polypeptide with two cytochrome b_6 hemes, a 20-kDa FeS-Rieske protein, two other smaller polypeptides, and bound plastoquinol.

The aspects of the organization of the components of the thylakoid membranes discussed in Sections V and VI are summarized in Fig. 11.22 (Anderson, 1981, 1987). Figure 11.22*a*

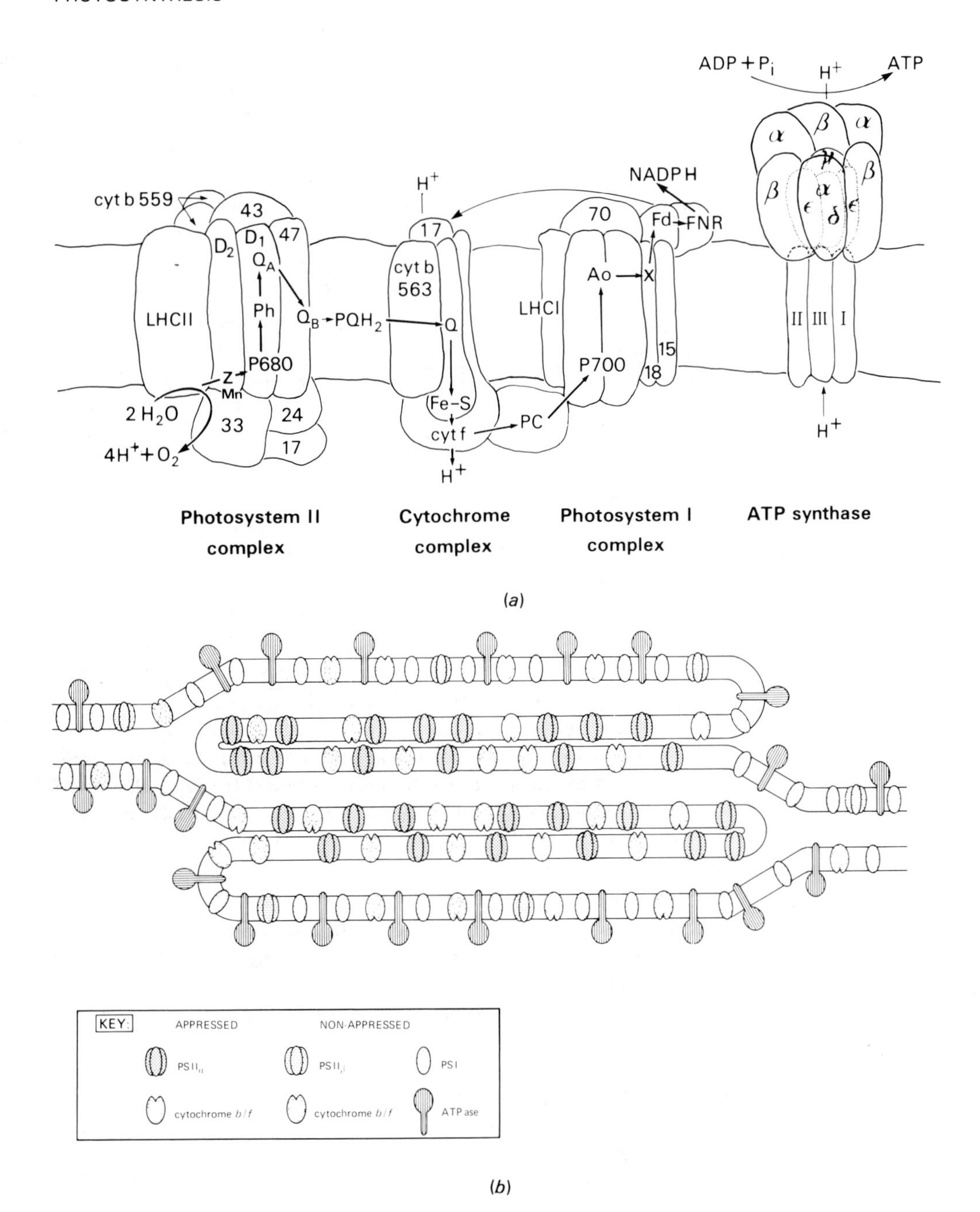

Fig. 11.22 (*a*) Arrangement of the supramolecular protein complexes and mobile electron transport carrier in thylakoid membranes. From J. M. Anderson, *Photosynthesis,* p. 275, with permission of Elsevier Science Publishers, Amsterdam, 1987. (*b*) Possible static representation of the lateral heterogeneity in the distribution of the supramolecular thylakoid complexes between appressed and nonappressed thylakoids. From J. M. Anderson, *FEBS Letters,* 124(1), 1981, with permission of the author.

shows the organization of the complexes; Fig. 11.22*b* shows their organization in relation to the thylakoid structural arrangement.

VII. COORDINATION BETWEEN PSI AND PSII

The efficient operation of the two systems, PSI and PSII, requires smooth coordination. Although the systems are likely to be excited equally by white light, shade favors PSI. Excess energy absorbed by either system would be dissipated as heat and lost to the system.

Unbalanced excitation of PSI is known as *state I* and that of PSII as *state II*. Apparently, these imbalances are corrected by what have been called *state transitions*. These are regulatory effects by which the flow of energy between the two systems is altered. The mechanism or mechanisms are still not entirely understood. Conceivably, the amount of energy normally transferred from PSII to PSI could be changed. Some role of Mg^{2+} or Na^{+} in the amount of energy transferred from PSII to PSI is suspected (Wong and Govinjee, 1979). Alternatively, the proportion of the light absorbed by the two photosystems could be altered, an effect referred to as a change in *optical cross section*.

There is some evidence that the optical cross section changes. Barley mutants lacking LHC II fail to show state transitions (Canaani and Malkin, 1984), which suggests that LHC II is responsible for these transitions. Since PSI and PSII are in separate locations in the membrane, a change in the position of the movable portion of LHC II could favor the transfer of energy to either PSI or PSII. The phosphorylation and dephosphorylation of LHC II affect its distribution, phosphorylation favoring a position closer to PSI (Bennett, 1983; Kyle et al., 1983). The phosphorylation is catalyzed by a membrane-bound Mg^{2+}-dependent kinase, whereas dephosphorylation is catalyzed by a phosphatase. Presumably, the phosphorylated LHC II moves laterally to approach PSI in the unstacked region, thereby favoring the transfer of energy from LHC II to PSI (Larsson et al., 1983). This could be the result of electrostatic repulsion. In contrast, a phosphatase dephosphorylates LHC II, favoring its displacement toward PSII. The phosphorylation cycle is thought to be regulated by the redox state of plastoquinone, acting as a sensor of the energy distribution (Allen et al., 1981). Reduction of plastoquinone, resulting from the electron transport following the photoactivation of PSII, would favor the kinase, whereas the phosphatase would be stimulated by oxidation of plastoquinone.

SUGGESTED READING

General Reading

Cramer, W. A., and Knaff, D. B. (1990) *Energy Transduction in Biological Membranes*, Chapters 5 and 6. Springer-Verlag, New York.

Lawlor, D. W. (1987) *Photosynthesis: Metabolism, Control and Physiology*. Longman Scientific, London, and Wiley, New York.

Other

Hunter, C. N., van Grondelle, R., and Olsen, J. D. (1989) Photosynthetic antenna proteins: 100 ps before photochemistry starts. *Trends Biochem. Sci.* 14:72–76.

Knaff, D. B., and Kämpf, C. (1987) Substrate oxidation and NAD^{+} reduction by phototrophic bacteria. In *Photosynthesis* (Amesz, J., ed.), pp. 199–211. Elsevier, New York.

Ort, D. R. (1985) Energy transduction in oxygenic photosynthesis: an overview of structure and mechanisms. In *Encyclopedia of Plant Physiology*, New Series, Vol. 19, pp. 165–196. Springer-Verlag, Berlin.

Rutherford, A. W. (1989) Photosystem II, the water splitting enzyme. *Trends Biochem. Sci.* 14:227–232.
Sauer, R. (1986) Photosynthetic light reactions—physical aspects. In *Encyclopedia of Plant Physiology*, New Series, Vol. 19, pp. 85–97. Springer-Verlag, Berlin.
Staehelin, L. A. (1986) Chloroplast structure and supramolecular organization of photosynthetic membranes. In *Encyclopedia of Plant Physiology*, New Series, Vol. 19, pp. 1–84. Springer-Verlag, Berlin.

REFERENCES

Allen, J. F., Steinbach, K. E., and Arntzen, K. E. (1981) Chloroplast protein phosphorylation couples plastoquinone redox state to distribution of excitation energy between photosystems. *Nature (London)* 291:25–29.
Anderson, J. M. (1981) Consequences of spatial separation of photosystem 1 and 2 in thylakoid membranes. *FEBS Lett.* 124:1–10.
Anderson, J. M. (1987) Molecular organization of thylakoid membranes. In *Photosynthesis* (Amesz, J., ed.), pp. 273–297. Elsevier, New York.
Anderson, J. M., and Anderson, B. (1982) The architecture of the photosynthetic membrane: lateral and transverse organization. *Trends Biochem. Sci.* 7: 288–292.
Arnon, D. I. (1961) Changing concepts of photosynthesis. *Bull. Torrey Bot. Club* 88:215–259.
Bengis, C., and Nelson, N. (1977) Subunit structure of the chloroplast PSI reaction center. *J. Biol. Chem.* 252:4564–4569.
Bennett, J. (1983) Regulation of photosynthesis by reversible phosphorylation of the light-harvesting chlorophyll *a/b* protein. *Biochem. J.* 212:1–13.
Blakenship, R. E., and Prince, R. C. (1986) State redox potentials and Z scheme of photosynthesis. *Trends Biochem. Sci.* 10:382–383.
Boska, M., Sauer, K., Buttner, W., and Babcock, G. T. (1983) Similarity of EPR signal II rise and $P680^+$ decay kinetics in Tris-washed chloroplast II preparations as a function of pH. *Biochim. Biophys. Acta* 722:327–330.
Canaani, O., and Malkin, S. (1984) Distribution of light excitation in an intact leaf between the two photosystems of photosynthesis. Changes in absorption cross-section following state 1-state 2 transitions. *Biochim. Biophys. Acta* 766:513–524.
Clayton, R. K. (1980) *Photosynthesis: Physical Mechanisms and Chemical Patterns*. Cambridge Univ. Press, London.
Cramer, W. A., Widger, W. R., Herrmann, R. G., and Trebst, A. (1985) Topography and function of thylakoid membrane proteins. *Trends Biochem. Sci.* 10:125–129.
Cramer, W. A., Black, M. T., Widger, W. R., and Girivin, M. E. (1987) Structure and function of photosynthetic cytochrome b-c_1, and b_6-f complexes. In *The Light Reaction* (Barber, J., ed.), pp. 447–494. Elsevier, New York.
Debus, R. J., Barry, B. A., Babcock, G. T., and McIntosh, L. (1988) Site directed mutagenesis identified specific tyrosine residues shown to be redox components of photosystem II. *Biophys. J.* 53:270a.
Deisenhofer, J., Epp, O., Miki, K., Huber, R., and Michel, H. (1984) X-ray structure analysis of a membrane-protein complex. *J. Mol. Biol.* 180:385–398.
Deisenhofer, J., Michel, H., and Huber, R. (1985) Structural basis of photosynthetic light reactions in bacteria. *Trends Biochem. Sci.* 10:243–248.
Doring, G., Renger, G., Vater, J., and Witt, H. T. (1967) Properties of the photoactive chlorophyll-a_{II} in photosynthesis. *Z. Naturforsch. Teil B* 1139–1143.
Dunahay, T. G., Staehelin, L. A., Siebert, M., Ogilvie, P. D., and Berg, S. P. (1984) Structural, biochemical and biophysical characterization of four oxygen-evolving photosystem II preparations from spinach. *Biochim. Biophys. Acta* 764:179–193.
Dutton, P. L. (1986) Energy transduction in an oxygenic photosynthesis. In *Encyclopedia of Plant Physiology*, New Series, Vol. 19, pp. 197–237. Springer-Verlag, Berlin.
Duysens, L. N. M., Amesz, J., and Kamp, B. M. (1961) Two photochemical systems in photosynthesis. *Nature (London)* 190:510–514.

Emerson, R., Chambers, R., and Cederstrand, C. (1957) Some factors influencing the long-wave limit of photosynthesis. *Proc. Natl. Acad. Sci. U.S.A.* 43: 133–143.

Glazer, A. N. and Melis, A. (1987) Photochemical reaction centers, structure, organization, and function. *Annu. Rev. Plant Physiol.* 38:11–45.

Goodenough, U. W. and Staehelin, L. A. (1971) Structural differentiation of stacked and unstacked chloroplast membranes. Freeze etch electron microscopy of wild-type and mutant strains of *Chlamydomonas. J. Cell Biol.* 48:594–619.

Goodenough, U. W., Armstrong, J. J., and Levine, R. P. (1969) Photosynthetic properties of ac-31, a mutant strain of *Chlamydomanas reinhardi* devoid of chloroplast membrane stacking. *Plant Physiol.* 44:1001–1012.

Haehnel, W. (1984) Photosynthetic electron transport in higher plants. *Annu. Rev. Plant Physiol.* 35:659–693.

Hauska, G. A., Hurt, E., Gabellini, N., and Lockau, W. (1983) Comparative aspects of quinol-cytochrome *c*/plastocyanin oxidoreductases. *Biochim. Biophys. Acta* 726:97–133.

Jones, C. W. and Vernon, L. P. (1969) Nicotinamide photoreduction in *Rhodospirillum rubrum* chromatophores. *Biochim. Biophys. Acta* 180:144–164.

Klimov, V. V., Dolan, E., Shaw, E. R., and Ke, B. (1980) Interaction between the intermediary electron acceptor (pheophytin) and a possible plastoquinone-iron complex in photosystem II reaction centers. *Proc. Natl. Acad. Sci. U.S.A.* 77:7227–7231.

Knaff, D. B., and Kampf, C. (1987) Substrate oxidation and NAD^+ reduction by prototrophic bacteria. In *Photosynthesis* (Amesz, J., ed.), pp. 199–212. Elsevier, New York.

Kok, B., Forbush, B., and McGloin, M. (1970). Cooperation of charges in photosynthetic O_2 evolution. I. A linear four step mechanism. *Photochem. Photobiol.* 11:457–475.

Kyle, D., Staehelin, L. A., and Arntzen, C. J. (1983) Lateral mobility of the light harvesting complex in chloroplast membranes controls excitation energy distribution in plants. *Arch. Biochem. Biophys.* 222:527–541.

Lam, E., Ortiz, W., Mayfield, S., and Malkin, R. (1984) Isolation and characterization of a light harvesting chlorophyll *a*/*b* protein complex associated with PSI. *Plant Physiol.* 74:650–655.

Larsson, U. K., Jergil, B., and Anderson, B. (1983) Changes in lateral distribution of the light-harvesting chlorophyll *a*/*b*-protein complex induced by its phosphorylation. *Eur. J. Biochem.* 136:25–29.

Losada, M., Trebst, A. V., Ogata, S., and Arnon, D. I. (1960) Equivalence of light and adenosine triphosphate in bacterial photosynthesis. *Nature (London)* 186:753–760.

Malkin, R. (1987) Photosystem I. In *The Light Reaction* (Barber, J., ed.), pp. 495–560. Elsevier, New York.

Malkin, R., Chain, R. K., Kraichoke, S., and Knaff, D. B. (1981) Studies of the function of the membrane bound iron-sulfur center of the photosynthetic bacterium *Cheomatium vinosum. Biochim. Biophys. Acta* 637:88–91.

Murata, N., and Miyao, M. (1985) Extrinsic membrane proteins in photosynthetic oxygen-evolving complex. *Trends Biochem. Sci.* 10:122–124.

Myers, J., and French, C. S. (1960) Evidence from action spectra for a specific participation of chlorophyll *b* in photosynthesis. *J. Gen. Physiol.* 43:723–736.

Namba, O., and Satoh, K. (1987) Isolation of photosystem II reaction center consisting of D1 and D2 polypeptides and cytochrome *b*-559. *Proc. Natl. Acad. Sci. U.S.A.* 84:109–112.

Ort, D. R. (1986) Energy transduction in oxygenic photosynthesis: an overview of structure and mechanisms. In *Encyclopedia of Plant Physiology,* New Series, Vol. 19, pp. 165–196. Springer-Verlag, Berlin.

Park, R. B. (1966) Thin section of $KMnO_4$ fixed *Spinacea oleracea* chloroplast, ×19,500. In *The Chlorophylls* (Vernon, L. P., and Seely, G. R., eds.), pp. 283–311. Academic Press, New York.

Parson, W. W. (1987) The bacterial reaction center. In *Photosynthesis* (Amesz, J., ed.), pp. 43–62. Elsevier, New York.

Pierson, B. K., and Olson, J. M. (1987) Photosynthetic bacteria. In *Photosynthesis* (Amesz, J., ed.), pp. 21–42. Elsevier, New York.

Sauer, K. (1986) Photosynthetic light reactions. In *Encyclopedia of Plant Physiology*, New Series, Vol. 19, pp. 85–97. Springer-Verlag, Berlin.

Sprague, S. G., and Varga, A. R. (1986) Topography, composition and assembly of photosynthetic membranes. In *Encyclopedia of Plant Physiology*, New Series, Vol. 19, pp. 603–631. Springer-Verlag, Berlin.

Staehelin, L. A. (1986) Chloroplast structure and supramolecular organization of photosynthetic membranes. In *Encyclopedia of Plant Physiology*, New Series, Vol. 19, pp. 11–83. Springer-Verlag, Berlin.

Stanier, R. Y. (1961) Photosynthetic mechanisms in bacteria and plants: development of a unitary concept. *Bacteriol. Rev.* 25:1–17.

Stemler, A., and Radmer, R. (1975) Source of photosynthetic oxygen in bicarbonate-stimulated Hill reaction. *Science* 190: 457–458.

Trebst, A. (1986) The topology of the plastoquinone and herbicide peptides in photosystem II in the thylakoid membrane. *Z. Naturforsch. Teil C* 41:240–245.

Trebst, A. V., Tsujimoto, H. Y., and Arnon, D. I. (1958) Separation of light and dark phases in photosynthesis of isolated chloroplasts. *Nature (London)* 182:351–355.

Vierling, E., and Alberte, R. S. (1983) P700 chlorophyll *a*-protein. Purification, characterization and antibody preparation. *Plant Physiol.* 72:625–633.

Whitmarsh, J., and Ort, D. R. (1984) Stoichiometries of electron transport complexes in spinach chloroplasts. *Arch. Biochem. Biophys.* 23:378–389.

Wong, D. and Govinjee (1979) Antagonistic effects of mono- and divalent cations on polarization of chlorophyll fluorescence in thylakoids and changes in excitation energy transfer. *FEBS Lett.* 97:373–377.

Zuber, H., Brunischolz, R., and Sidler, W. (1987) Structure and function of light-harvesting pigment-protein complexes. In *Photosynthesis* (Amesz, J., ed.), pp. 233–272. Elsevier, New York.

Chapter 12

Energy Transduction: Oxidative Phosphorylation and Photophosphorylation

The major function of the electron transport chains is to provide energy for the synthesis of ATP from ADP and P_i. ATP can be synthesized by either the oxidative reactions in mitochondria or bacteria (i.e., oxidative phosphorylation) or the reactions of photosynthetic systems (i.e., photophosphorylation). As we saw in Chapters 10 and 11, the electron transport chain functions in either oxidative or photosynthesizing systems as a series of redox reactions. There is every indication that the mechanism of the coupling between electron transport and phosphorylation is also very similar in both systems. Section I deals with the reactions in mitochondria and Section II with photosynthetic systems. Finally, the mechanism of coupling is discussed in more detail.

I. ENERGY TRANSDUCTION IN MITOCHONDRIA

The involvement of mitochondria in both phosphorylation and ion transport is illustrated in the experiment of Fig. 12.1 (Chance, 1965). In this experiment, the respiration of isolated guinea pig kidney mitochondria is monitored by an electrode sensitive to O_2 in the medium. The trace in the figure corresponds to a plot of the oxygen concentration (ordinate) with time (abscissa). Some respiration occurs without the addition of substrate, as indicated by the slight downward trace at the left-hand part of the graph; the mitochondria are presumably oxidizing an endogenous substrate. Addition of succinate (arrow 1) accelerates the respiration. Subsequent addition of ADP (arrow 2) to the suspension that already contains inorganic phosphate (P_i) produces a burst of respiration that coincides with oxidative phosphorylation. When all the ADP is used up, the respiration returns to a lower rate. Similar bursts of respiration occur after additions of Ca^{2+} (arrows 3 and 4), which is translocated into the mitochondrial lumen and precipitated as calcium

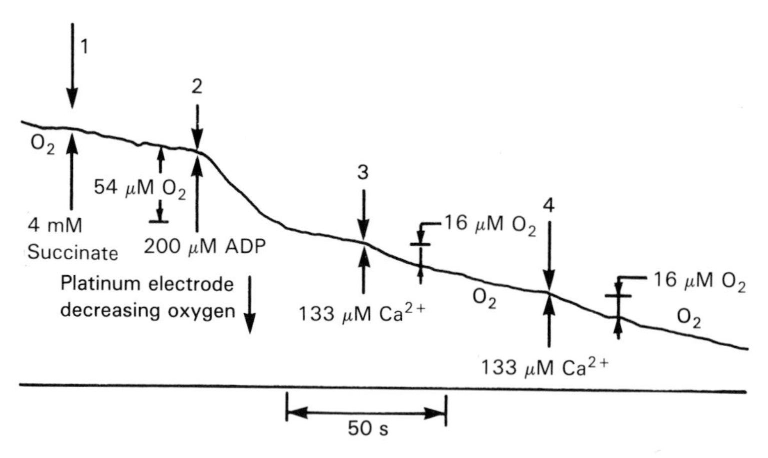

Fig. 12.1 Consumption of oxygen as a function of time, showing the increment in oxidation in the presence of ADP or Ca^{2+}. Guinea pig kidney mitochondria and an assay medium containing 4 mM phosphate were used. From B. Chance, *Journal of Biological Chemistry,* 240:2729–2748, 1965, with permission of the American Society of Biological Chemists, Inc.

phosphate. The rate of respiration before the addition of ADP divided by the rate of respiration during phosphorylation, the *respiratory control ratio,* has been taken as an index of the degree of coupling in mitochondrial preparations.

Part A of this section concerns oxidative phosphorylation primarily; part B is concerned with the coupling to active transport of ions in mitochondria.

A. Oxidative Phosphorylation

The maximal phosphorylative yields of electron transport can be predicted from thermodynamic considerations (see below). However, these arguments cannot establish the actual yield, which can be determined directly in well-coupled mitochondria. The amount of ATP formed from ADP and P_i is carefully measured and the oxygen taken up by the system is estimated. Each oxygen g-atom represents an electron pair traversing the cytochrome chain. It has become customary to express the phosphorylative yield as a P/O ratio: the moles of inorganic phosphate esterified per oxygen atom taken up. Oxygen uptake can be conveniently assessed with an oxygen electrode (as done in the experiment depicted in Fig. 12.1). Phosphorylation can be evaluated by measuring the disappearance of P_i or ADP or the appearance of ATP. The incorporation into ATP of the radioactive $^{32}P_i$ gives a reliable and convenient estimate of ATP synthesized. ATP can also be assayed by using biochemical reactions that are coupled to the hydrolysis of ATP. One of the most sensitive is the luciferin-luciferase reaction using enzymes purified from the firefly tail. The amount of light generated from the system is a function of the ATP concentration.

It is also possible to indirectly estimate the ATP formed. When small amounts of ADP are added, phosphorylation proceeds until virtually all of the ADP has been used to synthesize ATP. Therefore, the assumption can be made that the ATP formed corresponds to the ADP added. This is the method used in the interpretation of the experiment of Fig. 12.1.

Phosphorylation Sites The yield of ATP synthesized per oxygen taken up will depend on the substrate used. Substrates that require NAD^+ reduction for their oxidation, such as several of the tricarboxylic acid intermediates, isocitrate, α-ketoglutarate, and malate, involve all the redox couples of respiration. The oxidation of α-ketoglutarate has one additional substrate-level phosphorylation not involving the cytochrome chain. In practice, the fatty acid β-hydroxybutyric

Table 12.1 P/O Ratios in Isolated Mitochondria

Substrate	P/O	Reference
β-Hydroxybutyrate	2.4–2.5	*a*
Succinate	1.7	*a*
Ascorbate	0.88	*b*
Cytochrome *c*	0.61–0.68	*b*

[a]Copenhaver and Lardy (1952).
[b]Lehninger (1955).

acid is most commonly used. At least in liver mitochondria, this substrate is converted quantitatively into acetoacetate in a reaction that reduces one NAD. The P/O ratio of β-hydroxybutyrate oxidation is listed in Table 12.1 (Copenhaver and Lardy, 1952); the value is between 2 and 3. Since the mitochondria are never perfectly coupled, the higher figures are generally taken more seriously. The results are then consistent with the formation of three ATP for every electron pair traversing the chain.

The contribution of the various portions of the chain can be examined in more detail using other substrates. Succinic dehydrogenase is a flavoprotein, and the oxidation of succinate does not involve NAD. The oxidation of succinate therefore skips one of the steps of the electron transport chain (see Fig. 10.6, p. 288). Interestingly, with succinate as the substrate, electrons enter the chain well below those for an NAD-linked substrate and the P/O ratio is approximately 2 (Table 12.1). The results in Fig. 12.1 show that the phosphorylation of 200 μM ADP results in the uptake of 108 μg-atoms of oxygen. In this case the P/O ratio is 1.8. These results suggest that one phosphorylative step occurs in the span of the chain preceding CoQ (i.e., NAD–FP–CoQ) and the other two in the span below CoQ (i.e., CoQ–cyt *b*–cyt c_1–cyt *c*–cyt *a*–cyt a_3–O_2).

Although no natural substrate interacts with the chain span below CoQ, it has been found that ascorbate reduces cytochrome *c* nonenzymatically, resulting in P/O ratios that approach 1 (Table 12.1) (Lehninger, 1955). Added reduced cytochrome *c* can also be oxidized by the system, also yielding a P/O of about 1 (Table 12.1). Therefore, the second phosphorylation must occur in the span between CoQ and cytochrome *c* and the third between cytochrome *c* and oxygen, as shown in Fig. 12.2.

In very tightly coupled mitochondria, little oxidation takes place in the absence of ADP.

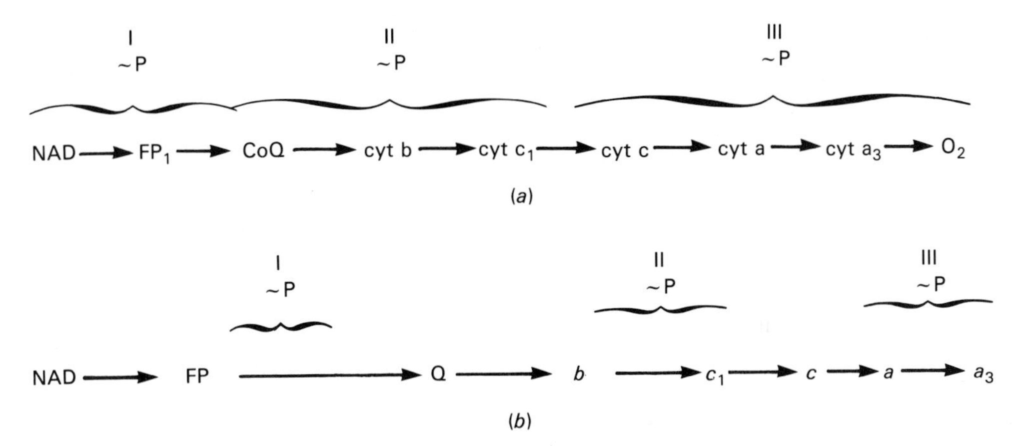

Fig. 12.2 Diagram summarizing (*a*) the sequence of electron transport components and (*b*) the location of the phosphorylative sites.

ADP is one of the substrates necessary for the oxidation to proceed. Since ADP is involved only at the phosphorylative sites, its absence should block electron flow at these sites. Accordingly, the components preceding the block should be more reduced and those following the block more oxidized. The electron transport chain should exhibit crossover points similar to those caused by a block in the chain by a specific inhibitor. In addition, since the oxidative phosphorylation reactions are reversible, the hydrolysis of ATP can reverse the flow of electrons. In this case, the components that precede the phosphorylative site in the course of normal electron transport should become more reduced; those that follow should become more oxidized. Therefore, the additions of ADP or ATP under the appropriate conditions aid in pinpointing the location of the phosphorylative sites.

Crossover points brought about by ADP can be observed by comparing the proportion of reduced molecules for each of the components of the cytochrome chain in two specific metabolic states. In one state (state 4), all components necessary for oxidative phosphorylation are present except for ADP. In the other (state 3), the conditions are the same except that ADP has been added. Therefore the mitochondria do not produce ATP in state 4 whereas they do in state 3. When all the ADP is used up, the mitochondria undergo a transition from state 3 to state 4. The metabolic states have been numbered for convenience in discussion (Chance and Williams, 1956). Three other states have been described and numbered [oxygenated mitochondria without added ADP or substrate (state 1), state 1 to which ADP has been added (state 2), and the complete system in the absence of oxygen (state 5)], but these are not discussed in this chapter.

The proportion of reduced components in the two key metabolic conditions is shown in Table 12.2 (Chance and Williams, 1956). There is a dramatic pyridine nucleotide reduction on transition from state 3 to state 4, which indicates that a phosphorylation point should occur after NAD. Flavoprotein and cytochrome *c* also become reduced to some extent. It is only at cytochrome *a* that oxidation is recognizable; therefore, there is an additional crossover point between cytochromes *c* and *a*.

Most of the components above the cyt *c*–cyt *a* crossover do not become fully reduced as they do when an inhibitor is appropriately used. This has been taken to suggest the possibility of a third crossover point, between cytochromes *b* and *c*, which is demonstrable by the addition of small amounts of azide to alter the steady-state proportion of reduced and oxidized components. However, the evidence for branching of electron transport in this region, the possible complex effect of azide (which is assumed to block electron transport from cytochrome a_3 to oxygen but might also uncouple), complicates the interpretation of this result.

As discussed above, the reverse of the phosphorylative reactions, i.e., the hydrolysis of ATP, should reduce components at the phosphorylation sites in the direction opposite to normal electron transport. For example, NAD could be reduced by a flow of electrons "upstream" from flavoprotein in a reaction requiring the energy of ATP hydrolysis. This does in fact occur, since oxidation of succinate can reduce pyridine nucleotides despite the fact that succinate is

Table 12.2 Oxidation–Reduction State of Respiratory Carriers in Aerobic Steady State in the Presence (State 3) or Absence (State 4) of ADP

State	Steady state (% reduction)				NAD + NADP (%)
	a	*c*	*b*	FP	
3 (+ ADP)	<4	6	16	20	53
4 (− ADP)	0	14	35	40	>99

Source: Chance and Williams (1956), with permission; copyright 1956 John Wiley & Sons.

"downstream" from NAD. The energy obtained from the succinate oxidation has been harnessed to the reduction of pyridine nucleotide. Oligomycin, which blocks the formation of ATP, has no effect; apparently the energy can be used without the formation of ATP. Under anaerobic conditions, i.e., in the absence of oxygen, the hydrolysis of ATP can also drive the electron transport chain backward, presumably by a reaction that is the reverse of phosphorylation. In agreement with this interpretation, oligomycin blocks this reaction.

Figure 12.3*a* represents an experiment in which the addition of ATP induces the reduction of pyridine nucleotide at the expense of the oxidation of the cytochromes and flavoproteins. In this experiment, Na_3S is used to block the cytochrome chain terminally at the cytochrome a_3 site. The two curves correspond to the changes in optical density at wavelengths that allow monitoring of the redox state of NAD (curve 1) and cytochrome *c* (curve 2) with time. The optical density changes have been calibrated in terms of NADH formed (upper ordinate) or expressed in terms of cytochrome oxidation (lower ordinate). In this kind of experiment several respiratory carriers, including flavoprotein, are oxidized. Typical results are tabulated as Fig. 12.3*b*. The results support the location of a phosphorylation site between the flavoprotein and NAD.

Other crossover points can be studied after isolating the relevant part of the chain by means of inhibitors. A block between cytochromes *b* and *c* with antimycin permits the demonstration of an ATP-induced oxidation of cytochrome *a* and reduction of cytochrome *c*. A slow reduction of cytochrome *b*, while components closer to the oxygen are oxidized, suggests a possible third point of interaction.

The conclusions reached in experiments in which the reactions are driven backward agree with those obtained for the transition from state 3 to state 4.

Insight into the location of steps in the electron transport chain responsible for oxidative phosphorylation can be obtained by the two approaches presented: the yield of ATP with various substrates and the crossover points revealed in the presence or absence of ADP or ATP. In fact, we have seen that one ATP is probably synthesized at each site. However, it is obvious that the synthesis of ATP from ADP and P_i can be coupled to the redox couples of the electron transport chain only when enough energy (i.e., ΔG) is available. The availability of this energy can be estimated directly by measuring the redox potential for the appropriate redox reactions. The redox potentials of individual mitochondrial components such as individual cytochromes seem to depend on their physical environment, e.g., whether they are in the mitochondrial structure or have been solubilized. Therefore thermodynamic parameters are not likely to be meaningful unless they are determined under conditions similar or identical to those of intact systems, if possible while the components are still present in the mitochondrial membrane. The redox state of the system can be varied chemically by adding precise amounts of oxidants or reductants, and the redox potential of the system as a whole can be measured. As long as the system is anaerobic (i.e., reducing equivalents cannot be oxidized by oxygen), the system is *closed*. Equilibration of the electron carriers would ideally then result in the entire system being poised at the same potential (E_h). As discussed in Chapter 6, the ΔG and ΔE of a half-cell compared to a standard value correspond to Eqs. (12.1) and (12.2), respectively.

$$\Delta G = \Delta G^\circ + RT \ln \frac{[\text{red}]}{[\text{ox}]} \tag{12.1}$$

$$\Delta E = \Delta E^\circ + RT \ln \frac{[\text{ox}]}{[\text{red}]} \tag{12.2}$$

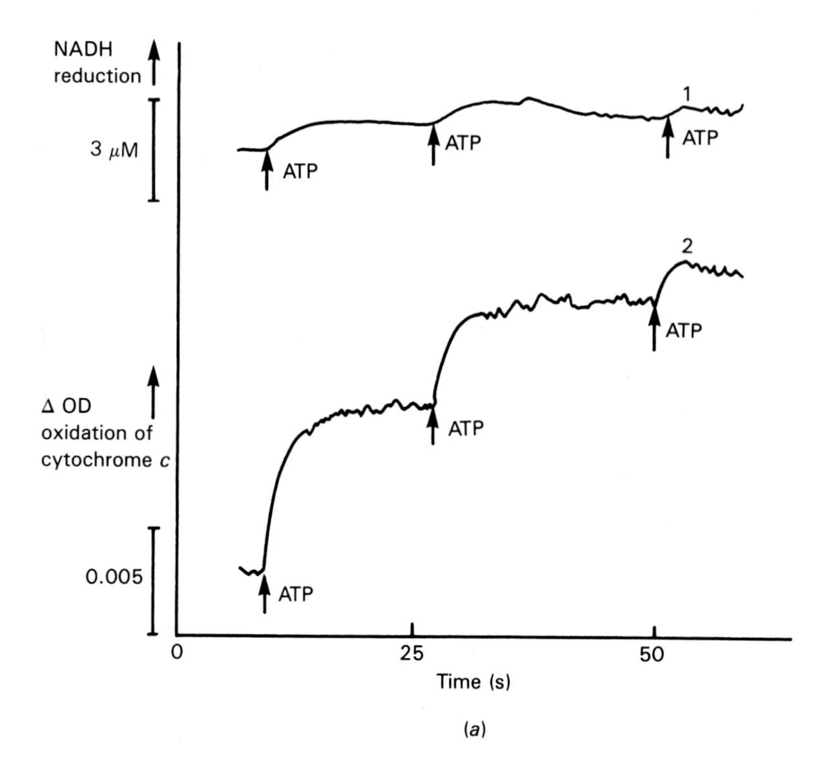

(a)

ATP-electron transfer efficiency in oxidation of cytochromes and flavoprotein of pigeon heart mitochondria

Mitochondria treated with 360 μM Na_3S; ATP, 5.6 μM; mannitol-sucrose-Tris medium (pH 7.4); temperature, 26°C.

Component	Cyto-chrome a_3* (+a)	Cyto-chrome a*	Cyto-chrome c	Flavo-protein	Total oxida-tion found	$\frac{ATP}{e}$
λ (nm)	445–460	605–630	550–540	460–510		
ΔΣ (cm^{-1} mM^{-1} (1-electron)	90	16	19	5.5		
Δc (μM) (1-electron)	0.17	0.36	0.36	0.63	1.52	3.7

(b)

Fig. 12.3 Reversal of electron transport on addition of ATP. The cytochrome c is oxidized and the pyridine nucleotide is reduced in sulfide-inhibited pigeon heart mitochondria in the absence and presence of ATP. The mitochondria were in a mannitol-sucrose medium, buffered with Tris at pH 7.4 and a temperature of 26°C. From B. Chance and C. Hollunger, *Journal of Biological Chemistry,* 236:1577–1584, 1961, with permission of the American Society of Biological Chemists, Inc.

In practice, the redox potential can be measured in the presence of low molecular weight mediators (see below) with two electrodes: a *reference electrode* (e.g., calomel electrode) and an *indicator electrode* (e.g., platinum electrode). The indicator electrode responds to the redox components of the system. Basically, each electrode acts as a half-cell. For the calomel electrode the reaction of the half-cell is

$$Hg_2Cl_2\,(\text{solid}) + 2e^- \rightleftharpoons 2Hg\,(\text{solid}) + 2Cl^- \tag{12.3}$$

For the platinum electrode the reaction corresponds to

$$Pt^{2+} + 2e^- \rightleftharpoons Pt \tag{12.4}$$

Since it is immobilized in the mitochondrial membrane, the electron transport chain cannot interact directly with the platinum electrode. Therefore, catalytic amounts of synthetic carriers (electron *donors-acceptors*) that can shuttle electrons from the mitochondria to the indicator electrode must be used.

For each level of oxidation-reduction of each component, there will be a corresponding ratio [ox]/[red]. As shown by Eq. (12.2), when this ratio is unity ΔE becomes $\Delta E°$, the half reduction potential (ΔE_{m}).

The concentrations of oxidized or reduced components are determined either from their light absorption or, in the case of the iron-sulfur proteins, from the electron spin resonance of the reduced form (see below). As discussed in Chapter 6, it is customary to express redox potentials of the half-cell by comparing them to the H_2 electrode half-cell, not the calomel half-cell. One can readily be converted to the other. The redox potentials are customarily expressed as

$$E_h = E_m + \frac{RT}{nF} \ln \frac{[\text{ox}]}{[\text{red}]} \tag{12.5}$$

where $E_{\mathrm{h}} = \Delta E$ for the reaction

$$\text{ox} + H_2 \rightleftharpoons \text{red} + 2H^+ \tag{12.6}$$

In Fig. 12.4 (Wilson et al., 1983), components of the electron transport chain are displayed as a function of their redox potential (E_{h} and $\Delta E°$). The various redox potentials seem to fall into three groups that are at least approximately equipotential. However, between the equipotential groups there is about 250 mV, which corresponds to a ΔG of -5.7 kcal/mol. Since the ΔG for the synthesis of ATP is approximately 10 kcal and two electrons are needed to go through the cytochrome chain per phosphorylation site, there seems to be enough energy for the production of ATP when two electrons pass from one equipotential group to another. The redox potentials of three components (cyt a_3, cyt b_{t}, and a flavoprotein) appear to depend on whether the mitochondria are uncoupled (lower value) or coupled and in the presence of ATP (higher value). In the original studies, the results were interpreted as suggesting that the three components are directly involved in transduction of the redox reactions to the synthesis of ATP. However, other interpretations are possible. For example, the ox/red ratio could have been altered by reversed passage of electrons at the expense of energy from ATP hydrolysis, as discussed above. Nevertheless, it seems likely that these components are either involved somehow in the coupling or very close to the coupling site.

The order of the electron transport carriers shown previously and the data summarized in Fig. 12.4 (Wilson et al., 1973) support the location of the three phosphorylation sites shown by the jumps in potentials in the figure, schematically expressed in Fig. 12.2*b*.

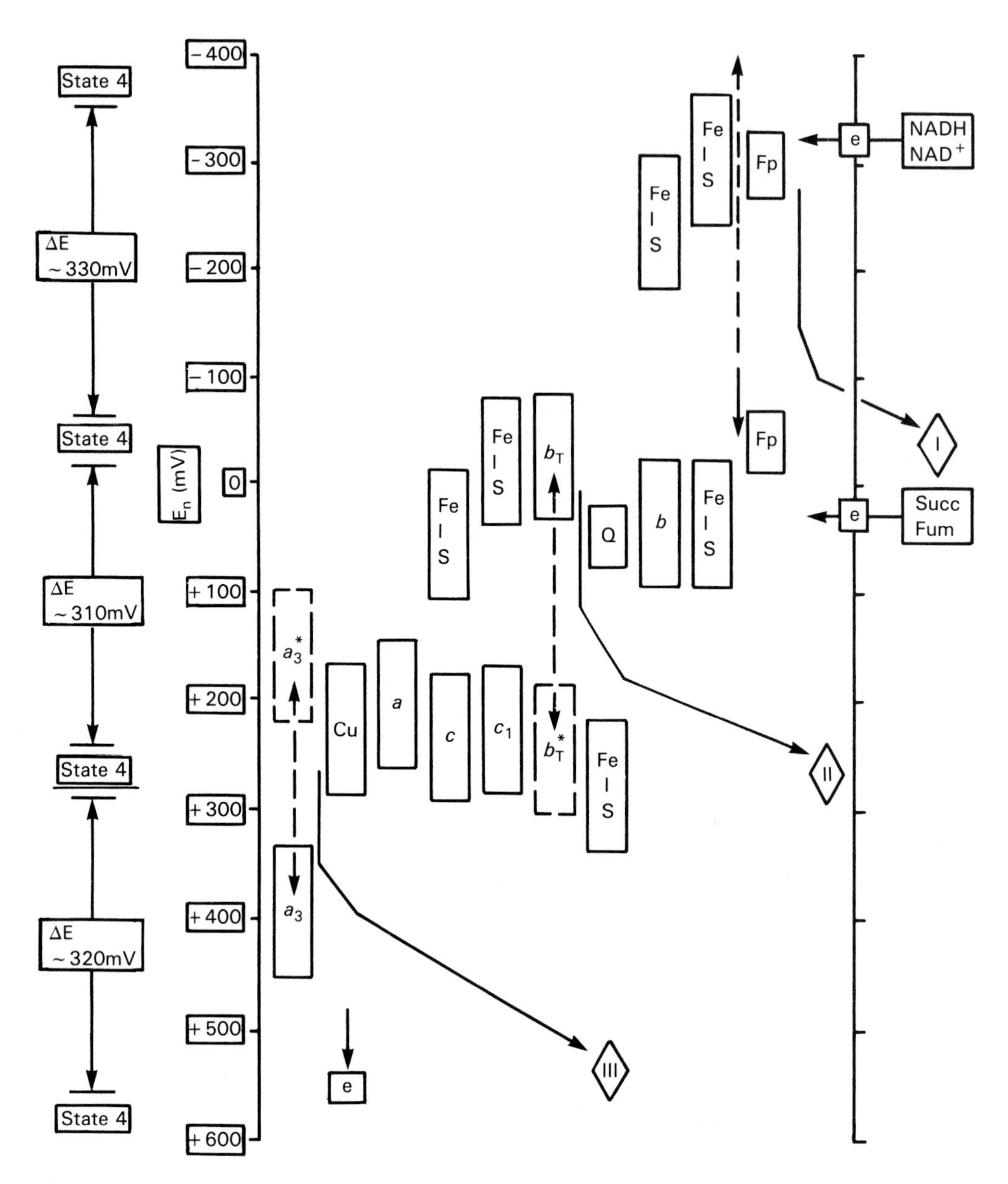

Fig. 12.4 Thermodynamic profile of the oxidation-reduction components of the respiratory chain of pigeon heart mitochondria. Each component is represented by a rectangle, which is centered on its half-reduction potential at pH 7.2 and extends from the potential at which the component is 9% reduced to the potential at which it is 91% reduced. The positions of the rectangles are not intended to indicate the sequence of electron transfer. On the left side are indicated the E_h values of the isopotential groups in pigeon heart mitochondria under state 4 conditions with succinate and glutamate as substrate. Two values are indicated for cytochromes b_T and a_3, one measured for uncoupled mitochondria and the other measured for coupled mitochondria in the presence of excess ATP. From D. P. Wilson, et al., *Current Topics in Bioenergetics,* 5:234–265, 1973, with permission of Academic Press.

Proton Production and Asymmetry Some of the redox reactions produce H^+ and others take up H^+, as shown in Eqs. (12.7) to (12.10), where S represents a substrate:

$$NAD^+ + SH_2 \rightarrow NADH + H^+ + S \tag{12.7}$$

$$H^+ + FP_{ox} + NADH \rightarrow FPH_2 + NAD^+ \tag{12.8}$$

$$2\,\text{cyt}\,b^{3+} + FPH_2 \rightarrow 2\,\text{cyt}\,b^{2+} + FP + 2H^+ \tag{12.9}$$

$$2H^+ + 2\,\text{cyt}\,a_3^{2+} + \tfrac{1}{2}O_2 \rightarrow 2\,\text{cyt}\,a_3^{3+} + 2H_2O - 3H^+ \rightarrow + 3H^+ \tag{12.10}$$

An H^+ is produced whenever the reducing equivalent carries H (containing one e^-) and reduces a component that carries electrons in their reduced form. Conversely, H^+ is taken up when the reducing equivalents are transferred from an electron-carrying component to an H-carrying form. When electron pairs traverse the electron transport chain from dehydrogenases, no net production of H^+ takes place. As many are produced by the reactions as are taken up. However, it is conceivable that some of the components are located on different faces of the mitochondrial membrane; i.e., the membrane is asymmetric. Therefore, one side could release H^+ and the other take it up, resulting in a net flow of H^+ during the passage of electrons. In fact, Chapter 10 gives evidence for asymmetry in the mitochondrial components. However, the proton movement linked to the electron transport chain is likely to proceed by other, still unknown mechanisms. No H carriers are available at the transhydrogenase (Anderson et al., 1981; Rydstrom, 1977), site I (Ohnishi, 1979), or cytochrome-*c* oxidase steps (Wikstrom and Krab, 1982), all involved in proton movements. Regardless of the model, it is difficult to visualize a mechanism that would allow the transfer of as many as four H^+ per site, as reported by some investigators (Reynafarjee et al., 1976). This question has been recently addressed in detail for the cytochrome c-oxidase reaction (Babcock and Wikström, 1992). However, if the H^+ fluxes are a secondary result of ion transport (Tedeschi, 1981), none of these considerations are relevant.

The well-known chemiosmotic model is discussed along with other models in Section III of this chapter. Basically, it proposes that a flux of H^+ in the direction of the proton electrochemical gradient, generated by electron transport, is coupled to the phosphorylation of ADP.

The steps in which the cytochromes are involved correspond to the transfer of a single electron per molecule of cytochrome. By contrast, redox reactions involving pyridine nucleotides, flavoproteins, or CoQ occur in two-electron steps. Perhaps in the normal mechanism of electron transport, two molecules of cytochrome always have to interact with one of the hydrogen carriers. If one electron is transferred at a time through the cytochromes, a semireduced state must exist in the immediately preceding component (e.g., CoQ).

The synthesis of a high-energy phosphate bond requires the energy available from the passage of two electrons (see above). In addition, the reduction of one atom of oxygen requires two electrons. For these reasons, the idea that two electrons are transferred simultaneously is favored. Nevertheless, one should not dismiss the possibility that the energy available from one of the two electrons could be stored to be used in phosphorylation when a second electron traverses the cytochrome chain.

B. Transport of Ions

We have discussed how the addition of certain ions, such as Ca^{2+}, to mitochondrial suspensions in the presence of a substrate and oxygen can result in a burst of respiration (Fig. 12.1). Coincident with this burst of respiration, Ca^{2+} is taken up by the mitochondria. Apparently the oxidative machinery of the mitochondria has been coupled to the ion translocation. Energy-dependent uptakes have been demonstrated for a number of divalent cations (e.g., Mg^{2+}, Mn^{2+}, and Sr^{2+}) and monovalent cations (e.g., K^+, Na^+, or Li^+). The translocations can be supported by oxidation or alternatively by the hydrolysis of ATP. Interestingly, oligomycin blocks ATP-dependent uptake but not the uptake supported by oxidation. In mitochondria and submitochondrial particles, oligomycin blocks the phosphorylation of ADP and the hydrolysis of ATP. On the other hand, uncouplers of oxidative phosphorylation prevent the translocation. The results suggest that part of the mechanism linking electron transport and ion transport is in common with oxidative phosphorylation.

The experiment shown in Table 12.3 (Brierley et al., 1962) illustrates the uptake of Mg^{2+} by beef heart mitochondria. The uptake in the complete system is shown in line 2. The uptake of Mg^{2+} requires energy, as shown by the need for a substrate (line 5) and by the inability to transport in the presence of an uncoupler (line 8) or the inhibitor of electron transport, antimycin (line 9). Transport and oxidative phosphorylation compete as the uptake is reduced in the presence of ADP (line 6) or ADP plus hexokinase and glucose (line 7) (the latter to regenerate ADP by phosphorylating glucose). However, oligomycin, which blocks phosphorylation, fails to affect the transport (line 10), as discussed in more detail later. In the experiments of Table 12.3, Mg^{2+} is accompanied by P_i. In fact, accumulation of a cation is generally accompanied by the passage of an anion, maintaining electric neutrality; alternatively, an internal cation must leave. The uptake of a divalent cation in the presence of P_i results in precipitation of the salt. The anion is, therefore, functioning in a dual capacity, accompanying the cation and acting as a trapping agent. Under these conditions, massive amounts of the salt can be accumulated. However, the anion need not be phosphate; arsenate, acetate, and many others will do. Without removal of the accumulated ions by precipitation (as in the case of acetate), however, the uptake is not as marked.

The uptake of cations is sometimes accompanied by a countermovement of H^+. The precise magnitude of the exchange depends on the conditions. For example, in the absence of an excess of penetrating anion, two H^+ appear in the medium per Ca^{2+}. This ratio can be lowered considerably by the presence of acetate or some other anion to which the membranes are permeable, and in some cases no protons are ejected.

Do the steps in the electron transport chain provide the energy for ion transport and for the phosphorylation of ADP? Several experiments suggest that this is so. As in the case of phosphorylation, this question can be analyzed by providing the respiratory chain with electron donors that are oxidized by different parts of the chain. For example, the contribution of the terminal portion of the chain can be examined by blocking the rest of the chain with antimycin.

Table 12.3 P_i and Mg^{2+} Accumulation by Isolated Heart Mitochondria

Conditions	P_i[a]	Mg^{2+}[a]
1. Mitochondria alone	30	50
2. Complete system	1010	1800
3. Mg^{2+} omitted	10	35
4. P_i omitted	10	95
5. Substrate omitted	10	75
6. ADP added (10 μmoles)	130	205
7. ADP + hexokinase system	30	85
8. Complete + dinitrophenol (0.3 μmoles)	8	75
9. Complete + antimycin (6 μg)	40	75
10. Complete + oligomycin (12 μg)	980	1800

Reprinted with permission from G. P. Brierley, et al., *Proceedings of the National Academy of Science,* 48:1928–1935, 1962.

[a]In micromoles per milligram protein.

The reducing equivalents are then provided by ascorbate in the presence of the dye *N,N,N′*-tetramethylphenylenediamine (TMPD) acting as a carrier. Only one phosphorylation site is available under these conditions, yet the system is capable of supporting the uptake of Mg^{2+} phosphate or Ca^{2+} phosphate using ascorbate plus TMPD or ATP (Brierley and Murer, 1964). This finding is in agreement with results of other experiments using submitochondrial particles coupled only at a third site (Penniston et al., 1966). Amytal partially inhibits the rapid accumulation of Ca^{2+} in experiments in which NAD-linked substrates and flavoprotein-linked substrates are used together. Therefore, the NAD-flavin site is probably also involved in transport. In agreement with this view, the respiratory responses of the system to Ca^{2+}, including crossover points, are very similar to those obtained by adding ADP to the system (Chance and Hollunger, 1961).

The results suggest that the transport of ions involves the same sites in the electron transport chain as oxidative phosphorylation. There is, however, a significant difference between the two. Whereas oxidative phosphorylation is blocked by the antibiotic oligomycin, the transport of ions supported by a respiratory substrate is unaffected. Oligomycin is known to combine with the ATP synthase complex (F_0F_1), suggesting that ion transport does not involve this complex when the energy is available from electron transport.

II. PHOTOPHOSPHORYLATION

As we saw in Chapter 11, the electron transport chain and the phosphorylative reactions in photosynthetic systems are very similar to those of the mitochondrial system. Analogous diagrammatic schemes can be designed, as shown in Fig. 12.5 (Hall, 1976). Figure 12.5 shows not only the various electron transport components and their position in the photosynthetic electron transport chain but also inhibitors, electron acceptors, and donors. Through these latter electron carriers, part of the chain can be functionally isolated. For example, with methylviologen (MV) or ferricyanide (FeCy) the electron flow that normally reduces NADP can be bypassed. Similarly, benzidine and ascorbate can bypass the reactions that normally release oxygen. In fact, as we saw in Chapter 11, ascorbate as an electron donor and indophenol blue as an acceptor were used to separate the two photosynthetic systems (PSI and PSII).

The relationship between electron transport and phosphorylative reactions can be studied in the chloroplasts by methodology analogous to that used in the study of mitochondria. Figure 12.6 (Hall, 1976) shows a record of oxygen release in an experiment in which ferricyanide is used as an electron acceptor. It is seen that light in the presence of ferricyanide increases the amount of O_2 released. When ADP is added (state 3), the amount of O_2 released increases. When the ADP is used up, the O_2 released returns to the state 4 value (no ADP). The P/O ratios calculated from these results approach 2. However, such high ratios can be calculated only after correcting for the basal electron transport in the absence of ADP, a correction that may not be justified. In the experiment of Fig. 12.6, without this correction, the ADP/O ratio would be slightly above 1 rather than 1.7. The ATP/2e ratios above unity, generally between 1.1 and 1.3, have been calculated (Horton and Hall, 1968; Winget et al., 1965). In addition to determining phosphorylative yield, it is possible to calculate photosynthetic control (PC) ratios analogous to respiratory control (RC) ratios in mitochondria by dividing the oxygen released in state 3 by that released in state 4.

If the components associated with PSI or PSII with the appropriate electron acceptors and donors are isolated, it can be shown that the portions of the electron transport chain appropriately marked in Fig. 12.5 are each associated with the synthesis of ATP. These results suggest one such site per span.

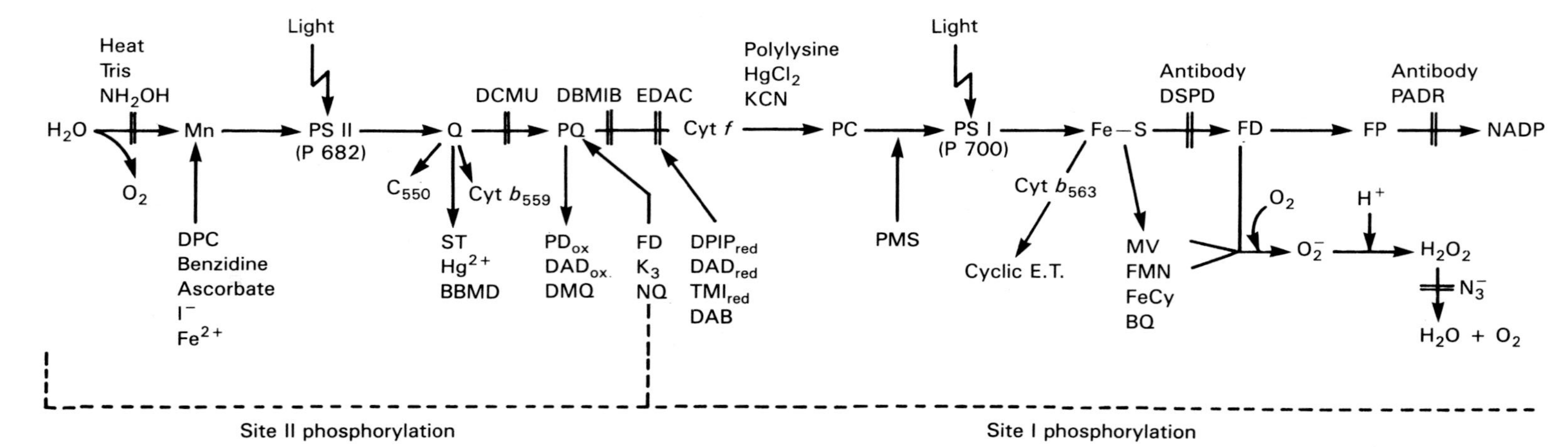

Fig. 12.5 Scheme for noncyclic electron transport and phosphorylation. From Hall (1976), with permission. The abbreviations are as follows. Electron donors: DPC, diphenyl carbazide; FD, ferrodoxin; K_3, vitamin K_3 (menadione); NQ, napthoquinone; DAD, diaminodurene; TMI, tetramethylindamine; DAB, diaminobenzidine; PMS, phenazine methosulfate. Electron acceptors: ST, silicotungstate; BBMD, benzyl-α-bromomalodinitrile; DAD_{ox}, diaminodurene; DMQ, dimethyl-*p*-benzoquinone; MV, methyl viologen; FeCy, ferricyanide; BQ, benzoquinone. Inhibitors: DCMU, 3-(3,4-dichlorophenyl)-1,1-dimethylurea; Tris, tris(hydroxymethylaminomethane); DBMIB, 2,5-dibromo-3-methyl-6-isopropyl-*p*-benzoquinone; EDAC, 1-ethyl-3-(3-dimethylaminopropyl)carbodiimide; DSPD, disalicylidene propane diamine; PADR, phosphoadenosine diphosphate ribose. Reprinted with permission from D. O. Hall, *The Intact Chloroplast.* Copyright © 1976 Elsevier Science Publishers, Amsterdam.

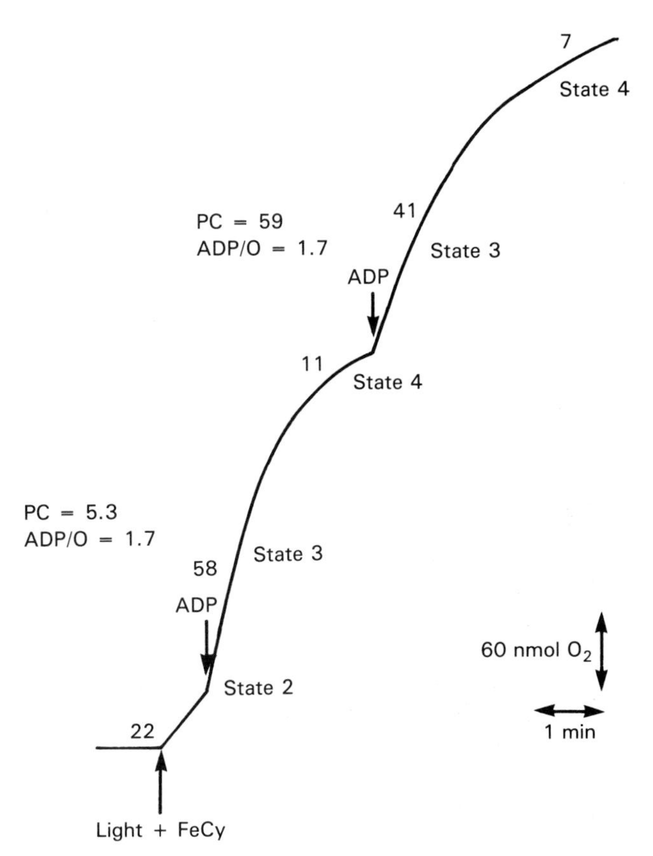

Fig. 12.6 Record of oxygen release over time. The electron acceptor is ferricyanide (FeCy). Reprinted with permission from D. O. Hall, *The Intact Chloroplast.* Copyright © 1976 Elsevier Science Publishers, Amsterdam.

The results for photophosphorylation appear analogous to those for oxidative phosphorylation. For this reason the two are thought to involve the same mechanisms, as discussed in Section III.

III. MECHANISMS OF COUPLING

The mechanisms by which the passage of electrons through the cytochrome chain is coupled to either phosphorylation or the translocation of ions are not well understood. Years of debate have centered on two basic proposals.

One view, the *chemical intermediate model,* assumes that phosphorylation or the transport of ions is coupled to oxidation by the formation of high-energy chemical intermediates (e.g. high phosphate group transfer potential). These high-energy intermediates provide the energy required for the synthesis of ATP or ion transport. In this model, the synthesis of ATP could take place when the ion translocation is run in reverse, in the direction of the electrochemical gradient. However, this would not be a primary feature of the model. The possibility will be discussed in Chapter 15 and is well documented for a number of cases involving ion transport (e.g., Na^+, K^+-ATPase and Ca^{2+}-ATPase). The *chemiosmotic hypothesis* of Mitchell (1966) proposes the formation of a proton electrochemical gradient by the coupling of electron transport to the efflux of H^+. The passage of H^+ in the direction of its electrochemical gradient is coupled

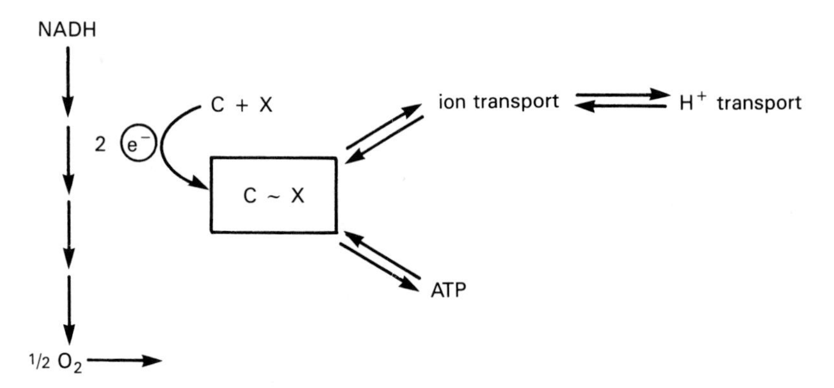

Fig. 12.7 Diagram summarizing the possible interpretation of the data. A non-chemiosmotic model.

to the phosphorylation of ADP. The H^+ flow coupled to electron transport is outward in intact mitochondria and inward in thylakoid vesicles. The transducing membranes are assumed to be extremely impermeable to ions so that the electrochemical gradient is not passively dissipated. Translocations of ions are assumed to occur mostly by electrically neutral exchanges in which external ions are exchanged for internal ions of the same charge (*antiport*) or oppositely charged ions are simultaneously transferred in the same direction (*symport*).

The crucial difference between these two alternatives can be seen by comparing Fig. 12.7, representing the chemical intermediate model, and Fig. 12.8, representing the chemiosmotic model.

The chemical intermediate model can be thought of as shown in Eq. (12.11):

$$AH_2 + B + C \rightleftharpoons A \sim C + BH_2 \tag{12.11a}$$

$$A \sim C + P_i \rightleftharpoons C \sim P + A \tag{12.11b}$$

$$C \sim P + ADP \rightleftharpoons ATP + C \tag{12.11c}$$

Basically, the oxidation of AH_2 results in the formation of $A \sim C$, and in a subsequent step C is replaced by phosphate, which can then be transferred to ADP. A and B would be components of the electron transport chain. An uncoupler would conceivably act by complexing with C. This model is analogous to substrate-level phosphorylation, which is relatively well understood

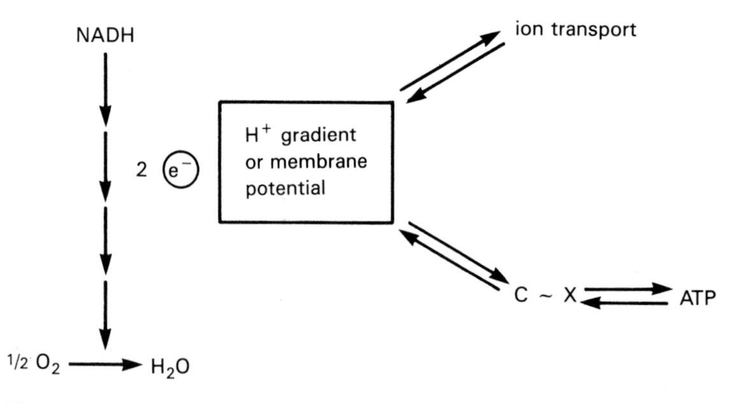

Fig. 12.8 Diagram summarizing the possible interpretation of the data. A chemiosmotic model.

(see Chapter 6). However, as we shall see in Chapters 14 and 15, high-energy states may involve particular conformations of protein molecules and not conventional "high-energy" intermediates (see also Dunn, 1980; Ryrie and Jagendorf, 1972). Chemical intermediate models equivalent to Eq. (12.11) but with additional steps have been proposed to accommodate all the data, but since they differ only in detail from that depicted in the equations they will not be considered here (Slater, 1971). There seems to be little point in arguing details when the available information is still sketchy. The evidence indicates, for example, that in the mitochondrial membrane energy is required not for the synthesis of ATP but for its release (Boyer et al., 1973). Similarly, in vitro, purified ATP synthase can form ATP from bound ADP and medium P_i without an energy input (Feldman and Sigman, 1982).

Formally, the chemiosmotic model has the advantage of apparent simplicity (Mitchell, 1967); since the high-energy state corresponds to the electrochemical gradient of H^+, only two sets of events have to take place. The proton transport coupled to the electron transport is responsible for producing the gradient (see Section I,A), and the passage of the proton (presumably through part of the ATP synthase complex) is coupled to the phosphorylation of ADP.

The system can be operated as an ATPase or as a phosphorylative mechanism, as shown for mitochondria in Fig. 12.9. The operation of the cytochrome chain to produce an H^+ ion gradient or a membrane potential is represented in Fig. 12.9*a*. The operation of the ATP synthase either in phosphorylation (heavy lines) or as an ATPase (dashed lines) is shown in Fig. 12.9*b*. The pH differential or an internal negative potential resulting from the expulsion of H^+ during oxidation would drive the H^+ inward for phosphorylation. The functioning of the system in the transport of cations, in this example Ca^{2+}, is shown in Fig. 12.9*c*. Figure 12.9 represents the chemiosmotic hypothesis as proposed in mitochondria. In the case of the thylakoid vesicles, all the events would have the opposite polarity, since the H^+ transport occurs inwardly.

The possible role of the chemiosmotic model can be evaluated through two basic questions: (1) Can the flux of H^+ in the direction of its electrochemical gradient produce ATP from ADP and P_i? (2) During normal electron transport, is there an H^+ electrochemical gradient sufficient to play a role in the various transducing systems (mitochondria, thylakoid vesicles, and bacteria)?

The answer to the first question is yes for thylakoid vesicles, as shown by the classic experiment of Jagendorf and Uribe (1966) summarized in Table 12.4. In this experiment leaky chloroplasts, probably equivalent to thylakoid vesicles, were maintained in the dark or in the

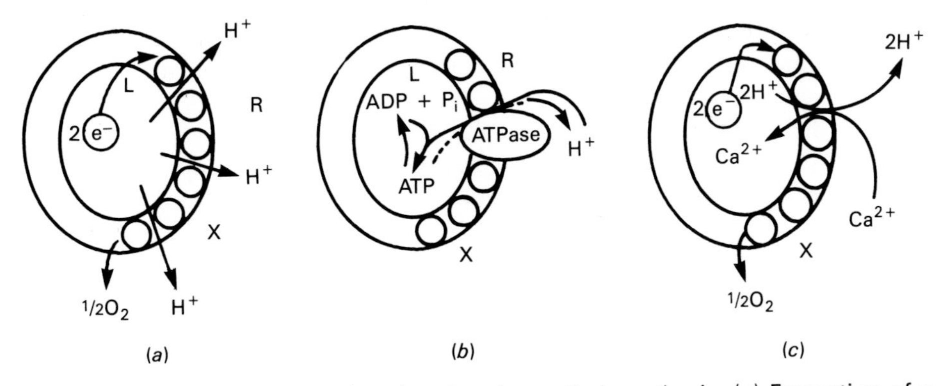

Fig. 12.9 Diagram representing the chemiosmotic hypothesis. (*a*) Formation of a membrane potential or a pH gradient by passage of the electrons through the cytochrome chain. (*b*) ATPase functioning either to hydrolyze ATP (dashed line) or to phosphorylate ADP (solid line). (*c*) Exchange of cations for H^+.

Table 12.4 Formation of ATP by Acid-Base Transitions[a]

Reaction mixture (1)	pH of acid (2)	ATP (estimated by luciferase assay) Total (3)	Net synthesis (4)	ATP (estimated by phosphomolybdate extraction) Total (5)	Net synthesis (6)
a. Complete	3.8	141	129	166	163
b. Complete	7.0	12	—	3	—
c. PO_4 omitted	3.8	12	—	—	—
d. ADP omitted	3.8	4	—	3	—
e. Mg^{2+} omitted	3.8	60	48	48	45
f. Chloroplasts omitted	3.8	7	—	3	—

Reprinted with permission from A. T. Jagendorf and E. Uribe, *Proceedings of the National Academy of Science,* 55:170–177, 1966.

[a]The chloroplast fragments were first exposed to acid, then to alkaline medium (pH 8). The results are given as micromoles of ATP per milligram of protein.

presence of the appropriate inhibitor. The chloroplasts were first equilibrated with organic acid and then shifted to an alkaline medium. Presumably the inside would initially be acidic, as it would be with illumination; the proton flow is in the direction opposite to that in mitochondria. The external alkalinization would then provide the appropriate gradient for an H^+ efflux. Item a corresponds to the experimental determination with the complete system. The amount of ATP formed was measured either with the luciferin-luciferase reaction (column 4) or by measuring the amount of phosphate incorporated into ATP (column 6), and the two methods show that ATP was formed by this procedure. On the other hand, ATP was not synthesized when the initial incubation was at pH 7.0 (item b), or when any of the necessary components were left out (such as PO_4, item c; ADP, item d; Mg^{2+}, item e; or chloroplasts, item f). Analogous demonstrations are available for submitochondrial particles or reconstituted vesicular preparations containing only ATP synthase, F_0F_1 (e.g., see Sone et al., 1977). The latter experiments also establish that F_0F_1 and no other protein is necessary for the synthesis of ATP.

Even though the experiments just described support the chemiosmotic hypothesis, experiments with intact mitochondria do not. In isolated mitochondria, a small alkaline shift in the medium induces ATP synthesis (Malenkova et al., 1982). Since in mitochondria the H^+ efflux induced by electron transport is the reverse of that in the thylakoids, these experiments are at variance with the present chemiosmotic model.

The answer to the second question is also not clear. The electrochemical potential gradient (the *proton-motive force* of Mitchell) contains two terms, one referring to the H^+ concentration gradient and the other to the electrical potential across the transducing membrane, Eq. (12.12):

$$\Delta G = nRT \ln \frac{(H^+)_i}{(H^+)_o} + nF\,\Delta\psi \tag{12.12a}$$

$$\Delta G = -2.3\,nRT\,\Delta pH + nF\,\Delta\psi \tag{12.12b}$$

In these equations the subscripts i and o refer to location inside or outside the lumen, F is the Faraday, and $\Delta\psi$ corresponds to the membrane potential. If two H^+ are presumed to be used up per ATP formed, $n = 2$; however, other stoichiometries are possible and at present $n = 3$ is favored by many.

The two terms have been evaluated in a number of studies. The mitochondrion will be used for the discussion of the technique used to calculate the ΔpH. Analogous reasoning can be applied to the thylakoid vesicles.

In metabolizing mitochondria in which the interior is alkaline in relation to the outside, the ΔpH has been evaluated most commonly by measuring the distribution of a weak acid (HA). The acid is assumed to permeate the inner mitochondrial space rapidly and primarily in its undissociated form. With these assumptions, at equilibrium,

$$[\mathrm{HA}]_i = [\mathrm{HA}]_o \tag{12.13}$$

The dissociation constant would be the same for the two phases:

$$\frac{[\mathrm{H}^+]_i[\mathrm{A}^-]_i}{[\mathrm{HA}]_i} = \frac{[\mathrm{H}^+]_o[\mathrm{A}^-]_o}{[\mathrm{HA}]_o} = K_d \tag{12.14}$$

so that

$$[\mathrm{H}^+]_i[\mathrm{A}^-]_i = [\mathrm{H}^+]_o[\mathrm{A}^-]_o \tag{12.15}$$

$$\Delta\mathrm{pH} = \log_{10}\frac{[\mathrm{A}^-]_i}{[\mathrm{A}^-]_o} \tag{12.16}$$

Therefore, in mitochondria, the ΔpH can be readily calculated from the distribution of the weak acid. In the case of the thylakoid vesicles, the inside is acid in relation to the medium and the analogous technique uses weak bases. For illuminated chloroplasts at steady state, the values calculated by a similar technique suggest a ΔpH as high as 3 to 3.5 (Pick et al., 1974). In fact, little phosphorylation occurs below a ΔpH of 2.5. In effect, this would allow for 8.4 kcal, which would not suffice to synthesize one ATP if two H^+ were transferred. However, it is generally thought now that more than two H^+ are transferred per ATP synthesized. In contrast, with mitochondria, the ΔpH under physiological conditions is generally approximately 0.5 (Addanki et al., 1968), not enough to play a significant role (although sometimes values as high as 1 have been reported).

Calculations of membrane potentials follow a different rationale. In a number of cells it is possible to estimate the electric potential across the plasma membrane by indirect techniques using the distribution (i.e., the ratio of concentrations) of a cation essentially completely dissociated at biological pH values. Commonly, these techniques have been applied to mitochondria or chloroplasts. Equation (12.17) shows the free energy available from the distribution of a monovalent cation per mole transferred.

$$\Delta G = RT \ln\frac{[\mathrm{C}^+]_i}{[\mathrm{C}^+]_o} + \Delta\Psi F \tag{12.17}$$

At equilibrium $\Delta G = 0$, so Eq. (12.18), the Nernst equation, is applicable.

$$\Delta\Psi = \frac{RT}{F}\ln\frac{[\mathrm{C}^+]_i}{[\mathrm{C}^+]_o} \tag{12.18}$$

Where it is possible to calculate the potential from the distribution of a cation, the procedure assumes that (1) the system is in equilibrium in relation to the cation used as a probe, (2) the ion distribution has not disturbed the system, and (3) the cation has distributed solely as the result of the potential across the membrane. In other words, no other mechanism should be responsible for the distribution.

Presumably, the same reasoning should apply for the thylakoid vesicles. However, in this case the potential across the membrane would be positive inside and an anion would have to be used as a probe.

In the case of the thylakoid vesicles, the distribution shows that the potential is negligible in magnitude, at least at steady state. In the case of mitochondria, there are preparations that phosphorylate in the absence of a significant potential as calculated with this method. In other preparations, however, it is possible to calculate a substantial membrane potential, as high as -200 mV. However, there is evidence that a process other than the potential itself is responsible for the distribution, that is, a transport system involving an H^+/C^+ stoichiometric exchange (for a review, see Tedeschi, 1981). Microelectrode impalement of giant mitochondria also shows that phosphorylation takes place in the absence of a significant membrane potential (Campo et al., 1984). In addition, the protons that have to be pumped out in the absence of ion transport to produce a potential across the membrane of -200 mV can be easily calculated (see Eq. (16.6) of Chapter 16) to be approximately 1×10^{-6} moles per g protein, as done by Mitchell (1966). When checked experimentally a significant H^+ efflux has never been observed in the absence of ion transport (e.g., Archbold et al.; 1979).

These findings are by no means the only evidence that the distribution of cation probe molecules does not reflect a membrane potential in mitochondria or that the apparent $\Delta\psi$ measured with such probes does not have a role in energy coupling. For example, the presence or absence of respiration has no effect on the rate constant of the efflux of cationic probe molecules; it affects only the rate constant of the influx (Skulskii et al., 1983). Both should be affected (in opposite ways) if a membrane potential were responsible for distribution. Furthermore, in at least some experiments, the membrane potential and the *proton-motive force* measured by conventional means remain essentially the same when both respiration and phosphorylation are decreased in parallel by the use of inhibitors (Mandolino et al., 1983).

Alternatives If we discarded the chemiosmotic model, we would be left with a major dilemma. The problem remains of explaining the synthesis of ATP in the experiments in which an H^+ flux has been imposed. In addition, the failure of the chemiosmotic mechanism to explain the data for mitochondria would seem to leave us without a viable model. However, several models have now been proposed that depend on localized effects (e.g., see Dilley and Schreiber, 1984; Hong and Junge, 1983; Kagawa, 1984; Westerhoff et al., 1984). Furthermore, some alternatives of the chemical intermediate hypothesis may still apply. It is also possible to consider elements of both the chemical intermediate and chemiosmotic models in a similar framework.

In Chapters 13 to 15 we will see that in the case of the Na^+,K^+-ATPase or the Ca^{2+}- ATPase proteins, the passage of ions in the direction of the gradient permits the synthesis of ATP. This would be analogous to the experiments in which H^+ fluxes have been induced and shown to result in ATP synthesis. In addition, these two transport ATPases can synthesize ATP without a gradient. When the purified ATPases are exposed to a high concentration of the salt they transport (Ca^{2+} or Na^+), they can produce ATP from ADP and P_i. Apparently, these ionic concentrations can force the enzyme into a high-energy configuration. It is therefore entirely possible to synthesize ATP by delivery of the appropriate ion directly to the appropriate site of an ATPase. Conceivably, the electron transport chain could do exactly that, delivering H^+ to the ATP synthetase. Figure 12.10 (Senior, 1979) illustrates a model that can operate either with or without a proton gradient. The proton directly delivered to the ATP synthase (step I) produces the release of ATP (by changing the conformation and hence the dissociation constant) in step II. This apparently corresponds to the energy-requiring step. The simultaneous attachment of ADP and P_i (step II) results in the release of H^+ (step III), presumably by decreasing the binding constant for H^+. The synthesis of bound ATP is shown in step IV.

As already mentioned, there is considerable evidence that the formation of bound ATP does not require a significant amount of energy (Boyer et al., 1973; Feldman and Sigman, 1982, 1983). Bound ATP, (ATP_b), can be present in the presence of an uncoupler (Boyer et al., 1973; Feldman and Sigman, 1983) and furthermore, in preparations of F_1 alone, at very low concentrations the equilibrium constant for $[(ATP)_b]/[ADP][P_i]$ is approximately 0.6, corresponding to a ΔG of 0.3 kcal. Similarly, there is evidence for conformational changes of ATP synthase dependent on energy input (e.g., Reynafarjee et al., 1976) or the presence of ATP (Dilley and Schreiber, 1984), and these changes result in different binding constants of nucleo-

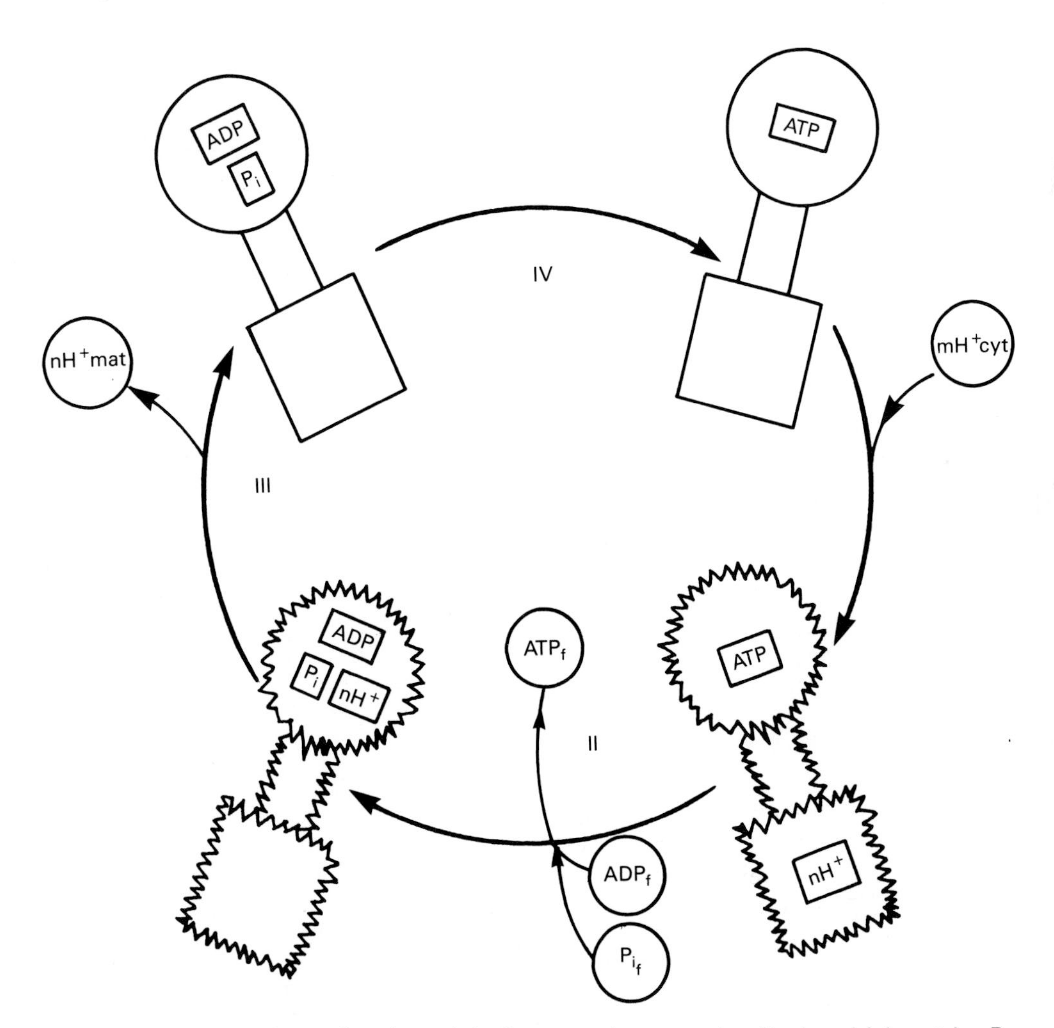

Fig. 12.10 The conformational model: *Cyt,* cytoplasm; *mat,* mitochondrial matrix. Reprinted from Senior (1979), by courtesy of Marcel Dekker, Inc.

tides. The synthesis of ATP by isolated ATP synthase induced by a shift to an alkaline medium supports a conformational model (Blumenfeld et al., 1987), although not the details of Fig. 12.10.

IV. THE ATP SYNTHASE

The ATP synthases or proton-ATPases, F_0F_1, is a complex of many polypeptides and can function either forward in oxidative phosphorylation or photophosphorylation or backward, i.e., to hydrolyze ATP.

A. Structure of F_0F_1

The outline of the structure derived primarily from electron micrographs is shown in Fig. 12.11. The ATP synthase subunits are highly conserved in organisms ranging from *E. coli* and thermophilic bacteria to the so-called higher eukaryotes. In addition, there is a high degree of homology between subunits α and β. The mechanism of action of the F_1 subunits is probably identical, since recombination of the various components from different sources produces hybrid complexes that are functional.

The polypeptide composition of the water-soluble F_1, which corresponds to the globular projection and part of the stalk in Fig. 12.11, is shown in Table 12.5 (Senior, 1988) for *E. coli* and beef heart mitochondria. With the exception of the ε subunit of beef heart mitochondria, which is distinct, the subunits from the two different systems appear to be very similar. However, the nomenclature, i.e., the Greek letter denoting each polypeptide, differs. The components are present in the stoichiometry of $\alpha_3\beta_3\gamma\delta\varepsilon$.

A combination of cryoelectron microscopy and image reconstruction techniques (Chapter 1, I, A) allows the three-dimensional reconstruction of negatively stained and unstained F_1. The complex appears to be roughly spherical with six elongated protein densities (the α and β subunits) and a central aqueous cavity. Three distinct conformations were found (Gogol et al., 1989b). The use of the Fab′ fragment of monoclonal antibodies to the various subunits allowed further localization of the components (Gogol et al., 1989a). Fab′ is a single antigen-binding chain of the antibody obtained after pepsin digestion of the whole molecule followed by reduction

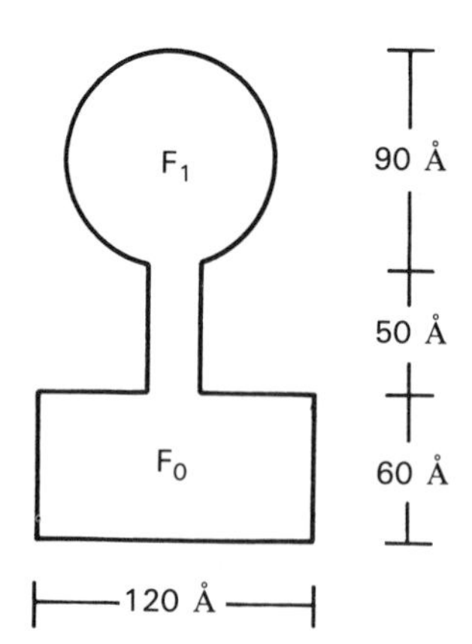

Fig. 12.11 Idealized drawing of the ATP synthase complex from rat liver mitochondria showing the approximate dimensions of F_1, F_0, and the stalk. Redrawn from Soper et al. (1979).

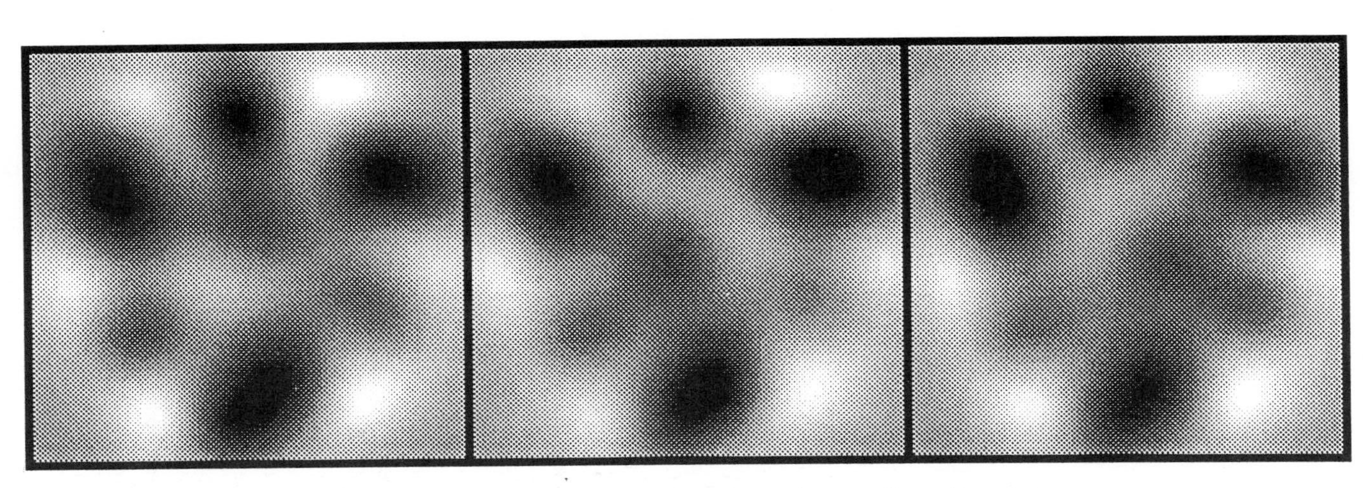

Fig. 12.12 Averaged images of F_1-complex after sorting into three classes. From Gogol et al. (1990), with permission.

of the disulfide bond. The anti-α Fab', for some reason, oriented the complex in a uniform manner, simplifying the analysis. The alternation of the α and β subunits was confirmed. A central density more centrally located in the cavity appears to be linked to a β subunit and is thought to represent one of the smaller subunits of F_1. Three different configurations were found. These are shown in Fig. 12.12. The frequency of occurrence of the various configurations (A,B,C) are tabulated in Table 12.6. The F_1 was tagged with both anti-α and anti-ε Fab', the latter to mark the location of one of the β subunits. As shown, the presence of ATP and Mg^{2+} which initiate the hydrolysis of ATP converts the majority of the images to the A conformation. Mg^{2+}, ADP, and P_i had a somewhat lesser effect on the conformation, whereas the chelator EDTA (to insure the absence of divalent cations) and ATP, which do not sustain hydrolysis, are without effect.

The F_0 portion is probably mostly embedded in the inner mitochondrial membrane. Its composition is shown in Table 12.7 (Senior, 1988). The *E. coli* F_0 proteins *a*, *b*, and *c* are present in the stoichiometry $a_1b_2c_n$, where *n* has been estimated to be between 10 and 12. Many other components have been shown in mitochondria: factor B, 11–15 kDa; F_6, 8–9 kDa; subunit 8, 6–8 kDa; an uncoupler binding protein, 30 kDa; and the inhibitor polypeptide, 9.6 kDa. The inhibitor peptide probably functions in blocking the ATPase activity in favor of the synthase activity. Subunit 6 of mitochondrial F_0 of 25 to 29 kDa is homologous to *a* of *E. coli*, whereas subunit 9, an 8-kDa proteolipid, is homologous to *c*.

Table 12.5 Subunit Composition of *E. coli* and Beef Heart Mitochondrial F_1

	E. coli	Beef heart mitochondria
Subunit composition	$\alpha_3\beta_3\gamma\delta\varepsilon$	$\alpha_3\beta_3\gamma\delta\varepsilon$
Total molecular size[a]	381,000	371,000
α subunit		
No. of residues	513	509
Molecular size	55,200	55,164
β subunit		
No. of residues	459	480
Molecular size	50,155	51,595
γ subunit		
No. of residues	286	272
Molecular size	31,428	30,141
δ subunit/OSCP[b]		
No. of residues	177	190
Molecular size	19,328	20,967
ε subunit/δ subunit[b]		
No. of residues	138	146
Molecular size	14,920	15,065
ε subunit[b]		
No. of residues		50
Molecular size		5,652

From A. E. Senior, *Physiological Reviews,* 68(1):177–231.

[a]Molecular size given in daltons.

[b]Note that *E. coli* δ is analogous to mitochondrial oligomycin sensi-

Table 12.6 Classification of ECF_1 images

	% in class		
Ligand(s) added	**A**	**B**	**C**
EDTA + ATP	30	27	43
	36	22	42
Mg^{2+} + ATP	67	4	29
	62	8	30
Mg^{2+} + ADP + P_1	52	5	43
	57	3	40

Reprinted with permission from E. P. Gogol, *Proceedings of the National Academy of Science,* 87:9585–9589, 1990.

The binding of F_1 to F_0 involves the δ subunit in *E. coli* (i.e., OSCP in eukaryotes) and probably ε.

The *b* subunit of F_0 is involved in binding F_1. The *b* dimer is largely hydrophobic, and the NH_2 terminal is thought to be embedded in the membrane with at least part of the molecule in the stalk of F_0F_1.

The hydrophobicity profile of the various subunits, their reactivity with various reagents, and analyses of mutants have suggested certain structural arrangements (see Senior, 1988). Models of subunit *a* suggest that certain amino acid side chains may form a "proton wire," i.e., may provide anionic groups that can pass H^+ from one to the other in a continuum.

Subunit *c* also contains residues that are likely to be embedded in the membrane. Subunit *c* has been shown to be involved in proton conduction in artificial bilayer systems. Since mutations impairing proton conduction occur in similar regions of *a* and *c*, it has been suggested that these regions are juxtaposed. The idea is emerging that protons can be passed from *a* (the

Table 12.7 Subunits of F_0

	E. coli	Beef heart mitochondria
Subunit *a*		
No. of residues	271	226
Molecular size[a]	30,285	24,816
Subunit *b*		
No. of residues	156	
Molecular size	17,202	
Subunit *c*		
No. of residues	79	75
Molecular size	8,264	7,402
A6L/aapl		
No. of residues	Not present	66
Molecular size		7,965
Other subunits	Not present	Likely
Stoichiometry	$a_1b_2c_n$	Unknown (c_n)

From A. E. Senior, *Physiological Reviews,* 68(1):177–231. Copyright © 1988 the American Physiological Society.

[a]Molecular size given in daltons.

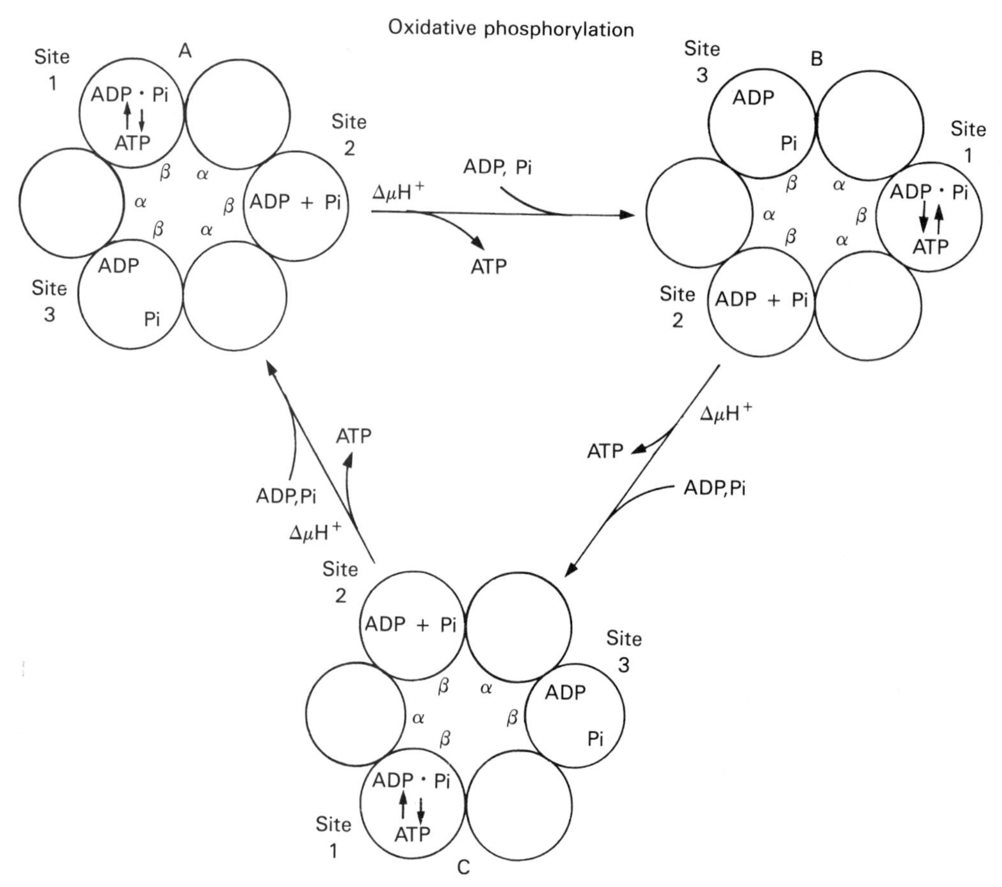

Fig. 12.13 ATP synthesis: proposed cyclical mechanism using all three catalytic sites of F_1. (A) catalytic site 1 is a high-affinity site, undergoing reversible synthesis of ATP from tightly bound ADP + P_i. Site 2 is a catalytic site of intermediate affinity, which must be occupied by ADP + P_i, and site 3 is a loose catalytic site. Energy from the proton gradient causes synchronous binding affinity changes at each site to give state B. (B) Site 1 has become lowest-affinity site, number 3, and ATP has been released and replaced by ADP and P_i from the medium. Site 2 has become high-affinity site, number 1, and reversible ATP synthesis has commenced. Site 3 has become an intermediate-affinity site, now number 2. Another proton gradient-induced binding-affinity change will release ATP and switch sites again to give the situation in state C. From Senior (1988), with permission.

presumed proton wire) to *c*, where *c* can act as a *protonophore*, or side group that can ferry protons back and forth, out of the membrane.

B. Catalytic Activity of F_1

Each F_1 complex contains six nucleotide binding sites. Three of these exchange rapidly with ATP added to the medium and are presumed to be catalytic sites. The other three exchange only slowly and might be regulatory sites.

Binding studies with isolated F_1 show that each α or β subunit binds ATP or ADP. The binding site of the β subunit exchanges rapidly and affinity labeling of F_1 with nucleotide analogs labels mostly β, although there is a suggestion that α may also be involved, perhaps at the interface between the two.

Binding affinity for substrates and products and one catalytic site was found to be profoundly affected by binding at a second site by ADP and P_i, which appeared to be needed for net ATP synthesis or release at the first site.

One of the models presently considered proposes that ADP and P_i tightly bound at one catalytic site can form ATP. However, binding of ADP and P_i to a second catalytic site is obligatory for release of ATP. As discussed previously, the energy for the release would come from the proton gradient. The model involves *alternating* catalytic sites; i.e., each β subunit acts in the binding of ADP and P_i synthesis and release of ATP alternatingly. Figure 12.13 (Senior, 1988) incorporates these features in a model in which three alternating sites are present (see Cross, 1981). In this respect, it is of considerable interest that hybrid F_1 containing one inactive β subunit along with two normal β subunits is inactive.

SUGGESTED READING

Boyer, P. D. (1989) A perspective of the binding change mechanism for ATP synthesis. *FASEB J.* 3:2164–2178.

Cramer, W. A., and Knaff, D. B. (1990) *Energy Transduction in Biological Membranes*, Chapter 8. Springer-Verlag, New York.

Penefsky, H. S., and Cross, R. L. (1991) Structure and mechanisms of F_0F_1-type ATP synthases and ATPases. *Adv. Enz.* 64:173–214.

Senior, A. E. (1988) ATP synthesis by oxidative phosphorylation. *Physiol. Rev.* 68:177–231.

Senior, A. E. (1990) The proton-translocating ATPase of *Escherichia coli*. *Annu. Rev. Biophys. Biophys. Chem.* 19:7–41.

Tzagaloff, A. (1982) *Mitochondria*, Chapters 6–9. Plenum Press, New York.

Special Aspects

Cox, G. B., and Gibson, F. (1987) The assembly of F_1F_0- ATPase in *Escherichia coli*. *Curr. Top. Bioenerg.* 15:163–174.

Ferguson, S. J. (1985) Fully localised chemiosmotic or localised proton flow pathway in energy coupling? A scrutiny of experimental evidence. *Biochim. Biophys. Acta* 811:47–95.

Ferguson, S. J. (1986) Towards a mechanism for the ATP synthase of oxidative phosphorylation. *Trends Biochem. Sci.* 11:100–101.

Junge, W. (1982) Electrogenic reactions and proton pumping in green plant photosynthesis. *Curr. Top. Membr. Transport* 16:431–465.

Kaback, H. R. (1982) Membrane vesicles, electrochemical ion gradients, and active transport. *Curr. Top. Membr. Transport* 16:393–430.

Kagawa, Y. (1984) Proton motive ATP synthesis. In *Bioenergetics* (Ernster, L., ed.), pp. 149–186. Elsevier, New York.

Tedeschi, H. (1981) The transport of cations in mitochondria. *Biochim. Biophys. Acta* 639:157–196; see pp. 162–170.

Weber, G. (1972) Addition of chemical and osmotic energies by ligand protein interactions. *Proc. Natl. Acad. Sci. U.S.A.* 69:3000–3003.

REFERENCES

Addanki, S., Dallas, F., Cahill, F. S., and Sotos, J. F. (1968) Determinations of intramitochondrial pH and intramitochondrial extramitochondrial pH gradient of isolated heart mitochondria by the use of 5,5-dimethyl 1-2,4 oxazolinedione. *J. Biol. Chem.* 243:2337–2348.

Anderson, W. M., Fowler, W. T., Pennington, R. M., and Fischer, R. R. (1981) Immunochemical characterization and purification of bovine heart mitochondrial pyridine nucleotide transhydrogenase. *J. Biol. Chem.* 259:1888–1895.

Archbold, G. P. R., Farrington, C. L., Lappin, S. A., McKay, A. M., and Malpress, F. H. (1979) Oxygen pulse curves in rat liver mitochondrial suspensions: Some observations and deductions. *Biochem. J.* 180:161–174.

Babcock, G. T., and Wikström, M. (1992) Oxygen activation and the conservation of energy in cell respiration, *Nature* 356:301–309.

Blumenfeld, M., Goldfeld, G., Mikoyan, V. D., and Soloyev, I. S. (1987) ATP synthesis by isolated coupling factors from chloroplasts during acidic and alkaline pH shifts. [translated in *Mol. Biol.* 21:268–274 (1987)] *Molekulyarnaya Biologiya* 21:323–329.

Boyer, P. D., Cross, R. L., and Momsen, W. (1973) A new concept in energy coupling in oxidative phosphorylation based on a molecular explanation of the oxygen exchange reactions. *Proc. Natl. Acad. Sci. U.S.A.* 70:2837–2839.

Brierley, G. P., and Murer, E. (1964) Ion accumulation in heart mitochondria supported by reduced cytochrome *c*. *Biochem. Biophys. Res. Commun.* 14:437–442.

Brierley, G. P., Bachman, E., and Green, D. E. (1962) Active transport of inorganic phosphate and magnesium ions by beef heart mitochondria. *Proc. Natl. Acad. Sci. U.S.A.* 48:1928–1935.

Campo, M. L., Bowman, C. L., and Tedeschi, H. (1984) Assays of ATP synthesis using single giant mitochondria. *Eur. J. Biochem.* 141:1–4.

Chance, B. (1965) The energy-linked reaction of calcium with mitochondria. *J. Biol. Chem.* 240:2729–2748.

Chance, B., and Hollunger, G. (1961) The interaction of energy and electron transfer reactions in mitochondria. VI. The efficiency of the reaction. *J. Biol. Chem.* 236:1577–1584.

Chance, B., and Williams, G. R. (1956) The respiratory chain and oxidative phosphorylation. *Adv. Enzymol.* 17:65–134.

Copenhaver, J. H., Jr., and Lardy, H. A. (1952) Oxidative phosphorylation: pathways and yield in mitochondrial preparations. *J. Biol. Chem.* 195:225–238.

Cross, R. L. (1981) The mechanism and regulation of ATP synthesis by F_1F_0-ATPases. *Annu. Rev. Biochem.* 50:681–714.

Dilley, R. A., and Schreiber, U. (1984) Correlation between membrane-localized protons and flash-driven ATP formation in chloroplast thylakoids. *J. Bioenerg. Biomembr.* 16:173–193.

Dunn, S. D. (1980) ATP causes a large change in conformation of the isolated β-subunit of *Escherichia coli* F1 ATPase. *J. Biol. Chem.* 255:11857–11860.

Feldman, R. I., and Sigman, D. S. (1982) The synthesis of enzyme bound ATP by soluble chloroplast factor 1. *J. Biol. Chem.* 257:1676–1683.

Feldman, R. I., and Sigman, D. S. (1983) The synthesis of ATP by membrane bound ATP synthetase complex from medium P under completely uncoupled conditions. *J. Biol. Chem.* 258:12178–12183.

Gogol, E. P., Aggeler, R., Sagermann, M., and Capaldi, R. A. (1989a) Cryoelectron microscopy of *Escherichia coli* F_1 Adenosinetriphosphatase decorated with monoclonal antibodies to individual subunits of the complex. *Biochem.* 28:4717–4724.

Gogol, E. P., Lücken, U., Bork, T., and Capaldi, R. A. (1989b) Molecular architecture of *Escherichia coli* F_1 Adenosinetriphosphatase. *Biochem.* 28:4709–4716.

Gogol, E. P., Johnston, E., Aggeler, R., and Capaldi, R. A. (1990) Ligand-dependent structural variation in *Escherichia coli* F_1ATPase revealed by cryoelectron microscopy. *Proc. Natl. Acad. Sci.* 87:9585–9589.

Hall, D. O. (1976) The coupling of photophosphorylation to electron transport in isolated chloroplasts. In *The Intact Chloroplast* (Barber, J., ed.), pp. 135–170. Elsevier, New York.

Hong, Y. Q., and Junge, W. (1983) Localized or delocalized protons in photophosphorylation. On the accessibility of the thylakoid lumen for ions and buffers. *Biochim. Biophys. Acta* 722:197–208.

Horton, A. A., and Hall, D. O. (1968) Determining stoichiometry of photosynthetic phosphorylation. *Nature (London)* 218:386–388.

Jagendorf, A. T., and Uribe, E. (1966) ATP formation caused by acid-base transition of spinach chloroplasts. *Proc. Natl. Acad. Sci. U.S.A.* 55:170–177.

Kagawa, Y. (1984) A new model of protonmotive ATP synthesis: acid-base cluster hypothesis. *J. Biochem.* 95:295–298.

Lehninger, A. L. (1955) Oxidative phosphorylation. *Harvey Lect.* 49:176–215.

Malenkova, I. V., Kuprin, S. P., Davydov, R. M., and Blumenfeld, L. A. (1982) pH-jump-induced ADP phosphorylation in mitochondria. *Biochim. Biophys. Acta* 682:179–183.

Mandolino, G., De Santis, A., and Melandri, B. A. (1983) Localized coupling in oxidative phosphorylation by mitochondria from Jerusalem artichoke (*Helianthus tuberosus*). *Biochim. Biophys. Acta* 723:428–439.

Mitchell, P. (1966) Chemiosmotic coupling in oxidative and photosynthetic phosphorylation. *Biol. Rev.* 41:445–502.

Ohnishi, T. (1979) Mitochondrial iron-sulfur Flavodehydrogenases. In *Membrane Proteins in Energy Transduction* (Capaldi, R. A., ed.), pp. 1–87; see pp. 58–80. Marcel Dekker, New York.

Penniston, J. T., Zande, H. V., and Green, D. E. (1966) Mitochondrial particles resolved for ion translocation. I. Preparation and properties of a particle coupled only at the phosphorylation site III of the electron transfer chain. *Arch. Biochem. Biophys.* 113:507–511.

Pick, U., Rottenberg, H., and Avron, M. (1974) The dependence of photophosphorylation in chloroplast on pH and external pH. *FEBS Lett.* 48:32–36.

Reynafarjee, B., Brand, M. D., and Lehninger, A. L. (1976) Evaluation of the H^+ site ratio of mitochondrial electron transport from rate measurements. *J. Biol. Chem.* 251:7442–7451.

Rydstrom, J. (1977) Energy linked nicotinamide nucleotide transhydrogenases. *Biochim. Biophys. Acta* 463:155–184.

Ryrie, I. J. and Jagendorf, A. T. (1972) Correlation between a conformational change in the coupling factor protein and the high energy state in chloroplasts. *J. Biol. Chem.* 247:4453–4459.

Senior, A. E. (1979) The mitochondrial ATPase. In *Membrane Proteins in Energy Transduction* (Capaldi, R. A., ed.), pp. 233–278. Marcel Dekker, New York.

Senior, A. E. (1988) ATP synthesis by oxidative phosphorylation. *Physiol. Rev.* 68:177–231.

Skulskii, I. A., Saris, N. E. L., and Glusunov, V. V. (1983) The effect of the energy state of mitochondria on the kinetics of unidirectional cation fluxes. *Arch. Biochem. Biophys.* 226:337–346.

Slater, E. C. (1971) The coupling between energy-yielding and energy-utilizing reactions in mitochondria. *Q. Rev. Biophys.* 4:35–71.

Sone, N., Yoshida, M., Hirata, H., and Kagawa, Y. (1977) Adenosine triphosphate synthesis by electrochemical proton gradient in vesicles reconstituted from purified adenosine triphosphatase and phospholipids of thermophilic bacterium. *J. Biol. Chem.* 252:2756–2960.

Soper, J. W., Decker, G. L., and Peterson, P. L. (1979) Mitochondrial ATPase complex. A dispersed cytochrome divalent oligomycin-sensitive preparation from rat-liver containing molecules with a tripartite structure. *J. Biol. Chem.* 254:11170–11176.

Tedeschi, H. (1981) The transport of cations in mitochondria. *Biochim. Biophys. Acta* 639:157–196; see pp. 162–170.

Westerhoff, H. V., Melandri, B. A., Venturoli, G., Azzone, G. F., and Kell, D. B. (1984) Mosaic protonic coupling hypothesis for free energy transduction. *FEBS Lett.* 165:1–5.

Wikstrom, M., and Krab, K. (1982) Proton translocation of cytochrome oxidase. *Curr. Top. Membr. Transport* 16:303–321.

Wilson, D. F., Dutton, P. L., and Wagner, M. (1973) Energy transducing components in mitochondrial respiration. *Curr. Top. Bioenerg.* 5:234–265.

Winget, D. G., Izawa, S., and Good, N. E. (1965) The stoichiometry of photophosphorylation. *Biochem. Biophys. Res. Commun.* 21:438–443.

Chapter 13

The Cell Membrane: Transport and Permeability

The integrity of the cell depends in a fundamental way on the presence of the cell membrane. The membrane prevents the loss of internal components and metabolites. Its ability to exclude some solutes and to permit the passage of others leads to the maintenance of an internal environment that differs from the medium external to the cell. The low concentration of Ca^{2+} in the cytoplasm permits its use as a sensitive signal for a variety of physiological regulatory effects. Similarly, the creation of an Na^+ electrochemical gradient provides a reservoir of energy that can be used for the transport of other solutes such as amino acids and sugars. The membrane controls the entrance of substrates used in either energy metabolism (e.g., glucose) or the synthesis of cell components (e.g., amino acids). The present chapter is directed primarily to this problem. Chapters 14 and 15 discuss the transport of ions.

Enzymes act to increase the rates of reactions by lowering the activation energy of the reaction and thereby allowing more molecules to react at a given time. As seen in Chapters 7 and 8, the activity of enzymes can be increased or decreased by effectors. The cell may also be able to increase either the rate of synthesis of an enzyme or its rate of breakdown (Chapter 9). Thus, in the cell there are various ways to regulate the rate of each reaction.

In addition, the activity of an enzyme can be regulated by its accessibility to its substrate and necessary cofactors. Below a saturating concentration of substrate, the rate of a reaction can be controlled by varying the concentration of the reactants. The lower the frequency of the encounter between reactants and enzyme, the lower the reaction rate. The availability of substrate molecules could then control the rate of the reaction. At one extreme, the impermeability of a membrane could prevent a reaction entirely. At the other extreme, if the substrate accumulated well beyond its extracellular concentration, the rate of the reaction could be increased, provided the concentration was not beyond the saturation level for the enzyme. The capacity of the cell to accumulate substances is illustrated dramatically in bacteria, where the accumulation is sufficient to alter the dry weight of the cells, albeit under artificial conditions. Generally, however, even in bacteria the accumulation occurs at high levels only when the enzyme system of the cell is unable to break down the substrate. This may occur because the organism is a mutant lacking the appropriate enzyme present in the wild type. Alternatively, the molecule that is accumulated may be similar in structure to the natural substrate but sufficiently different not to be metabolized. This chapter addresses the mechanisms of control of the concentration of substrates or substrate analogs by processes that occur at the cell membrane.

I. CHARACTERIZATION OF TRANSPORT

Passage of a solute from one biological compartment to another may involve passage of the molecule through a biological membrane. This process need not differ from ordinary diffusion. The passage from, say, compartment 1 to compartment 2 is proportional to the concentration of solute in the first compartment $[S_1]$, as shown in Eq. (13.1).

$$J_{1\rightarrow2} = k[S_1] \tag{13.1}$$

In this equation, k is a constant related to the diffusion coefficient (diffusion coefficient/thickness of the membrane). Equation (13.1) is a commonsense equation, a way of expressing Fick's law. The probability of the molecules going through a barrier must obviously increase with the number of molecules present. The flux (J) can be measured experimentally simply by measuring the passage from 1 to 2 ($J_{1\rightarrow2}$) of a radioactive isotope before the concentration in the second compartment $[S_2]$ is sufficiently high to produce a significant flux in the opposite direction. The passage in the opposite direction ($J_{2\rightarrow1}$) is given by an identical relationship. The net passage is given by the differences between the two fluxes.

$$J_{2\rightarrow1} = k[S_2] \tag{13.2}$$

$$J_{1\rightarrow2} - J_{2\rightarrow1} = k([S_2] - [S_1]) = \frac{dS_1}{dt} \tag{13.3}$$

Equation (13.3) predicts an equilibrium ($dS_1/dt = 0$) when $[S_1] = [S_2]$. It also predicts that the net passage due to diffusion alone will be in the direction of the concentration gradient.

The chemical potentials (see Chapter 6) for the components in compartments 1 and 2 are, respectively,

$$\mu_1 = \mu_0 + RT \ln [S_1] \tag{13.4}$$

$$\mu_2 = \mu_0 + RT \ln [S_2] \tag{13.5}$$

At equilibrium $\mu_1 - \mu_2 = 0$ and $[S_1] = [S_2]$.

It is interesting to consider that, at least in theory, when μ_1 is not equal to μ_2 the net chemical potential could be used to drive another system to perform work. Conversely, energy would have to be expended to transfer the solute against the concentration gradient. As we shall see, this does take place in a number of cases.

The constant k will depend on the nature of the solute and the properties of the biological membrane in question. The rate constant will also depend directly on the surface area exposed to the exchanges. The constant k corrected for surface area (A) is the permeability constant (P).

$$\frac{k}{A} = P \tag{13.6}$$

Equation (13.3), together with energetic and other considerations, allows a number of predictions. If a substance is assumed to pass through a biological membrane by dissolving in the membrane material and diffusing to the other compartment, the net passage of solute should be directly proportional to the concentration gradient. The passage should be largely independent of the presence of other substances. A steady state should be reached eventually in the form of a true equilibrium at which $[S_1] = [S_2]$. In addition, the small difference in properties between isomers should mean that there will be little difference in their rate of transfer. For example, transfer of the D form of a compound should generally be about the same as transfer of the L form.

These simple expectations are frequently not realized. The rate of penetration may not be a simple linear function of the concentration of the penetrant, as is the case in the transport of some amino acids. Figure 13.1*a* (Christensen et al., 1963) represents the uptake of L-*tert*-leucine by Ehrlich ascites tumor cells (a line of mouse peritoneal cells) as a function of concentration of the amino acid. With increasing concentration the rate of uptake does not increase as predicted. In addition, L-*tert*-leucine is taken up more rapidly than the D isomer, as shown in Fig. 13.1*c*. Furthermore, the steady-state level reached is greater for the L isomer. The membrane's ability to distinguish between isomers is known as *stereospecificity*.

Other apparent anomalies have also been observed. Where many amino acids are used simultaneously, they interfere with each other as if they were competing for entry.

Where a single amino acid is used, the internal concentration eventually reaches a steady-state level, as shown in Fig. 13.1*c*. However, this steady-state level does not correspond to a thermodynamic equilibrium. The internal concentration is higher than the external concentration. In fact, in this experiment the ratio between the internal and external concentrations approaches 50 for the L isomer. In other systems, it may be greater. For example, the *E. coli* β-galactoside transport system is capable of attaining ratios as high as 400. Any accumulation against a chemical or an electrochemical gradient requires the expenditure of energy.

Some of the findings may be explained without changing the diffusion model significantly. The saturation effect shown in Fig. 13.1*a* could be the result of passage of the amino acid through a restrictive channel that could admit only a certain number of molecules at a time. However, most of the other observations cannot be explained by such a simple concept. We would not expect, for example, that a simple channel would be able to distinguish between an L and a D isomer without introducing a number of complexities into the model. Neither would we expect accumulation against a chemical or electrochemical gradient unless the model were extensively modified.

The transport of many substances, charged and uncharged, may be explained with a rather simple model by considering the possibility that passage through the membrane is *mediated* by a *transporter* or *carrier* molecule. The term "transporter" is preferred in this discussion, since the term "carrier" has been used in the past to indicate that the transporter is capable of diffusing from one membrane interface to the other, which is not likely to take place. The translocation through a transporter mechanism requires reversible binding of the transported molecule to specific sites in the transporter molecule, followed by transfer of the sites to the opposite interface of the membrane and release of the ligand. We discussed reactions of this kind when examining

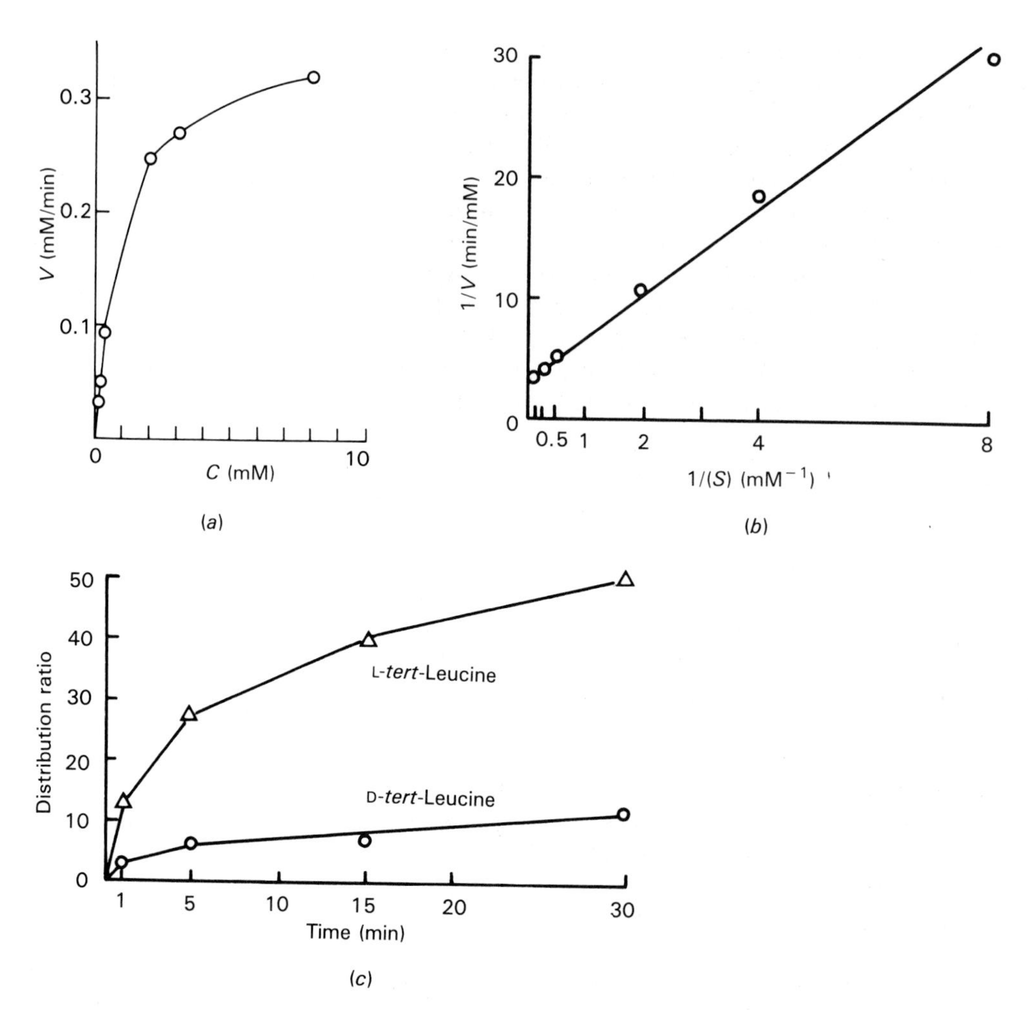

Fig. 13.1 (*a*) Kinetics of the transport of L-*tert*-leucine. (*b*) Lineweaver-Burk plot of L-*tert*-leucine concentrations against its rate of uptake. See text for details. V_m, 3.0 μmol ml^{-1} min^{-1}; K_m, 1.0×10^{-3} M. (*c*) Time course of the uptake of the iosmers of *tert*-leucine by Ehrlich ascites tumor cells. The cells were incubated as 4% suspensions in 1 mM L- or D-*tert*[1 − ^{14}C]leucine. The distribution ratio represents the ratio of radioactivity per kilogram of cellular water to the radioactivity of the suspending solution. From H. N. Christensen, et al., *Biochemica et Biophysica Acta,* 78:206–213. Copyright © 1963 Elsevier Science Publishers, Amsterdam.

the properties of enzymes. The model reactions describing the general patterns of enzyme-substrate interactions could also be used here as shown by

$$\underset{\text{outside phase}}{S_o} + \left[T \underset{2}{\overset{1}{\rightleftharpoons}} \underset{\text{membrane phase}}{TS_o} \overset{3}{\rightarrow} T \right] + \underset{\text{inside phase}}{S_i} \qquad (13.7)$$

In Eq. (13.7), T represents the transporter, which is restricted to the membrane phase (indicated by the brackets); S represents the substance being transported; and the subscripts *o* and i refer to the outside and inside of the cell, respectively, where the accumulation occurs against a concentration gradient. A necessary energy-requiring step could correspond to any one of the steps of Eq. (13.7). The transporter itself or other molecular elements involved in the reaction between substrate and transporter could have the requisite stereospecificity. As in the case of enzymes, the number of transporter molecules would necessarily be limited, and the kinetics of the transfer should therefore exhibit saturation. This saturation could explain the constancy of the rate of transfer with large increases in substrate concentration (Fig. 13.1*a*).

Except for the localization of the system in a membrane and the fact that the reaction corresponds to a vectorial displacement rather than catalysis, the model formally corresponds to that discussed for enzyme reactions (Chapter 7). In fact, the equations that are applicable are the same, and the rate of the transfer should depend on the concentration of substrate in a completely analogous manner, as shown in Eq. (13.8).

$$V = \frac{dS_i}{dt} = \frac{V_m(S_o)}{K_m + (S_o)} \qquad \text{or} \qquad \frac{1}{V} = \frac{K_m}{V_m(S_o)} + \frac{1}{V_m} \tag{13.8}$$

As shown in Fig. 13.1*a*, when the substrate concentration increases, the velocity of the reaction approaches a limiting value (V_m). As for the action of enzymes, and as shown in Eq. (13.8), a plot of $1/V$ as a function of $1/S_o$, a Lineweaver-Burk plot, should yield the constants K_m and V_m. The close agreement of data with the model is shown by the type of analysis of Fig. 13.1*b*, where, for the transport of L-*tert*-leucine into cultured tumor cells, the V_m is 3 mol ml^{-1} min^{-1} and the K_m is 1×10^{-3} M.

The K_m of Eq. (13.8) depends on the various rate constants of the reactions that are depicted in Eq. (13.7) (see Christensen 1975, pp. 115–119).

Many other features of transport suggest analogies with enzyme reactions. For example, as noted above, in the presence of analogs the kinetics exhibit the characteristics of competitive inhibition. Furthermore, some of the transport processes have been found to be blocked by chemicals (for instance, mercurials), which have been shown to react with proteins.

The representation of Eq. (13.7) does not consider a mechanism for the expenditure of energy necessary to transfer solute against an electrochemical gradient. Such a mechanism could correspond to phosphorylation of the transporter molecule followed by hydrolysis of the phosphate ester, as proposed for the transport of Na^+ (Chapters 14 and 15). However, the mechanism could be entirely different. The coupling to the energy-expending system will be discussed later. The kinetics of the uptake are frequently not precisely those predicted by Eq. (13.8). The rate of the reaction may not approach a maximal value asymptotically but may continue to rise slightly and steadily with increasing substrate concentration. Figure 13.2 shows the results for the rate of uptake of alanine by Ehrlich ascites tumor cells at various external concentrations (Oxender and Christensen, 1963). An explanation for this observation may be that more than one process is operative. It is conceivable, for example, that the amino acid exchanges at a rate directly proportional to the difference between the external and internal concentrations (i.e., by diffusion) and simultaneously by the carrier-mediated process represented by

$$\begin{aligned} \frac{dS}{dt} &= \text{carrier rate} + \text{diffusional rate} \\ \frac{dS_i}{dt} &= V + k([S_o] - [S_i]) \end{aligned} \tag{13.9}$$

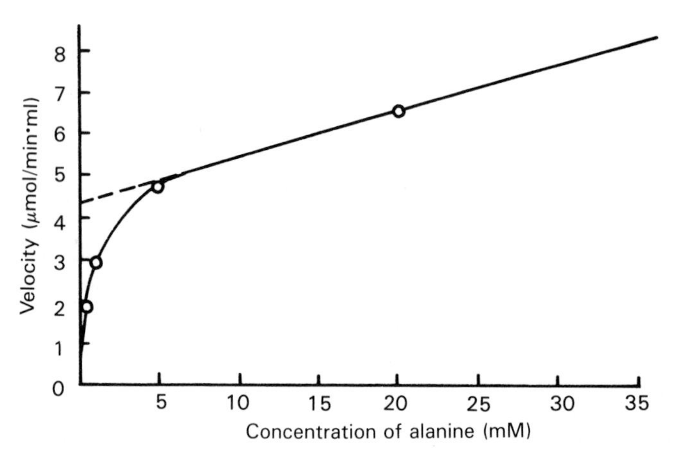

Fig. 13.2 Uptake of labeled alanine during 1 min as a function of external alanine concentration. Reprinted with permission from D. L. Oxender and H. N. Christensen, *Journal of Biological Chemistry,* 238:3691. Copyright © 1963 American Society of Biological Chemists, Inc.

Equation (13.9) can be integrated simply, allowing the determination of V and k. The line of Fig. 13.2 was arrived at by evaluating the appropriate constants and using the integrated form of Eq. (13.9).

In the example of the permease system of β-galactosides in *E. coli,* the accumulation can be against an extremely steep gradient. When a mutant is selected that is unable to utilize the sugar when it is transported (i.e., galactosidase-less mutant), the transport system can maintain a gradient up to 400-fold, and in exceptional cases a 100,000-fold accumulation has been claimed (Vorisek and Kepes, 1972) for galactose transport.

Where transport is against a concentration gradient, since the passive influx is negligible, Eq. (13.9) reduces to

$$\frac{dS_i}{dt} = V - kS_i \tag{13.10}$$

In both Eqs. (13.9) and (13.10) the assumption is made that the efflux is directly proportional to the internal concentration of the substrate, in harmony with the assumption that the efflux is the result of a diffusional process. However, for a number of substrates, the mediation of the transporter is also suspected for passage in the opposite direction.

The amino acid transport system discussed above and the β-galactoside transport system of bacteria clearly accumulate solute against a concentration gradient. Other translocation systems also follow saturation kinetics and exhibit stereospecificity. In many cases, such as the transport of sugars in human red blood cells and mammalian muscle, the transfer is not against a concentration gradient. For the latter systems, a mechanism such as that proposed in Eq. (13.7) would explain the kinetics. In addition, no energy expenditure is required. The process involved in these cases is known as *facilitated diffusion*. Similarly, exchange of a molecule in the medium for another in the cytoplasm is known as *exchange diffusion*. A number of systems using these two mechanisms have been tabulated by Stein (1986).

II. MOLECULAR MECHANISMS OF TRANSPORT

The molecular mechanism responsible for the transport of solutes across a biological membrane from one phase to the other can be envisioned in a variety of ways. However, all indications support the view that the transporters are analogous to enzymes. In recent years, the protein nature of the transporters has been verified by their isolation and reconstitution in artificial membrane systems.

At one time or another, several models have been suggested for the molecular mechanism of transport as represented in Fig. 13.3. The transporter-substrate complex could be transferred from one interface of the membrane to the other by diffusion of the complex through the membrane (model 1). A larger molecule spanning the membrane could also transfer the solute, perhaps by rotation or some other movement (model 2). However, the movement need not involve a large portion of the molecule but simply a small portion (not shown) or a conformational movement of the molecule or molecules constituting the pore structure (model 3). All these models involve movement of either the whole transporter or a portion of the transporter molecule (Fig. 13.3*a*). In contrast, fixed groups capable of binding the substrate arranged through a membrane pore (Fig. 13.3*b*) could also transfer the solute.

The mechanisms represented by the two sets of general models—mobile (*a*) or fixed (*b*)—can be distinguished experimentally (Fig. 13.4). Let us examine the situation in which the substrate does not normally accumulate against a concentration gradient. On reaching a steady state (here, a true equilibrium), the inside and outside concentrations are the same. The inward flux is precisely the same as the outward flux. A competitor should decrease the flux by making the transporter less available to the substrate. If the transporter is mobile, the competitor, present on only one side, would naturally block the passage of substrate in a single direction (i.e., if the competitor is outside, it should affect only the inward flux). The flux in the opposite direction should remain unaffected. The inhibition would, therefore, be asymmetric. This situation could result in temporary accumulation against a gradient. Some of these ideas are represented in Fig. 13.4*a* and are applicable to any of the mobile transporter models; model 1 was chosen arbitrarily in this diagram. The reasoning would be identical if we were to consider the situation in which the interior of the cell is preloaded with a competitor. On resuspension of the cells in a solution of the substrate, accumulation of substrate should take place in an inward direction against a concentration gradient and the efflux of substrate would be inhibited. Note that if the substrate is labeled with a radioactive isotope, the competitive inhibitor can be the unlabeled substrate molecule. The counterflow of a substrate against a concentration gradient involves the dissipation of chemical potential due to the flow of a second substance in the direction of its own concentration gradient. Counterflow is also known as *trans facilitation* (trans indicating that the competitor is on the other side of the membrane from the substrate). This phenomenon is apparently also aided by the fact that the rate of movement of the transporter is considerably higher when it is loaded than when it is empty (Gorga and Lienhard, 1981) (see below).

For the case of a *fixed* carrier (Fig. 13.4*b*), both the inward and outward fluxes should be equally affected and therefore an inhibition and not a trans facilitation should take place.

Many experiments provide in essence the same information, for example, in the monosaccharide transport of red blood cells (Rosenberg and Wilbrandt, 1958) discussed in Chapter 6 in relation to the bioenergetics of the process. Figure 6.3 shows trans facilitation in human blood cells.

There are many other indications that the binding sites involved in transport alternate between the two faces of the membrane. For example, cytochalasin B (CB) binds to the glucose transporter of red blood cells only at the inner surface, whereas 4,6-*O*-ethylidene-D-glucose (EG) binds only to the outer surface. In the intact cell CB has no effect on the binding of EG, since CB does not enter the red blood cell. However, it does interfere with the binding of EG when leaky ghosts are used (Gorga and Lienhard, 1981).

This observation also suggests that the mechanism involves some conformational rearrangement of the transporter molecule, since the intrinsic fluorescence of the glucose transporter

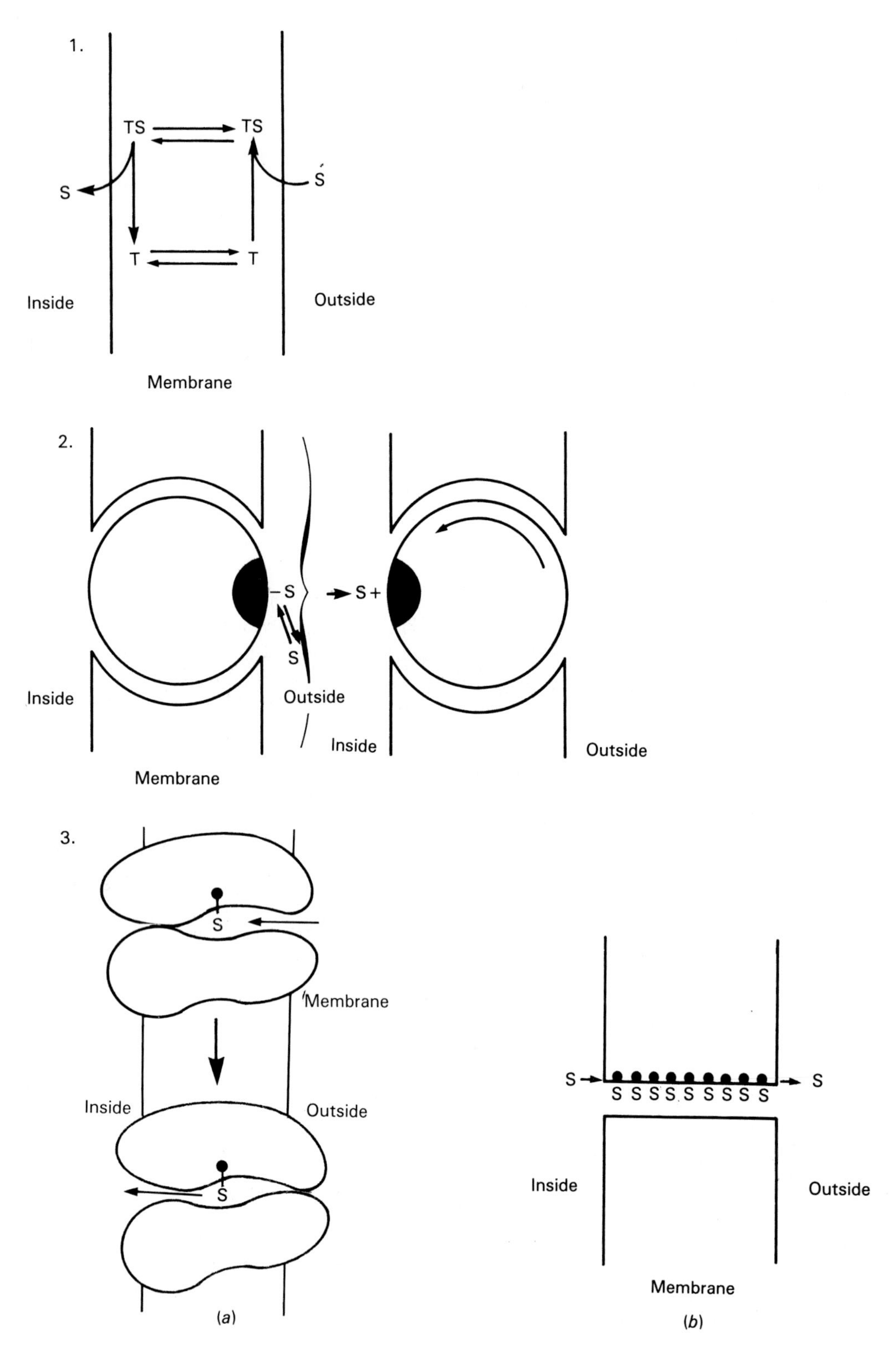

Fig. 13.3 Diagrammatic representation of carrier models. Circle, carrier; S, substrate. (*a*) Mobile transporters, (*b*) fixed transporters.

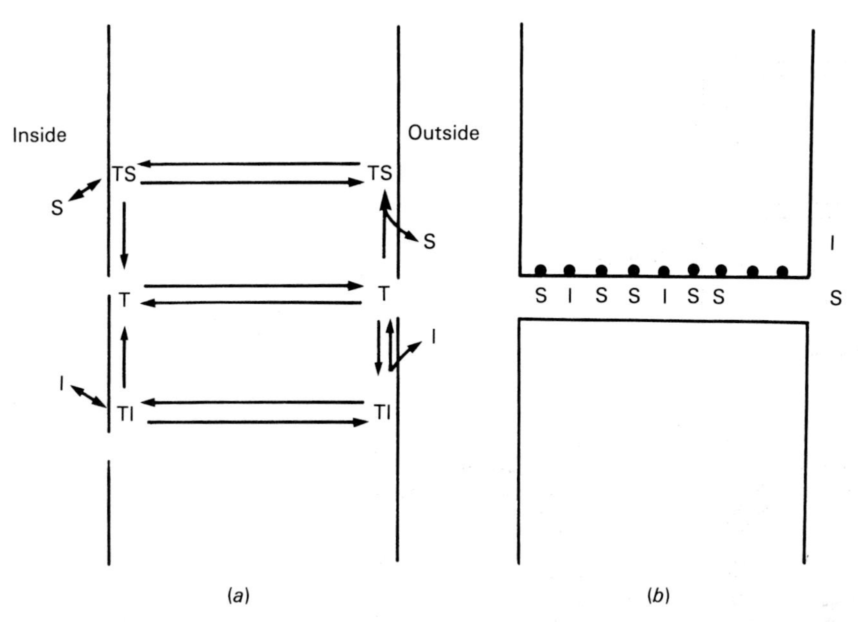

Fig. 13.4 Diagrammatic representation of transporter models and their behavior in relation to competitive inhibitors. T, Transporter; S, substrate; I, inhibitor; filled circles indicate binding sites. (*a*) Mobile transporter; (*b*) fixed transporter.

changes with binding to the substrate (Gorga and Lienhard, 1981). A conformational change is also demonstrated by changes in its inactivation by chemical treatment. The sugar transport of the red blood cell can be inactivated by 1-fluoro-2,4-dinitrobenzene (FDNB), which reacts with free amino groups. The rate of inactivation varies with the sugar bound to the carrier site. Deoxyglucose, for example, increases the rate of inactivation by FDNB fivefold. These results suggest that the FDNB-sensitive groups in the carrier are exposed differently, depending on the transported molecule bound to them (Krupka, 1971). Similarly, the galactose-binding protein of *E. coli* shows an increase in intrinsic fluorescence and in electrophoretic mobility when attached to its substrate (Boos et al., 1972).

The evidence presently favors a mobile transporter mechanism. However, a number of factors, some already examined, argue against a rotational model (Fig. 13.3, model 2) or any model requiring a major movement of the transporter. In Chapter 2 we saw that a number of intrinsic proteins span the membrane asymmetrically so that specific groups are found exclusively on the outside or the inside phase. This is not what we would expect if proteins had the capacity to move or rotate in the membrane. It could be argued that transport proteins have unique properties, but all the available evidence suggests that proteins involved in transport behave like any other intrinsic protein (see also Chapter 14). Trypsin treatment of red blood cells does not interfere with glucose transport, whereas trypsin treatment of inside-out vesicles prepared from the red blood cells blocks the transport (Baldwin et al., 1980). This finding points to the conclusion that the trypsin-sensitive portion of the glucose transporter is always located on the side of the membrane that normally faces inside the cell.

These considerations make models 1 and 2 of Fig. 13.3 highly unlikely. One of the models that appears most likely is model 3, in which the groups binding the substrate are made available in alternation to two different phases (see Chapter 15). Model 3 shows two protein molecules (i.e., a dimer) forming a water-filled channel. Two separate chains of the same protein molecule traversing the membrane could also form a similar channel.

III. VECTORIAL ENZYMES

We have seen that the mechanisms of transport may involve the formation of a complex with a transporter molecule itself, behavior analogous to that of an enzyme. However, the transporter does not catalyze a chemical reaction and does not release a product that differs from the original substrate. Rather it mediates the translocation of the substrate across the cell membrane and the substrate remains unchanged. In principle, a process could involve both transfer through the cell membrane and catalysis. In this case, the reaction would differ from one catalyzed by an enzyme in solution: the action of this kind of enzyme has a direction; it is *vectorial*. A vectorial reaction is thought to be involved in the simultaneous translocation and phosphorylation of glucose and related monosaccharides in *E. coli, Salmonella typhimurium,* and *Bacillus subtilis*. The sugars are translocated from the external medium into the cell, where they are released in a phosphorylated form. A similar process is thought to take place in yeast (e.g., Van Steveninck, 1972) and in *Aerobacter aerogenes* (Kelker and Anderson, 1972). Simultaneous translocation and enzyme action are by no means restricted to microorganisms. They are thought to take place, for example, in the intestinal transport of sugars, where sucrose is hydrolyzed and the products of the reaction, fructose and glucose, are released into cells.

The most direct evidence for a role of vectorial reactions in translocation comes from studies of the phosphoenolpyruvate phosphotransferase reaction in vesicles obtained from *E. coli* membranes. The enzyme is responsible for the transfer of phosphate from phosphoenolpyruvate to sugar molecules. The experiment is represented in Fig. 13.5 (Kaback, 1968), where the ordinate represents the amount of sugar taken up as a function of time. The interior of the vesicles was preloaded with [^{14}C]glucose, and [^{3}H]glucose and phosphoenolpyruvate were added to the outside. The phosphorylation and the transfer of the phosphate showed a preference for the outside [^{3}H]glucose. The phosphorylation did not precede the transfer, since the vesicles were unable to take up external glucose-6-phosphate.

In some bacteria, the phosphorylation translocation of sugars involves the phosphoenolpy-

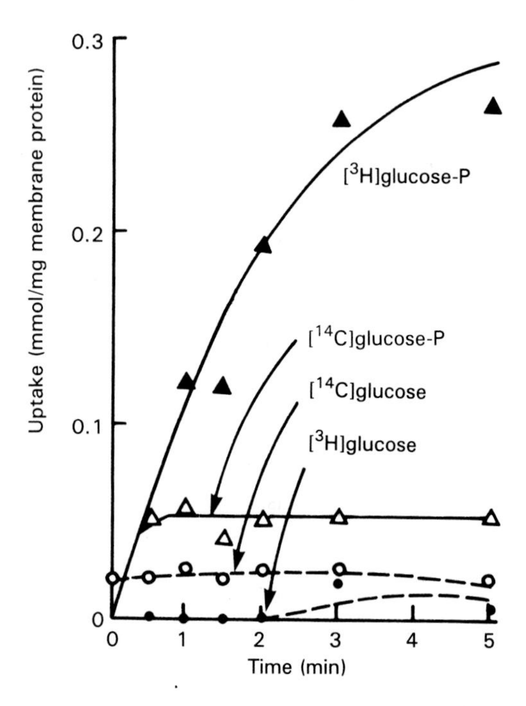

Fig. 13.5 Uptake and phosphorylation of [^{3}H]glucose by *E. coli* membranes previously loaded with [^{14}C]glucose. Reprinted with permission from H. R. Kaback, *Journal of Biological Chemistry,* 243:3711–3724. Copyright © 1968 American Society of Biological Chemists, Inc.

ruvate phosphotransferase reaction as in the example of Fig. 13.5. In this reaction sugars are phosphorylated by the transfer of phosphate from phosphoenolpyruvate (PEP). The reaction occurs in two steps, one in which the phosphate is transferred from PEP to a protein (HPr) and a second in which it is transferred from the protein to the sugar. This can be represented as

$$\text{PEP} + \text{HPr} \xrightleftharpoons{\text{enzyme I, Mg}^{2+}} \text{pyruvate} + \text{PHPr} \quad (13.11a)$$

$$\text{PHPr} + \text{sugar} \xrightleftharpoons{\text{enzyme II}} \text{sugar-6-phosphate} + \text{HPr} \quad (13.11b)$$

The protein represented by the symbol HPr is small (molecular weight 9400), and it is the histidine residue that is phosphorylated.

The phosphotransferase system of bacteria may well account for the first reaction of many metabolized sugars for which no active kinase has been found. The apparent absence of an appropriate kinase has been one of the most intriguing questions in the physiology of some microorganisms (Wood, 1966). There are many indications that the phosphotransferase system plays a role in the translocation of sugars and that they are accumulated in the phosphorylated form. Mutants of *S. typhimurium* lacking any of the components of the system fail to accumulate sugars (Simoni et al., 1967). However, it should be recognized that the mutants may be pleiotrophic (Kennedy and Scarborough, 1967) and have more than one deficiency. In some cases, the substrate is accumulated as a phosphorylated intermediate, as in the case of β-glucosides (Fox and Wilson, 1968; Wang and Morse, 1968). It has been proposed that the PEP system is involved in the transport of β-glycosides in *E. coli*. The protein HPr is deficient in treated *E. coli* cells that have impaired β-galactoside transport, and the addition of HPr reestablishes function. On the other hand, the finding that *E. coli* mutants deficient in enzyme I can transport β-galactosides makes this interpretation unlikely (Asensio et al., 1963).

IV. STRATEGIES FOR THE ISOLATION OF TRANSPORT PROTEINS

Our insights into a biological system cannot be considered sufficient until all molecular components and their characteristics are recognized. The ultimate test of our understanding consists of reestablishing biological activities in the test tube after putting the component molecules back together. A number of transport proteins have been isolated, and with many of them it has been possible to reconstitute transport by recombining them with artificial membranes (Hokin, 1981).

Isolation of a transport protein requires disruption of the membrane and some means of recognizing the presence of the molecules through the various fractionation procedures. Detergents must be used because the transporters are intrinsic proteins held to the membrane structure by hydrophobic bonds.

Recognition is not always a simple matter. Since the transport proteins must bind either substrates or inhibitors, they may be recognized by this property. The task is simplest when a specific inhibitor that can be labeled is strongly bound. The anion transporter of the red blood cell (Chapter 15) has been recognized in this fashion. The usefulness of this approach can be much extended with the use of modified substrates that can be cross-linked at the sites of attachment, perhaps in a light-dependent fashion (*photoaffinity labeling*).

Since the ion transport is sometimes associated with the hydrolysis of ATP, so that the ATPase activity should depend on the presence of the appropriate ion (e.g., Ca^{2+}-ATPase,

Na^+,K^+-ATPase), the various fractions can be assayed by their ability to hydrolyze ATP in the presence of the ions in question.

A general method that in principle could be very useful is the separation of the transport proteins through their property of binding to either substrate or inhibitor. The latter (e.g., glucose, ATP) are covalently bound to a column and will bind the transporter but allow other proteins through (this is known as *affinity* chromatography). An excess of the free substrate or inhibitor can then be used to combine with the transport protein and free it from the column in a process known as elution.

A group of related transport proteins with ATPase activity has been identified. This superfamily of *traffic ATPases* are also referred to as the *ATP-binding casette proteins* (ABC) (Hyde et al., 1990). It includes the multidrug resistance P-glycoprotein (MDR), the cystic fibrosis gene product (CPTR), and a variety of prokaryotic permeases. CPTR probably functions as a Cl^--conducting channel (Anderson et al., 1991; Kartner et al., 1991).

Many years of work by many people have brought us close to unraveling one of the most intriguing problems of cell physiology: how a cell can control its integrity by controlling the flow of material in and out of its internal compartments. In this chapter, we have dealt primarily with the translocation of nonionic components. The next two chapters (Chapters 14 and 15) discuss the systems responsible for the transfer of ions in the cell.

SUGGESTED READING

Christensen, H. N. (1975) *Biological Transport,* Chapters 1, 4–8. Benjamin, Reading, Mass.

Macey, R. I. (1987) Mathematical models of membrane transport processes. In *Membrane Physiology,* 2d ed. (Andreoli, T. E., Hoffman, J. F., Fanestil, D. D., and Schulz, S. G., eds.), pp. 111–132. Plenum Medical Book Company, New York.

Schafer, J. A., and Andreoli, T. E. (1987) Principles of water and nonelectrolyte transport across membranes. In *Membrane Physiology,* 2d ed. (Andreoli, T. E., Hoffman, J. F., Fanestil, D. D., and Schulz, S. G., eds.), pp. 177–190. Plenum Medical Book Company, New York.

Schulz, S. G. (1987) Ion-coupled transport of organic solutes across biological membranes. In *Membrane Physiology,* 2d ed. (Andreoli, T. E., Hoffman, J. F., Fanestil, D. D., and Schulz, S. G., eds.), pp. 283–294. Plenum Medical Book Company, New York.

Stein, W. D. (1986) *Transport and Diffusion Across Cell Membranes,* Chapters 1, 2, 4, and 5. Academic Press, New York.

REFERENCES

Anderson, M. P., Rich, D. P., Gregory, R. J., Smith, A. E., and Welsh, M. J. (1991) Generation of cAMP-activated chloride currents by expression of CFTR. *Science* 251:679–682.

Asensio, C., Avigad, G., and Horecker, B. L. (1963) Preferential galactose utilization in a mutant strain of *E. coli. Arch. Biochem. Biophys.* 103:299–309.

Baldwin, J. M., Lienhard, G. E., and Baldwin, S. A. (1980) The monosaccharide transport system of the human erythrocyte. *Biochim Biophys. Acta* 599:699–714.

Boos, W., Gordon, A. S., Hall, R. E., and Price, H. D. (1972) Transport properties of the galactose-binding protein of *Escherichia coli*. Substrate induced conformational change. *J. Biol. Chem.* 247:917–924.

Christensen, H. N. (1975) *Biological Transport.* Benjamin, Reading, Mass.

Christensen, H. N., Clifford, J. B., and Oxender, D. L. (1963) Stereospecificity of the transport of *tert*-leucine. *Biochim. Biophys. Acta* 78:206–213.

Fox, C. F., and Wilson, G. (1968) The role of a phosphoenolpyruvate-dependent kinase in β-glucoside catabolism in *Escherichia coli. Proc. Natl. Acad. Sci. U.S.A.* 59:988–995.

Gorga, F. R., and Lienhard, G. E. (1981) Equilibria and kinetics of ligand binding to the human erythrocyte glucose transporter. Evident for an alternative conformational model. *Biochemistry* 20:5108–5113.

Hokin, L. E. (1981) Reconstitution of carriers in artificial membranes. *J. Membr. Biol.* 60:77–93.

Hyde, S. C., Emsley, P., Hartshorn, M. J., Mimmack, M. M., Gileadi, U., Pearce, S. R., Gallagher, M. P., Gill, D. R., Hubbard, R. E., and Higgins, C. (1990) Structural model of ATP-binding proteins associated with cystic fibrosis, multidrug resistance and bacterial transport. *Nature* 346:362–365.

Kaback, H. R. (1968) The role of the phosphoenolpyruvate-phosphotransferase system in the transport of sugars by isolated membrane preparations of *Escherichia coli*. *J. Biol. Chem.* 243:3711–3724.

Kartner, N., Hanrahan, J. W., Jensen, T. J., Naismith, A. L., Sun, S., Ackerley, C. A., Reyes, E. F., Tsui, L. C., Rommens, J. M., Bear, C. E., and Riordan, J. R. (1991) Expression of the cystic-fibrosis gene in nonepithelial invertebrate cells produce a regulated anion conductance. *Cell* 64:681–691.

Kelker, N. E., and Anderson, R. L. (1972) Evidence for vectorial phosphorylation of *n*-fructose by intact cells of *Aerobacter aerogenes*. *J. Bacteriol.* 112:1441–1443.

Kennedy, E. P., and Scarborough, G. A. (1967) Mechanism of hydrolysis of *o*-nitrophenyl-β-galactoside in *Staphylococcus aureus* and its significance for theories of sugar transport. *Proc. Natl. Acad. Sci. U.S.A.* 58:225–228.

Krupka, R. M. (1971) Evidence for a carrier conformational change associated with sugar transport in erythrocytes. *Biochemistry* 10:1143–1148.

Oxender, D. L., and Christensen, H. N. (1963) Distinct mediating systems for the transport of neutral amino acids by the Ehrlich cell. *J. Biol. Chem.* 238:3686–3699.

Rosenberg, T., and Wilbrandt, W. (1958) Uphill transport induced by counterflow. *J. Gen. Physiol.* 41:289–296.

Simoni, R. D., Levinthal, L. M., Kundig, F. D., Kunding, W., Anderson, B., Hartman, P. E., and Roseman, S. (1967) Genetic evidence for the role of a bacterial phosphotransferase system in sugar transport. *Proc. Natl. Acad. Sci. U.S.A.* 58:1963–1970.

Stein, W. D. (1986) *Transport and Diffusion across Cell Membranes*, pp. 339–343. Academic Press, New York.

Van Steveninck, J. (1972) Transport and transport associated phosphorylation of galactose in *Saccharomyces cerevisiae*. *Biochim. Biophys. Acta* 274:575–583.

Vorisek, J., and Kepes, A. (1972) Galactose transport in *Escherichia coli* and the galactose binding protein. *Eur. J. Biochem.* 28:364–372.

Wang, R. J., and Morse, M. L. (1968) Carbohydrate accumulation metabolism in *Escherichia coli*. I. Description pleiotrophic mutants. *J. Mol. Biol.* 32:59–66.

Wood, W. A. (1966) Carbohydrate metabolism. *Annu. Rev. Biochem.* 35:521–558.

Chapter 14

The Cell Membrane: Transport of Ions

As discussed in Chapter 6, active transport of a solute can be powered by the flow of another solute in the opposite direction. In animal cells the active influx of sugars and amino acids results from the flow of Na^+ in the opposite direction. In contrast, the active transport of Ca^{2+} and the coupled transport of Na^+ and K^+ or of H^+ and K^+ have been shown to be powered by ATP hydrolysis (that is, the transporter functions as an ATPase). These transport reactions are outlined in Eqs. (14.1) to (14.3). In these equations, the location of the component, inside or outside, is indicated by i or o, respectively. These are the only active transport systems coupled to ATP hydrolysis in the plasma membrane of animal cells.

$$2K_o^+ + 3Na_i^+ + MgATP_i \rightleftharpoons MgADP_i + P_i + 3Na_o^+ + 2K_i^+ \qquad (14.1)$$

$$2Ca_i^{2+} + MgATP_i \rightleftharpoons MgADP_i + P_i + 2Ca_o^{2+} \qquad (14.2)$$

$$nK_o^+ + nH_i^+ + MgATP_i \rightleftharpoons MgADP_i + P_i + nH_o^+ + nK_i^+ \qquad (14.3)$$

Each of the three proteins responsible for the ATP-powered active transport of these ions has been isolated (Kyte, 1971; McLennan, 1969; Sachs et al., 1976) and it has been possible to reconstitute the Ca^{2+} and the Na^+- K^+ transport systems by recombining these proteins with artificial lipid membranes (for a review, see Hokin, 1981). The various transport ATPases resemble each other strikingly. All are polypeptides of approximately 900 to 1200 amino acids and their activity involves phosphorylation in an aspartate residue. The Ca^{2+}- and Na^+,K^+-ATPases have been studied in more detail and appear to share many other similarities, including the portions of the molecules associated with the phospholipid portion of the membrane and the location within the polypeptide of the phosphorylated residue. In the membrane, the Na^+,K^+-

ATPase corresponds to two polypeptides: a larger polypeptide (α) of 94 to 106 kDa and a smaller sialoglycoprotein (β) of 41 to 52 kDa (Kyte, 1974), which may have a role in the transport and membrane assembly of the α-subunit (McDonough et al., 1990). The two polypeptides are present in equimolar amounts, but it is generally agreed that the α component corresponds to the transporter. Certainly, the ATPase activity is in the α subunit, since the β subunit can be removed without changing the ATPase activity (Freytag, 1983). The Ca^{2+}-ATPase lacks the smaller polypeptide.

The Na^+-K^+ transport system is associated with almost all cells that have been studied, including those of specialized tissues such as the kidney and the rectal gland of elasmobranchs that function in salt transport. The Ca^{2+}-ATPase functions to maintain a low cytoplasmic Ca^{2+} concentration. One seems to be associated with the plasma membrane, while another distinct ATPase functions to sequester Ca^{2+} in the sarcoplasmic reticulum of muscle cells (see Chapter 18). Similar Ca^{2+}-sequestering systems are likely to be present in other contractile systems. In contrast, the H^+,K^+-ATPase is the means by which acid is accumulated by the gastric mucosa (Sachs et al., 1976). A similar ATPase in the fungus *Neurospora* (Scarborough, 1980) is responsible for the high membrane potential of its cells; here the transport is *electrogenic*.

This chapter treats primarily the functional significance and characteristics of the Na^+,K^+-ATPase transport system. Chapter 15 will discuss the characterization of the proteins responsible for the transport of cations, together with possible models of transport. A protein that functions in the exchange of anions will be discussed later.

I. FUNCTIONAL SIGNIFICANCE OF THE Na^+-K^+ TRANSPORT SYSTEM

In all cells there seems to be some mechanism for controlling the concentration of ions in the internal medium. A high internal concentration of K^+ seems to be the rule but is not invariably the case. A low concentration of Na^+ is also generally the case, although again there are exceptions. The range of internal K^+ concentration in vertebrates is generally narrow, 100 to 200 mM. In freshwater organisms, the internal level of K^+ is frequently much smaller, 15 to 30 mM, but this is remarkably high compared to the external environment, in which K^+ may be present in only trace amounts. The capacity to control the internal concentration is such that certain organisms can grow at extreme conditions of salinity and still concentrate K^+ selectively despite the high Na^+ concentration of the medium and the relative absence of K^+. *Halobacterium salinarium* concentrates K^+ in the face of an external NaCl concentration of 4 M and can attain the remarkable internal K^+ concentration of 4 M (Christian and Waltho, 1962). Examples of the concentrations of Na^+ and K^+ in various organisms are listed in Table 14.1 (Steinbach, 1963).

The reasons for the universality of these ion distributions are still a subject for speculation. As we shall see in Chapter 16, the cell membrane's resting and action potentials depend largely on this ion concentration imbalance between the inside and the outside of the cell. These potentials play a role not only in the signal conduction of nerves and muscle in higher organisms but also in the responses of protists to their environment. Furthermore, the internal concentration of cations has to be controlled (in this case by pumping out Na^+) to maintain cell volume within physiological limits. The Na^+ tends to accumulate inside (Tosteson and Hoffman, 1960) to neutralize the negative charges of the macromolecules trapped in the cytoplasm. Inhibition of the Na^+ pumping activity therefore leads to osmotic swelling. There is no doubt that the exclusion of Na^+ plays a role in cell function and integrity in at least animal cells. However, it is difficult to argue that these needs are universal. Electrical events may not be invariably required, and the rigid walls of plant cells and microorganisms are effective in limiting swelling. Other reasons

Table 14.1 Selected Values for Na^+ and K^+

	Erythrocytes (mmol/kg wet wt.)		Plasma or serum (mmol/kg wet wt.)	
Animal	**Na**	**K**	**Na**	**K**
Human	11	91	138	4.2
Beef	70	25	142	4.8
Sheep (average)	82	11	160	4.8
Dog	106	5	150	4.4
Rabbit	16	99	158	4.1
Elephant seal	95	7	142	4.5
Rat	12	100	151	5.9
Duck	7	112	141	6.0
Chicken	18	119	154	6.0
Dolphin (mammal)	13	99	153	4.3
Fish (mackerel)	—	—	183	10.0
Frog	—	—	105	4.8
Reptile (turtle)	—	—	140	4.6

Plant	Na	K
Asparagus	1	50
Beet, leaves	60	130
Beet, roots	30	75
Lettuce	6	60
String beans	1	60
Broccoli	7	75
Celery	50	90
Spinach	30	120
Yucca, leaves	19	70
Yucca, stems	98	50
Vetch	40	156
Salt grass (sea shore)	70	45
Salt bush	23	63
Rye, tops	4	120
Clover, tops	20	150

Reprinted by permission from H. B. Steinbach, *Comparative Biochemistry,* Vol. IV:677–720. Copyright © 1963 Academic Press.

may therefore exist for the universality of the control of the cell's internal environment. Evidence accumulated over the years indicates that K^+ is generally required for growth (Steinbach, 1963). Potassium ions are required for protein synthesis in a number of unrelated organisms or preparations (Table 14.2). This requirement probably explains the effect of K^+ on growth. In addition, K^+ is required for maximal activity of a number of enzymes concerned primarily with other functions (see Table 14.3) (Lubin, 1964).

The K^+ requirement for protein synthesis is elegantly shown in an experiment with a mutant strain of *E. coli* that is incapable of accumulating K^+ (Fig. 14.1) (Lubin and Ennis, 1964). Other experiments of this kind have been carried out with a *B. subtilis* mutant. The cells are incubated in a medium containing leucine labeled with ^{14}C. The ordinate in Fig. 14.1 represents the labeled leucine incorporated by the bacteria. In this experiment, leucine incorporation serves as an indicator of protein synthesis. Clearly, the K^+ inside the cells plays a role in the incorporation of the amino acid tested. A similar demonstration has been carried out with mammalian tumor cells in tissue culture in which the K^+ transport has been inhibited by the drug amphotericin B.

From these experiments, it is clear that K^+ accumulation has a fundamental role in protein synthesis. In addition, it activates a number of metabolic enzymes.

Table 14.2 Dependence of Protein Synthesis on K^+

System	Reference
Liver extract	Sachs, H. *J. Biol. Chem.* 228:23 (1957)
Pancreas extract	Gazzinelli, G., and Dickman, S. R. *Biochim. Biophys. Acta* 61:980 (1962)
Sea urchin egg	Hultin, T. *Exp. Cell Res.* 25:405 (1961)
E. coli intact mutant or extract	Lubin, M. *Nature (London)* 213:415 (1967)
Bacillus subtilis intact mutant	Lubin, M. *Fed. Proc. Fed. Am. Soc. Exp. Biol.* 23:994 (1964)
Sarcoma 180, intact, pharmacological manipulation	Lubin, M. *Nature (London)* 213:451 (1967)

II. COUPLING OF ATP HYDROLYSIS TO THE TRANSPORT OF Na^+

The active transport of Na^+, K^+, Ca^{2+}, or H^+ in the plasma membrane is coupled to the hydrolysis of ATP. For practical reasons, the Na^+, K^+-ATPases studied most intensely have been those of giant nerve cell axons and red blood cells. As we shall see, they have obvious advantages. In the internal medium of these cells, as in many others, a high K^+ concentration and a low Na^+

Table 14.3 Enzymes Strongly Activated by K^+

Enzyme	Tissue or organism
Glycolysis and related pathways	
Fructokinase	Liver
6-Phosphofructokinase	Yeast
Pyruvate phosphokinase	Animal tissues, yeast
Phosphotransacetylase	*Clostridium*
Malic enzyme	*Lactobacillus*
Acetyl-CoA synthetase	Heart
Biosynthesis of essential small molecules	
Inosine-5′-phosphate dehydrogenase	*Aerobacter*
S-Adenosylmethionine synthetase	Yeast, liver
Deoxyguanylate kinase	*E. coli*
Pantothenate synthetase	*E. coli*
AICAR transformylase	Liver
Carbamoyl-phosphate synthetase	Liver
Protein synthesis	
Tyrosine-activating enzyme	Pancreas, liver
Phosphodiesterase	*E. coli*
Miscellaneous	
5′-Adenylic acid deaminase	Erythrocytes
Aldehyde dehydrogenase	Yeast
Glycerol dehydrogenase	*Aerobacter*
β-Galactosidase	*E. coli*
Tryptophanase	*E. coli*
Glutamylcysteine synthetase	Wheat
Glutathione synthetase	Wheat
D-Alanyl-D-alanine synthetase	*Streptococcus*
L-Threonine dehydrase	Liver

Martin Lubin, "Cell Potassium and the Regulation of Protein Synthesis," in *The Cellular Functions of Membrane Transport* by Hoffman, Ed., © 1964, p. 197. Reprinted by permission of Prentice-Hall, Inc., Englewood Cliffs, NJ.

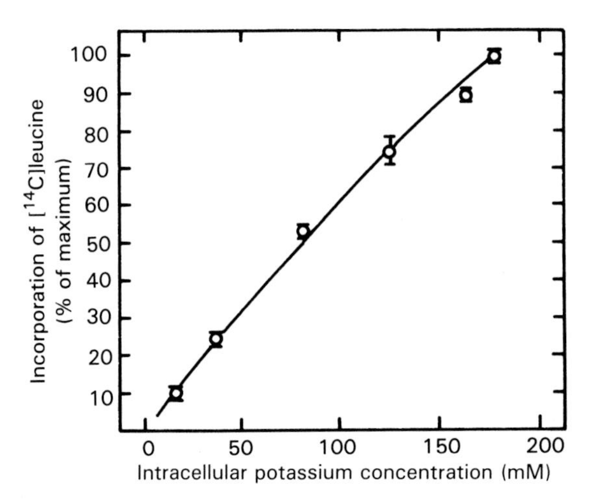

Fig. 14.1 Incorporation of [^{14}C]leucine into protein at various levels of K^+ at 37°C. Reprinted with permission from M. Lubin and H. Ennis, *Biochimica et Biophysica Acta,* 80:614–631. Copyright © 1964 Elsevier Science Publishers, Amsterdam.

concentration are maintained in the presence of high Na^+ and low K^+ concentrations in the external medium. Here we will consider the evidence for an active transport mechanism that functions as an ATPase.

Giant axons of certain invertebrates are ideal experimental material for the study of ionic transfers and electrical phenomena. Their size greatly facilitates manipulation (see Fig. 14.2*a*) (Caldwell et al., 1960), for example, in the internal injection of compounds with micropipettes. In addition, considerable data can be collected from a single experiment, since these axons are sturdy enough to allow prolonged observations. With the squid axon, a number of incisive experiments have explored a whole spectrum of problems related to ion transfer and excitability, including that of the coupling of Na^+ transport to energy expenditure.

When $^{22}Na^+$ is injected into the nerve fiber, the exit of Na^+ can be followed by placing the axon in unlabeled medium and simply measuring the level of $^{22}Na^+$ in the external medium. Oxidative metabolism can be blocked with certain inhibitors (e.g., CN^-). The role of high-energy phosphates in Na^+ transport can be tested directly by injecting these compounds into the axon after blocking the metabolic reactions that would regenerate them (see Fig. 14.2*a*). Results of such an experiment are shown in Fig. 14.2*b* and *c*. The addition of 2 mM cyanide to the system considerably reduces the Na^+ efflux from the axon (arrow 1, Fig. 14.2*b* or *c*). However, injections of ATP (arrow 4) increase the loss of $^{22}Na^+$ significantly, whereas the injection of breakdown products of ATP is ineffective (arrow 2). The results with arginine phosphate, another phosphate ester with a high phosphate group-transfer potential, are similar (Fig. 14.2*c*). In the experiments shown, the effects of the ATP or arginine phosphate are transient, but they can be prolonged for several hours if higher concentrations are used.

Although injections of ATP or arginine phosphate into the axon are effective, their addition to the medium is not. Thus the system is asymmetric, as might be expected in a transport system that transports in only one direction. In both experiments, after about 4 h, the cyanide is washed off and the high rate of Na^+ loss (arrow 4) returns.

From these experiments, it is relatively simple to calculate the Na/P ratio in much the same way as done for the mitochondrial system. The ratio was found to be about 0.7 using either arginine phosphate or ATP. These results agree with the notion that an ATP-powered transport is taking place in these axons. Later experiments (Mullins and Brinley, 1967) actually analyzed

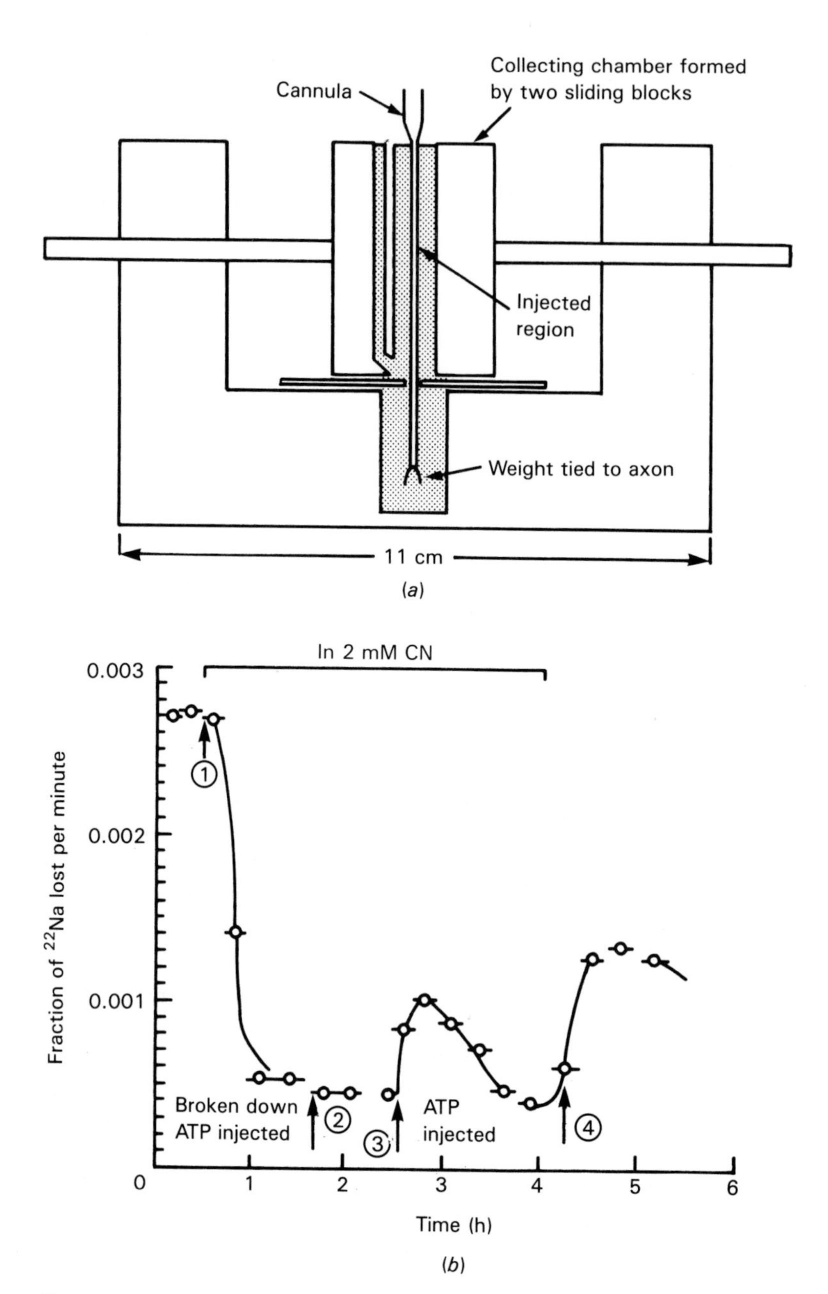

Fig. 14.2 (*a*) System for measuring ^{22}Na efflux in giant axon. (*b*) ^{22}Na efflux; the effect of CN^- and ATP.

the amount of ATP present in metabolically blocked, internally perfused giant axons; the results were about the same.

Similar experiments have been carried out with intact red blood cells or isolated membranes. Erythrocytes are generally sturdy and, of course, are available in large quantities. Analysis of the internal medium or the suspending solution can be carried out readily after separating the cells or subcellular preparations from the suspension fluid by centrifugation or filtration. The

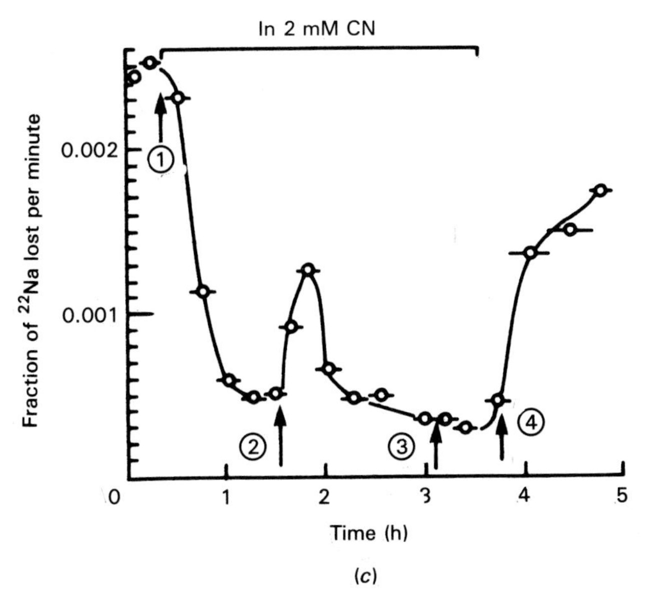

Fig. 14.2 (*Continued*) (*c*) ^{22}Na efflux; the effect of CN^- and arginine phosphate. Reprinted with permission from P. C. Caldwell, et al., *Journal of Physiology,* 152:561–590. Copyright © 1960 The Physiological Society, Oxford, England.

erythrocytes can be lysed (*hemolyzed*) by hypotonic conditions or the use of detergents. The intact cells are shaped like slightly concave plates. When suspended in a hypoosmotic medium, they swell to a spherical shape so that the surface membrane is stretched. At this point, they become extremely permeable and lose their internal contents, including hemoglobin; what is left are the *erythrocyte ghosts*. Because of their high permeability, the ghosts tend to equilibrate with the external medium. Interestingly, the empty sacs can at least partially regain their low permeability under the appropriate conditions, as if their membranes were capable of "resealing." Because of this property, it is possible to vary the internal contents of the cells not by injection but by hemolysis in media containing the desired components. This procedure is illustrated in Fig. 14.3 (Whittam, 1962), which shows the good correlation between the Na^+ composition inside the resealed ghosts (ordinate) and that of the suspension medium used for hemolysis (abscissa). Comparable results have been shown for other solutes contained in the hemolysis medium.

Ghost fragments are more amenable to biochemical studies, since the absence of permeability barriers avoids unnecessary complications in experimental design (for example, changes in concentration of ions due to the presence of a permeability barrier). A number of other cell membrane preparations, such as brain and kidney microsomes, are similar to these ghost fragments.

Incorporation of ATP into the ghosts gives results very similar to those obtained with the giant axon. Such an experiment is represented in Fig. 14.4. Curve 1 shows the results obtained with ATP, and curve 2 shows the inhibition of the ATP effect by the use of an inhibitor of the Na^+ transport system, ouabain (G-strophanthin, a glycoside used as a cardiac drug). Table 14.4 (Hoffman, 1962) compares the Na^+ exit in the presence of ATP (row 1) with that occurring in its absence (row 5). The effects of several other compounds of high phosphate group transfer potential (rows 2–4) are also shown. From the results, it can be concluded that ATP could serve as the source of energy for the translocation of Na^+ in the two systems discussed.

The effect of ouabain is a useful one, since this drug seems to inhibit active transport of

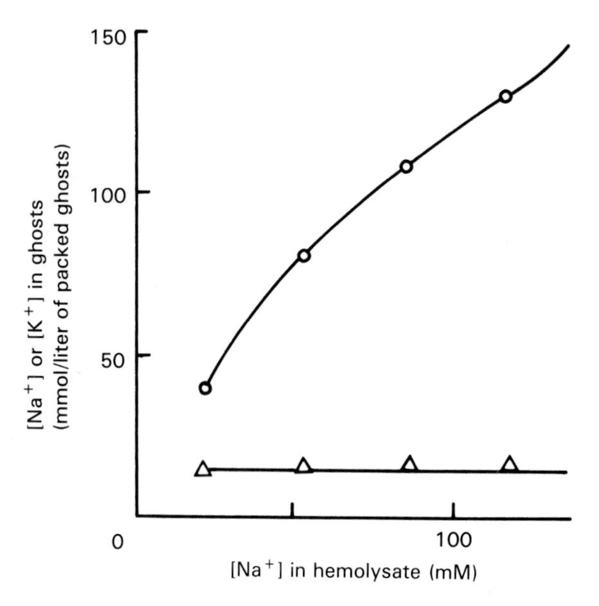

Fig. 14.3 Ionic composition of resealed ghosts as a function of the composition of the hemolysate. (○) Na^+, (△) K^+. From Whittam (1962). Reprinted by permission from *Biochemistry Journal*, 84:110–118, copyright 1962 The Biochemical Society, London.

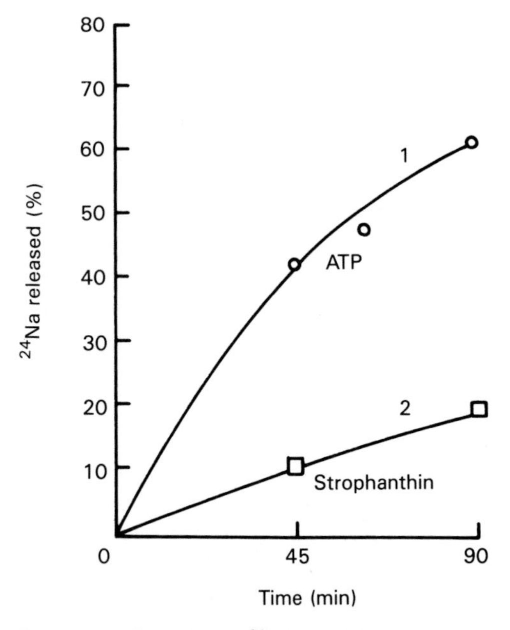

Fig. 14.4 Release of ^{24}Na induced by ATP (curve 1) and its inhibition by ouabain (curve 2). From J. Hoffman, *Circulation*, 26:1201–1213, 1962, by permission of the American Heart Association, Inc.

Table 14.4 Influence of Different Incorporated Nucleotides on the Activation of Transport in Depleted Ghosts

Incorporated substrate	^{24}Na released in 80 min (%)	
	Alone	+Ouabain
1. ATP	50.6	22.8
2. ITP	27.6	22.2
3. GTP	25.9	25.5
4. UTP	33.4	33.2
5. Control[a]	29.1	24.2

From J. Hoffman, *Circulation,* 26:1201–1213, 1962, by permission of the American Heart Association, Inc.
[a]No substrates were added to the control.

Na^+ in many different kinds of cells and organisms. With it, the active transport of Na^+ can easily be distinguished from nonspecific exchanges, such as those brought about by a change in permeability. A parallel change would be expected for an ATPase involved in transport; other unrelated ATPases are not likely to be affected. The ATPases of both nerve and erythrocyte ghosts appear to be sensitive to the inhibitor. At least in squid axon, which has been studied in this respect, cardiac glycosides are effective only when added to the outside; their injection into the axon does not inhibit Na^+ transport. This is a polarity opposite to that of ATP, which is effective only on the inside.

III. THE Na^+- K^+ TRANSPORT ATPase SYSTEM

The experiments just discussed show that the energy available from the hydrolysis of ATP can be used for the transport of Na^+ (and K^+). It is conceivable that the site of the hydrolysis could be different from that of the translocation. We would expect a transport system to be in the membrane, and the ATPase has been localized in the membrane in, for example, erythrocyte ghosts that have lost virtually all of their internal contents. The experiment of Fig. 14.4 is representative of these membranes. Similar evidence has also been obtained in neurons, where the individual nerve cell sheaths can be mechanically isolated by microdissection, as shown in Fig. 14.5 (Cummins and Hyden, 1962). The activity of the material isolated in this way is shown in Table 14.5 (Cummins and Hyden, 1962). It is interesting to see that in this case ouabain is capable of inhibiting the membrane ATPase entirely. These results leave little doubt that the transport ATPase is present in the cell membrane. Many other membrane preparations from widely different tissues including plants (Lai and Thomson, 1971) have been shown to contain Na^+- and K^+-dependent ATPases.

These experiments substantiate the idea that the ATPase system corresponds to the Na^+-K^+ transport system, and they have opened the way for a more detailed examination of this transport. The maintenance of a high internal concentration of K^+ and a low internal concentration of Na^+ may be most simply explained by a model such as that represented in Fig. 14.6 (Glynn, 1957). Here a high phosphate group transfer potential form of a carrier, Y, in the inner phase (step 1) permits the translocation of Na^+ to the external phase (steps 2–4) against an electrochemical gradient. In this process Na^+ complexes with the carrier Y (step 2) and, after movement of the complex NaY to the external surface (step 3), the NaY complex dissociates (step 4). The transition from a high-energy form of the carrier (Y) to a low-energy form (X) (step 5) fulfills

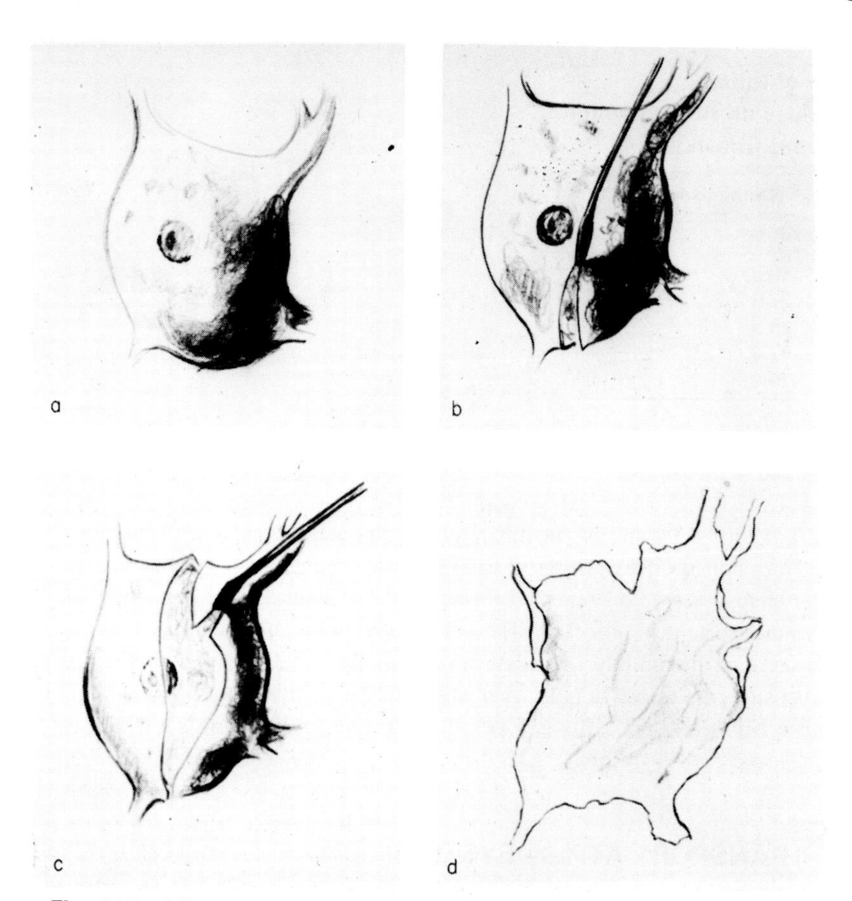

Fig. 14.5 Microsurgical procedure to obtain nerve cell membrane. (*a*) Whole nerve cells; (*b*) initial incision; (*c*) the incision; (*d*) final membrane preparation. Reprinted with permission from J. Cummins and H. Hyden, *Biochimica et Biophysica Acta,* 60:271–283. Copyright © 1962 Elsevier Science Publishers, Amsterdam.

Table 14.5 Effect of Ouabain on the ATPase of Nerve Cell Membrane[a]

Concentration of ouabain (M)	ATPase activity (pmol ATP hydrolyzed per membrane/h)
0	0.4
2×10^{-6}	0.2
2×10^{-5}	0

Reprinted with permission from J. Cummins and H. Hyden, *Biochimica et Biophysica Acta,* 60:271–283. Copyright © 1962 Elsevier Science Publishing, Amsterdam.

[a]Composition of saline was: NaCl, 66 mM; KCl, 33 mM; $MgCl_2$, 5 mM; Tris, 25 mM (pH 8.0); final volume, 6 µl. 15 pmol of radioactive ATP added (250 counts/min). Total of 10 experiments consisting of 8–10 membranes each.

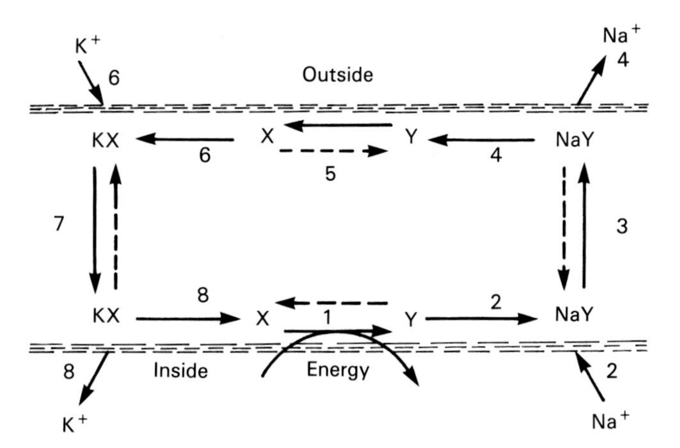

Fig. 14.6 Shaw's hypothesis of a K^+ carrier that is converted to an Na^+ carrier by the expenditure of energy. Reprinted with permission from *Progress in Biophysics and Molecular Biology,* 8:241–307, I. M. Glynn, copyright 1957 Pergamon Journals, Ltd.

the thermodynamic requirement of an energy expenditure. It also presents K^+ with a carrier for its translocation to the internal phase (steps 6–8). The K^+ transport does not require energy, since it is in the direction of its electrochemical gradient.

The cyclic operation of this mechanism would then provide linked Na^+ and K^+ translocation with the movement of two ions in opposite directions coupled to an energy expenditure. The passage of the complexed moiety of the transporter (either NaY or KX) inside the membrane could be by diffusion of the transporter from one membrane interface to the other. As discussed in Chapter 15, it is more likely that the transporter alternatively exposes its reactive group first to one interface and then to the other through a conformational rearrangement. The form of the energy expended is not specified in Fig. 14.6; however, the evidence discussed for the Na^+-K^+ transport shows that ATP hydrolysis serves as the energy source in this case.

A mechanism involving the hydrolysis of ATP could function in the manner illustrated in Fig. 14.7. Figure 14.7 specifies a phosphorylated form of the carrier as the transporter of Na^+

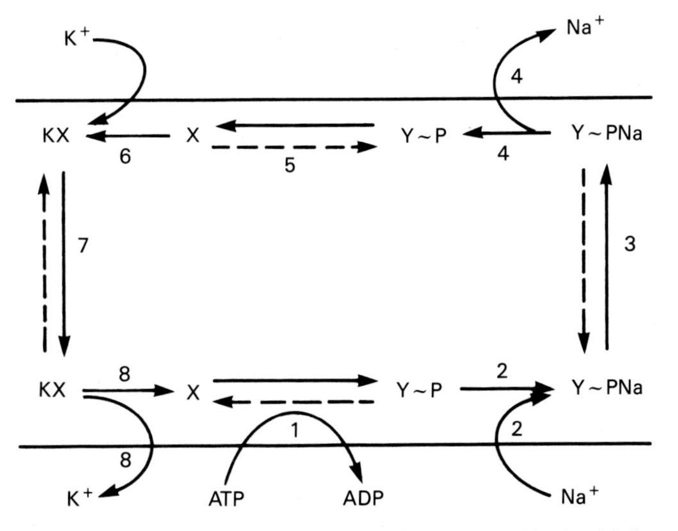

Fig. 14.7 Shaw's hypothesis modified to specify the high-energy intermediate as a phosphorylated form.

(step 1) and the dephosphorylated carrier (formed in step 5) as the transporter of K^+. Otherwise, it does not differ from the model of Fig. 14.6. As shown in Fig. 14.7, the transport system, and hence the energy-expending step 1, cannot operate without the presence of Na^+ and K^+. Hence an ATPase involved in Na^+-K^+ transport must depend on the concentrations of these two ions, Na^+ to facilitate the removal of Y~P from the inside surface and K^+ for the return of the molecule of X from the outside surface to the phosphorylation site. In addition, since cardiac glycosides inhibit the transport, any ATPase activity reflecting this transport must be inhibited. ATPases from many systems have been found to fulfill these requirements (see Table 14.6) (Bonting and Caravaggio, 1963), and today it is generally assumed that they correspond to the Na^+-K^+ transport system. The results, from a preparation of erythrocyte ghosts, are presented in Fig. 14.8*a* and *b* (Post et al., 1960). Similar data are available for other tissues (see Table 14.6 and Bonting et al., 1961).

The dependence of the membrane ATPase on Na^+ and K^+ is represented in Fig. 14.8*a* and *b*. The total osmotic pressure of the solution was maintained constant by varying the proportion of the two cations. Figure 14.8*a* represents the experiment over a wide range of Na^+ concentrations (0–100 mM) with K^+ varied accordingly from 20 to 140 mM. The experiment essentially provides an approximate constant (apparent K_m = 24 mM) for Na^+, as indicated by the half-maximal concentration of Na^+. This is arrived at readily, since the K^+ effect is maximal even at 20 mM. Figure 14.8*b* shows the variation of activity with K^+ concentration, which is varied over a much narrower range (0–25 mM). Accordingly, Na^+ is varied at a high concentration (120–145 mM), where its effect is constant and maximal. This provides an apparent K_m of 3 mM for K^+.

The dependence of the Na^+ transport on Na^+ concentration is shown in Fig. 14.8*c*. The constant for this process (K_m = 20 mM) is approximately the same as that for ATPase. Thus the ATPase depends on the Na^+ and K^+ concentrations as required by the model of Fig. 14.7. The apparent Michaelis-Menten kinetics of Fig. 14.8 are probably coincidental, since the ATPase generally shows sigmoidal kinetics (Robinson, 1970); nevertheless, the details of the kinetics do not alter the interpretation.

One of the significant features of the model of Fig. 14.7 is the asymmetry of the system. In this model, the Na^+ combines with the energized carrier at the inner surface of the membrane,

Table 14.6 Comparison of Cation Fluxes and ATPase Activities[a]

Tissue	Temperature (°C)	Cation flux (10^{-14}mol cm^{-2} s^{-1}) (average)	Na^+,K^+-ATPase activity (10^{-14}mol cm^{-2} s^{-1})	Ratio	Total ATPase activity (10^{-14}mol cm^{-2} s^{-1})	Ratio
Human erythrocytes	37	3.87	1.38 (±0.36; 4)	2.80	2.64 (±0.66; 4)	1.47
Frog muscle	17	985	530 (±94; 4)	1.86	3.220 (±610; 4)	0.31
Squid giant axon	19	1,200	400 (±79; 5)	3.00	1,790 (±370; 5)	0.67
Frog skin	20	19,700	6,640 (±1100; 4)	2.97	15,300 (±4250; 4)	1.29
Toad bladder	27	43,700	17,600 (±1640; 15)	2.48	96,400 (±4120; 15)	0.45
Electric eel, noninnervated membrane, Sachs organ	23	86,100	38,800 (±4160; 3)	2.22	42,400 (±3920; 3)	2.03
				2.56 ±0.19		1.04 ±0.27

From S. L. Bonting and L. L. Caravaggio, *Archives of Biochemistry and Biophysics,* 101:37–46, 1963, with permission.
[a]ATPase activities are shown as means; numbers in parentheses are the standard error and number of determinations.

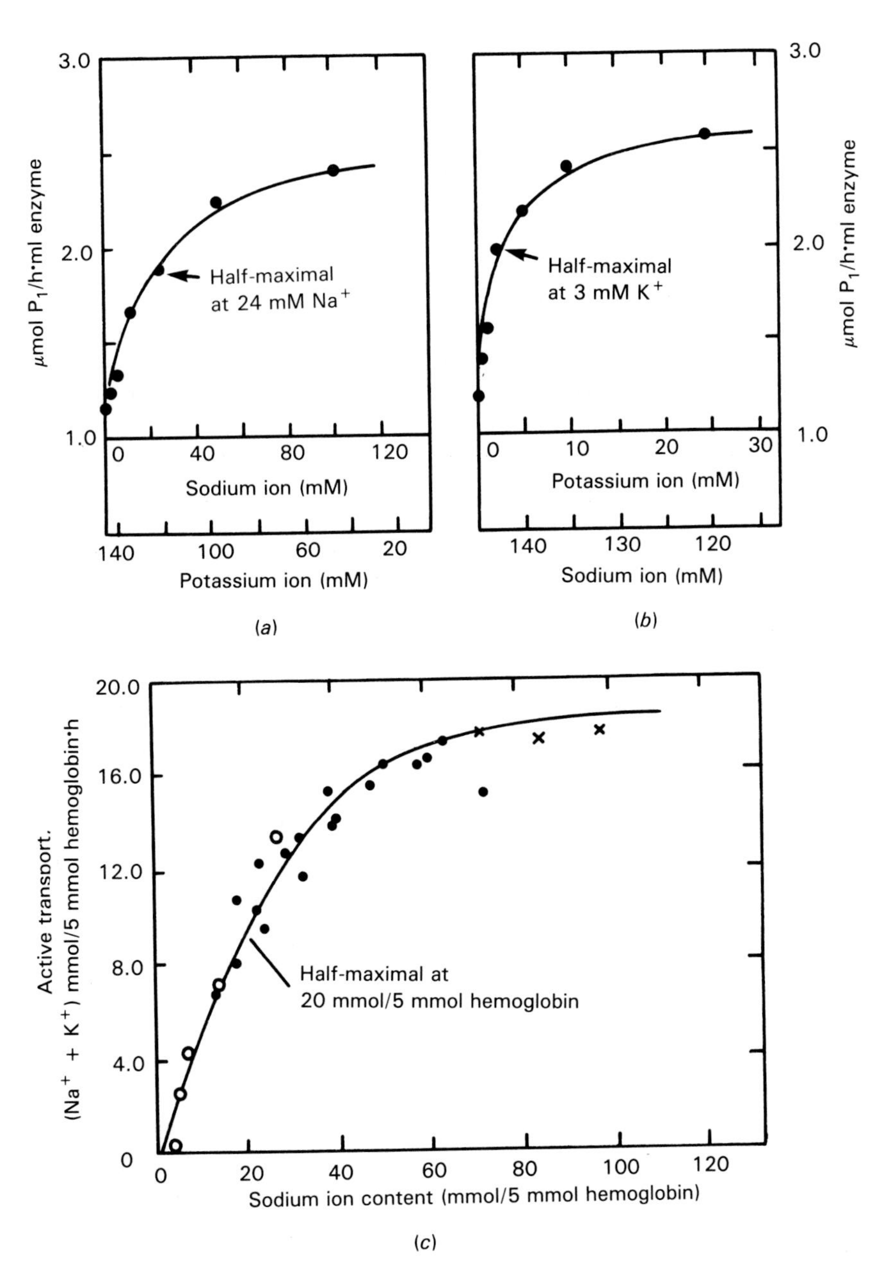

Fig. 14.8 (*a* and *b*) Effect of varying Na^+ and K^+ concentrations on the ATPase activity of a preparation of human erythrocyte membranes. Isotonicity was maintained by maintaining Na^+ and K^+ constant. (*c*) Influence of cell Na^+ content on rate of active transport in intact cells. The millimoles of hemoglobin correspond to millimolar concentrations. The sum of the active transport rates of sodium plus potassium in samples taken at 2-h intervals is plotted against the mean Na^+ content. External K^+ was always more than 30 mM and was not rate limiting. The different symbols indicate three different experiments. Passive transport corrections were less than 7% of the corresponding active rates. All the results plotted for the last experiment were arbitrarily multiplied by 1.5 to obtain a smooth continuation of the curve obtained in the first experiments with fresher cells. This was done because cells stored in the cold for a long period have regularly had a lower pumping capacity than fresher cells. Reprinted by permission from R. L. Post, et al., *Journal of Biological Chemistry,* 235:1789–1802. Copyright © 1960 American Society of Biological Chemists, Inc.

whereas the K^+ combines with a different form of the transporter at the outer surface. Experiments already discussed do show some asymmetry. As mentioned, ouabain inhibits the reactions only when placed on the outside of the cells; injected into axons, it is ineffective. The opposite is true of ATP; it must be present internally to be utilized by the system. The asymmetry of ATP hydrolysis is consistent with the model (step 1 of Fig. 14.7).

The asymmetric localization of the components of the transport system on either side of the membrane is also brought out by many other studies. Compounds unable to penetrate the cell are capable of blocking the ATPase activity (Ohta et al., 1971). On the other hand, antibodies to pig kidney ATPase block the Na^+ efflux only when placed inside the red blood cell (by hemolysis and resealing) (Jorgensen et al., 1973) and have no effect on the outside. A similar antibody blocks the Na^+-dependent phosphorylation of membrane preparations but not the K^+-dependent hydrolysis (Atkinson et al., 1971).

We have seen that it is possible to vary the internal as well as the external environment of resealed red blood cell ghosts (see Fig. 14.3 and Table 14.7). Therefore, whether the ATPase system is asymmetric in relation to the Na^+ and K^+ can be tested in a forthright manner. The internal and external environments can be varied in relation to the Na^+ and K^+.

The results of such experiments are shown in Table 14.7 (Whittam, 1962). The amount of ATP hydrolyzed is measured by following the increase in P_i. Since the system is rather impure, the ATPase activity that is irrelevant to transport must be ignored. This is done by using ouabain in parallel experiments; the P_i liberated in the presence of ouabain, and presumably due to nonspecific ATP hydrolysis, is subtracted from the total formed in its absence. This difference, shown in column 5 of Table 14.7, represents the ATPase sensitive to ouabain and therefore presumed to be associated with transport. This view is reasonable, since the effect of ouabain has approximately the same constants in relation to either the transport or the ATPase activity. In addition, we have seen that ouabain blocks the ATPase of the neuronal membranes (Table 14.5) as well as the ATP-energized Na^+ translocation of erythrocyte ghosts (Fig. 14.4). The variation in external K^+ is shown in column 4 of Table 14.7, part A. The ATPase activity (column 5) depends heavily on the external K^+ concentration. On the other hand, the external

Table 14.7 Dependence of Na^+ and K^+ on Ouabain-Sensitive (0.1 mM) ATPase

	(1)	(2) Cations inside (mM)	(3)	(4) Ouabain sensitive	(5)
Main solute (140–150 mM)	Na^+	K^+	$Na^+ + K^+$	K^+ in medium (mM)	ΔP_i liberated (mM/liter ghosts per hour)
A 1. NaCl	a. 149	18	167	0	0.4
	b.			10	1.5
2. Choline chloride	c. 115	20	135	0	0.5
	d.			10	1.4
3. KCl	e. —	—	—	150	1.9
B 4. NaCl	a. 100	75	175	10	1.9
	b. 83	14	97	10	1.7
	c. 41	79	120	10	1.2
5. Choline chloride	a. 80	16	96	10	1.6
	b. 84	82	166	10	1.6
	c. 14	81	95	10	1.0

Source: Whittam (1962). Reprinted by permission from *Biochemistry Journal,* 84:110–118, copyright © 1962 The Biochemical Society, London.

Na^+ level does not seem to matter, since KCl or choline chloride medium (items 2, 3, and 5) is as effective as Na^+ media (items 1 and 4).

The results obtained by varying the internal medium are very different (Table 14.7). The internal K^+ concentration seems to have little or no effect on the ATP hydrolysis (e. g., compare items 5a with 5b and 4a with 4b). However, the internal concentration of Na^+ becomes critical. From lines 4 and 5, it is clear that decreasing the internal Na^+ (column 1) leads to a decrease in the hydrolysis of ATP (column 5).

The "transport" ATPase is therefore dependent on the internal Na^+ and on the external K^+ as outlined by the model of Fig. 14.7.

It is thus clear that in a number of cell systems the transport of Na^+ and K^+ can be coupled to the hydrolysis of ATP. The results are consistent with the idea that the outflow of Na^+ is linked to the inflow of K^+ by some mechanism that is outlined in the model of Fig. 14.7.

The ATPase that shows this characteristic Na^+-K^+ dependence and a cardiac glycoside sensitivity seems to be associated in some way with the cell membrane. It would be of great interest to define the nature of this association and the molecular organization of the ATPase more precisely. A complete analysis is not possible from what is known at present; nevertheless, a good deal is known and will be discussed in the next chapter. The information available will first be examined from the point of view of the Na^+,K^+-ATPase and then discussed from the broader point of view of models.

SUGGESTED READING

Cross, R. L., and Taiz, L. (1990) Gene duplication as a means of altering H^+/ATP ratio during the evolution of F_0 F_1 ATPase and synthases. *FEBS Lett.* 259:227–229.

Hokin, L. E. (1981) Reconstitution of "carriers" in artificial membranes. *J. Membr. Biol.* 60:77–93.

Karlish, S. J. D. (1989) The mechanism of cation transport by the Na^+, K^+ ATPase. In *Ion Transport* (Keeling, D. and Benham, C., ed.), pp. 19–34, Academic Press, New York.

Kyte, J. (1981) Molecular considerations relevant to the mechanism of active transport. *Nature (London)* 292:201–204.

Pedersen, P. L., and Carafoli, E. (1987) Ion motive ATPases. Part I. Ubiquity, properties and significance to cell function. *Trends Biochem. Sci.* 12:146–150.

Pedersen, P. L., and Carafoli, E. (1987) Ion motive ATPases. Part II. Energy, coupling and work output. *Trends Biochem. Sci.* 12:186–189.

Stekhoven, F. S., and Bonting, S. L. (1981) Transport adenosine triphosphatase: properties and functions. *Physiol. Rev.* 61:1–76.

REFERENCES

Atkinson, A., Gatenby, A. D., and Lowe, A. G. (1971) Transport ATPase-subunit structure analyzed. *Nature (London) New Biol.* 233:145–146.

Bonting, S. L., and Caravaggio, L. L. (1963) Studies on sodium potassium activated adenosine triphosphatase. V. Correlation of enzyme activity with cation flux in six tissues. *Arch. Biochem. Biophys.* 101:37–46.

Bonting, S. L., Simon, K. A., and Hawkins, N. M. (1961) Studies on sodium-potassium-activated adenosine triphosphatase. I. Quantitative distribution in several tissues of the cat. *Arch. Biochem. Biophys.* 95:416–423.

Caldwell, P. C., Hodgkin, A. L., Keynes, R. D., and Shaw, T. I. (1960) The effects of injecting "energy-rich" phosphate compounds on the active transport of ions in the giant axons of Loligo. *J. Physiol. (London)* 152:561–590.

Christian, J. H. B., and Waltho, J. A. (1962) Solute concentrations within cells of halophilic and non-halophilic bacteria. *Biochim. Biophys. Acta* 65:506–508.

Craig, W. S., and Kyte, J. (1980) Stoichiometry and molecular weight of the minimum asymmetric unit of canine renal sodium and potassium ion-activated adenosine triphosphatase. *J. Biol. Chem.* 255:6262–6269.

Cummins, J., and Hyden, H. (1962) Adenosine triphosphate levels and adenosine triphosphatases in neurons, glia and neuronal membranes of the vestibular nucleus. *Biochim. Biophys. Acta* 60:271–283.

Freytag, J. W. (1983) The (Na^+,K^+)ATPase exhibits enzymic activity in the absence of the glycoprotein subunit. *FEBS Lett.* 159:280–289.

Glynn, I. M. (1957) The ionic permeability of the red cell membrane. *Prog. Biophys. Mol. Biol.* 8:241–307.

Hoffman, J. F. (1962) Cation transport and structure of the red-cell plasma membrane. *Circulation* 26:1201–1213.

Hokin, L. E. (1981) Reconstitution of "carriers" in artificial membranes. *J. Membr. Biol.* 60:77–93.

Jorgensen, P. L., Hansen, O., Glynn, I. M., and Cavieres, J. O. (1973) Antibodies to pig kidney (Na^+-K^+)-ATPase inhibit the sodium pump in human red cells provided they have access to the inner surface of the cell membrane. *Biochim. Biophys. Acta* 291:795–800.

Kyte, J. (1971) Purification of the sodium- and potassium-dependent adenosine triphosphatase from canine renal medulla. *J. Biol. Chem.* 246:4157–4165.

Kyte, J. (1974) Properties of the two polypeptides of sodium- and potassium-dependent adenosine triphosphatase. *J. Biol. Chem.* 247:7642–7649.

Lai, Y. F., and Thomson, J. E. (1971) The preparation and properties of an isolated plant membrane fraction enriched in (Na^+-K^+)-stimulated ATPase. *Biochim. Biophys. Acta* 233:84–90.

Lubin, M. (1964) Cell potassium and the regulation of protein synthesis. In *The Cellular Functions of Membrane Transport* (Hoffman, J. F., ed.), pp. 193–209. Prentice-Hall, Englewood Cliffs, N. J.

Lubin, M., and Ennis, H. L. (1964) On the role of intracellular potassium in protein synthesis. *Biochim. Biophys. Acta* 80:614–631.

McDonough, A. A., Geering, K., and Farley, R. A. (1990) The sodium pump needs its β subunit. *FASEB J.* 4:1598–1605.

McLennan, D. H. (1969) Purification and properties of an adenosine triphosphatase from sarcoplasmic reticulum. *J. Biol. Chem.* 245:4508–4515.

Mullins, L. J., and Brinley, F. J., Jr. (1967) Some factors influencing sodium extrusion by internally dialyzed squid axons. *J. Gen. Physiol.* 50:2333–2355.

Ohta, H., Matsumoto, J., Nagano, K., Fujita, M., and Nakao, M. (1971) The inhibitions of Na, K-activated adenosine triphosphatase by a large molecule derivative of *p*-chloromercuribenzoic acid at the outer surface of the human red cell. *Biochem. Biophys. Res. Commun.* 42:1127–1133.

Post, R. L., Merritt, C. R., Kinsolving, C. R., and Albright, C. D. (1960) Membrane adenosine triphosphatase as a participant in the active transport of sodium and potassium in the human erythrocyte. *J. Biol. Chem.* 235:1796–1802.

Robinson, J. D. (1970) Interactions between monovalent cations and the (Na^+K^+)-dependent adenosine triphosphatase. *Arch. Biochem. Biophys.* 139:17–27.

Sachs, G., Chang, H. H., Rabon, E., Schackman, R., Lewin, M., and Saccomani, G. (1976) A nonelectrogenic H^+ pump in plasma membranes of hog stomachs. *J. Biol. Chem.* 251:7690–7698.

Scarborough, G. A. (1980) Proton translocation catalyzed by the electrogenic ATPase in the plasma-membrane of neurospora. *Biochemistry* 19:2925–2931.

Steinbach, H. B. (1963) Comparative biochemistry of the alkali metals. In *Comparative Biochemistry* (Florkin, M., and Mason, H. S., eds.), Vol. 4, Part B, pp. 677–720. Academic Press, New York.

Tedeschi, H. (1974) *Cell Physiology: Molecular Dynamics*. Academic Press, New York.

Tosteson, D. C., and Hoffman, J. F. (1960) Regulation of cell volume by active cation transport in high and low potassium sheep red cells. *J. Gen. Physiol.* 44:169–194.

Whittam, R. (1962) The asymmetrical stimulation of a membrane adenosine triphosphatase in relation to active cation transport. *Biochem. J.* 84:110–118.

Chapter 15

Transport of Ions: Mechanisms and Models

Examining simple models and alternatives sometimes can provide insights into biological processes. This approach has proved very useful in sorting out the data on ion transport and their possible interpretation, and it provides the perspective of this chapter.

Section I examines data obtained in a study of Na^+,K^+-ATPase and uses for discussion the model represented in Fig. 15.1 based on the experiments presented in Chapter 14. This model undoubtedly will require extensive modification and elaboration, but it is a useful summary. Very similar data are available from studies of the Ca^{2+}-ATPase, and a similar model could also be drawn for the transport of Ca^{2+}. Section II examines some of the characteristics of the phosphorylation of ADP by inorganic phosphate, catalyzed by transport ATPases in the absence of ionic gradients. These phenomena may reveal some new features of the ATPases and perhaps have some bearing on our understanding of the synthesis of ATP by the ATP synthase of mitochondria, chloroplasts, and bacteria. Section III concentrates on possible molecular mechanisms of ion transport and discusses the information gained from knowledge of the amino acid sequences.

I. COUPLING BETWEEN ATP HYDROLYSIS AND TRANSPORT

The evidence reviewed in Chapter 14 unmistakably links the transport of Na^+ and K^+ to the hydrolysis of ATP. As suggested in step 1 of Fig. 15.1, the coupling between the translocation of the ions and the hydrolysis of ATP may result from the required phosphorylation of the transporter molecule. The incubation of membrane preparations with ATP labeled with ^{32}P in its terminal position labels the membranes. The phosphate, and not the whole ATP molecule, is incorporated since [^{14}C]ATP does not label the membranes. Table 15.1 (Post et al., 1965)

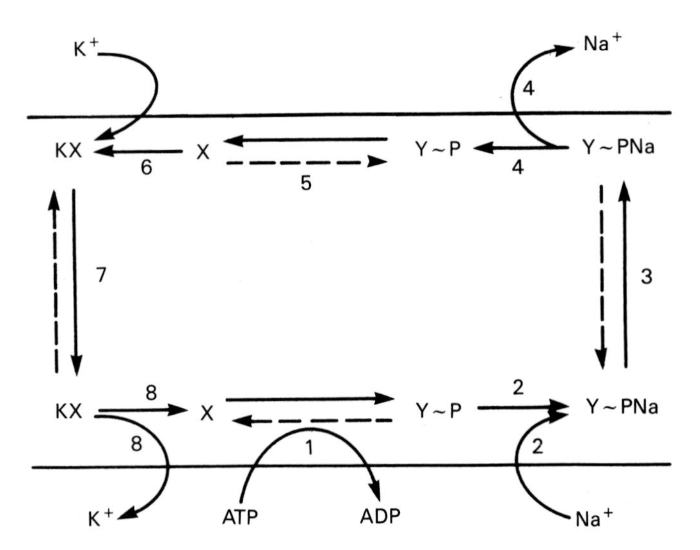

Fig. 15.1 Early model of the functioning of the Na^+,K^+-ATPase. Step 1 corresponds to the phosphorylation of the ATPase indicated as Y; step 2 the binding of Na^+ to Y ~ P; step 3 the movement of the binding group from the cytoplasmic side of the membrane to the outside; step 4 the release of Na^+; step 5 the hydrolysis of Y ~ P to form a different form of Y, X; step 6 the binding of K^+; step 7 its displacement to the cytoplasmic side of the membrane; and step 8 its release into the cytoplasm.

summarizes the incorporation of ^{32}P into kidney plasma membranes as a function of the cation present in the medium. When Na^+ is present, the incorporation is highest, 97 pmol/mg protein, compared to the incorporation in its absence (between 14 and 29 pmol). Even in the presence of Na^+ the incorporation may seem rather small. This is because the Na^+,K^+-ATPase is a minor component of the cell membrane (see below). Much higher values can be obtained for membrane fragments containing the Ca^{2+}-ATPase, which represents a very large proportion of the total protein of the sarcoplasmic reticulum. Actually, in both cases the amount of ^{32}P incorporated corresponds to one per ATPase molecule. In step 1 of the model the phosphorylation of X produces Y ~ P. Different letters, X and Y, are used to denote the two forms because they have

Table 15.1 Effect of Monovalent Cations on Labeling of Kidney Membranes After Incubation with Mg^{2+} and [^{32}P]ATP

Addition	Labeling (pmol ^{32}P/mg protein)
None	26
Li^+	20
Na^+	97
K^+	16
NH_4^+	14
Rb^+	18
Cs^+	14
$Tris^+$	19

very different properties: Y is able to bind Na^+ (step 2) and transfer it to the external membrane interface (step 3), from which it is released (step 4). On the other hand, X, which is regenerated by the hydrolysis of Y ~ P (step 5), binds K^+ (step 6), transfers it to the internal membrane interface (step 7), and releases it to the cell's interior (step 8).

This scheme suggests that the formation of Y ~ P requires the presence of Na^+, as shown by the results in Table 15.1. The phosphorylation is related to transport (Fig. 15.2), since a large portion of the Na^+-dependent phosphorylation is inhibited by ouabain (Post et al., 1965). In Chapter 14 we saw that ouabain is an inhibitor of the Na^+,K^+-ATPase. The scheme also predicts that K^+ would favor the hydrolysis of Y ~ P as shown by the experiment represented in Fig. 15.3 (Post et al., 1965), in which the membranes first labeled with [^{32}P]ATP have been incubated in the presence of K^+. Although the ^{32}P is released even in the absence of K^+, the release is sharply accelerated when K^+ is present. The rates of phosphorylation and dephosphorylation are comparable to those of the ATPase activity (e.g., Kyte, 1974), which in turn correspond very closely to the moles of ions being transported, as shown in Table 14.6 of Chapter 14.

The nature of the phosphorylated ATPase has also been examined in relation to its sensitivity to ADP. The increased hydrolysis favored by K^+ also appears in the experiment of Fig. 15.4 (curve 1) (Post et al., 1969). In this experiment the addition of ADP (curve 2) has no effect, suggesting that the phosphorylated ATPase is no longer in a high-energy form. However, the results are different when the ATPase has first been treated with *N*-ethylmaleimide (NEM), which reacts with sulfhydryl groups. Treatment of the ATPase with NEM blocks the ATPase activity but not the phosphorylation. As shown in Fig. 15.5 (Post et al., 1969), the NEM-treated ATPase is not sensitive to K^+ but is sensitive to ADP. These results suggest that the ATPase may be present in two distinct forms, a form with a high phosphate group transfer potential and a low-energy, K^+-sensitive form. This scheme is consistent with the following reactions:

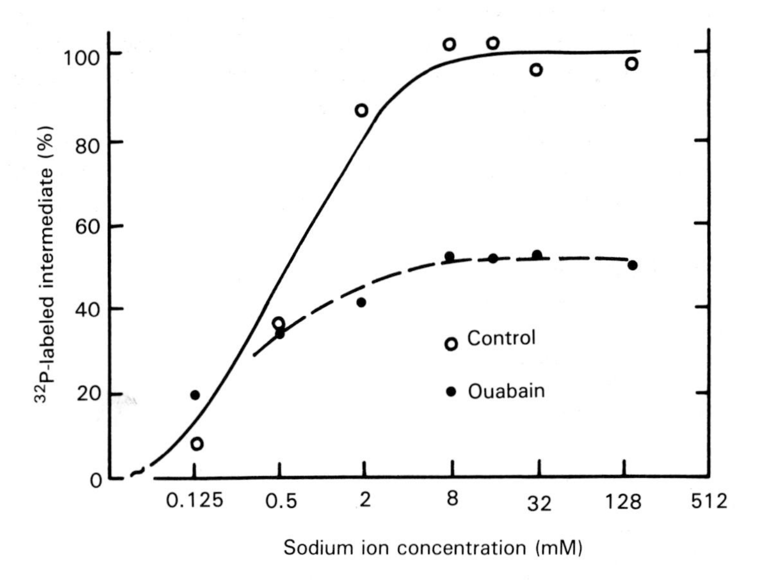

Fig. 15.2 Effect of ouabain on the sensitivity of the ^{32}P-labeled intermediate to the concentration of sodium ion. The concentration of ouabain was 2.5×10^{-4} M, and that of Mg-ATP was 0.1 mM. Incubation was for 12 s at 23°C. The results are the average of two experiments. Reprinted with permission from R. L. Post, et al., *Journal of Biological Chemistry,* 240:1437–1445. Copyright © 1965 American Society of Biological Chemists, Inc.

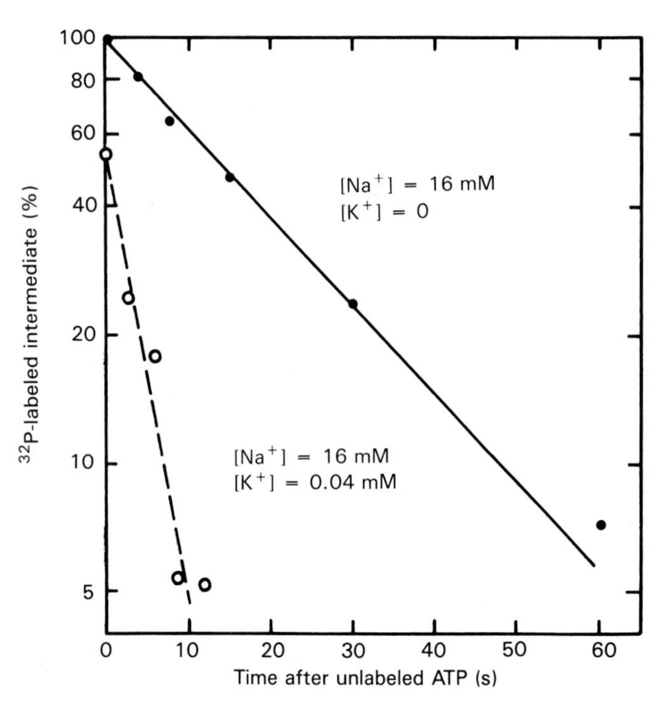

Fig. 15.3 Influence of K^+ on the rate of breakdown of the ^{32}P-labeled intermediate. Kidney membranes were stirred with 0.04 mM Mg-ATP labeled with ^{32}P for 2 min at 8.5°C in the presence of 16 mM Na^+ in a volume of 1.0 ml. (●) K^+ absent; (○) K^+ present at 0.04 mM. Then 0.1 ml of 20 mM unlabeled (Tris) ATP was added to reduce the specific activity of the labeled ATP to 2% of its initial value. After the time intervals on the horizontal axis the reaction was stopped with acid. The solid line indicates exponential disappearance with a time constant of 21 s. The dashed line is similar with a time constant of 4 s. Reprinted with permission from R. L. Post, et al., *Journal of Biological Chemistry,* 240:1437–1445. Copyright © 1965 American Society of Biological Chemists, Inc.

$$Na^+ + E_1 + ATP \rightleftharpoons E_1 \sim P{\cdot}Na^+ + ADP \tag{15.1}$$

$$Na^+ E_1 \sim P \underset{}{\overset{NEM}{\rightleftharpoons}} E_2\text{-}P + Na^+ \tag{15.2}$$

$$E_2\text{-}P + K^+ \rightleftharpoons E_2\text{-}PK \tag{15.3}$$

$$E_2\text{-}PK \rightleftharpoons E_2\text{-}PK \tag{15.4}$$

where E represents the transporter molecule. The subscripts are used to distinguish the various molecular configurations of the enzyme; E_1 and E_2 correspond to the Y and X of Fig. 15.1, respectively.

The models represented by these equations do not require a change in binding constants for transport to occur. However, all transport systems known have been shown to have this feature (see Table 15.5, p. 411).

The estimates of size of the molecule together with estimates of turnover number of the transport ATPase [i.e., moles of product × (moles of enzyme × minutes)$^{-1}$] permit a number of interesting approximations. The turnover number was calculated to be about 12,000 based on the phosphate hydrolyzed × (minutes)$^{-1}$ × (moles of ^{32}P intermediate)$^{-1}$. Since 1 mmol of P_i per hour is hydrolyzed from the ATP by 1 liter of cells, if we assume that there are 1.1 ×

10^{13} cells per liter, there must be about 80 sites per cell. Assuming that the volume of each transporter molecule is 3.2×10^{-19} cm^3, the total volume of transporter per cell is $80 \times (3.2 \times 10^{-19}) = 2.6 \times 10^{-17}$ cm^3. Since the red blood cell surface area is about 1.55×10^{-6} cm^2 and its thickness is approximately 10 nm, the volume of the membrane is about 1.55×10^{-12} cm^3. Therefore, the transport ATPase occupies about 0.002% of the membrane volume. Estimates based on ouabain binding sites correspond to a much higher proportion (about 0.04%). In any case, it seems that an extremely tiny portion of the cell membrane is involved. This is apparently true of almost all membranes, although in some the proportion may be higher; this is the case for the Ca^{2+}-ATPase, which is the major protein present in the sarcoplasmic reticulum. At least in the red cell ghost, the ATPase has been found by a cytochemical electron microscopic method (Charnock et al., 1972) to be distributed evenly throughout the membrane surface.

The phosphorylated portion of the Na^+,K^+-ATPase corresponds to a carboxyl phosphate (i.e., an acylphosphate). This conclusion is based on the reactivity of the phosphate group when the transporter molecule is treated with hydroxylamine, acylphosphatase, or molybdate or on its

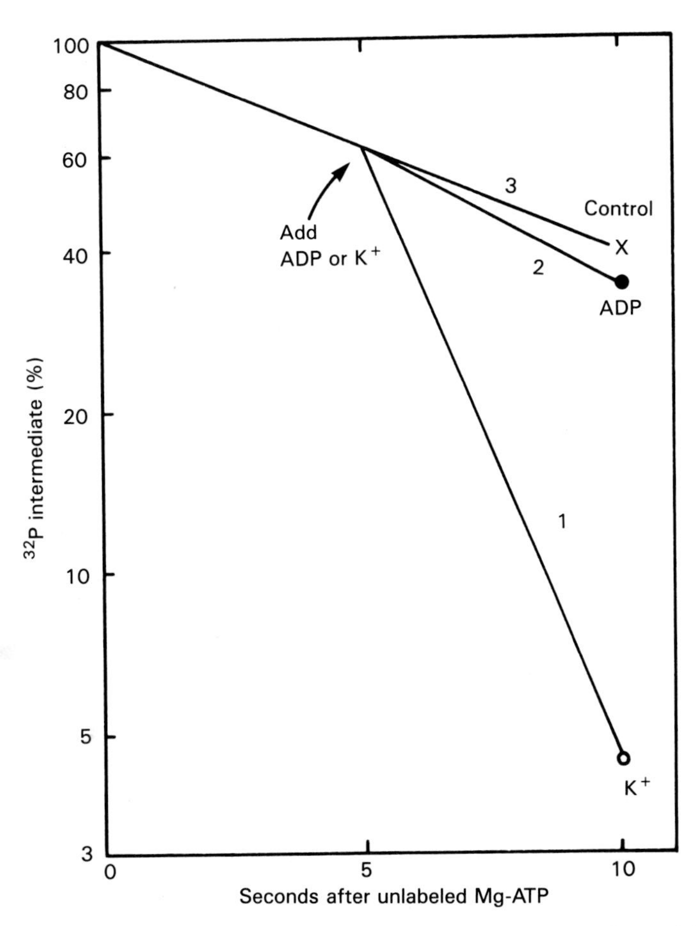

Fig. 15.4 Sensitivity of the phosphorylated intermediate of the native enzyme to ADP and K^+. The ATPase was labeled with [^{32}P]ATP. At zero time the radioactivity of the ATP was chased using a 100-fold excess of unlabeled ATP. From Post et al. (1969). Reproduced from *The Journal of General Physiology*, 1969, 54:3065–3265, by copyright permission of the Rockefeller University Press.

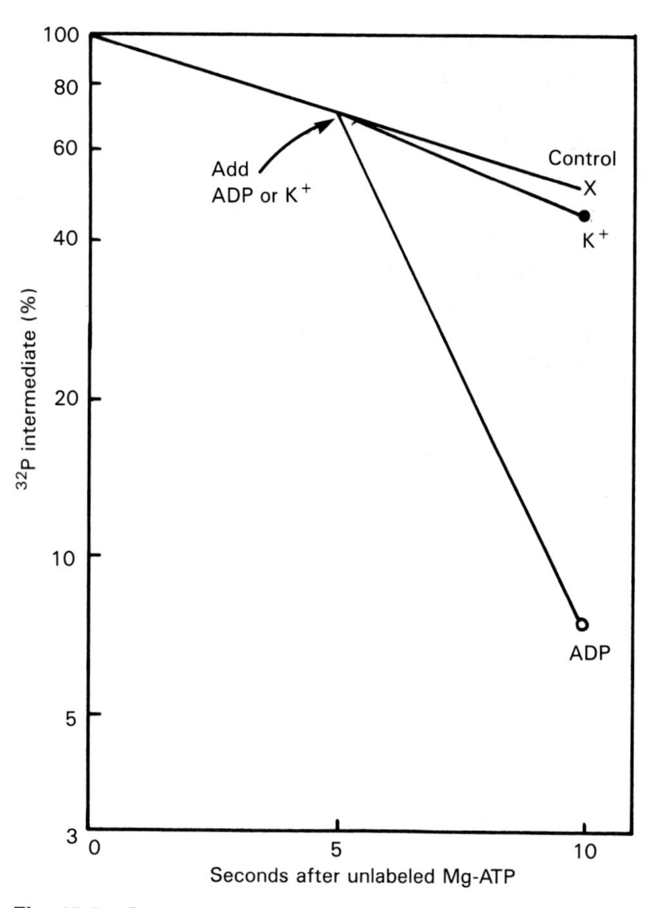

Fig. 15.5 Sensitivity of the phosphorylated intermediate of the Na^+,K^+-ATPase to ADP and K^+ after treatment with *N*-ethylmaleimide. From Post et al. (1969). Reproduced from *The Journal of General Physiology*, 1969, 54:3065–3265, by copyright permission of the Rockefeller University Press.

ability to be transferred to alcohols when in acidic solutions (Hokin and Dahl, 1972; Whittam and Wheeler, 1972).

Proteins have three kinds of carboxyl groups: the glutamyl group, the aspartyl group, and the carboxyl terminal of the protein. The phosphorylated site of the ATPase is an aspartate residue (Nishigaki et al., 1974; Post and Kume, 1973), although there have been some indications of the involvement of a glutamyl residue (Kahlenberg et al., 1968). The aspartyl residue was implicated by first phosphorylating the ATPase with [^{32}P]ATP. Then the denatured phosphoprotein was digested with pronase. After prolonged digestion, the ^{32}P was found in a tripeptide. Analog model phosphoaspartyl and phosphoglutamyl tripeptides were synthesized and compared to the authentic tripeptide. The properties of the latter were found to correspond closely to those of prolylphosphoaspartyllysine.

II. SYNTHESIS OF ATP BY TRANSPORT ATPases

As we saw in Chapter 6, when ion pumps are run in reverse, ATP can be synthesized from ADP and P_i, at least in the case of Na^+,K^+-ATPase (Garrahan and Glynn, 1967) and Ca^{2+}-ATPase (Kanazawa et al., 1971; Makinose, 1972). These findings have certain implications related to the model of Fig. 15.1. ATP can be synthesized only if the phosphorylated form of Y (Y ~ P) is a high-energy form (high phosphate group transfer potential); i.e., the ΔG for its hydrolysis is sufficiently low to support the synthesis of ATP. However, as described above, the evidence indicates that the usual phosphorylated form of the ATPase is hydrolyzed with the addition of K^+ but not ADP. As already noted, a possible explanation is that there are two phosphorylated forms of the transporter molecule, a high-energy form involved in the transport of Na^+ and a low-energy form that interacts with K^+ (Y-P). Furthermore, since Y-P is a low-energy form it should be possible to phosphorylate the molecule with P_i in the absence of Na^+, and this was found to be the case (Post et al., 1975; Schoot et al., 1977; Sen et al., 1969). In the formulation of Fig. 15.1, Y$\sim$P would then be the precursor of Y-P. The Na^+,K^+-ATPase is present in two forms, E_1 and E_2, which differ in conformation, as shown by various means such as exposure of regions of the molecule at the membrane surface to tryptic digestion (see below). The Y$\sim$P and Y-P would then correspond, respectively, to the phosphorylated forms of E_1 and E_2 (Jorgensen and Petersen, 1979). Digestion of the enzyme phosphorylated with ATP and that phosphorylated with P_i produces identical electrophoretic patterns (Bontig et al., 1979; Siegel et al., 1969).

However, it is not necessary to have an ion gradient to synthesize ATP in the case of either the Na^+,K^+-ATPase (Post et al., 1974) or the Ca^{2+}-ATPase (Knowles and Racker, 1975). The in vitro synthesis is carried out in two steps. First, the ATPase is phosphorylated; we saw that this can be done in the case of the Na^+, K^+-ATPase by incubation with P_i. Then ATP is synthesized when ADP is added in the presence of a high concentration of Na^+. Obviously, this proceeds only for a single turnover.

The sequence of events can perhaps be understood best by examining the reactions in some detail. If K^+ is ignored, the reactions would be as shown in Eqs. (15.5) to (15.7):

$$E_2 + P_i \rightleftharpoons E_2\text{-P} \tag{15.5}$$

$$E_2\text{-P} + Na^+ \rightleftharpoons Na^+ \, E_1 \sim P \tag{15.6}$$

$$Na^+ \, E_1 \sim P + ADP \rightleftharpoons Na^+ + E_1 + ATP \tag{15.7}$$

These reactions represent the reverse of the normal sequence of active transport, which would proceed from Eq. (15.7) to Eq. (15.5). The passage from Eq. (15.5) to Eq. (15.7) would be highly improbable unless the Na^+ concentration was raised sufficiently, as predictable from the law of mass action. However, as discussed more fully in Section III, the phosphorylation of the ATPase by ATP—presumably reaction (15.7) run from right to left—decreases the binding constant of the cation and the effect is reversible. It has been proposed that when the binding of one component (e. g., the phosphorylation) to a protein capable of undergoing conformational change affects the binding of another (e. g., the cation), the inverse will be true and the nature of the enzyme-phosphate bond will thereby be affected by the binding of the cation (Weber, 1972, 1974). Presumably, the binding of Na^+ would then convert the low-energy bond into a high-energy bond.

The possibility of obtaining ATP from the reverse of ion transport can be explained by the considerations discussed in this section. The high Na^+ present on the outside of the cell will permit the formation of the high-energy phosphate [reaction of Eq. (15.6)]. The phosphorylation of ADP removes the phosphate and the enzyme can be used again for another cycle of phosphorylation. Under normal conditions, the reactions would run from right to left, since the higher affinity of the ATPase sites for Na^+ on the inside surface will favor the reaction that makes use of ATP.

The possibility that the transport ATPases can function in the synthesis of ATP, albeit under artificial conditions, suggests the idea of considering the ATP synthases (F_1) of mitochondria, bacteria, or chloroplasts using the same model, i.e., as H^+ transport ATPases that are run in reverse during oxidative phosphorylation or photophosphorylation. The cytochrome chain would be responsible for either maintaining an electrochemical gradient for H^+ (i.e., the chemiosmotic hypothesis; see Chapter 10) or delivering H^+ to the ATPase.

III. MODELS OF ION TRANSPORT

In the simplest approach, it seems that the transport of an ion could take place by a process including the following steps: (1) the solute first complexes with specific binding sites on the transporter molecule; (2) the binding sites then move from one interface to the other and the ion is released.

The precise stoichiometry depends on the transport system. For Ca^{2+} transport across the sarcoplasmic reticulum membrane, two Ca^{2+} are transported per ATP hydrolyzed—presumably one turnover of the Ca^{2+} pump. For the Na^+/K^+ pump we saw that approximately three Na^+ are transported outward and two K^+ are transported inward per ATP hydrolyzed.

A system represented by this mechanism would carry out the net transport of the ion until a steady state is reached. Since the binding constants at the two interfaces would be the same, the concentrations in the two compartments separated by the membrane would be the same. These considerations ignore the possible involvement of a membrane potential to avoid complications unnecessary for our present argument.

Active transport, i.e., transport against an electrochemical gradient, could take place when two other conditions are met:

1 The affinity of the binding group for the ions differs at the two interfaces; it should be less at the interface between the membrane and the compartment at which the accumulation occurs.

2 The series of reactions involved in the transport have a suitable decrease in free energy; i.e., they are coupled to a reaction that can supply the required energy, in this case the hydrolysis of ATP.

In this model at steady state (not an equilibrium), the concentration will not be the same in the two compartments; it will be greater in the compartment with the lower binding constant.

Models capable of carrying active transport can be constructed without postulating a change in binding constants. However, all transport systems known have been shown to have this feature (see Table 15.5, p. 411). For simplicity, in the present discussion we assume that the transport of all ions occurs by the same basic process. This approach is not unreasonable since, as we saw in Chapter 13, there is considerable evidence that the transport functions are analogous for the Na^+,K^+-ATPase (Kyte, 1971), Ca^{2+}-ATPase (MacLennan, 1970), and the H^+,K^+-ATPase (Sachs et al., 1976) and that the properties of these molecules are very closely related (Kyte, 1981). The anion transporter, which is best known from studies of the red blood cell, also has some similarities to the other ion transporters; however, not enough is known about this transporter to generalize with any degree of confidence. What is the evidence supporting these models? That the ion must bind to the transporter molecule is clear from either direct measurements or the kinetics of the transport process. Some of these have been discussed in Chapter 14 for the Na^+,K^+-ATPase. Movement of binding sites from one membrane interface to the other is shown in the case of anion transport. There is also evidence of transporter movement for choline (Gorga and Lienhard, 1981) and glucose (Deves and Krupka, 1979) transport and for the adenine nucleotide transport of mitochondria (see below). Counterflow (Chapters 6 and 13) is most easily explained if movement of this kind takes place.

The anion transport protein of the red blood cell normally serves to exchange anions, physiologically, Cl^- and HCO_3^- during the passage of the red blood cell through the body (Sachs et al., 1975). The process involves the interaction between anion and transporter binding sites, since the rate of transport reaches a limiting value when the anion concentration is increased (Gunn et al.,

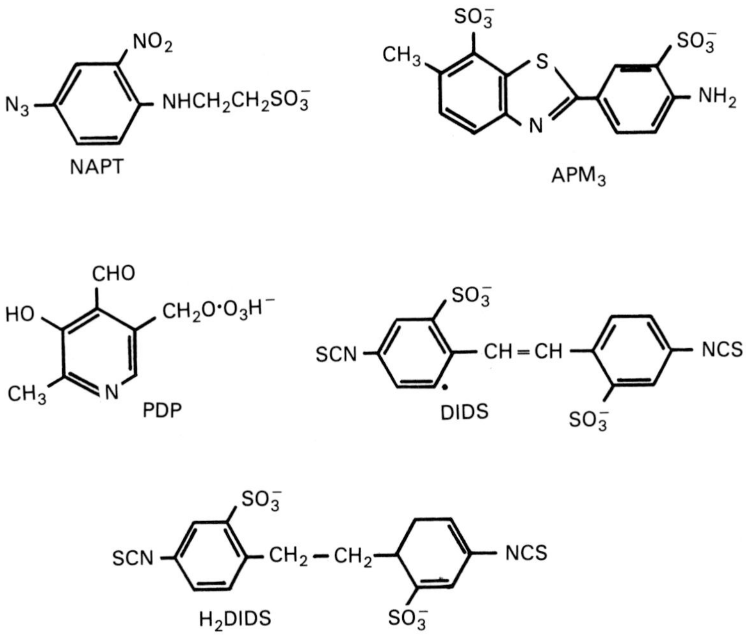

NO_2
N_3 — $NHCH_2CH_2SO_3^-$
NAPT

SO_3^- CH_3 S N SO_3^- NH_2
APM_3

CHO
HO $CH_2O{\cdot}O_3H^-$
CH_3 N
PDP

SO_3^-
SCN — CH=CH — NCS
DIDS SO_3^-

SO_3^-
SCN — CH_2-CH_2 — NCS
H_2DIDS SO_3^-

Fig. 15.6 Structures of inhibitors of the anion exchanger.

1973). Furthermore, several different anions can be exchanged by the anion transporter and, when present simultaneously, they compete for the transporter. The transporter is located across the cell membrane and therefore has portions that are exposed to the outer medium and others that are exposed to the cytoplasm. The rapid turnover of the transport suggests that the anion is transferred through a small distance, perhaps indicating that only a small portion of the transporter molecule moves. That movement of the transport sites takes place is shown in several kinds of experiments. Some experiments show that sites involved in the transport appear to have accumulated on the inner face by accumulating anions in the cell's interior.

An experimental test of this question requires recognition of the sites involved in transport. 4,4′-Diisothiocyano-2,2′-disulfonate (DIDS) has been found to block anion transport in several cells. DIDS, whose structure is shown in Fig. 15.6, reacts with amino groups as shown in Eq. (15.8).

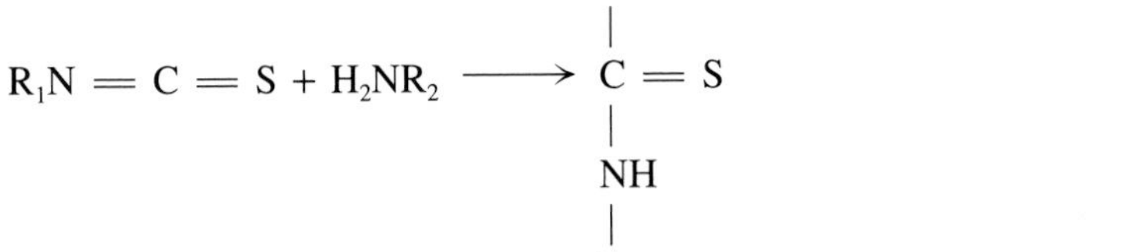

$$S = C = N - (C_6H_3SO_3^-) - C(^3H)(H) - C(^3H)(H) - (C_6H_3SO_3^-) - N = C = N = R_1$$

$$R_1N = C = S + H_2NR_2 \longrightarrow R_1\text{-}NH\text{-}C(=S)\text{-}NH\text{-}R_2 \qquad (15.8)$$

Under the conditions of the experiments it seems to react mostly with the transporter, at the external face of the plasma membrane. The experiment represented in Fig. 15.7 (Cabantchik and Rothstein, 1976) shows the reactivity of DIDS with the anion transporter. When the DIDS is labeled with a radioactive atom, in this case ^3H, it is possible to follow its fate simply by measuring the radioactivity through various fractionating procedures. Figure 15.7 represents a sodium dodecyl sulfate (SDS) gel of the proteins of the membrane after their reaction with [^3H]DIDS. The upper portion corresponds to the various peptide bands that can be seen after staining. The upper track of bands shows the staining with Coomassie Blue, which is a general stain for all proteins. The lower track represents the staining with periodic acid-Schiff (PAS), which stains carbohydrate, in this case glycoproteins. The graph below shows the radioactivity of the various fractions obtained from the gel; the radioactivity appears as a single peak

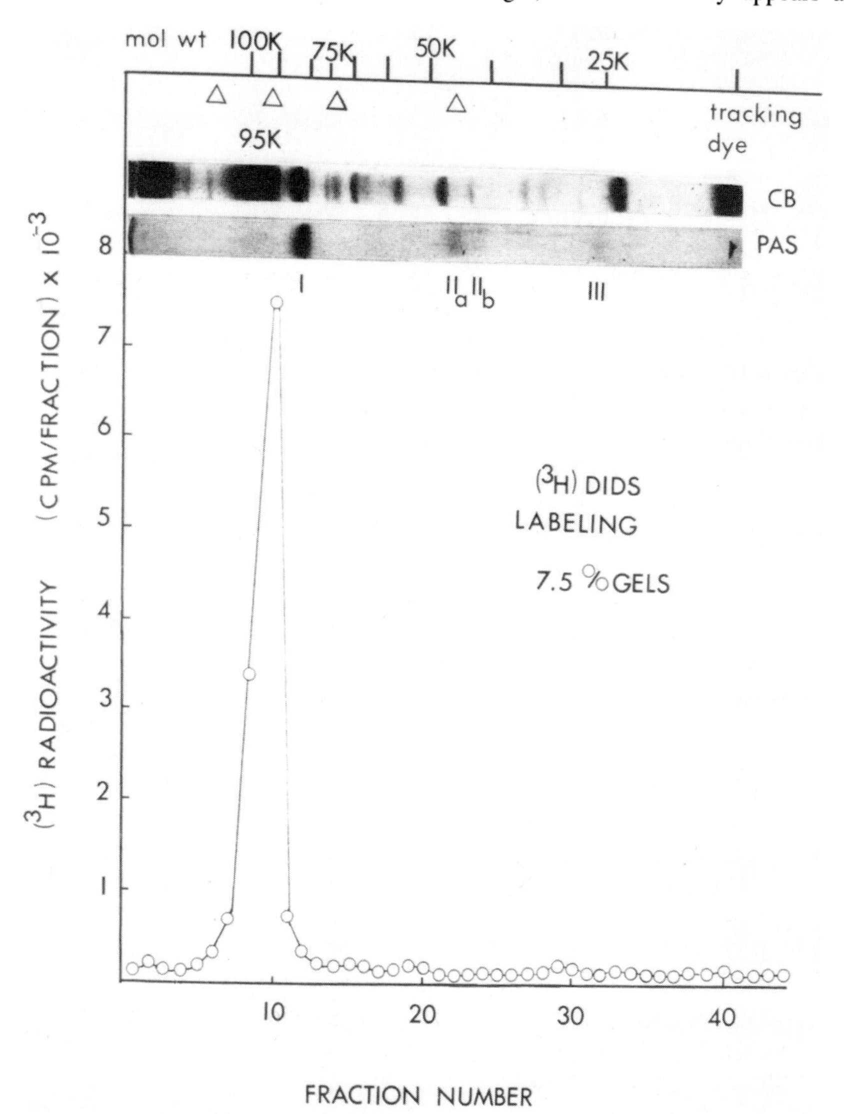

Fig. 15.7 Labeling and staining profiles of ghost protein isolated from [^3H]DIDS-labeled ghosts. From Cabantchik and Rothstein (1976), with permission.

corresponding to a single band, showing that the DIDS is likely to bind specifically only to one protein species.

Figure 15.8*a* shows the effect of DIDS on the transport, represented as percent inhibition (right ordinate), and the binding of the DIDS to the cells (left ordinate). The abscissa corresponds to the concentration of DIDS. Figure 15.8*b* compares sites bound by DIDS and the inhibition of the exchange of anion transport using the data of Fig. 15.8*a*. The two correspond closely, suggesting that DIDS binding can serve to identify the sites involved in the anion transport.

The presence of excess anions only in the cell's interior should force a redistribution of any mobile transporter sites so that their concentration is higher at the inner face of the plasma membrane, where they would be bound to the anions. Because the number of transporting sites is small, this redistribution would be recognizable as a decrease in the number of groups binding to DIDS, which reacts only with external sites. In the experiment of Table 15.2 (Rothstein et al., 1976) the anion chosen was pyridoxal phosphate (PDP), which exchanges slowly. Therefore, when inside the cell, it can be washed from the outside without substantial loss from the interior. After this internal loading, the transport was blocked by the addition of the anion transport inhibitor dipyridamole (which does not block the binding of DIDS).

Item a of Table 15.2, column 2, corresponds to the control, to which neither PDP nor dipyridamole was added. The binding of DIDS is taken as 100%. The presence of PDP in the outside medium interfered with the DIDS binding (item b), possibly because the two molecules bind to the same sites. The effect was reversible, since after removal of the PDP from the external medium (item c) the binding returned. In item d, DIDS was added to the cells incubated for 20 h with PDP. Not unexpectedly, the binding of DIDS was depressed comparable to that shown in item b because of the presence of PDP. The binding of DIDS was unaffected by the presence of the anion transport inhibitor dipyridamole in all these cases (items a–d, column 3). However, after the outside was washed free of the PDP, the binding of DIDS decreased (item e, column 3). Presumably the presence of PDP on the inside caused a redistribution of the transporter sites so that now more of them were facing the inside interface.

Recruitment of the transport sites for the red blood cell anion exchange transporter has also been demonstrated using the nuclear magnetic resonance (NMR) signal of $^{35}Cl^-$ (Falke et al., 1984a). These experiments took advantage of the competitive inhibitors of anion transport 4,4′-dinitrostilbene-2,2′-disulfonate (DNDS) and *p*-nitrobenzene sulfonate (PNBS), which, like DIDS, can interact with the transporter binding sites only at the external surface of the red blood cells or red blood cell ghosts. The binding of $^{35}Cl^-$ can be recognized because it broadens the NMR signal. In Fig. 15.9 the lower line corresponds to the signal from free $^{35}Cl^-$ and the upper line to the broadened signal in the presence of red blood cell ghosts. The line broadening can be expressed quantitatively as the increase in the distance at the half-height of the spectrum, as done for Fig. 15.10 (p. 400) (Falke et al., 1984b). Part of the line broadening can be shown to correspond to the binding to the anion transporter, since it is diminished by the addition of DNDS. Figure 15.10*a* shows the broadening as a function of increasing additions of leaky red blood cell ghosts (expressed as ghost protein). The upper line shows the total broadening, whereas the lower line shows the broadening in the presence of a saturating concentration of DNDS. The difference between the two lines corresponds to the signal from the transport sites. Figure 15.10*b* shows the line broadening produced by various preparations.

DNDS acts as a competitive inhibitor of Cl^- by binding to transporter sites only at the outside surface. The NMR study compares six different preparations (Fig. 15.10). The first three shown in the figure correspond to systems with intact membranes that respond only to the external Cl^- for differing reasons (in the red blood cells the Cl^- is bound by the hemoglobin; the other two are collapsed membranes containing little Cl^-). The last three are leaky membranes

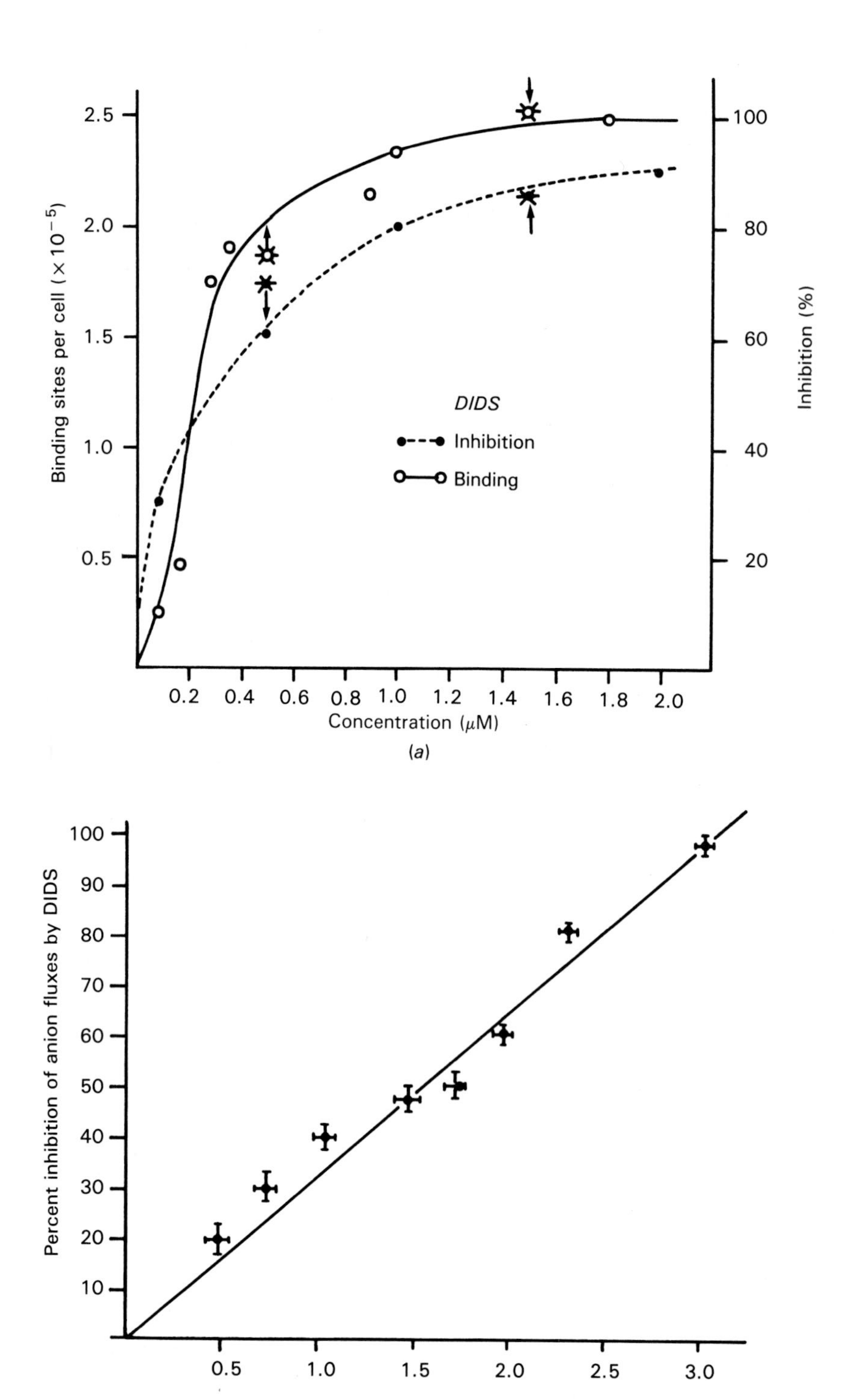

Fig. 15.8 (*a*) [^{3}H]DIDS binding and the inhibition of sulfate transport. (*b*) Relationship between [^{3}H]DIDS binding and its effect on sulfate transport. From Z. I. Cabantchik and A. Rothstein, *Journal of Membrane Biology,* 15:207–226, with permission. Copyright © 1976 Springer-Verlag, Heidelberg.

Table 15.2 Reduction of [^{3}H]DIDS Binding to Outer Surface of Cell by Pyridoxal-5′-phosphate (PDP) Present on Either Side of the Membrane

(1) PDP treatment	[^{3}H]DIDS binding to cells (2) Dipyridamole absent (control)	(3) Dipyridamole present
a. None	100	100
b. Outside	11	12
c. Outside—washed	83	85
d. Outside and inside	10	9
e. Outside and inside—washed	89	55

Reprinted with permission from A. Rothstein, et al., *Proceedings of the Federation of American Societies for Experimental Biology*, pp. 6–9, 1976.

and the line broadening corresponds to both internal and external Cl^-. The DNDS, which can only react with the external binding sites, has an effect on both the internal and the external Cl^- line broadening. This could only happen if the DNDS had recruited the binding sites to the external membrane face.

Other experiments on the nucleotide transport system of mitochondria have suggested that the binding sites of the transporter molecules move (Klingenberg and Burchholtz, 1973). Bongkrekate is an inhibitor of the transport. The binding of bongkrekate and that of ADP to mitochondria mutually enhance each other, and an interaction of the two results in irreversible binding. The results suggest that the ADP transporter becomes trapped at the interface by bongkrekate.

The evidence for movement in other transporter molecules is indirect. However, it is clear that conformational changes take place, and these changes obviously must involve small movements of the molecule. Although the examples of this section concentrate on Na^+,K^+-ATPase, similar data are available for other transport systems. These include not only the analogous Ca^{2+}-ATPase (e.g., Jenkins and Tanner, 1975) but also the glucose transport powered by Na^+ influx (Peerce and Wright, 1984) and the transport of at least one amino acid (Wright and Peerce, 1984). As mentioned above, two forms of Na^+,K^+-ATPase, usually called E_1 and E_2, have been shown to exist. The two have corresponding phosphorylated forms, E_1' and E_2', which differ in their intrinsic fluorescence spectra (Karlish and Yates, 1978), their susceptibility to trypsin (Castro and Farley, 1979; Jorgensen, 1975; Koepsell, 1979), and the availability of their sulfhydryl groups (Hart and Titus, 1973). Since there is no rotational or translational movement of the transporter molecules across the membrane (Kyte, 1974), it is most likely that

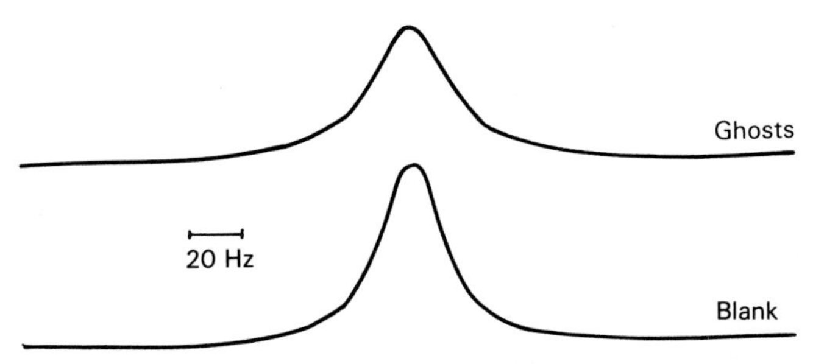

Fig. 15.9 Effect of red blood cell ghost membranes on $^{35}Cl^-$ NMR spectrum. Reprinted with permission from J. J. Falke, et al., *Journal of Biological Chemistry*, 259:6472–6480. Copyright © 1984 American Society of Biological Chemists, Inc.

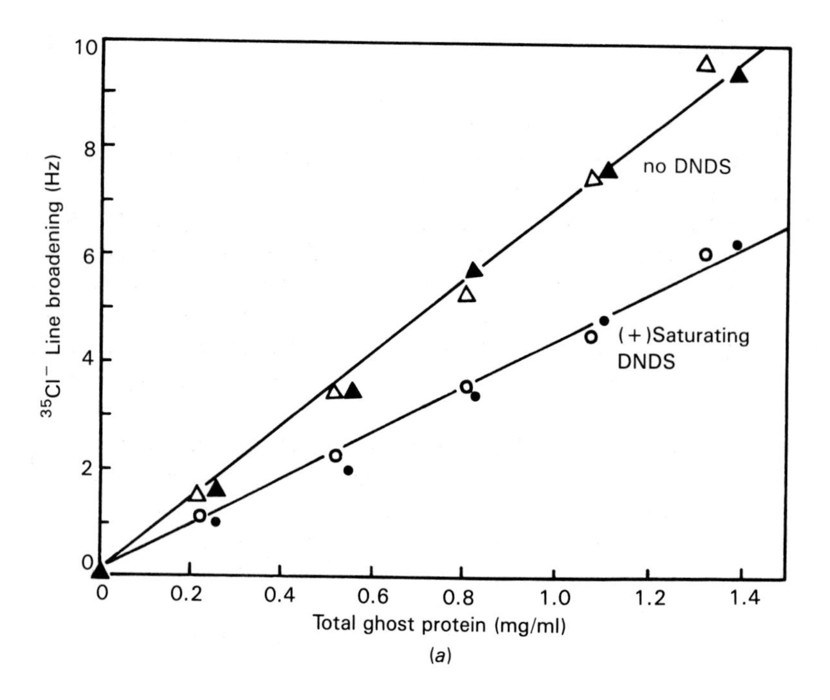

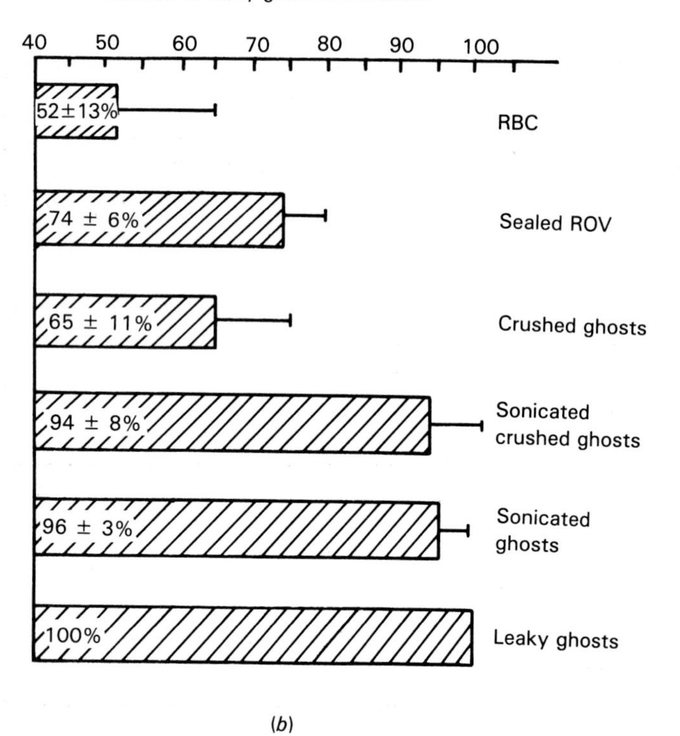

Fig. 15.10 (*a*) Relationship between $^{35}Cl^-$ line broadening and ghost membrane concentration. The samples contained leaky ghost membranes with (○, ●) or without (△, ▲) 1 mM DNDS. (*b*) Sidedness of the red cell membrane $^{35}Cl^-$ line broadening. ROV, Right side out vesicle; RBC, red blood cell. Reprinted with permission from J. J. Falke, et al., *Journal of Biological Chemistry,* 259:6472–6480. Copyright © 1984 American Society of Biological Chemists, Inc.

these conformational changes represent changes in the shape of the transporter to shuttle its binding groups between the two interfaces. Evidence for a conformational change is also available from circular dichroism studies of Na^+,K^+-ATPase.

Nearly all biological molecules are optically active due to their lack of symmetry. In an optically active solution, the absorbance of left circularly polarized light (A_L) is different from the absorbance of right circularly polarized light (A_R). After passing through the sample, each light component is still circularly polarized, but the radii of the circles traced by their electric vectors are now different; this is circular dichroism (CD). In CD, the differential absorption between left and right circularly polarized light, i.e., $A_L - A_R$, is measured at various wavelengths.

CD has been largely applied empirically with model polypeptides to determine how the data can be interpreted in terms of molecular conformation. A simple example is shown in the case of myoglobin (Greenfield and Fasman, 1969). Figure 15.11*a* shows the CD spectra of poly-L-lysine in various conformations: curve 1, the α-helical configuration; curve 2, the β-helical configuration; and curve 3, the configuration of a random coil. Figure 15.11*b* corresponds to the CD spectrum of myoglobin. The points are calculated assuming 68.3% α, 4.7% β, and 27% random coil and closely match the line that represents the experimental results.

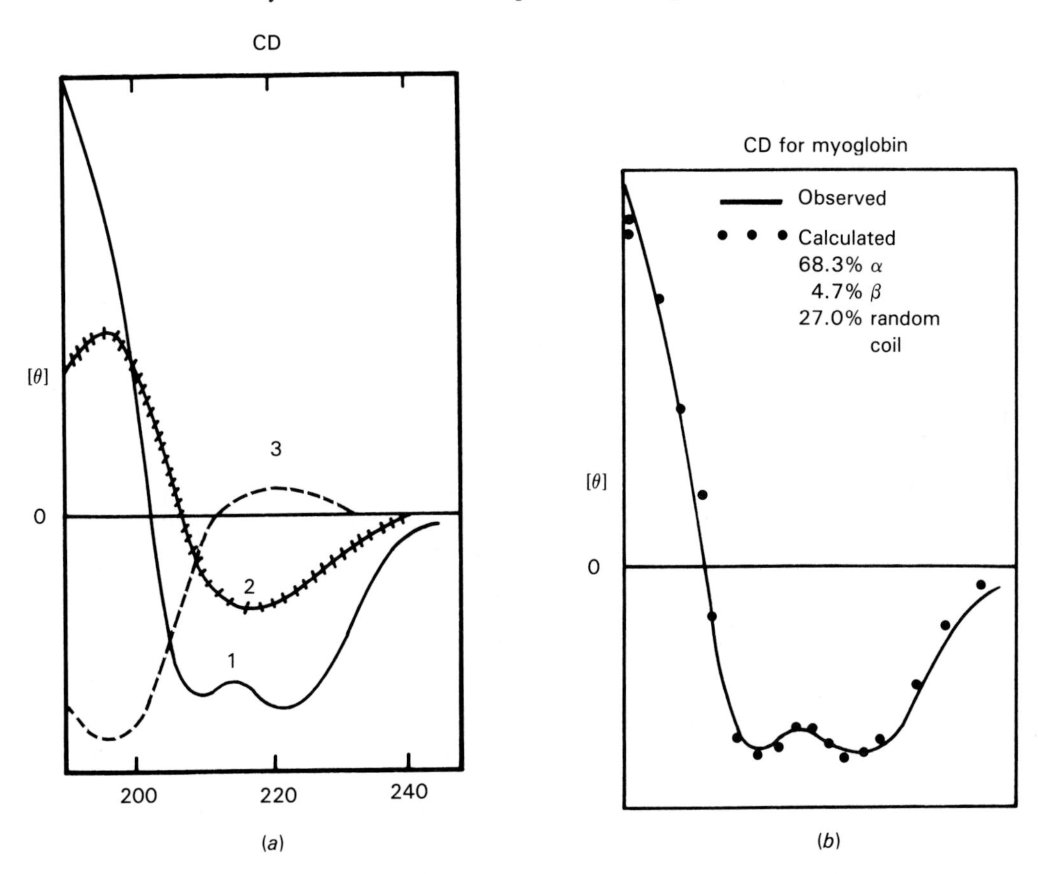

Fig. 15.11 (*a*) CD spectrum for poly-L-lysine in the α, β, and random conformations. (*b*) Observed CD curve for myoglobin and the curve calculated from the poly-L-lysine data of part (*a*). From Greenfield and Fasman (1969). Reprinted with permission from *Biochemistry* 8:4108–4115, copyright 1969 American Chemical Society.

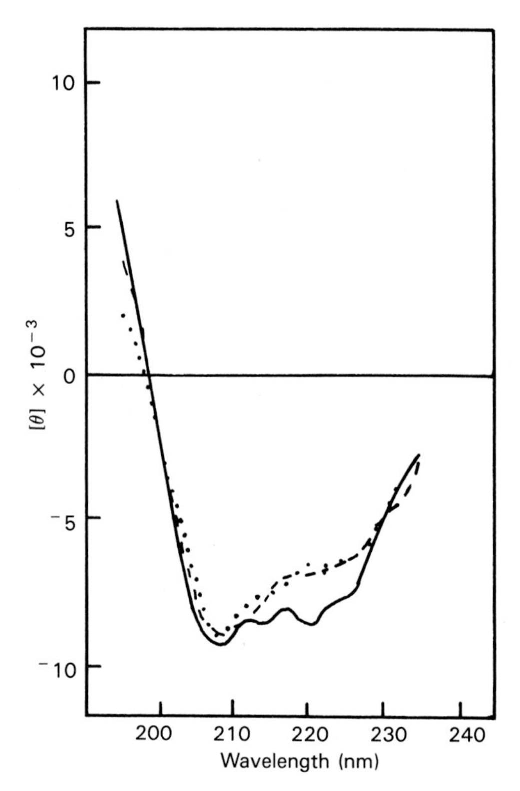

Fig. 15.12 CD spectra of the Na^+,K^+-ATPase showing reversibility of conformational change. Sample A (dashed line), Na^+,K^+-ATPase in the presence of 10 mM KCl. Sample B (solid line), as in sample A plus 90 mM NaCl. Sample C (dotted line), as in sample B but in the presence of 40 mM KCl. Reprinted with permission from T. J. Gresalfi and B. A. Wallace, *Journal of Biological Chemistry,* 259:2622–2655. Copyright © 1984 American Society of Biological Chemists, Inc.

Gresalfi and Wallace (1984) have examined the CD spectra of purified and membrane-attached Na^+,K^+- ATPase in its E_1 and E_2 forms obtained by introducing either Na^+ or K^+. The spectra for the peptide backbone (190–240 nm), shown in Fig. 15.12, were consistent with extensive conformational differences between E_1 and E_2. The changes appear to be reversible when the ion composition is altered. The CD changes are interpreted by comparison with results for model polypeptides, as summarized in Table 15.3 (Gresalfi and Wallace, 1984).

The involvement in transport of rearrangements within the ATPase is also shown by x-ray diffraction studies of packed membranes from the sarcoplasmic reticulum containing Ca^{2+}-ATPase (Blasie et al., 1985). The interpretation of the results is shown in Fig. 15.13. A significant portion of the ATPase juts into the cytoplasmic phase, as also shown by three-

Table 15.3 Calculated Secondary Structures

	No flattening corrections						Flattening corrections			
Sample	α[a]	β	T	R	NRMSD	Sum	α	β	T	R
Na^+	0.27	0.18	0.17	0.38	0.050	0.91	0.25	0.31	0.09	0.35
K^+	0.20	0.28	0.14	0.37	0.078	0.99	0.18	0.38	0.09	0.35
Cholate	0.27	0.19	0.15	0.39	0.042	0.86	0.25	0.34	0.07	0.35
No salt	0.29	0.19	0.17	0.35	0.050	0.88	0.28	0.32	0.09	0.31
No salt—solubilized	0.19	0.41	0.08	0.32	0.058	1.08				
Na^+—solubilized	0.18	0.41	0.09	0.33	0.081	1.03				
K^+—solubilized	0.17	0.45	0.06	0.32	0.073	1.10				

Reprinted with permission from T. J. Gresalfi and B. A. Wallace, *Journal of Biological Chemistry,* 259:2622–2655. Copyright © 1984 American Society of Biological Chemists, Inc.

[a]α, helix; β, sheet; T, turn; R, random coil.

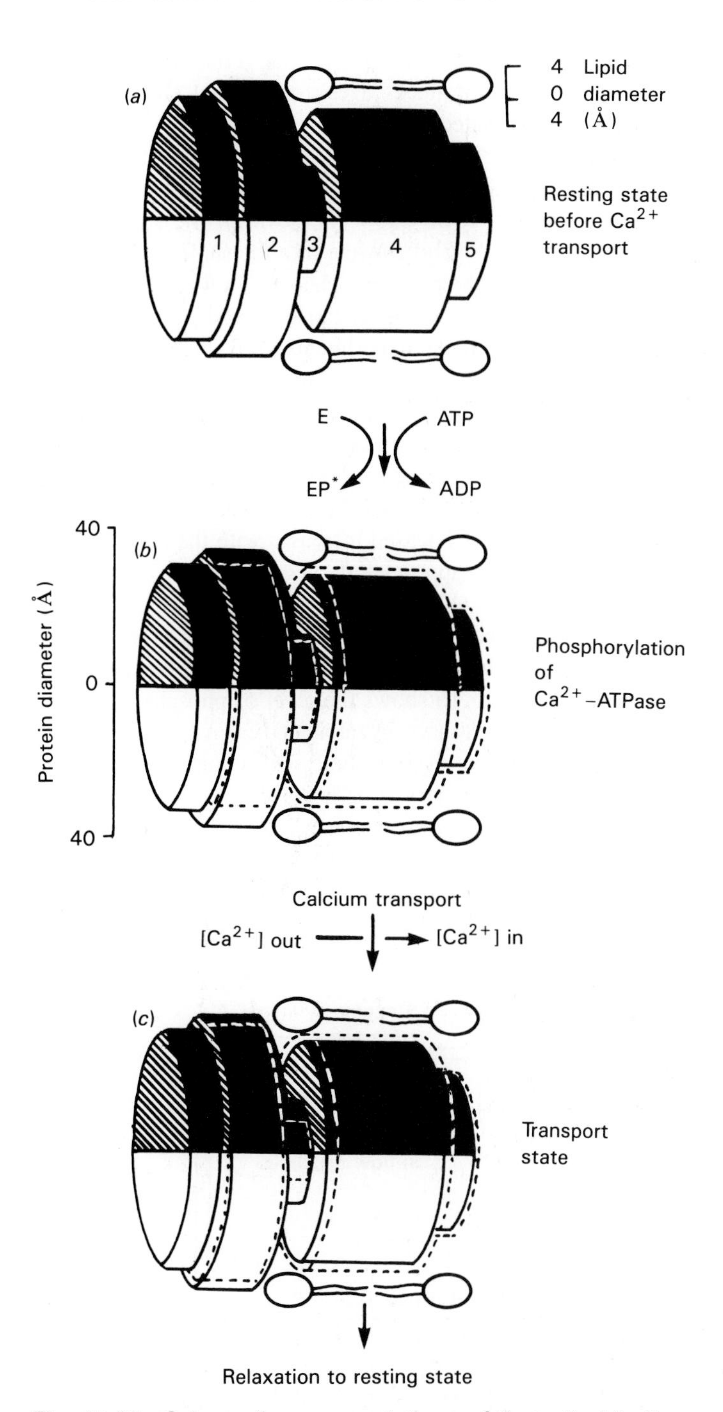

Fig. 15.13 Schematic representations of the cylindrically averaged (about the normal to the membrane plane) SR membrane protein structure at relatively low resolution (~ 2.9 nm). The changes in conformation are indicated by the dashed lines. From Blasie et al. (1985). Reproduced from the *Biophysical Journal*, 1985, 48:9–18, by copyright permission of the Biophysical Society.

dimensional reconstruction of negatively stained crystals in sarcoplasmic reticulum membranes (Taylor et al., 1986). Activation of the ATPase produces a conformational change with a displacement of the structure into the bilayer indicated by the dashed lines.

Other experiments (Ikemoto, 1976) provide the evidence for both the binding and the change in affinity of the model. The experiments discussed here were carried out with a stop-flow apparatus (Fig. 15.14) using a purified preparation of Ca^{2+}-ATPase from the sarcoplasmic reticulum. The stop-flow apparatus delivers reactants and enzyme into the same chamber with very rapid mixing in relation to the time course of the reaction. Then the flow is stopped, also very rapidly. The light absorption of the contents of the chamber can be recorded, a process that for very rapid reactions requires an oscilloscope. These experiments used the Ca^{2+} indicator Arsenazo III, which changes color when Ca^{2+} is bound. The record of Fig. 15.15 represents the light absorption with time. The two sets of panels differ in the time scale, set I showing fast changes with intervals corresponding to 50 ms and set II showing slower changes with intervals representing 5 s. The downward deflections reflect increases in the concentration of Ca^{2+}. The concentration of ATP added is shown at the left in the records. In the control (IA and IA′), no ATP was added and no Ca^{2+} was released. The Ca^{2+} released increases with the concentration of ATP added (compare B and D) until the system appears saturated (compare D and E), as would be expected from the fact that the amount of ATPase is finite. As shown by the longer time scale in set II, the release is temporary; eventually the Ca^{2+} is bound again, presumably when all the ATP is hydrolyzed. The results show that the ATPase binds Ca^{2+} and that activation by ATP reversibly decreases the binding. Figure 15.16 shows the level of phosphorylation of the enzyme (curve 1) compared to the Ca^{2+} release (curve 2) calculated from Fig. 15.15. The two panels represent identical results plotted on two different time scales coincident with the phosphorylation of the ATPase, suggesting that phosphorylation is responsible for the change. Similar data are available for other transport systems, such as Na^{+},K^{+}-ATPase (Masui and

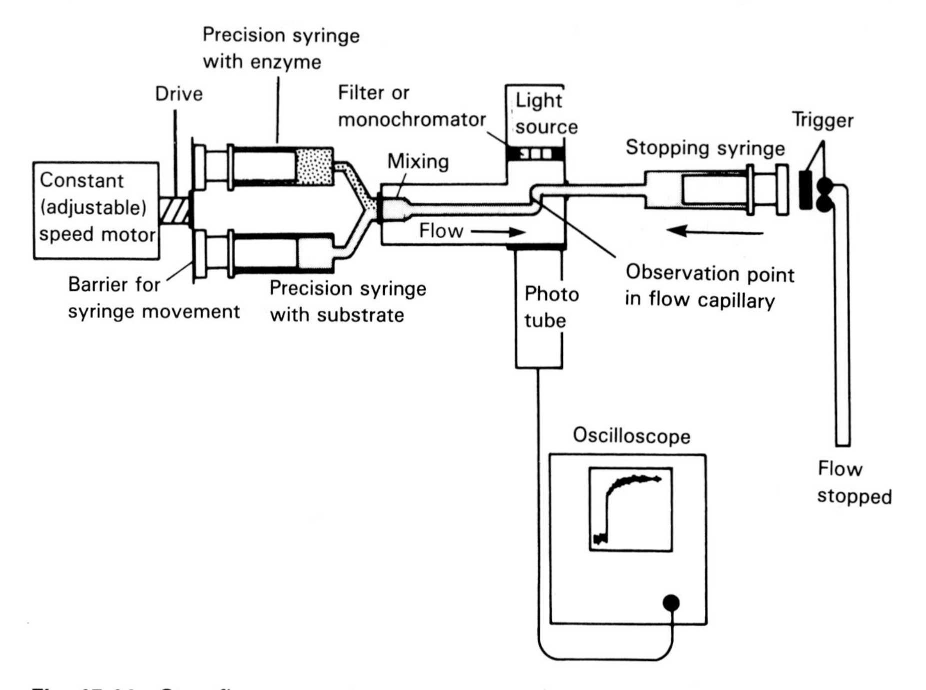

Fig. 15.14 Stop-flow apparatus.

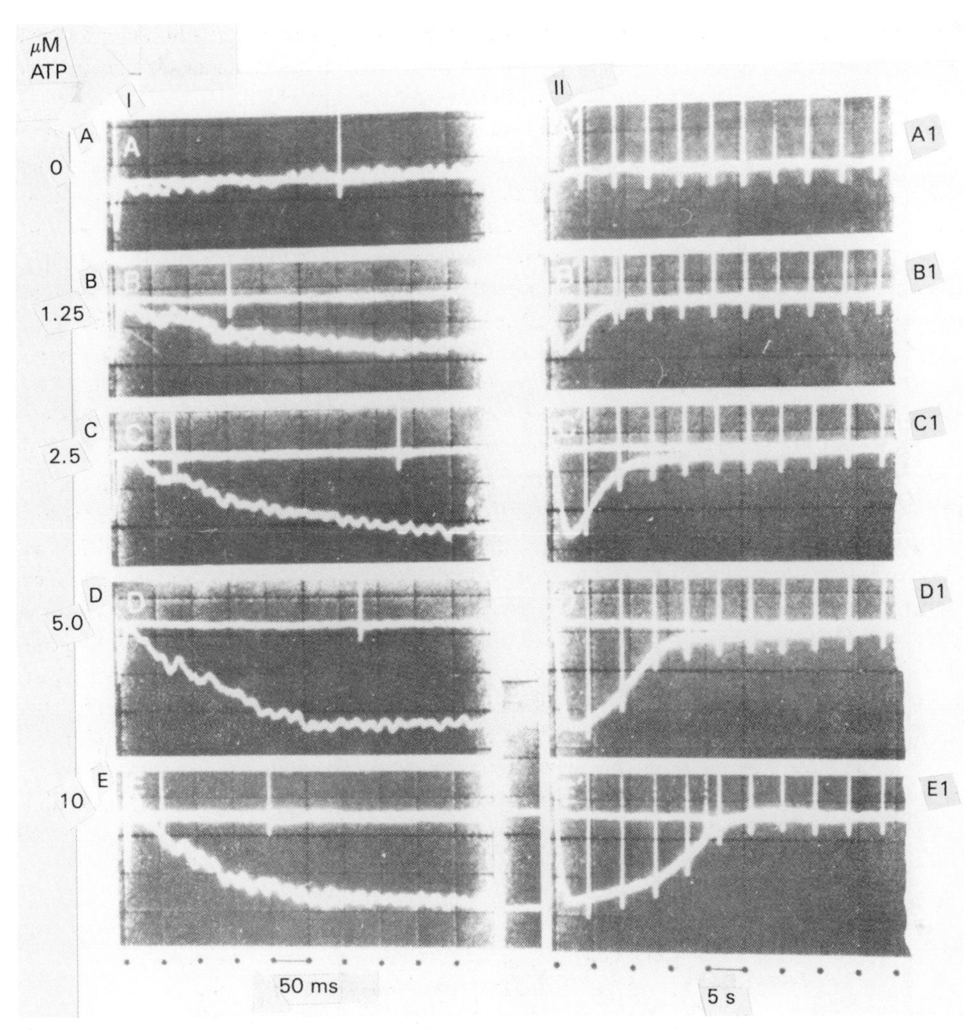

Fig. 15.15 ATPase-coupled changes in Ca^{2+} binding to purified Ca^{2+}-ATPase of the sarcoplasmic reticulum. From Ikemoto (1976), with permission.

Homareda, 1982; Yamaguchi and Tonomura, 1980). The binding constants on the two sides of the membrane for different transport systems are shown in Table 15.4 (Tanford, 1983).

The experiments discussed so far indicate that during transport the binding sites present in the transporter protein move from one interface to the other. The transport process seems to involve precise stoichiometry. As mentioned, two Ca^{2+} are transported per ATP hydrolyzed. Furthermore, two Ca^{2+} are bound per phosphorylated transporter molecule (Inesi et al., 1980). In the case of active transport, the affinity of the binding groups of the transporter for the ligand decreases when the transporter molecule is phosphorylated and this lower affinity should represent the state of the transporter on the side with the higher concentration at steady state. A model of active transport involving ion binding sites and shuttling of ions across the plasma membrane is consistent with the data.

However, these considerations leave unresolved how the binding sites can move from one interface to the other without a major movement of the transporter. Since the sites of the

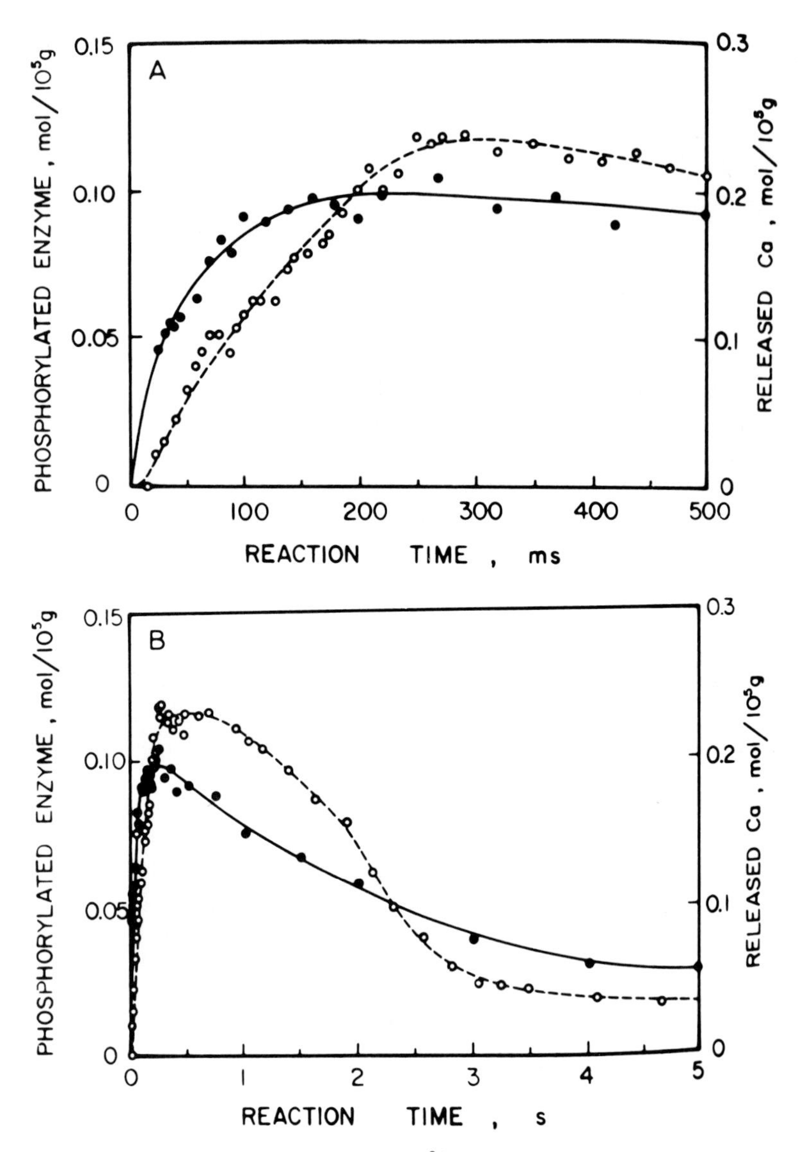

Fig. 15.16 Relationship between Ca^{2+} release and rebinding and the formation and decay of the phosphorylated intermediate. (○) Ca^{2+} release; (●) P in enzyme. Reprinted with permission from N. Ikemoto, *Journal of Biological Chemistry,* 251:7275–7277. Copyright © 1976 American Society of Biological Chemists, Inc.

transporter molecule interacting with chemicals at the two membrane interfaces differ, it follows that the transporter molecule does not flip or rotate. Furthermore, the Na^+,K^+-ATPase continues to function in the hydrolysis of ATP even when anchored at one interface with an antibody (Kyte, 1974). These difficulties could be resolved by proposing that the binding sites do not traverse the whole membrane thickness but rather move over much shorter distances. This would be possible if the binding sites were inside a channel traversing the membrane.

There are, in fact, many indications that the transporters can act as channels. The Na^+,K^+-ATPase incorporated into planar bilayers of phosphatidylethanolamine and phosphatidylserine increases their electrical conductance, and this high-conductance state was found to be sensitive

Table 15.4 Binding Constants for Transported Ions

Protein	Ion	Binding constant, K_{eq} (M^{-1})	
		Uptake side	Discharge side
SR Ca^{2+} pump	Ca^{2+}	10^7–10^8	300
Na^+ pump	Na^+	4×10^3	<20
Chloroplast F_0F_1	H^+	$>10^8$	$<10^6$
Na^+/Ca^{2+} exchange	Ca^{2+}	2×10^5–10^6	400

Source: Tanford (1983). Reproduced, with permission, from the *Annual Review of Biochemistry,* Volume 52, © 1983 by Annual Reviews Inc.

to ouabain and vanadate (Last et al., 1983). Similarly, addition of purified Ca^{2+}-ATPase to a bilayer (in this case a bilayer made of oxidized cholesterol) changes the conductivity of the bilayer (Shamoo and MacLennan, 1975). The measurement can be carried out by placing the bilayer on an opening separating two different compartments containing salt solutions, in this example Ca^{2+} salts, and passing a current between the two. Results of such an experiment are illustrated in Fig. 15.17, which shows that the conductivity increases very rapidly when the ATPase is added. The effect is specific, since it can be prevented with $HgCl_2$. As shown in Fig. 15.18, $HgCl_2$ inhibits both the ATPase activity and the Ca^{2+} transport activity. The ability to increase the conductance of lipid bilayers to ions is common to a group of antibiotics, which have been called *ionophores*. Analogously, this capacity has been called ionophoric ability. In addition to its ionophoric behavior, Ca^{2+}-ATPase, the anion transporter, also behaves like a channel and may therefore be consistent with the model proposed. In this case the selectivity of

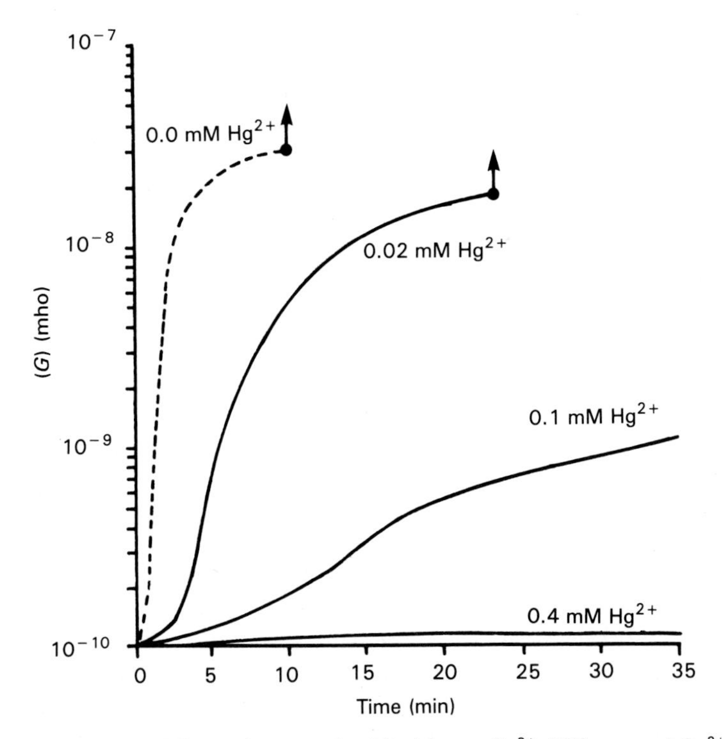

Fig. 15.17 Effect of mercuric chloride on Ca^{2+}-ATPase and Ca^{2+} transport activities. From A. E. Shamoo and D. J. MacLennan, *Journal of Membrane Biology,* 25:65–74, with permission. Copyright © 1975 Springer-Verlag, Heidelberg.

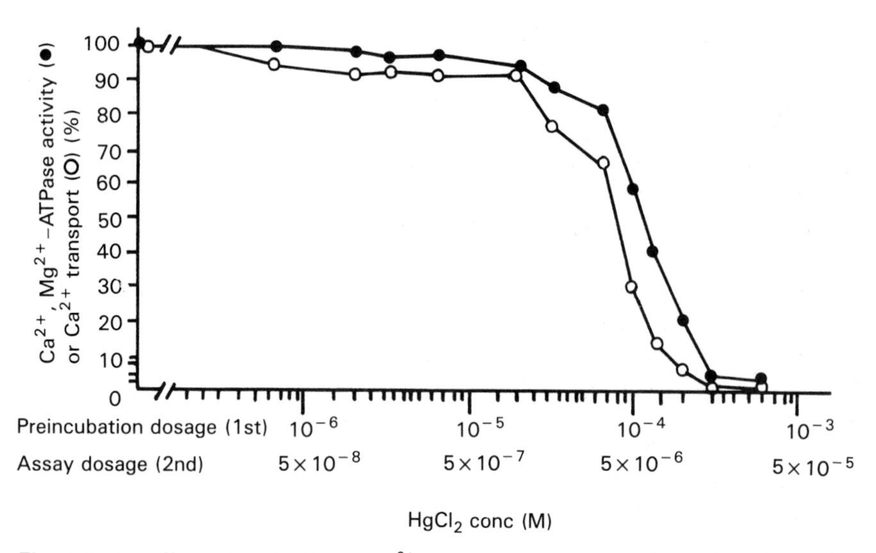

Fig. 15.18 Effect of solubilized Ca^{2+}-ATPase on conductance of a bilayer at 50 mV in the presence and absence of Hg^{2+}. From A. E. Shamoo and D. J. MacLennan, *Journal of Membrane Biology,* 25:65–74, with permission. Copyright © 1975 Springer-Verlag, Heidelberg.

the transport system appears to depend on the size of the molecule, suggesting a channel 0.8 to 0.9 nm in diameter (Giebel and Passow, 1960).

After cleavage of the Ca^{2+}-ATPase by incubation of sarcoplasmic reticulum membranes with trypsin, various fragments of the molecule have been tested for ionophoric properties and other characteristics. Limited digestion produces two Ca^{2+}-ATPase segments of 55 and 45 kDa. The 45-kDa fragment seems to be embedded in the lipid portion of the membrane. Further digestion produces two fragments of 30 and 20 kDa from the 55-kDa piece. The ATP and Mg^{2+} binding as well as the phosphorylation correspond to the 30-kDa fragment. All the segments have ionophoric activity, but the 20-kDa piece has some specificity for Ca^{2+}. The conductivity change induced by this fragment is sensitive to ruthenium red and $HgCl_2$, both inhibitors of the transport.

In summary, it appears that the transport of ions proceeds by binding the ions to specific sites. These sites are probably present in a channel of the transporter that traverses the membrane. The translocation is associated with some movement of the binding sites, so that in effect the sites are exposed first to one and then to the other side of the membrane. The presence of a channel open at all times cannot be part of an active transport mechanism, i.e., a mechanism capable of transporting against an electrochemical gradient. Such a channel would allow passive flow in the direction of the gradient, resulting in its dissipation. For this reason, the models generally considered propose alternating access (see Fig. 15.19) in which a small conformational change (in this case a rotation) exposes the binding sites first to the water phase on one side of the membrane and then to the water phase on the other side (Tanford, 1983). A movement of the binding groups as represented in the diagram could also account for a change in the affinity for the transported ion. During the working cycle of the alternating access pump, the ion is unavailable for exchange. The ion in the transporter molecule is said to be *occluded*. Occlusion suggests the presence of an intermediate position of the binding sites, apart from their location at either the uptake or the discharge site (see Glynn and Karlish, 1990).

As discussed in Chapter 2, computer analysis of membrane protein hydrophobicity (Kyte and Doolittle, 1982) based on the amino acid sequence permits predictions of the way the protein is likely to be arranged in the membrane. With these predictions and other information, it is now possible to speculate about which part of the protein may be involved in the various functions of transport, in particular the channel function.

The deduction of the amino acid sequence of Ca^{2+}- ATPase has been possible after isolating, cloning, and sequencing the cDNA (Brandl et al., 1986) (see Chapter 1 for a discussion of these procedures). A possible arrangement of the protein in relation to the membrane derived from hydrophobicity plots (see Chapter 2) is shown in Fig. 15.20. In this model, 10 peptide sectors are presumed to traverse the membrane rather than the 8 shown in the similar model of Chapter 2. The α and β chains are shown in the cytoplasmic domain, based on their sensitivity to trypsin when in vesicles. Their functions (see above) are shown at the top. The stalk portion is rich in acidic residues (indicated by circles), and there are some in the hydrophobic region as well. These areas of the chains may cluster to produce a channel. The negatively charged residues could provide the Ca^{2+} binding sites. It is conceivable that the transition between E_1 and E_2 corresponds to movement of the Ca^{2+} binding groups past a barrier in the stalk into a channel of transmembrane helices in what would basically be a gating step. This movement might occur by a rotational movement of the stalk helices with an increase in the amount of material entering the bilayer, a view supported by the x-ray diffraction data discussed above (Blasie et al., 1985). The helices in the bilayer together could provide an ion exchange channel and complete the transfer of the Ca^{2+}. The model shown assumes that a channel can be formed by a single protein molecule.

The properties common to at least some of the transport systems are summarized in Table 15.5. Most of the transports considered in this table are against the electrochemical gradient of the transported substrate. For all transport systems the transporter binds the transported substrate. Furthermore, the transporters have been shown to undergo a conformational change. Channel behavior has been shown at least under some conditions for some of the transporters. As discussed, the alternating-access model of transport (see Fig. 15.19) requires channel-like behavior.

Cogent arguments have been made that either a monomer (Kyte, 1981) or an oligomer (Klingenberg, 1981) is able to form a channel. Several polypeptides are known to form the walls of a pore; e.g., the sodium channel of nerve tissue involves three separate polypeptides (Hartshone and Catterall, 1984). A channel between adjoining cells, the *connexon*, is formed from

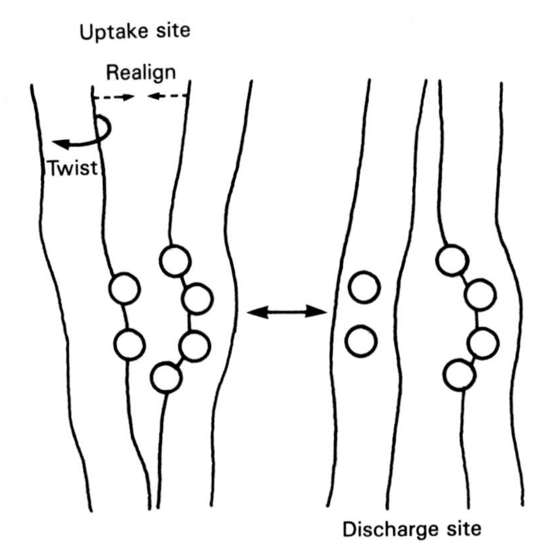

Fig. 15.19 Representation of the alternate-access model of transport. The structures represent polypeptide chains traversing the phospholipid bilayer of the plasma membrane. The circles indicate the binding sites of the transported ion. The closeness of the binding groups on the left accounts for the high-affinity binding, the separation on the right for the decrease in affinity. The slight rotation of the polypeptides accounts for the access of the binding sites from either the uptake site (left) or the discharge site (right).

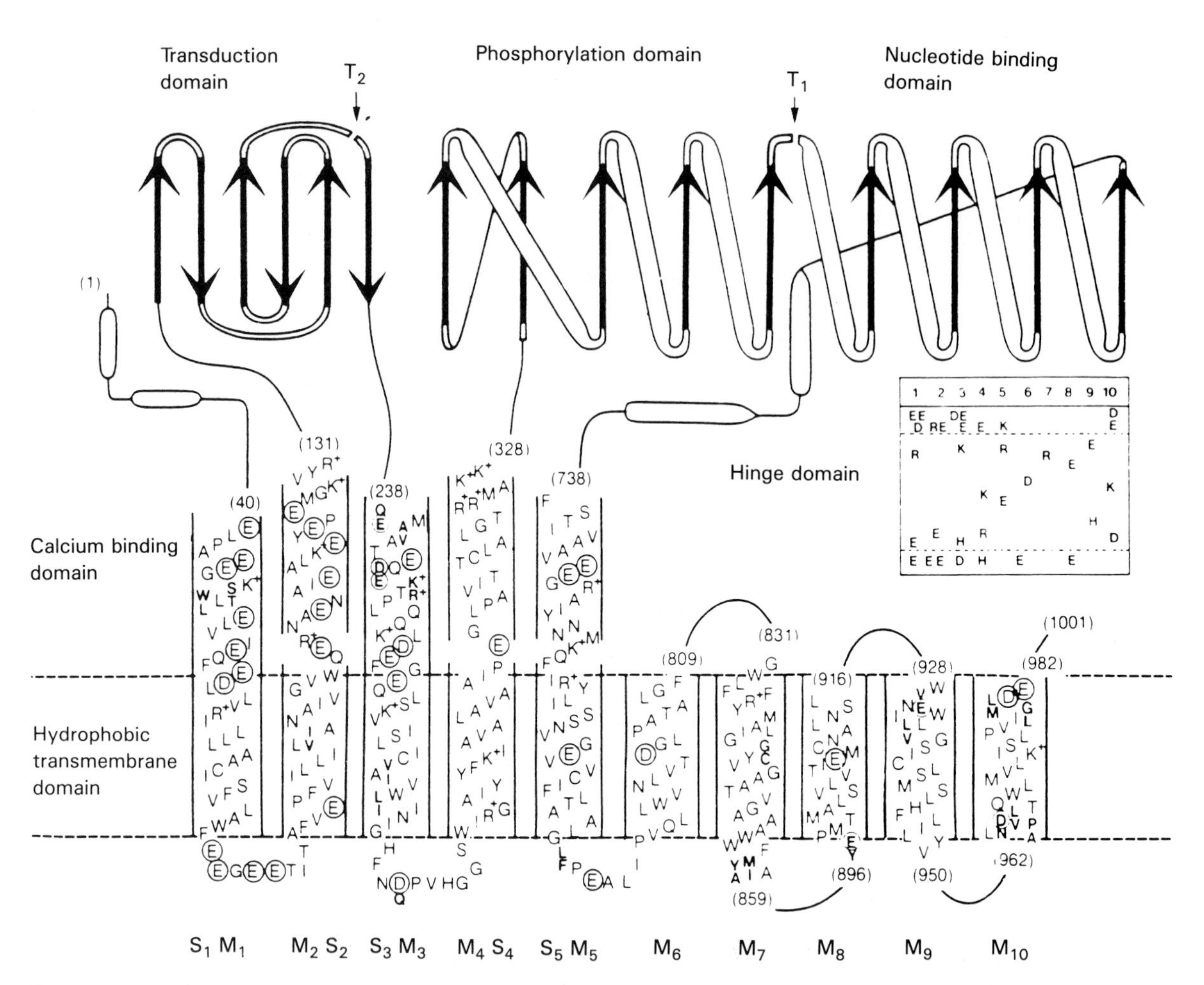

Fig. 15.20 Structural diagram of the Ca^{2+}-ATPase molecule based on predicted domain structure and hydrophobicity plots. In the α-helical segments of the stalk and the transmembrane region the individual residues are shown on an α-helical net. The scale of this section of the diagram is larger than that of the globular domains. The folding of the latter is indicated schematically in accordance with the general principles established for antiparallel or parallel β-sheet domains. The connections between individual β-strands and β-helices are arbitrary. The inset shows the location of the negative charges, where the numbers correspond to those of the transmembrane helices. Reprinted with permission from Brandl, et al., *Cell,* 44:597–607, copyright 1986 by Cell Press.

Table 15.5 Summary of the Properties of Some Transport Systems[a]

Ion or solute transported	Active transport	Binding to transporter	Channel properties	Conformational change or movement of transporter
Anion exchanger	No	Yes[b]	Yes[c]	Yes[d]
Na^+, K^+ (ATPase)	Yes[e]	Yes (Table 15.3)	Yes[f]	Yes[g]
Ca^{2+} (ATPase)	Yes	Yes (Table 15.3)	Yes[h]	Yes[i]
H^+ (ATP synthase)	Yes	Yes (Table 15.3)	Yes[j]	Yes (See Chapter 12)
Na^+-glucose cotransporter	Yes	—	—	Yes[k]
Na^+-amino acid cotransporter	Yes	—	—	Yes[l]

[a]"Yes" indicates that the phenomenon has been observed for the corresponding transporter.
[b]Falke et al. (1984a).
[c]Giebel and Passow (1960).
[d]Falke et al. (1984b).
[e]Yamaguchi and Tonomura (1980).
[f]Last et al. (1983).
[g]Jorgensen (1975); Karlish and Yates (1978); Koepsell (1979).
[h]Shamoo and MacLennan (1975).
[i]Imamura et al. (1984).
[j]Tanford (1983).
[k]Peerce and Wright (1984).
[l]Wright and Peerce (1984).

several polypeptides (see Chapter 2). A study of Na^+,K^+-ATPase presents evidence that a single α-β complex can be active (Craig, 1982). Other data, however, indicate that it is the α portion of the complex that is responsible for the ATPase activity, since the β subunit can be selectively removed without removing the ATP-hydrolytic activity (Freytag, 1983). However, some studies suggest involvement of the β subunit in association with the α subunit in channel formation (Shamoo and Myers, 1976). It has also been suggested that the β subunit is needed for the transport and membrane assembly of the α-β complex (McDonough et al., 1990).

The experiments implicating a single complex were carried out by fractionating the various oligomers of the purified enzyme dispersed with the nonionic detergent octaethylene glycol dodecyl ether in glycerol gradients. When most of the detergent is removed, the complexes remain in solution, presumably because of a thin layer of detergent on their hydrophobic surfaces. The various fractions were assayed for ATP-hydrolyzing activity and then cross-linked briefly with glutaraldehyde, a procedure that in effect "freezes" the configuration of the complexes. The cross-linked complexes can then be examined with SDS-gel electrophoresis (which fractionates mostly on the basis of molecular weight). The maximal activity was that of a single α-β complex. Figure 15.21 (Craig, 1982) shows the peaks corresponding to each oligomeric fraction (as determined by subsequent SDS-gel electrophoresis). The dashed line shows the Na^+,K^+-ATPase activity. Since the α-β monomer retains the properties of the transporting unit (e.g., the K_m in relation to the ion concentrations), it has been presumed to be functionally equivalent to a transporting unit, and hence one α-β unit must have a channel-like activity.

The size of the transporting unit in the reconstituted system was estimated by the technique of radiation inactivation and was consistent with one α-β combination (Karlish and Kempner, 1984). In this technique (Kempner and Schlegel, 1979) the size of a macromolecule is estimated, at least approximately, by its sensitivity to radiation. The larger the target, the greater the chance of inactivation. The results of these experiments with Na^+,K^+-ATPase should be regarded with caution, since the target size, approximately 200 kDa, is only marginally lower than the size of

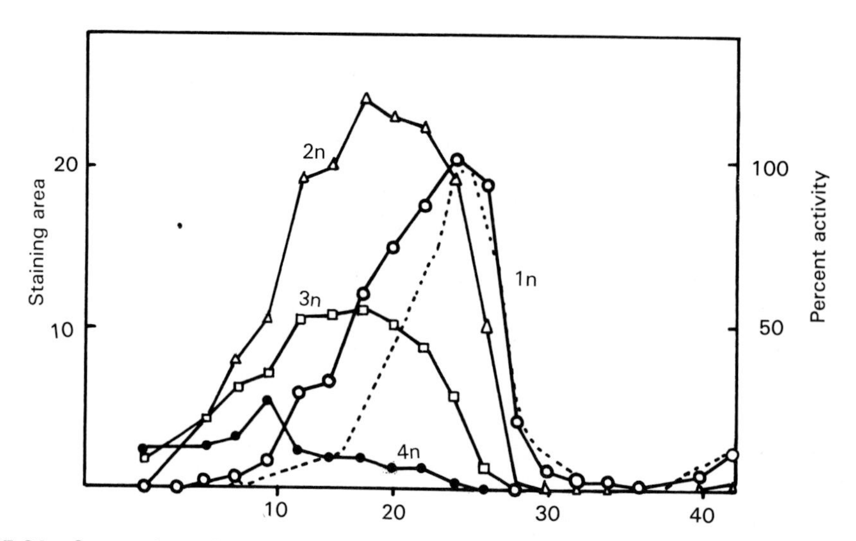

Fig. 15.21 Separation of extract on a linear glycerol gradient of 10–30%. The broken line indicates Na^+,K^+-ATPase activity, n indicates one αβ complex. The number of n was assessed after cross-linking with glutaraldehyde. From Craig (1982). Reprinted with permission from *Biochemistry* 21:5707–5717. Copyright 1982 American Chemical Society.

a dimer of α. Furthermore, results of such techniques applied to Ca^{2+}-ATPase (which lacks the β subunit but is otherwise very similar) suggest that the functional unit is a dimer (Falke et al., 1984a; Hymel et al., 1984). There are also indications that a dimer is involved in the Cl^- channels of *Torpedo* in a reconstituted system of channels extracted from the electroplax (Miller and White, 1984).

SUGGESTED READING

Brandl, C. J., Green, N. M., Korczak, B., and MacLennan, D. H. (1986) Two Ca ATPase genes: homologies and mechanistic implication of deduced amino acid sequences. *Cell* 44:597–607.

Karlish, S. J. D. (1989) The mechanism of cation transport by the Na^+,K^+-ATPase. In *Ion Transport* (Keeling, D., and Benham, C., eds.), pp. 19–34. Academic Press, New York.

Pedersen, P. L., and Carafoli, E. (1987) Ion motive ATPases. I. Ubiquity, properties and significance to cell function. *Trends Biochem. Sci.* 12:146–150.

Pedersen, P. L., and Carafoli, E. (1987) Ion motive ATPases. II. Energy coupling and work output. *Trends Biochem. Sci.* 12:186–189.

Stein, W. D., and Lieb, W. R. (1986) *Transport and Diffusion Across Cell Membranes*, Chapter 6, pp. 475–612. Academic Press, New York.

Tanford, C. (1984) The sarcoplasmic reticulum calcium pump. Localization of free energy transfer to discrete steps of the reaction cycle. *FEBS Lett.* 166:1–7.

Complete Review

Schuurmans Stekhoven, F., and Bonting, S. L. (1981) Transport adenosine triphosphatases: properties and functions. *Physiol. Rev.* 61:1–76.

Special Aspects

Hokin, L. E. (1981) Reconstitution of carriers in artificial membranes. *J. Membr. Biol.* 60:77–93.

Hokin, L. E., and Dahl, J. L. (1972) The sodium-potassium adenosinetriphosphatase. In *Metabolic Pathways* (Greenberg, D. M., ed.), Vol. 6, pp. 269–315. Academic Press, New York.

Jorgensen, P. L. (1983) Principal conformations of the α-subunit and ion translocation. *Curr. Top. Membr. Transport* 19:377–401.

Kyte, J. (1981) Molecular considerations relevant to the mechanism of active transport. *Nature (London)* 292:201–204.

Lauger, P. (1984) Channels and multiple conformational states: interrelations with carriers and pumps. *Curr. Top. Membr. Transport* 21:309–326.

Post, R. L. (1979) A perspective on sodium and potassium ion transport adenosine triphosphatase. In *Cation Flux Across Biomembranes* (Mukohata, Y., and Packer, L., eds.), pp. 3–20. Academic Press, New York.

Rothstein, A., Grinstein, S., Ship, S., and Knauf, P. A. (1978) Asymmetry of functional sites of the erythrocyte anion transport protein. *Trends Biochem. Sci.* 3:126–128.

Shamoo, A. E., and Murphy, T. J. (1979) Ionophores and ion transport across natural membranes. *Curr. Top. Bioenerg.* 9:147–177.

REFERENCES

Barzilay, M., and Cabantchik, Z. I. (1979a) Anion transport in red blood cells. II. Kinetics of reversible inhibition by nitroaromatic sulfonic acids. *Membr. Biochem.* 2:255–281.

Barzilay, M., and Cabantchik, Z. I. (1979b) Anion transport in red blood cells. III. Sites and sidedness of inhibition by high-affinity reversible binding probes. *Membr. Biochem.* 2:297–322.

Blasie, J. K., Herbette, L. G., Pascolini, D., Skita, V., Pierce, D. H., and Scarpa, A. (1985) Time resolved x-ray diffraction of sarcoplasmic reticulum membrane during active transport. *Biophys. J.* 48:9–18.

Bontig, S. I., Schuurmans Stekhoven, F. M. A. H., Swarts, H. G. P., and dePont, J. J. H. H. M. (1979) The low-energy phosphorylated intermediate of Na^+,K^+-ATPase. In *Na,K-ATPase, Structure and Kinetics* (Skou, J. C., and Norby, J. G., eds.), pp. 317–330. Academic Press, New York.

Brandl, C. J., Green, N. M., Korczak, B., and MacLennan, D. H. (1986) Two Ca^{2+} ATPase genes: homologies and mechanistic implications of deduced amino acid sequences. *Cell* 44:597–607.

Cabantchik, Z. I., and Rothstein, A. (1976) Membrane proteins related to anion permeability of human red blood cells. *J. Membr. Biol.* 15:207–226.

Castro, J., and Farley, R. A. (1979) Proteolytic fragmentation of the catalytic subunit of the sodium and potassium adenosine triphosphatase. *J. Biol. Chem.* 254:2221–2228.

Charnock, J. S., Trebilcock, H. A., and Casley-Smith, J. R. (1972) Demonstration of transport adenosine triphosphatase in the plasma membranes of erythrocyte ghosts by quantitative electron microscopy. *J. Histochem. Cytochem.* 20:1069–1080.

Craig, W. S. (1982) Monomer of sodium and potassium ion activated adenosine triphosphatase displays complete enzymatic function. *Biochemistry* 21:5707–5717.

Deves, R., and Krupka, R. M. (1979) The binding and translocation steps in transport as related to substrate structure. A study of the choline carrier of erythrocyte. *Biochim. Biophys. Acta* 557:469–485.

Falke, J. J., Pace, R. J., and Chan, S. I. (1984a) Chloride binding to anion binding sites of band 3. *J. Biol. Chem.* 259:6472–6480.

Falke, J. J., Pace, R. J., and Chan, S. I. (1984b) Direct observations of the transmembrane recruitment of band 3 transport sites by competitive inhibitors. *J. Biol. Chem.* 259:6481–6491.

Freytag, J. W. (1983) The (Na^+,K^+)ATPase exhibits enzymic activity in the absence of the glycoprotein subunit. *FEBS Lett.* 159:280–284.

Garrahan, P. J., and Glynn, I. M. (1967) The incorporation of inorganic phosphate into adenosine triphosphate by reversal of the sodium pump. *J. Physiol. (London)* 192:237–256.

Giebel, O., and Passow, H. (1960) Die permeabilitat der erythrocytenmembran fur organische anionen. *Pfluegers Arch.* 271:378–388.

Glynn, I. M., and Karlish, S. J. D. (1990) Occluded cations in active transport, *Annu. Rev. Biochem.* 59:171–205.

Gorga, F. R., and Lienhard, G. E. (1981) Equilibria and kinetics of ligand binding to the human erythrocyte glucose transporter. Evidence for an alternating conformation model for transport. *Biochemistry* 20:5108–5113.

Greenfield, N., and Fasman, G. D. (1969) Computed circular dichroism spectra for the evaluation of protein conformation. *Biochemistry* 8:4108–4116.

Gresalfi, J., and Wallace, B. A. (1984) Secondary structural composition of the Na/K-ATPase E_1 and E_2 conformers. *J. Biol. Chem.* 259:2622–2655.

Gunn, R. B., Dalmark, M., Tosteson, D. C., and Wieth, J. O. (1973) Characteristics of chloride transport in human red blood cells. *J. Gen. Physiol.* 61:185–206.

Hart, N. M., and Titus, E. O. (1973) Sulfhydryl groups of sodium-potassium transport adenosine triphosphatase. *J. Biol. Chem.* 248:4674–4681.

Hartshone, R. P., and Catterall, W. A. (1984) The sodium channel from rat brain. Purification and subunit composition. *J. Cell Biol.* 259:1667–1675.

Hokin, L. E., and Dahl, J. L. (1972) In *Metabolic Pathways* (Hokin, L. E., ed.), Vol. 6, pp. 269–315. Academic Press, New York.

Hymel, L., Maurer, A., Berenski, C., Jung, C. Y., and Fleischer, S. (1984) Target size of calcium pump protein from skeletal muscle sarcoplasmic reticulum. *J. Biol. Chem.* 259:4890–4895.

Ikemoto, N. (1976) Behavior of Ca^{2+} transport sites linked with the phosphorylation reaction of ATPase purified from the sarcoplasmic reticulum. *J. Biol. Chem.* 251:7275–7277.

Imamura, Y., Saito, K., and Kawakita, M. (1984) Conformational change of Ca^{2+}, Mg^{2+} adenosine triphosphatase of sarcoplasmic reticulum upon binding of Ca^{2+} and adenyl-5′-yl-imidodiphosphate as detected by trypsin sensitivity analysis. *J. Biochem.* 95:1305–1313.

Inesi, G., Kurzmack, M., Coan, C., and Lewis, E. (1980) Cooperative calcium binding and ATPase activation in sarcoplasmic reticulum vesicles. *J. Biol. Chem.* 255:3025–3031.

Jenkins, R. E., and Tanner, M. (1975) The major human erythrocyte membrane protein. *Biochem. J.* 147:393–399.

Jorgensen, P. L. (1975) Purification and characterization of (Na^+,K^+)-ATPase. V. Conformational changes in the enzyme. Transitions between the Na-form and the K-form studied with tryptic digestion as a tool. *Biochim. Biophys. Acta* 401:399–415.

Jorgensen, P. L., and Petersen, J. (1979) Protein conformations of the phosphorylated intermediates of purified Na^+,K^+-ATPase studied with tryptic digestion and intrinsic fluorescence as tools. In *Na^+,K^+-ATPase Structure and Kinetics* (Skou, J. C., and Norby, J. G., eds.), pp. 143–155. Academic Press, New York.

Kahlenberg, A., Galsworthy, P. R., and Hokin, L. E. (1968) Studies of the characterization of the sodium-potassium transport adenosinetriphosphatase. *Arch. Biochem. Biophys.* 126:331–342.

Kanazawa, T., Yamada, S., Yamamoto, T., and Tonomura, Y. (1971) Reaction mechanism of the Ca^{2+}-dependent ATPase of sarcoplasmic reticulum from skeletal muscle. *J. Biochem.* 70:95–123.

Karlish, S. J. D., and Kempner, E. S. (1984) Minimal functional unit for transport and enzyme activities of Na^+K^+-ATPase as determined by radiation inactivation. *Biochim. Biophys. Acta* 776:288–298.

Karlish, S. J. D., and Yates, D. W. (1978) Tryptophan fluorescence of (Na^+,K^+)-ATPase as a tool for study of the enzyme mechanism. *Biochim. Biophys. Acta* 527:115–130.

Kempner, E. S., and Schlegel, W. (1979) Size determination of enzyme by radiation inactivation. *Anal. Biochem.* 92:2–10.

Klingenberg, M. (1981) Membrane protein oligomeric structure and transport function. *Nature (London)* 290:449–454.

Klingenberg, M., and Burchholtz, M. (1973) On the mechanism of bongkrekate effect on the mitochondrial adenine-nucleotide carrier as studied through the binding of ADP. *Eur. J. Biochem.* 38:346.

Knowles, A. F., and Racker, R. (1975) Formation of adenosine triphosphate from P_i and adenosine triphosphate by purified Ca^{2+}-adenosine triphosphatase. *J. Biol. Chem.* 250:1949–1951.

Koepsell, H. (1979) Conformational changes of membrane-bound (Na^+,K^+)-ATPase as revealed by trysin digestion. *J. Membr. Biol.* 48:69–94.

Kyte, J. (1971) Purification of the sodium- and potassium-dependent adenosine triphosphatase from canine renal medulla. *J. Biol. Chem.* 246:4157–4165.

Kyte, J. (1974) The reactions of sodium and potassium ion activated adenosine triphosphatase with specific antibodies. *J. Biol. Chem.* 249:3652–3660.

Kyte, J. (1981) Molecular considerations relevant to the mechanism of active transport. *Nature (London)* 292:201–204.

Kyte, J., and Doolittle, R. F. (1982) A simple method for displaying the hydropathic character of protein. *J. Mol. Biol.* 157:105–132.

Last, T. A., Gantzer, M. L., and Tyler, C. D. (1983) Ion-gated channel induced in planar bilayers by incorporation of (Na^+,K^+)-ATPase. *J. Biol. Chem.* 258:2399–2404.

MacLennan, D. H. (1970) Purification and properties of an adenosine triphosphate from sarcoplasmic reticulum. *J. Biol. Chem.* 245:4508–4518.

Makinose, M. (1972) Phosphoprotein formation during osmo-chemical energy conversion in the membrane of the sarcoplasmic reticulum. *FEBS Lett.* 25:113–115.

Masui, H., and Homareda, H. J. (1982) Interaction of sodium and potassium ions with Na^+,K^+-ATPase. I. Ouabain-sensitive alternative binding of three Na^+ or two K^+ to the enzyme. *J. Biochem.* 92:193–217.

McDonough, A. A., Geering, K., and Farley, R. A. (1990) The sodium pump needs its β subunit. *FASEB J.* 4:1598–1605.

Miller, C., and White, M. M. (1984) Dimeric structure of single chloride channels from *Torpedo* electroplax. *Proc. Natl. Acad. Sci. U.S.A.* 81:2772–2775.

Nishigaki, I., Chen, F. T., and Hokin, L. E. (1974) Studies on the characterization of the sodium-potassium transport adenosine triphosphatase. *J. Biol. Chem.* 249:4911–4916.

Peerce, B. E., and Wright, E. M. (1984) Sodium induced conformational changes in the glucose transporter of intestinal brush borders. *J. Biol. Chem.* 259:14105–14112.

Post, R. L., and Kume, S. (1973) Evidence for an aspartate phosphate residue at the active site of sodium and potassium ion transport adenosine triphosphatase. *J. Biol. Chem.* 248:6993–7000.

Post, R. L., Sen, A. K., and Rosenthal, A. S. (1965) A phosphorylated intermediate in adenosine triphosphate-dependent sodium and potassium transport across kidney membranes. *J. Biol. Chem.* 240:1437–1445.

Post, R. L., Kume, S., Tobin, T., Orgutt, B., and Shu, A. K. (1969) Flexibility of an active center in sodium plus potassium adenosine triphosphatase. *J. Gen. Phys.* 54:306s–326s.

Post, R. L., Taniguchi, K., and Toda, G. (1974) Synthesis of adenosine triphosphate by Na^+,K^+-ATPase. *Ann. N.Y. Acad. Sci.* 242:80–91.

Post, R. L., Toda, G., and Rogers, F. N. (1975) Phosphorylation by inorganic phosphate of sodium plus potassium ion transport adenosine triphosphatase. Four reactive states. *J. Biol. Chem.* 250:691–701.

Rothstein, A., Cabantchik, Z. I., and Knauf, P. (1976) Mechanism of anion transport in red blood cells: role of membrane proteins. *Fed. Proc.* 35:3–10.

Sachs, G., Chang, H., Rabon, E., Schackman, R., Lewin, M., and Saccomani, G. (1976) A nonelectrogenic H^+ pump in plasma membranes of hog stomach. *J. Biol. Chem.* 251:7690–7698.

Sachs, J. R., Knauf, P. A., and Dunham, P. B. (1975) Transport through red cell membranes. In *The Red Blood Cell* (Surgenor, D. M., ed.), 2d ed., Vol. 2, pp. 613–705. Academic Press, New York.

Schoot, B. M., Schoots, A. F. M., dePont, J. J. H. H. M., Schuurmans Stekhoven, F. M. A. H., and Bonting, S. L. (1977) Studies on (Na^+-K^+) activated ATPase. XVI. Effects of *N*-ethylmaleimide on overall and partial reactions. *Biochim. Biophys. Acta* 483:181–192.

Sen, A., Tobin, T., and Post, R. L. (1969) A cycle for ouabain inhibition of sodium- and potassium-dependent adenosine triphosphatase. *J. Biol. Chem.* 244:6596–6604.

Shamoo, A., and MacLennan, D. H. (1975) Separate effects of mercurial compounds on the ionophoric and hydrolytic functions of the $(Ca^{2+}\ Mg^{2+})$-ATPase of sarcoplasmic reticulum. *J. Membr. Biol.* 25:65–74.

Shamoo, A. E., and Myers, M. (1976) Na^+-dependent ionophore as part of the small polypeptide of the $(Na^+$ and $K^+)$-ATPase from eel electroplax membrane. *J. Membr. Biol.* 19:163–178.

Siegel, G. J., Koval, G. J., and Albers, R. W. (1969) Sodium-potassium-activated adenosine triphosphatase. *J. Biol. Chem.* 244:3264–3269.

Steck, T. L., Ramos, B., and Strapazon, E. (1976) Proteolytic dissection of band 3, the predominant transmembrane polypeptide of the human erythrocyte membrane. *Biochemistry* 15:1115–1161.

Tanford, C. (1983) Mechanism of free energy: coupling in active transport. *Annu. Rev. Biochem.* 52:379–409.

Taniguchi, K., and Post, R. L. (1975) Synthesis of adenosine triphosphate and exchange between inorganic phosphate and adenosine triphosphate in sodium and potassium ion transport adenosine triphosphatase. *J. Biol. Chem.* 250:3010–3018.

Taylor, K. A., Dux, L., and Martonosi, A. (1986) Three-dimensional reconstruction of negatively stained crystals of Ca^{2+}-ATPase from muscle sarcoplasmic reticulum. *J. Mol. Biol.* 187:417–427.

Weber, G. (1972) Ligand binding and internal equilibria in proteins. *Biochemistry* 11:864–878.

Weber, G. (1974) Addition of chemical and osmotic energies by ligand protein interactions. *Ann. N.Y. Acad. Sci.* 227:486–496.

Whittam, R., and Wheeler, K. P. (1972) The sodium-potassium adenosinetriphosphatase. In *Metabolic Pathways* (Greenberg, D. M., ed.), Vol. 6, pp. 269–315. Academic Press, New York.

Wright, E. M., and Peerce, B. E. (1984) Identification and conformational changes of the intestinal proline carrier. *J. Biol. Chem.* 259:14993–14996.

Yamaguchi, M., and Tonomura, Y. (1980) Binding of monovalent cations to Na^+,K^+-dependent ATPase purified from porcine kidney. *J. Biochem.* 88:1365–1375.

Chapter 16

Signaling in Biological Systems: Excitation and Conduction

Signals between cells and within cells are necessary for most physiological functions. These signals are composed of chemical and electrical events of some complexity.

In one way or another, signaling frequently makes use of the electrochemical gradient across the cell membrane. This is the case not only for excitable cells, the primary topic of this chapter, but also for other cells. For example, the entry of Ca^{2+} from either the medium or internal stores, driven by the Ca^{2+} electrochemical gradient, has a role in the release of secretory products in either exocrine or endocrine secretory cells. This mechanism is also used for the release of neurotransmitters, which transmit signals between nerve cells. Electrical signals are also coupled to the release of Ca^{2+} needed for the contraction of muscle or less organized contractile systems (see Chapter 18). Resting membrane potentials are present in most cells, and even in some plant cells dynamic electrical changes take place on stimulation (discussed later in this chapter). The precise physiological role of these electrical events in plants is still not clear.

Nerve cells (and muscle cells) are specialized for conducting signals through the transmission of an electrical event (the *nerve impulse*) on stimulation. A variety of stimuli are effective in initiating impulses. Perhaps the most direct way to elicit a nerve impulse in the laboratory is by means of an electrical current of an intensity above a critical level (the *threshold*). Once initiated, the nerve impulse can be propagated without loss of intensity over the entire length of the nerve cell, which can be 1 m or more long.

Although cells generally have similar underlying mechanisms for conducting impulses, the speed of nerve conduction varies enormously from organism to organism and from one conducting system to another. Some mammalian nerve fibers can conduct with speeds as high as 100 m/s (over 200 miles per hour!). Others are slower, and the speed may be as low as 0.1 m/s.

The response to stimulation of nerves can be relatively direct and readily observable. One stimulating event in a motor nerve may result in one contraction of a striated muscle. The response from other cells may also be much more complex and subtle. This is particularly true where nerve cells interact, as in ganglia or in the central nervous system, or where the effector cell responds in a complex manner (e.g., smooth muscle does not respond with a single contraction). The analysis of such complex effects is beyond the scope of this discussion.

I. NEURONS: UNITS OF CONDUCTION

Many kinds of cells can propagate an electrical event or an electrical impulse. However, nerve cells (*neurons*) function primarily to transmit impulses.

Some of the types of neurons encountered in different biological systems are shown in Fig. 16.1 (Bullock and Horridge, 1965). In general, different shapes are related to different functions

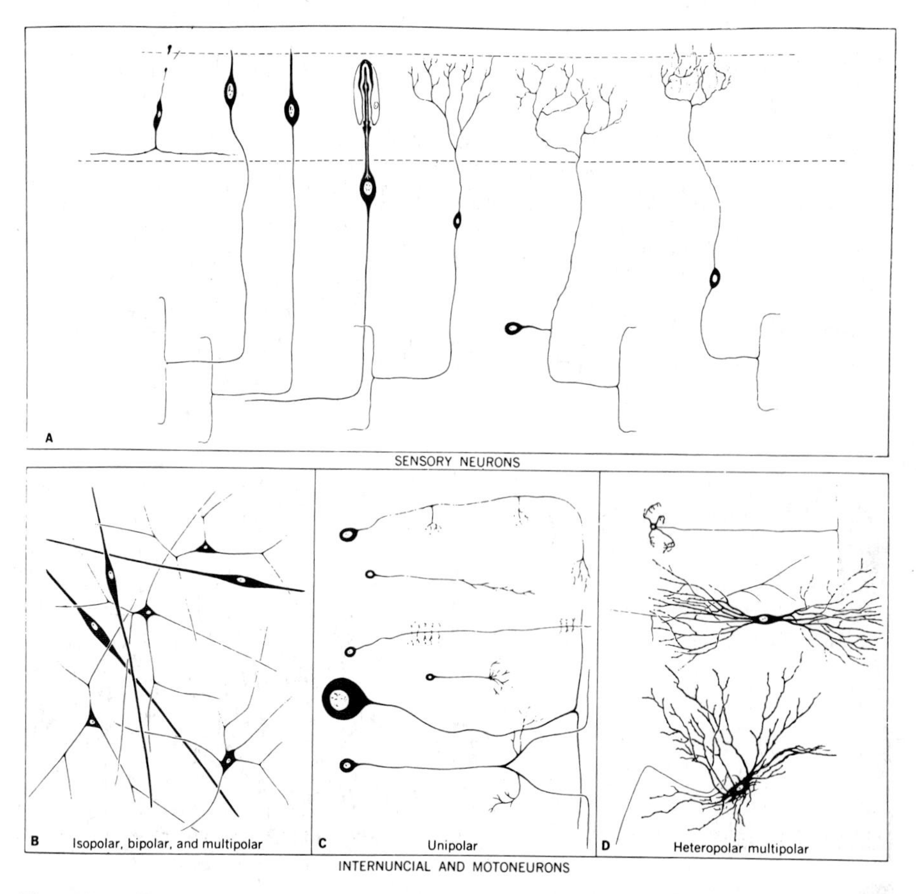

Fig. 16.1 Types of neurons based on the number and differentiation of processes. (*a*) Sensory neurons of several systems. (*b*) Isopolar, bipolar, and multipolar neurons from the nerve net of medusa. (*c*) Unipolar neurons, the predominant type in invertebrates. (*d*) Heteropolar, multipolar neurons, the predominant type in the central nervous system of vertebrates. From *Structure and Function in the Nervous Systems of Invertebrates.* By Theodore Holmes Bullock and G. Adrian Horridge. Copyright © 1965 by Theodore Holmes Bullock and G. Adrian Horridge. Reprinted by permission of W. H. Freeman and Company.

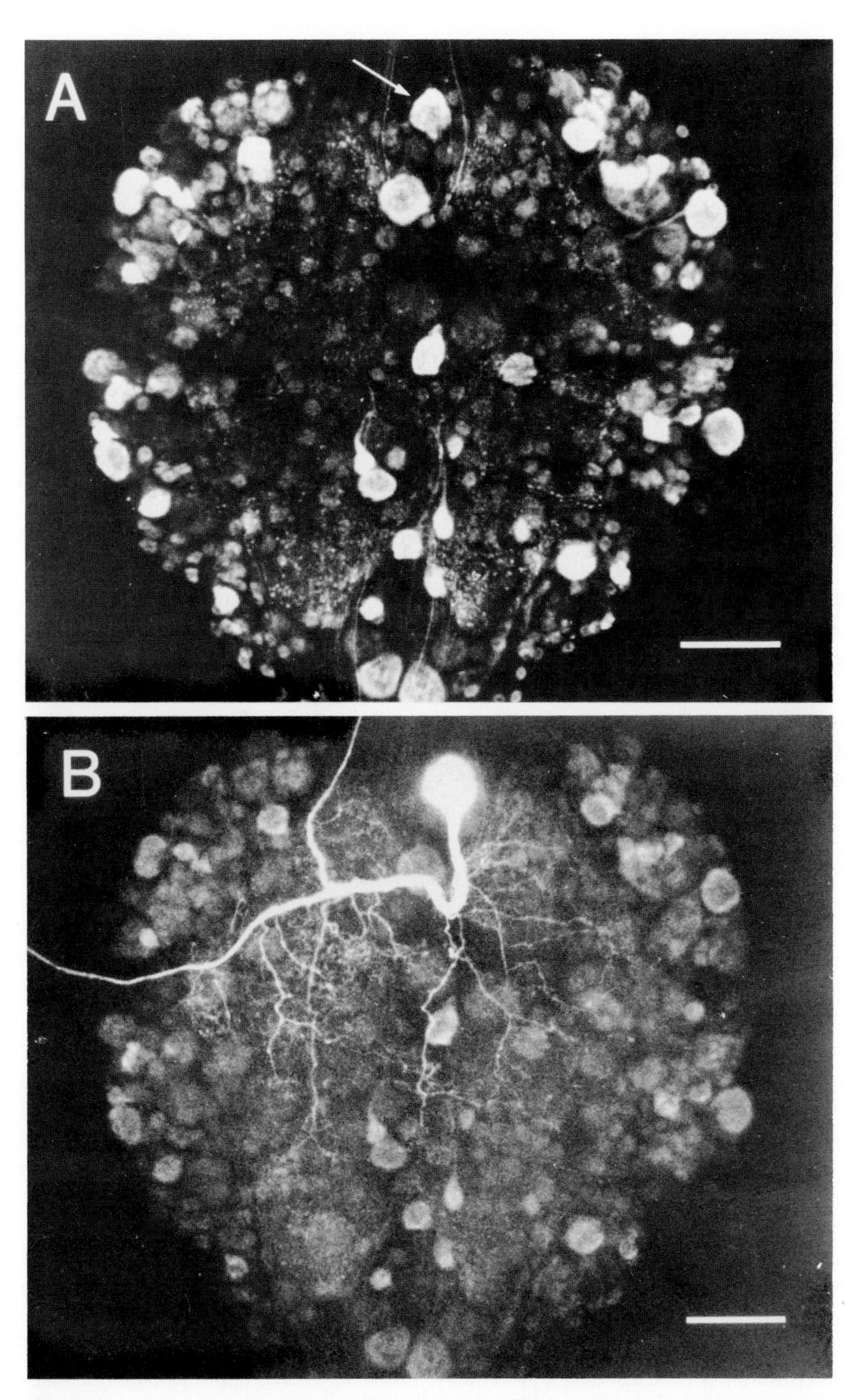

Fig. 16.2 Leech segmental ganglion stained with an antiserum directed against the neuropeptide FMRFamide. Primary antiserum is visualized by indirect immunofluorescence. (A) Demonstration of several cell bodies with an antigen sensitive to the antibody. (B) The same preparation but after intracellular injection of the water-soluble dye Lucifer Yellow. The arrow in (A) indicates the neuron (the penile evertor motor neuron) that is microinjected with Lucifer Yellow. Courtesy of Ronald C. Calabrese.

of these neurons, i.e., where they receive inputs and send their output signals. Many neurons have a characteristic stellate shape. The cell has many small branches, the *dendrites,* and a long process, the *axon*. Generally, the cell bodies are present in ganglia or in the central nervous system. The axon, a nerve fiber, is the portion that conducts the nerve impulse over long distances.

Neurons assume many shapes and forms. The size and morphology of the cell body (*soma*) of neurons vary widely. *Globulus* cells of invertebrates can be smaller than 3 μm in diameter. However, neurons can also be huge; for example, in gastropods they can be larger than 800 μm and therefore visible with the naked eye. The variation in size and shape within the same organism is surprising. Generally, the smaller cells have very little cytoplasm and larger cells have a good deal of it. The size of the cell is only roughly correlated with the size of the axon that originates from it.

Figure 16.2 shows a leech segmental ganglion stained by immunofluorescence using an antibody to the neurotetrapeptide FMRFamide (Phe-Met-Arg-Phe-NH_2) (Kuhlman et al., 1985). The neurons, which occupy specific positions, differ not only in size or morphology but also in their capacity to interact with the antibody. In Fig. 16.2B, the neuron that is indicated in Fig. 16.2A by the arrow has been microinjected with the water-soluble dye Lucifer Yellow. Diffusion of the dye inside the cell permits tracing the various processes corresponding to the same cell.

At specialized junctions, the *synapses,* a nerve impulse in one cell can be communicated to another neuron. Synapses are usually on the surface of the dendrites or on the cell body but may occasionally be on the axon of the cell receiving the impulse. There are also synaptic connections between nerve fibers and effector organs, such as muscle (smooth, cardiac, and striated) and glands. The mechanism of synaptic transmission may be quite different depending on the synapse, as discussed later in the chapter. In some synapses, the transmission is electrical through specialized junctions (the *gap junctions*), resembling the transmission in the nerve fibers themselves (*electrical synapses*). In others, the *chemical synapses,* the transmission is carried out by the release of a neurotransmitter, which then interacts with the postsynaptic cell and may have an excitatory or an inhibitory effect.

In invertebrates, the cell bodies are on the outside layer of the ganglia (the *rind*), whereas axons, synapses, and dendrites are in the *core* of the ganglia. In the vertebrate central nervous system, the cell bodies are in the *gray matter* and the axons predominantly in the *white matter*. In the vertebrate nervous system, in portions that are considered more primitive, the gray matter is frequently on the inside of the tissues (e.g., in the cerebellum); in higher brain regions, the gray matter is on the outside (e.g., in the cerebral cortex).

Many invertebrates and some primitive vertebrates have giant axons. In some organisms each giant axon is formed from a single cell. In others, it is formed by the fusion of axons from many separate cells. Giant axons such as those of the squid can be as much as 1 mm in diameter. Their size and hardiness when excised have made them a favorite experimental preparation.

In ganglia and in tissues of the central nervous system, specialized cells (*neuroglia*) form sheaths around one or more cells and also act as packing between the cells. They probably have an important maintenance role. In vertebrates, an insulating layer of structured lipoprotein, called *myelin,* surrounds the larger axons. The myelin layer is interrupted every millimeter or so by deep constrictions or breaks, called the *nodes of Ranvier,* where impulse conduction takes place. In some axons, the covering is simpler and may consist of a single layer of glial cells.

In most vertebrate nerves, the axons are held together in bundles that are enclosed in sheaths. A single bundle of nerves may contain fibers from neurons with very distinct functions. For example, some of the axons may be from *motor* neurons, which control muscle contraction, and others may be from *sensory* neurons, which transmit information from a receptor. Similar arrangements can also occur in invertebrates, but these animals usually have far fewer nerve fibers.

In most cells there is an electrical potential difference (the *resting potential*) between the internal cytoplasm and the external environment. Generally the inside is negative relative to the outside; the cell membrane is said to be *polarized*. In excitable cells, this potential is poised to permit the release of a *nerve impulse* at a speed unmatched by most other cellular events. This nerve impulse, once initiated, is propagated along the axon without decrement.

For purposes of discussion, we shall regard a nerve impulse as the discharge of the stored (resting) electrical potential (a *depolarization*). As we shall see, it is actually much more complex than this and is due to a transient reversal of polarization (the inside becomes positive).

Resting potentials and changes therein can be examined fruitfully in terms of ionic gradients, ionic channels, and ionic movements across the membrane, although some of the mechanisms underlying these events still remain obscure.

II. IONIC ORIGINS OF THE RESTING POTENTIAL

When two different concentrations of the same salt come in contact with each other, there is a net flow of both ions from the area of high concentration to that of low concentration. If the mobilities of the two component ions differ significantly, the cation and the anion tend to separate. However, the actual amount of separation is limited because the electrical potential set up by the separation opposes the diffusional forces by accelerating the slower ion. This tendency of one of the component ions to get ahead of the other results in the production of a potential (diffusion or *junction potential*). The charge of the more dilute phase corresponds to that of the more mobile ion, which moves ahead of the oppositely charged ion.

For the passage of 1 equivalent of charge from one phase (phase 1) to the other (phase 2), the change in free energy is

$$\Delta G = t_+ RT \ln \frac{\overset{+}{a}_2}{\overset{+}{a}_1} + t_- RT \ln \frac{\overset{-}{a}_2}{\overset{-}{a}_1} \tag{16.1}$$

where a refers to the activity of the ion in question and generally can be approximated by the concentration. The subscript 1 or 2 refers to the phase, t is the transference number defined in terms of the mobility of the ions (u)

$$t_+ = \frac{u_+}{u_+ + u_-} \tag{16.1a}$$

$$t_- = \frac{u_-}{u_+ + u_-} \tag{16.1b}$$

and the subscript signs (+ and −) refer to the charge of the ion.

As discussed in a previous chapter, $\Delta\Psi$, the electrical potential, corresponds to $\Delta G/zF$. The diffusion potential $\Delta\Psi_d$ will therefore take the form

$$\Delta\Psi_d = t_+ \frac{RT}{zF} \ln \frac{\overset{+}{a}_2}{\overset{+}{a}_1} + t_- \frac{RT}{zF} \ln \frac{\overset{-}{a}_2}{\overset{-}{a}_1} \tag{16.2}$$

Since in each phase the concentration of the cation must be virtually the same as that of the anion, $\overset{+}{a}_2 = \overset{-}{a}_2$ and $\overset{+}{a}_1 = \overset{-}{a}_1$. For monovalent ions, the relationship can therefore be written in a simpler form:

$$\Delta\Psi_d = (t_+ + t_-) \frac{RT}{F} \ln \frac{a_2}{a_1} \tag{16.3}$$

Accordingly, Eq. (16.3) can be represented as

$$\Delta\Psi_d = (1 - 2t_-) \frac{RT}{F} \ln \frac{a_2}{a_1} \tag{16.4}$$

When the mobility of one of the ions is much greater than that of the other, only the faster ion needs to be considered. Supposing that u approaches 0, then t approaches 0 [Eq. (16.1b)] and $1 - 2t$ approaches 1, so that $\Delta\Psi_d$ is now independent of the mobilities. This point becomes rather significant in relation to biological potentials. Where the mobility of one ion surpasses that of the other, Eq. (16.4) becomes (16.5), the familiar *Nernst equation,* which we have encountered before.

$$\Delta\Psi_d = \frac{\mathrm{RT}}{\mathrm{zF}} \ln \frac{a_2}{a_1} \tag{16.5}$$
$$= -58 \text{ mV} \log_{10} \frac{a_2}{a_1} \qquad \text{for} \quad T = 20°\text{C}$$

The resting potentials of most cells are governed by the principles that have been outlined. Generally, the permeability of the membrane to K^+, equivalent to the mobility, and hence t is far greater than the permeability to the other ions present; therefore Eq. (16.5) can be used and a in the equation becomes the concentration of K^+. For example, in frog sartorius muscle the permeability to K^+ is 100 times greater than to Na^+, and in the freshwater alga *Nitella* it is 52 times greater.

In some freshwater cells, such as the algae *Nitella* and *Chara,* the internal concentrations of both Na^+ and K^+ are much higher than those in the surroundings. However, in most complex multicellular organisms, the total ionic concentrations of intra- and extracellular fluids are nearly equal. Nevertheless, there are steep gradients of ions, notably K^+ (high on the inside) and Na^+ (high on the outside).

The K^+ of the cell's interior tends to leak out and is replenished by the transport activities of the cell membrane. In muscle and nerve, the Na^+ tends to flow into the cell and has to be pumped out. In the steady state, the internal ionic composition is maintained by the balance between the movement of ions in the direction of the electrochemical gradient and the active transport of the ions against the electrochemical gradient, so that the internal $[K^+]$ remains high and the internal $[Na^+]$ low (see Chapter 14).

Clearly, the cell membrane plays a fundamental role in the maintenance of the resting potential. However, as already discussed, a potential difference between two phases does not require the presence of a membrane. Furthermore, when one of the ionic components is restrained (e.g., is part of the cell structure), the mobile ions follow a Donnan distribution and the potential may still be considerable (see Collins and Edwards, 1971) even in the absence of a membrane.

The ionic compositions of the internal and external media of some cells have been studied extensively. The values for frog sartorius muscle are represented in Table 16.1 (Conway, 1957). It is possible to predict the magnitude of the potential between the external and internal phases, the resting potential, by means of Eq. (16.5), remembering the higher mobility of K^+. At 18°C

Table 16.1 Intrafiber Composition[a]

Constituent	Muscle	Plasma	Concentration in fiber water
K	83.8	2.15	124
Na	23.9	103.8	3.6
Ca	4.0	2.0	4.9
Mg	9.6	1.2	14.0
Cl	10.7	74.3	1.5
HCO_2 (including Ba-soluble CO_2)	11.6	25.4	12.4
Phosphate	5.3	3.1	7.3
Sulfate	0.3	1.9	0.4
Phosphocreatine	23.7		35.2
Carnosine	11.0		14.7
Amino acids	6.8	6.9	8.8
Creatine	5.3	2.1	7.4
Lactate	3.1	3.3	3.9
ATP	2.7		4.0
Hexose monophosphate	1.7		2.5
Glucose	0.5	3.9	
Protein	1.5	0.6	2.1
Urea	1.6	2.0	2.0
Water (g/kg)	800	954	
Interspace water (g/kg)	127		

[a]Frog muscle and plasma (moles per kilogram), except where noted.

this value is calculated to be 102 mV with the inside negative. The potential can be measured directly by inserting an electrode with a tip of microscopic dimensions (a microelectrode) into the muscle fibers. With an external concentration of about 2.5 mM K^+, the measured value is 80 to 92 mV, which is not very different from the calculated value.

It is possible to study the resting potential over a wide range of external K^+ concentrations. This has been done by equilibrating the muscle fiber in various concentrations of KCl or of a K^+ salt of a nonpermeable anion (e.g., acetate). The results of such experiments yield the relationship represented in Fig. 16.3 (Conway, 1957). The straight line (line A) corresponds to a slope of 57 mV for a 10-fold change in K^+ concentration. The value predicted from Eq. (16.5) is 58 mV. Thus, the equation has very good predictive value.

Similar results are obtained with nerve and other cells as well. For example, the ionic composition of *Chara* (cell sap) and that of its environment (pond water) are shown in Table 16.2 (Gaffey and Mullins, 1958). The resting potential, measured directly, is -181 mV. Since the permeability to K^+ far outstrips that to Na^+ or Cl^- (Table 16.3), Eq. (16.5) can be used; the potential calculated from these values is -184 mV, in good agreement with the actual measurement.

Clearly, the results indicate that in general the differential distribution of K^+ between the intracellular and extracellular phases is responsible for the resting potential.

III. DYNAMICS OF THE MEMBRANE POTENTIAL

The resting potential of nerve and muscle fibers was shown to be a reflection of the differential mobility of ions between the internal and external phases of the fibers. The resting potential, therefore, critically depends on the permeability properties of the cell membrane.

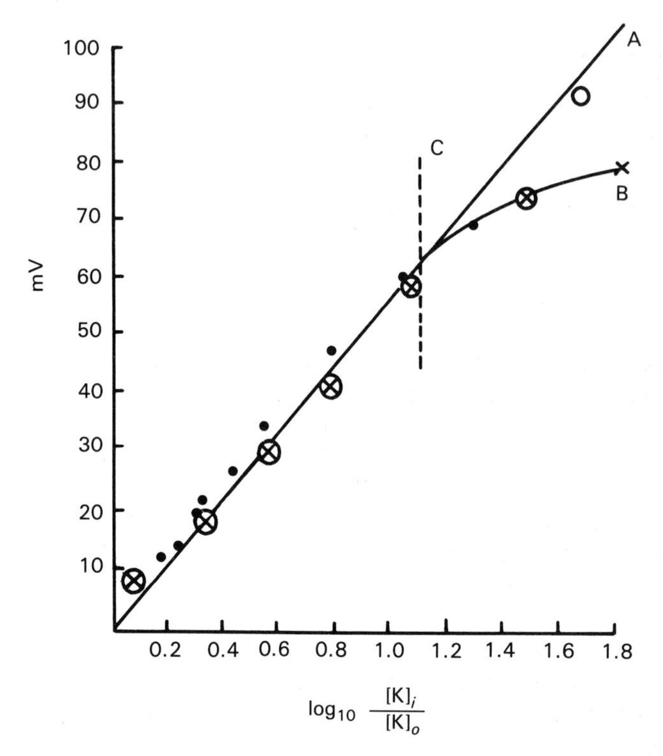

Fig. 16.3 Mean E_m values of frog sartorii immersed in Ringer-Barkan fluid with addition of external KCl. (●) Averages of results after overnight immersion at 0–3°C, then brought to room temperature. (⊗) Immediate results from isotonic mixtures with acetate ion replacing chloride and bicarbonate. (○) Average of observations taken immediately with Ringer-Conway fluid containing 2.5 mM K^+. The vertical line C at 1.11 on the abscissa represents the level of 10 mM K^+ in the external fluid. Line A corresponds to the theoretical expectation. Reprinted with permission from E. J. Conway, *Physiological Review,* 37:84–132.

The permeability of cells can be studied by following the entry or exit of the ions with time. In addition, the electrical resistance of the membrane is related to the permeability of the membrane to ions. This resistance can be determined by measuring the voltage produced by a current pulse between a microelectrode inserted into the cytoplasm and a reference electrode in the medium. An analogous but more complex method allows calculation of the resistance of the membrane by the passage of alternating currents through cell suspensions. Resistance measurements obtained with a number of cells are shown in Table 16.4, expressed per unit area

Table 16.2 Steady-State Concentrations of Ions in Sap of *Chara*

Ion	Concentration (mM) ± SE	
	Pond water	Sap
K^+	0.046	65 ± 1.9
Na^+	0.15	66 ± 1.9
Cl^-	0.04	112 ± 1.5

Reprinted with permission from C. T. Gaffey and L. J. Mullins, *Journal of Physiology,* 144:505–524.

Table 16.3 Resting Ion Fluxes in *Chara*

Ion	External conc. (mM)	Influx/mM ($pmol/cm^2$ s)
K^+	1.4	2.0
Na^+	10.0	0.07
Cl^-	35.0	0.05

Reprinted with permission from C. T. Gaffey and L. J. Mullins, *Journal of Physiology,* 144:505–524.

Table 16.4 Membrane Resistances

Material	Resistance (Ω cm^2) At rest	Resistance (Ω cm^2) With activity	Reference
Nitella	10^5	5×10^2	*a*
Loligo axon	10^3	2.5×10	*b*
Chara	2×10^5	2.0×10^3	*c*

[a]Cole and Curtis (1938).
[b]Cole and Curtis (1939).
[c]Gaffey and Mullins (1958).

(square centimeter) of membrane. The resistance of either the external medium or the cytoplasm is much lower than that of the cell membrane. The resistance measurements therefore indicate the presence of a specialized structure of high electrical resistance and hence low permeability to ions at the surface of the cells.

Many experiments support the view that the cell membrane has an important role in excitation. Most of the axoplasm can be squeezed out from one end of a cut giant axon and be replaced with an artificial medium, such as a KF solution. Yet the excitability and the potentials of the axon remain undisturbed as long as the membrane remains undamaged. The importance of phospholipid components of the membrane is shown by treatment of axons with phospholipase C, an enzyme that hydrolyzes lecithin and abolishes the resting potential (Tobias, 1958).

A diagrammatic representation of the potential difference across the membrane (+ on the outside and − inside), the resting potential, is shown in Fig. 16.4*a*. A nerve impulse or *action potential* involves in part a depolarization, represented in Fig. 16.4*b*. Nerve conduction corresponds to this depolarizing event and the propagation of depolarization.

The action potential can be initiated by an electric current at the site of the negatively

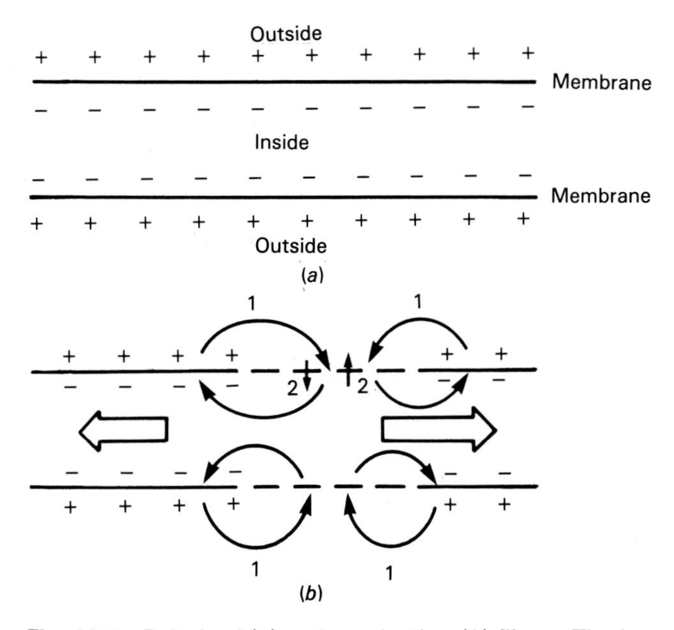

Fig. 16.4 Polarized (*a*) and conducting (*b*) fibers. The large arrows represent the direction of the propagation after artificial depolarization. The small arrows indicate the passage of ions in the longitudinal direction (1) and across the cell membrane (2).

charged electrode (*cathode*) placed at the outer surface of the nerve fiber. These events can be observed most readily with appropriate electrical amplification and recording equipment such as amplifiers and an oscilloscope.

A stimulus must be above a minimal value (the *threshold*) to elicit an action potential. It is possible to study in detail what happens to the membrane potential with stimuli that vary in intensity above and below threshold (Fig. 16.5*a*) (Hodgkin, 1939). The results for squid axon represent the potentials recorded at the cell surface at the site of an external cathode (negatively charged electrode, curves above) and at the anode (positively charged electrode, curves below the axis). The ordinate is in units relative to the action potential taken as unity. At very low intensity, the pattern is the same at the two stimulating electrodes. Naturally, the potentials are opposite in sign. There will be an enhancement of the potential difference between the inside and the outside at the anode. The system is said to be *hyperpolarized*. Under the cathode, a partial depolarization will take place. However, at higher intensities (curves 6, 7, 8, etc.) the curves representing the potential at the cathode change in shape. There is a depolarization beyond the direct electrode effect that is greater in both magnitude and duration. The responses of the nerve to these depolarizations can be shown by subtracting the direct effect of the stimulus (which mirrors the anodal response) from the total depolarization. This difference is shown in Fig. 16.5*b*. When the stimulus produces a sufficiently high depolarization, the potential is unstable and can give rise to an action potential (e.g., curves 10–12). The action potential is self-sustained, since it is independent of the input of the stimulating current, and it is also self-propagated. It turns out to be the same in magnitude at all points along the fiber.

A propagated wave of depolarization (or action potential) can be recorded from a nerve or muscle following each stimulation above threshold, provided that the interval between the stimuli is greater than the *refractory period*. This is the period during and immediately after an action potential when new action potentials cannot be elicited because the channels that allow Na^+ to enter are inactivated (see Section III,C). An action potential recorded from an electrode inside the squid axon is shown in Fig. 16.6. The zero level on the scale represents the point at which there is no potential difference between the inside and the outside of the fiber; i.e., there is no potential difference across the membrane. The initial level (-70 mV) is the resting potential, where the inside is negative in relation to the outside. The upward swing that follows is the action potential or spike.

In addition to involving a depolarization, the action potential reverses the polarity of the fiber (Fig. 16.6). Then the resting potential is rapidly reestablished after a period of hyperpolarization (Hodgkin, 1951) and the nerve can be stimulated again. Therefore, a process must exist that repolarizes the fibers very rapidly. A discussion of the ionic basis for the depolarization phenomenon and the repolarization follows. Analogous events take place in muscle.

The depolarization underlying the nerve impulse (the action potential or spike) causes a flow of current from the depolarized areas to the adjacent polarized areas. This current flow depolarizes the polarized region, setting up an action potential there. In this way an action potential is propagated, or conducted, along an axon. The heavy arrows of Fig. 16.4 represent the direction of propagation of the action potential, and the broken line represents the change in membrane resistance. Normally, the action potential is conducted in a single direction since it originates from the cell body. Because the portion behind the depolarized area is in the refractory state, the impulse cannot travel backward.

A. Ionic Basis of Depolarization

Depolarization might be partially explained by proposing that the permeability of the membrane to all ions increases during the action potential so that the flux of ions causes the electrochemical

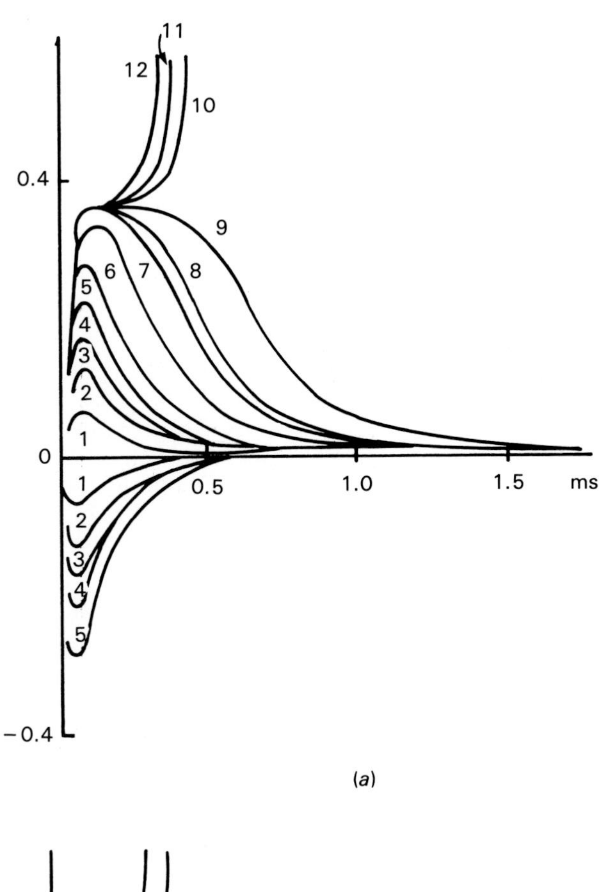

(a)

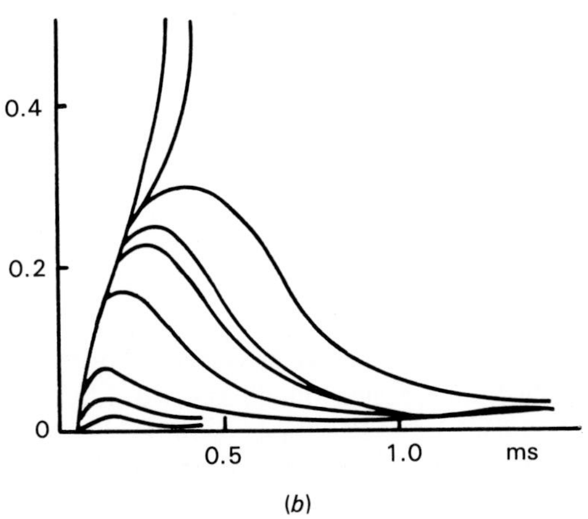

(b)

Fig. 16.5 (a) Electrical changes at stimulating electrode produced by shocks with relative strengths, successively from above, 1.00 (upper 6 curves), 0.96, 0.85, 0.71, 0.57, 0.43, 0.21, −0.21, −0.43, −0.57, −0.71, and −1.00. The ordinate scale gives the potential as a fraction of the propagated spike, which was about 40 mV in amplitude. The 0.96 curve is thicker than the others because the local response had begun to fluctuate very slightly at this strength. The width of the line indicates the extent of the fluctuation. (b) Responses produced by shocks with strengths, successively from above, 1.00 (upper 5 curves), 0.96, 0.85, 0.71, and 0.57; obtained from curves in (a) by subtracting anodic changes from corresponding cathodic curves. Two of the anodic curves necessary for this analysis were recorded but are not shown in (a). Ordinate, as in (a). From A. L. Hodgkin, *Proceedings of Royal Society Series B.,* 148:1–37, with permission. Copyright © 1958 The Royal Society, London.

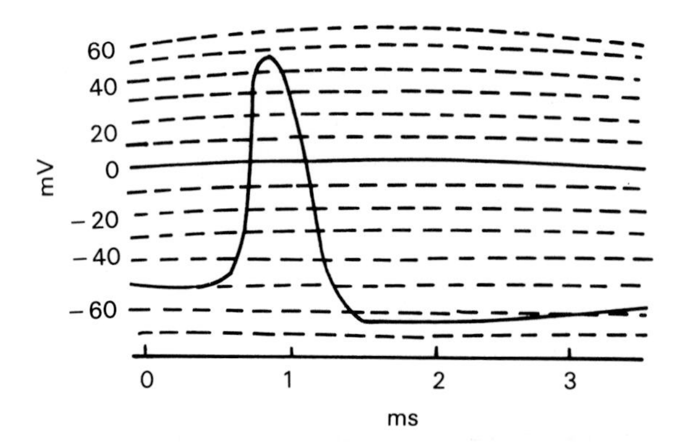

Fig. 16.6 Resting and action potential in giant squid axon at 18.5°C. Reprinted with permission from A. L. Hodgkin, *Biological Review of Cambridge Philosophical Society,* 26:339–409.

gradient to collapse. Measurements of resistance do support this view (see Table 16.4). The resistance during the peak of the action potential is at best a small fraction of the resting resistance. However, an increase in the permeability to all ions would cause the membrane to move toward 0 mV and would not account for the overshoot, which is typically to +50 mV. Moreover, experiments in which different external ions are used show that the entry of Na^+ alone accounts for the depolarization. The number of moles of ion required for a given change in potential is given by

$$\text{Moles} = (C\,\Delta\Psi_m)F \tag{16.6}$$

where C is the membrane capacitance in farads, or amount of charge (coulombs) across the membrane per volt, $\Delta\Psi_m$ is the maximum change in membrane potential during the rising phase of the impulse, and F is Faraday's constant (96,500 coulombs per mole of monovalent ion). Since the action potential of the axon corresponds to about 100 mV and the membrane capacitance is about 1.5 $\mu F/cm^2$, the flux of Na^+ cannot be less than 1.6×10^{-12} mol/cm^2. This can only be a minimum value, since it ignores the possibilities of accompanying leakage of K^+ or entrance of Na^+ during the falling phase of the action potential.

The quantity of Na^+ taken up per stimulus is so small that it cannot be detected; however, it can be calculated when the total uptake resulting from repeated stimuli is divided by the number of impulses. In the giant axon of the squid, *Loligo,* 3.5×10^{-12} mol/cm^2 of Na^+ is taken up per nerve impulse. From these considerations it seems likely that the entry of Na^+ suffices to account for the action potential. This conclusion is reinforced by the observation that Na^+ in the medium is required to achieve the action potential. Table 16.5 summarizes similar results obtained for other systems.

These results imply that the membrane permeability to Na^+ during the rising phase of the action potential is greater than the permeability to other ions, and the relationship between action potential and $[Na^+]$ should be quantitatively predictable from Eq. (16.5) using the appropriate $[Na^+]$. This is shown more clearly by Eq. (16.7), where $[Na^+_{ext}]$ is the external Na^+ level, which is changed experimentally, and $[Na^+_{st}]$ is the normal value in the extracellular medium.

$$\Delta\Psi_{ext} - \Delta\Psi_{st} = (58\text{ mV})\log_{10}\frac{[Na^+_{st}]}{[Na^+_{ext}]} \tag{16.7}$$

In Eq. (16.7) it is assumed that the internal Na^+ level is not changed by this manipulation of the external medium.

Table 16.5 Na^+ and K^+ Exchanges During Nerve Excitation

Material	(1) Na^+ influx/impulse (10^{-12} mol/cm^2)	(2) K^+ efflux/impulse (10^{-12} mol/cm^2)	Reference
Carcinus maenas	—	1.7	*a*
Carcinus maenas	—	2.5	*b*
Sepia officinalis	—	3.4	*c*
Sepia officinalis	3.7	4.3	*d*
Sepia officinalis	3.8	3.6	*e*
Loligo forbesi	3.5	3.0	*e*
Loligo pealli	4.5	—	*f*
Loligo pealii	4.4	—	*g*

[a]Hodgkin and Huxley (1947).
[b]Keynes (1951a).
[c]Weidmann (1951).
[d]Keynes (1951b).
[e]Keynes and Lewis (1951).
[f]Rothenberg (1950).
[g]Grundfest and Nachmansohn (1950).

Results for different tissues are shown in Fig. 16.7 (Hodgkin, 1951). Curve 1 of each graph represents the resting potential, which is little affected by the Na^+ concentration. Curve 2 represents the results calculated from Eq. (16.7). The predictions are close, although significant deviations do occur for the squid giant axon. These deviations may result from the approximations assumed in deriving Eqs. (16.5) and (16.7). All the results seem to support the idea that Na^+ is responsible for carrying the depolarizing current during the rising phase of the action potential.

Action potentials also take place in freshwater algae such as *Nitella* and *Chara*. The functional significance of these potentials is not clear, although they seem to be related to the movements of the cytoplasm. Since the pond water is almost free of ions, the action potential can be carried only by the efflux of an internal anion. An examination of the dependence of the action potential on the external concentration of Cl^- (analogous to that of Fig. 16.7) is shown in Fig. 16.8 (Gaffey and Mullins, 1958), in which the broken line (curve 2) represents the theoretical 58-mV slope. Curve 1 represents the resting potential, which remains unaffected by changing $[Cl^-]$. As curve 2 shows, the magnitude of the action potential (shown by the circles) conforms well to the expectations of theory. In these experiments choline chloride was used, since choline enters only very slowly.

The efflux of Cl^- during the action potential can be calculated from Eq. (16.6), as done previously for the squid axon. Considering the action potential to be about 200 mV and the capacitance to be about 1 μF/cm^2, the minimal amount necessary to carry the action potential is 2×10^{-12} mol of Cl^- per impulse. The measured efflux is about $10{,}000 \times 10^{-12}$ mol per impulse, well in excess of the calculated value. This excess may result from simultaneous efflux of K^+ and Cl^- during the prolonged experimental period required to follow the slow action potential of *Chara*.

For nerve and muscle, the reversal of polarity and the Na^+ permeation can be summarized as in Figs. 16.9 and 16.10. The resting potential is again represented by the + and − charges. The arrow shown perpendicular to the surface indicates the Na^+ influx.

As mentioned, the depolarization of the action potential is followed almost immediately by repolarization of the axon (Fig. 16.6). In fact, the whole cycle of depolarization and repolarization generally takes place in 1 ms or so. Repolarization would be accomplished most rapidly by removal of the excess positive charge that has entered the nerve or muscle cell. This could be done most simply by a rapid efflux of K^+ in the direction of the electrochemical gradient. The efflux of K^+ shown in Table 16.5 (column 2) is of the same order of magnitude as that of Na^+, as required by the fact that the two should represent equal but opposite phenomena. The changes

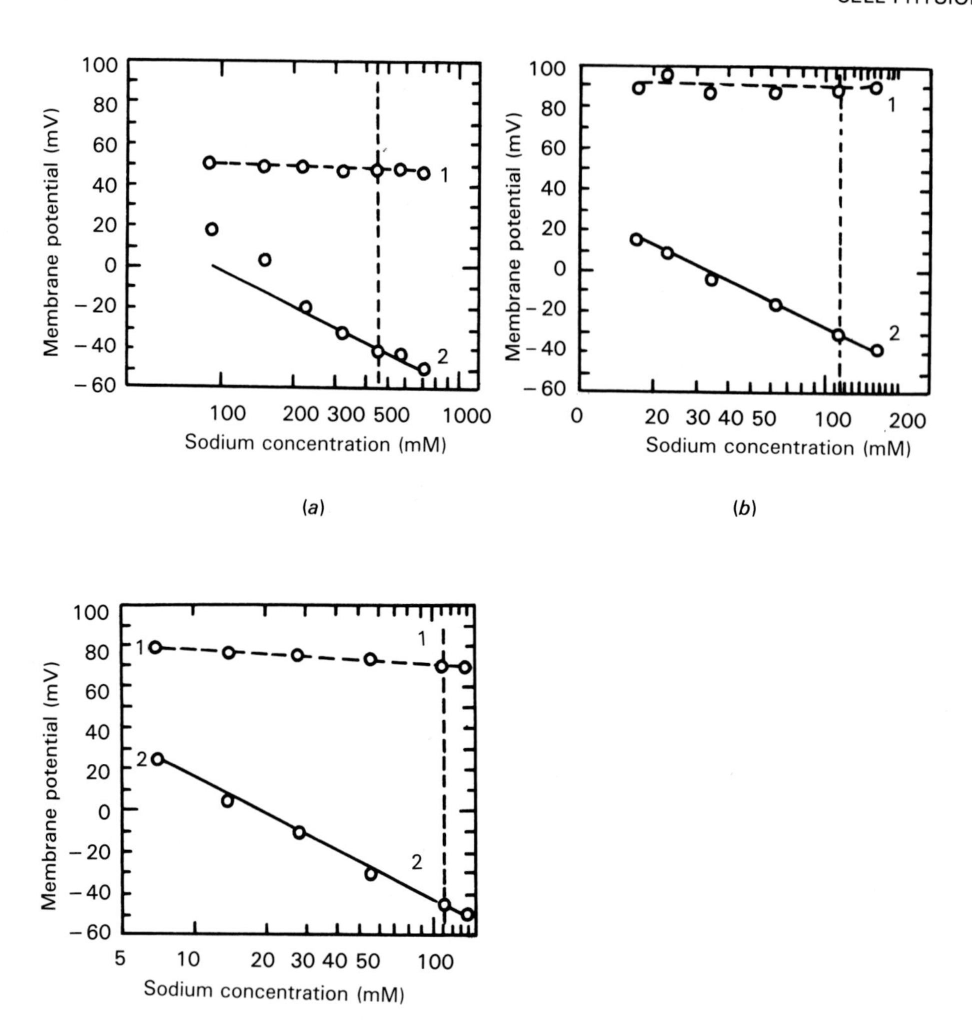

Fig. 16.7 Relation between sodium concentration in external solution and potential difference across resting and active membrane. (*a*) Squid giant axon (from work of Hodgkin and Katz); (*b*) frog sartorius muscle (from work of Nastuk and Hodgkin); (*c*) frog myelinated nerve (from work of Huxley and Stampfli). Abscissa: sodium concentration on logarithmic scale (dashed line shows concentration in Ringer's fluid or seawater). Ordinate: potential difference across membrane (outside potential minus inside potential) at rest (1) and at crest of action potential (2). The solid line is drawn with a slope of 58 mV for a tenfold change in sodium concentration. The points in these curves were obtained by adding the original author's values for the resting potential or for the reversed potential difference across the active membrane. Reprinted with permission from A. L. Hodgkin, *Biological Review of Cambridge Philosophical Society,* 26:339–409.

in permeability for specific ions are likely to be the result of opening and closing of protein-lined channels (Keynes, 1972), as discussed in Section III,C.

Resting and action potentials seem to be governed by similar principles in all these conducting systems, even if the depolarizing action potential of *Chara* is carried by an outward flow of Cl^-. The values listed in Table 16.6 show that the efflux of K^+ is well in excess of the Cl^- efflux and can account for the recovery in *Chara*.

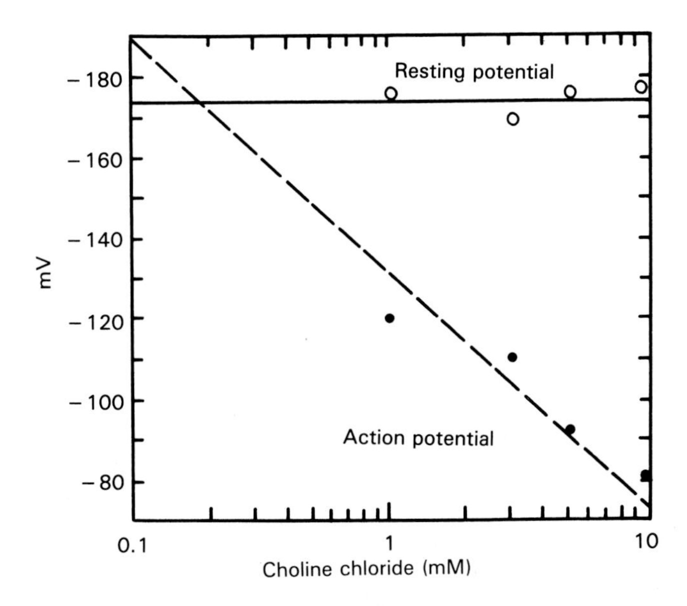

Fig. 16.8 Magnitude of the resting and action potentials in various concentrations of choline chloride and sucrose solutions. The sucrose concentration at any point is 200 mM − 2 × choline chloride concentration. The dashed line has a slope of 58 mV for a tenfold change in Cl^- concentration. Recording is with a microelectrode in the sap. Semilog scale. Reprinted with permission from C. T. Gaffey and L. J. Mullins, *Journal of Physiology,* 144:505–524. Copyright © 1958 The Physiological Society, Oxford, England.

Both Na^+ and K^+ can shift across the nerve membrane with action potentials with little change in ionic concentrations. We have seen that many stimuli are needed to detect any change. The K^+ loss associated with one stimulus corresponds to about 4×10^{-12} mol/cm^2 (Table 16.5). Since the axon is about 500 μm in diameter, the loss per liter of axoplasm is approximately 1 $\times$ 10^{-5} mol, less than 1 part in 10,000. Although the loss is small, eventually work must be carried out to restore the internal K^+ and Na^+ levels. In Chapter 14 we discussed a transport system responsible for pumping Na^+ out and pumping K^+ in. It is this system that is involved in the maintenance of the internal ionic environment.

Although the Na^+ currents generally underlie the action potential in most animal tissues, there are exceptions. Crustacean muscle depends on Ca^{2+} for conduction. This is most readily demonstrable in muscles in which the K^+ channels have been blocked (e.g., with TEA). The involvement of Ca^{2+} fluxes was demonstrated with ^{45}Ca using giant muscle fiber from a barnacle in which the internal Ca^{2+} was reduced. During induced action potentials the Ca^{2+} was found to correspond to 2–6 pmoles/μF where 0.5 are needed to depolarize the membrane by 100mV (Hagiwara and Naka, 1964). As demonstrated later, most cells appear to have Ca^{2+}-channels, which play significant roles.

Table 16.6 Na^+, K^+, and Cl^- Exchanges During Excitation in *Chara*

Material	Na^+ efflux/impulse (mol/cm^2 × 10^{-8} ± SE)	K^+ efflux/impulse (mol/cm^2 × 10^{-8} ± SE)	Cl^- efflux/impulse (mol/cm^2 × 10^{-8} ± SE)
Chara	0	3.1 ± 1	0.9 ± 1

B. Membrane Mechanisms: Voltage Clamping

The model of the behavior of the membrane of an axon can be outlined simply:

1 The unequal distribution of K^+ is responsible for the resting potential.

2 A small depolarization beyond a critical value leads to a change in permeability to Na^+, which in turn leads to Na^+ flow into the fiber. This inflow is responsible for the reversal of the resting potential. The initial depolarization could be the result of an electrical stimulus or the depolarization of an adjoining area.

3 The immediate recovery of the resting potential is the consequence of a brief increase in the permeability to K^+, causing movement of K^+ in the direction of the electrochemical gradient.

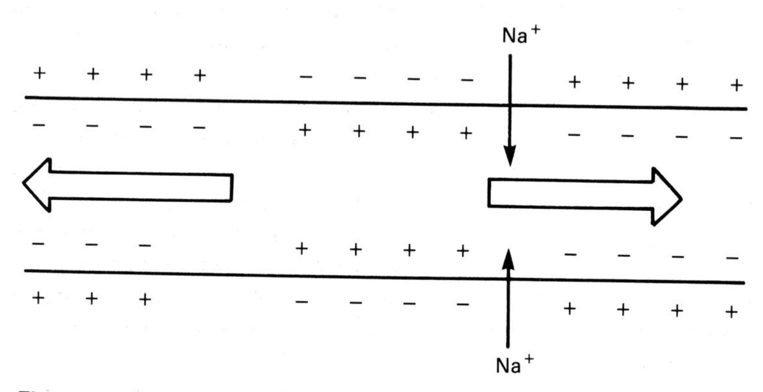

Fig. 16.9 Diagram representing the membrane potential during the action potential (center) and at rest.

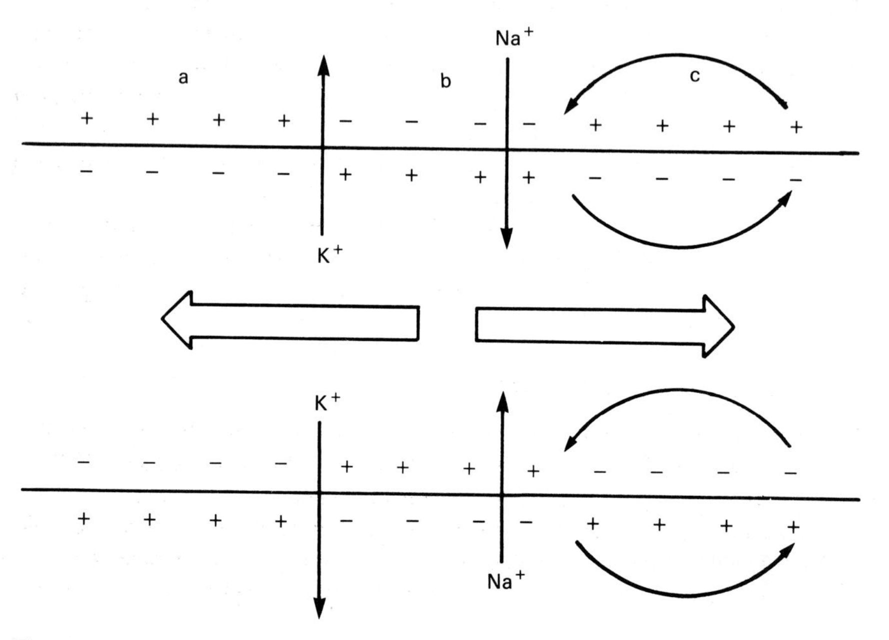

Fig. 16.10 Diagram representing the ionic basis of propagation of the action potential followed by recovery.

If this model is correct, it seems possible to manipulate the electrical potentials of the axons and to test these premises independently. The potential difference between the inside and the outside can be set at any level by means of an external electrode, a microelectrode implanted in an axon, and appropriate electronics. In the *voltage clamp* technique, the membrane potential is kept constant at any desired level by passing a current equal and opposite to that generated by the flow of ions across the membrane.

With this method, it has been possible to test whether the potentials per se trigger the changes in flux necessary for the polarization-depolarization cycles. Increases in the potential difference (hyperpolarization) cause an inflow of current. This is expected from the passive resistive properties of the membrane. A small decrease in the potential causes an outflow of current, as expected. A decrease in the potential difference beyond a critical value has a very different effect (Fig. 16.11) (Hodgkin, 1958). There is initially a large inward flow of current (dashed line, Fig. 16.11*a*); however, it is quickly followed by an outward current (full line), which in the absence of a clamp would repolarize the nerve.

The inward current should correspond to the Na^+ influx that would take place during depolarization (in the absence of a clamp). Experimental tests of this point show that this current is, in fact, critically dependent on the presence of Na^+; replacement of the Na^+ in the external fluid by choline (which does not penetrate) blocks this event completely (Fig. 16.11*b*). The outward current that follows is likely to be carried by K^+, which normally would repolarize the nerve. A test of whether K^+ is involved can be carried out by measuring the current flow and the K^+ efflux simultaneously. The results of this experiment are shown in Fig. 16.12 (Hodgkin and Huxley, 1953). The efflux of K^+ and the outflow of current correspond quantitatively.

These potentials can be illustrated in a rather convenient form as shown in Fig. 16.13*a* (Narahashi et al., 1964). The inflowing current (the Na^+ current) and the outflowing current (the K^+ current) can be plotted as a function of the voltage at which the membrane potential is clamped. The current can be corrected for leakage by assuming that the leakage current is a linear function of the potential (dashed line) and subtracting it from the total current.

A meaningful application of this system is shown in Fig. 16.13*b*, where the Na^+ passage is blocked by the toxic drug tetrodotoxin. Tetrodotoxin is a toxin extracted from certain organs of the puffer fish (the poison of the *fugu* fish of James Bond fame). The Na^+ current is completely blocked, whereas the K^+ current remains unaffected.

In contrast to tetrodotoxin, tetraethylammonium interferes with the passage of K^+ and hence with the recovery of the resting potential after stimulation (not shown).

The results we have examined support the idea that the phenomenon of excitation can be explained entirely by underlying ionic currents. However, the ionic currents must be a reflection of events occurring in the structure of the membrane, and these events are beginning to be studied in detail.

C. Molecular Mechanisms and Channels

The lipid portion of the membrane is not likely to allow the passage of ions at the high rates that have been measured. The activation energy is of the order of 250 kJ/mol (Parsegian, 1969). In contrast, passage through channels would lower the activation energy, for K^+, to about 20 kJ/mol (Frankenhaeuser and Moore, 1963). Furthermore, it is difficult to visualize the lipid components being regulated to vary the passage of ions in response to a membrane potential. For these reasons, the presence of protein-lined channels has been considered for some time.

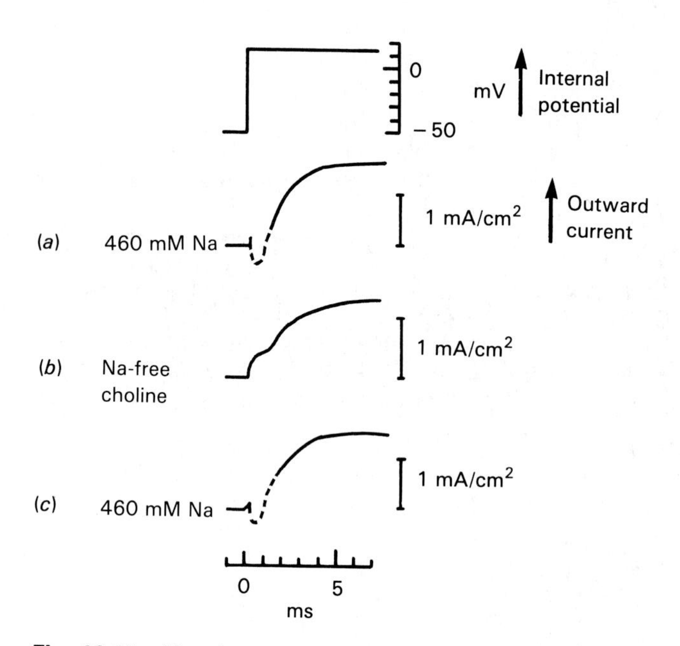

Fig. 16.11 Membrane currents associated with depolarization of 65 mV in presence and absence of external sodium ions. The change in membrane potential is shown at the top; the lower three records give the membrane current density; 11°C, outward current and internal potential shown upward. From A. L. Hodgkin, *Proceedings of Royal Society Series B,* 148:1–37, with permission. Copyright © The Royal Society, London.

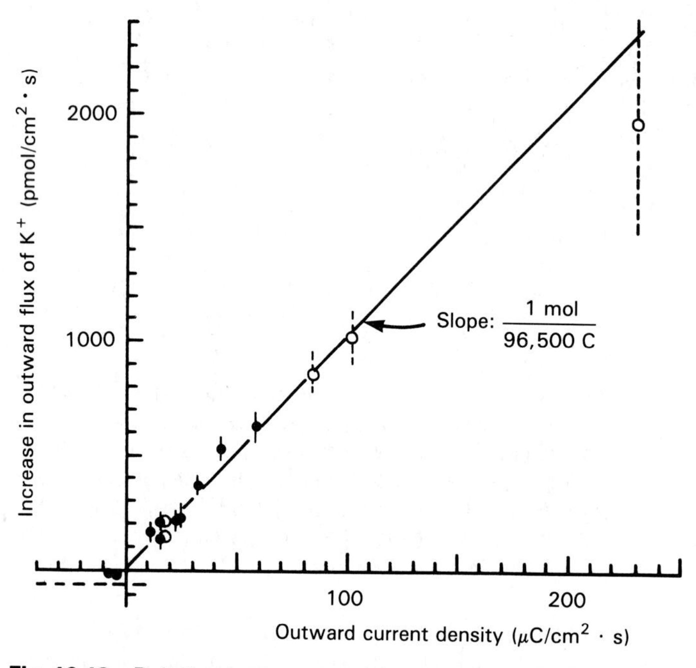

Fig. 16.12 Relation between membrane current density and potassium efflux when a *Sepia* axon is depolarized. The axon was depolarized by an applied current for periods of 60 to 600 s. Vertical lines show $\pm 2 \times$ SE; the horizontal line is drawn at a level corresponding to complete suppression of the average resting efflux. From A. L. Hodgkin and A. F. Huxley, *Journal of Physiology,* 121:403–414, with permission. Copyright © 1953 The Physiological Society, Oxford, England.

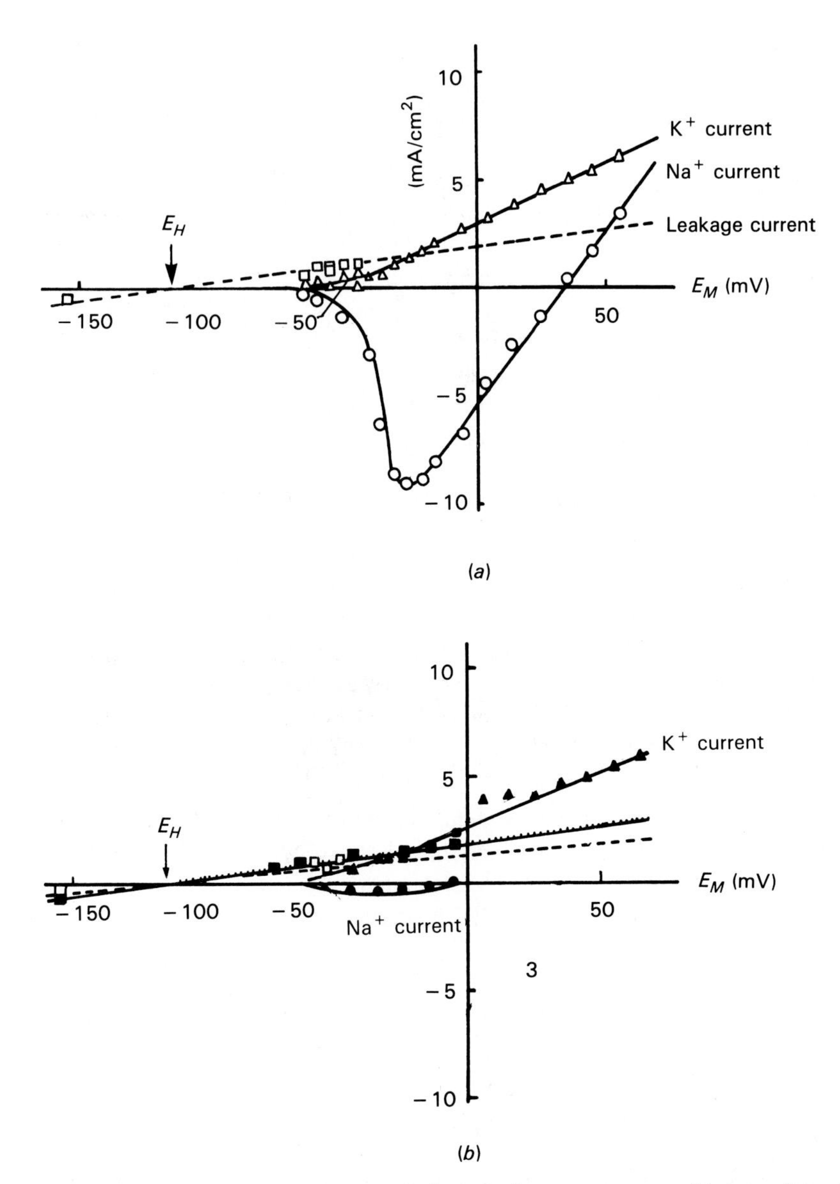

Fig. 16.13 (*a*) Current-voltage relations before treatment with tetrodotoxin. Circles refer to peak Na^+ current corrected for leakage current, triangles refer to steady-state K^+ current corrected for leakage current, and squares refer to leakage current; *I*, designated component of the membrane current (inward direction negative); E_M, membrane potential; E_H, holding potential. (*b*) Current-voltage relations during treatment with tetrodotoxin 3×10^{-5} g/ml. Reproduced from the *Journal of General Physiology*, 47:965–974, 1964, by copyright permission of the Rockefeller University Press.

Membrane channels are integral proteins that span the lipid bilayer. They are generally present as oligomers, and they operate to regulate the passage of ions needed for the propagation of the electrical impulses in excitable cells and for postsynaptic responses to neurotransmitters, which are discussed later. In the latter case, the receptors are coupled to ion channels, which they induce to open when they bind neurotransmitters. Ion channels also have an important role in hormone secretion, visual transduction, transepithelial ion transport, and the activation of contractile mechanisms.

Ion channels display specificity for certain ions, saturation kinetics in relation to ion concentration, competitive inhibition by analogs, and conformational changes (to go from the open to the closed configuration)—all characteristic of enzyme activity. However, they also exhibit behavior that differs from that of enzymes. The rate of ion transport is orders of magnitude higher than enzyme turnover, and the temperature dependence of the ion passage is very low (Latorre and Miller, 1983). As we shall see, proteins that have the appropriate characteristics have been isolated and reconstituted into bilayers.

Channels have been studied in their native state through a relatively new technique (patch clamping) by which a small heat-polished pipette is sealed against the cell membrane. In effect, a small patch of membrane has been isolated and can be electrically studied independent of the rest of the membrane. This tight seal has been referred to as a *giga* seal (*giga* meaning a billion, referring to the fact that the pipette and the sealed patch together have a resistance in the range of 10^9 to 10^{11} Ω). When the voltage is clamped, a recording of the current permits the study of a few channels or even a single channel in the membrane (Hamill et al., 1981). The relationship between voltage and current can be studied over a wide range of clamped voltages. In the neurotransmitter-regulated channels of the synapse, discussed later in this chapter, the effects of agonists (activators) and antagonists (blockers) can be studied at the level of the simplest unit. The relationship between current and clamped voltage is shown in Fig. 16.14 (Hamill et al., 1981). The larger the voltage, the larger the currents in a linear relationship, reflecting a constant conductance (reciprocal of resistance). Note that at any given voltage, the levels of current in all openings are constant and therefore characteristic of those channels.

It is also possible to isolate the patch mechanically in either an inside-out or right-side out (i.e., outside-out) configuration or to rupture the patch, providing a direct connection between the cytoplasm and the inside of the pipette.

The records of Fig. 16.14 show that the individual channels open or close for variable periods of time. It should be possible then to examine the possible states in relation to the state of polarization of an excitable cell by using voltage clamping. Would the behavior of the channels correspond in some way to the high Na^+ or K^+ conductance at the appropriate voltage?

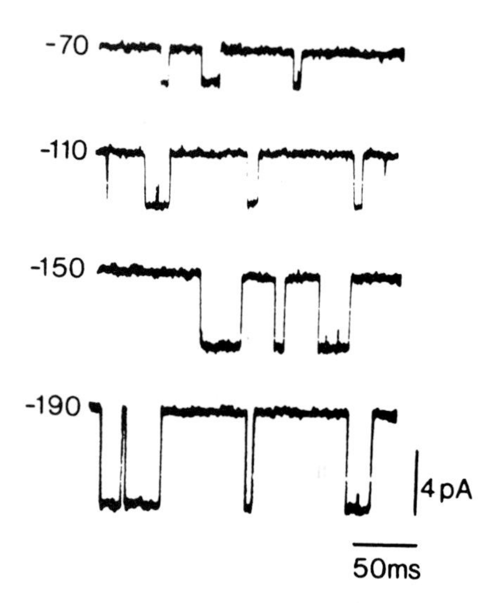

Fig. 16.14 Control of voltage and of ionic environment in the pipette after formation of a giga seal. Single-channel current recordings. From O. P. Hamill, et al., *European Journal of Physiology,* 391:85–100, with permission. Copyright © 1981 Springer-Verlag, Heidelberg.

The ability to open or close in response to voltage changes across the membrane is known as voltage gating. Voltage gating is presumably the result of conformational changes in the proteins constituting the channels. Many different kinds of channels that differ in ionic specificity have been recognized. Most of these are voltage gated.

The voltage dependence of the conductance under voltage clamp conditions is represented at the top of Fig. 16.15 in an excitable cell in tissue culture, a bovine chromaffin cell.

The individual channels that are specific for Na^+ open in response to a depolarization. Three different states have been reported: resting states (R), which are present at hyperpolarizing voltages (no ions go through); open states (O), which allow the passage of Na^+ and which have opened as the result of an initial depolarization; and inactivated states (I), which are closed and cannot be opened with further depolarization. The latter correspond to the state during the refractory period.

The Na^+ current represented in Fig. 16.15 can be generated by the behavior of the channels shown under the curve and summarized by the schemes shown at the right in the figure. Figure 16.15*a* corresponds to a long-maintained opening; more channels open in the rising current phase and close shortly afterward. This effect could result from a fast transition to the open state (O) followed by a slow transition to both the resting (R) and inactivated (I) states. Figure 16.15*b* shows how short-lived open states could generate the same pattern. In this case, the rate of opening and returning to the resting state must be high and the rate of conversion to the inactive state low; otherwise repeated opening and closing would not be possible. The open state would nevertheless predominate when the current passage is maximal. Similarly, the pattern could be explained by Fig. 16.15*c*, in which the opening is short-lived as the result of slow transitions from the open to the resting state by rapid inactivation. The last case appears to be correct (Aldrich et al., 1983).

The appropriate behavior for the K^+ channels has also been shown using patch clamping of internally perfused giant squid axons. Since the repolarization of the axon after the action potential depends on an increase in K^+ permeability, we would expect the frequency of opening the K^+-specific channels to increase with depolarization. The results of Fig. 16.16 (Conti and Neher, 1980) show exactly this behavior. In Fig. 16.16*a* the current is shown with increasing depolarization from the bottom of the figure to the top. The voltages at which the patches are clamped are listed at the left of each record. The frequency of opening is minimal when the axon is polarized (see lower two records). It increases with depolarization, and then the channels close again. The increased conductance (i.e., current/voltage) is an indication of the number of channels that are open. Figure 16.16*b* compares the variances observed experimentally (the points) to those calculated from theoretical considerations. The agreement between the points and the line shows that the observed current changes correspond to random open-close transitions of the same channels.

IV. ELECTROGENIC PUMPS

We have seen that the resting membrane potential is the result of the mobility of ions (usually K^+) through the cell membrane. Since the passage of Na^+ and K^+ through cell membranes is in part controlled by active transport, it is possible that transport events also become reflected in a membrane potential. In fact, hyperpolarizations have been observed in a number of cases (e.g., snail ganglion cells and striated muscle) during periods of rapid Na^+ extrusion. These potentials have been considered to be the result of the transport mechanisms, i.e., electrogenic pumps. In *Neurospora* hyphae, part of the resting potential across the membrane responds to K^+ concentration and part responds rapidly and reversibly to interference with metabolism by

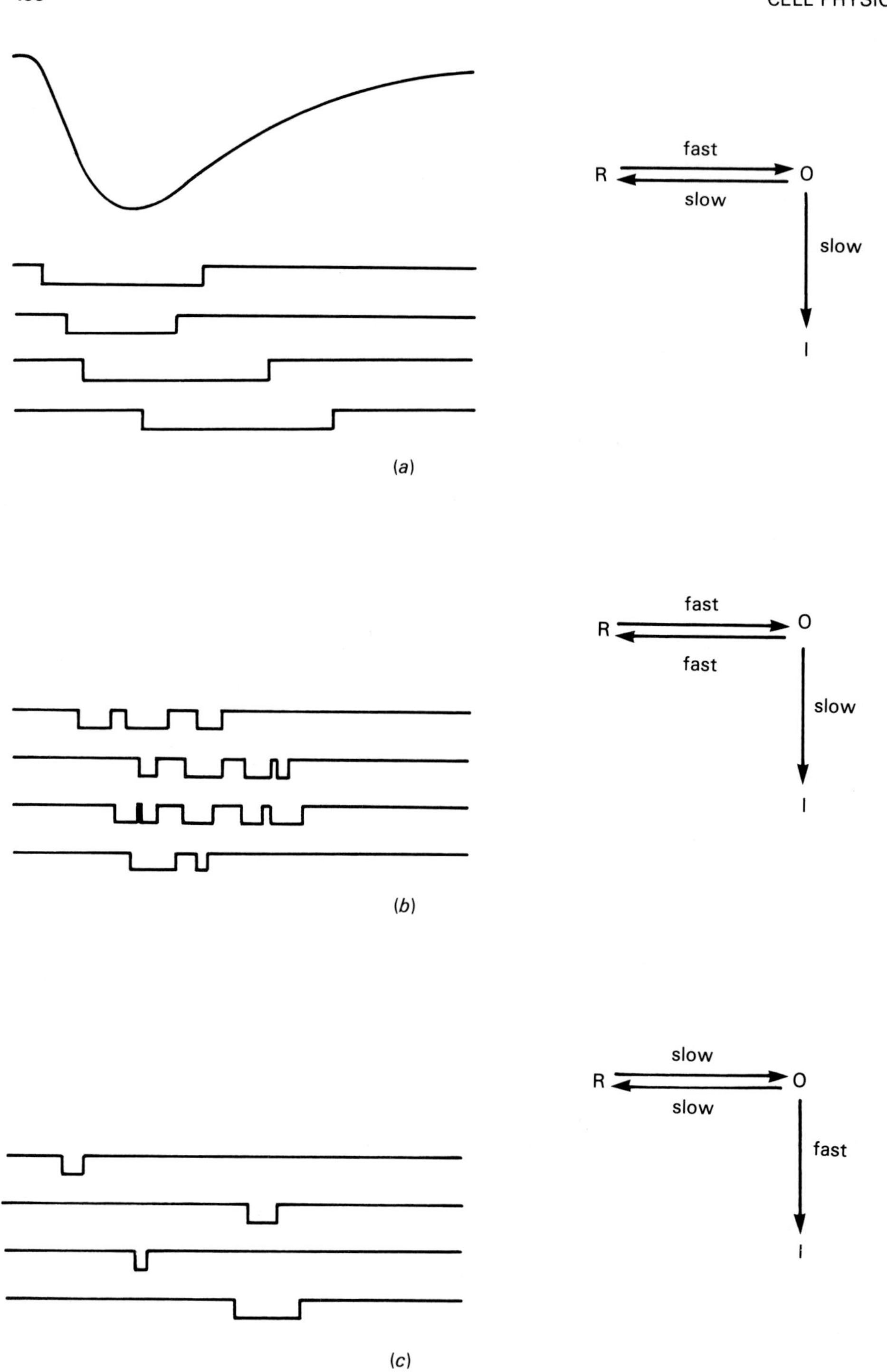

Fig. 16.15 Three possibilities for sodium channel gating that predict identical macroscopic sodium currents but different single-channel behavior. From R. W. Aldrich, *Trends in Neuro-Science,* 9:82–85, with permission. Copyright © 1986 Elsevier Science Publishers, England.

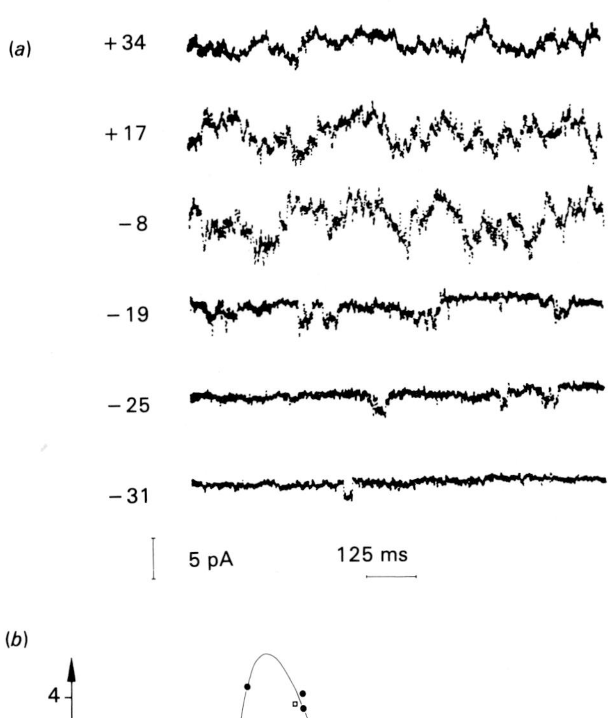

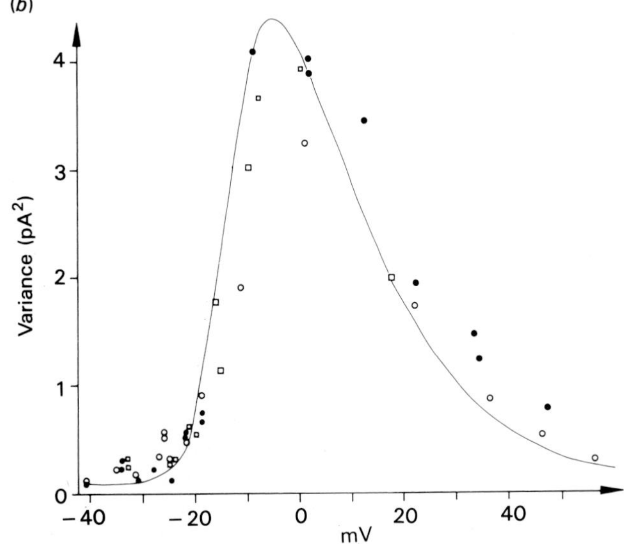

Fig. 16.16 (*a*) Recordings of patch current at different membrane potentials. From Conti and Neher (1980). (*b*) Variance as a function of voltage. The points are experimental points and the line has been calculated from the observed probability of opening on the assumption that there are seven channels in the patch. Reprinted by permission from *Nature* 285:140–143, copyright 1980 Macmillan Magazines Ltd.

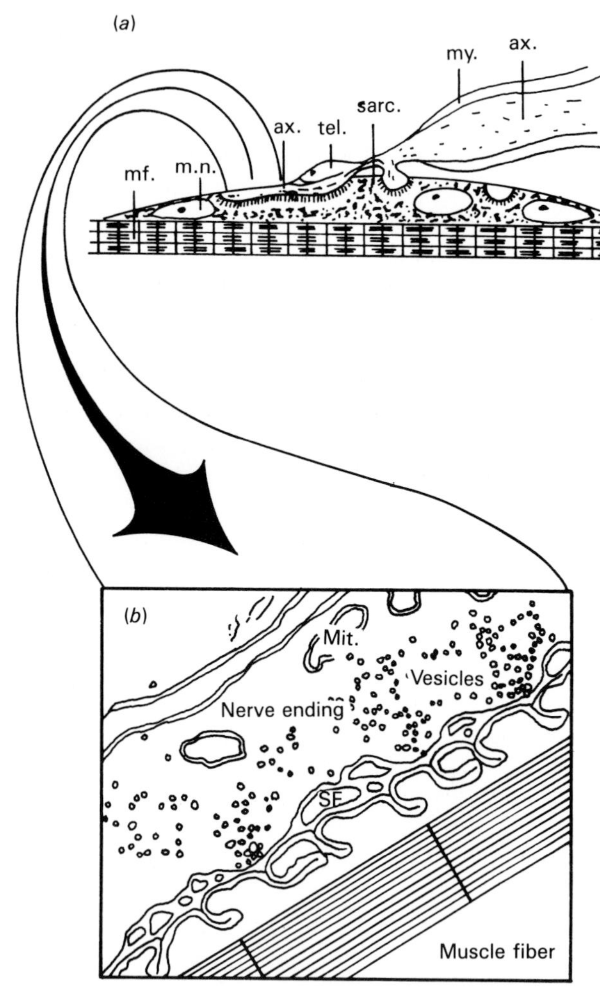

Fig. 16.17 (*a*) Diagram of a motor end plate showing axoplasm (ax) and myelin (my) of motor nerve, and saclike terminals (arrows) lying in gutterlike depressions of the mitochondrion-rich muscle sarcoplasm (sarc). The terminals are protected by teloglia (tel). Muscle nuclei (m.n.) and myofibrils (mf) are also diagrammed. (*b*) Tracing of an electron micrograph of a portion of the nerve terminal similar to that between the arrows in (*a*). Note the highly folded postsynaptic membrane extending into the muscle sarcoplasm, the fingerlike projections of teloglia, and the numerous vesicles and mitochondria (Mit.) in the terminal cytoplasm. Reprinted with permission from V. P. Whittaker, *Proceedings of the National Academy of Sciences,* 60:1082, 1084, 1968.

inhibitors of respiration such as azide or carbon monoxide (Slayman, 1965). In this case (Slayman and Tatum, 1965), an electrogenic pump is the major component. Similar results were obtained with plant cells.

V. TRANSMISSION OF EXCITATION BETWEEN CELLS

Excitable cells can communicate with each other, generally through specialized regions of contact, the synapses. In vertebrates the synaptic connections form a network of great complexity. It has been estimated that there are 4×10^{11} synapses per gram of guinea pig cerebral cortex.

Many of the complexities of neural behavior are thought to be a reflection of the organization of neuronal networks and the properties of the synapses. The convergence of different terminals can produce complex effects, since a variety of excitatory and inhibitory influences are summed and integrated.

The contact between nerve and striated muscle, the *neuromuscular junction*, is represented diagrammatically in Fig. 16.17 (Whittaker, 1968). At this junction the nerve terminal expands into a baglike arrangement; nerve and muscle are separated by a space. Transmission occurs by the release of a transmitter, acetylcholine, which induces a depolarization of the specialized portion of the muscle membrane present at the junction, the "end plate." Presumably the transmitter attaches to specific receptor sites on the end plate. The application of acetylcholine to the junction with a micropipette generates postjunctional potential changes. The postjunctional potentials induced by the action potential of the nerve fiber can be blocked by certain drugs that compete with acetylcholine for attachment to the active sites. Curare, the poison used on arrow tips by South American Indians, acts in this manner. At rest, there is an occasional release of small packets or quanta of acetylcholine from the undisturbed cells. This release produces the small postjunctional potentials (*miniature end-plate potentials*) (Fatt and Katz, 1952; Miledi, 1966). The size of these potentials can be calculated to correspond to the release of 1000 to 10,000 molecules of acetylcholine. It has been suggested that they correspond to the release of the contents of acetylcholine-containing vesicles that are visible with the electron microscope (for a discussion, see Auerback, 1972). These vesicles have been isolated (Whittaker and Sheridan, 1965) and shown to contain acetylcholine.

Graded end-plate potentials can be produced by applying electrical currents locally to a nerve-muscle preparation in which the nerve and the muscle are rendered unresponsive with tetrodotoxin (Katz and Miledi, 1967a, 1967b). Conversely, the end-plate potential can be blocked by hyperpolarization of the nerve terminal by means of a current pulse delivered locally by a microelectrode. Thus the critical event for release of acetylcholine is depolarization. Apparently, it operates through the opening of voltage-gated Ca^{2+} channels. External Ca^{2+} is required for neurotransmitter release.

Connections between nerve cells have similar specialized regions. The part of the cell through which incoming events influence the synapse is called the *presynaptic terminal*. The presynaptic terminal stimulates the *postsynaptic cell*. The synapse may be a tight connection in which the cell membranes are closely apposed with only a small space between the gap junctions (Pappas and Bennett, 1966), which have been discussed in Chapter 2. In these cases transmission occurs by purely electrical events through these specialized low-resistance areas. At other junctions, there is a considerable space between the membranes of the two cells and the transmission is thought to require a neurotransmitter. Figure 16.18 shows an electron micrograph of a section of the spinal cord of the toadfish in which the two kinds of synaptic ending are present side by side. The two types of synapses are represented diagrammatically in Fig. 16.19 (Whittaker, 1968).

By inserting a stimulation electrode into one kind of cell (i.e., presynaptic or postsynaptic) and a recording electrode into the other, one can follow the various events. The results confirm the presence of two distinct kinds of synapses. In gap synaptic junctions, both depolarizations and hyperpolarizations can be transmitted from one cell to the other in either direction (Bennett et al., 1967). Consequently, presynaptic cells can be stimulated from a postsynaptic action potential. The gap junction synapses also lack neurotransmitter vesicles, in agreement with the notion that in these cases the model of transmission is electrical.

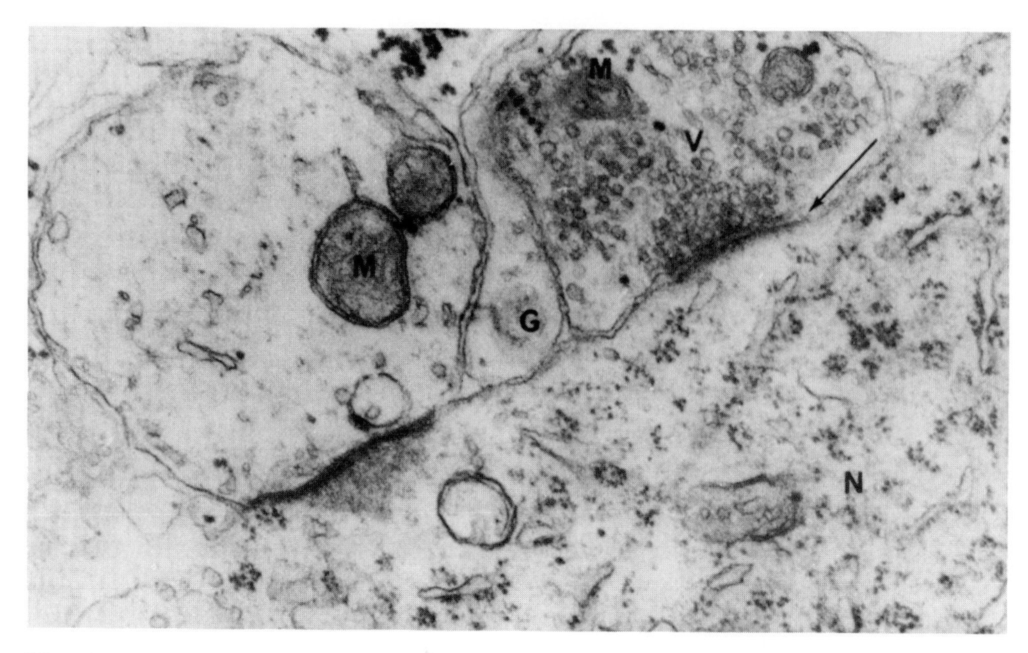

Fig. 16.18 Electron micrograph of a section from the spinal cord of the toadfish. Profiles of two synaptic endings can be seen separated by a glial cell process (G). The two synaptic endings form contact on a neuronal cell body (N). In the chemically transmitting synapse vesicles (V) are clustered close to the presynaptic membrane. The pre- and postsynaptic membranes are distinctly separated by a 20-nm space (arrow) at the chemically transmitting synapse. At the electrotonic synapse (at the left) the apposing membranes are very close and no space is discernible. M, Mitochondria. ×28,000. From Pappas and Bennett (1966), with permission.

In the synapses with a large space between pre- and postsynaptic membranes, the presynaptic terminals exhibit numerous vesicles suggesting the involvement of a neurotransmitter. This idea is supported by the observation that experimentally induced postsynaptic action potentials cannot stimulate presynaptic cells. In contrast, the postsynaptic potentials can be evoked electrically even when the action potentials are blocked by tetrodotoxin (Katz and Miledi, 1967a) (which blocks the Na^+ channels) or by tetraethylammonium (which blocks the passage of K^+ needed for repolarization) (Katz and Miledi, 1967c), indicating that the presynaptic electrical events are not directly involved and can be bypassed. Small spontaneous postsynaptic depolarizations have been observed and are thought to reflect a continuous subthreshold release of packets of neurotransmitter. Certain pharmacological agents can also block these synapses by preventing neurotransmitter binding. Much of this evidence suggests a similarity to neuromuscular transmission.

Although acetylcholine is a synaptic neurotransmitter, many other chemicals have been implicated. At some junctions the transmitters may be noradrenaline, dopamine, and 5-hydroxytryptamine or serotonin (the *monoamines*). Some amino acids, e.g., glutamate, γ-aminobutyric acid, and glycine, are thought to act at synapses as neurotransmitters. The two latter compounds are generally found to have an inhibitory influence, whereas glutamate is usually excitatory. Some of the drugs that produce hallucinations (e.g., LSD and mescaline) and some tranquilizers

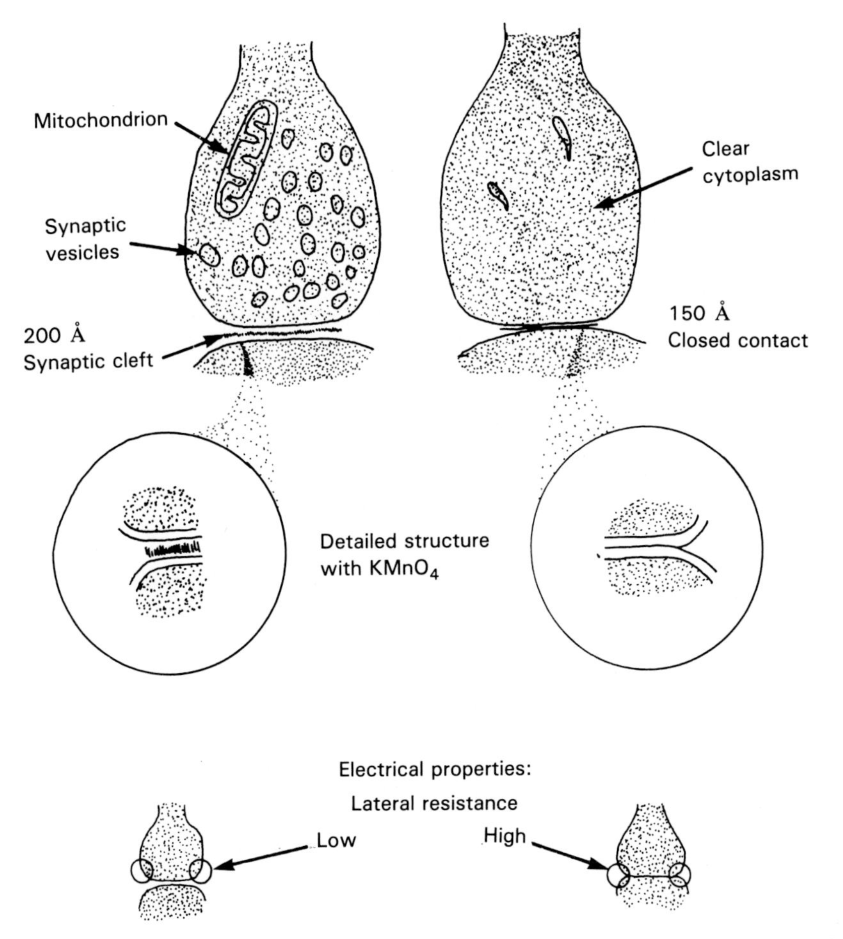

Fig. 16.19 Characteristics of open, chemical synapse (left) and closed, electrical synapse (right). Note in the latter synapse the absence of synaptic vesicles and cleft and the invasion of the postsynaptic elements by action currents, which in the chemical synapse are short-circuited by the low-resistance cleft. Reprinted with permission from V. P. Whittaker, *Proceedings of the National Academy of Sciences,* 60:1082, 1084, 1968.

(e.g., thalidomide) are thought to produce their effects by their structural similarity to neurotransmitters. There is some evidence that different synapses on the same cell may involve different transmitters (Gerschenfeld et al., 1967).

The mechanism for the release of neurotransmitter vesicles or secretory vesicles depends on the influx of Ca^{2+} through voltage-sensitive calcium channels (Katz, 1969). The details of the release mechanism are not known. However, it is thought to involve in some way the protein synapsin I (Browning et al., 1985), which is associated with the neurotransmitter vesicles of the neurons and is likely to be attached to cytoskeletal elements.

The most likely mechanism for the release of the vesicles involves the phosphorylation of synapsin I in a Ca^{2+}- calmodulin-dependent manner, resulting in a decrease in the affinity of the synapsin for the vesicles. In fact, microinjection of dephosphorylated synapsin I decreases the evoked output of neurotransmitter, whereas injection of Ca^{2+}- calmodulin-dependent kinase has the opposite effect (Llinas et al., 1985).

Neurotransmitters depolarize the postsynaptic membrane by a mechanism in which the increase in membrane conductance depolarizes the membrane. As might be expected from our previous examination of depolarization, this process involves the opening of channels. The

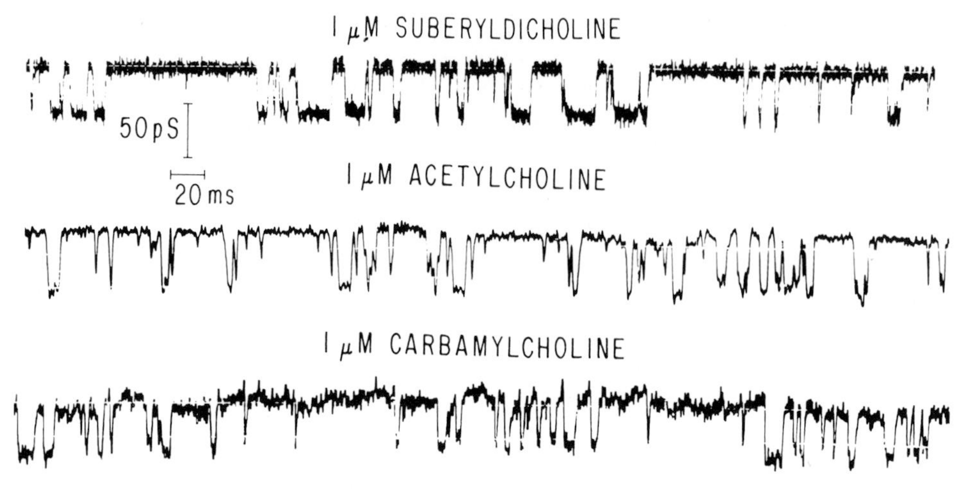

Fig. 16.20 Single acetylcholine receptor channel currents activated by different cholinergic agonists in planar lipid bilayers. Reproduced from *Biophysics Journal,* 45:165–174, 1984, by copyright permission of Rockefeller University Press.

opening depends on binding of the neurotransmitters; i.e., the channels act as receptors and are induced to open. Postsynaptic depolarization has been studied with patch clamping techniques primarily in cells in cell culture or freshly dissociated neurons, since the native structures are enveloped in protective sheets.

One of the most versatile techniques involves the reconstitution of extracted channel proteins in bilayers or in the lipid at the tip of a patch pipette. This approach has been used extensively with many proteins including Na^+ and K^+ channels and acetylcholine-activated channels.

The receptors have been isolated most commonly from the electric organ of *Torpedo*. The acetylcholine-activated channels increase the permeability of the membrane to a variety of ions when they bind acetylcholine as well as other agonists. The four distinct glycoproteins are present as a pentamer (Montal et al., 1986) with the composition $\alpha_2\beta\gamma\delta$.

The activation of single channels reconstituted into a bilayer is shown in Fig. 16.20 (Montal et al., 1984).

SUGGESTED READING

Catterall, W. A. (1988) Structure and function of voltage-sensitive ion channels. *Science* 242:50–61.

Hille, B. (1989) Voltage-gated sodium channels since 1952, in *Ion Transport* (Keeling, D., and Benham, C., eds.), pp. 57–71. Academic Press, New York.

Hille, B. (1991) *Ionic Channels of Excitable Membranes,* 2d ed., Chapters 1–6, Sinauer Associates Inc., Sunderland, MA.

Kandel, E. R., and Schwartz, J. H. (1985) *Principles of Neural Science,* 2d ed., Chapters 2–8, pp. 14–90. Elsevier, New York.

Kuffler, S. W., Nicholls, J. G., and Martin, A. R. (1984). *From Neuron to Brain: A Cellular Approach to the Function of the Nervous System,* Chapters 4–7 and 9, pp. 99–186 and 207–240. Sinauer Associates, Sunderland, Mass.

McGeer, P. L., Eccles, J. C., and McGeer, E. G. (1987) *Molecular Biology of the Mammalian Brain,* 2d ed. Chapters 1–5, pp. 1–174. Plenum, New York.

Shepherd, G. M. (1988) *Neurobiology,* 2d ed., Chapters 1–8, pp. 1–176. Oxford Univ. Press, New York.

Special Aspects

Catterall, W. A., Gonoi, T., and Costa, M. (1985) Sodium channels in neural cells: molecular properties and analysis of mutants. *Curr. Top. Membr. Transport* 23:79–100.

Horn, R. (1984) Gating of channels in nerve and muscle: a stochastic approach. *Curr. Top. Membr. Transport* 21:53–97.

Montal, M. (1990) Molecular anatomy and molecular design of channel proteins, *FASEB J.* 4:2623–2635.

REFERENCES

Aldrich, R. W. (1986) Voltage-dependent gating of sodium channels: towards an integrated approach. *Trends Neurosci.* 9:82–86.

Aldrich, R. W., Corey, D. P., and Stevens, C. F. (1983) A reinterpretation of mammalian sodium channel gating based on single channel recording. *Nature (London)* 306:438–441.

Auerback, A. (1972) Transmitter release at chemical synapses. In *Structure and Function of Synapses* (Pappas, G. D., and Purpura, D. P., eds.), pp. 137–159. Raven, New York.

Bennett, M. V. L., Nakajima, Y., and Pappas, G. D. (1967) Physiology and ultrastructure of electrotonic junctions. I. Supramedullary neurons. *J. Neurophysiol.* 30:161–179.

Browning, M. D., Huganir, R., and Greengard, P. (1985) Protein phosphorylation and neuronal function. *J. Neurochem.* 45:11–23.

Bullock, T. H., and Horridge, G. A. (1965) *Structure and Function in the Nervous Systems of Invertebrates*, Vol. 1. (Whitaker, D. M., Emerson, R., Kennedy, D., and Beadle, G. W., eds.). Freeman, San Francisco.

Cole, K. S., and Curtis, H. J. (1938) Electrical impedence of *Nitella* during activity. *J. Gen. Physiol.* 22:37–64.

Cole, K. S., and Curtis, H. J. (1939) Electrical impedence of squid giant axon during activity. *J. Gen. Physiol.* 22:649–670.

Collins, E. W., Jr., and Edwards, C. (1971) Role of Donnan equilibrium in the resting potentials in glycerol extracted muscle. *Am. J. Physiol.* 221:1130–1133.

Conti, F., and Neher, E. (1980) Single channel recording of K currents in squid axons. *Nature (London)* 285:140–143.

Conway, E. J. (1957) Nature and significance of concentration relations of K^+ and Na^+ ions in skeletal muscle. *Physiol. Rev.* 37:84–132.

Fatt, P., and Katz, B. (1952) Spontaneous subthreshold activity of motor nerve endings. *J. Physiol (London)* 117:109–128.

Frankenhaeuser, B., and Moore, L. E. (1963) The effect of temperature on the sodium and potassium permeability changes in myelinated nerve fibers of *Xenopus laevis*. *J. Physiol. (London)* 169:431–437.

Gaffey, C. T., and Mullins, L. J. (1958) Ion fluxes during the action potential in *Chara*. *J. Physiol. (London)* 144:505–524.

Gerschenfeld, H. M., Ascher, P., and Tauc, L. (1967) Two different excitatory transmitters acting on a single molluscan neurone. *Nature (London)* 213:358–359.

Grundfest, H., and Nachmansohn, D. (1950) Increased sodium entry into squid giant axons during activities at high frequencies and during reversible inactivation of cholinesterase. *Fed. Proc. Fed. Amer. Soc. Exp. Biol.* 9:53.

Hagiwara, S., and Naka, K. I. (1964) The initiation of the spike potential in barnacle muscle fibers under low intracellular Ca^{2+}. *J. Gen. Physiol.* 48:141–161.

Hamill, A. M., Marty, A., Neher, H., Sakman, B., and Sigworth, F. J. (1981) Improved patch-clamp techniques for high resolution current recording from cells and cell free membrane patches. *Eur. J. Physiol.* 391:85–100.

Hodgkin, A. L. (1939) The subthreshold potentials in a crustacean nerve fibre. *Proc. R. Soc. London. Ser. B* 126:87–121.

Hodgkin, A. L. (1951) The ionic basis of electrical activity in nerve and muscle. *Biol. Rev. Cambridge Philos. Soc.* 26:339–409.

Hodgkin, A. L. (1958) Ionic movements and electrical activity in giant nerve fibres. *Proc. R. Soc. London Ser. B* 148:1–37.

Hodgkin, A. L., and Huxley, A. F. (1947) Potassium leakage from an active nerve fibre. *J. Physiol. (London)* 106:341–367.

Hodgkin, A. L., and Huxley, A. F. (1953) Movement of radioactive potassium and membrane current in a giant axon. *J. Physiol. (London)* 121:403–414.

Hodgkin, A. L., and Katz, B. (1949) The effect of sodium ions on the electrical activity of the giant axon of the squid. *J. Physiol. (London)* 108:33–77.

Huxley, A. F., and Stämfli, R. (1951) Effects of potassium and sodium on resting and action potentials of single myelinated fibres. *J. Physiol. (London)* 112:496–508.

Katz, B. (1969) *The Release of Neurotransmitter Substances,* Sherrington Lectures, Vol. 10. Thomas, Springfield, Ill.

Katz, B., and Miledi, R. A. (1967a) A study of synaptic transmission in the absence of nerve impulses. *J. Physiol. (London)* 192:407–436.

Katz, B., and Miledi, R. (1967b) Tetrodotoxin and neuromuscular transmission. *Proc. R. Soc. London Ser. B* 167:8–22.

Katz, B., and Miledi, R. (1967c) The release of acetylcholine from nerve endings by graded electric pulses. *Proc. R. Soc. London Ser. B* 167:23–28.

Keynes, R. D. (1951a) Leakage of radioactive potassium from stimulated nerve. *J. Physiol. (London)* 113:99–114.

Keynes, R. D. (1951b) The ionic movements during nervous activity. *J. Physiol. (London)* 114:119–150.

Keynes, R. D. (1972) Excitable membranes. *Nature (London)* 239:29–32.

Keynes, R. D., and Lewis, P. R. (1951) The sodium and potassium content of cephalopod nerve fibers. *J. Physiol. (London)* 114:151–182.

Kuhlman, J. R., Li, C., and Calabrese, R. L. (1985) FMRF-amide-like substances in the leech: immunocytochemical localization. *J. Neurosci.* 5:2301–2309.

Latorre, R., and Miller, C. (1983) Conduction and selectivity in potassium channels. *J. Membr. Biol.* 71:11–30.

Llinas, R., McGuiness, T. L., Leonard, C. S., Sugimori, M., and Greengard, P. (1985) Intraterminal injection of synapsin-I or calcium calmodulin dependent protein kinase-II alters neurotransmitter release at the squid giant synapse. *Proc. Natl. Acad. Sci. U.S.A.* 82:3035–3039.

Miledi, R. (1966) Miniature synaptic potentials in squid nerve cells. *Nature (London)* 212:1240–1242.

Montal, M., Labarca, P., Fredkin, D. R., and Isla, B. A. (1984) Channel properties of the purified acetylcholine receptor from *Torpedo californica.* Reconstituted in planar lipid bilayer membranes! *Biophys. J.* 45:165–174.

Montal, M., Aholt, R., and Labarca, P. (1986) The reconstituted acetylcholine receptor. In *Ion Channel Reconstitution* (Miller, C., ed.). Plenum, New York.

Narahashi, T., Moore, J. W., and Scott, W. R. (1964) Tetrodotoxin blockage of sodium conductance increase in lobster giant axons. *J. Gen. Physiol.* 47:965–974.

Nastuk, W. L., and Hodgkin, A. L. (1950) The electrical activity of single muscle fibres. *J. Cell Comp. Physiol.* 35:39–73.

Pappas, G. D., and Bennett, M. V. L. (1966) Specialized junctions involved in electrical transmission between neurons. *Ann. N.Y. Acad. Sci.* 137:495–508.

Parsegian, A. (1969) Energy of an ion crossing a low dielectric membrane: solutions to four relevant static problems. *Nature (London)* 221:844–846.

Rothenberg, M. A. (1950) Studies on permeability in relation to nerve function. II. Ionic movements across axonal membranes. *Biochim. Biophys. Acta* 4:96–114.

Slayman, C. L. (1965) Electrical properties of *Neurospora crassa* respiration and intracellular potential. *J. Gen. Physiol.* 49:93–116.

Slayman, C. L., and Tatum, E. L. (1965) Potassium transport in *Neurospora*. II. Measurements of steady-state potassium fluxes. *Biochim. Biophys. Acta* 102:149–160.

Tobias, J. M. (1958) Experimentally altered structure related to function in the lobster axon with an extrapolation to molecular mechanism in excitation. *J. Cell. Comp. Physiol.* 52:89–125.

Weldman, S. (1951) Electrical characteristics of *Sepia* axons. *J. Physiol. (London)* 114:372–381.

Whittaker, V. P. (1968) Synaptic transmission. *Proc. Natl. Acad. Sci. U.S.A.* 60:1081–1091.

Whittaker, V. P., and Sheridan, M. N. (1965) The morphology and acetylcholine content of isolated cerebral cortical synaptic vesicles. *J. Neurochem.* 12:363–372.

Chapter 17

Mechanochemical Coupling: Movement in Various Systems

Whether we are concerned with single cells, multicellular organisms, or populations, biological motion is of fundamental importance. Movement plays obvious roles in feeding, avoidance, digestion, respiration, circulation, and reproduction. Contractile proteins are likely to be involved in the shape of a cell and its changes. The continuous flow of the cytoplasm of plant cells may well be analogous to circulation in a multicellular organism. The movement and rearrangement of cells are fundamental to morphogenesis.

Motility may be the result of the action of special structures, such as cilia, flagella, or muscle fibers. It also takes place in the cytoplasm of cells, in which the contractile machinery, not readily apparent, may involve the assembly and dissociation of contractile units.

Regardless of details, the displacement of matter will require the performance of work. In the living organism, metabolic and photosynthetic events generally make energy available in a chemical form, such as ATP or some other compound of high phosphate group transfer potential. In producing movement, the hydrolysis of high-energy compounds is coupled to the mechanical events.

I. HIGH-ENERGY PHOSPHATE AND MOVEMENT

The hydrolysis of high-energy phosphate, generally ATP, is involved in motility. ATP has frequently been implicated in experiments in which most of the soluble components of cells or contractile structures were extracted either with cold glycerol solutions (in a procedure known as *glycerination*) or in more recent procedures with detergents. The extraction leaves the contractile apparatus intact. The addition of ATP, normally extracted along with the other soluble components, induces contraction (e.g., Szent-György, 1949; Summers and Gibbons, 1971).

In the case of muscle, either the force generated or the amount of work performed can be readily measured. The muscle can be attached to a lever. When the system is connected to an appropriate transducer, the tension generated is recorded by measuring the current generated by the transducer. Where work has to be measured, the muscle can be allowed to shorten and lift a weight or shorten against a force exerted by the apparatus. In addition to a role of ATP in contraction of extracted muscle, the direct involvement of ATP hydrolysis in the contraction of intact striated muscle has been shown. In muscle, phosphocreatine is usually present at higher concentrations than ATP and acts as a high-energy phosphate reserve. In the reaction of Eq. (17.1), catalyzed by creatine phosphotransferase, creatine phosphate (Cr ~ P) transfers the phosphate from creatine (Cr) to ADP.

$$\text{Cr} \sim \text{P} + \text{ADP} \rightarrow \text{ATP} + \text{Cr} \tag{17.1}$$

An involvement of ~ P, i.e., a compound of high phosphate group transfer potential, is shown by the decrease in the level of phosphocreatine when the synthesis of new ATP is blocked by adding an uncoupler of oxidative phosphorylation, 2,4-dinitrophenol (DNP). In the experiment of Fig. 17.1 (Cain et al., 1962), the frog's rectus abdominis muscle is stimulated electrically and contracts against a constant load. The amount of phosphocreatine hydrolyzed is directly proportional to the amount of work performed (displacement × mass).

To test for involvement of ATP requires blocking creatine phosphotransferase, which can be inhibited by 1-fluoro-2,4-dinitrobenzene (FDNB). Table 17.1 shows the constancy of Cr ~ P in a system in which metabolism is blocked by DNP in the presence of FDNB. In these experiments, matched pairs of muscles from the same animal were used. All were treated with DNP and all except the last with FDNB. One muscle served as control and was allowed to rest, whereas the other carried out work as tabulated in column 3. Column 4 shows the difference in Cr ~ P concentration between control and experimental muscles. The difference (indicated by

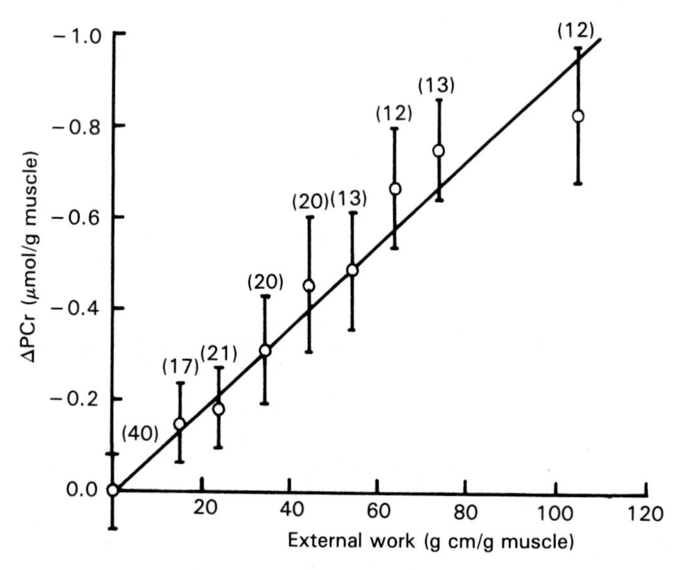

Fig. 17.1 Relationship of phosphocreatine breakdown to the amount of external work performed by frog rectus abdominis muscle contraction once or twice against a constant load to different degrees and for different times at 0°C. From Cain et al. (1962). Reprinted by permission from *Nature* 196:214–217, copyright © 1962 Macmillan Magazines Ltd.

Table 17.1 Production of Inorganic Phosphate (P_i) without Change in Phosphocreatine (PCr) After Three Small Contractions at 0°C in Frog Rectus Abdominis Muscles Pretreated with Dinitrophenol (DNP) Plus Fluorodinitrobenzene (FDNB)[a]

(1) Type of experiment	(2) Pairs of muscles	(3) External work (g cm/g muscle)	(4) ΔPCr (μmol/g muscle)	(5) ΔP_i (μmol/g muscle)
Control minus contractions	27	81 ± 5	−0.10 ± 0.08	—
Control minus contractions	12	79 ± 8	−0.17 ± 0.08	+1.23 ± 0.48
Control minus control	4	0	0.17 ± 0.17	—
Control minus contractions (DNP but no FDNB)	4	76 ± 8	−1.02 ± 0.28	—

Source: Cain et al (1962). Reprinted by permission from *Nature* 196:214–217, copyright © 1962 Macmillan Magazines Ltd.

[a] The results are expressed as differences.

the box) is small unless FDNB is left out (last experiment). Although the hydrolysis of Cr ~ P is blocked by FDNB, the contractions produce inorganic phosphate (column 5, boxed value), suggesting that hydrolysis of some other high-energy phosphate, presumably ATP, is taking place. Table 17.2 shows that the amount of ATP is indeed decreased during contraction to produce ADP and AMP. The reactions involved can be expressed as

$$\text{ATP} \rightleftharpoons \text{ADP} + \text{P}_i + \text{work} \tag{17.2}$$

$$2\text{ADP} \rightleftharpoons \text{ATP} + \text{AMP} \tag{17.3}$$

Not all the energy is expended as work; a significant portion is lost in the form of heat.

ATP hydrolysis also provides the energy for ciliary and flagellar motion. Figure 17.2 (Brokaw, 1967) represents an experiment in which the beating of flagella of glycerinated sea urchin sperm is observed. The frequency of beat (Fig. 17.2*a*) is shown as a function of ATP concentration. The higher the ATP concentration, the greater the frequency. At a sufficiently low concentration, the flagella do not beat at all. Breakdown of the added ATP is clearly involved, since the beat frequency is linearly related to the ATP hydrolyzed (Fig. 17.2*b*). The contractile machinery functions as an ATPase.

These experiments demonstrate that the energy expenditure for these two kinds of movement

Table 17.2 Breakdown of Adenosine Triphosphate (ATP) to Form Adenosine Diphosphate (ADP) and Adenosine Monophosphate (AMP) During Contraction of Frog Rectus Abdominis Muscles after Treatment with Fluorodinitrobenzene (FDNB)[a]

	ATP (μmol/g muscle)	ADP (μmol/g muscle)	AMP (μmol/g muscle)
Single contraction			
Control	1.25	0.64	0.10
After one contraction	0.81	0.90	0.24
Change ± SE for nine pairs	−0.44 ± 0.046	+0.26 ± 0.023	+0.14 ± 0.027
Double contraction			
Control	1.24	0.61	0.07
After two contractions	0.59	0.88	0.41
Change ± SE for three pairs	−0.65 ± 0.045	+0.27 ± 0.051	+0.34 ± 0.037

Source: Cain et al. (1962). Reprinted by permission from *Nature* 196:214–217, copyright © 1962 Macmillan Magazines Ltd.

[a]External work ≏ 100 g cm per gram of muscle per contraction.

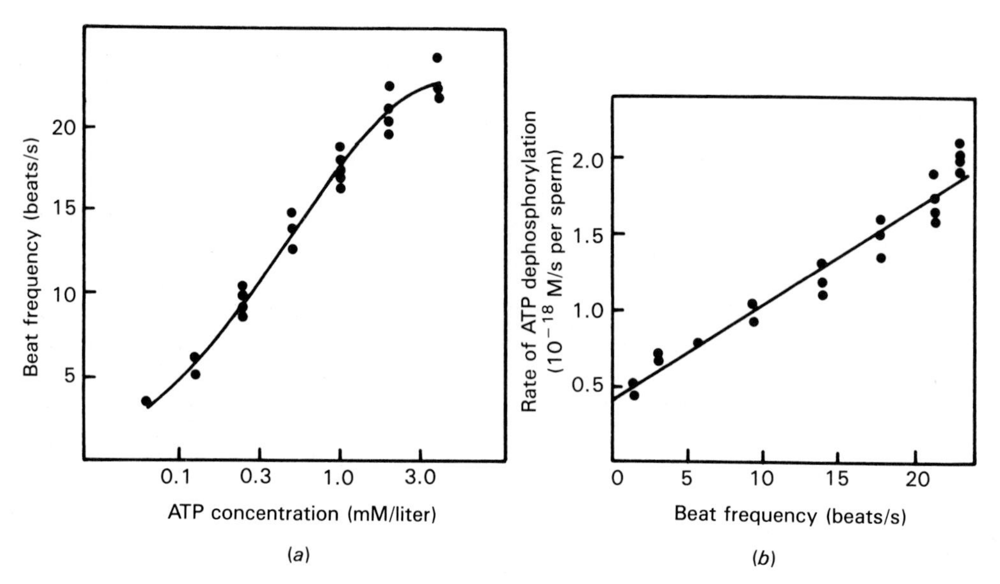

Fig. 17.2 (*a*) Effect of ATP concentration on beat frequency of glycerinated sea urchin spermatozoa. Each point represents the average of measurements of 20 spermatozoa. (*b*) Dephosphorylation of ATP by sperm suspensions at various beat frequencies obtained by varying the ATP concentration. Each point represents a single measurement of the rate of ATP dephosphorylation. The line was obtained by the method of least squares. From Brokaw (1967), *Science* 156:76–78, copyright 1967 by the AAAS.

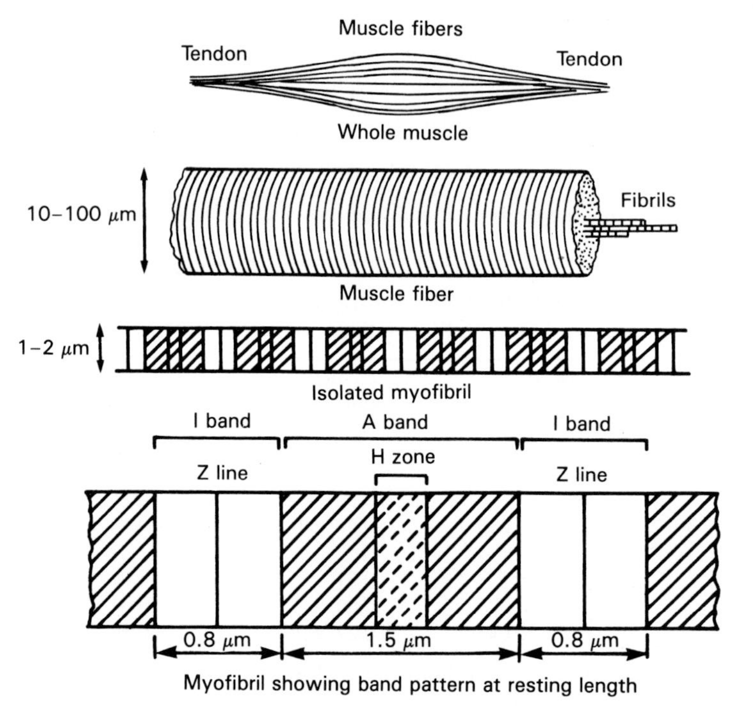

Fig. 17.3 Structure of striated muscle at different levels of organization; dimensions shown are those for rabbit psoas muscle. From H. E. Huxley, *The Cell,* Vol. 4(1):365–481, with permission. Copyright © 1960 Academic Press.

is provided by the hydrolysis of ATP. Unfortunately, this conclusion says little about the mechanism of the movement. The best understood events in cell movement are those involved in the contraction of striated muscle, where the structured elements are fixed and regular, and for this reason they are discussed first.

II. CONTRACTION IN STRIATED MUSCLE

The regularity of structure of striated muscle approaches that of a paracrystalline state; contraction events and structural states can be directly correlated. Striated muscle shortens on electrical, mechanical, or chemical stimulation; when loaded, it can perform work.

Striated muscle is made up of longitudinal elements. A diagrammatic view of vertebrate striated muscle at different levels of organization is shown in Fig. 17.3 (Huxley, 1960). The smallest functional element is the myofibril. The repeating unit of the myofibril is the sarcomere, which extends from Z line (or disk) to Z line. The structure of the myofibril shown in the figure is based on observations with both the light microscope (Fig. 17.4) (Hanson and Huxley, 1955) and the electron microscope (Fig. 17.5) (Huxley, 1960). As shown, the striations are the result of the presence of dark or anisotropic (A) bands and light or isotropic (I) bands. Anisotropy and isotropy refer to behavior in relation to polarized light. The isotropic bands transmit incident polarized light at the same velocity regardless of the light's direction. The anisotropic bands transmit light at different velocities depending on its direction; the bands are birefringent.

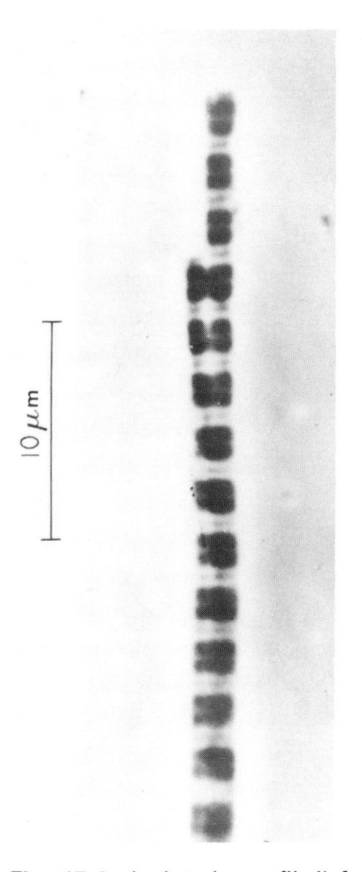

Fig. 17.4 Isolated myofibril from rabbit psoas muscle (glycerinated) in phase-contrast illumination with positive contrast. From Hanson and Huxley (1955), with permission.

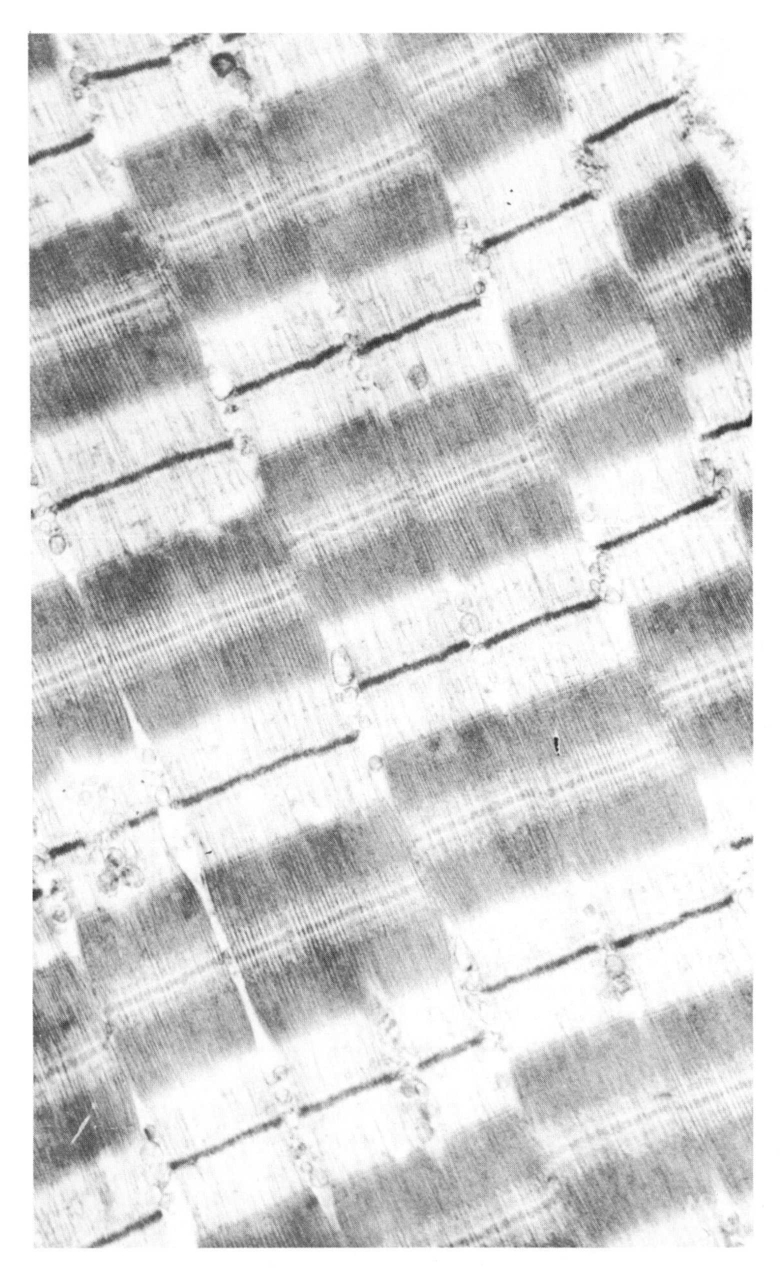

Fig. 17.5 Low-power electron micrograph of section of rabbit psoas muscle; the pattern of A, I, A, and H bands may be seen much more clearly than in light microscope pictures, and the longitudinal filaments that make up the contractile material are also visible. Granules and/or vesicles lie in between the myofibrils. ×20,700. From Huxley (1960), with permission.

The changes occurring in isolated muscle fibers during contraction and stretching may well approximate the events occurring in the living muscle. Observations with the light microscope can provide some insight or, at least, permit posing the problem in a more meaningful manner. The results obtained with the interference microscope are shown in Fig. 17.6 (Huxley and Niedergerke, 1954). In this figure, the sarcomere lengths under the different conditions are shown at the left. The sarcomere length is about 2 to 3 μm at rest. The A band remains largely unchanged by stretching (Fig. 17.6) or contraction (not shown here) of the muscle fiber. However, the I band is wider when the fiber is stretched, and it is narrowed when the myofibril contracts. A model capable of explaining the basic organization of the muscle fibril should be able to explain these observations.

First, some knowledge of the components of the fibers and how they are put together is necessary. Extraction of the fibers by different procedures can provide a good deal of information. The myofibrils are composed largely of proteins. Since the solubility properties of the major protein components are fairly well known, it is possible to extract selectively one protein at a

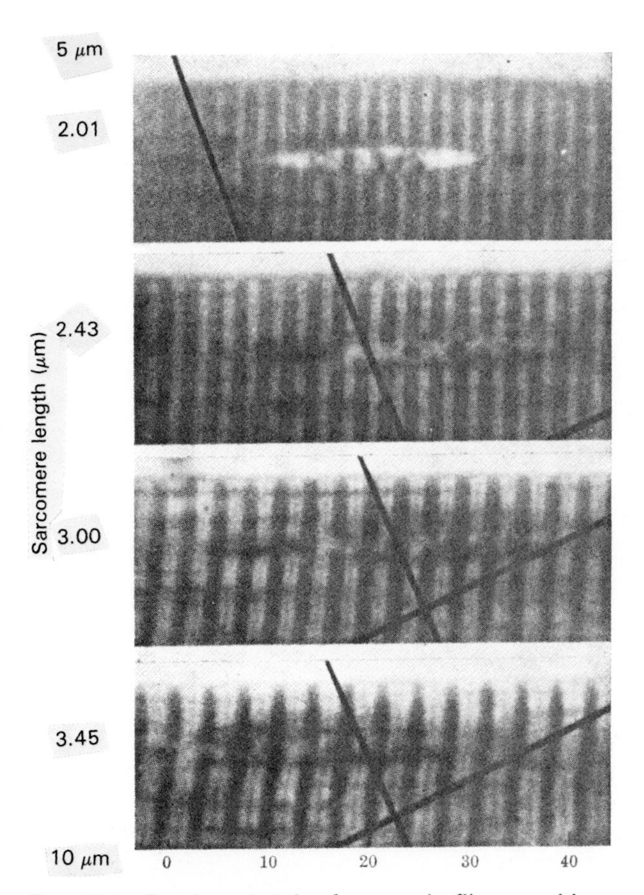

Fig. 17.6 Passive stretch of a muscle fiber, positive contrast (A bands dark). Sarcomere lengths indicated beside the photographs. Almost all the change of length is in the I bands (light). Muscle fibers during isotonic tetanus are not shown. The results are similar to those obtained with contraction. From Huxley and Niedergerke (1954). Reprinted by permission from *Nature* 173:971–973, copyright © 1954 MacMillan Magazines Ltd.

time. Myosin makes up about 55%, actin about 20%, tropomyosin about 5%, and troponin about 3 to 4% of the fiber.

We have seen that myofibrils extracted with glycerol can still contract when ATP is added to the medium. The glycerinated system seems to be a suitable system to analyze, since most irrelevant components have probably been removed. Myosin can be differentially extracted with 0.6 M KCl, 0.01 M pyrophosphate, and 10 mM $MgCl_2$. The material treated in this fashion loses its A bands (Fig. 17.7) (Hanson and Huxley, 1955). Treatment with KI, on the other hand, removes the actin and the I band simultaneously. These results indicate that the A band corresponds predominantly to myosin and the I band to actin. It is likely that other components are also present. For example, the fact that the organization of the remaining components is not disrupted when the bands are extracted speaks for the persistence of some other component.

Electron micrographs of longitudinal sections of muscle show that the sarcomeres are made up of thick and thin fibers that are interdigitated (Fig. 17.8*a*) (Huxley, 1960). The thick filaments are predominantly in the A bands and the thin filaments in the I bands. On the basis of the

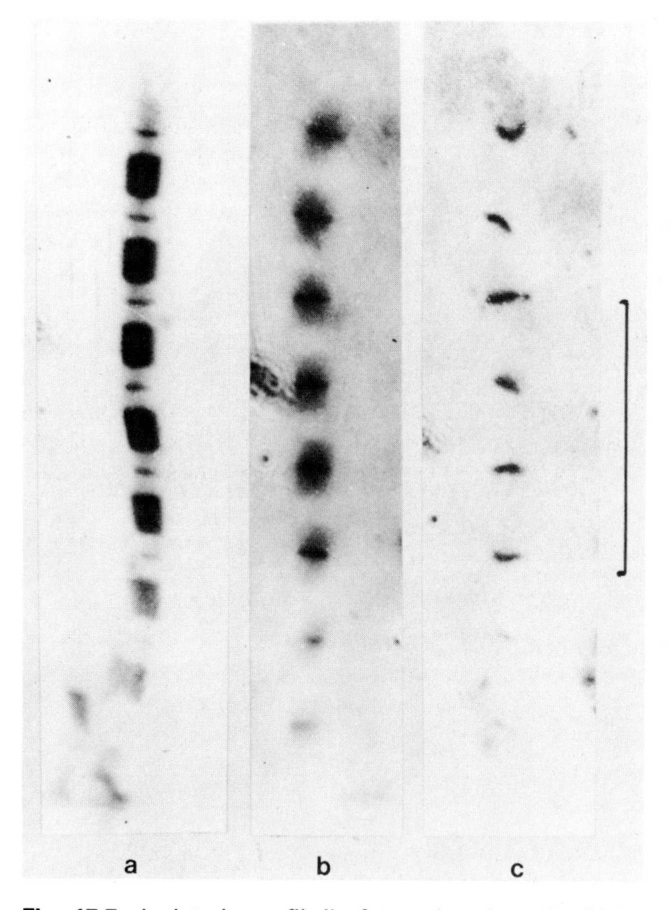

Fig. 17.7 Isolated myofibrils from glycerinated rabbit psoas muscle in phase contrast illumination. (*a*) Intact; (*b*) after actin extraction; (*c*) after myosin extraction. A large amount of the secondary material disappears when actin extraction takes place, leaving behind the Z zones connected together by some residual backbone substance. From Hanson and Huxley (1955), with permission.

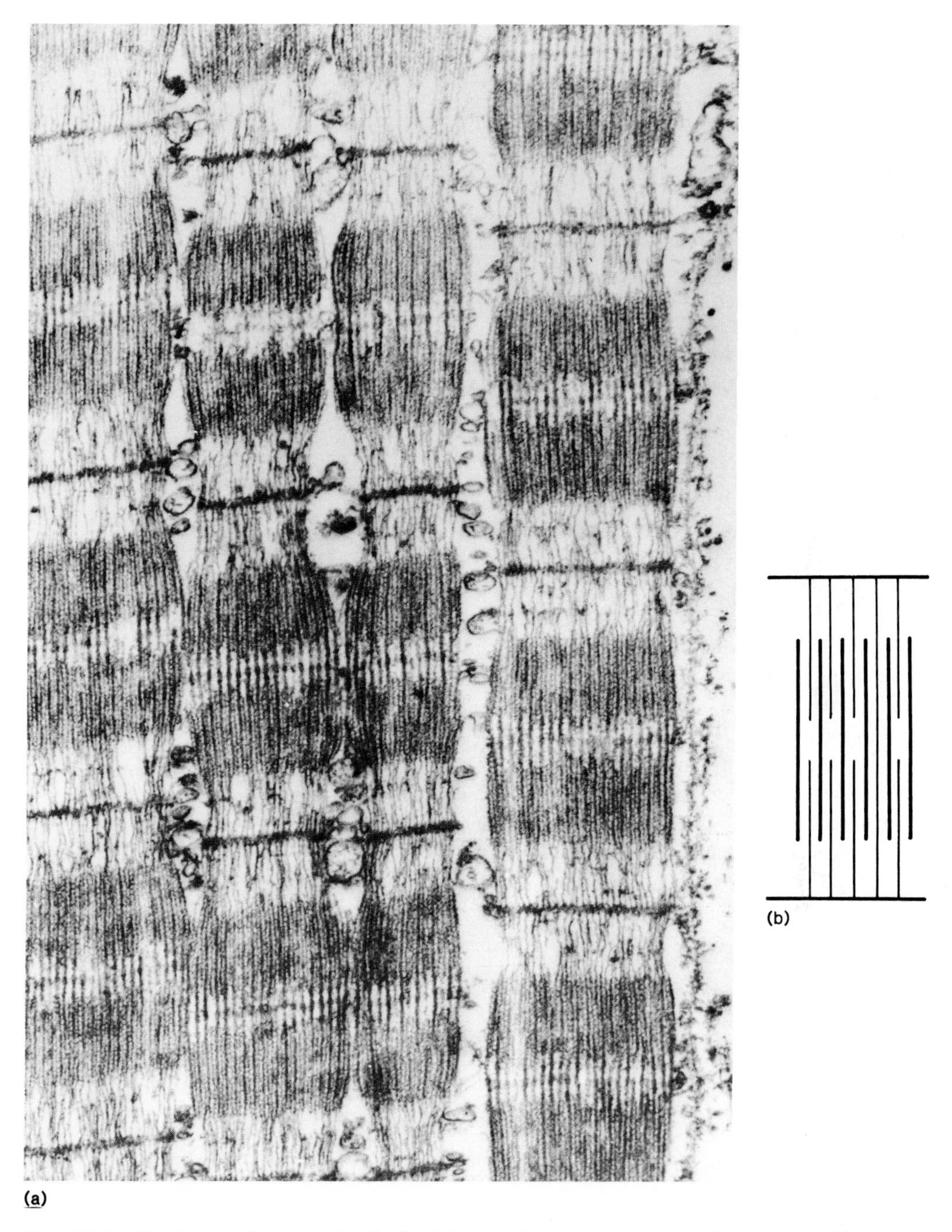

Fig. 17.8 Electron micrograph of ultrathin section through part of a muscle fiber; the hexagonal arrays of filaments in different myofibrils and in different sarcomeres have their axes oriented in a variety of directions. From Huxley (1960), with permission.

differential disappearance of the bands with the extraction procedures, we can assign myosin to the thick fibers and actin to the thin fibers. Such sarcomeres can be represented diagrammatically as shown in Fig. 17.8*b*.

The interdigitating structure of the sarcomere fibers suggests that contraction could take place if the thin filaments slid over the thicker filaments, shortening the sarcomere. The I band would then be shortened with no change in width of the A band, as was observed. This proposal is known as the sliding-filament model of muscle contraction. Such effects should be visible with the electron microscope in sections perpendicular to the long axis of the fiber. Figures 17.9 (Huxley, 1960) and 17.10 (Cartsen et al., 1961) show that electron microscopic observations of thick sections are consistent with this view. Figure 17.9*a* illustrates a stretched muscle and Fig. 17.9*b* a resting muscle. Figure 17.10 shows a preparation that was fixed at rest (Fig. 17.10*a*) or in the contracted state where the myofibril is held at constant length (Fig. 17.10*b*). Figure 17.10*c* shows a contracted myofibril. The sliding is thought to correspond to a rearrangement of the bonds between thick and thin filaments at the bridges visible with the electron microscope (Fig. 17.11) (Huxley, 1960).

Many new studies have revealed additional complexities. Nevertheless, the events that now need to be described in terms of molecular rearrangements and forces may be sufficiently known to permit the construction of realistic molecular models.

III. CILIA AND FLAGELLA

Motile cell processes are known as cilia and flagella. Generally, they are called cilia when they are short and numerous and flagella when they are long in relation to cell size and few in number. Although the motion of flagella is frequently undulant and that of cilia pendular, this is not always the case.

Cilia may form a composite of many shafts. The shafts may have their own individual membranes (as in the compound cilia of some ctenophores and protozoans). More rarely (in the ctenophore *Beroe*) they are enclosed by the same membrane.

Cilia and flagella are fundamental to the motility and feeding activity of many unicellular organisms. In more complex organisms they play basic roles in respiration, circulation, digestion, and reproduction.

Fig. 17.9 Changes in band pattern at different muscle lengths as seen in the electron microscope (thick sections) oriented for sectioning so as not to foreshorten band lengths. (*a*) Stretched muscle showing long I bands and H zone. (*b*) Rest length muscle, showing decrease in length of I bands and H zone and constancy of length of A band. From Huxley (1960), with permission.

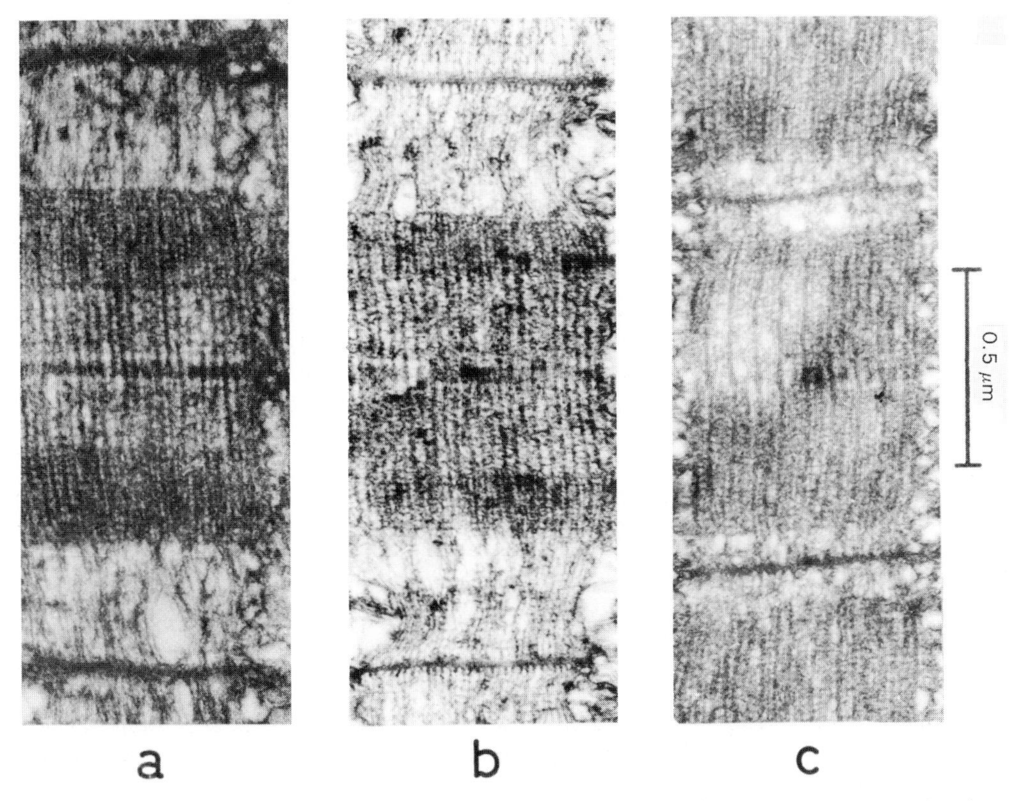

Fig. 17.10 Electron micrographs of (*a*) sarcomere at rest, (*b*) sarcomere contracting but held at constant rest length (isometric contraction), and (*c*) sarcomere shortening 20%. The insert on the right shows the orientation of the myofibril in relation to the sectioning glass knife. From Carlsen et al. (1961), with permission.

The processes are usually attached to specialized structures, the basal bodies, which can vary considerably in structure. Throughout completely unrelated phyla, the internal arrangement of the component parts of cilia and flagella is amazingly constant, and it is likely that the basic mechanism responsible for motility may be the same.

All cilia and flagella so far studied have nine pairs of longitudinal tubules, the doublets, which are arranged around two central tubules. The central tubules are not always present and may not be essential. The component tubules of the doublets differ and are known as A and B to distinguish them (see Fig. 17.12). Together, the tubular components of a cilium or a flagellum are known as the *axoneme*. Coarse fibers are also present in vertebrate sperm, but their function is likely to be only indirectly related to motion. The ubiquity of the peripheral tubules suggests that they are involved in movement.

Figure 17.12 (Johnson, 1985) summarizes the structure of an axoneme of a cilium. It represents a reconstruction based on electron micrographs of intact and disrupted cilia. The structures labeled D correspond to dynein complexes, which are attached to the A tubule of a doublet and are in a position to interact with the B tubule of an adjacent doublet.

As we saw previously, flagellar motion is powered by the hydrolysis of ATP. ATPase activity has also been demonstrated with the electron microscope by precipitating the phosphate produced by the hydrolysis of ATP with a heavy metal. ATPase activity shown by these studies appears to be localized in the peripheral tubules (Nelson, 1958).

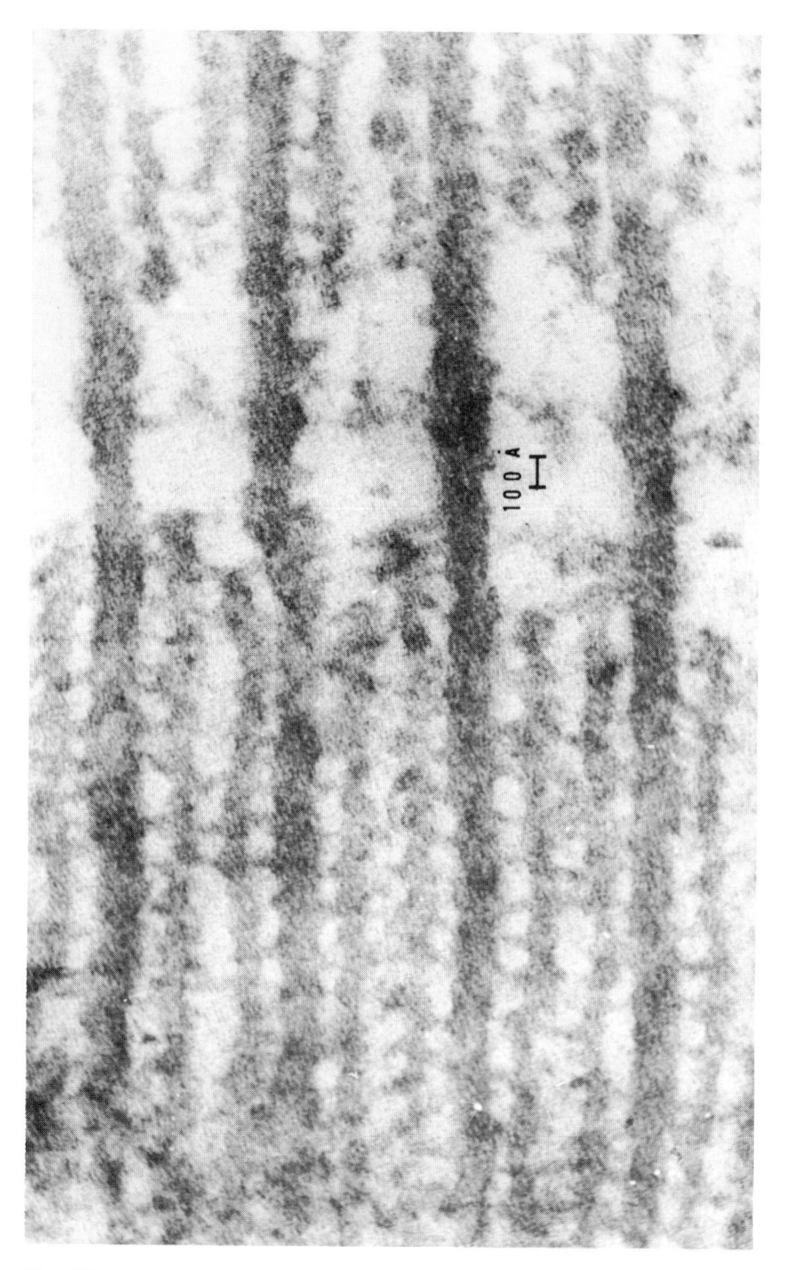

Fig. 17.11 Highly magnified electron micrograph through A band of myofibril from rabbit psoas muscle, showing the system of cross connections between the large and small filaments. ×418,000. From Huxley (1960), with permission.

The study of the cilia of *Tetrahymena* has also proved very useful in the further localization of the ATPase. From this organism it is possible to extract a ciliary protein fraction with ATPase activity. The extraction removes the "arms" of the tubules, i.e., the dynein complex (Fig. 17.13) (Gibbons, 1963).

Figure 17.13*a* shows a cross section of normal cilia isolated from *Tetrahymena* and fixed with OsO_4. A diagram of this view is shown in Fig. 17.13*b*. Figure 17.13*c* represents a

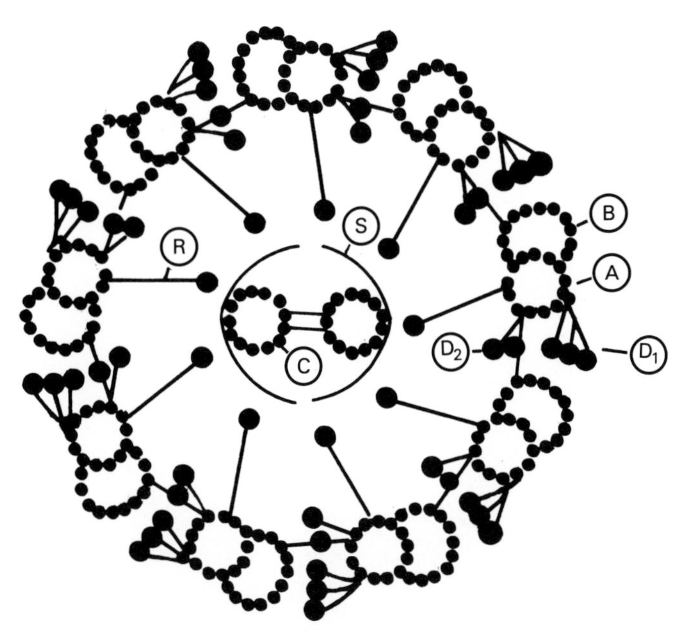

Fig. 17.12 Cross-sectional view of a ciliary axoneme represented schematically. D_1, Outer dynein arm; D_2, inner dynein arm; A, A tubule; B, B tubule; C, central tubule; R, radial spoke; S, central sheath. From Johnson (1985). Reproduced, with permission, from the Annual Review of Biophysical Chemistry, Volume 14, © 1985 by Annual Reviews Inc.

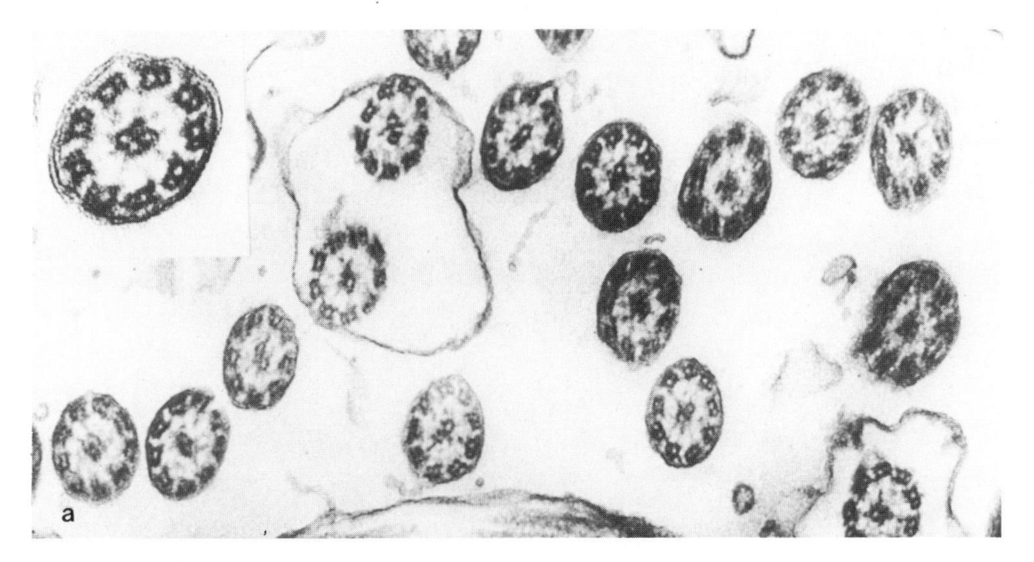

Fig. 17.13 (*a*) Freshly isolated cilia, ×55,250 (inset, ×85,000).

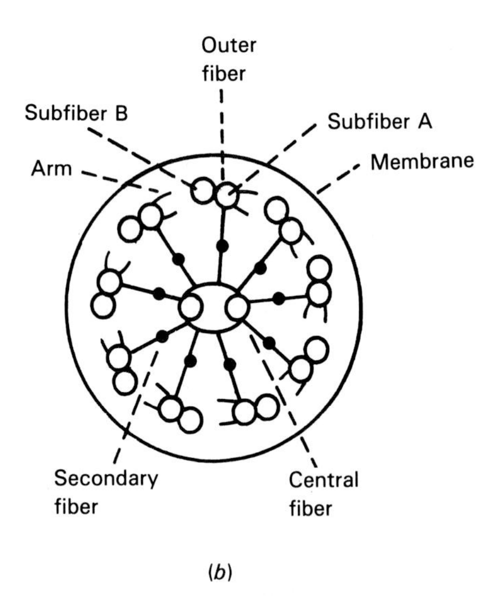

(*b*)

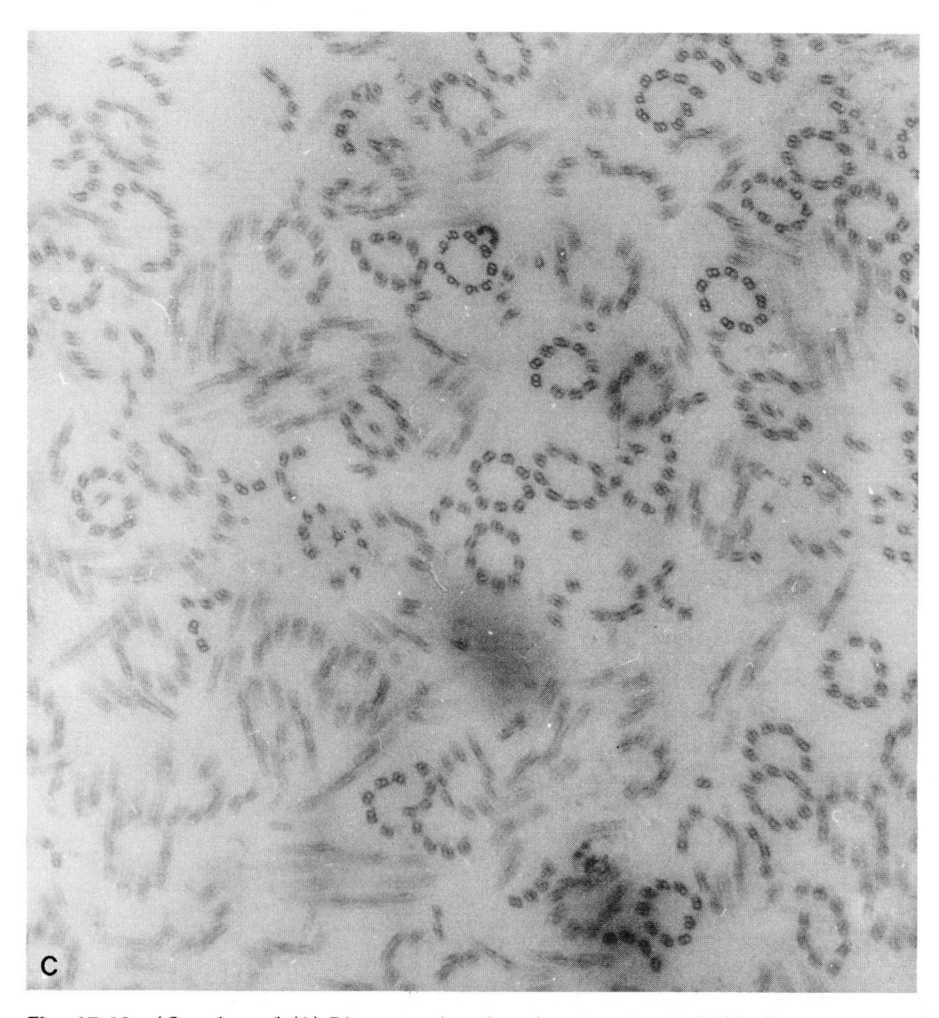

Fig. 17.13 (*Continued*) (*b*) Diagram showing the structures visible in a cross section of a cilium corresponding to (*a*). (*c*) Insoluble residue after dialyzing digitonin-extracted cilia against Tris-EDTA solution (×39,000).

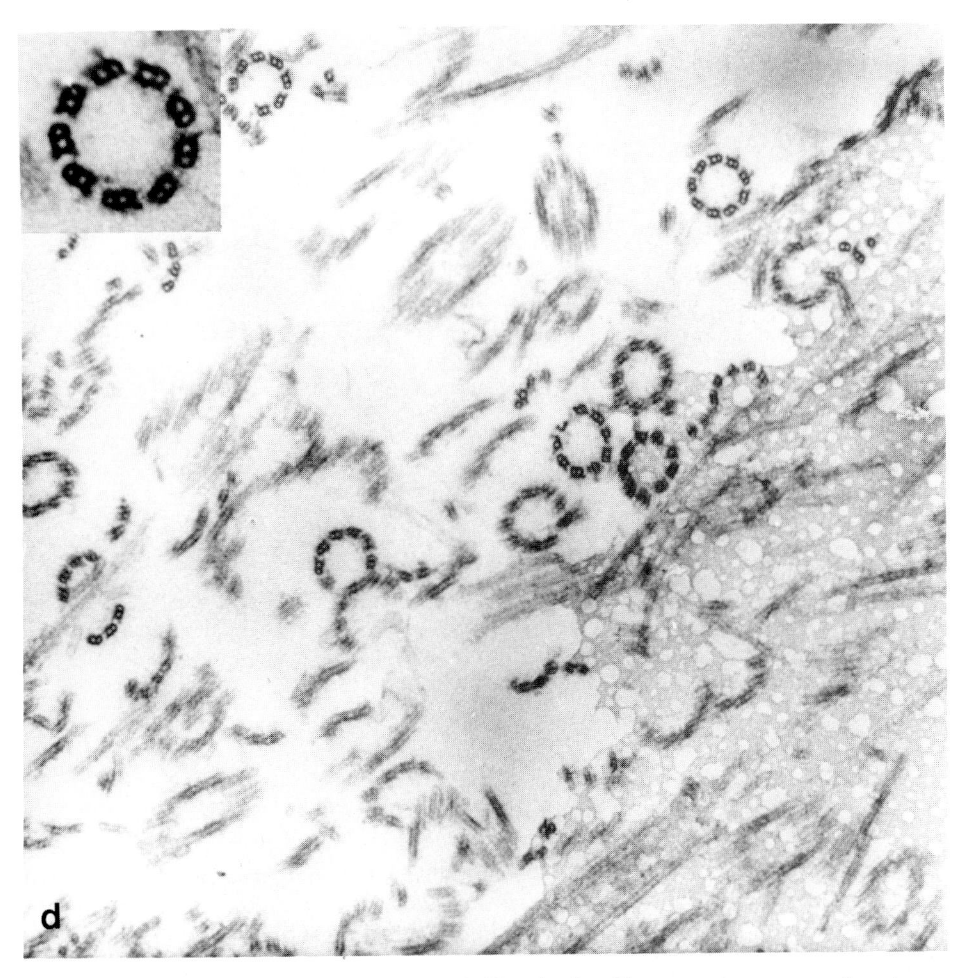

Fig. 17.13 (*Continued*) (*d*) Reconstituted cilia obtained by restoring magnesium to a preparation of digitonin-extracted cilia that had been dialyzed against Tris-EDTA solution (×40,000). Inset shows a portion of the same micrograph (×80,000). Without dialyzing, only the membrane is missing. From Gibbons (1963), with permission.

preparation extracted with digitonin and then dialyzed against a solution containing a chelator (presumably to remove Mg^{2+} and soluble components); the arms have largely disappeared. The removal of the arms (Fig. 17.13*c*) coincides with the virtual disappearance of the ATPase activity, which remains in the extract. In contrast, when the Mg^{2+} and soluble factors are replaced after the digitonin extraction, the typical ciliary structure with arms is reconstituted (Fig. 17.13*d*).

One would think that during motion the cilium must bend. Observations with the electron microscope (Horridge, 1965; Satir, 1968) suggest that the tubules in cilia do not buckle or become deformed. They may be sliding in relation to each other. The sliding out of a tubule preferentially on one side of the cilium could produce a bending in that direction, as shown in Fig. 17.14 (Satir, 1968). Detailed analyses of the motion of flagella and cilia tend to agree with a mechanism based on sliding filaments (Bryan and Wilson, 1971; Warner and Mitchell, 1981).

A sliding-filament mechanism is supported more directly by a variety of experiments carried out with isolated components containing the axonemes prepared from the flagella of sea urchin

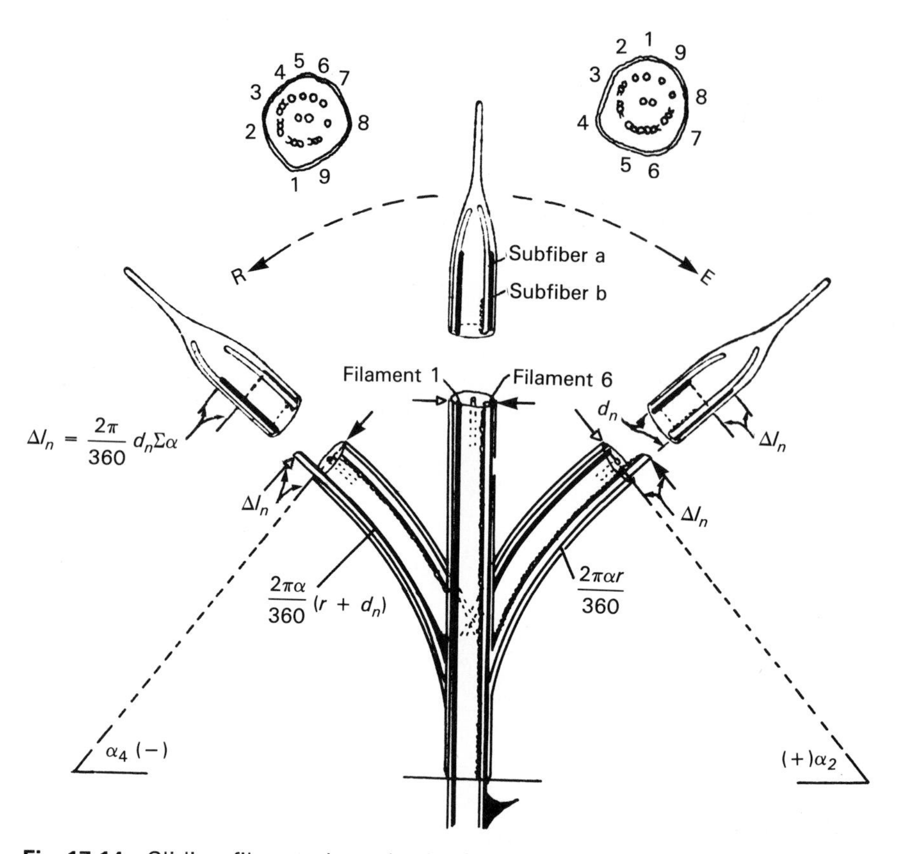

Fig. 17.14 Sliding-filament hypothesis of ciliary motility. Behavior of two doublet peripheral filaments (1 and 6) at the tip and base is illustrated when a schematized cilium is bent to either side of a straight position (center). In the neutral position subfibers b of the filaments end together at one level at the tip (i. e., are equally long) while subfibers a continue onward as naked singlet microtubules to different termination points (i. e., not equally long). The arrows with open and solid arrowheads mark equal shaft lengths from the basal plate (base line) on projections of the two filaments in the plane of the diagram (plane of beat). When the cilium bends, a circular arc arises at the base. A cross section through the tip at a level where some of the filaments are doublet and some are singlet is shown for both E and R cilia at the top of the diagram. The eye indicates that the view in both cases is from the abfrontal (effective) side of the cell. Axis tilt is neglected. From Satir (1968). Reproduced from *The Journal of Cell Biology*, 1968, 39:77–94, by copyright permission of the Rockefeller University Press.

(*Tripneustes gratilla*) sperm. The sea urchin sperm axonemes treated with trypsin remain largely intact, although they have probably become detached from their normal point of attachment. Addition of ATP causes the axonemes to dissociate. The process can be followed by dark-field light microscopy (in which the light reflected by the specimen is observed). The axonemes elongate by a process in which the tubules appear to be extruded. These observations are shown in Fig. 17.15*a* and *b* (Summers and Gibbons, 1971). In Fig. 17.15*a* the successive photographs represent progressive changes. The markers show a stationary position in the field. Apparently, part of the axoneme is stuck to the slide and does not move from its position next to the lower stationary marker. However, some of the filaments move in relation to the upper marker. This result can be most readily explained by postulating that the tubules are sliding in relation to each other.

The electron micrograph of a similar preparation from *Tetrahymena* is shown in Fig. 17.16 (Warner and Mitchell, 1981). The tubules, originally aligned, slid in relation to each other in the direction of the arrow after the addition of MgATP. This micrograph also implicates the dynein cross-bridges, which form the connection between the doublets.

IV. MOVEMENT IN OTHER SYSTEMS

In some systems no special structure is clearly associated with movement, possibly because the structures are formed transiently. The apparent variety of processes makes even a simple tabulation of events a difficult task (Allen, 1967).

We have to explain, for example, the sudden directed migration of particles involving a displacement of 10 to 30 μm in a few seconds. This *saltatory* motion may be interrupted as suddenly as it was started. While it takes place, the particle is sometimes displaced at a constant rate. It is particularly difficult to fit such an event into a simple and familiar model, since adjacent particles may be affected in entirely different ways.

Rotation of intracellular inclusions such as chloroplasts and nuclei may occur in the absence of cytoplasmic flow (Kamiya and Kuroda, 1956). In a number of systems, cytoplasm particles may flow in opposite direction as in the filamentous pseudopods of foraminifera or the stalks of *Acetabularia* (Kamiya, 1962). Here, obviously, the motion of the cytoplasm cannot be the consequence of a pressure gradient, since in the absence of rigid tubes we would expect the pressure to be transmitted in the same fashion in the different channels.

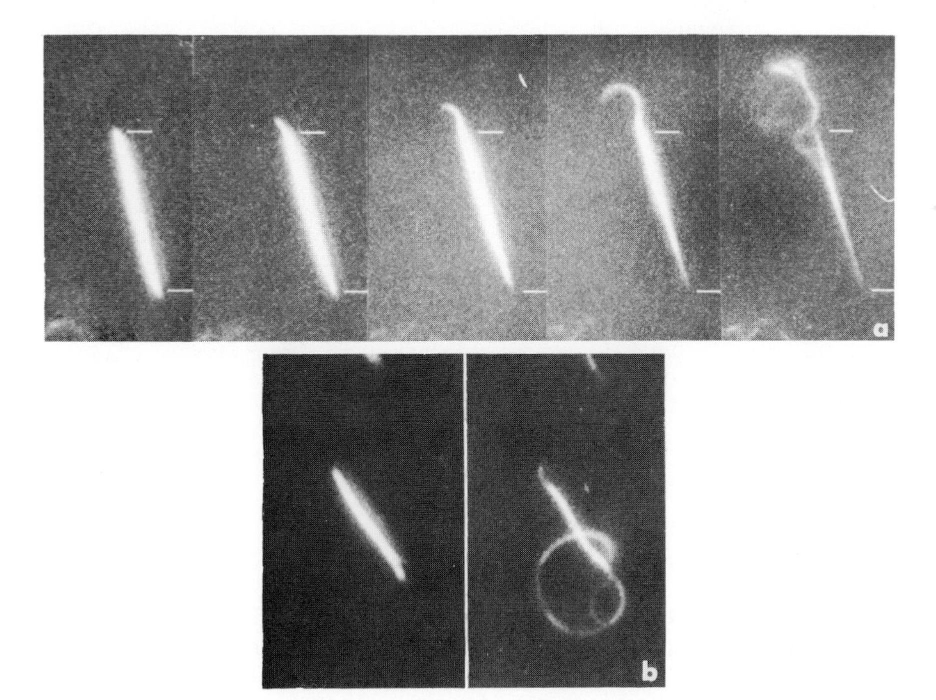

Fig. 17.15 Dark-field light micrographs of trypsin-treated axonemes after the addition of ATP. ×1300. (*a*) Each successive micrograph was taken after 12- to 30-s intervals. From Summers and Gibbons (1971). (*b*) Initial and final frames in an experiment similar to that of (*a*). In this preparation, a group of tubules slide down the attached tubules and loop around, forming a circle of tubules. The final figure is more than three times longer than the original fragment. From Summers and Gibbons (1971), with permission.

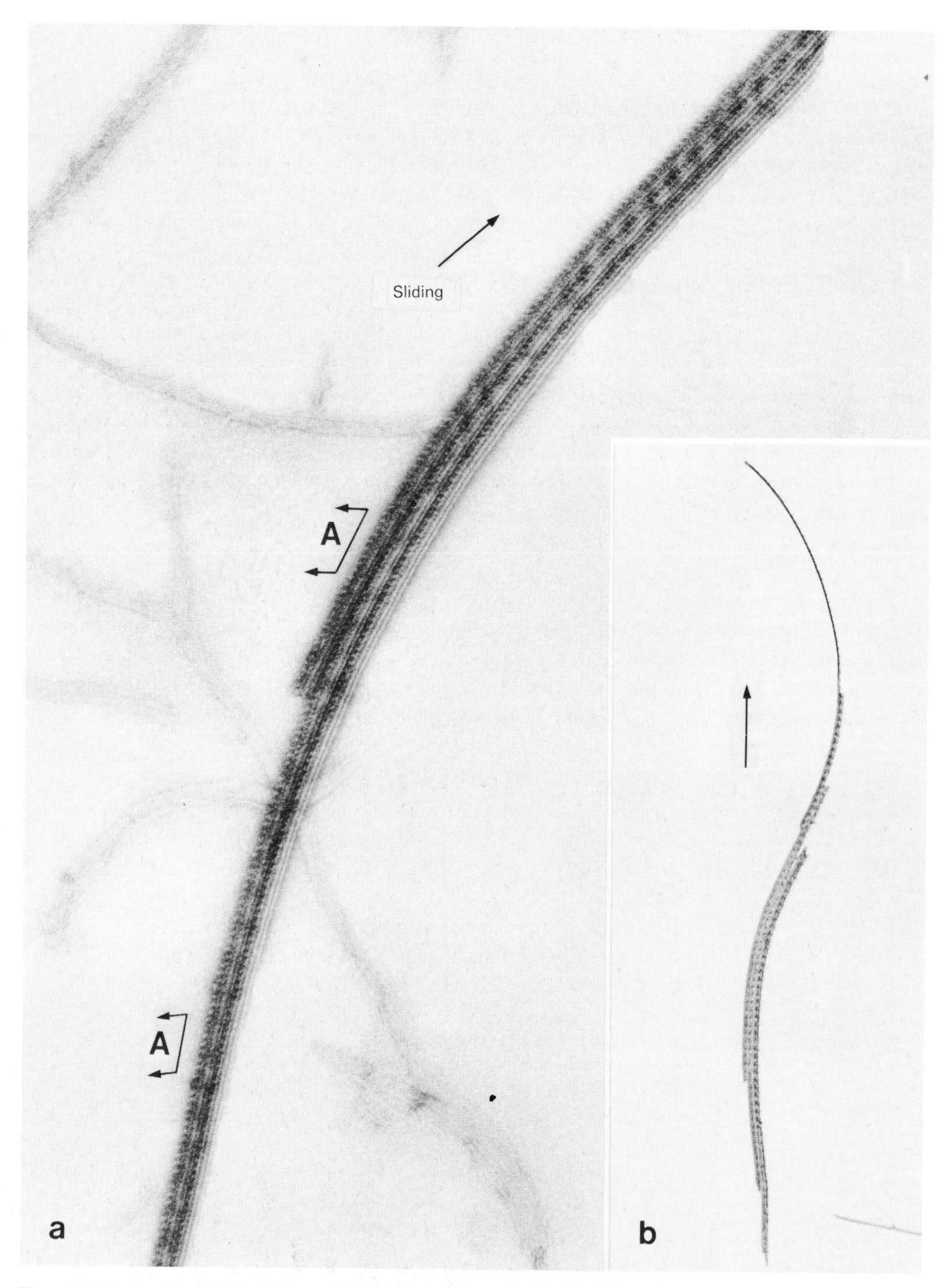

Fig. 17.16 Isolated *Tetrahymena* cilia reactivated with 0.1 mM MgATP to cause sliding disintegration. Typical sliding figures are recognized as partially overlapping doublets cross-bridged by the dynein arms. Free dynein arms are polarized and tilt uniformly toward the base of the cilium (bracketed arrows) away from the direction of active sliding. (*a*) ×51,000; (*b*) ×1,200. From Warner and Mitchell (1981). Reproduced from the *Journal of Cell Biology*, 1981, 89:35–44, by copyright permission of the Rockefeller University Press.

The nature of motility is vectorial, since it must displace mass. It is difficult to see how any motion could occur without linear structures to impose a direction to the displacement. Not surprisingly, linear elements frequently containing actin or tubulin have been found to be associated with cell movement. In many cases, these molecules have been shown to be present in linear structures by *immunofluorescence*. In immunofluorescence, antibodies conjugated to fluorescent dyes are used to identify or trace proteins. Antibodies to the appropriate purified protein (e.g., tubulin or actin) may be labeled. The structures to which they bind appear fluorescent when viewed with the appropriate microscope. In indirect immunofluorescence, the antibody is not labeled and a second fluorescent antibody, capable of binding to the first (a different species can be used to produce this second antibody), is used to locate the antigen.

Most of the detailed information has been derived from flattened cells in culture (e.g., see Osborn and Weber, 1977; Weber et al., 1975), since the structures are in the same focal plane. In mouse 3T3 fibroblasts (Fig. 17.17*a*) microtubular bundles seem to radiate from the nucleus. On the other hand, actin bundles seem to be distributed longitudinally (Fig. 17.17*b*). The pattern differs significantly with the cell type, and the intricacies of detail and arrangement of the individual fibers can be examined only with the electron microscope. The emergence of the use of high-voltage electron microscopy of thick sections, the tilting of sections, and computer-aided image processing provides a wealth of detail (see Chapter 1).

The arrangement of fibers in the cytoplasm has been seen as a three-dimensional network, the so-called *cytoskeleton*. Since the system is not rigid and is changing from moment to moment with cell movement, this name is really not appropriate. A more detailed discussion of movement in various cells is presented below.

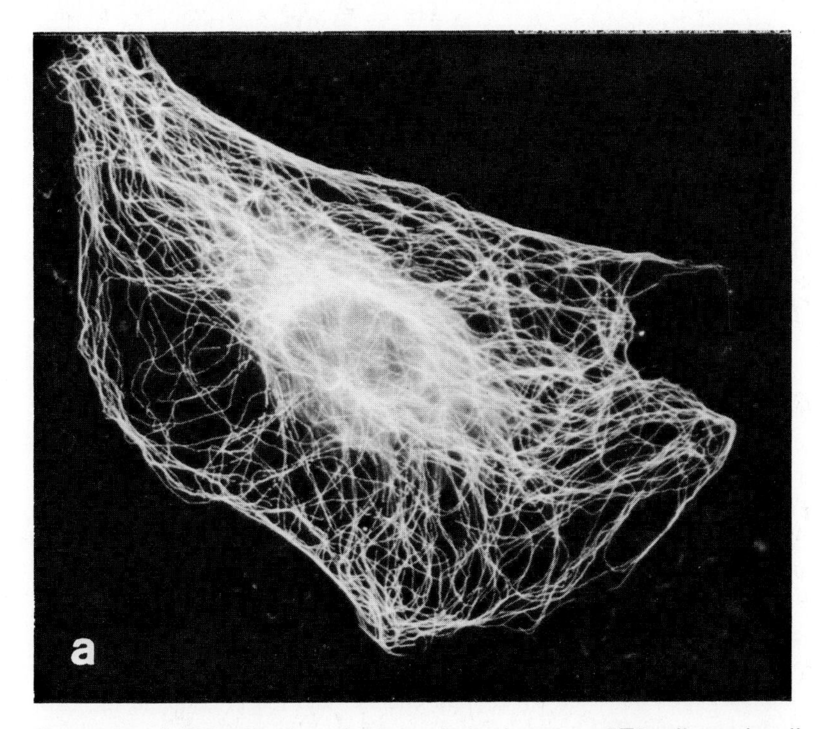

Fig. 17.17 (*a*) Distribution of microtubules in mouse 3T3 cells as visualized by immunofluorescence. Microtubules appear as an intricate network of fine fibers radiating from the region of the nucleus. From Weber (1976), with permission.

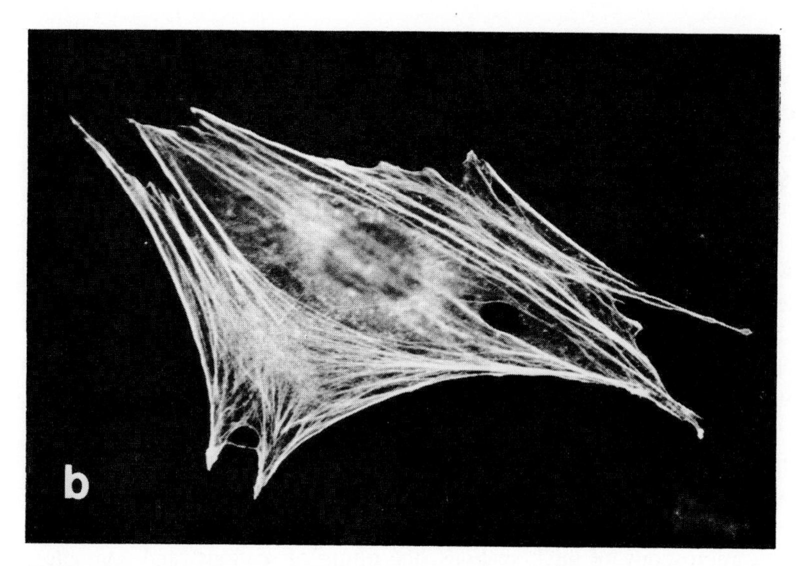

Fig. 17.17 (*Continued*) (*b*) Distribution of microfilaments in mouse 3T3 cells as visualized by immunofluorescence. Microfilaments form bundles that lie parallel to the long axis of the cell. From Osborn and Weber (1977), with permission.

A. Plant Cells

The cytoplasm of some plant cells flows between an ectoplasmic static layer and the central vacuole in a rotational manner (cyclosis) as shown in Fig. 17.18 (Hayashi, 1964). A pair of indifferent zones in which there are no chloroplasts separates the two opposing streams.

A velocity profile of particles present in the cytoplasm of the internodal cells of *Nitella* is very revealing (Fig. 17.19) (Kamiya and Kuroda, 1956). The bulk of the endoplasm flows at a constant rate regardless of the location. The ectoplasm does not flow at all. These results differ from those found for slime mold, and they suggest that the motion is the result of some interaction in a region between endoplasm and ectoplasm. In experiments in which the chloroplasts have been detached or removed, it is possible to observe longitudinally arranged fibers (Kamiya and Kuroda, 1957). These fibers disappear when the motion is arrested by the passage of an electrical current, and they return after motion is reestablished.

B. Slime Molds

Acellular slime molds are easy to manipulate and give a good rate of cytoplasmic flow. The cytoplasm flows in channels (endoplasm). The direction of flow can reverse rhythmically. Streaming cytoplasm appears more fluid and less gel-like than the surrounding cytoplasm (ectoplasm). The two cytoplasmic states are dynamic, and transitions from one form to the other may occur rapidly.

The system behaves as if it were in a state of tension, since cutting a channel produces a spurt of material. Movement can be interpreted as an increase in tension at some site accompanied by a weakening of gel-like structures where new channels are formed. The movement of the endoplasm appears to be completely passive. This has been demonstrated in experiments in which two separate chambers are connected by a small channel and the pressure in the two chambers can be readily manipulated. The normal cytoplasmic flow shows the velocity gradient represented in Fig. 17.20 (Kamiya and Kuroda, 1958). The pattern of flow is identical whether the flow is endogenous (Fig. 17.20*a)* or imposed by an external pressure (Fig. 17.20*b)*. The results can therefore be interpreted in terms of models in which contraction occurs at a site other than the endoplasm and the passive flow is a result of the increase in pressure.

Since the hydrolysis of ATP is likely to be involved in motion and fibrillar structures must be involved for directional movement, it may be particularly important to examine the distribution of ATPase activity and of fibers. Both approaches are fruitful. The ATPase activity visualized by the precipitation of inorganic phosphate formed from ATP hydrolysis appears in the ectoplasm and, more specifically, where the fibers are located.

Slime molds have remarkable properties, and various experimental manipulations can be carried out that are difficult or impossible with other organisms. The cytoplasmic threads can be hung up and shown to flow in the direction of gravitational pull to form a droplet. In this case, little work is performed to maintain tension (Wohlfarth-Bottermann, 1964). On the other hand, the flow can oppose the pull of gravity, so the system must be performing work. Fibers appear only when work is performed. The endoplasmic fibers are oriented in the direction of

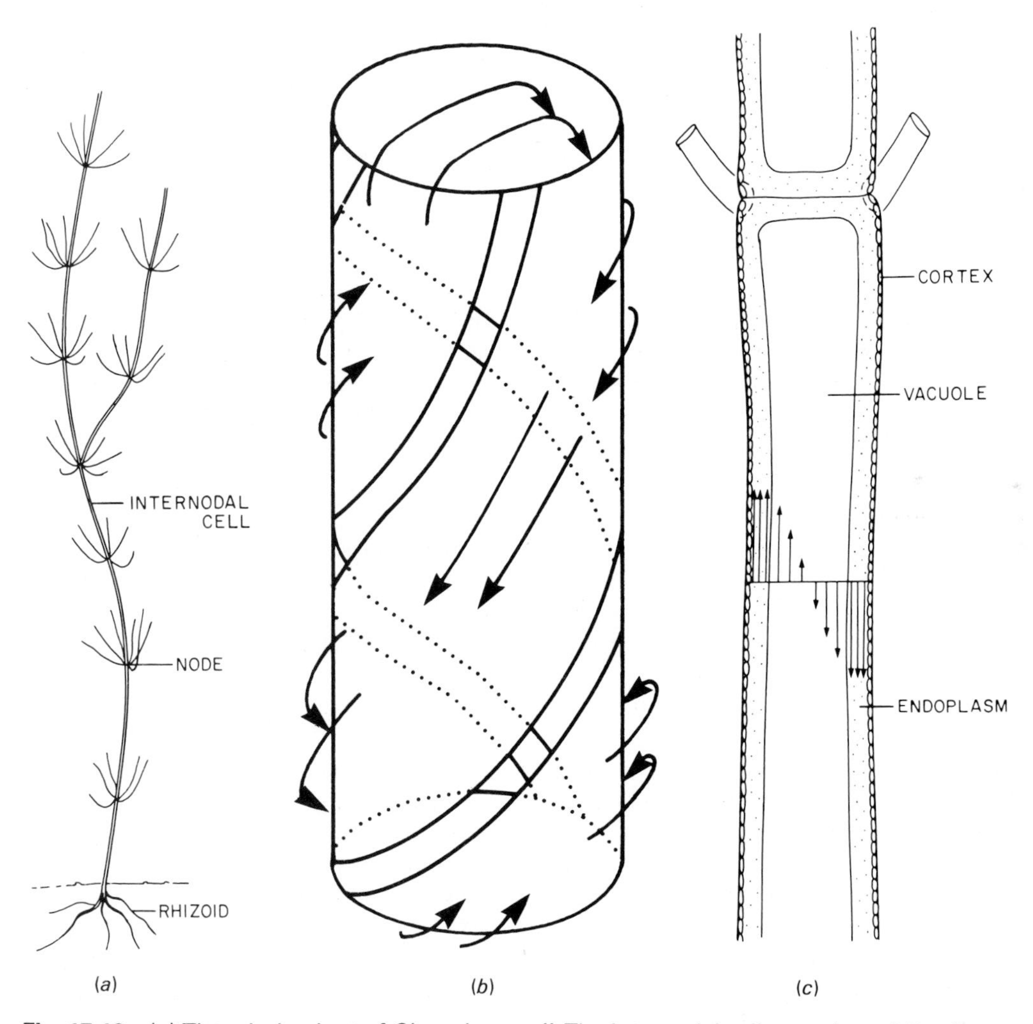

Fig. 17.18 (*a*) The whole plant of *Chara braunnii.* The internodal cells are about 0.5 to 2 cm long. (*b*) Diagram illustrating the flow in the intact internodal cell. (*c*) Longitudinal section of (*b*). In part from T. Hayashi, *Primitive Motile Systems in Cell Biology,* with permission. Copyright © 1964 Academic Press.

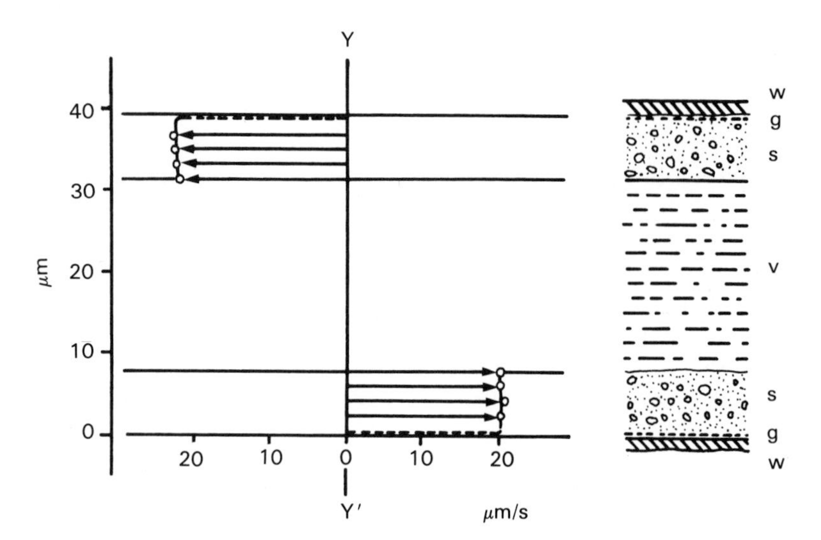

Fig. 17.19 Velocity distribution of the rotational streaming in a rhizoid cell of *Nitella flexilis*. Part of the rhizoid cell as seen under the microscope is shown on the right; w, cell wall; g, plasmagel layer (ectoplasm); s, flowing plasmasol (endoplasm); v, vacuole. From N. Kamiya and K. Kuroda, *Botanical* Magazine, 69:544–554, with permission. Copyright © 1956 Botanical Society of Japan, Tokyo.

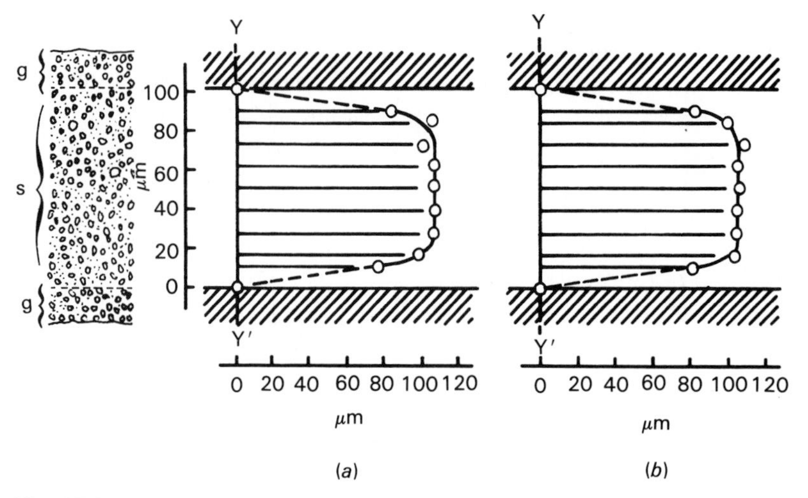

Fig. 17.20 Velocity distributions of the protoplasmic streaming in a strand of the plasmodium. Circles show positions of the granules that were in cross sections *YY'* 3 s ago. (*a*) Spontaneous streaming under natural condition. (*b*) Artificially induced streaming when pressure difference of 15 mm of water was established between the two ends of the strand 5 mm long. Microscopic view of part of the strand is shown on the left; g, plasmagel forming the wall of the capillary tube; s, plasmasol involved in streaming. From N. Kamiya and K. Kuroda, *Protoplasma,* 49:1–4, with permission. Copyright © 1958 Springer-Verlag, Heidelberg.

flow, whereas the ectoplasmic fibers, with which ATP hydrolysis is associated, are perpendicular to the direction of flow. Results consistent with this are obtained in the study of birefringence of the threads. When fibrous material is oriented all in one direction, the threads act as a polarizer and polarize light. This birefringence can be detected with the appropriate polarizer inserted in the microscopic system.

Tension could be developed by contraction of the fibers that are anchored to the ectoplasm and laid across the endoplasmic tube, producing the pressure described above and the flow in the channels of the more fluid endoplasm. The mechanism of the contraction could be similar to the sliding-filament mechanism. Perhaps filaments temporarily oriented in a particular direction slide in relation to each other.

C. Amoeboid Motion

Amoeboid movement has intrigued biologists since the first microscopic observations and has been the focus of many studies. Amoebas attached to a substratum advance by the formation of pseudopods and the retraction of their posterior mass, in a sense walking on the substratum, using their pseudopods as appendages. The endoplasmic cytoplasm flowing in a pseudopod is central to the more stationary ectoplasm. An amoeba in motion can be represented as shown in Fig. 17.21 (Allen, 1961). The cytoplasm of an amoeba, such as *A. proteus,* moves as represented in the figure. The lines across the cytoplasm represent velocity profiles. The nomenclature of the figure is that used by R. D. Allen.

The mechanism by which the cytoplasm moves has been the subject of considerable speculation over the years. Analysis of the data (Allen, 1961) favors a model in which contractile tension develops at the anchored, rigid rim of the ectoplasmic tube, formed from everting endoplasm at the fountain zone (in front). In addition, when amoebas are broken inside a capillary, flow occurs in the front in the absence of the ectoplasmic tube, in agreement with the frontal contraction model. In addition the heat released is greater at the front end of the amoeba. In contrast, suction at the rear end of an amoeba does not interfere with the flow in the pseudopods (Allen et al., 1971).

It should be noted that some amoebas may move by an entirely distinct process. For example, the giant herbivorous amoeba *Pelomixa palustris* moves in a way that is not consistent with the idea that a frontal contraction takes place and that may involve a contraction at its posterior end.

D. Cells in Culture

The movement of cells in culture has been studied, generally with fibroblasts. In all likelihood, the behavior would be the same in situ. The cells move when they become attached to the substratum. The sites of attachment have been recognized using light microscopy, in particular interference reflection microscopy (Izzard and Lochner, 1980). In this method, the rays of light reflected by two closely apposed parallel planes (for example, the slide and the plasma membrane on the underside of the cell) interfere and appear gray or gray-black. Two types of attachments have been observed, so-called *close contact* in which the membrane is approximately 30 nm away from the slide (appearing as gray in Fig. 17.22*a*) and *focal contact* in which it is approximately 10 to 15 nm away from the slide (appearing as black in Fig. 17.22*a*). Matching the image of the same cell obtained by differential interference microscopy (Fig. 17.22*b*) reveals that the points of contact (mostly focal contacts) coincide with the apparent attachment of microfilament bundles (Izzard and Lochner, 1980), the stress fibers. After glycerination of these cells, bundles contract when ATP is added, either while still in the cytoplasm or when separated out from the rest of the cell by microdissection with a laser beam (Isenberg et al., 1976). Contact

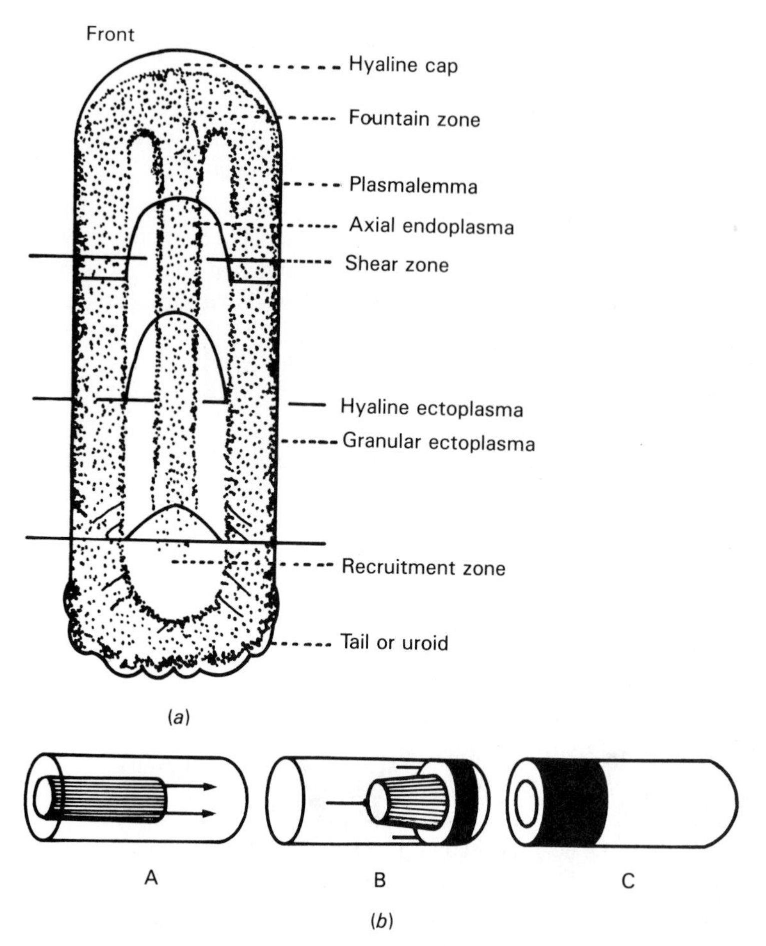

Fig. 17.21 (*a*) Diagram of an amoeba introducing the terminology and pattern of cytoplasmic streaming with average velocity profiles from the data of Allen and Roslansky. (*b*) Diagram illustrating the fountain zone contraction theory. A solid cylinder of endoplasm (A) is drawn with straight lines to indicate that it is cytoplasm in the extended condition. In B the same cylinder of endoplasm is followed into the fountain zone, where it begins to shorten (represented by folded lines) just before and as it becomes everted to form a hollow ring of ectoplasm (C). Actually, the endoplasm that becomes everted is thinner than represented in the diagram, as the shear zone has been omitted for simplicity. From R. D. Allen, *Experimental Cell Research Supplements,* 8:17–31, with permission.

structures must be continuously remade and broken down during movement. However, they are not likely to be involved in the actual mechanism of movement. Stress fibers and associated structures are eliminated in chick fibroblasts by the microinjection of antibodies to brush border myosin (Hörner et al., 1988), whereas cell movement remains.

E. Tubules and Cytoplasmic Transport

The Microtubules Tubules resembling those present in cilia and flagella, the *microtubules,* have been identified in the cytoplasm of many cells. A number of observations have suggested a direct or indirect involvement of these microtubules in movement within the cytoplasm.

(a)

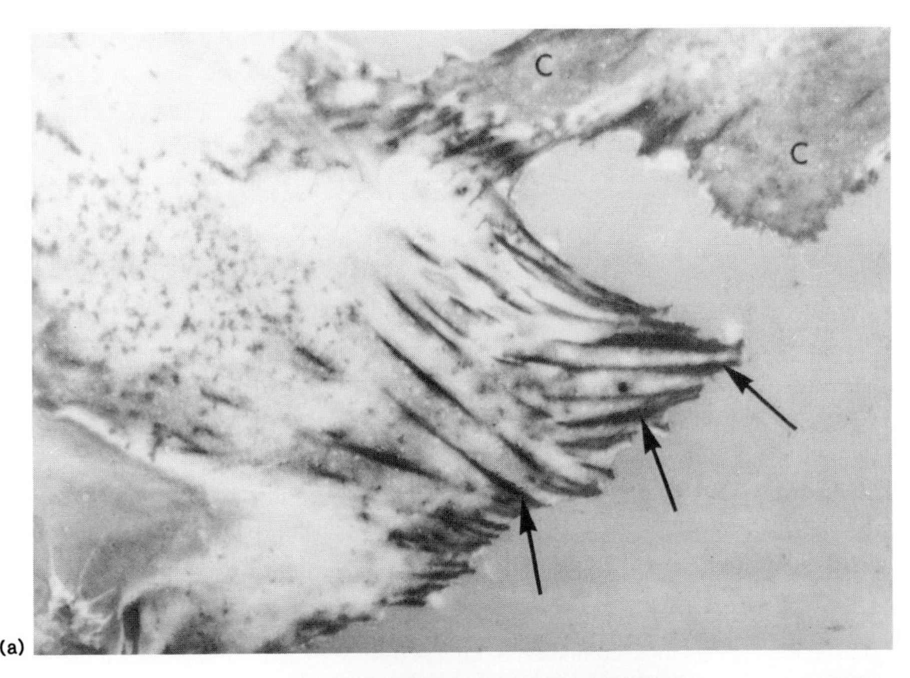

(b)

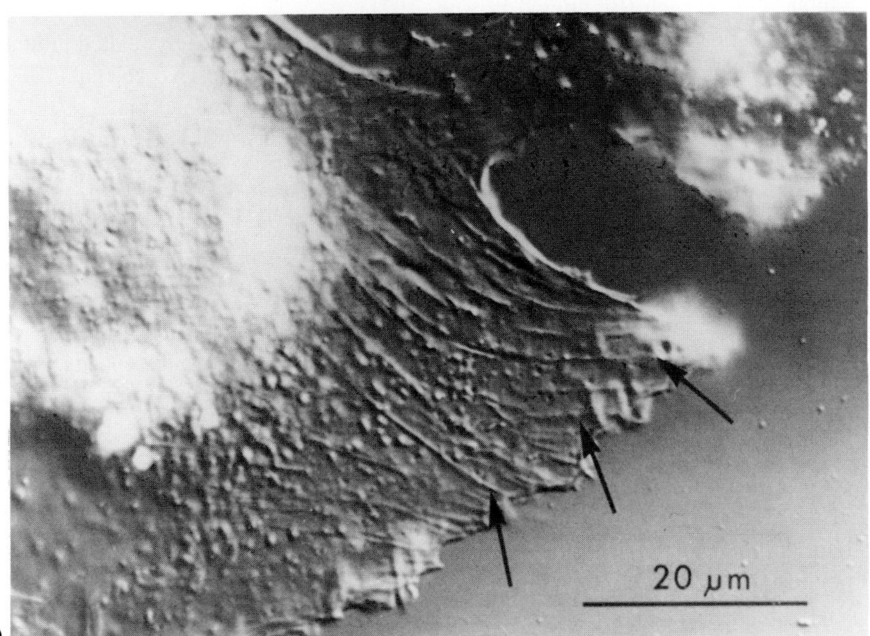

Fig. 17.22 Paired light micrographs of the same Balb/c 3T3 fibroblasts viewed by interference reflection microscopy (top) and differential interference contrast (bottom). The interference reflection microscope provides information on the distance between the cell surface and substrate (a glass coverslip) by treating this space as a thin film, which will generate an interference pattern in reflected light in the same way as gasoline on a wet road surface. The black, elongate *focal contacts* (arrows, top), located at the periphery of the cell, are the stronger substrate adhesions. The actomyosin-containing *stress fibers* in the cytoplasm terminate at thin distal ends in the focal contacts (matching arrows, bottom). The second type of substrate adhesion, the *close contact*, is present under the cell entering the upper-right corner (c, top). It is broad in area and involved in the spreading of the cell margin across the substrate. Courtesy of C. S. Izzard.

Microtubules, generally straight cylinders 25 to 36 nm in diameter, have been found in many cells. Frequently subfilaments 4 to 5 nm in diameter have been demonstrated in these cylinders (Porter, 1966). The tubules may be hollow, although central cores have been described for some.

Microtubules are present in the mitotic spindle, which has been estimated to contain as many as 5,000 to 10,000 in *Haemanthus catherinae*. They have also been found in neurons, in various other cells (Porter, 1966), in cytoplasmic spiky projections, and adjacent to the walls of some plant cells.

Colchicine prevents chromosomes from separating, apparently by binding to the spindle protein. This observation led to the idea that colchicine might be used as a marker of microtubule subunits (the *tubulins*) in various systems. By using ^{3}H-labeled colchicine to identify the molecules through fractionation procedures, proteins isolated from several sources were shown to have very similar properties (Adelman et al., 1968). These molecules are the building blocks of microtubules in different systems. Isolated subunits can be made to reconstitute microtubules. The tubulins from different organisms and tissues are similar in molecular weight and amino acid composition. They bind to colchicine and the antitumor drug vinblastine, and they react similarly to common antitubulin antibodies (e.g., Dales, 1972; Donges and Roth, 1972; Fulton et al., 1971). Tubulin appears to be composed of dimers formed by two similar but not identical subunits of about 55 kDa each (Bryan and Wilson, 1971).

Recent studies indicate that tubulin actually represents a family of closely related proteins, the *isotubulins*. As many as 17 isotubulins may be present in nerve tissue (George et al., 1981), and their presence or absence is strongly influenced by the developmental state (Dahl and Weibel, 1979; Denoulet et al., 1982; Gozes and Littauer, 1978). Interestingly, a single neuron may have as many as nine isotubulins (Gozes and Sweadner, 1981) and there are indications that their location within the cell is specific. A much smaller number of isotubulins has been detected in liver cells, where their presence does not seem as dependent on developmental state (Donges and Roth, 1972).

The microtubules of cilia and flagella are involved in cell movement. It would be indeed surprising if the microtubules in the cytoplasm of other cells were not responsible in some way for cell movement. Some of the evidence, however, is only indirect and depends on the ability of colchicine and vinblastine to block cell motility and simultaneously interfere with microtubules. The effect of colchicine seems to be relatively specific, but vinblastine reacts with a number of cell components. In at least some cases, there is more direct evidence of a microtubular involvement. The effects of these two drugs are summarized in Table 17.3, which suggests a very general role of microtubules in motility.

The role of microtubules in the transport of materials has been explored extensively in the *axopodia* of heliozoans and foraminifera. Axopodia are slender cytoplasmic processes radiating

Table 17.3 Effect of Colchicine or Vinblastine in Cell Motility

Tissue	Blocking agent	Effect on microtubules or oriented fibers	References
Chick nerve endings	Vinblastine	Yes	*a*
Crayfish neurons	Vinblastine	Yes	*b*
Hypothalamic neurons	Colchicine	No	*c*
Cultured hamster kidney cells	Colchicine	Yes	*d*

[a]Feit et al. (1971).
[b]Fernandes et al. (1971).
[c]Flament-Durand (1972).
[d]Goldman (1971).

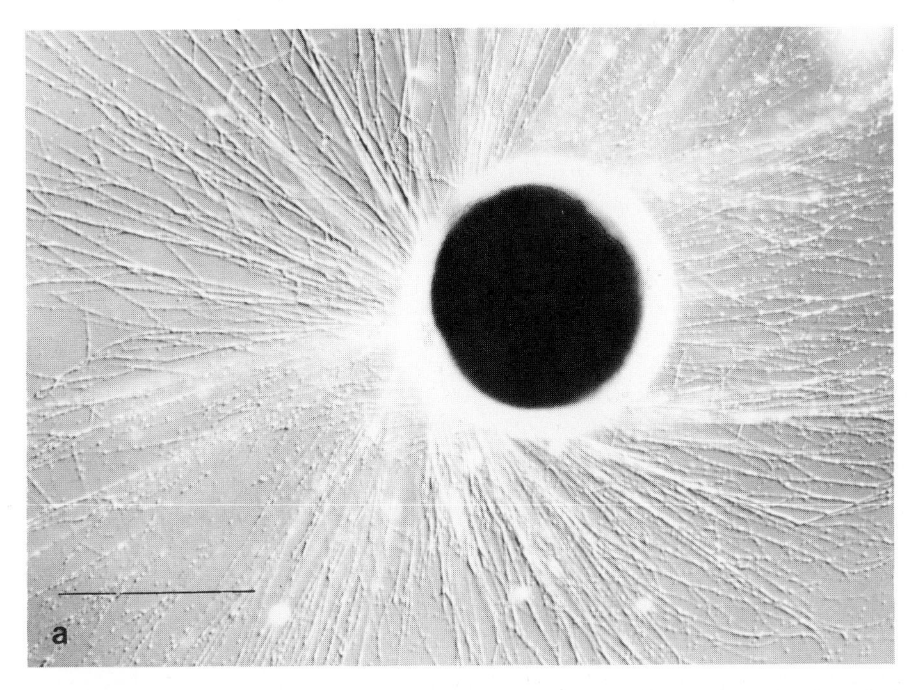

Fig. 17.23 (*a*) *Allogromia laticollaris*, a typical foraminiferan. This single-celled organism consists of a spherical cell body that contains the nucleus and an elaborate network of interconnected pseudopods. The pseudopodial networks may become quite expansive, with the main trunks attaining lengths of nearly 1 cm. Food particles, such as bacteria and diatoms, bind to the pseudopodial membrane and are transported along the outside of the plasma membrane until they accumulate near the cell body. The pseudopodial movements are driven by movements of the cytoskeletal microtubules. Bidirectional intracellular transport of organelles occurs throughout the network. This transport, as well as the cell surface transport of food particles, occurs only along the cytoplasmic microtubules. ×350; bar = 100 μm. From Travis and Allen (1981), with permission.

from the main cytoplasmic mass. They are a few micrometers thick, but they can extend as far as half a millimeter (Fig. 17.23*a)* (Travis and Allen, 1981). The material within the axopodia exhibits cytoplasmic streaming. The internal structure of the axopodia, the *axoneme,* contains longitudinally oriented microtubules (Fig. 17.23*b* and *c)* (Travis and Bowser, 1988). The electron micrographs also show cross-bridges between microtubules and vesicles. The interaction between microtubules and vesicles is discussed in Chapters 4 and 18. In the axoneme, cytoplasmic particles may stream independently and even in opposite directions (MacDonald and Kitching, 1967). A direct involvement of microtubules is indicated most clearly by experiments in which keratocytes were studied with light microscopy and video enhancement techniques (Allen et al., 1981). Motion of particles was observed in linear elements visible in the thin parts of the cells. Although individual microtubules and even microtubular bundles may be below the level of resolution (see Chapter 1), they can be perceived with the light microscope as structures in the range of 100 to 200 nm forming linear arrays. That these linear arrays correspond to microtubules was shown in the following manner. The cells were lysed and fixed under continuous observation to ensure that no structural rearrangement took place. Then they were labeled with the appropriate antitubulin antibodies (Fig. 17.24) (Hayden et al., 1983). The labeled linear arrays clearly corresponded to those visible with differential interference optics.

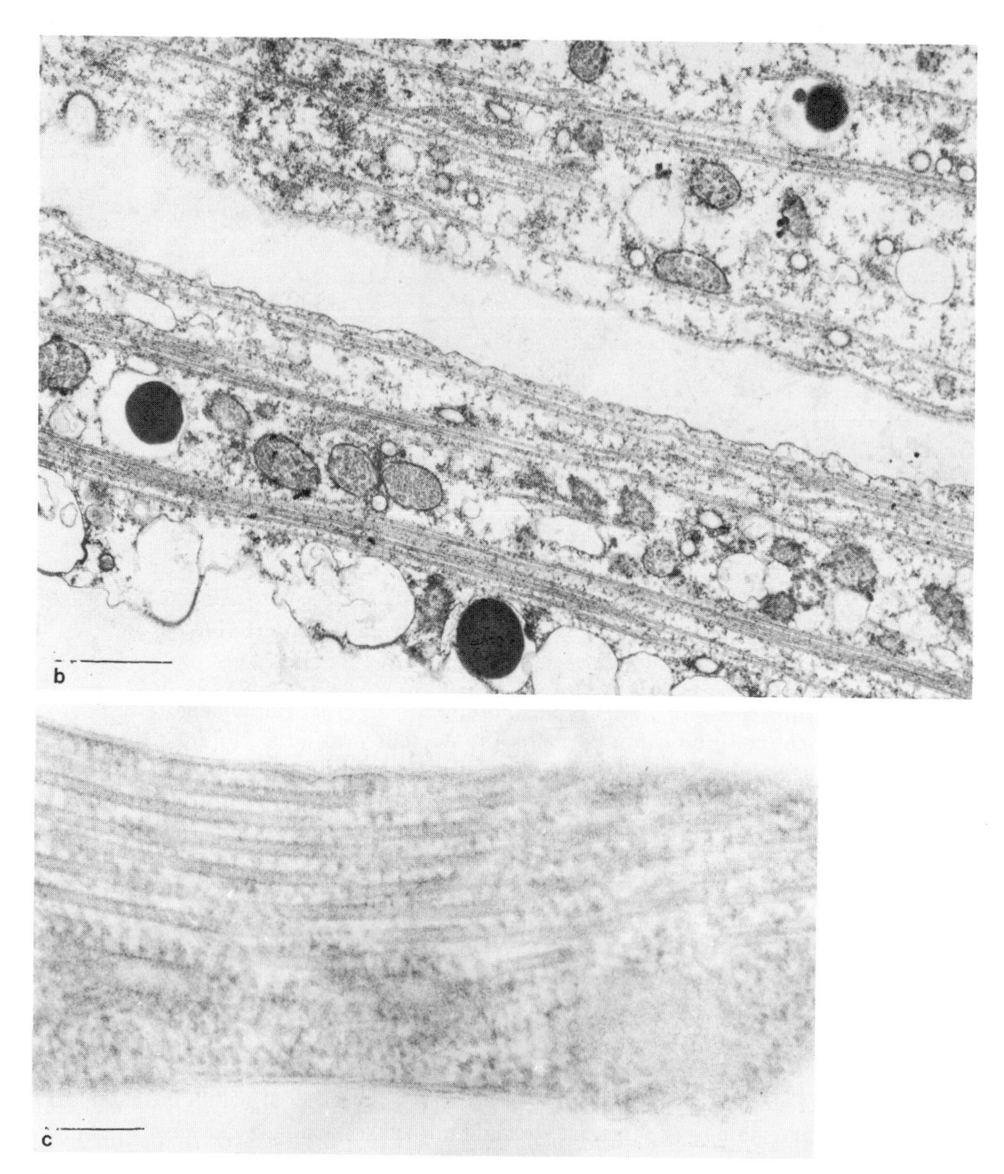

Fig. 17.23 (*Continued*) (*b*) Conventional thin-section transmission electron micrograph through a foraminiferan pseudopod showing the close association between the microtubules and the transported organelles. Mitochondria, coated vesicles, and other membranous organelles move along the microtubules. In this figure, several organelles appear to be linked to the microtubules by cross-bridge structures. ×20,000; bar = 1 µm. Unpublished micrograph from J. L. Travis. (*c*) A two-step lateral translation of a thin-section electron micrograph of a foraminiferan pseudopod. This technique enhances the periodic cross-bridge structures that link adjacent microtubules. Note that similar side arms serve to attach the microtubules to the plasma membrane. ×80,000; bar = 0.2 µm. From Travis and Bowser (1988), with permission.

Cytoplasmic microtubules, like their counterparts in the axonemes of cilia and flagella, are polar assemblies of α- and β-tubulin, which alternate in the formation of *protofilaments* (i.e., the individual threads). There are generally 13 protofilaments per microtubule. One end of the microtubule differs from the other end functionally and in molecular terms, since the front and the back ends have different properties. In vitro, tubulin polymerizes to form microtubules. The rate of growth differs, however. The end that grows faster is called the plus end; the slower growing end is the minus end. The direction of movement of different motors (see Chapter 18, IV,B) in relation to the two ends also differs. The polarity of microtubules can be demonstrated experimentally by two different approaches. Dynein attaches to microtubules with its arms directed toward the MOTC. As discussed in more detail in Chapter 18, dynein is one of the "motors" associated with movement. Tubules are also capable of attaching to more tubulin subunits thereby forming incomplete tubules. These have been found to form arms with the appearance of right- or left-handed hooks. In cross section, right-handed hooks indicate that the positive end faces up.

Even at steady state there is a rapid exchange of subunits between microtubules and the soluble tubulin, in vitro and in intact cells (Saxton et al., 1984). A variety of *microtubule-associated proteins* (MAPs) have been found that suppress microtubular dynamics (e.g., Bré and Karsenti, 1990) and can serve as nucleation centers for polymerization. Regulation of the polymerization has very important roles in cell structure and function, and at least some of this regulation results from the physiologically controlled phosphorylation of MAPs.

Inside cells, the assembly is highly regulated spatially and temporally. For example, the microtubules that will form the mitotic apparatus (see next section) must assemble in a particular location (that is, at the two poles) and at the appropriate time (i.e., at the beginning of mitosis) to create the proper structure. Conversely, its disassembly must be organized. The same regulation should take place at other sites of microtubular assembly.

The assembly of microtubules and the polarity itself depend on the *microtubules-organizing centers* (MTOCs) which serve to nucleate the microtubular assembly and appear to be responsible for the number of *protofilaments* (i.e., individual threads) per microtubule (generally 13 in vivo). The MTOCs consist of diverse structures, such as the basal bodies of cilia and flagella and in animal cells, the area of the centrioles. The role of MTOCs in nucleation is still obscure.

As mentioned above, microtubules are assembled from α- and β-tubulin. A third kind of tubulin, γ has been found. However, γ-tubulin is not a component of microtubules. It is present, however, at other locations, for example, at the centrosome (i.e., MTOC) region of the mitotic spindle. It has been suggested that the γ-tubulin is associated with MTOCs and that it imposes the polarity of microtubules by providing a link between the MTOC and β-tubulin by binding to both (e.g., Oakley, 1992).

The Axon As we saw in Chapter 16, neurons are unique cells functionally and anatomically. The axons extend from the cell body, sometimes for several meters. Since synthetic and assembly processes take place in the cell body, the distant regions can be supplied only by a special transport system. Conversely, to be removed from the periphery, continuous transport in the direction of the cell body is needed. Movement from the cell body out into the axon is referred to as *anterograde*. Movement in the opposite direction is called *retrograde*.

In recent years a good deal of progress has been made in the study of axonal movement (see Vallee and Bloom, 1991).

When neuronal cell bodies are labelled by radioactive amino acids (e.g., by injection into a ganglion), labelled proteins are detected along the axon as a function of time and distance. The proteins can be separated by SDS-PAGE electrophoresis (see Chapter 1, II,B). This approach has shown that different proteins travel at different rates (see Grafstein and Forman, 1980).

Rapid anterograde transport (approximately 20 to 400 mm per day) is responsible for the movement of vesicle, endoplasmic reticulum, synaptic vesicles, and plasma membrane components. Slow movement (0.1 to 4 mm per day), on the other hand, concerns the movement of cytoskeletal elements and cytoplasmic enzymes of intermediate metabolism. Retrograde transport resembles fast movement and seems to correspond to movement of vesicles in the endocytotic pathway (see Chapter 3, I).

Fast axonal transport involves microtubules. Virtually all microtubules are oriented with the plus end toward the axon terminal and the minus end toward the cell body (e.g., Heidemann et al., 1981). Because of this innate molecular asymmetry, transport of materials along the microtubules in one direction is different (structurally and kinetically) than transport in the other direction. Gliding in either direction along single axonal microtubules has been observed (Allen et al., 1985), and two major mechanochemical transducing enzymes or so-called motors, dynein

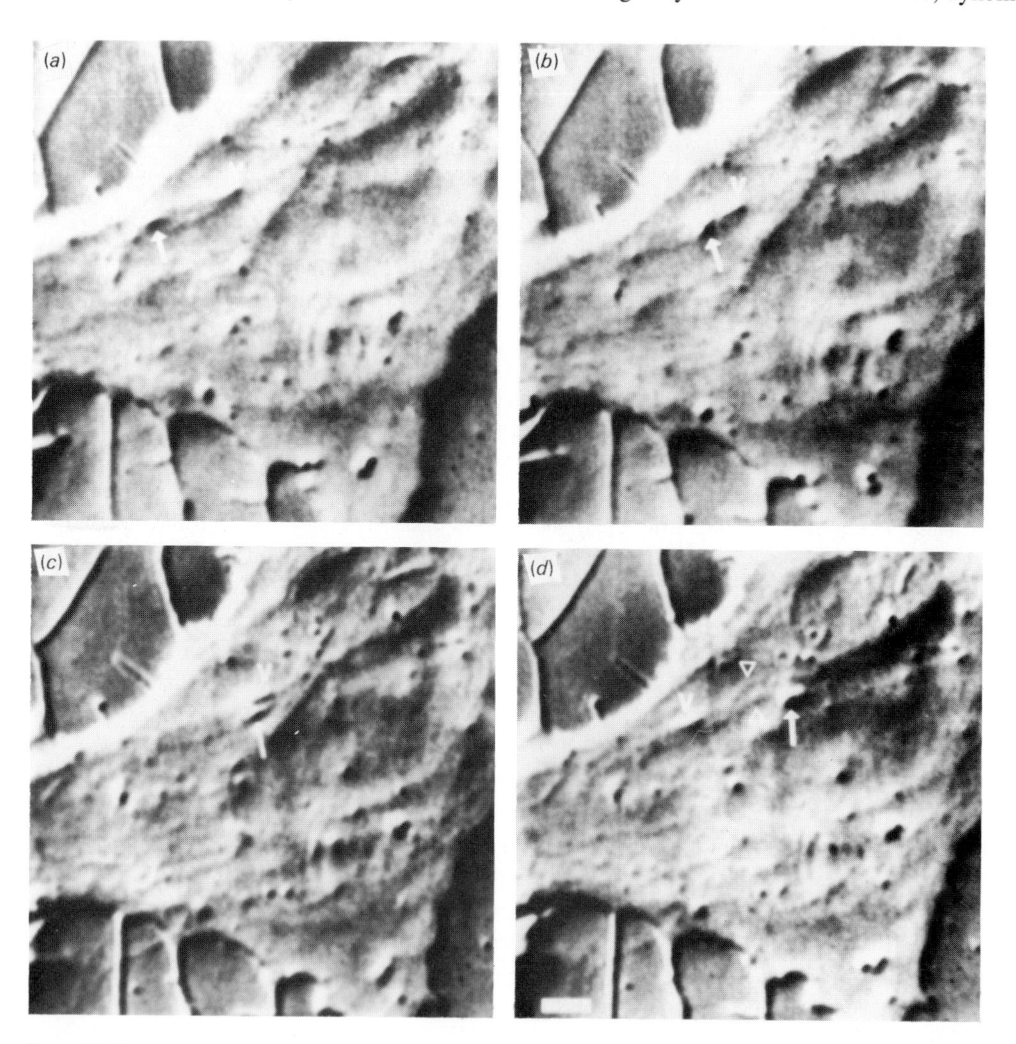

Fig. 17.24 Demonstration of movement involving linear elements (*a–d*) and demonstration that the linear elements correspond to microtubules in keratocytes. (*a*) One particle in motion (arrow) approached a stationary particle (arrowhead). (*b*) The particles collided and motion stopped. (*c*) The particle in motion moved to the side of the stationary particle. (*d*) Both particles moved in opposite directions.

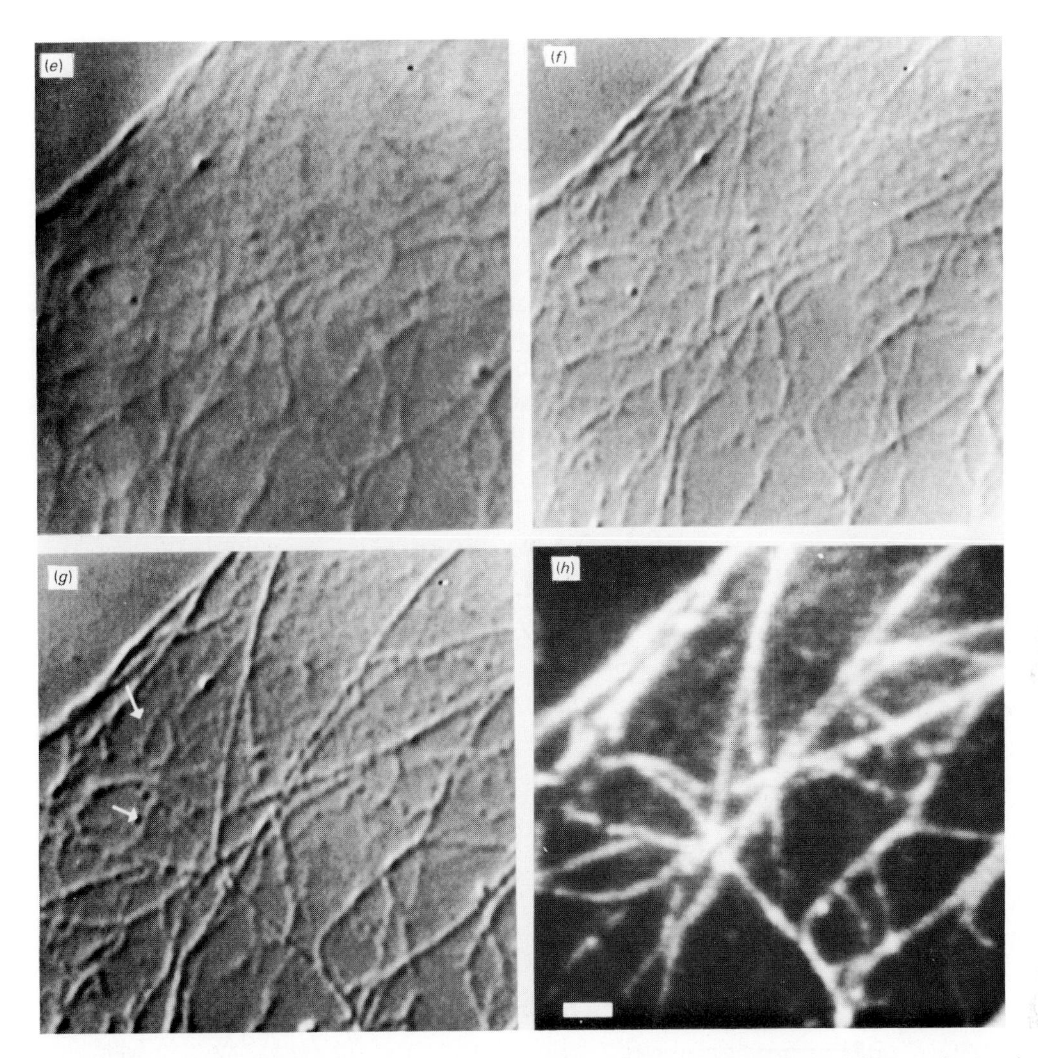

Fig. 17.24 Demonstration of movement of vesicle along a microtubule. Video-enhanced images using differential interference contrast (AVEC-DIC) and immunofluorescence. (*a*) and (*b*) show the position of the vesicle at different times. (*c*) to (*g*) demonstrate that the linear elements (LEs) are microtubules. (*c*) Images of three cells showing the LEs present and partially exhibiting movement along the LE; (*d*) image of cell after lysis in microtubule stabilizing solution (Hayden et al., 1983, with permission).

and kinesin are responsible for transport in a single direction. The involvement of microtubules and these motors will be discussed in more detail in Chapter 18 (IV,B).

Movements in the Mitotic Spindle The complexity, precision, and drama (it has all but a surprise ending) of cell division have fascinated many investigators from the very first days of the study of cells. Microtubules are heavily involved in many of the processes. This section can only cover some of the salient points. The first part will discuss the general pattern of mitosis, followed by some of the details of chromosome movement.

Mitosis follows a specific choreography that differs somewhat in detail from one kind of cell to another (for details see e.g., Hyams and Brinkley, 1989). The cells enter mitosis (Fig. 17.25), that is, the M phase from the G_2 phase. At *prophase,* the chromatin which is present in a diffuse form at interphase and has duplicated during the preceding S-phase (the phase before G_2) to form sister chromatids, condenses to form chromosomes which remain attached at the centromere (see below). The microtubules, disassembled from the cytoskeleton, begin assembling to form an *aster* at the *microtubule organizing centers* (MTOC) corresponding to the *centrioles* in animal cells. The centrioles have originated from the duplication of the original centriole pair (the *centrosome*) and eventually arrange themselves at opposite ends of the cells at the *poles.* At *prometaphase,* the nuclear envelope breaks up into many vesicles and the microtubules of the nascent *spindle* extend into the nuclear region. Some of the microtubules attach to *kinetochores,* the kinetochore microtubules (K-MT in Fig. 17.26). The kinetochores are specialized structures formed from the *centromeres,* a specific portion of each chromosome.

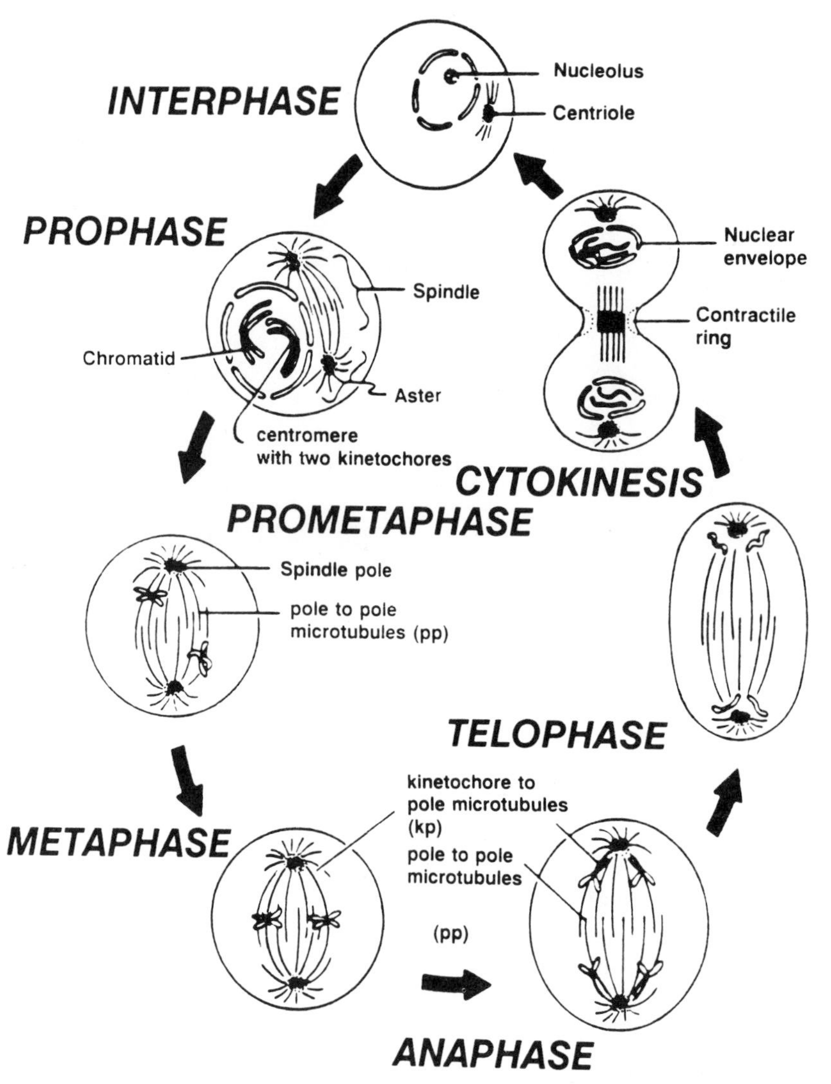

Fig. 17.25 Diagrammatic representation of the processes occurring during mitosis.

The polar microtubules (P-MT in Fig. 17.26) overlap with microtubules originating from the opposite pole in the spindle equator. Other microtubules remain in the original aster and may play a role during the movement of the centrioles in cell elongation during anaphase (see below). At *metaphase*, the chromosomes become aligned in the *metaphase plate*. The splitting of the centromeres initiates *anaphase* in which the kinetochore microtubules shorten and the sister chromatids move in opposite directions. At anaphase, the spindle elongates, when the poles of the spindle are pulled apart and the P-MT elongate by polymerization. During *telophase*, the separated daughter chromatids arrive at the poles and the K-MT microtubules dissociate, whereas the P-MT microtubules elongate some more. The nuclear envelope then reforms, and the chromatids expand from the condensed configuration. At *cytokinesis*, the two newly formed cells separate in a process which includes the contraction of the *contractile ring*, based on an actomyosin system (Schroeder, 1973; Mabuchi and Okuno, 1977).

During some of these stages, microtubules and kinetochores are connected dynamically through the plus end of the microtubules connecting the kinetochores to the poles (K-MT) (see e.g., Euteneur and McIntosh, 1981). The two kinetochores of each sister chromatid are attached to K-MT that are connected to opposite poles. Tubulin is thought to be continuously incorporated at the kinetochore during metaphase. During anaphase, when the K-MT shorten and the kinetochores are moved toward the poles, tubulin is lost from the kinetochore region. These exchanges have been followed by injecting fluorescently labelled tubulin into dividing cells (see below).

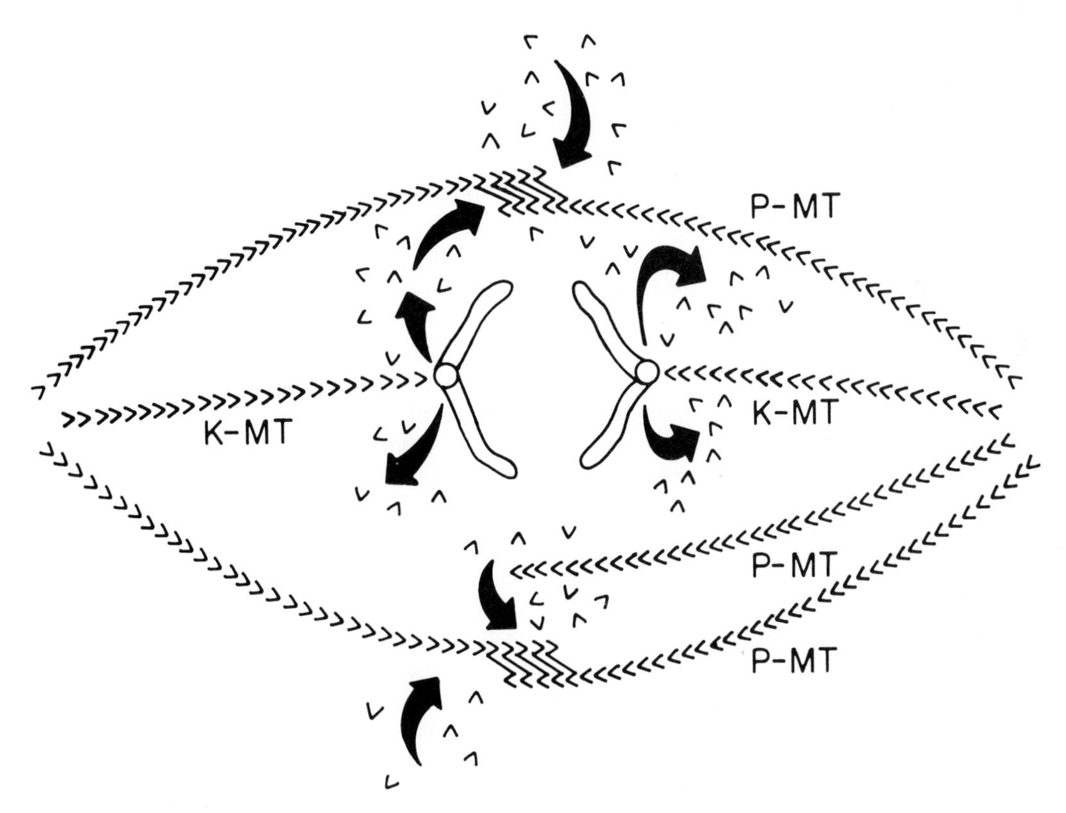

Fig. 17.26 Diagrammatic representation of events during anaphase. Only one chromosome pair is shown. K-MT corresponds to the kinetochore microtubules, whereas P-MT corresponds to the polar microtubules. The arrowheads represent the tubulin subunits that form the microtubules. The heavy arrows indicate the direction of the polymerization of the P-MTs or the K-MTs. The lines between the oppositely oriented P-MTs represent cross-links.

The details of mitosis are far more complex, and this section can only outline the major aspects. There are a number of questions that may be asked. How do the microtubules attach to the kinetochores? Using same-cell correlative video-DIC light microscopy, immunofluorescence, and electron microscopy, Rieder and Alexander (1990) studied the attachment of chromosomes which were initially positioned many micrometers from the polar region in early prometaphase. Attachment occurred when a single microtubule became associated with one of the kinetochores. The kinetochore then moved poleward along the surface of the microtubule, suggesting a mechanism of movement similar to the other microtubule based motor-driven movements as discussed for the axon (see also below). Before attachment, the kinetochore lacks microtubules (shown by EM and immunofluorescence) so that it cannot serve as a nucleation site. Another important question is the mechanism of alignment of the chromosomes in the metaphase plate. This is thought to be the result of a balance between the pull by the K-MTs' attached kinetochores connected to opposite poles. The force appears to be proportional to the length of the attached microtubule, tending to keep the chromosome on the equatorial region. What initiates the anaphase movements? Ample evidence suggests that anaphase is triggered by a sudden increase in the Ca^{2+} concentration release by vesicles in the spindle (see Hepler, 1980, 1987).

The events of anaphase have two components. At anaphase A, the K-MTs shorten and the sister chromatids are pulled to opposite poles (Gobsky et al., 1988). In contrast, at anaphase B, the poles move apart and the P-MTs lengthen (Masuda and Cande 1987, Pickett-Heaps, 1986). The two are clearly distinct processes since they can take place independently. For example, chloral hydrate blocks anaphase B but has no effect on chromosome movement.

Movements of the kinetochores in anaphase could be generated by several possible mechanisms. Apparently, the actomyosin system is not involved since antimyosin or antiactin applied to isolated spindles, lysed cells, or microinjected into cells does not interfere with anaphase movements (Sakai et al., 1976) (whereas they block cytokinesis). Of several possible alternatives, two of these are considered the most likely. One involves protein *motors* for axonal transport and is discussed more fully in the next chapter (Chapter 18, IV,B). The other corresponds to the disassembly of the microtubules.

Two motors with opposite polarity have been demonstrated on kinetochores of isolated chromosomes (e.g., Hyman and Mitchison, 1991). Experiments using different concentrations of ATP-γ-S, an ATP analog that can phosphorylate but cannot be hydrolyzed, were able to vary the direction of the movement, suggesting a regulation of the two motors by phosphorylation. In addition, the presence of two motors with opposite polarity suggests a possible role of the kinetochore in a variety of chromosome movements indicated in Fig. 17.25. A plus-end motor (i.e., acting away from the poles) could be involved at or before metaphase. The minus-end motor (i.e., acting toward the poles) would be involved in the anaphase A chromosome movement. In support of this position, antibody to one of the proteolytic fragments of dynein was found to block the process (Sakai et al., 1976). Since no K-MTs remain on the equatorial side of the chromosomes as the kinetochores are moved toward the poles, the K-MTs must depolymerize at the kinetochore. Such a depolymerization has been demonstrated (Gorbsky et al., 1988) using fluorescent tubulin (derivatized with the dye X-rhodamine) injected into cells. This procedure introduces an homogeneous labelling of the microtubules. A photobleached spot on a K-MT remains stationary while the microtubule shortens. This could only happen if the K-MT microtubules were depolymerizing at the kinetochore end.

The evidence strongly supports the involvement of a motor in the movement of the kinetochores. However, the depolymerization could provide the needed energy by itself as demonstrated with model systems (Coue et al., 1991). In this study, lysed and extracted *Tetrahymena* cells were used. Microtubules were polymerized from ordered arrays of basal bodies so that the minus end is fixed. They could be depolymerized simply by perfusing with a tubulin-free medium. Attached chromosomes introduced in the system were found to move in the direction of the basal bodies upon depolymerization. The movement appears not to require ATP (e.g., it occurred in the presence of orthovanadate which blocks ATP hydrolysis or apyrase, an enzyme which hydrolyses ATP), and under appropriate conditions it exhibited constant velocity.

Anaphase B spindle elongation involves the P-MTs. In this case, isolated spindles from diatom were shown to incorporate fluorescently labelled tubulin at the spindle midzone (Masuda and Cande, 1987). However, the actual elongation requires ATP hydrolysis and might involve a motor in a mechanism resembling the sliding of axonemes in disrupted cilia or flagella (discussed earlier in this chapter, III). An overlap between P-MTs originating from the two different poles has been shown with electron microscopy (McDonald et al., 1979). Cross-bridges between P-MTs have been shown in isolated mitotic spindles using a colloidal gold-labelled monoclonal antibody to flagellar dynein (Hirokawa et al., 1985). In contrast to cilia or flagella (see Fig. 15.26), in this case the sliding tubules are antiparallel. However, dynein cross-links are possible with either parallel or antiparallel configuration (Warner and Mitchell, 1981).

V. CONCLUDING REMARKS

The examples presented make it clear that motility occurs in a variety of ways that may be based on fundamentally different molecular processes. On the other hand, as we shall see, there are a number of similarities in the behavior of the macromolecules extracted from these motile systems. For this reason, a number of investigators have used as their working hypothesis the idea that the different motile systems are different facets of similar molecular events.

SUGGESTED READING

Gibbons, I. R. (1981) Cilia and flagella of eukaryotes. *J. Cell Biol.* 91:107s–124s.

Lackie, J. M. (1986) *Cell Movement and Cell Behavior,* Chapters 1–3. Allen & Unwin, London.

Kamiya, N. (1981) Physical and chemical basis of cytoplasmic streaming. *Annu. Rev. Plant Physiol.* 32:205–236.

Kamiya, N. (1986) Cytoplasmic streaming in giant algal cells: a historical survey of experimental approaches. *Bot. Mag. (Tokyo)* 99:441–467.

Lackie, J. M. (1986) *Cell Movement and Cell Behavior,* Chapters 1–3. Allen & Unwin, London.

McIntosh, J. R., and Pfarr, C. M. (1991) Mitotic motors. *J. Cell Biol.* 115:577–585.

Sawin, K. E., and Scholey, J. M. (1991) Motor proteins in cell division. *Trends in Cell Biol.* 1:123–129.

Shroer, T. A., and Sheetz, M. P. (1991) Functions of microtubule-based motors. *Annu. Rev. Physiol.* 53:629–652.

Vallee, R. B., and Bloom, G. S. (1991) Mechanisms of fast and slow axonal transport. *Annu. Rev. Neuroscie.* 14:59–92.

REFERENCES

Adelman, M. R., Borisy, G. C., Shelanski, N. L., Weisenberg, R. C., and Taylor, E. W. (1968) Cytoplasmic filaments and tubules. *Fed. Proc. Fed. Am. Soc. Exp. Biol.* 27:1186–1193.

Allen, R. D. (1961) A new theory of ameboid movement and protoplasmic streaming. *Exp. Cell Res. Suppl.* 8:17–31.

Allen, R. D. (1967) Diversity and characteristics of cytoplasmic movement. *Neurosci. Res. Program., Bull.* 5:329–332.

Allen, R. D., Francis, D., and Zeh, R. (1971) Direct test of the positive pressure gradient theory of pseudopod extension and retraction in amoebae. *Science* 174:1237–1240.

Allen, R. D., Allen, N. S., and Travis, J. L. (1981) Video-enhanced contrast, differential interference contrast (AVEC-DIC) microscopy: a new method capable of analyzing microtubule-related motility in the reticulopodial network of *Allogromia laticollaris*. *Cell Motil.* 1:291–302.

Allen, R. D., Metuzals, J., Tasaki, I., Brady, S. T., and Gilbert, S. P. (1982) Fast axonal transport in squid giant axon. *Science* 218:1127–1129.

Allen, R. D., Weiss, D. G., Hayden, J. H., Brown, D. T., Fujiwake, H., and Simpson, M. (1985) Gliding movement of and bidirectional transport along single native microtubules from squid axoplasm: evidence for an active role of microtubules in cytoplasmic transport. *J. Cell Biol.* 100:1736–1752.

Bré, M. H., and Karsenti, E. (1990) Effect of brain microtubule-associated protein on microtubule dynamics and nucleating activity of centrosomes. *Cell Motil. Cytosk.* 15:88–98.

Brokaw, C. J. (1967) Adenosine triphosphate usage by flagella. *Science* 156:76–78.

Bryan, J., and Wilson, L. (1971) Are cytoplasmic microtubules heteropolymers? *Proc. Natl. Acad. Sci. U.S.A.* 68:1762–1766.

Cain, D. F., Infante, A. A., and Davies, R. E. (1962) Chemistry of muscle contraction: adenosine triphosphate and phosphorylcreatine as energy supplies for single contractions of working muscle. *Nature (London)* 196:214–217.

Carlsen, F., Knappeis, G. G., and Buchtal, F. (1961) Ultrastructure of the resting and contracted striated muscle fiber at different degrees of stretch. *J. Biophys. Biochem. Cytol.* 11:95–117.

Coue, M., Lombillo, V. A., and McIntosh, J. R. (1991) Microtubule depolymerization promotes particle and chromosome movement in vitro. *J. Cell Biol.* 112:1165–1175.

Dahl, J. L., and Weibel, V. J. (1979) Changes in tubulin heterogeneity during postnatal development of rat brain. *Biochem. Biophys. Res. Commun.* 86:822–828.

Dales, S. (1972) Concerning the universality of a microtubule antigen in animal cells. *J. Cell Biol.* 52:748–754.

Denoulet, P., Jentet, C., and Gros, F. (1982) Tubulin microheterogenicity during mouse liver development. *Biochem. Biophys. Res. Commun.* 105:806–813.

Donges, S., and Roth, E. A. (1972) Serological similarities of microtubule protein. *Naturwissenschaften* 59:372.

Euteneur, V., and McIntosh, J. R. (1981) Structural polarity of kinetochore microtubules in Ptk1-cells. *J. Cell Biol.* 89:338–345.

Fulton, C., Kane, R. E., and Stephens, R. E. (1971) Serological similarities of flagellar and mitotic microtubules. *J. Cell Biol.* 50:762–773.

George, H. J., Misra, L., Field, D. J., and Lee, C. (1981) Polymorphism of brain tubulin. *Biochemistry* 20:2402–2409.

Gibbons, I. R. (1963) Studies on the protein components of cilia from *Tetrahymena pyriformis*. *Proc. Natl. Acad. Sci. U.S.A.* 50:1002–1010.

Goldberg, D. J., Harris, D. A., Lubit, B. W, and Schwartz, J. H. (1980) Analysis of the mechanism of fast axonal transport by intracellular injection of potentially inhibitory macromolecules, evidence for a possible role of actin-filaments. *Proc. Natl. Acad. Sci. U.S.A.* 77:7448–7452.

Gorbsky, G. J., Sammak, P. J., and Borisy, G. G. (1988) Microtuble dynamics and chromosome motion visualized in living anaphase cells. *J. Cell Biol.* 106:1185–1192.

Gozes, I., and Littauer, U. Z. (1978) Tubulin microheterogeneity increase with rat brain maturation. *Nature (London)* 276:411–413.

Gozes, I., and Sweadner, K. J. (1981) Multiple tubulin forms are expressed by a single neurone. *Nature (London)* 294:477–480.

Grafstein, B., and Forman, D. S. (1980) Intracellular transport in neurons. *Physiol. Rev.* 60:1167–1283.

Hanson, J., and Huxley, H. E. (1955) The structural basis of contraction in striated muscle. *Symp. Soc. Exp. Biol.* 9:228–264.

Hayashi, T. (1964) Role of the cortical gel layer in cytoplasmic streaming. In *Primitive Motile Systems in Cell Biology* (Allen, R. D., and Kamiya, N., eds.), pp. 19–29. Academic Press, New York.

Hayden, J. H., Allen, R. D., and Goldman, R. D. (1983) Cytoplasmic transport in keratocytes; direct visualization of particle translocation along microtubules. *Cell Motil.* 3:1–19.

Heidemann, S. R., Landers, J. M., and Hamburg, M. A. (1981) Polarity orientation of axonal microtubules. *J. Cell Biol.* 91:661–665.
Hepler, P. K. (1980) Membranes in the mitotic apparatus of barley cells. *J. Cell Biol.* 86:490–499.
Hepler, P. K., and Callahan, D. A. (1987) Free calcium increases during anaphase in stamen hair cells of *Tradescantia*. *J. Cell Biol.* 105:2137–2143.
Hirokawa, N., Takemura, R., and Hisanaga, S.-I. (1985) Cytoskeletal architecture of isolated mitotic spindle with special reference to microtubule-associated proteins and cytoplasmic dynein. *J. Cell Biol.* 101:1858–1870.
Höner, B., Citi, S., Kendrick-Jones, J., and Jockusch, B. M. (1988) Modulation of cellular morphology and locomotory activity by antibodies against myosin. *J. Cell Biol.* 107:2181–2189.
Horridge, G. A. (1965) Macrocilia with numerous shafts from the lips of the ctenophore *Beroe*. *Proc. R. Soc. London Ser. B* 162:351–364.
Huxley, A. F., and Niedergerke, R. (1954) Structural changes in muscle during contraction. *Nature (London)* 173:971–973.
Huxley, H. E. (1960) Muscle cells. In *The Cell* (Brachet, J., and Mirsky, A. E., eds.), Vol. 4, Part 1, pp. 365–481. Academic Press, New York.
Hyams, J. S., and Brinkley, B. R. (1989) *Mitosis,* pp. 1–350. Academic Press, Orlando, Fl.
Hyman, A. A., and Mitchison, T. J. (1991) Two different microtubule-based motor activities with opposite polarities in kinetochores. *Nature* 351:206–211.
Isenberg, G., Rathke, P. C., Hulsman, N., and Franke, N. N. (1976) Cytoplasmic actomyosin fibrils in tissue culture cells. Direct proof of contractility by visualization of ATP-induced contraction in fibrils isolated by laser microbeam dissection. *Cell Tissue Res.* 166:427–443.
Izzard, C. S., and Lochner, L. R. (1980) Formation of cell-to-substrate contacts during fibroblast motility: an interference-reflection study. *J. Cell Sci.* 42:81–116.
Johnson, K. A. (1985) Pathway of the microtubule-dynein ATPase and the structure of dynein: a comparison with actomyosin. *Annu. Rev. Biophys. Biophys. Chem.* 14:161–188.
Kamiya, N. (1962) Protoplasmic streaming. In *Handbuch der Pflanzenphysiologie* (Ruhland, W., ed.), Vol. 17, Part 2, pp. 979–1035. Springer-Verlag, Berlin.
Kamiya, N., and Kuroda, K. (1956) Velocity distribution of the protoplasmic streaming in *Nitella* cells. I. Velocity distribution of protoplasm in a rhizoid cell. *Bot. Mag.* 69:544–554.
Kamiya, N., and Kuroda, K. (1957) Cell operation in *Nitella*. II. Behavior of isolated endoplasm. *Proc. Jpn. Acad.* 33:201–205.
Kamiya, N., and Kuroda, K. (1958) Studies on the velocity distribution of the protoplasmic streaming in the myxomycete *Plasmodium*. *Protoplasma* 49:1–4.
Mabuchi, I., and Okuno, M. (1977) The effect of myosin antibody on the division of starfish blastomeres. *J. Cell Biol.* 74:251–263.
MacDonald, A. C., and Kitching, J. A. (1967) Axopodial filament of heliozoa. *Nature (London)* 215:99–100.
Masuda, H., and Cande, W. Z. (1987) The role of tubulin polymerization during spindle elongation. *Cell* 49:193–202.
Nelson, L. (1958) Cytochemical studies with the electron microscope. I. Adenosine triphosphatase in rat spermatozoa. *Biochim. Biophys. Acta* 27:634–641.
Osborn, M., and Weber, K. (1977) The detergent-resistant cytoskeleton of tissue culture cells includes the nucleus and the microfilament bundle. *Exp. Cell Res.* 106:339–349.
Pickett-Heaps, J. D. (1986) Mitotic mechanisms: an alternative view. *Trends in Biochem. Sci.* 11:504–507.
Porter, K. R. (1966) Cytoplasmic microtubules and their function. Principles of biomolecular organization. *CIBA Found. Symp. 1965,* pp. 308–345.
Rieder, C. L., and Alexander, S. P. (1990) Kinetochores are transported poleward along a single astral microtubule during chromosome attachment to the spindle in newt lung cells. *J. Cell Biol.* 110:81–95.
Sakai, H., Mabuchi, I., Shimoda, S., Kuriyama, R., Ogawa, K., and Mohri, H. (1976) Induction of chromosome motion in the glycerol-isolated mitotic apparatus: nucleotide specificity and effects of antidynein and myosin sera on motion. *Dev. Growth Differ.* 18:211–219.

Satir, P. (1968) Studies on cilia. III. Further studies on the cilium tip and a "sliding filament" model of ciliary motility. *J. Cell Biol.* 39:77–94.

Saxton, W. M., Stemple, D. L., Leslie, R. J., Salmon, E. D., Zavortink, M., and McIntosh, J. R. (1984) Tubulin, dynamics in cultured mammalian cells. *J. Cell Biol.* 99:2175–2186.

Schroeder, T. E. (1963) Actin in dividing cells: contractile ring filaments bind heavy meromyosin. *Proc. Natl. Acad. Sci. U.S.A.* 70:1688–1692.

Smith, S. J. (1988) Neuronal cytomechanics: the actin-based motility of growth cones. *Science* 242:708–715.

Stephens, R. E. (1968) Reassociation of microtubule protein. *J. Mol. Biol.* 33:517–519.

Summers, K. E., and Gibbons, I. R. (1971) Adenosine triphosphate-induced sliding of tubules in trypsin-treated flagellae of sea urchin sperm. *Proc. Natl. Acad. Sci. U.S.A.* 68:3092–3096.

Szent-György, A. (1949) Free energy relations and contraction of actomyosin, *Biol. Bull.* 96:140–161.

Travis, J. L., and Allen, R. D. (1981) Studies on the motility of foraminifera. I. Ultrastructure of the reticulopodial network of *Allogromia laticollaris* (Arnold). *J. Cell Biol.* 90:211–221.

Travis, J. L., and Bowser, S. S. (1988) Optical approaches to the study of foraminifera. *Cell Motil. Cytoskel.* 10:226–236.

Vallee, R. B., and Bloom, G. S. (1991) Mechanisms of fast and slow axonal transport. *Annu. Rev. Neurosci.* 14:59–92.

Warner, F. D., and Mitchell, D. R. (1978) Structural conformation of ciliary dynein arms and the generation of sliding forces in *Tetrahymena* cilia. *J. Cell Biol.* 76:261–272.

Warner, F. D., and Mitchell, D. R. (1981) Polarity of dynein-microtubule interactions *in vitro:* cross linking between parallel and antiparallel microtubules. *J. Cell Biol.* 89:35–44.

Waters, J., Cole, R., and Rieder, C. L. (1991) A kinetic analysis of centrosome separation during mitosis in newt pneumocytes *J. Cell Biol.* 115:347a.

Weber, K. (1976) Biochemical anatomy of microfilaments in cells in tissue culture using immunofluorescence microscopy. In *Contractile Systems in Non-Muscle Tissues* (Perry, S. V., ed.), pp. 51–66. Elsevier/North-Holland, New York.

Weber, K., Pollack, R., and Bibring, T. (1975) Antibody against tubulins: the specific visualization of cytoplasmic microtubules in tissue culture cells. *Proc. Natl. Acad. Sci. U.S.A.* 72:459–465.

Weisenberg, R. C. (1972) Microtubule formation in vitro in solutions containing low calcium concentrations. *Science* 177:1104–1105.

Wohlfarth-Bottermann, K. E. (1964) Differentiations of the ground cytoplasm and their significance for the generation of the motive force of ameboid movement. In *Primitive Motile Systems in Cell Biology* (Allen, R., and Kamiya, N., eds.), pp. 79–109. Academic Press, New York.

Chapter 18

Mechanochemical Coupling: Molecular Basis

At times the extraction and purification of the molecular components of a system lead to a deeper understanding of its intimate mechanisms. Ideally, it should be possible to reassemble the various component molecules and reestablish function. Only then can the significance of the molecular elements of the system be certain and perhaps lead us to decipher the mechanism of the action at the molecular level. The experimental approach to contraction need not be different from the studies carried out with enzymes and multienzyme complexes (Chapter 7). Conformational rearrangements of the molecules accompany many enzymatic reactions. Similarly, movement, perhaps in the form of a conformational change, is likely to play a role in the transport across cell membranes. Where contraction occurs, the conformational change may be much greater, having been magnified by a variety of means. Although in principle the mechanisms may not be much different from those involved in enzyme action, the task of elucidating their nature may be more complex. In most cases intermediate compounds are not readily evident, and more ephemeral structural states may perhaps be involved.

I. MOLECULAR BASIS OF CONTRACTION IN STRIATED MUSCLE

The study of the molecular mechanism of contraction requires some indication of activity. Since the system transduces chemical energy into mechanical energy (contraction), it could be possible to examine preparations for their capacity to contract. However, in some respects it may be more revealing to examine activities that must be associated with mechanical activity, e.g., the Ca^{2+} dependence of ATPase that is characteristic of the myofibril system or the capacity of components to complex with other macromolecules. Several other useful assays will be described.

A. Myosin and Actin

The myofibrils contain protein associated with contraction; 55% of this is myosin, 20% actin, 5% tropomyosin, and 3 to 4% troponin. We will have to focus our attention on these components. Two of these, actin and myosin, can be recombined to form *actomyosin*. The complex can be made into a fiber by releasing it into distilled water or into a dilute solution. The fiber shortens in the presence of ATP. Because the actomyosin molecules are randomly distributed, the shortening occurs in all directions and the thread cannot displace a weight. But when actomyosin is appropriately oriented, by spreading it on a surface, for example, it will contract only longitudinally on the addition of ATP and is thereby capable of lifting weights. This experiment is shown in Fig. 18.1 (Hayashi, 1952). The contractions are slow—it takes several minutes to reach the maximal level—in contrast to the very fast contraction of intact muscle. This slowness may be the result of the slow diffusion of ATP into the relatively thick fibers. It is also possible that not all the appropriate components of the system are present. During the contraction, ATP is hydrolyzed, as estimated from parallel experiments carried out under the same conditions.

Events thought to correspond only indirectly to the primary events of contraction have also been found useful in studying actomyosin. For example, it has been found that the viscosity of the complex decreases on addition of ATP and then returns to normal, indicating dissociation and reformation of a complex. The viscosity depends on the shape of the macromolecules in solution. The longer the molecule, the higher the viscosity. The dissociation of the rodlike actomyosin produces a decrease in viscosity, since the axial ratios (i.e., the length to the width) of actin and myosin are much lower than the axial ratio of actomyosin. As we shall see, the

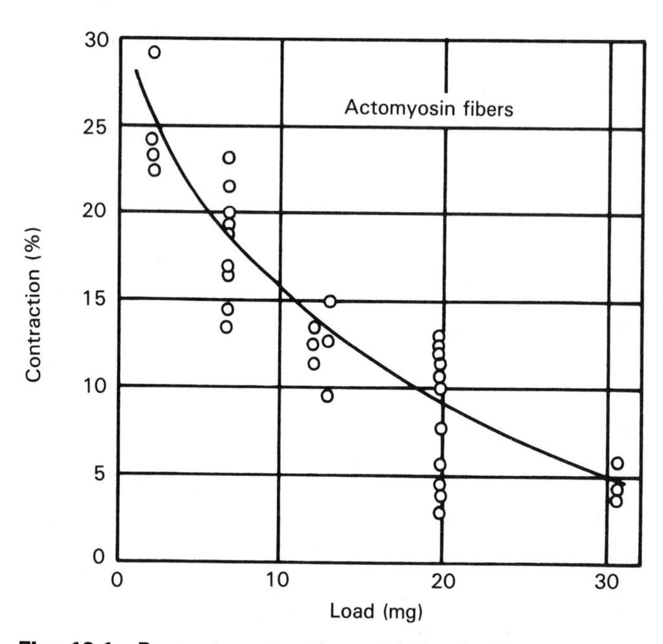

Fig. 18.1 Percent contraction with load of actomyosin fibers. Each point represents a 15-minute contraction. From Hayashi (1952). Reproduced from *The Journal of General Physiology*, 1952, 36:139–151, by copyright permission of the Rockefeller University Press.

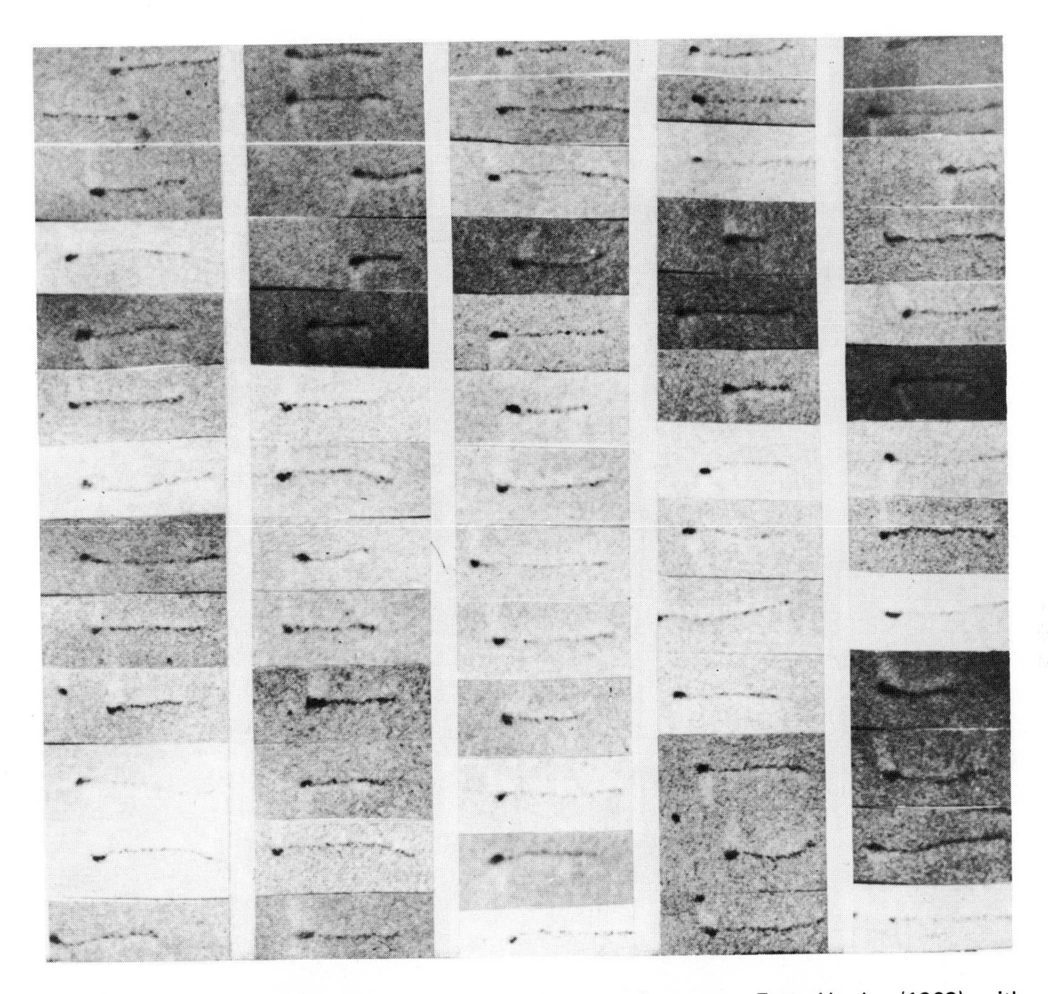

Fig. 18.2 Myosin monomers seen with the electron microscope. From Huxley (1963), with permission.

dissociation of actomyosin and its subsequent reassociation from actin and myosin are likely to be part of the contractile mechanism. Actomyosin gels decrease in volume on addition of ATP (the superprecipitation reaction), a change that has also been used as an index of contractility.

Myosin can be seen with the electron microscope after negative staining. It is a long molecule, about 160 nm in length and 20 to 40 nm in diameter. A single myosin unit has a molecular mass of about 500 kDa and contains two heavy chains of about 230 kDa and four light chains of 16 to 20 kDa. Electron microscopic views of the monomers are shown in Fig. 18.2 (Huxley, 1963). When either the pH or the ionic strength is reduced, myosin monomers associate (i.e., polymerize) into thick filaments remarkably similar to the thick filaments of the myofibrils. As we saw in Chapter 17, there is evidence that the thick filaments of striated muscle are made up of myosin. In the fibers that polymerize from myosin monomers, the tail ends of the molecules cling together and the head pieces jut out. The molecules align in the filament in such a way that some of the head pieces face in one direction while others face in the other

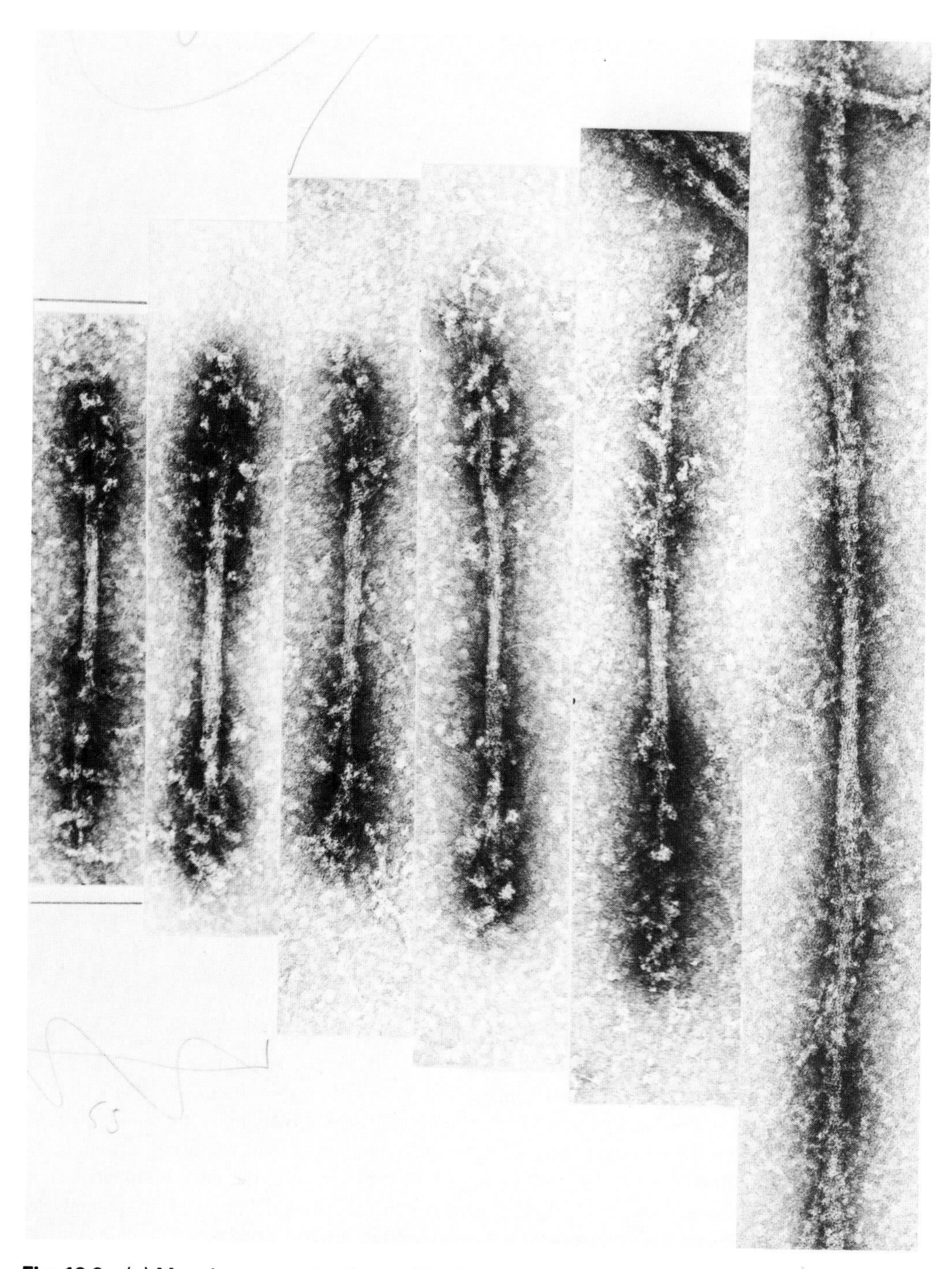

Fig. 18.3 (*a*) Myosin aggregates formed in vitro as seen with the electron microscope after negative staining.

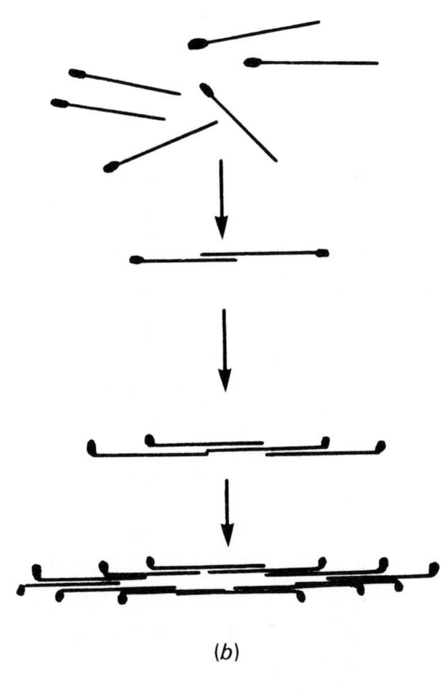

Fig. 18.3 *(Continued)* (*b*) Scheme showing mechanism of formation of fibrous aggregates from myosin monomers. From Huxley (1963), with permission.

direction, forming a symmetrical composite. Figure 18.3*a* (Huxley, 1963) shows such polymerized filaments. Figure 18.3*b* is a diagrammatic representation of how these filaments are probably formed.

The head portion of the individual myosin molecules at the amino-terminal appears to correspond to bridges between thick and thin filaments seen with the electron microscope in sections of sarcomeres. The head portions can be isolated after partial proteolytic digestion of the myosin as heavy meromyosin (HMM). Their ability to then combine with actin supports the view that they correspond to bridges observed with the electron microscope. The HMM fragment can be further cleaved into two smaller fragments, S1 and S2.

The myosin head pieces seen with the electron microscope after shadowing appear to be formed by two subunits (Slayton and Lowry, 1967). A model incorporating our present knowledge is shown in Fig. 18.4 (Stebbins and Hyams, 1979).

The addition of HMM (or S1) to actin produces an appearance of many neatly arranged arrowheads. This arrowhead configuration could occur only if the HMM fragments were specifi-

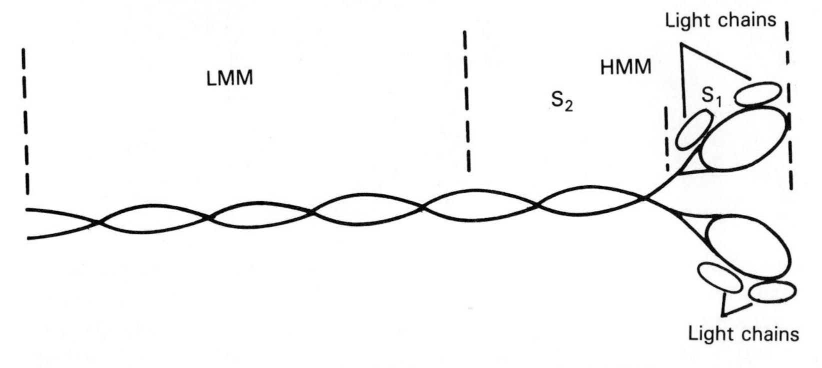

Fig. 18.4 Structure of myosin. Light meromyosin (LMM) and heavy meromyosin (HMM) and the subfragments of HMM, S1 and S2, are shown. The broken lines indicate the cleavage sites of the proteolytic digestion. From H. Stebbins and J. S. Hyams, *Cell Motility,* with permission. Copyright © 1979 Longman Group Ltd., Edinburgh.

cally oriented by the thin filaments. It is therefore considered to be a specific test for the presence of actin filaments and indicates that the actin itself has polarity. The ends of the actin have been named after the appearance of the arrowheads, i.e., pointed (P) and barbed (B). The actin with oriented arrowheads, as seen in the negatively stained preparation shown in Fig. 18.5, is referred to as *decorated*. The two views shown in the figure represent electron micrographs at slightly different tilts, and they permit visualization of the arrangement in three dimensions using a simple standard or homemade stereoscope.

Actin is a smaller molecule than myosin; it has a molecular mass of 42 kDa. It is globular, and G-actin is about 5.5 nm in diameter. In this conformation, 1 mol of actin contains 1 mol of ATP and 1 mol of divalent cation, Ca^{2+} or Mg^{2+}. G-actin polymerizes in a process in which ATP is hydrolyzed to form a long thread of F-actin, a double helix 7 to 8 nm in diameter (Fig. 18.6) (Hanson and Lowy, 1963). The ADP derived from the ATP and the divalent cation remain attached to the actin. The divalent cation is thought to be involved in regulation of the polymerization rate.

Since actomyosin threads formed from the combination of actin and myosin contract, the two components and the complex must have properties that play a fundamental role in contraction. Myosin, for example, hydrolyzes ATP. We have seen that ATP is probably the energy source for contraction. In fact, there is a good correlation between the speed of contraction of various striated muscles and their ATPase activity (Maddox and Perry, 1966).

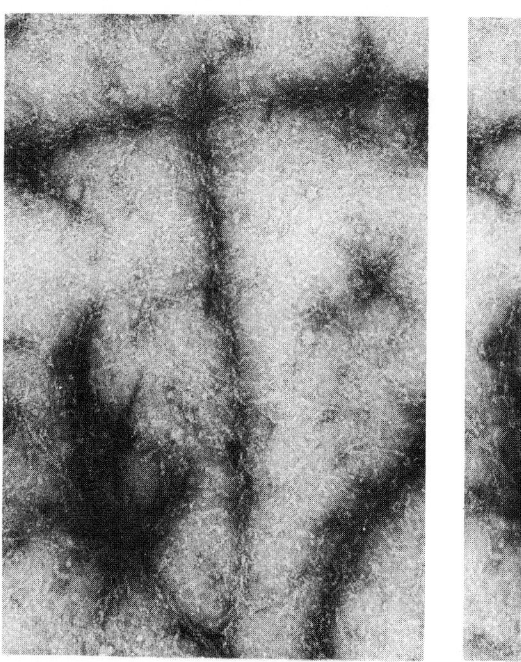

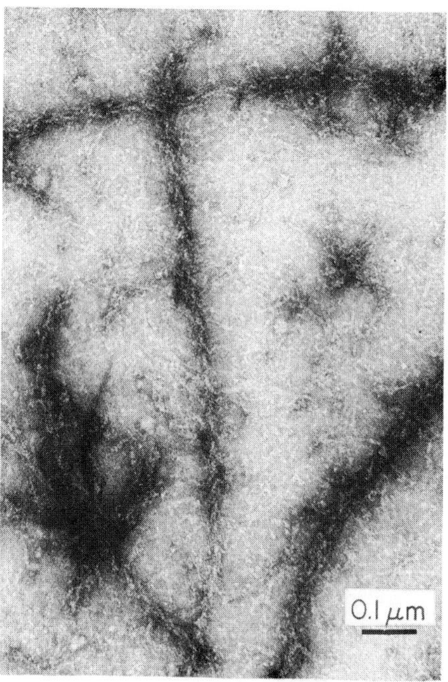

Fig. 18.5 Rabbit actin complexed with heavy meromyosin (HMM), negatively stained with 1% aqueous uranyl acetate. ×70,240. Stereophotographs at +6° and −6° tilt. For three-dimensional depth, view with a standard stereoscope or a simple homemade prism stereoscope [see E. G. Gray and R. A. Willis, *J. Cell Sci.* 3:309-326 (1968)]. Toward upper part of filament note clear demonstration of arrowhead orientation in relation to the filament. Electron microscopy by Barry S. Eckert and S. M. McGee-Russell, Department of Biological Sciences, S.U.N.Y, at Albany.

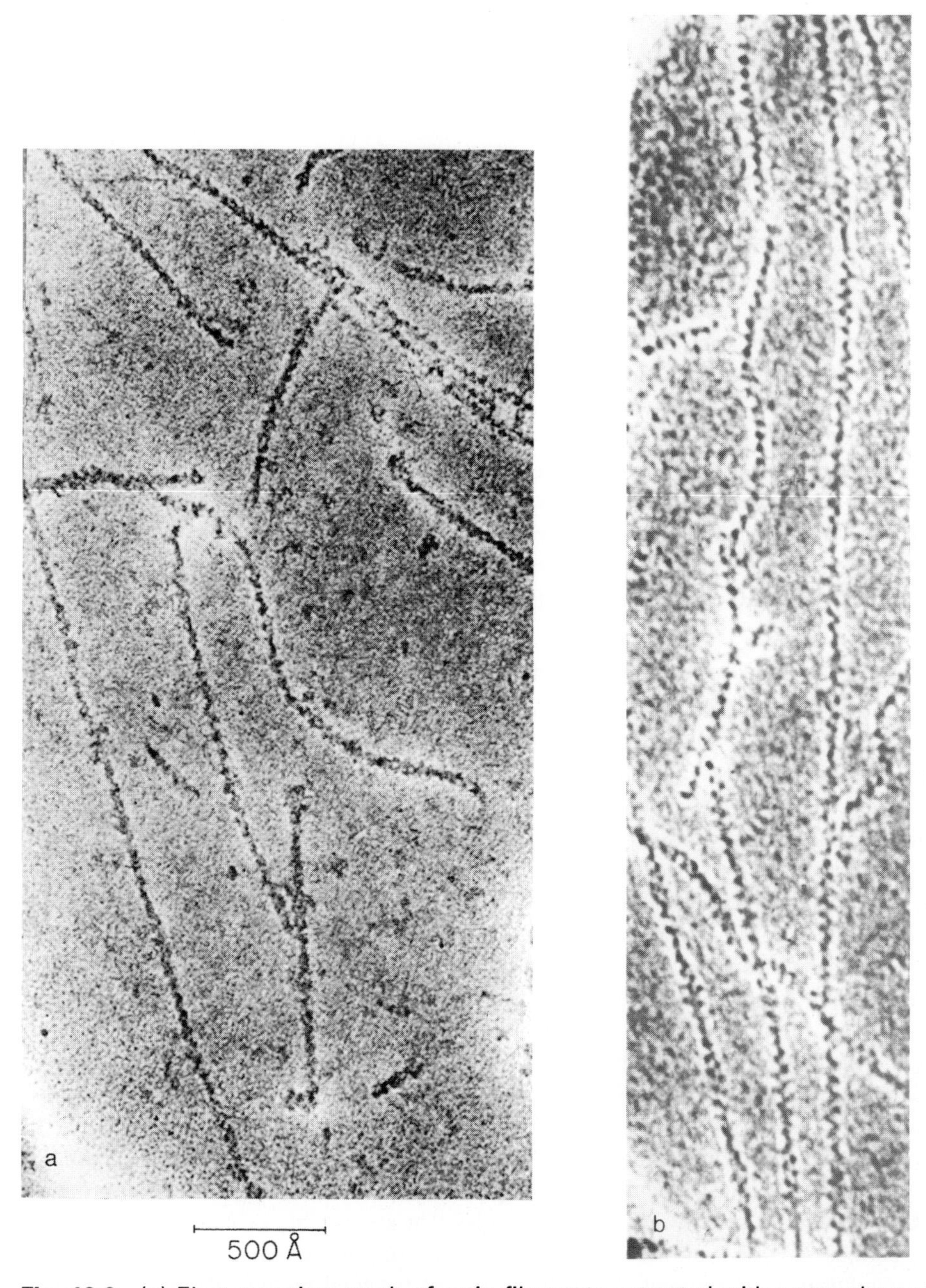

Fig. 18.6 (*a*) Electron micrograph of actin filaments extracted with water-glycerol in the presence of Mg-ATP and EDTA. Negative staining with uranyl acetate. ×36,125. (*b*) Electron micrograph similar to that in (*a*). However, the preparation corresponds to F-actin. The electron micrograph permits counting the number of globular units per turn of the two-stranded helix. ×44,625. From Hanson and Lowy (1963), with permission.

The events accompanying contraction, according to the sliding-filament model, must be cyclic. At least four events must take place: (1) binding between actin and myosin, (2) some structural change that pulls the actin filaments from the two sides of the sarcomere toward each other, (3) coupling of the hydrolysis of ATP to the appropriate structural changes, and (4) breaking of the bond between actin and myosin. Since a sarcomere shortens more than the distance between side chains or cross-bridges, the cyclic breaking and remaking of bonds must occur over a distance corresponding to several cross-bridges; in frog muscle the shortening corresponds to 30% of the sarcomere length, or about 370 nm, whereas the myosin projections are approximately 40 nm apart. Let us examine some of the properties of the actomyosin system with these steps in mind.

As mentioned, it is possible to fragment the myosin molecule by proteolytic digestion. The individual fragments can be separated out and tested for various properties. The ATPase activity resides in only a portion of the myosin molecule, the heavy meromyosin associated with the thicker part of the molecule (Jones and Perry, 1966; Mueller and Perry, 1962). A basic role of the interaction between actin and myosin predicted by the sliding-filament model is supported by the observation that actin and myosin have to be complexed for shortening to occur. It is interesting to note that the complexing ability also resides in the myosin fragment that has the ATPase activity. However, the reactive sites are likely to be different for these two activities, since thiol reagents affect the two activities differently (Bárány and Bárány, 1959; Kaldor et al., 1964).

B. Tropomyosin, Paramyosin, and Troponin

Although the ATPase is in the myosin molecule, little ATP is hydrolyzed by myosin alone (Mommaerts and Green, 1954) despite the appropriate Ca^{2+} and Mg^{2+} concentrations. However, in the presence of actin, the Mg^{2+}-ATPase activity is increased as much as 20-fold (Hasselbach, 1952) and approaches the level found in intact muscle (Perry, 1956). Myofibril contractility requires the presence of Ca^{2+}. In vertebrates and arthropods, the Ca^{2+} sensitivity of actomyosin apparently depends on the presence of two proteins, tropomyosin and troponin. Troponin inhibits the Mg^{2+}-actomyosin ATPase, and this inhibition is reversed by Ca^{2+}. Tropomyosin is involved in holding troponin in an appropriate configuration along the actin filament (Ebashi et al., 1972). These two components seem to act in a manner analogous to that of allosteric regulatory subunits. The arrangement of troponin and tropomyosin along the thin filaments is shown by exposing sections of muscle to the appropriate antibodies covalently attached to fluorescent dyes (Endo et al., 1966; Pipe, 1966). In some other organisms, such as mollusks, brachiopods, and some worms (Lehman et al., 1972), a light chain of myosin seems to be directly involved in this control mechanism (Kendrick-Jones et al., 1972). In annelids (e.g., the sandworm *Nereis*), both mechanisms are present (Lehman et al., 1972).

Tropomyosin is a fibrous molecule 40 nm long (molecular mass 68 kDa). A similar protein, *paramyosin,* is found in certain invertebrate muscles that are capable of maintaining tension over long periods of time with a minimum expenditure of energy (*catch muscle*). In some muscles paramyosin may represent as much as 30% of the total protein. Paramyosin forms threads 133 nm long and 2 nm in diameter. Intact muscle fibers probably corresponding to paramyosin are intermeshed with thin filaments that probably correspond to actin (Hanson and Lowy, 1961). This arrangement is thought to have some role in contraction and the catch phenomenon.

C. Cross-Bridges and Contraction

It would seem from the discussion in Section A that a small portion of the myosin filament, the globular projection, combines with actin in the formation of cross-bridges. This portion of the molecule is also responsible for the hydrolysis of ATP. As we have seen, the reaction must be cyclic and, during contraction, association and dissociation involving several cross-bridges must occur. The isometric tension (i.e., tension at constant length) developed during contraction at various sarcomere lengths (Gordon et al., 1966) correlates well with the degree of overlap of thick and thin filaments, which in turn allows the maximum number of cross-bridges. Figure 18.7*a* (Gordon et al., 1966) shows a diagrammatic representation of the overlap in filaments based on studies with the electron microscope. Figure 18.7*b* is a graphical summary of the results. The tension developed is expressed as a percentage of the maximal value (ordinate). The sarcomere length, from Z line to Z line, is represented on the abscissa. The numbered arrows correspond to the numbers in the diagram (Fig. 18.7*a*). Comparison of two diagrams shows that maximal tension is developed when the overlap of the cross-bridges is maximal. The tension falls when the myofibrils are stretched beyond this length and fewer bridges are in register with the corresponding point of attachment in the actin fibers. The tension also drops sharply when the myofibril is shorter than this optimal value; apparently in this case the thin filaments overlap (Fig. 18.7*a*).

There are two objections to a model that depends on the formation of cross-bridges between actin and myosin. First, the spacing between thick and thin filaments increases with increasing contraction, and it is difficult to visualize a bridge that varies in length depending on the space

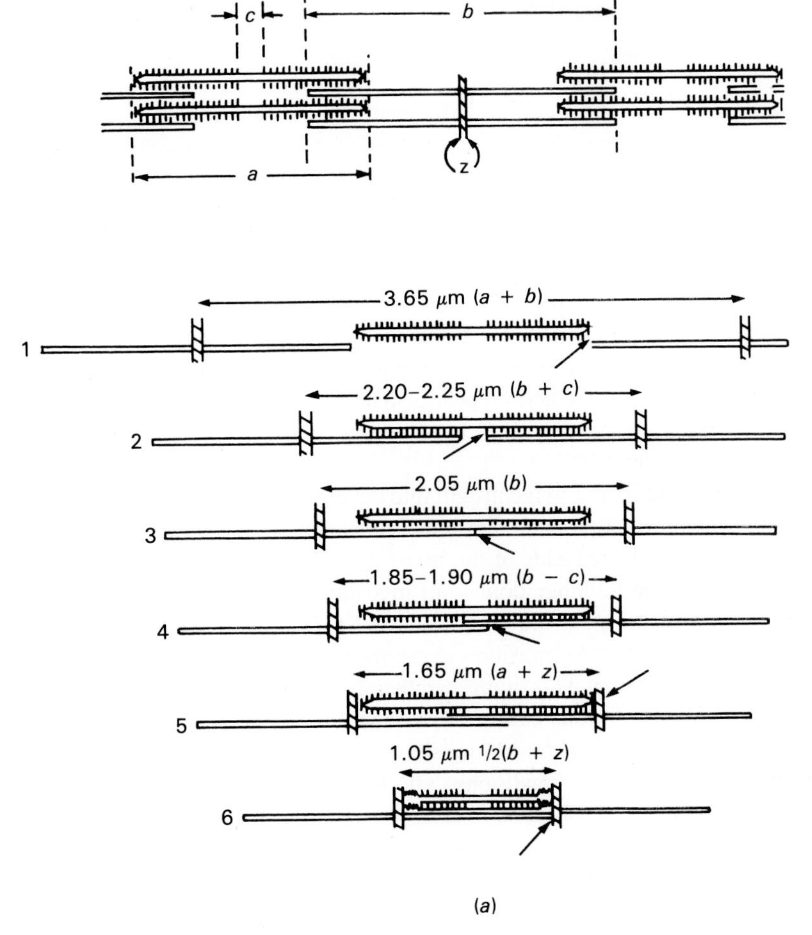

Fig. 18.7 (*a*) Schematic diagram of filaments. The cross-bridges are represented by the small lines on the thick filaments.

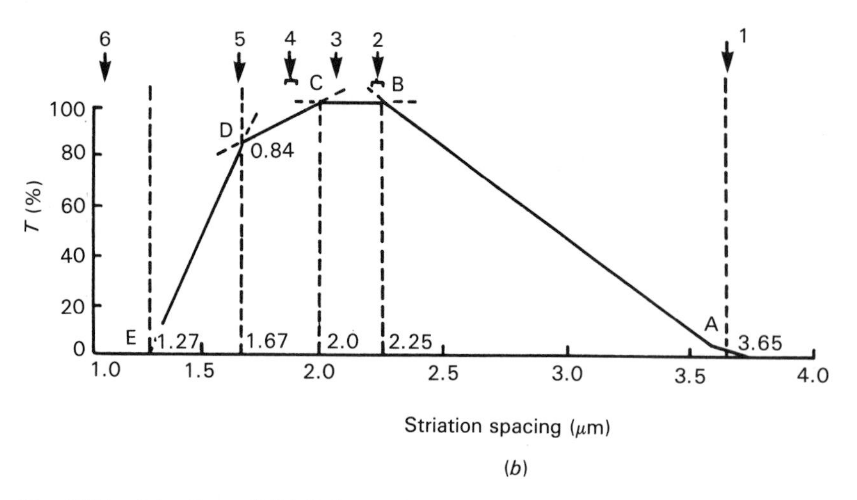

Fig. 18.7 (*Continued*) (*b*) Schematic summary of results. *T* = percent tension. The arrows along the top are placed opposite the striation spacings at which the critical stages of overlap of filaments occur, numbered as in (*a*). From A. M. Gordon, et al., *Journal of Physiology*, 184:170–192, with permission. Copyright © 1966 The Physiological Society, Oxford, England.

that has to be spanned. Second, the actin and myosin molecules are not in register to permit the formation of cross-bridges. F-actin forms a helix with a pitch of 36.5 nm. The myosin projections have a repeat of 42.9 nm. In fact, if one cross-bridge were formed at one point, the next match would be 550 nm away.

However, x-ray diffraction data indicate that during contraction the period corresponding to actin remains the same, whereas myosin, while conserving a shorter repeat unit of 14.3 nm, acquires repeat units with a pitch of 72 nm. The results can be most readily explained by assuming that the myosin heads are mounted on an arm that is flexible at least at two points (Fig. 18.8*a*) (Huxley, 1969). The arm can swing to fit the variable distance between filaments and can adjust its repeat distance to fit the actin repeat distance (Fig. 18.8*b*). In such a model the motive force could be generated in the small head region, possibly by a motion of the component units (Fig. 18.9) (Huxley, 1969).

Although the experiments described provide insight into the events that result in contraction, more information is needed about what produces the actual movement of the sliding molecules. This could take place by a tilting of the heads of the myosin projections (Fig. 18.9).

A conformational change in the S2 hinge domain of myosin (see Figs. 18.8 and 18.12), a helix-to-coil transition (i.e., a melting), could shorten the S2 region and also produce force (Flory, 1956; Harrington, 1971). After the transition to the more unfolded form, the myosin is more susceptible to proteolytic digestion. Transitions can therefore be detected by following the sensitivity to exposure to a protease, in this case α-chymotrypsin.

Figure 18.10 (p. 499) (Ueno and Harrington, 1986) shows the results of proteolytic digestion of glycerinated muscle under a variety of conditions. The muscle proteins were analyzed by sodium dodecyl sulfate (SDS) gel electrophoresis after appropriate incubation and termination of the proteolysis with a chymotrypsin inhibitor. The results represent the muscle proteins in their native state (lane a); after incubation in the presence of the chymotrypsin in the absence of ATP (rigor solution), a medium that produces maximally cross-linked muscle fibers (lanes b and c); in the presence of ADP but not ATP, a medium that corresponds to the relaxed state (lanes d and e); and in an activating medium containing ATP and an ATP-generating system (phosphocreatine and creatine phosphokinase), which is a contraction medium (lanes f and g).

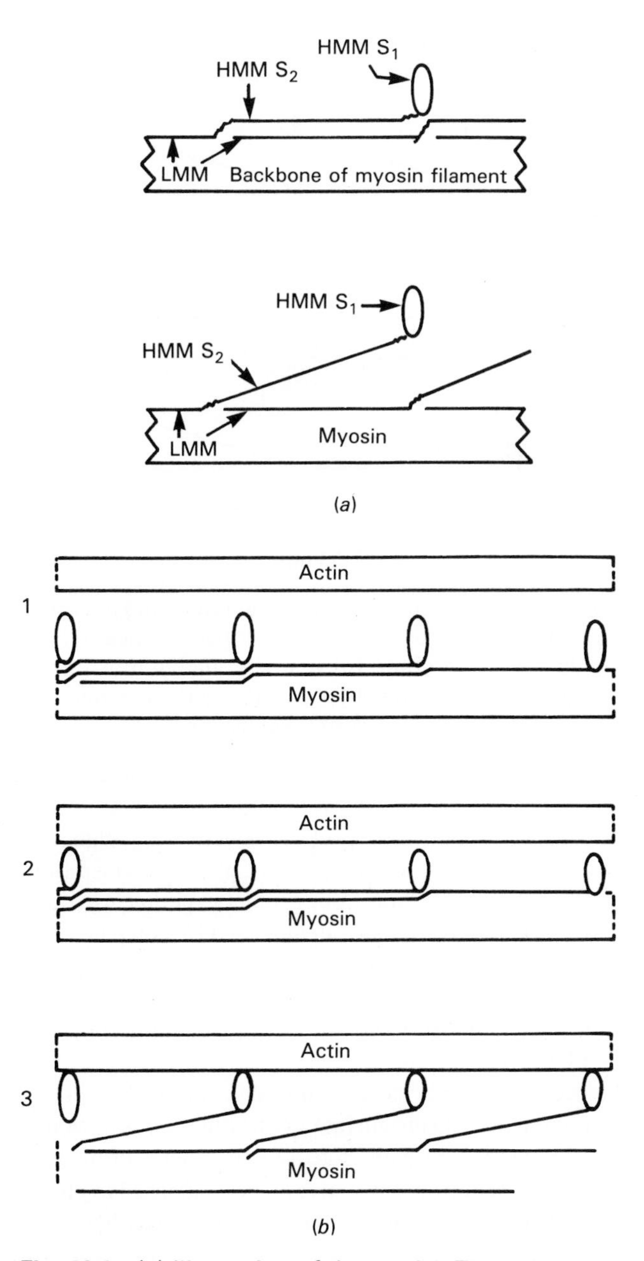

Fig. 18.8 (*a*) Illustration of the model. The points where bending can occur are indicated by the wavy line. (*b*) Diagram showing relative positions of filaments and cross-bridges at two different interfilament spacings, (1) 25 nm and (2) 20 nm, corresponding to sarcomere lengths of 2.0 and 3.1 μm in frog sartorius muscle. During contraction or in rigor, the cross-bridges could attach to the actin filaments by bending at two flexible junctions, as shown in (3). From Huxley (1969), *Science* 164:1356–1366, copyright 1969 by the AAAS.

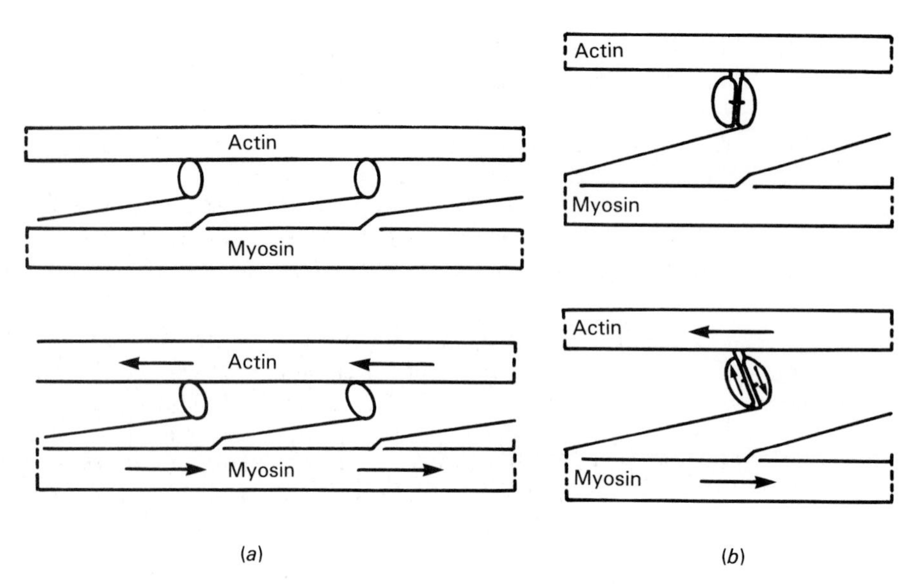

Fig. 18.9 Possible mechanisms for producing relative sliding movement by tilting of cross-bridges. (*a*) If separation of filaments is maintained by electrostatic force balance, tilting must give rise to movement of filaments past each other. (*b*) A small relative movement between two subunits of myosin could give rise to a large change in tilt by the mechanism shown. From Huxley (1969), *Science* 164:1356–1366, copyright 1969 by the AAAS.

The fragments generated in the activating solution (lane g) were isolated and chromatographed in lane h. As shown in Fig. 18.10 and summarized in the diagram at the bottom, cleavage products representing the S2 region were generated almost exclusively in the activating solutions.

Figure 18.11*a* (Ueno and Harrington, 1986) shows the rate constant of the cleavage (left ordinate and filled circles) and the isometric force generation from parallel experiments (right ordinate and dotted line) at various Mg-ATP concentrations. Figure 18.11*b* compares the cleavage rate constant (left ordinate and filled circles) with the fraction of active cross-bridges calculated from the work of others (Arata and Shimizu, 1981) (right ordinate and dashed line). The conformational change, isometric force, and number of cross-bridges functioning correlate well. The model is summarized in Fig. 18.12 (p. 501).

This evidence strongly supports the occurrence of a change in myosin conformation in the hinge region during contraction. However, it provides no insight into whether the conformational change is responsible for movement. Studies carried out with proteolytic fragments of myosin do not support such a role. The S_1-fragment representing a single myosin head attached to glass, supports the movement of actin (Yoyoshima et al., 1987) and can generate force (Kishino and Yanagida, 1988).

D. Attachment of the Thin Filaments: α-Actinin and Vinculin

The contractile event, expressed in the shortening of the sarcomere, requires that the thin actin filament be attached to the Z disks. The latter are probably mainly made up of a rod-shaped protein of about 95 kDa, α-actinin. The location of the α-actinin in the Z disk has been demonstrated with immunofluorescence using anti-α-actinin antibodies (Lazarides and Burridge, 1975; Masaki et al., 1967). Since α-actinin cross-links in vitro with actin, it is presumed to constitute the attachment site for the thin filaments.

In smooth muscle another protein, *vinculin,* of 130 kDa (Geiger, 1979) seems to play a similar role in connecting the actin filaments to the dense plaques (Geiger et al., 1980).

II. MECHANISMS IN OTHER MUSCLES

We have seen that much of the information available for striated muscle points to contraction occurring as a consequence of the sliding of actin filaments in relation to myosin filaments. Some clues are presented as to how this might occur at the molecular level. As previously

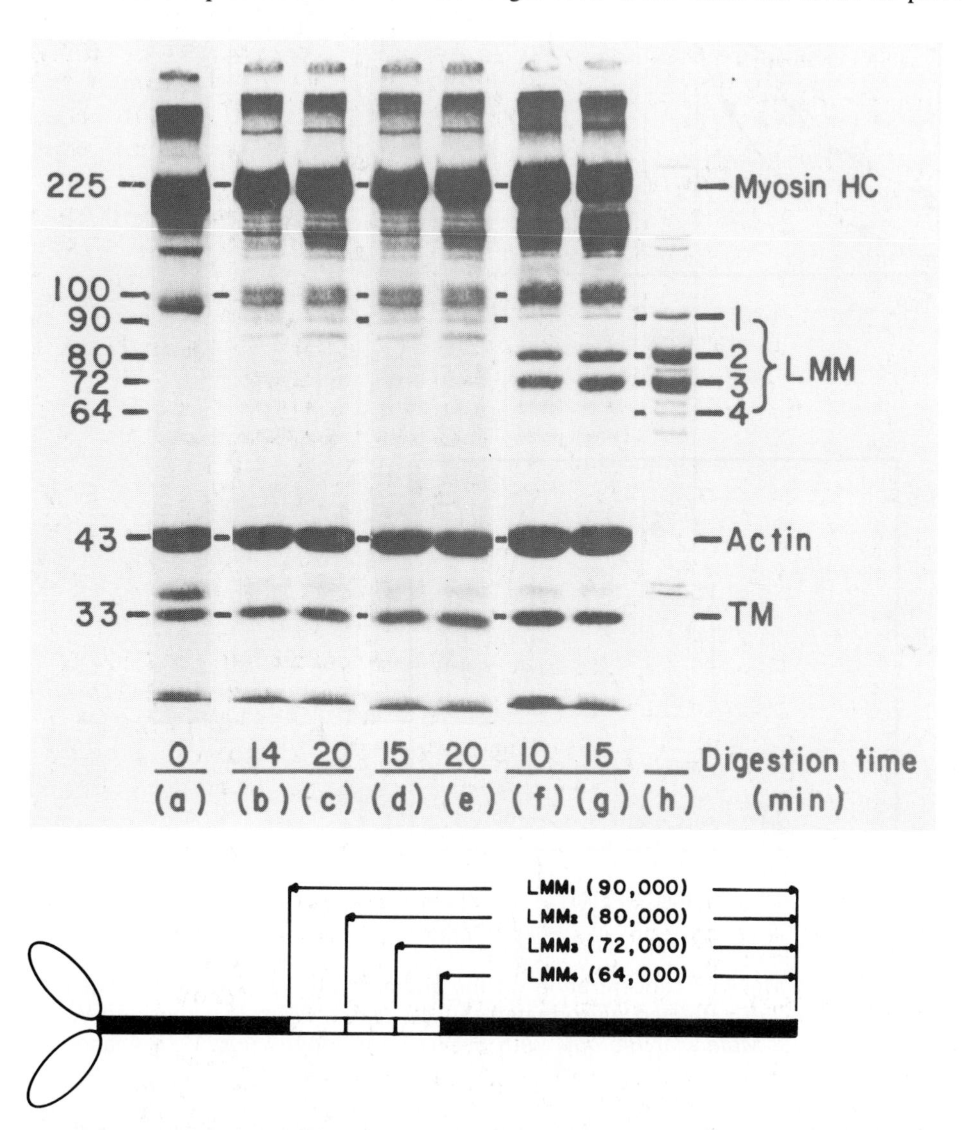

Fig. 18.10 Electrophoresis of muscle proteins on SDS gels after α-chymotrypsin digestion of glycerinated rabbit psoas muscle. (a) Undigested fiber, (b and c) incubation in rigor medium, (d and e) relaxation solution (rigor solution plus ADP), and (f and g) activating solution (see text). (h) LMM subfragments isolated from (g). From H. Ueno and W. F. Harrington, *Journal of Molecular Biology,* with permission. Copyright © 1986 Academic Press.

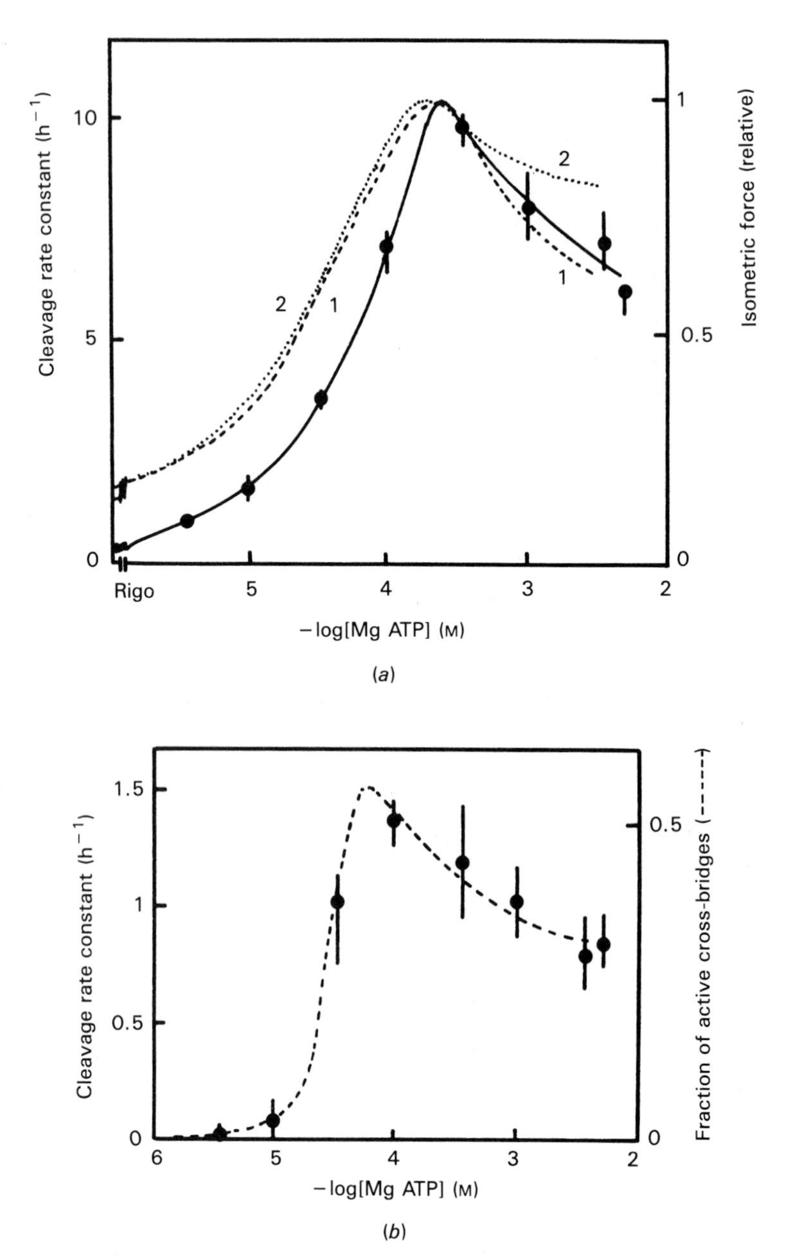

Fig. 18.11 Effect of Mg ATP concentration on the chymotryptic cleavage rate constant and isometric tension of activated glycerinated muscle fibers. From H. Ueno and W. F. Harrington, *Journal of Molecular Biology,* with permission. Copyright © 1986 Academic Press.

discussed, the striations are the consequence of the presence of fibers of myosin in the thick filaments of the denser anisotropic bands and of actin in the thin filaments spanning the less dense isotropic bands. The thin filaments interdigitate with the thick filaments of the A bands. The sarcomeres are in register so that the A or I bands of one sarcomere are adjacent to those of a sarcomere located next to it.

Striated muscle is found in vertebrates and a number of arthropods. Presumably the same contractile mechanism operates in all these striated muscles, although certain details may differ.

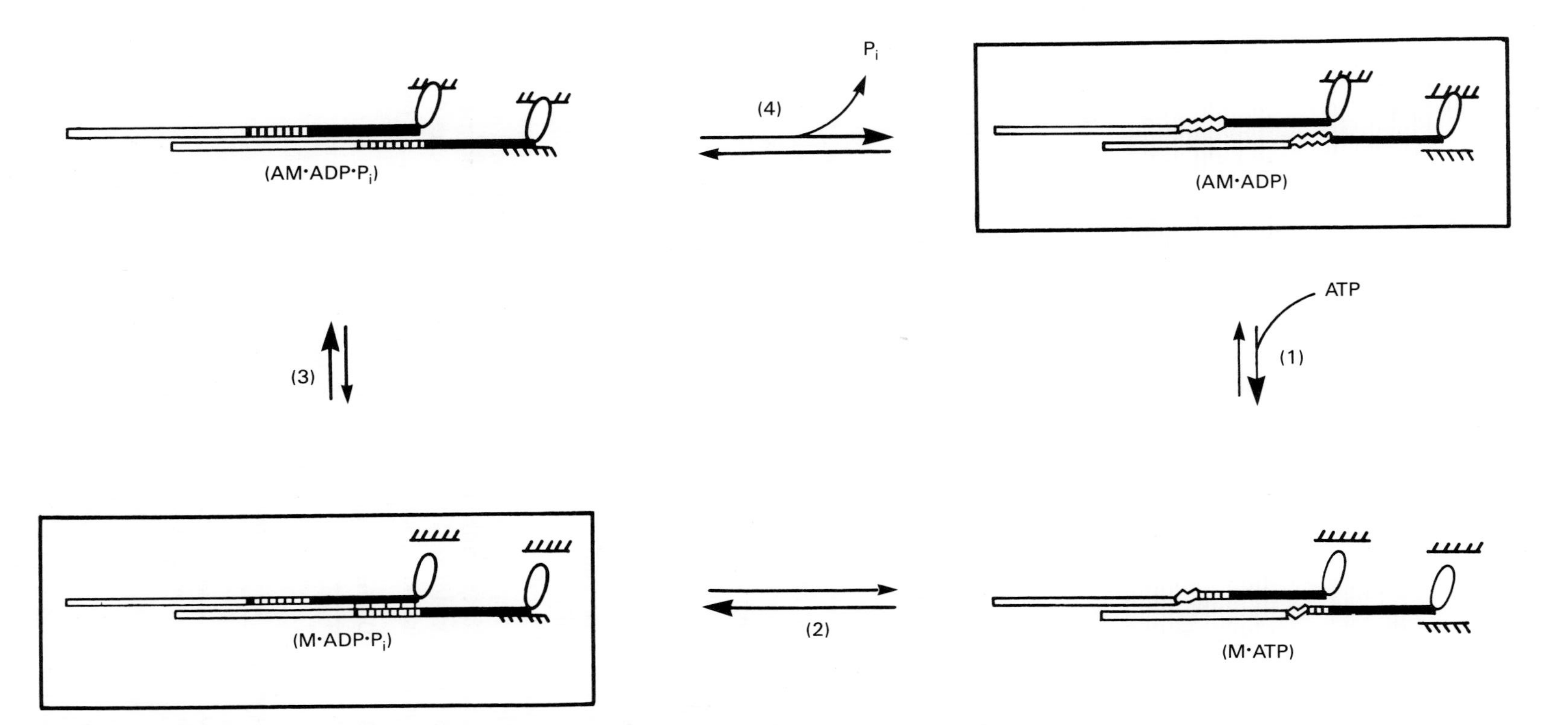

Fig. 18.12 Schematic representation of a contractile mechanism. This simplified scheme indicates an active cross-bridge cycle. From H. Ueno and W. F. Harrington, *Journal of Molecular Biology,* with permission. Copyright © 1986 Academic Press.

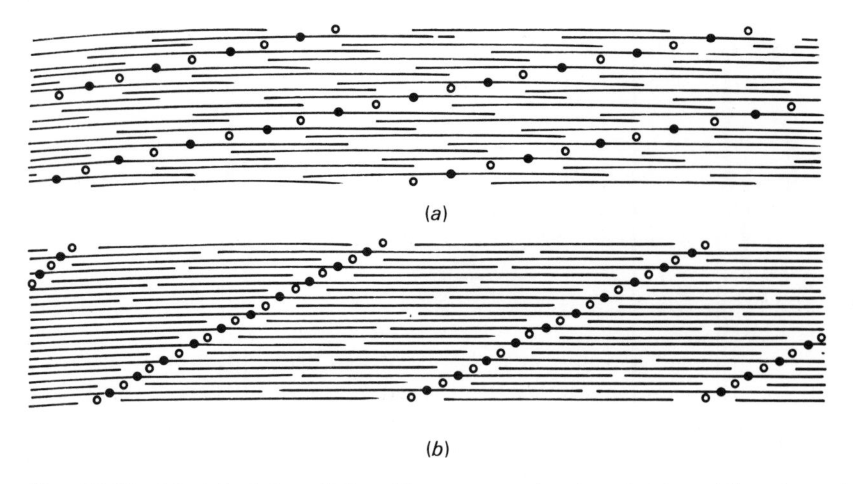

Fig. 18.13 Model of the sliding-filament mechanism in the obliquely striated body wall muscle of the earthworm. (*a*) Relaxed state; (*b*) contracted state. Note that the thick filaments slide not only relative to thin filaments but also relative to each other. From H. G. Heumann and E. Zebe, *Zeitschrift fuer Zellforschung und Mikroskopische Anatomie,* 78:131–150, with permission. Copyright © 1967 Springer-Verlag, Heidelberg.

Other muscles are arranged differently from striated muscle. Some, such as the body wall muscle of the earthworm, are obliquely striated (Fig. 18.13) (Heumann and Zebe, 1967); others, such as smooth muscles, have no apparent striations at all. Do these muscles contract in a different manner? The answer is not entirely clear. Smooth or obliquely striated muscles generally have some properties that distinguish them from striated muscle. For example, although striated muscle generally contracts to about 80% of its resting length, some smooth muscles can contract to 30% of their resting length (Hasselbach and Ledermair, 1958; Winton, 1926). Are these manifestations of fundamental differences or do they represent minor modifications of the same basic mechanism? Close examination of at least some of the cases indicates that usually it is not necessary to propose a different mechanism. Where shortening is extreme, the change in length could take place only if the filaments did not abut against a barrier such as a Z disk or it would become necessary for the thick filaments to slide relative to other thick filaments, as apparently does occur. An example of extreme contraction, that of the body wall muscle of the earthworm, is shown in Fig. 18.13. Different types of atypically striated or nonstriated muscle are listed in Table 18.1 (Rüegg, 1968). The table summarizes the type of muscle, the Z line structure, the proteins in the thin and thick filaments, and where these muscles are present.

Table 18.1 The Diversity of Smooth Muscle

		Proteins of myofilaments		
Type of muscle	Z-line structure	Thin filaments	Thick filaments	Examples
Helical smooth or obliquely striated	Z-column or dense bodies	Actin	Myosin and tropomyosin A (paramyosin)	Earthworm body wall, oyster yellow adductor
Invertebrate smooth, type I (paramyosin muscle)	Dense bodies	Actin	Myosin, very much paramyosin	Muscle anterior byssal retractor, oyster opaque adductor
Invertebrate smooth, type II (classical smooth)	Dense bodies	Actin	Myosin	Pharynx retractor, snail
Vertebrate smooth muscle	Dense bodies	Actin	Thick fibers difficult to see, no paramyosin	Uterus, taenia coli, chicken gizzard

From J. C. Rüegg, *Symposia of the Society for Experimental Biology No. XXII: Aspects of Cell Motility,* with permission. Copyright © 1968 Academic Press.

All the systems that have been studied seem to have both actin and myosin and may well work by a similar mechanism, although differences must occur. Not surprisingly, the properties of the component molecules are different. This is particularly true of myosin, which differs depending on the system from which it has been isolated. Evidence as to whether sliding actually occurs or whether the thick and thin filaments actually correspond to actin and myosin is not always readily available. In addition, the thick filaments have been difficult to demonstrate in vertebrate smooth muscle by means of electron microscopic techniques.

III. MOVEMENT IN CILIA AND FLAGELLA

In Chapter 17 we saw that the bending of cilia or flagella may occur through a mechanism analogous to the sliding filament of striated muscles. Nevertheless, the details of the mechanism are likely to differ significantly. The elements that make up the microtubules of cilia or flagella have been isolated and studied. The tubules are made up of *tubulin* α and β, which are present in cilia, flagella, and the cytoplasm. Dynein, responsible for the ATPase activity, corresponds to the cross-bridges in the tubular doublets, as shown in Fig. 17.12 (Chapter 17). Tubulin seems to be analogous to actin and dynein to myosin. However, tubulin differs from actin, and dynein does not correspond to myosin. The details of the interactions between the two proteins making up the microtubular system of cilia or flagella are still not well understood.

Dynein is made up of several polypeptides (Johnson, 1985). In *Tetrahymena* cilia and in *Chlamydomonas* flagella the dynein is three-headed, whereas in sea urchin sperm flagella the dynein is double-headed. A comparison of myosin and dynein is shown in Fig. 18.14 (Johnson, 1985). The base of the dynein is anchored to one microtubule, the so-called A subfiber, by ionic interactions. The head, which contains the ATP binding site, is free to interact with the B subfiber of the adjacent doublet, as shown in Chapter 17, Fig. 17.12. Although this arrangement is entirely analogous to the one in actomyosin, the dynein heads are much larger.

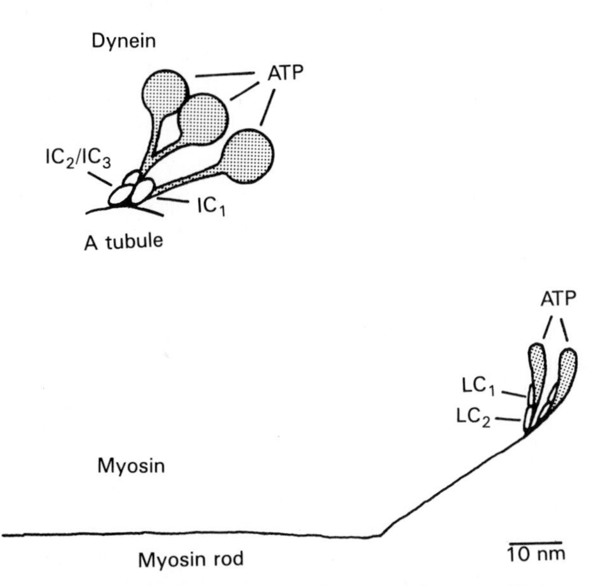

Fig. 18.14 Structures of dynein and myosin. The structure of *Tetrahymena* 22S dynein is schematically compared to that of myosin, which is drawn to the same scale. The dynein structure combines information obtained from the three dynein sources (see text). IC_n, Dynein intermediate chains; LC_n, myosin light chains. From Johnson (1985). Reproduced, with permission, from the *Annual Review of Biophysics and Biophysical Chemistry*, Vol. 14, copyright © 1985 by Annual Reviews Inc.

Dynein forms a so-called rigor bond with the B subfiber and is released by the addition of ATP. The rate of this dissociation suggests that the microtubule-dynein complex may operate in essentially the same way as the actomyosin complex, with the same cycle of dissociation, movement of bridge, ATP hydrolysis, formation of bond, and generation of force as outlined for actomyosin.

IV. MOVEMENT IN THE CYTOPLASM

Movement apparently also takes place in the absence of obvious contractile structures. Perhaps the proper molecular arrangement could be assembled when needed and then disassembled (e.g., mitotic apparatus) or could actually be labile and present only for a short while. In the slime mold the flow of the cytoplasm reverses direction continuously (Chapter 17). Fibers could produce contraction of the kind observed in slime molds if arranged across the flowing cytoplasmic channel at one location and could then be disassembled while another contractile structure is formed elsewhere.

The idea that the underlying mechanisms of motility are similar is reinforced by the fact that microtubules or fibers have been found in several contractile systems and in some cases actomyosin and actin molecules have been isolated.

A. Microfilaments and Microfilament Bundles

As mentioned, filaments and tubules are present in the cytoplasm. Not only tubules but also filaments are prominent in neurons (*neurofilaments*). These are 8 to 10 nm in diameter (Davidson and Taylor, 1960) and appear to be formed of globular subunits. Filaments also appear in many other cell systems. Generally the microfilaments fall into two groups. Some of the fibers range from 5 to 7 nm in diameter. Others, such as the neurofilaments, are in the range of 8 to 10 nm. The 5- to 7-nm fibers correspond in size to actin filaments. The microfilaments that do not correspond to actin may well have a structural role. There is extensive evidence that the neurofilaments form a network with microtubules. They form cross-bridges, possibly through the so-called *microtubule-associated proteins* (MAPs) (Leterrier et al., 1982).

This section examines the possible role of the actomyosin system of cells other than muscle and then the role of actin-binding proteins.

The Actomyosin System In *Nitella* and *Chara,* as we saw in Chapter 17, the pattern of flow of the cytoplasm suggests that the motile force is produced in the boundary between the stationary cortex and the outer edge of the cytoplasm. Fibers in this zone have been recognized with the light microscope (Ishikawa et al., 1969; Kamitsubo, 1966, 1972) and the electron microscope (Nagai and Rebhun, 1966). The fibrils are 5 to 6 nm thick, which suggests that they are actin filaments. Immunofluorescence techniques (Williamson and Toh, 1979) and HMM decoration (Korn and Hammer, 1988) confirm that actin is in fact involved. In *Nitella* myosin has been isolated (Kato and Tonomura, 1977) suggesting that an actomyosin contractile system is present.

The mechanism of cytoplasmic flow implied by the boundary model described above for *Nitella* and *Chara,* together with the location of the filaments in the cortex, suggests that whereas actin is in the cortex, myosin is most likely to be present in the moving endoplasm. This model is supported by experiments in which cortex and endoplasm are separated out by centrifugation of these very large cells. Centrifugation at low speed collects the endoplasm in the centrifugal end of the cell, whereas the cortex remains in place. The two components are differentially treated and then reassembled by centrifugation in the reverse direction (Chen and Kamiya, 1975; Nagai and Kamiya, 1977). *N*-Ethylmaleimide (NEM) is known to interfere with the F-actin-activated ATPase activity of myosin. In the separation-reassembly experiments, treatment of the cortex with NEM did not interfere with streaming after reconstitution. However, treatment of the endoplasm blocked the streaming. Similarly, cytochalasin B, which reacts with actin, blocked streaming when the cortex alone was treated. These observations seem to be in agreement with the model previously proposed for motility in this system, discussed in Chapter 17. Another experiment supports this interpretation. The ectoplasmic fibers were isolated by cutting open the cell and washing away the cytoplasm. The oriented fibers and the chloroplasts stayed behind. Fluorescent beads 0.7 μm in diameter were coated with HMM and then placed on these fibers, on which they proceeded to move unidirectionally along the array of filaments. This motion required the presence of ATP and did not take place when the HMM was inactivated (Sheetz and Spudich, 1983a). Therefore, all indicators support the models that were proposed.

The structural versatility and the dynamics of all cells are in large part the result of dynamic interactions of the components of the cytoskeleton. Some of these interactions are based on the system of microtubules and associated proteins discussed in the next section (see also Chapter 17, IV, E). Other interactions require actin and actin-associated proteins, and these are addressed in this section.

Actin and Actin-Binding Proteins Actin is ubiquitous in cells. Its polymerization is thought to play a significant role not only in cytoplasmic movement, but also in the structure and mechanical properties of the cytoplasm. Several proteins interact with actin to influence its polymerization and the structure of actin complexes. These will be discussed first, followed by a discussion of other actin-binding proteins.

F-actin is a polar molecule. One end is referred to as pointed (P) and the other as barbed (B) in reference to its appearance in electron micrographs when complexed to S1. Both ends can add monomers; however, the growth is generally biased in favor of the B end. Polymerization from a monomer has a delay followed by accelerated growth, reflecting the need for nucleation.

At least four different groups of proteins bind actin (Pollard and Cooper, 1986). The *capping proteins* generally bind to the B end of the filaments and thereby interfere with the binding of monomers at that end. They can then grow with their barbed ends attached to particles or surfaces coated with capping proteins such as villin or severin. The capping proteins in the presence of Ca^{2+} favor the polymerization of actin monomer by facilitating nucleation to form short filaments, possibly by severing longer segments.

Capping may stabilize filament length by stopping *treadmilling,* i.e., the addition of monomers at one end and disassembly at the other end, a phenomenon that takes place in the intact cell. By facilitating the formation of short segments, they favor the sol state and not the gel state and hence presumably favor the flow of cytoplasm.

Some low molecular weight proteins, the *severing proteins,* bind to actin monomers and sever longer actin filaments to produce shorter ones.

Various *bundling proteins,* which are present as dimers, cross-link actin filaments to form bundles. Generally three to five actin molecules are cross-linked by one bundling protein molecule. Bundling proteins play a role in intact cells; in the brush border of the intestinal microvilli, for example, *fimbrin* and *villin* cross-link actins to form bundles. In addition, cross-linking favors the formation of a gel. The role of these actin-binding proteins is summarized in Fig. 18.15.

Much of the recent information on the actin-based interactions has been derived by primary amino acid sequence data of the binding proteins, determined by using recombinant DNA techniques (see Chapter 1, II,A). The search for common sequences has allowed the recognition of actin-binding domains. Many of these actin-binding proteins mediate interactions by binding to yet other proteins. Many of the actin-binding proteins can bind directly to polyphosphoinosides

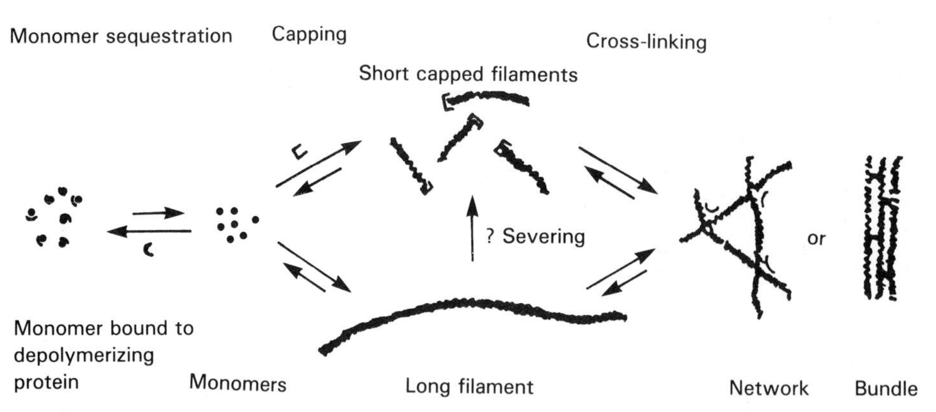

Fig. 18.15 Regulation of actin assembly by three classes of actin-binding proteins. Class I cross-links filaments into networks or bundles. Class II caps an end of the filament and may sever preformed filaments. Class III inhibits polymerization by binding to actin monomers. From S. W. Craig and T. D. Pollard, *Trends in Biochemical Sciences,* with permission. Copyright © 1982 Elsevier Science Publishers, England.

(i.e., a phospholipid component of the membrane); others to integral membrane proteins, possibly implicating the actin system in the regulation of both structure and metabolism. For example, profilin inhibits the hydrolysis of PIP_2 by some phospholipases C, and therefore has a potential regulatory role on the production of second messenger of the inositol system (Chapter 5, V).

The known actin-binding proteins share a 270 amino acid region which is presumed to be the actin-binding domain (see Hardwig and Kwiatkowski, 1991). Subfragments of these domains bind actin (e.g., a 17 kDa fragment of *Dictyostelium* ABP-120, Bresnick et al., 1990). These molecules generally are rod-shaped, and apart from at least one actin-binding domain, the rest of the rod contains tandems of structural repeats of variable length, presumably to distance the actin from specific binding sites to other proteins that are also present. For example, spectrin has binding sites in the rod region for calmodulin, ankyrin, and band 4.1 (Morrow, 1989). Many of these proteins also contain domains which have homology to sequences of the Ca^{2+}-binding proteins (e.g., Vandekerhove, 1990) suggesting a regulatory role of Ca^{2+}. Such a region is at the carboxy-terminal of α-actinin (Noegel et al., 1987) and of the α-spectrin (Wasenius et al., 1990), and at the amino-terminal region of fimbrin (DeArruda et al., 1990).

Except for fimbrin/plastin, the actin-binding proteins are thought to be in the form of dimers in antiparallel subunit chains so that the actin-binding domains, generally present at the amino-terminal are at both ends of the composite rods (Fig. 18.16). The subunits are either side by side (e.g., spectrin, α-actinin, and ABP-120), or end to end (e.g., filamin). Spectrin is formed by α and β chains that self-associate in overlapped antiparallel alignment to form heterodimers 100 nm in length; these associate end to end to form tetramers (see Fig. 18.16) (Hartwig and Kwiatkowski, 1991).

The assembly of cytoskeletal elements containing actin has been examined in detail in the red blood cells (see Bennett, V., 1990 and Anderson and Marchesi, 1985) and in intestinal microvilli. In the red blood cell, a meshwork of spectrin tetramers interacting with F-actin is attached to the plasma membrane through two high affinity associations (see Chapter 2, Fig. 2.5 for a graphical representation): a binding to ankyrin which attaches to band 3 (anion exchanger) protein, and a binding to band 4.1 attached to glycophorin, an association favored by polyphosphoinositides.

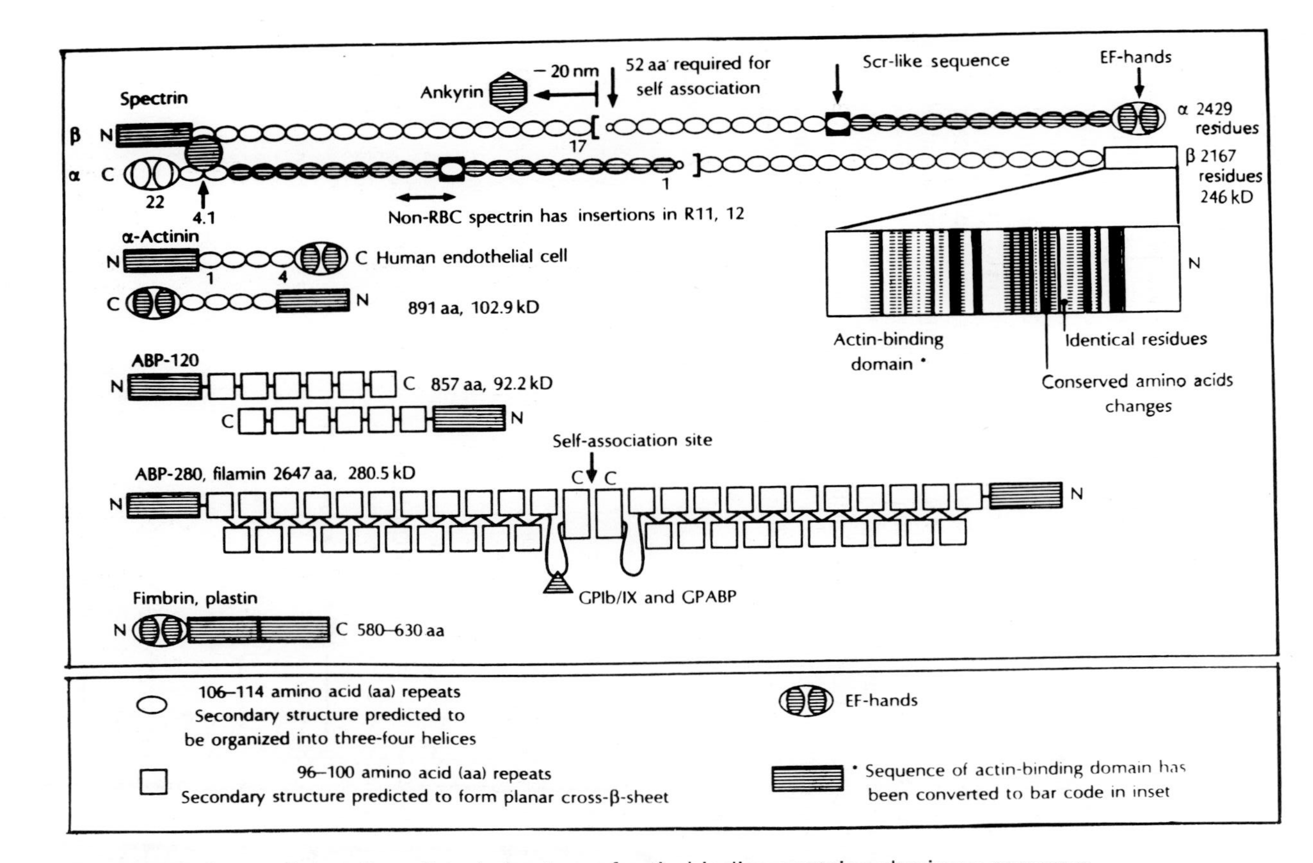

Fig. 18.16 Comparison of predicted structure of actin-binding proteins sharing a common actin-binding domain. From J. H. Hartwig and D. J. Kwiatkowski, *Current Opinion in Cell Biology*, 3:87–97. Copyright © 1991 Current Biology Ltd.

In the brush border of microvilli, a myosin I complex containing calmodulin subunits attaches to actin, while at the carboxy-terminal it binds to acidic phospholipids (Hayden et al., 1990). The complex also contains the bundling proteins villin and fibrin. These four proteins can assemble in vitro to form the microvillar core (Coluccio and Bretscher, 1989). A model of this association is shown in Fig. 18.17.

The polarity of cells may also result from interactions with actin (Nelson and Hammerton, 1989, 1990; McNeil et al., 1990). In kidney epithelial cells in culture, the integral transport protein Na^+, K^+-ATPase is randomly distributed. However, when cells contact other cells the ATPase becomes associated with only one membrane sector, the basal-lateral membrane. This happens when this protein associates with fodrin and ankyrin which in turn associate with the cell adhesion molecule uvomorulin (Nelson et al., 1990). The association of uvomorulin with extracellular or cell surface components at the baso-lateral membranes could impose this distribution. In support of this view, transfection of fibroblast to express uvomorulin redistributes the Na^+, K^+-ATPase and fodrin to sites of uvomorulin-mediated cell to cell contacts (McNeil et al., 1990).

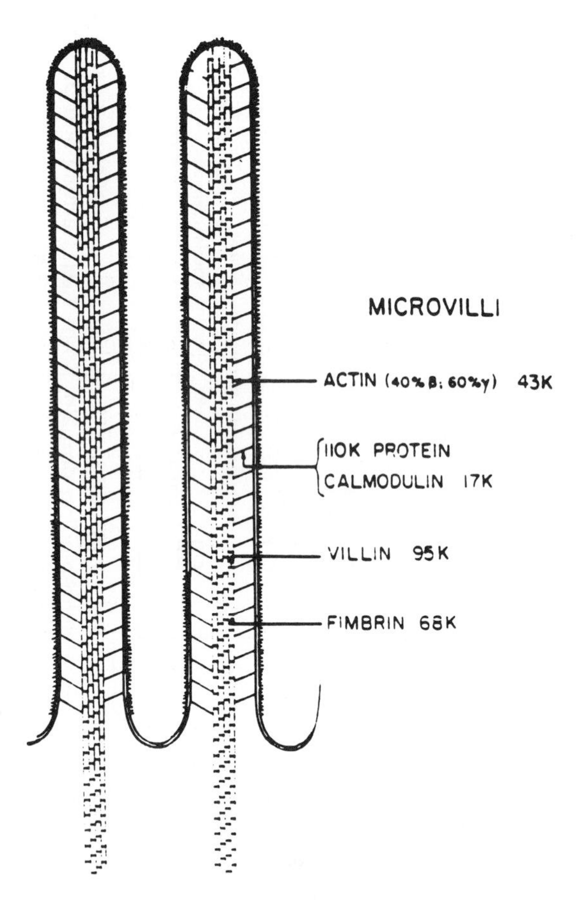

Fig. 18.17. Model of the molecular organization of the microvillus cytoskeleton. The microvilli are approximately 100 nm in diameter. The 100 kDa-calmodulin 17 kDa is a myosin I. Reproduced from the *Annual Review,* copyright © 1991 by Annual Reviews Inc.

Myosins There are at least three different kinds of myosin. The myosin of striated muscle just discussed (this chapter, I,A) has been called conventional myosin, or myosin II, and there are closely related forms in smooth muscle and cells other than muscle, nonmuscle myosin II. As previously discussed, muscle myosin is double-headed and composed of two approximately 200 kDa heavy chains with an amino-terminal globular domain. A single unit of the latter (S_1), which corresponds to one head, is responsible for the ATPase activity and furthermore, has all the machinery needed to generate force (Toyoshima et al., 1987; Kishino and Yanagida, 1988). These general properties are shared by the other myosins II. In addition, the amino acid sequence of the rod part, responsible for the formation of filaments, contains a 28 amino acid repeat sequence which gives rise to a 14.5 repeat (McLachlan and Karn, 1982). However, differences in these tail regions are responsible for many divergent properties of the various myosins II.

Phosphorylation of the myosin light chains has a regulatory function in striated muscle, increasing isometric tension (Persichini et al., 1985). Smooth muscle and nonmuscle myosin assembly depends on the phosphorylation of the light regulatory subunits for their ability to assemble in filaments; in addition, the phosphorylated form has a high ATPase activity (see Trybus, 1991). In at least some cases, phosphorylation has been shown to be in response to physiological stimuli (Devreotes et al., 1987). A regulation of myosin and actin filaments assembly by phosphorylation-dephosphorylation of the myosin light chains has been demonstrated in living fibroblast cells by injection of a protein phosphatase into human fibroblasts (Fernandez et al., 1990). The microinjection resulted in a disassembly of the actin network as seen by immunofluorescence. After long incubations, the cell's actin returned to the original distribution. Neutralization of the phosphatase by microinjection of the corresponding antibody prevented the reorganization. In contrast to this function in vertebrate cells, in lower eukaryotes such as the cellular slime mold *Dictyostelium discoideum,* phosphorylation of the tail inhibits assembly, and the assembly of myosin II of *Acanthamoeba castellanii* is not affected by phosphorylation (see Trybus, 1991).

Myosins I are present in protozoans and at least some vertebrate cells such as in the microvilli of the intestinal brush border. Five different kinds of myosin I have been isolated and studied. As their striated muscle counterpart, they have ATPase activity and can also generate force on actin filaments (Albanesi et al., 1985; Mooseker et al., 1989; Collins et al., 1990). However, they contain a single chain, are single-headed, and cannot form filaments. Different kinds of myosins I vary considerably in the length of the tail region and have distinct amino acid sequences. Some have binding domains for actin, membranes (e.g., Miyata et al., 1989), and calmodulin.

The association of myosin I with membranes and actin suggests that myosin I may be able to support pseudopod extension, membrane ruffling, and the movement of cellular organelles or vesicles on actin filaments. The idea that it is involved in vesicular transport is supported by the inhibition of the movement of endogenous vesicles on actin filaments by antimyosin I and not antimyosin II in *Acanthamoeba* cell extracts (Adams and Pollard, 1989). When absorbed to lipid surfaces, myosin I can also move actin filaments along the surface (Zot and Pollard, 1991). The tail-region actin-binding site allows cross-linking with actin to form gels which, in the presence of ATP when phosphorylated by special kinases (Fujisaki et al., 1985), condense. This phenomenon suggests a mechanochemical role that could play a role in phagocytosis and the extension of pseudopods.

Insights in the relative roles of myosin I and II in vivo are provided by experiments in which the expression or the function of one of these myosins is blocked. In practice it has been easiest to block myosin II (e.g., in *Dictyostelium,* there is a single gene corresponding to this myosin, whereas there are five for myosin I). *Dictyostelium* cells lacking myosin II, either produced by genetic manipulation (e.g., DeLozanne and Spudich, 1987) or using antisense RNA (which by hybridizing with the mRNA blocks its translation) (e.g., Knecht and Loomis, 1987), are unable

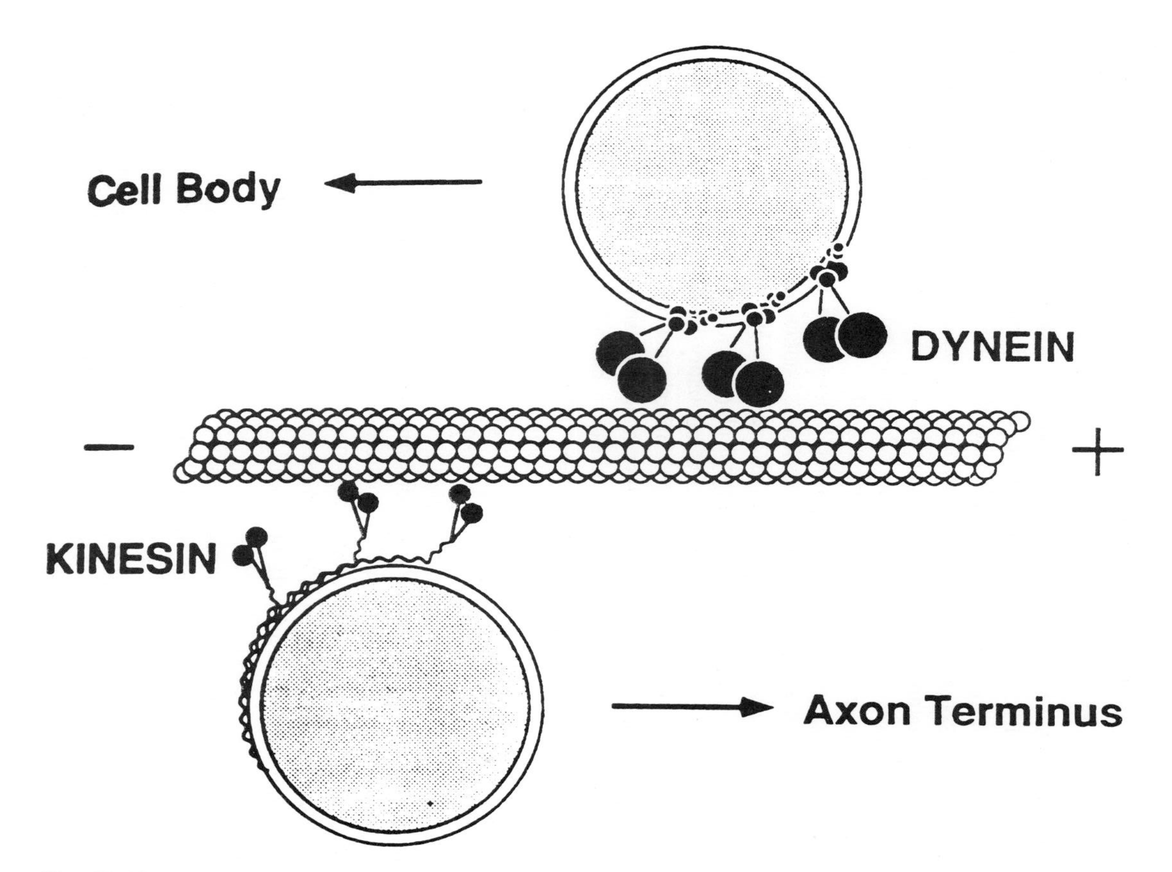

Fig. 18.18. Axonal transport. Organelles are depicted using either kinesin for anterograde transport or cytoplasmic dynein for retrograde transport. Vallee and Bloom (1991). Reproduced from the *Annual Review of NeuroSciences,* Volume 14, copyright © 1991 by Annual Reviews Inc.

to carry out cytokinesis and receptor capping and to maintain cortical tension (as determined by deformability) (Pasternak et al., 1989). Capping of certain receptors is a sign of mobility of the receptor, shown by an accumulation due to cross-linking to multivalent ligands (e.g., concanavalin A). However, cell locomotion, formation of pseudopods, and phagocytosis remain unaffected. A similar strategy was used by injecting an excess of antimyosin II (Sinard and Pollard, 1989) which slows down but does not stop motility. Therefore, in these cells the former properties are likely to depend on myosin II and the latter on myosin I.

In the microvilli of the intestinal brush border, myosin I laterally links actin filament bundles and the plasma membrane (Matsudaira and Burgess, 1979; Coluccio and Bretscher, 1989). Microvilli are not motile, although an active tension may be needed to maintain structure. Myosin I of microvilli can support movement on actin filaments (Mooseker et al., 1989) and the movement of actin on myosin adsorbed to a surface (Collins et al., 1990). A domain at the carboxy-terminal of the tail region of microvillar myosin I can bind to membranes, whereas an amino-terminal domain has been shown to bind calmodulin (Conzelman and Mooseker, 1987; Coluccio and Bretscher, 1988).

Three distinct monoclonal antibodies to chicken brush border myosin were injected into chicken fibroblasts (Höner et al., 1988). The antibodies were to three distinct epitopes of the tail region of the molecule. The microinjection resulted in the loss of stress fibers, change in shape, and increased locomotory activity. Other changes were consistent with the interpretation that there was an increase in fluidity in the cells.

In addition to myosin I and II other myosins may also exist. Two novel myosins (perhaps of the myosin I family) have been found in *Drosophila*, and possibly a new family of myosin has been identified in mice and yeast (see Hammer, 1991).

Obviously the interactions between the myosins and actin are very complex and many questions still remain unanswered.

B. Microtubules

Like actin filaments, microtubules are ubiquitous in cells and are thought to play a role in both structure and movement. The general role of microtubules in movement was briefly described in Chapter 17 (IV,E). Aside from their role in chromosome movement, microtubules are likely to be associated with Golgi stacks (Allan and Kreis, 1987), endoplasmic reticulum (Franke, 1971), mitochondria (Lindén et al., 1989), and lysosomes (Swanson et al., 1987). Chapter 4 examined intracellular transport and discussed evidence for an involvement of microtubules (e.g., III,B). Chapter 17 examined the role of microtubules in axonal movement and in cell division (IV,E) and the dependence of microtubular transport on motors was noted. This section will take a closer look at the currently known motors and their possible mode of action.

Current studies have so far suggested two major kinds of microtubule-associated motors, the dynein and the kinesin classes. Kinesin drives the transport of organelles from the minus to the plus end (see Fig. 18.18). In neurons and undifferentiated cells, this corresponds to the movement from the cell nucleus to the periphery (axon in the case of neurons). Kinesin can also translocate microtubules linearly in a minus direction when attached to a surface. Dynein drives movement in the opposite direction, translocating particles toward the minus end and moving microtubules in the plus direction (Vale et al., 1986). The role of these motors is summarized in Fig. 18.18. The two motors are also distinguishable by their different sensitivities to inhibitors or ATP-analogs as well as direction and rate of movement along microtubules. Categorizing motors according to directionality of movements can be too simplistic. A recently discovered kinesin-like motor protein, *claret*, moves particles toward the minus end, and resembles cytoplasmic dynein in its sensitivity to inhibitors (Walker et al., 1990). In addition, a dynein motor of *Reticulomyxa* has been reported to provide bidirectional organelle movement in a permeabilized cell system (Schliwa et al., 1991). However, it is not yet clear whether it is the microtubules in the lysed cell model that are moving relative to one another. This is an important consideration since dynein can effect bidirectional microtubule sliding in the ciliary axoneme.

Apart from directional motion some cytoplasmic dynein and kinesin can produce torque, causing rotation of microtubules coincident to linear translocation. This is the case for *claret* (Walker et al., 1990) and ciliary dynein from *Tetrahymena* (Vale and Toyoshima, 1988).

The Kinesins Movement of particles on single microtubules can be followed using DIC and enhanced contrast video technology (Chapter 17 IV,E). The microtubules of extruded axoplasm from giant squid axons support the fast movement of particles on microtubules in either direction (see Chapter 17, IV,E), and not surprisingly, giant axons were used in the early studies. The transport was found to require ATP (Brady et al., 1982, 1985). The nonhydrolyzable ATP analog 5′adenylimidodiphosphate (AMPPNP) abolished motility and also seemed to facilitate the attachment of organelles to microtubules (Lasek and Brady, 1985). This observation suggested the presence of a motor protein trapped on microtubules by crosslinking, a state similar to the rigor found in the absence of ATP in cilia, flagella, and striated muscle. Isolation procedures were therefore devised to isolate microtubules in the presence of AMPPNP (Brady, 1985; Vale et al., 1985*a*). Extracts of these preparations, enriched in a polypeptide of 100–130 kDa, were shown to increase the level of ATPase activity of chick brain preparations. This polypeptide, which was found later to be an ATPase, could be tested for motor activity readily; for example, when adsorbed to glass in the presence of ATP, it propels microtubules along the glass, and the motility can be observed with the appropriate light-microscope technology. In a different kind of assay, synthetic microspheres can be coated with the protein and tested on oriented microtubules. Because of its involvement in movement this protein was called *kinesin*. When tested on isolated microtubules assembled on a centrosome (with the plus end out) using the microbead assay, the beads were found to move away from the centrosome (i.e., in the anterograde axonal direction, toward the plus end) (Vale et al., 1985b).

Kinesin was found to be widely distributed in different tissues. The protein is tetrameric, consisting of two copies of a 120 kDa and two copies of a 65 kDa complex (e.g., Bloom et al., 1988). Like other motors, kinesin was found to be an elongated molecule with a pair of globular heads at one end (e.g., Hirokawa et al., 1989). cDNA corresponding to *Drosophila* kinesin has been isolated. Like myosin II (Fig. 18.4) or dynein from cilia and flagella (Fig. 18.14), kinesin was found to have a globular portion at the amino-terminal.

Truncated forms of kinesin with progressively less tail region were produced by genetic manipulation of *E. coli* using vectors that encoded variable portions of the tail region. When tested in an assay system, it was found that the head portion was sufficient to produce motility (Yang et al., 1990).

A minimum of seven proteins of the kinesin family have recently been identified by determining the amino acid sequence through the techniques involving the isolation of cDNA and PCR amplification (Endow, 1991) (Chapter 1, II). Once the sequence was known, database searches for amino acid homologies were then carried out. The head portion is at the amino-terminal and contains the ATP and the microtubule binding sites (e.g., Yang et al., 1989), further evidence that it corresponds to the motor domain of the proteins. This portion is very similar for all these proteins. In contrast, the sequences of the tail portion have little in common. Hybridization of PCR-amplified cDNA produced from degenerate primers to the *Drosophila* polytene chromosomes (where the genome is amplified) suggests the existence of as many as 25 members of the kinesin family (Endow and Hatsumi, 1991). The identification of so many possible kinesins may indicate that each may be required for different functions.

Now that the amino acid sequence of at least some of the kinesins is known, some more precise questions can be asked. What determines directionality? Since *claret,* which has reverse directionality, has the motor region in the carboxy-end of the molecule instead of the amino-end, it is conceivable that directionality is determined by the position of the motor region.

Cytoplasmic dynein Cytoplasmic dynein was isolated from microtubules and found to bind microtubules in an ATP-dependent manner (Paschal et al., 1987). It was originally named microtubular associated protein 1C (MAP 1C), since its homology to dynein was not immediately recognized. The protein had ATPase activity, markedly stimulated by binding to microtubules. When tested with the microtubule-gliding assay (Paschal and Valee, 1987), it was found to be a motor. When tested on isolated axonemes of *Chlamydomonas reinhardtii,* a biflagellate single-celled alga, the axonemes were found to move toward the plus end (so that the force exerted by the motor was toward the minus end). As indicated earlier, this polarity is opposite that of kinesin.

MAP 1C was identified as cytoplasmic dynein from its biochemical and physical properties. Perhaps most significantly, scanning transmission electron microscopy (STEM) (Vallee et al., 1988) revealed the same morphology and molecular mass as the two-headed forms of ciliary and flagellar dynein. The native protein has two 410 kDa heavy chains, three 74 kDa subunits, and one subunit each of low molecular weight subunits ranging from 53 to 59 kDa (Vallee et al., 1988). The protein has been found in a variety of cells and tissues.

Mechanisms of movement What is the mechanism of transduction of the microtubule-based motors? Several models have been proposed (Block et al., 1990; Schnapp et al., 1990). Obviously an ATP-dependent cycling of cross-bridges between motors and microtubule is necessary. However, the details are still unknown. The possibility of controlling the protein structure through amino acid substitutions and the studies through movement assays may well shed light on this question.

Association of Motors with Vesicles The motors were found to be located on the organelle surfaces (Langford et al., 1987). In the native system, binding of the motor to the organelle is likely to require an integral membrane protein, since phospholipid vesicles or trypsinized organelles do not move on microtubules (Gilbert et al., 1985; Vale et al., 1985b). Specific binding to such motor receptors could determine the sorting out of vesicles accompanying intracellular transport (Chapter 4). A role of microtubules in this transport is supported by the observations that microtubules are associated with the endoplasmic reticulum (Franke, 1971), Golgi vesicles (Allan and Kreis, 1987), and lysosomes (Swanson et al., 1987).

Calculations of the energy required for movement (Sheetz and Spudich, 1983b) show that only a few of these motor molecules need to be attached to the organelles, and electron microscopic observations suggest only five cross-bridges between organelles and microtubules (Langford et al., 1987).

C. Intermediate Filaments

Intermediate filaments (IFs) are cytoskeletal fibers generally 8 to 10 nm in diameter. They are therefore intermediate in thickness between the thinner F-actin and the thicker microtubules. In most cells the IFs form a basketlike array around the nucleus but also reach the periphery of the cell. They are present at specialized junctions (see Chapter 4, V,D), and they are prominent throughout the length of axons.

Strictly speaking, IFs are not directly associated with movement. However, they will be discussed in this section. We have seen how difficult it is to separate a discussion of movement from that of structure (e.g., for the case of microvilli whose cytoskeleton is in part composed of actin and myosin). Furthermore, one of the roles of neurofilaments is thought to be the maintenance of the caliber of large myelinated axons, a factor with obvious implications in intra-axonal transport. The complexity and variety of IF proteins and tissue-specific expression suggest that they have an important role. They are most likely to play a role in specialized cell functions and in maintaining cell structure. However, a general housekeeping role seems unlikely since at least some cultured cells function well and even divide in the absence of IFs (although lamins, the IFs of the nucleus are present) (see Steinert and Roop, 1988).

IFs are unique among the cytoskeletal structure in that their protein subunits are fibrous (for a review on IF, see Steinert and Roop, 1988). As previously discussed, F-actin and microtubules are formed from the polymerization of globular subunits. The IF proteins form long filaments of high-tensile strength. Four types of IF are recognized, spanning a size range between 40 to 130 kDa, and they have been classified from their amino acid sequence.

Type I, the keratins (acid, basic, or neutral) are heteropolymers formed from an equal number of subunits from each subgroup. They are found predominantly in epithelial cells (besides hair and nails). There are at least 19 different types of IF proteins in human epithelia.

Type II proteins include *vimentin* (mesenchyme), *desmin* (striated and smooth muscle), and *glial fibrillar acidic proteins* (astrocytes and Schwann cells). These IF proteins assemble spontaneously in vitro to form homopolymers and heteropolymers. Type III proteins are neurofilament proteins of neurons, and in vertebrates the type III IF proteins represent three different polypeptides. Type IV are nuclear IF proteins, the *lamins,* that form a highly organized network, the *nuclear laminae*.

The central portions of all IF proteins (the so-called rod portions) of approximately 310 amino acids are highly homologous and in an α-helical configuration (see Fig. 18.19 for a diagrammatic representation). Association of the central region of two fibrous subunits produces a double-stranded coil (i.e., a dimer), a 3 nm *protofilament* (Fig. 18.20b). Two of these arrange in parallel to form a tetramer with two globular domains at each end, the 4.5 nm protofilament (Fig. 18.20c). These further arrange in a staggered manner (Fig. 18.20d) to produce a striated appearance when viewed with the EM.

Phosphorylation of specific domains in the IF proteins controls assembly. For example, vimentin (Inagachi et al., 1987) is disassembled by phosphorylation whereas phosphorylation of desmin inhibits assembly (Geisler and Weber, 1988). In fact, a dynamic remodeling of the cytoskeleton during neuronal growth and the establishment of directionality is thought to depend on the phosphorylation of specific domains of neurofilament proteins (Nixon and Sihag, 1991).

V. TRIGGERING OF CONTRACTION

The events that trigger contraction are reasonably well understood for vertebrate skeletal muscle. Several steps are interposed between the excitation of the nerve and contraction. The action potential of the nerve is transmitted to the muscle. The nerve terminals release acetylcholine. A specialized structure on the muscle fiber, the *end plate,* responds to the arrival of acetylcholine by depolarizing, giving rise to the *end-plate potential* (Chapter 16). The end-plate potential in turn triggers an action potential in the muscle, which results in contraction by processes discussed in this section. The sequence of events in other motile systems discussed here is less well understood. The coupling between stimulus events and motility in the various motor systems is the topic of this section.

Calcium is necessary to couple excitation to contraction; it has been implicated in muscle contraction from the earliest experiments. Heart muscle, for example, exhibits an action potential but fails to contract when Ca^{2+} is absent from its external medium (Mines, 1913). The situation is somewhat more complex in the striated muscle of the frog, which fails to conduct or contract when stimulated in a medium lacking Ca^{2+} (Ishiko and Sato, 1957). Apparently, the two

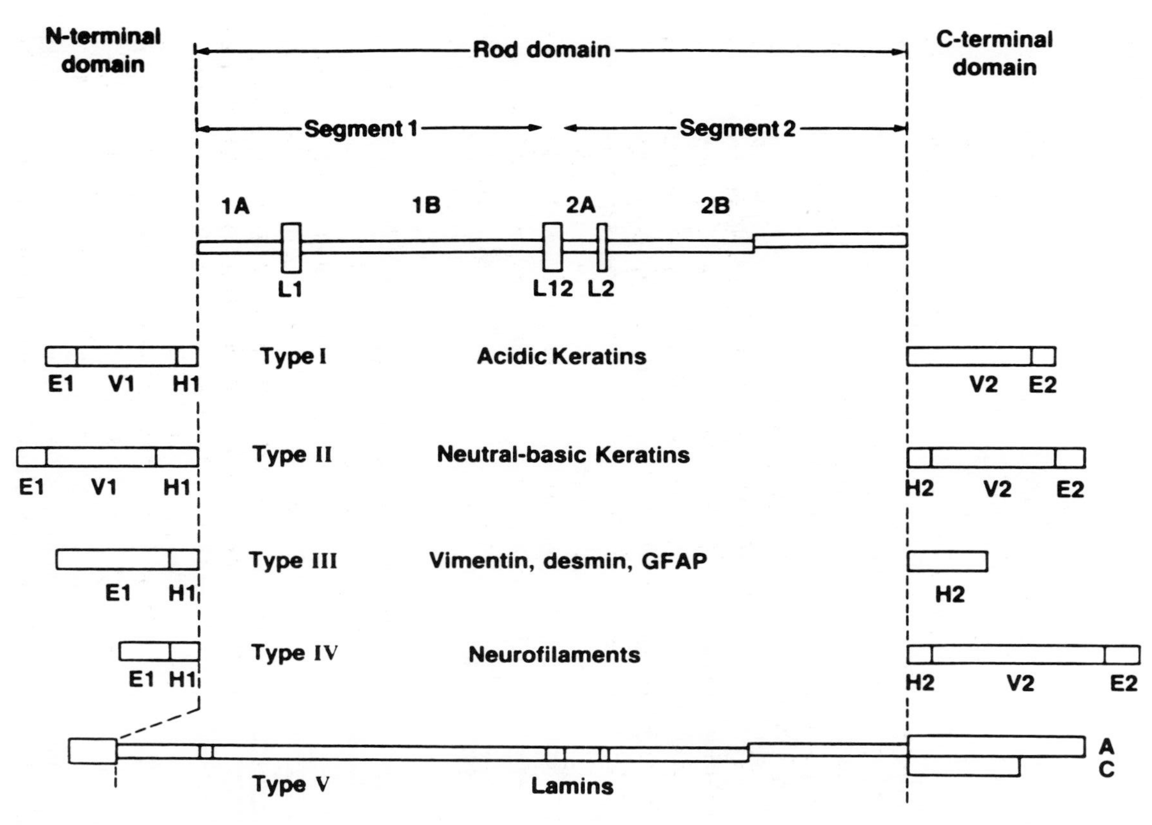

Fig. 18.19. Subdomains of IF chains. All central rod domains are flanked by end domains. For details see Steinert and Roop (1988). Steinert and Roop (1988). Reproduced from the *Annual Review of Biochemistry,* Volume 57, copyright © 1988 by Annual Reviews Inc.

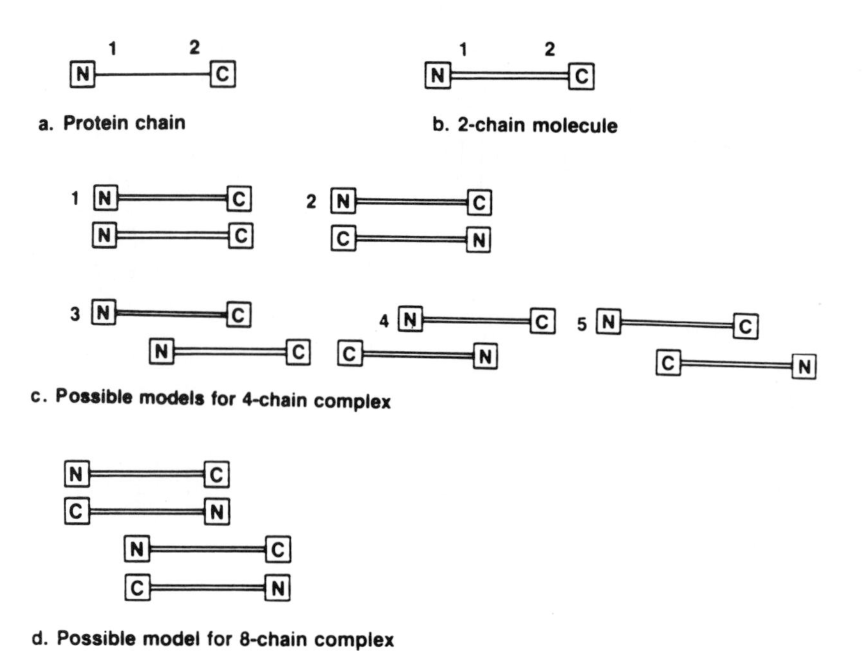

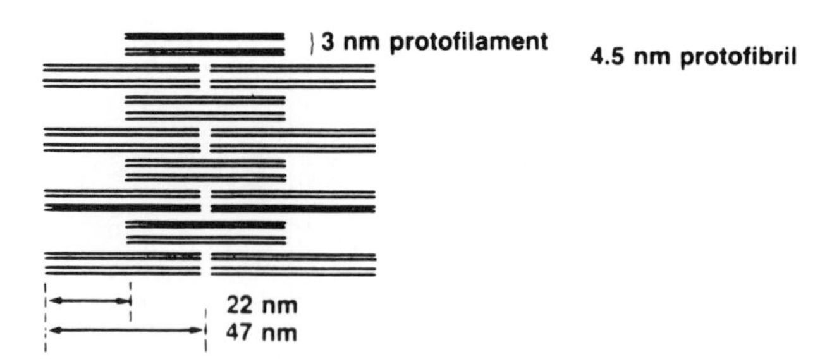

Fig. 18.20. IF chains and formation of oligomers. (a) IF chain, (b) two coiled-coil molecules in parallel and in register, (c) Proposed models for four chain complexes, (d) possible 8-chain complex and (e) lattice of an entire IF containing 32 chains / 47 nm. The axial banding at 22 and 47 nm is that observed with the EM and X-ray diffraction. Steinert and Roop (1988). Reproduced from the *Annual Review of Biochemistry,* Volume 57, copyright © 1988 by Annual Reviews Inc.

independent mechanisms, conduction and contraction, are affected by the lack of Ca^{2+}. When depolarization is induced by increasing the K^+ concentration in the medium, contraction still cannot be elicited without Ca^{2+} (Franck, 1960).

Although Ca^{2+} is necessary for contraction, muscle can contract in the absence of the action potential. Microinjection of Ca^{2+} into the fiber (Heilbrunn and Wiercinski, 1947; Portzehl et al., 1964) in concentrations as low as 0.3 to 1.5 μM (Portzehl et al., 1964) induces a contraction. In contrast, injections of Mg^{2+}, Na^+, K^+, ATP, AMP, arginine, and inorganic phosphate are ineffective (Caldwell and Walster, 1963; Heilbrunn and Wiercinski, 1947). The Ca^{2+} could then serve as a trigger for initiating contraction. This view is supported by the contraction elicited by delivering small amounts of Ca^{2+} with a micropipette on muscle fibers stripped of their surface membrane (skinned preparations) (Constantin et al., 1965). The contraction is limited to a small area on which the Ca^{2+} has been delivered, and it is followed soon by relaxation. The

Fig. 18.21 Tail myotome of guppy. The A and I bands and the Z line are labeled in the left corner at the bottom of the figure. At each Z line a *triad* [the central tubule (part of the T system) and the lateral sacs], is indicated by the triple arrows. The rest of the sarcoplasmic reticulum (single, large arrows) is probably continuous with the lateral sacs. The central tubule is convoluted and was originally present in and out of the section. It appears in this section as vesicles. From Franzini-Armstrong (1964), with permission.

requirement of Ca^{2+} for contraction, either in intact muscle or in stripped fibers, suggests that the release of Ca^{2+} into the fiber's interior induces contraction. Its subsequent removal by some special mechanism induces the contractile elements to relax.

The Ca^{2+} required for contraction does not originate from the medium external to the fiber. The influx of Ca^{2+} during activation is insufficient to provide enough for contraction (Franck, 1961; Winegrad, 1961). On the other hand, vesicles derived from the muscle's endoplasmic reticulum (the *sarcoplasmic reticulum*) can remove enough Ca^{2+} from the medium to induce isolated fibrils to relax after having been made to contract by the addition of ATP (Weber et al., 1963). Therefore the sarcoplasmic reticulum could provide the system for removal and release of the Ca^{2+} necessary for the relaxation-contraction cycles of functional striated muscle.

The structural arrangement of the internal membranes is well suited for a role in triggering contraction. It is composed of a system of transverse tubules continuous with the surface membrane (Fig. 18.21*a* and *b*) and a longitudinal system, the lateral sacs, that reach each sarcomere (Fig. 18.21*c*) (Peachey, 1965; Porter, 1961). The T tubules and the two associated cisternae at the sarcomere level constitute the *triad*. The continuity of the transverse system with the surface membrane has been demonstrated by electron microscopy (Fig. 18.21*a*) (Franzini-Armstrong, 1964). Similarly, the longitudinal and transverse systems are closely associated (Fahrenbach, 1965; Franzini-Armstrong, 1964, 1973). The continuity of the transverse tubules (T tubules) with the surface membrane is supported by observation of the passage of large

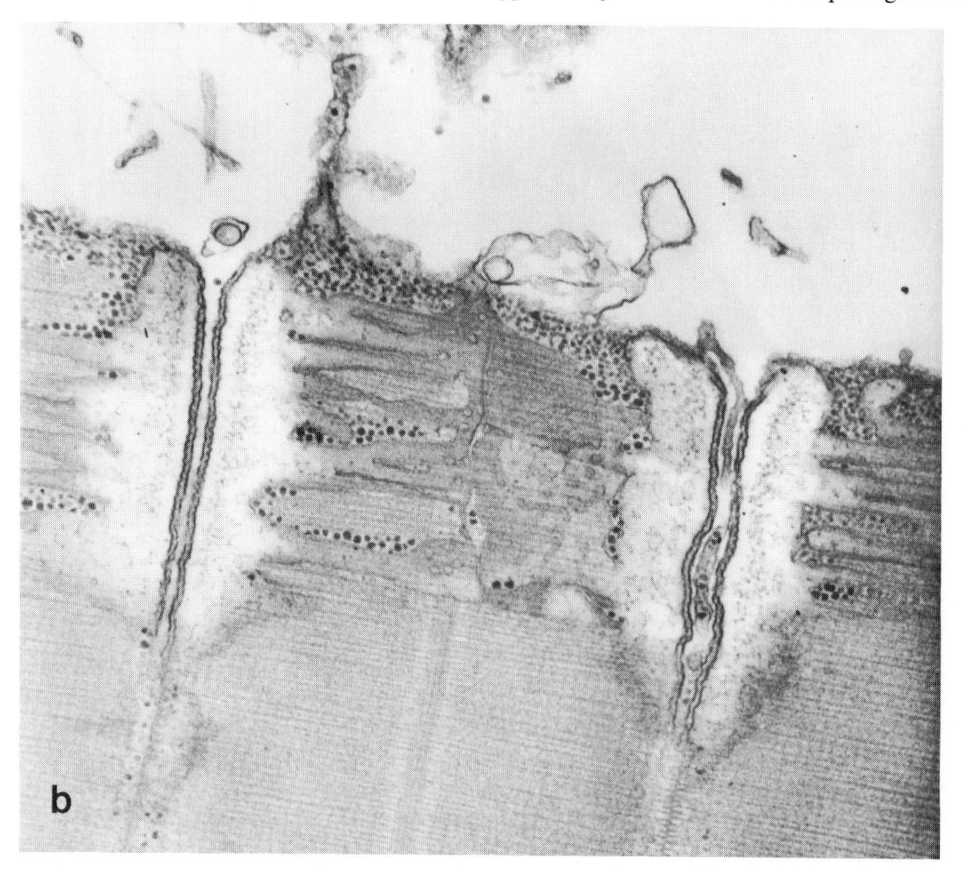

Fig. 18.21 (*Continued*) (*b*) Opening of the T-system to the outside showing continuity with the cell membrane (the *plasmalemma* or *sarcolemma*). White twitch fiber from the black mollie. ×50,000. From Franzini-Armstrong (1973), with permission.

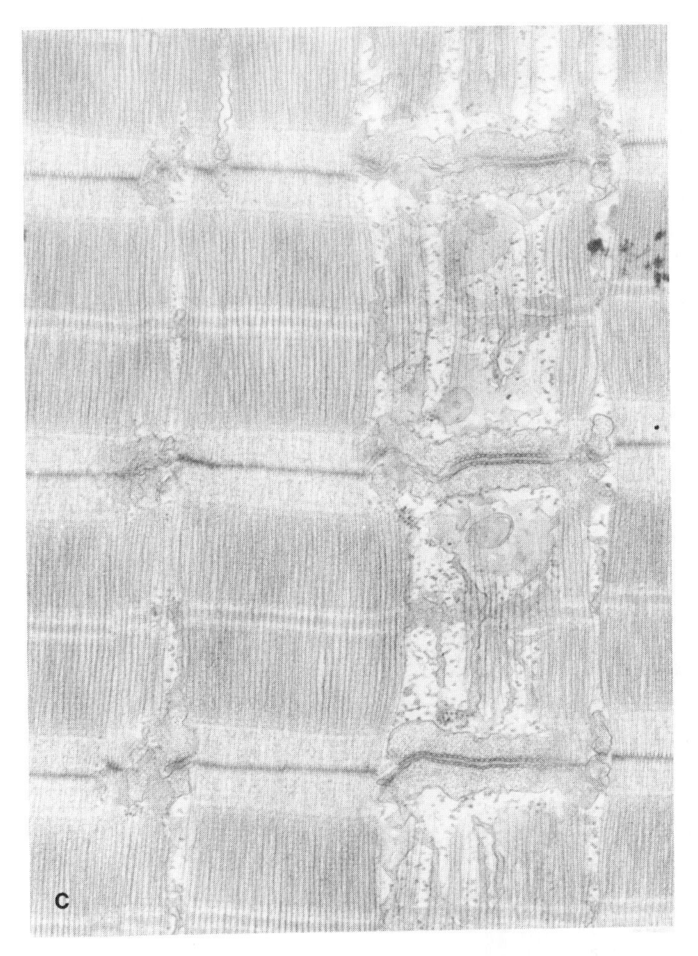

Fig. 18.21 *(Continued)* (*c*) Sarcoplasmic reticulum of frog sartorius muscle. From Peachey (1965), by copyright permission of the Rockefeller University Press.

molecules, such as ferritin (Huxley, 1969; Page, 1964) or albumin (Hill, 1964), into the transverse system from the medium. Three-dimensional reconstructions are shown in Fig. 18.22 (Peachey, 1965).

A recent reconstruction of the triad is shown in Fig. 18.23 (Franzini-Armstrong et al., 1987, Block et al., 1988). The T tubule (in the center of the image, toward the viewer) is shown with two terminal cisternae (TC). The lower one is shown occupied by *calsequestrin,* an acidic Ca^{2+}-binding protein of skeletal, cardiac, and smooth muscle (MacLennan et al., 1983). Calsequestrin has a molecular mass of 42 kDa and can bind approximately 50 Ca^{2+} per molecule. The surface of the TC shows projection of the Ca^{2+}-ATPase, visible also as intramembranous particles. The projections are lacking in the junctional portion of the triad. The large structures of four subunits in the junctional portion are the so called *junctional feet* (JF).

That the sarcoplasmic reticulum is involved in triggering a contraction is supported by experiments in which small portions of the muscle fibers are given electrical subthreshold stimuli (Huxley and Taylor, 1958; Huxley, 1969) at the location of the T tubules and experiments in which direct electrical stimulation is applied to fibers stripped of surface membranes (Constantin and Podolski, 1966; Csapo, 1959). In contrast to the effect of normal stimulation, the responses are local, are not propagated, and are graded depending on the strength of the stimulus. The effectiveness of the stimulation varies dramatically with the position of the electrode. In frog muscle, stimulation at the Z line is effective (Huxley, 1959; Huxley and Taylor, 1958), and in the crab and lizard, stimulation between the A and I bands (Huxley, 1959) is effective. These susceptible spots correspond to the location of the openings of the transverse tubules in these animals. The portions of the sarcomeres that contract correspond in distribution to the longitudinal vesicles. These experiments support a role of the sarcoplasmic reticulum in the control of muscle contraction and also implicate the T tubules in conducting depolarization.

In the proposed mechanism, Ca^{2+} would be released from the sarcoplasmic reticulum to produce a contraction, whereas it would be transported into the vesicles during relaxation. Tests of this premise have been carried out with several approaches. The Ca^{2+} can be detected in a variety of ways. It can be traced by using a radioactive isotope such as ^{45}Ca. Alternatively, its concentration can be estimated using Ca^{2+} indicators. These can be dyes such as murexide, which complexes with Ca^{2+} and whose light absorption varies with the degree of binding. Murexide was discussed previously in connection with a study of the interaction between Ca^{2+} and Ca^{2+}-transporting enzyme, the Ca^{2+}-ATPase (Chapter 14). A convenient indicator is the protein aequorin, extracted from the luminous jellyfish, which fluoresces when complexed with Ca^{2+} so that the light emitted is proportional to the concentration of Ca^{2+}.

The results obtained with these three approaches essentially agree. Muscle can be incubated

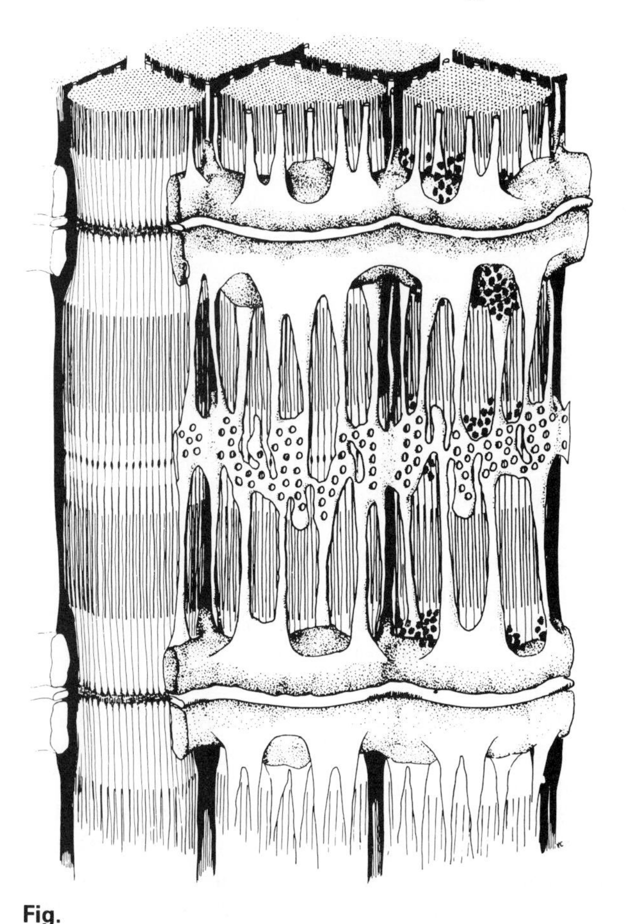

Fig. 18.22 Three-dimensional reconstruction of the sarcoplasmic reticulum of the frog. From Peachey (1965). Reproduced from *The Journal of Cell Biology*, 1965, 25:209–231, by copyright permission of the Rockefeller University Press.

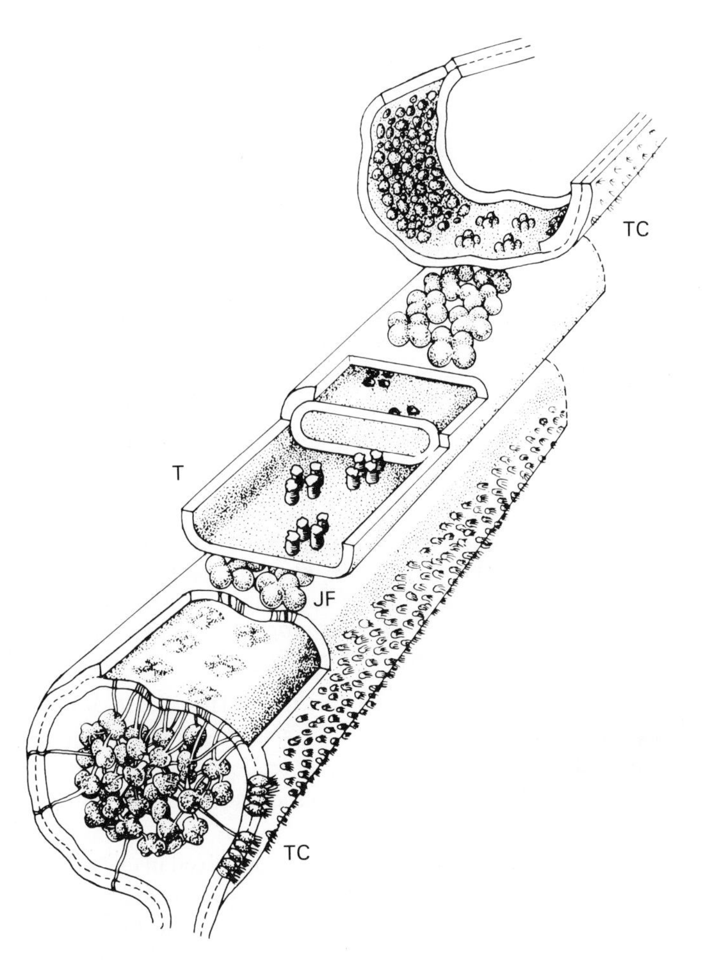

Fig. 18.23 Three-dimensional reconstruction of a triad. From Franzini-Armstrong et al. (1987) and Block et al. (1988). Reproduced from *The Journal of Cell Biology* (1988) 107:2587–2600, by copyright permission of the Rockefeller University Press.

in the presence of ^{45}Ca and then fixed after contraction or in the relaxed state (Winegrad, 1965a, 1965b). The position of the radioactivity viewed with autoradiography then serves to locate the Ca^{2+}. At rest, the ^{45}Ca was found in a position corresponding to the lateral sacs close to the I-band. During contraction it was found to shift to the A-band, where fiber overlap would be expected. The Ca^{2+} can also be followed using Ca^{2+} indicators. The experiments with murexide showed that the amount of free Ca^{2+} inside the muscle fiber increases with contraction (Jobis and O'Connor, 1966). In other experiments in which aequorin was injected into barnacle muscle fibers, the time course of the fluorescent emission preceded the contraction, and the decrease in fluorescence preceded the relaxation. The results are shown in Fig. 18.24 (Ashley and Ridgeway, 1970). Curve 1 indicates the changes in membrane potential that precede the release of Ca^{2+}, shown as light emission in trace 2. The isometric tension of the muscle, shown in trace 3 (the muscle is held at constant length), follows the release of Ca^{2+} and decreases after the decrease in internal Ca^{2+}. These experiments suggest that Ca^{2+} is indeed the trigger for muscle contraction.

The portion of the sarcoplasmic reticulum (SR) that intimately controls the Ca^{2+} release-sequestration cycles must correspond to the *triad*. Physiologically, excitation-contraction must depend on three processes that correspond to the structures. There must be a mechanism of

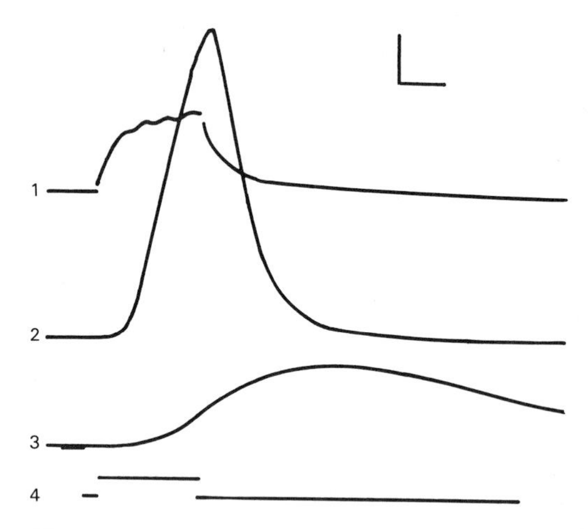

Fig. 18.24 Results of applying a single depolarizing pulse to a single barnacle muscle fiber. Curve 1, membrane potential; curve 2, light emission; curve 3, isometric tension; curve 4, stimulation marker. Calibration: ordinate, 20 mV (curve 1) or 5 g (curve 2); abscissa, 100 ms. From C. C. Ashley and E. B. Ridgeway, *Journal of Physiology,* 209:105–130, with permission. Copyright © 1970 The Physiological Society, Oxford, England.

coupling the depolarization of the cell membrane and the release of Ca^{2+}. The release of Ca^{2+} needed for contraction of the myofibril is most likely to involve Ca^{2+} channels of the junctional SR. Removal of the Ca^{2+} by the terminal cisternae involves the operation of Ca^{2+}-ATPase, which we discussed in part in Chapters 14 and 15. In addition, calsequestrin helps retain the Ca^{2+} in the sarcoplasmic reticulum by its capacity to bind it.

There is little question that the electrical currents in the T tubules are involved in excitation-contraction coupling. All the evidence suggests that the depolarization is equivalent to a conventional action potential. Voltage clamp experiments on isolated portions of muscle fibers have demonstrated an inward delayed Na^+ current that follows the current of the action potential as shown in Fig. 18.25, curve A (Mandrino, 1977). When the Na^+ in the medium is reduced (in this

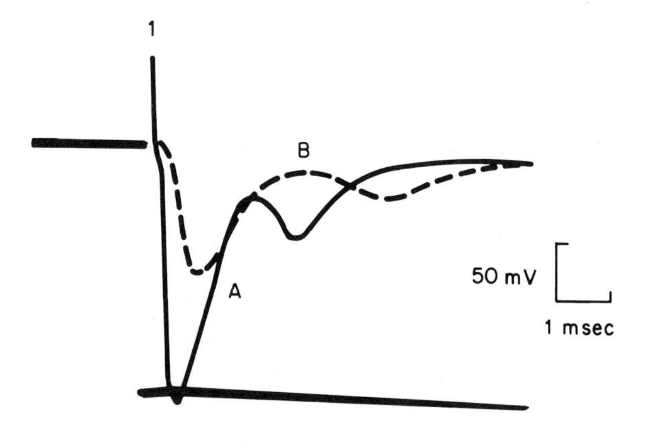

Fig. 18.25 Effects of decreasing the external sodium concentration on the inward currents. A, Record in normal Ringer's solution; B, record in a solution with 50% of the normal sodium concentration. The peak current is approximately 1×10^{-7} A. From M. Mandrino, *Journal of Physiology,* 269:605–625, with permission. Copyright © 1977 The Physiological Society, Oxford, England.

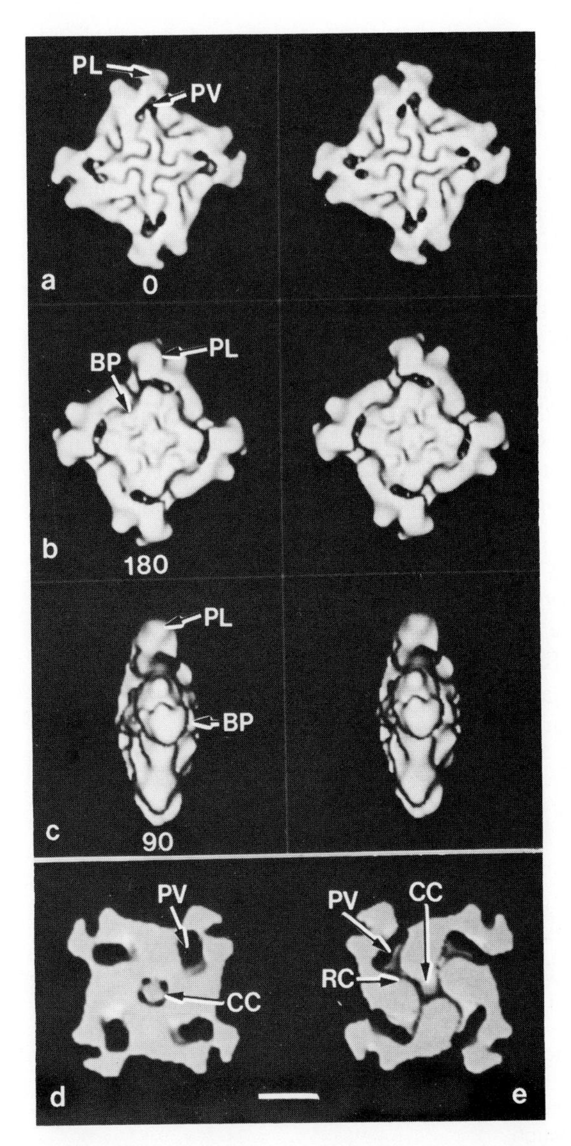

Fig. 18.26 Computer-generated surface representations of the three-dimensionally reconstructed junctional channel complex. (*a–c*) Stereo pairs of the reconstruction in various orientations related by rotation about a vertical axis. (*d* and *e*) The two complementary halves of the reconstruction after slicing it in half to reveal internal structural features. In (*e*) the view is from the interior of the channel toward the surface adsorbed to the grid (the "platform" side) and in (*d*) it is toward the T-tubule surface. Abbreviations: BP, base platform; PL, peripheral lobes; PV, peripheral vestibules; CC, central channel; RC, radial channels. Scale bar, 10 nm. From Wagenknecht et al. (1989), with permission.

case to half the normal amount), the delayed Na^+ current is also reduced, as shown in Fig. 18.25, curve B. In agreement with this observation, tetrodotoxin, which blocks Na^+ channels, also delays the current. The delayed current disappears when the T tubules are removed by treatment with glycerol. The presence of Na^+ channels in the T tubules is also supported by the localization of monoclonal antibodies to the Na^+ channels in the T tubules (Haimovich et al., 1987).

The mechanism of coupling between the T tubules and the terminal cisternae is still not clear. It is conceivable that the depolarization of the T tubules releases a second messenger that induces the release of Ca^{2+} from the TC, presumably through Ca^{2+} channels. This second messenger may be IP_3 or very low concentrations of Ca^{2+} (see Chapter 5). Both have been shown to induce the discharge of Ca^{2+} from the sarcoplasmic reticulum. A role of a second messenger is very likely (see Volpe et al., 1986), and at this time a strong argument can be made to support a role of either IP_3 or Ca^{2+}.

The closeness of the association between the terminal cisternae and the T tubules suggests the possibility of a direct mechanical connection between the two (see Caswell and Brandt, 1989). A molecular sensor in the T tubules in contact with the Ca^{2+} channels could induce a conformational change in the latter, initiating the Ca^{2+} release.

The Ca^{2+}-efflux channels correspond to the feet protein. The channels are blocked by the drug ryanodine (an alkaloid generally used as an insecticide). Binding of ryanodine provides a convenient way to recognize the channel protein during fractionation procedures. The ryanodine receptor was found to correspond to a single peptide of approximately 360 kDa. Figure 18.26 (Wagenknecht et al., 1989) shows an image reconstruction of the isolated protein from electron micrographs (see Chapter 1). When incorporated in bilayers, the ryanodine receptor was found to have channel activity (Hymel et al., 1988) expected of the Ca^{2+}-release channel (e.g., inhibition by ruthenium red and Mg^{2+}).

Many studies have suggested that Ca^{2+} activation also occurs in primitive motile systems. The freshwater protozoan *Spirostomum ambiguum*, for example, contracts when stimulated electrically. Release of Ca^{2+}, detected by the aequorin assay, coincides with the contraction. Removal of Ca^{2+} coincides with relaxation (Ettienne, 1970).

A role of Ca^{2+} is also indicated in the ciliary motion of *Paramecium* (Eckert, 1972) and in other systems as well, but the evidence is indirect. Some examples of the indirect evidence for Ca^{2+} involvement are shown in Table 18.2.

The initiation of chromosomal movement in the mitotic spindle also seems to be triggered by a release of Ca^{2+} from vesicles distributed in the mitotic spindle (Hepler, 1980; Hepler and Callahan, 1987).

Table 18.2 Possible Role of Ca^{2+} in the Motility of Various Cells

Organism or cell	Evidence	References
Vorticella	Ca^{2+} activation of contraction	*a*
Xenopus laevis	Cleavage of eggs requires Ca^{2+}	*b*
Spirostomum	Release of Ca^{2+} to activate contraction followed with aequorin	*c*
Fibroblasts	Ca^{2+} required for contraction; vesicular fraction inhibits contraction	*d*
Stentor	Initiation of contraction requires calcium	*e*
Leukocytes	Demonstration of the presence of a contractile protein that requires Ca^{2+} to contract	*f*
Amoeba	Ca^{2+} needed for contraction of ruptured cells	*g*
Physarum polycephalum	Ca^{2+}-requiring actinomysin	*h*
	Ca^{2+}-requiring vacuoles	*i*
	Release of Ca^{2+} with electrical charge	*j*

[a]Amos, W. B. *Nature (London)* 229:127 (1971).
[b]Baker, P. F., and Warner, A. E. *J. Cell Biol.* 53:579 (1972).
[c]Ettienne, E. M. *J. Gen. Physiol.* 56:168 (1970).
[d]Kinoshita, S., Andoh, B., and Hoffman-Berling, H. *Biochim. Biophys. Acta* 79:88 (1964).
[e]Huang, B., and Pitelka, D. R. *J. Cell Biol.* 57:704 (1973).
[f]Shibata, N., Tatsumi, N., Tanaka, K., Okamura, Y., and Senda, N. *Biochim. Biophys. Acta* 256:565 (1972).
[g]Taylor, D. L., Condeelis, J. S., Moore, P. L., and Allen, R. D. *J. Cell Biol.* 59:378–394 (1973).
[h]Kato, T., and Tonomura, Y. *J. Biochem.* 77:1127–1134 (1975).
[i]Kato, T., and Tonomura, Y. *J. Biochem.* 81:207–213 (1977).
[j]Ridgway, E. B., and Durham, A. C. H. *J. Cell Biol.* 69:223–226 (1976).

SUGGESTED READING

Adams, R., and Pollard, T. D. (1989) Prediction of common properties of particle translocation motors through comparison of myosin I, cytoplasmic dynein, and kinesin. In *Cell Movement,* Vol. 2, pp. 3–10. Liss, New York.

Amos, L. A., and Amos, W. B. (1991) *Molecules of the Cytoskeleton,* The Guilford Press, New York.

Brokaw, C. J., and Johnson, K. A. (1989) Perspectives, dynein induced microtubule sliding and force generation. In *Cell Movement,* Vol. I, pp. 191–198. Liss, New York.

Caswell, A. H., and Brandt, N. R. (1989) Does muscle activation occur by direct mechanical coupling of tranverse tubules to the sarcoplasmic reticulum? *Trends Biochem. Sci.* 14:161–165.

Ebashi, S. (1983) Regulation of contractility. In *Muscle and Nonmuscle Motility,* Vol. 1, pp. 217–232 (Stracher, A., ed.). Academic Press, New York.

Gibbons, I. R. (1981) Cilia and flagella of eukaryotes. *J. Cell Biol.* 91:107S–124S.

Gibbons, I. R. (1988) Dynein ATPases as microtubule motors. *J. Biol. Chem.* 263(31):15837–15840.

Gibbons, I. R. (1989) Microtubule-based motility: an overview of a fast moving field. In *Cell Movement,* Vol. 1, pp. 3–22. Liss, New York.

Hammer, J. A. III (1991) Novel Myosins. *Trends in Cell Biol.* 1:50–56.

Huxley, H. E. (1990) Minireview: Sliding filaments and molecular motile system, *J. Biol. Chem.* 265:8347–8350.

Huxley, H. E., and Kress, M. (1985) Cross bridge behavior during muscle contraction. *J. Muscle Res. Cell Motil.* 6:153–161.

Lackie, J. M. (1986) *Cell Movement and Cell Behavior,* Chapters 1–3. Allen & Unwin, London.

Martonosi, A. N. (1983) The regulation of cytoplasmic Ca^{2+} concentration in muscle and non-muscle cells. In *Muscle and Nonmuscle Motility,* Vol. 1, pp. 233–357 (Stracher, A., ed.). Academic Press, New York.

Pollard, T. D., Doberstein, S. K., and Zot, H. G. (1991) Myosin I. *Annu. Rev. Physiol.* 53:653–681.

Rüegg, J. C. (1986) Vertebrate smooth muscle. In *Calcium in Muscle Activation,* Chapter 8, pp. 201–238. Springer-Verlag, Berlin.

Satir, P. (1989) Mechanism of ciliary movement—what's new. *News Physiol. Sci.* 4:153–157.

Shroer, T. A., and Sheetz, M. P. (1991) Functions of microtubule-based motors. *Annu. Rev. Physiol.* 53:629–652.

Somlyo, A. P. (1984) Cellular site of calcium regulation. *Nature (London)* 309:516–517.

Somlyo, A. P., and Himpens, B. (1989) Cell calcium and its regulation in smooth muscle. *FASEB J.* 3:2266–2276.

Spudich, J. A. (1989) In pursuit of myosin function. *Cell Regul.* 1:1–11.

Steinert, P. M., and Roop, D. R. (1988) Molecular and cellular biology of intermediate filaments. *Annu. Rev. Biochem.* 57:593–626.

Vallee, R., and Bloom, G. S. (1991) Mechanism of fast and slow axonal transport. *Annu. Rev. NeuroScie.* 14:59–92.

Minireviews

Bretscher, A. (1991) Molecular aspects of microfilament structure and assembly. *Curr. Opin. Struct. Biol.* 1:281–287.

Endow, S. A. (1991) The emerging kinesin family of microtubule motor proteins. *Trends Biochem. Sci.* 16:221–225.

Hartwig, J. H., and Kwiatkowski, D. J. (1991) Actin-binding proteins. *Curr. Opin. Cell Biol.* 3:87–97.

Morrow, J. (1989) The spectrin membrane skeleton: emerging concepts. *Curr. Opin. Cell Biol.* 1:23–29.

REFERENCES

Adams, R. J., and Pollard, T. D. (1986) Propulsion of organelles isolated from *Acanthamoeba* along actin filaments by myosin I. *Nature* 322:754–756.

Albanesi, J. P., Fujisaki, H., Hammer, J. A. III, Korn, E. D., Jones, R., and Sheetz, M. P. (1985) Monomeric *Acanthamoeba* myosin I supports movement in vitro. *J. Biol. Chem.* 260:8649–8652.

Allan, V. J., and Kreis, T. E. (1987) A microtubule-binding protein associated with membranes of the Golgi apparatus. *J. Cell Biol.* 103:2229–2239.

Allen, R. D., Weiss, D. G., Hayden, J. H., Brown, D. T., Fujiwake, H., and Simpson, M. (1985) Gliding movement of and bidirectional organelle transport along single native microtubules from squid axoplasm: evidence for an active role of microtubules in cytoplasmic transport. *J. Cell Biol.* 100:1736–1752.

Amos, L. A. (1987) Kinesin from pig brain studied by electron microscopy. *J. Cell Sci.* 87:105–107.

Anderson, R. A., and Marchesi, V. T. (1985) Associations between glycophorin and protein 4.1 are modulated by polyphosphoinositides: a mechanism for membrane skeletal regulation. *Nature* 318:295–298.

Arata, T., and Shimizu, H. (1981) Spin-label study of actin-myosin-nucleotide interactions in contracting glycerinated muscle fibers. *J. Mol. Biol.* 151:411–437.

Ashley, C. C., and Ridgeway, E. B. (1970) On the relationship between membrane potential, calcium transient and tension in single barnacle muscle fibers. *J. Physiol. (London)* 209:105–130.

Bárány, M., and Bárány, K. (1959) Studies on "active centers" of L-myosin. *Biochim. Biophys. Acta* 35:293–309.

Beckerle, M. C., and Porter, K. R. (1982) Inhibitors of dynein activity block intracellular transport in erythrophores. *Nature (London)* 295:701–703.

Bennett, V. (1990) Spectrin: structural mediator between diverse plasma membrane proteins and the cytoplasm. *Curr. Opin. Cell Biol.* 2:51–56.

Block, B. A., Imagawa, T., Campbell, K. P., and Franzini-Armstrong, C. (1988) Structural evidence for direct interaction between the molecular components of the transverse tubule/sarcoplasmic reticulum junction in skeletal muscle. *J. Cell Biol.* 107:2587–2600.

Block, S. M., Goldstein, L. S. B., and Schnapp, B. J. (1990) Dead movement by single kinesin molecules studied with optical tweezers. *Nature* 348:348–352.

Bloom, G. S., Wagner, M. C., Pfister, K., and Brady, S. T. (1988) Native structure and physical properties of bovine brain kinesin, and identification of the ATP-binding polypeptide. *Biochem.* 27:3409–3416.

Brady, S. T. (1985). A novel brain ATPase with properties expected for the fast axonal transport motor. *Nature* 317:73–75.

Brady, S. T., Lasek, R. J., and Allen, R. D. (1982) Fast axonal transport in extruded axoplasm from squid giant axon. *Science* 218:1129–1131.

Brady, S. T., Lasek, R. J., and Allen, R. D. (1985) Video microscopy of fast axonal transport in extruded axoplasm: a new model for study of molecular mechanisms. *Cell Motil.* 5:81–101.

Bresnick, A., Warren, V., and Condeelis, J. (1990) Identification of a short sequence essential for actin binding by *Dictyostelium* ABP-120. *J. Biol. Chem.* 265:9236–9240.

Burton, P. R., and Fernandez, H. L. (1973) Delineation by lanthanum staining of filamentous elements associated with the surfaces of axonal microtubules. *J. Cell Sci.* 12:576–583.

Caldwell, P. C., and Walster, G. (1963) Studies on the microinjection of various substances into crab muscle fibers. *J. Physiol. (London)* 169:353–373.

Caswell, A. H., and Brandt, N. R. (1989) Does muscle activation occur by direct mechanical coupling of transverse tubules to the sarcoplasmic reticulum? *Trends Biochem. Sci.* 14:161–165.

Chen, J. C. W., and Kamiya, N. (1975) Localization of myosin in the internodal cell of *Nitella* as suggested by differential treatment with *N*-ethylmaleimide. *Cell Struct. Funct.* 1:1–9.

Collins, K., Sellers, J. R., and Matsudaira, P. (1990) Calmodulin dissociation regulates brush border myosin I (110-kD-calmodulin) mechanochemical activity in vitro. *J. Cell Biol.* 110:1137–1147.

Coluccio, L. M., and Bretscher, A. (1988) Mapping of the microvillar 100K-calmodulin complex: calmodulin-associated or -free fragments of the 110-kD polypeptide bind actin and retain ATPase activity. *J. Cell Biol.* 106:367–373.

Coluccio, L. M., and Bretscher, A. (1989) Reassociation of microvillar core proteins: making a microvillar core *in vitro*. *J. Cell Biol.* 108:495–502.

Constantin, L. L. and Podolosky, R. J. (1966) Evidence for the depolarization of the internal membrane system in activation of the frog semitendinosus muscle. *Nature (London)* 210:483–486.

Constantin, L. L., Franzini-Armstrong, C., and Podolsky, R. J. (1965) Localization of calcium-accumulating structures in striated muscle fibers. *Science* 147:158–159.

Conzelman, K. A., and Mooseker, M. (1987) The 110 kD protein-calmodulin complex of the intestinal microvilli is an actin activated MgATPase. *J. Cell Biol.* 105:313–324.

Craig, S. W. and Pollard, T. D. (1982) Actin binding proteins. *Trends Biochem. Sci.* 7:88–91.

Csapo, A. (1959) Studies on excitation-contraction coupling. *Ann. N.Y. Acad. Sci.* 81:453–467.

Davidson, P. F., and Taylor, E. W. (1960) Physical-chemical studies of proteins of squid nerve axoplasm, with special reference to the axon fibrous protein. *J. Gen. Physiol.* 43:801–803.

De Arruda, M. V., Watson, S., Leavitt, J., and Matsudaira, P. (1990) Fimbrin is a homologue of cytoplasmic phosphoprotein plastin and has domains homologous with calmodulin and actin gelation proteins. *J. Cell Biol.* 111:1069–1079.

DeLozanne, A., and Spudich, J. A. (1987) Disruption of *Dictyostelium* myosin heavy chain gene by homologous recombination. *Science* 236:1086–1091.

Devreotes, P., Fontana, D., Klein, P., Sherring, J., and Theibert, A. (1987) Trans-membrane signaling in *Dictyostelium*. *Methods Cell Biol.* 28:489–496.

Ebashi, S., Ohtsuki, I., and Mihashi, K. (1972) Regulatory proteins of muscle with special reference to troponin. *Cold Spring Harbor Symp. Quant. Biol.* 37:215–223.

Eckert, R. (1972) Bioelectrical control of ciliary activity. *Science* 176:473–481.

Endo, M., Nonomura, Y., Masaki, T., Ohtsuk, I., and Ebashi, S. (1966) Localization of native tropomyosin in relation to striation patterns. *J. Biochem. (Tokyo)* 60:605–608.

Endow, S. A. (1991) The emerging kinesin family of microtubule motor proteins. *Trends Biochem. Sci.* 16:221–225.

Endow, S. A., and Hatsumi, S. (1991) A multimember kinesin gene family in *Drosophila*. *Proc. Natl. Acad. Sci. U.S.A.* 88:4424–4427.

Ettienne, E. M. (1970) Control of contractility in *Spirostomum* by dissociated calcium ions. *J. Gen. Physiol.* 56:168–179.

Fahrenbach, W. H. (1965) Sarcoplasmic reticulum: ultrastructure of the triadic structure. *Science* 147:1308–1309.

Fernandez, A., Brautigan, D. L., Mumby, M., and Lamb, N. J. C. (1990) Protein phosphatase type-1, not type-2A, modulates actin microfilament integrity and myosin light chain phosphorylation in living nonmuscle cells. *J. Cell Biol.* 111:103–112.

Flory, P. J. (1956) Role of crystallization in polymers and proteins. *Science* 124:53–60.

Franck, G. B. (1960) Effects of changes in extracellular calcium concentration and potassium induced contracture of frog's skeletal muscle. *J. Physiol. (London)* 151:518–538.

Franck, G. B. (1961) Role of extracellular calcium ions in excitation-contraction coupling in skeletal muscle. In *Biophysics of Physiological and Pharmacological Actions*, Publ. 69, pp. 293–307. American Association for the Advancement of Science, Washington, D.C.

Franzini-Armstrong, C. (1964) Fine structure of sarcoplasmic reticulum and transverse tubular systems in muscle fibers. *Fed. Proc., Fed. Am. Soc. Exp. Biol.* 23:887–895.

Franzini-Armstrong, C. (1973) Membranous systems in muscle fibers. In *Structure and Function of Muscle* (Bourne, G. H., ed.), pp. 531–619. Academic Press, New York.

Franzini-Armstrong, C., Kenney, L. J., and Varriano-Marston, E. (1987) The structure of calsequestrin in triads of vertebrate skeletal muscle: a deep-etch study. *J. Cell Biol.* 105:49–56.

Fujisaki, H., Albanesi, J. P., and Korn, E. D. (1985) Experimental evidence for contractile activities of *Acanthoamoeba* myosin IA and IB. *J. Biol. Chem.* 260:1183–1189.

Fukui, Y., Lynch, T. J., Brzeska, H., and Korn, E. D. (1989) Myosin I is located at the leading edges of locomoting *Dictyostelium amoebae*. *Nature (London)* 341:328–331.

Geiger, B. (1979) A 130K protein from chicken gizzard: its localization at the termini of microfilament bundles in cultured chicken cells. *Cell* 18:193–205.

Geisler, N., and Weber, K. (1988) Phosphorylation of desmin *in vitro* inhibits formation of intermediate filaments: identification of three kinase A sites in the aminoterminal head domain. *EMBO J.* 7:15–20.

Gilbert, S. P., Slaboda, R. D., and Allen, R. D. (1985) Translocation of vesicles from squid axoplasm on flagellar microtubules. *Nature (London)* 315:245–248.

Gordon, A. M., Huxley, F., and Julian, F. J. (1966) The variation in isometric tension with sarcomere length in vertebrate muscle fibers. *J. Physiol. (London)* 184:170–192.

Griffith, L. M., and Pollard, T. D. (1978) Evidence for actin filament microtubule interaction mediated by microtubule-associated proteins. *J. Cell Biol.* 78:958–965.

Griffith, L. M., and Pollard, T. D. (1982) The interaction of actin filaments with microtubules and microtubule-associated proteins. *J. Biol. Chem.* 257:9143–9151.

Haimovich, B., Schotland, D. L., Fieles, W. E., and Barchi, R. L. (1987) Localization of sodium channel subtypes in adult rat skeletal muscle using channel-specific monoclonal antibodies. *J. Neurosci.* 7:2957–2966.

Hammer, J. A. III (1991) Novel myosins. *Trends in Cell Biol.* 1:50–56.

Hanson, J., and Lowy, J. (1961) The structure of the muscle fibres in the translucent part of the adductor of the oyster *Crassostrea angulata*. *Proc. R. Soc. London Ser. B* 154:173–196.

Hanson, J., and Lowy, J. (1963) The structure of F-actin filaments isolated from muscle. *J. Mol. Biol.* 6:46–60.

Harrington, W. F. (1971) A mechanochemical mechanism for muscle contraction. *Proc. Natl. Acad. Sci. U.S.A.* 68:685–689.

Hartwig, J. H., and Kwiatkowski, D. J. (1991) Actin-binding proteins. *Curr. Opin. Cell Biol.* 3:87–97.

Hasselbach, W. (1952) Die Uemwandlung von Actomyosin-ATPase in L-myosin ATPase durch Aktivatoren und die Resultierenden Aktivierungseffekte. *Z. Naturforsch.* 76:163–174.

Hasselbach, W., and Ledermair, O. (1958) Der Kontraktionscyclus der Isolierten contractiles Strukturen der Uterusmuskulatur und Seine Besonderheiten. *Pfluegers Arch. Gesamte Physiol.* 267:532–542.

Hayashi, T. (1953) Contractile properties of compressed monolayers of actomyosin. *J. Gen. Physiol.* 36:139–151.

Hayden, S. M., Wolenski, J. S., and Mooseker, M. S. (1990) Binding of brush border myosin to phospholipid vesicles. *J. Cell Biol.* 111:443–451.

Heilbrunn, L. V., and Wiercinski, F. J. (1947) The action of various cations on muscle protoplasm. *J. Cell Comp. Physiol.* 29:15–32.

Hepler, P. K. (1980) Membranes in the mitotic apparatus of barley cells. *J. Cell Biol.* 86:490–499.

Hepler, P. K., and Callahan, D. A. (1987) Free calcium increases during anaphase in stamen hair cells of *Tradescantia*. *J. Cell Biol.* 105:2137–2143.

Heumann, H. G., and Zebe, E. (1967) Über Feinbau und Funktionsweisse der Fasern aus dem Hautmuskelschlauch des Regenwurms *Lumbricus terrestris L. Z. Zellforsch. Mikrosk. Anat.* 78:131–150.

Hill, D. K. (1964) The space accessible to albumin within the striated fiber of the toad. *J. Physiol. (London)* 175:275–294.

Hirokawa, N., Pfister, K. K., Yorifuji, H., Wagner, M. C., Brady, S. T., and Bloom, G. S. (1989) Submolecular domains of bovine brain kinesin identified by electron microscopy and monoclonal antibody decoration. *Cell* 56:867–878.

Hollenbeck, P. J., and Chapman, K. (1986) A novel microtubule associated protein from mammalian nerve shows ATP sensitive binding to microtubule. *J. Cell Biol.* 103:1539–1545.

Höner, B., Citi, S, Kendrick-Jones, J., and Jockusch, B. M. (1988) Modulation of cellular morphology and locomotory activity by antibodies against myosin. *J. Cell Biol.* 107:2181–2189.

Huxley, F. (1959) Local activation of muscle. *Ann. N.Y. Acad Sci.* 81:446–452.

Huxley, F., and Taylor, R. E. (1958) Local activation of striated muscle fibres. *J. Physiol. (London)* 144:426–441.

Huxley, H. E. (1963) Electron microscope studies on the structure of natural and synthetic protein filaments from striated muscle. *J. Mol. Biol.* 7:281–308.

Huxley, H. E. (1969) The mechanism of muscular contraction. *Science* 164:1356–1366.

Hymel, L., Inui, M., Fleischer, S., and Schindler, H. (1988) Purified ryanodine receptor of skeletal muscle sarcoplasmic reticulum forms Ca^{2+}-activated oligomeric Ca^{2+} channels in planar bilayers. *Proc. Natl. Acad. Sci. U.S.A.* 85:441–445.

Inagaki, M., Nishi, Y., Nishizawa, K., Matsuyama, M., and Sato, C. (1987) Site-specific phosphorylation induces disassembly of vimentin filaments *in vitro*. *Nature* 328:649–652.

Ip, W., Hartzer, M. K., Pang, Y.-Y. S., and Robson, R. M. (1985) Assembly of vimentin *in vitro* and its implications concerning the structure of intermediate filaments. *J. Mol. Biol.* 183: 365–375.

Ishikawa, H., Bischoff, R., and Holtzer, H. (1969) Formation of arrowhead complexes with heavy meromyosin in a variety of cells. *J. Cell Biol.* 43:312–328.

Ishiko, N., and Sato, M. (1957) The effect of calcium ions on electrical properties of striated muscle fibres. *Jpn. J. Physiol* 7:51–63.

Jobis, F. F., and O'Connor, M. J. (1966) Calcium release and reabsorption in the sartorius muscle of the toad. *Biochem. Biophys. Res. Commun.* 25:246–252.

Johnson, K. A. (1985) Pathway of the microtubules-dynein ATPase and the structure of dynein: a comparison with actomyosin. *Annu. Rev. Biophys. Biophys. Chem.* 14:161–188.

Jones, J. M., and Perry, S. V. (1966) The biological activity of subfragments prepared from heavy meromyosin. *Biochem. J.* 100:120–130.

Kachar, B., Fujisake, H., Albanesi, J. P., and Korn, E. D. (1987) Soluble microtubule-based translocator in *Acanthamoeba castellanii* extracts. *Biophys. J.* 51:487a.

Kaldor, G., Gitlin, J., Westley, F., and Volk, B. W. (1964) Studies on the interaction of actin with myosin. *Biochemistry* 3:1137–1145.

Kamitsubo, E. (1966) Motile protoplasmic fibrils in Characeae. II. Linear fibrillar structure and its bearing on protoplasmic streaming. *Proc. Jpn. Acad.* 42:640–643.

Kamitsubo, E. (1972) Motile protoplasmic fibrils in cells of Characeae. *Protoplasma* 74:53–70.

Kato, T., and Tonomura, Y. (1977) Identification of myosin in *Nitella flexilis*. *J. Biochem.* 82:777–782.

Kendrick-Jones, J., Szentkiralyi, E. M., and Szent-Gyorgyi, A. O. (1972) Myosin-linked regulatory systems: the role of the light chains. *Cold Spring Harbor Symp. Quant. Biol.* 37:47–53.

Kersey, Y. M., Hepler, P. K., Palevity, B. A., and Wessels, N. K. (1976) Polarity of actin filaments in characean algae. *Proc. Natl. Acad. Sci. U.S.A.* 73:165–167.

Kishino, A., and Yanagida, T. (1988) Force measurement by micromanipulation of single actin filament by glass needles. *Nature* 334:74–76.

Knecht, D. A., and Loomis, W. F. (1987) Antisense RNA inactivation of myosin heavy-chain gene-expression in *Dictyostelium discoideum*. *Science* 236:1081–1086.

Langford, G. M., Allen, R. D., and Weiss, D. G. (1987) Substructure of side arms on squid axoplasmic vesicles and microtubules visualized by negative contrast electronmicroscopy. *Cell Motil.* 7:20–30.

Lazarides, E., and Burridge, K. (1975) α-Actinin: immunofluorescent localization of a muscle structural protein in non-muscle cells. *Cell* 6:289–298.

Lehman, W., Kendrick-Jones, J., and Szent-Gyorgyi, A. G. (1972) Myosin-linked regulatory systems: comparative studies. *Cold Spring Harbor Symp. Quant. Biol.* 37:319–330.

Leterrier, J. F., Liem, R. K. H., and Shelanski, M. L. (1982) Interactions between neurofilaments and microtubule-associated proteins: possible mechanism for intraorganellar bridging. *J. Cell Biol.* 95:982–986.

Lindén, M., Nelson, B. B., Loncar, D., and Leterrier, J.-F. (1989) Studies on the interaction between mitochondria and the cytoskeleton. *J. Bioeneg. Biomembr.* 21:507–518.

Lye, R. J., Porter, M. E., Scholey, J. M., and McIntosh, J. R. (1987) Identification of a microtubule-based cytoplasmic motor in the nematode, *C. elegans*. *Cell* 51:309–318.

MacLennan, D. H., Campbell, K., and Reithmeier, R. A. F. (1983) Calsequestrin. In *Calcium and Cell Function,* Vol. 4, pp. 151–172. Academic Press, New York.

Maddox, C. R., and Perry, S. V. (1966) Differences in the myosins of the red and white muscles of the pigeon. *Biochem. J.* 99:8p–9p.

Mandrino, M. (1977) Voltage clamp experiments on frog single skeletal muscle fibres: evidence for a tubular sodium current. *J. Physiol. (London)* 269:605–625.

Masaki, T., Endo, M., and Ebashi, S. (1967) Localization of 6S component of α-actinin at Z-band. *J. Biochem. (Tokyo)* 62(5):630–632.

Matsudaira, P. T., and Burgess, D. R. (1979) Identification and organization of the components in the isolated microvillus cytoskeleton. *J. Cell Biol.* 83:667–673.

McLachlan, A. D., and Karn, J. (1982) Periodic charge distributions in myosin rod amino acid sequence match cross-bridge spacing in muscle. *Nature* 299:226–231.

McNeil, H., Ozawa, M., Kempler, R., and Nelson, W. J. (1990) Novel function of the cell adhesion molecule uvomorulin as an inducer of cell surface polarity. *Cell* 62:309–316.

Mines, G. R. (1913) On the functional analysis of the action of electrolytes. *J. Physiol. (London)* 46:553–564.

Miyata, H., Bowers, B., and Korn, E. D. (1989) Plasma membrane association of *Acanthamoeba* myosin I. *J. Cell Biol.* 109:1519–1528.

Mommaerts, W. F. H. M., and Green, I. (1954) Adenosine triphosphatase systems of muscle. III. A survey of the adenosine triphosphatase activity of myosin. *J. Biol. Chem.* 208:833–843.

Mooseker, M. S., Conzelman, K. A., Coleman, T. R., Heuser, J. E., and Sheetz, M. P. (1989) Characterization of intestinal microvillar membrane disks: detergent resistant membrane sheets enriched in associated brush border myosin I (110K-calmodulin). *J. Cell Biol.* 109:1153–1161.

Mueller, H., and Perry, S. V. (1962) The degradation of heavy meromyosin by trypsin. *Biochem. J.* 85:431–439.

Murphey, D. B., and Tilney, L. G. (1974) The role of microtubules in the movement of pigment granules in teleost melanophores. *J. Cell Biol.* 61:757–779.

Nagai, R., and Kamiya, N. (1977) Differential treatment of *Chara* cells with cytochalasin B with special reference to its effect in cytoplasmic streaming. *Exp. Cell Res.* 108:231–237.

Nagai, R., and Rebhun, L. I. (1966) Cytoplasmic microfilaments in streaming *Nitella* cells. *J. Ultrastruct. Res.* 14:571–589.

Nelson, W. J., and Hammerton, R. W. (1989) A membrane-cytoskeleton complex containing Na^+, K^+-ATPase, ankyrin and fodrin in Madin-Darby canine kidney (MDCK) cells: implications from the biogenesis of epithelial cell polarity. *J. Cell Biol.* 108:893–902.

Nelson, W. J., Shore, E. M., Wang, A. Z., and Hammerton, R. W. (1990) Identification of membrane cytoskeletal, complex containing the cell adhesion molecule uvomorulin (E-cadherin), ankyrin and fodrin in Madin-Darby canine kidney epithelial cells. *J. Cell Biol.* 110:349–357.

Nixon, R. A., and Sihag, R. K. (1991) Neurofilament phosphorylation: a new regulation and function. *Trends in Neurosci.* 14:501–506.

Noegel, A., Witke, W., and Schleicher, M. (1987) Calcium-sensitive non-muscle α-actinin contains EF-hand structures and highly conserved regions. *FEBS Lett.* 221:391–396.

Ochs, S. (1971) Characteristics and a model for fast axoplasmic transport in nerve. *J. Neurobiol.* 2:331–345.

Page, S. (1964) The organization of the sarcoplasmic reticulum in frog muscle. *J. Physiol. (London)* 175:10P–11P.

Pasternak, C., Spudich, J. A., and Elson, E. L. (1989) Capping of surface receptors and concomitant cortical tension are generated by conventional myosin. *Nature* 341:549–551.

Peachey, L. D. (1965) The sarcoplasmic reticulum and transverse tubules of the frog's sartorius. *J. Cell Biol.* 25 (No. 3, Pt. 2):209–231.

Pepe, F. A. (1966) Some aspects of the structural organization of the myofibril as revealed by antibody-staining methods. *J. Cell Biol.* 28:505–525.

Perry, S. V. (1956) Relation between chemical and contractile function and structure of the skeletal muscle cell. *Physiol. Rev.* 36:1–76.

Persechini, A., Stull, J. T., and Cooke, R. (1985) The effect of myosin phosphorylation on the contractile properties of skinned rabbit skeletal muscle fibers. *J. Biol. Chem.* 260:7951–7954.

Pollard, T. D., and Cooper, J. A. (1986) Actin and actin-binding proteins. A critical evaluation of mechanisms and function. *Annu. Rev. Biochem.* 55:987–1035.

Pollard, T. D., and Weihing, R. R. (1974) Actin and myosin and cell movements. *CRC Crit. Rev. Biochem.* 2:1–65.

Porter, K. R. (1961) The sarcoplasmic reticulum. Its recent history and present status. *J. Biophys. Biochem. Cytol.* 10(2):219–226.

Portzehl, H. E., Caldwell, P. C., and Ruegg, J. C. (1964) The dependence of contraction and relaxation of muscle fibres from the crab *Maya squinado* on the internal concentration of free calcium ions. *Biochim. Biophys. Acta* 79:581–591.

Pratt, M. M. (1986) Homology of egg and flagellar dynein. *J. Biol. Chem.* 261:956–964.

Pryer, N. K., Wadsworth, P., and Salmon, E. D. (1986) Polarized microtubule gliding and particle saltations produced by soluble factors from sea urchin eggs and embryos. *Cell Motil.* 6:537–548.

Quinlan, R. A., Hatzfeld, M., Franke, W. W., Lustig, A., Schulthess, T., and Engel, J. (1986) Characterization of dimer subunits of intermediate filament proteins. *J. Mol. Biol.* 192:337–349.

Rüegg, J. C. (1968) Contractile mechanisms of smooth muscle. *Symp. Soc. Exp. Biol.* 22:45–66.

Schliwa, M. Shimizu, T., Vale, R. D., and Euteneur, U. (1991) Nucleotide specificities of anterograde and retrograde organelle transport in *Reticulomyxa* are undistinguishable. *J. Cell Biol.* 112:1199–1203.

Schnapp, B. J., Vale, R. D., Sheetz, M. P., and Reese, T. S. (1985) Single microtubules from squid axoplasm support bidirectional movement of organelles. *Cell* 40:455–462.

Schnapp, B. J., Crise, B., Sheetz, M. P., Reese, T. S., and Khan, S. (1990) Delayed start-up of kinesin-driven microtubule gliding following inhibition by adenosine 5′-[β,γ-imido]triphosphate. *Proc. Natl. Acad. Sci. U.S.A.* 87:10053–10057.

Sheetz, M. P., and Spudich, J. A. (1983a) Movement of myosin-coated fluorescent structures on actin cables. *Cell Motil.* 3:485–489.

Sheetz, M. P., and Spudich, J. A. (1983b) Movement of myosin-coated fluorescent beads on actin cables in vitro. *Nature (London)* 303:31–35.

Sinard, J. H., and Pollard, T. D. (1989) Microinjection into *Acanthamoeba castellanii* of monoclonal antibodies to myosin II slows but does not stop cell locomotion. *Cell Motil. Cytosk.* 12:42–52.

Slayter, H. S., and Lowry, S. (1967) Substructure of the myosin molecule as visualized by electron microscopy. *Proc. Natl. Acad. Sci. U.S.A.* 58:1611–1618.

Stebbins, H., and Hyams, J. S. (1979) In *Cell Motility* (Phillips, I. D. J., ed.), p. 192. Longman, London.

Steinert, P. M., and Roop, D. R. (1988) Molecular and cellular biology of intermediate filaments. *Annu. Rev. Biochem.* 57:593–626.

Swanson, J., Bushnell, A., and Silverstein, S. C. (1987) Tubular lysosomes, morphology and distribution within macrophages depend on the integrity of cytoplasmic microtubules. *Proc. Natl. Acad. Sci. U.S.A.* 84:1921–1925.

Toyoshima, Y. Y., Kron, S. J., McNally, E. M., Niebling, K. R., Toyoshima, C., and Spudich, J. A. (1987) Myosin subfragment-1 is sufficient to move actin filament *in vitro*. *Nature* 328:536–539 .

Travis, J. L., and Allen, R. D. (1981) Studies on the motility of the foraminifera. I. Ultrastructure of the reticulopodial network of *Allogromia latticollaris* (Arnold). *J. Cell Biol.* 90:211–221.

Trybus, K. M. (1991) Assembly of cytoplasmic and smooth muscle myosins. *Curr. Opin. Cell Biol.* 3:105–111.

Ueno, H., and Harrington, W. F. (1986) Local melting in the subfragment-2 region of myosin in activated muscle and its correlation with contractile force. *J. Mol. Biol.* 190:69–82.

Vale, R. D., Reese, T. S., and Sheetz, M. P. (1985a) Identification of a novel force generating protein, kinesin, involved in microtubule-based motility. *Cell* 42:39–50.

Vale, R. D., Reese, T. S., and Sheetz, M. P. (1985b) Different axoplasmic proteins movement in opposite directions along microtubules in vitro. *Cell* 43:623–632.

Vale, R. D., Schnapp, B. J., Mitchision, T., Steuer, A., Reese, T. S., and Sheetz, M. P. (1986) Different axoplasmic proteins generate movement in different directions along microtubules in vitro. *Cell* 43:623–632.

Vale, R. D., and Toyoshima, Y. Y. (1988) Rotation and translocation of microtubules in vitro induced by dyneins of *Tetrahymena* cilia. *Cell* 52:459–469.

Vallee, R., and Bloom, G. S. (1991) Mechanism of fast and slow axonal transport. *Annu. Rev. Neuroscie.* 14:59–92.

Vallee, R. B., Wall, J. S., Paschal, B. M., and Sheptner, H. S. (1988) Microtubule associated protein 1C from brain is a two-headed cytosolic dynein. *Nature* 332:561–563.

Vandekerkhove, J. (1990) Actin-binding proteins. *Curr. Opin. Cell Biol.* 2:41–50.

Volpe, P., Di Virgilio, F., Pozzan, T., and Salviati, G. (1986) Role of 1,4,5-triphosphate in excitation-contraction coupling in skeletal muscle. *FEBS Lett.* 197:1–4.

Wagenknecht, T., Grassucci, R., Frank, J., Saito, A., Inui, M., and Fleischer, S. (1989) Three-dimensional architecture of the calcium channel/foot structure of sarcoplasmic reticulum. *Nature (London)* 338:167–170.

Walker, R. A., Salmon, E. D., and Endow, S. A. (1990) The *Drosophila claret* segregation protein is a minus-end directed motor molecule. *Nature* 347:780–782.

Warrick, H. M., and Spudich, J. A. (1987) Myosin structure and function in cell motility. *Annu. Rev. Cell Biol.* 3:379–421.

Waszenius, V. M., Saraste, M., Salven, P., Eramaa, M., Holm, L., and Lehto, V. P. (1989) Primary structure of brain α-spectrin. *J. Cell Biol.* 108:79–93.

Weber, A., Hertz, R., and Reiss, I. (1963) On the mechanism of the relaxing effect of fragmented sarcoplasmic reticulum. *J. Gen. Physiol.* 46:679–702.

Williamson, R. E., and Toh, B. H. (1979) Motile models of plant cells and the immunofluorescent localization of actin in a motile *Chara* cell model. In *Cell Motility: Molecules and Organization* (Hatano, S., Ishikawa, H., and Sato, H., eds.), pp. 339–346. University Park Press, Baltimore.

Winegrad, S. (1961) The possible role of calcium in excitation-contraction coupling of heart muscle. *Circulation* 24:523–529.

Winegrad, S. (1965a) Autoradiographic studies of intracellular calcium in frog skeletal muscle. *J. Gen. Physiol.* 48:455–479.

Winegrad, S. (1965b) The location of muscle calcium with respect to the myofibrils. *J. Gen. Physiol.* 48:997–1002.

Winton, F. R. (1926) The influence of length on the responses of unstriated muscle to electrical and chemical stimulation and stretching. *J. Physiol. (London)* 61:368–382.

Yang, J. T., Laymon, R. A., and Goldstein, L. S. B. (1989) A three-domain structure of kinesin heavy chain revealed by DNA sequence and microtubule binding analysis. *Cell* 56:879–889.

Yang, J. T., Saxton, W. M., Stewart, R. J., Raff, E. C., and Goldstein, L. S. B. (1990) Evidence that the head of kinesin is sufficient for force generation and motility in vitro. *Science* 249:42–47.

Zot, H. G., and Pollard, T. D. (1991) Myosin I moves actin filaments over a pure lipid substrate. *Biophys. J.* 59:246a.

Index